VACUUM METALLURGY

VACUUM METALLURGY

Edited by
O. WINKLER
Executive Vice-President of Balzers Ltd. for High Vacuum Technique and Thin Films,
Balzers, Liechtenstein

AND

R. BAKISH
President, Bakish Materials Corporation
Englewood, N.J., U.S.A.

Adjunct Professor at the Fairleigh Dickinson University
Teaneck, N.J., U.S.A.

ELSEVIER PUBLISHING COMPANY
AMSTERDAM – LONDON – NEW YORK
1971

ELSEVIER PUBLISHING COMPANY
335 JAN VAN GALENSTRAAT
P. O. BOX 211, AMSTERDAM, THE NETHERLANDS

ELSEVIER PUBLISHING CO. LTD.
BARKING, ESSEX, ENGLAND

AMERICAN ELSEVIER PUBLISHING COMPANY, INC.
52 VANDERBILT AVENUE
NEW YORK, NEW YORK 10017

LIBRARY OF CONGRESS CARD NUMBER: 74–118258

STANDARD BOOK NUMBER: 0–444–40857–6

WITH 536 ILLUSTRATIONS AND 165 TABLES.

COPYRIGHT © 1971 BY ELSEVIER PUBLISHING COMPANY, AMSTERDAM
ALL RIGHTS RESERVED. NO PART OF THIS PUBLICATION MAY BE REPRODUCED, STORED IN A RETRIEVAL SYSTEM, OR TRANSMITTED IN ANY FORM OR BY ANY MEANS, ELECTRONIC, MECHANICAL, PHOTOCOPYING, RECORDING, OR OTHERWISE, WITHOUT THE PRIOR WRITTEN PERMISSION OF THE PUBLISHER,
ELSEVIER PUBLISHING COMPANY, JAN VAN GALENSTRAAT 335, AMSTERDAM

PRINTED IN THE NETHERLANDS

Preface

R. F. Bunshah, in his keynote address to the 1967 International Vacuum Metallurgy Conference, defined vacuum metallurgy as "the making, shaping and treating of metals and alloys under pressures ranging from sub-atmospheric to ultra high-vacuum, a study of their properties and applications". This definition indicates the breadth of the field. In the last fifteen years it has developed as an important industrial processing approach, and it is therefore not surprising that the various aspects of this technology have already been discussed at numerous vacuum metallurgy conferences. Up to now, however, a comprehensive treatment of the subject in depth has been lacking.

Diversification and specialization have proceeded at so great a pace in recent years that it now seems quite impossible for any one individual to have a complete command of the subject. It was therefore necessary to select, on an international basis, experts who would be competent to describe critically and as fully as possible, on the basis of their own results and experience, the state of the art in the various specialized fields. Even with this approach it was necessary to restrict the scope to essentials, particularly to those elements of the technology that have become important and permanent in industrial application.

A multi-author book can never be completely uniform in treatment: the individual style and approach of the authors inevitably shows through. Another problem is the fact that the period between completion of manuscripts and publication is relatively long for a field in such a state of rapid development. This should be borne in mind in attempting to assess the work critically.

Almost all metals in the Periodic Table nowadays belong either to the "vacuum metals", and are thus produced by vacuum methods, or to those that undergo vacuum processes at some stage in their manufacture. The aim of vacuum treatment is essentially the removal or reduction of harmful impurities during production or in the various processing phases, whether for increase of the yield, improvement of quality, or creation of quite new materials.

The first chapter considers the physical concepts involved and the way in which they affect the scope and possibilities of this technology. The thermodynamic principles of the equilibrium between metallic phases and a gas phase or metal–gas compounds are set out, in as far as these are of importance for vacuum metallurgy (chiefly for liquid metals). Subsequently, the factors which determine how quickly and how fully the possible final state is reached are dealt with. The theoretical and experimental work published in recent years on the kinetics of the vacuum metallurgical reactions forms an important basis for the rational further development of various processes.

Chapter II gives a brief survey of the present situation with regard to vacuum engineering. Though only a means to an end, it is nevertheless the prerequisite for all vacuum metallurgy techniques. For example, proposals far ahead of their time for degassing of steel were made more than 100 years ago but could not be carried out simply because no effective pumping systems were available. Fifty years ago, W. Rohn, pioneer in the field of vacuum melting, was much restricted in his possibilities for the same reason. The progress of vacuum metallurgy was in this way closely linked with the progress of vacuum technology.

The description of the various fields of vacuum metallurgy in Chapter III begins with a survey of the use of vacuum techniques in extractive metallurgy and refining of metals. It shows the great range of applications which have developed almost unnoticed. Many reduction processes, which make use of either the displacement of reaction equilibria or the increased volatility of one of the constituents under vacuum, have achieved technical significance. The same applies to the refining of metals by distillation separation, which in the case of lead and zinc for example is already carried out on a large scale. Also important are the chemical transport reactions and zone melting in vacuum, which are often the only possible methods to achieve high purities for refractory and very reactive metals.

The first part of Chapter IV deals with refining on a very large scale, namely, vacuum degassing of steels. Eighteen years ago the development of stream degassing by Bochumer Verein made vacuum metallurgy industrially significant for the first time. At present, steel degassing plants with a capacity of over 400 tons are in use. They are used not only for degassing but also for the synthesis of steel alloys with a narrow range of properties and compositions and for decarburization. This is one of the best examples of how an expensive technique can find economic justification in the improved and more uniform quality of the end product.

Vacuum degassing in the solid state, which is discussed in the second part of this chapter, is for the most part just an accompaniment to the execution of other vacuum processes (sintering, annealing, brazing). Only in a few cases is it an end in itself. The treatment is restricted therefore to the description of thermodynamic equilibria and phase diagrams of some important metal–gas systems and to the kinetics of solid state degassing and so provides a supplement to the first chapter.

Chapter V deals with the vacuum melting processes at present in use in which solid feed stock is simply converted to another form or simultaneously a refining, structural change and/or alloy formation is achieved. Of the so-called remelting processes (vacuum arc and electron beam melting), which were originally developed for the remelting of rare metals and refractory metals, vacuum arc melting is today used mainly for high alloy steels; it may in the future be used also for the processing of low alloy steels. This technique will always be considered whenever the possibilities of conventional metallurgy to improve quality or increase yield have been exhausted.

While vacuum induction melting is preferred for the preparation of controlled composition alloys, especially of reactive metals, and for controlled refining, the remelting processes are used particularly for structural improvement, the elimination of the anisotropy of the mechanical properties by means of favourable crystallization conditions and the resulting higher yield being the principal targets. However, since in electroslag melting at ambient pressure the crystallization conditions are similar, the vacuum remelting technique will be more and more restricted to those cases where removal of gases and volatile impurities is also of importance.

Chapter VI deals with the determination of the gas contents of metals. Particularly with metals of a high degree of purity, this is an important and often difficult task. If the gas determinations are conducted in vacuum, one is concerned with vacuum degassing processes on a small scale, analogous to those already described in a previous chapter. Vacuum gas analysis offers a greater range of application and often greater accuracy than other methods using an inert carrier gas at ambient pressure. Through automation of the actual analysis with digital output of results the use of this technique for process control will become increasingly easier.

A further possibility for the consolidation of metals and alloys, vacuum sintering, is described in Chapter VII. Its most important application at present is in the production of cemented carbides. The possibility of manufacturing finished components with particularly favourable structures and the unique variability of composition are the attractive features of this technique. Its future prospects seem promising, especially if the more important cost factors such as the price of raw materials, *e.g.* of alloyed steel powder, and the cost of the vacuum atmosphere develop favourably in comparison with inert gas or hydrogen atmospheres.

In Chapters VIII and IX vacuum heat treatment, vacuum brazing, and other vacuum techniques for joining components are considered. They are all relatively new methods. Vacuum annealing and vacuum brazing have become particularly important. Though initially developed for special materials and uses, they have since found wider technical application. In vacuum annealing installations, for example, the reduction of the cooling-down time by inert gas circulation and the possibility of quenching the charge in liquids under vacuum or inert atmosphere have significantly extended the range of applications. Here too, the better reproducibility of results, better surface quality, and improved mechanical properties more than compensate for the large investment involved. The same applies to vacuum brazing. Of the remaining joining methods, electron beam melting must be highlighted. This has reached a very high technical level in recent years, but nevertheless, owing to the competitive development of more conventional welding techniques, it has remained a specialized process.

Chapter X concerns the changes in solid surfaces brought about by the application of thin films by means of evaporation and sputtering. This in fact is one of the oldest vacuum processes, and it has found very diversified applications recently, *e.g.* in the manufacture of integrated circuits in the electronics industry. The ability to apply, in closely controlled thicknesses, thin films of refractory and other metals of almost any desired composition on non- and semi-conducting substrates was an important step in the technical progress of this field.

The final chapter (Chapter XI) presents the behaviour of metals at very low pressures. It deals with problems whose solutions were important in connection with the development of space travel. This work led to a whole new area of research and development in metal physics, the benefits of which are certain to spill over into other areas of technology.

In conclusion, a few remarks are necessary regarding the units in this book. Metric units are used throughout and in the form recommended in 1968 by the Council of the British Institution of Metallurgists*. In a few cases, British units are also given**. The unit "Torr" is still widely used in vacuum technology and has therefore been retained***, although several countries plan to adopt the "newton/m^2" or "pascal" in the near future.

Last but not least, the Editors wish to express their sincere thanks to the contributors to this volume for their willingness to meet deadlines and for the great care taken in preparing their manuscripts. Thanks are also due to Dr. Graff, who expertly translated the German manuscripts into English, and to all authors and publishers who have granted permission to reproduce diagrams and photographs.

Balzers, Liechtenstein O. WINKLER

Englewood, N. J., U.S.A. R. BAKISH

---

*   *Metals and Materials*, 2 (1968) 367
**   See also conversion table, p. 862
*** 1 Torr = 133.3 N/m^2 = 133.3 Pa

List of Contributors

R. Bakish, Bakish Materials Corporation, 171 Sherwood Place, Englewood, N. J. 07631, and Fairleigh Dickinson University, Teaneck, N. J., U.S.A.

D. T. Bourgette, Metals and Ceramics Division, Oak Ridge National Laboratory, P. O. Box X, Oak Ridge, Tenn. 37830, U.S.A.

H. C. Child, Birmingham Small Arms Ltd., Group Research Centre, Kitts Green, Birmingham B33 OLF, England

M. E. Harper, Steel Company of Australia Pty. Ltd., Sussex Street, Coburg, Victoria 3058, Australia

D. V. Keller, Department of Metallurgical Engineering, Syracuse University, Syracuse, N. Y. 13210, U.S.A.

R. Kieffer, Institut für Chemische Technologie anorganischer Stoffe, Technische Hochschule, Getreidemarkt 9, Wien-VI, Austria

† K. Köstlin, Hilti AG, FL 9494 Schaan, Liechtenstein

J. Krüger, formerly Institut für Metallhüttenwesen und Elektrometallurgie, Rhein.-Westf. Technische Hochschule, 51 Aachen, Intzestrasse 3, Germany, now with Lurgi Gesellschaft für Chemie und Hüttenwesen mbH, D 6000 Frankfurt am Main, Gervinusstrasse 17–19, Germany

E. Lassner, Metallwerk Plansee, Postfach 74, A–6600 Reutte–Tirol, Austria

A. Lawley, Department of Metallurgical Engineering, Drexel University, Philadelphia, Pa. 19104, U.S.A.

G. E. Oldfield, Jessop-Saville Ltd., Brightside Works, Sheffield S9, England

G. Paesold, Balzers Ltd. for High Vacuum Technique and Thin Films, FL 9496 Balzers, Liechtenstein

(continued overleaf)

---

† Deceased 19 February 1970

LIST OF CONTRIBUTORS

B. D. POWER, Edwards Vacuum Components, Manor Royal, Crawley, Sussex England

E. RITTER, Balzers Ltd. for High Vacuum Technique and Thin Films, FL 9496 Balzers, Liechtenstein

K. RÜTTIGER, Hoesch AG Hüttenwerke, D 4600 Dortmund, Postfach 902, Germany

H. D. SEGHEZZI, Hilti AG, FL 9494 Schaan, Liechtenstein

R. L. STEPHENSON, Department of Metallurgical Engineering, Syracuse University, Syracuse, N. Y. 13210, U.S.A.

S. S. WHITE, Technical Services Branch, Texas Instruments Inc., 34 Forest Street, Attleboro, Mass. 02703, U.S.A.

O. WINKLER, Balzers Ltd. for High Vacuum Technique and Thin Films, FL 9496 Balzers, Liechtenstein

Contents

Preface v

List of Contributors ix

Chapter I. Thermodynamics and Kinetics in Vacuum Metallurgy
by O. Winkler 1
1. Introduction 1
2. Thermodynamic principles of vacuum-metallurgical processes . . . 2
   2.1 The law of mass action 2
   2.2 Fundamental thermodynamic relationships 3
   2.3 Equilibria between a pure condensed phase and an ideal gas . 4
   2.4 Calculation of the temperature function of the equilibrium constants . 6
   2.5 Thermodynamics of solutions 7
   2.6 Solubility of gases 11
   2.7 Deoxidation by a reduction process or by evaporation of suboxides . 15
   2.8 Equilibria of metal vapours 22
   2.9 Reactions with refractory oxides and slags 24
   2.10 Refining in a vacuum 29
3. Kinetics of vacuum-metallurgical processes 30
   3.1 Kinetics of degassing 30
   3.2 Evaporation of metallic elements 74
   3.3 Evaporation losses of the basis metal during degassing and evaporation of alloying elements and impurities 76
   3.4 Reactions between melts, refractories and slags 82
Acknowledgement 89
List of symbols 90
References 92

Chapter II. Vacuum Engineering
by B. D. Power and M. E. Harper 95
1. Sources of gas in metallurgical process plant 95
   1.1 Processing pressure 95
   1.2 Effective pumping speed 96
   1.3 Initial gas content 97
   1.4 Leakage 98
   1.5 Virtual leakage 99
   1.6 Outgassing 99
   1.7 Gases released during the process 101
2. Pumps and pumping systems 102
   2.1 Water ring pumps 102
   2.2 Water-operated ejectors 104
   2.3 Oil-sealed rotary mechanical pumps 105
   2.4 Roots pumps (mechanical boosters) 108
   2.5 Steam ejectors 111

2.6	Vapour ejector and diffusion pumps	114
2.7	Pumps depending on gettering and ionization	120
2.8	Cryosorption pumps	123
2.9	Turbomolecular pumps	125
2.10	Systems for ultra-high vacuum (U.H.V.)	126

3. Pressure-measuring devices 130
   3.1 Mechanical gauges with atmospheric pressure reference 131
   3.2 Mechanical gauges with high vacuum reference 131
   3.3 Liquid "U" tube manometer. 132
   3.4 McLeod gauge 132
   3.5 Thermocouple gauge 133
   3.6 Pirani gauge 133
   3.7 "Alphatron" ionization gauge 134
   3.8 Cold cathode ionization gauge (Penning type) 135
   3.9 Hot cathode ionization gauge 135
   3.10 A note on partial pressure gauges (mass spectrometers) 137
4. Leak-proving and leak-finding 138
   4.1 Leak-proving methods 138
   4.2 Systematic leak-finding 140
   4.3 Some leak-sensing devices and search gases 140
Bibliography 143

Chapter III. Use of Vacuum Techniques in Extractive Metallurgy and Refining of Metals
by J. KRÜGER 145
1. Introduction 145
2. Vacuum technology in the preliminary stages of metallurgical reduction processes . . 147
   2.1 Physical processes 147
   2.2 Chemical processes 150
3. Vacuum-metallurgical reduction processes 160
   3.1 Carbothermic reduction processes 161
      Carbothermic reduction processes without volatilization of the metals to be reduced, 161; *Carbothermic reduction processes involving volatilization of the reduced metal,* 181
   3.2 Metallothermic reduction processes 185
      Metallothermic reduction processes without volatilization of the metals to be reduced, 185; *Metallothermic reduction processes with volatilization of the reduced metals,* 191
4. Refining of metals in a vacuum 212
   4.1 Separation of metals by distillation in a vacuum 213
      Discussion of the rate-determining steps, 213; *Methods of metal distillation,* 221; *Distillation processes in the metallurgy of lead,* 222; *Distillation processes in the metallurgy of the remaining low-melting-point heavy metals, in particular cadmium, zinc and tin,* 234; *Distillative separation and refining of rare-earth metals,* 241; *Distillative separation and refining of the alkali and alkaline-earth metals,* 243; *Distillative separation and refining of aluminium, gallium and indium,* 253; *Distillative separation of the iron-group metals,* 258; *Distillation of selenium and tellurium,* 259; *Distillation of heavy non-ferrous metal scrap,* 260
   4.2 Deoxidation and decarburization of non-ferrous metals in a vacuum . . . 262
      Decarburization of ferro-chromium in a vacuum, 262; *Deoxidation via suboxides,* 268
   4.3 Chemical transport reactions in a vacuum 279
      Refining of vanadium, niobium, tantalum, uranium, chromium and thorium, 281; *The refining and purification of titanium, zirconium and hafnium,* 282; *Refining of molybdenum,* 285
   4.4 Zone melting in a vacuum 286
      Basic principles in zone melting, 286; *Design of vacuum zone-melting apparatus,* 294; *Appraisal of the effects of zone-melting,* 298; *Examples for vacuum-zone-melting,* 298

List of symbols	320
Acknowledgement	321
References	321

Chapter IV. Vacuum Degassing
by K. Rüttiger (Section 1)
  H. D. Seghezzi and K. Köstlin (Section 2) 337
1. Vacuum degassing in the liquid state 337
   1.1 Introduction 337
       Historical survey, 337
   1.2 General points of view 347
       Aims of a vacuum treatment of liquid steel, 347; *Choice of pumping system, and dust problems*, 350; *Heat losses*, 353; *Refractory lining of the vessels for the vacuum treatment*, 356; *Insertion into the production schedule of steel works*, 363
   1.3 Ladle degassing processes 363
       Ordinary ladle degassing, 363; *Ladle degassing with injection of a purging gas*, 371; *Ladle degassing with stirring coil*, 374; *Ladle degassing with additional heating*, 377; *Refractory lining of the vacuum ladles*, 380
   1.4 Stream degassing processes 382
       Stream droplet degassing, 383; *Tap degassing*, 387; *Therm-I-Vac process*, 393
   1.5 Cycling and circulation degassing processes 396
       The DH process, 396; *The RH process*, 409; *Thermo-Flow process*, 421; *Transfer degassing*, 422
   1.6 Mould degassing processes 425
       Vacuum casting inside a vacuum tank, 425; *Vacuum casting without an outer vacuum chamber*, 434; *Ingot degassing*, 436
   1.7 Discussion of the vacuum processes 440
       Gas content and the kinetics of degassing, 441; *Cleanness*, 447; *Vacuum-metallurgical potentialities*, 448; *Incorporation in the production process*, 449; *Effect of degassing processes on the properties of steels*, 541
2. Vacuum degassing in the solid state 463
   2.1 Introduction 463
   2.2 Thermodynamic equilibria 466
       Introductory remarks, 466; *Binary systems*, 467; *Multicomponent systems having at least two non-metallic (gaseous) components*, 478
   2.3 Stationary states 481
   2.4 Kinetics of degassing 483
       Theory, 483; *Results of experiments*, 493
References 507

Chapter V. Vacuum Melting
by H. C. Child and G. E. Oldfield (Sections 1–3)
  R. Bakish (Section 4)
  A. Lawley (Section 5) 517
Introduction to Sections 1–3 517
1. Vacuum melting in resistance furnaces 518
2. Vacuum induction melting 519
   2.1 Design features of modern plant 519
   2.2 Process techniques 539
   2.3 Metallurgical effects 545
   2.4 Special requirements for precision casting furnaces 552
3. Vacuum arc melting 553
   3.1 Design features for modern plant 554
   3.2 Special designs of furnace 572
   3.3 Process techniques 575
   3.4 Metallurgical effects 580

4. Electron beam melting	593
4.1 Introduction	593
4.2 Equipment design as manifested by commercially produced systems	596
4.3 Melting and purification in electron beam furnaces	615
4.4 Electron beam processed materials	621
5. Crystal growing	633
5.1 Introduction	633
5.2 Methods of crystal growing	634
References	642

Chapter VI. Determination of the Gas Content of Metals by Vacuum Degassing Methods

by E. Lassner	649
1. Introduction	649
2. Technology of the vacuum extraction analysis	652
3. Apparatus	655
3.1 Furnaces	656
3.2 Pumping system	657
3.3 Analyser	657
4. Sources of error	659
4.1 Sources arising from the sample	660
4.2 Errors caused by the carrier melt or other factors in the furnace vessel	660
4.3 Errors caused by the apparatus	662
5. Accuracy and precision	663
6. Literature survey	666
References	668

Chapter VII. Vacuum Sintering

by R. Kieffer, G. Paesold and O. Winkler	673
1. Phenomena during sintering	673
1.1 Sintering of homogeneous systems	674
1.2 Sintering of heterogeneous systems	675
1.3 Fundamentals of the reactions	677
2. Production of sintered materials	681
2.1 Tantalum	682
2.2 Niobium	685
2.3 Tungsten and molybdenum	687
2.4 Beryllium	687
2.5 Hard metals and cemented carbides	688
2.6 Compound metals	692
2.7 Alnico sintered magnets	693
2.8 Stainless steel	693
3. High-vacuum sintering furnaces	694
3.1 Direct-resistance heating	694
3.2 Indirect-resistance heating	698
3.3 Induction furnaces	708
3.4 Vacuum sintering installations for the production of cemented carbides	710
3.5 Pressure sintering furnaces	718
References	718

Chapter VIII. Vacuum Heat Treatment

by S. S. White	721
1. The vacuum atmosphere	721
2. Types of furnaces	728
2.1 Hot wall furnaces	728
2.2 Cold wall furnaces	729
2.3 Other furnace types	735

3. Heat treatment processes carried out in vacuum		736
3.1 Hardening		736
3.2 Vacuum annealing		739
4. Conclusion		743
References		744

Chapter IX. Joining
by R. BAKISH (Sections 1 and 3)
   S. S. WHITE (Section 2) 745

1. Electron beam welding		745
1.1 Introduction		745
1.2 The process and its modifications		745
1.3 Equipment		762
1.4 Industrial applications and related topics		771
1.5 Conclusions		775
2. Vacuum brazing		775
2.1 Introduction		775
2.2 Characteristics of vacuum brazing		777
2.3 Wetting and spreading phenomena		778
2.4 Metallurgical reactions between filler and base metal		780
2.5 Metallurgical failure mechanisms in vacuum brazing		781
2.6 The strength of brazed joints		782
2.7 Materials that can be vacuum brazed		785
2.8 Vacuum brazing furnaces		789
2.9 Conclusion		790
3. Solid state bonding		790
3.1 Joining by diffusion bonding		791
3.2 Chemical vapor deposition welding		798
References		800

Chapter X. Vacuum Coating
by E. RITTER 803

1. Vacuum evaporation		803
1.1 Principle of the method		803
1.2 Substrate preparation		803
1.3 Vacuum required and the effect of residual gases		805
1.4 Methods of evaporation		805
1.5 Control of evaporation		808
1.6 Condensation		809
1.7 Operation of the processes		810
1.8 Applications		812
2. Cathode sputtering		814
2.1 Principle and modifications of the process		814
2.2 Materials which may be deposited by cathode sputtering		816
2.3 Sputtering practice		817
2.4 Applications		817
References		818

Chapter XI. Effect of High Vacuum on some Important Properties of Metals and Alloys
by D. T. BOURGETTE, D. V. KELLER AND R. L. STEPHENSON 821

1. Introduction		821
2. The space and vacuum environment		822
3. Space simulation		823
4. Metallic adhesion		825
4.1 Introduction		825
4.2 Adhesion strength data		827

4.3	Bulk dispersal mechanisms	831
4.4	Interfacial forces and material transfer during tensile tests	834
4.5	Coefficient of adhesion	835

List of symbols 837
5. Mechanical properties in vacuum 838

5.1	Interaction of materials with residual contaminants	838
5.2	Hardness	843
5.3	Ductility and ductile to brittle transition temperature	846
5.4	Tensile properties	849
5.5	Creep properties	852
5.6	Fatigue	855

List of symbols 858
References 858

Conversion Table 862

Author Index 863

Subject Index 880

Chapter I

Thermodynamics and Kinetics in Vacuum Metallurgy

O. WINKLER

1. Introduction

The thermodynamic principles of typical reactions occurring in vacuum-metallurgical processes will be considered in the first part of this chapter, and the phenomena and effects governing the course of these reactions in the second part.

The wide range of employment of a vacuum in metallurgical technology is such that it is not possible to discuss all thermodynamic and kinetic aspects in this chapter. This is not, indeed, regarded as necessary, as other authors in this book will deal with these particulars of the individual processes in the chapters that follow.

The vacuum-metallurgical processes are, in principle, almost always made up of heterogeneous reactions. It is only very rarely that equilibrium exists within individual or between different phases, as only quasi-stationary states are attained. Nonequilibrium between the phases is even created intentionally, in order to promote the desired material transport which leads to the removal of impurities or to a separation of components via the gas or vapour phase. Equilibrium, therefore, should exist only at the phase boundary. Darken[1] calls this state the "local equilibrium".

Although in most cases the problems involved are those of transport, the determination of the data which characterize the state that an enclosed system will finally attain is of a fundamental nature. Only on the basis of these data may we predict whether or not the desired reaction can occur at all or whether it is promoted by application of a vacuum, and which conditions must be fulfilled for a vacuum-metallurgical process. However, their determination is relatively simple because only stationary states need to be considered. In the last two decades the data of many important reactions have been determined and are available in tables [2-7].

Although a reaction may be hypothesized, it does not follow that it indeed takes place, *i.e.* at a sufficiently high reaction rate. The question of the quantities which determine the kinetics of a process is therefore at least of equal importance. These quantities not only decide the efficiency of a process or the size of the reactor for a required throughput, but also the relation of the main process to undesirable side reactions and losses. A typical example for such competing side reactions is found in the deoxidation of metal melts by carbon in an oxide-ceramic crucible, in which the elimination of the oxygen from the melt is limited by the supply of oxygen from the crucible material.

These questions require an analysis of the course of the reaction, which usually raises more problems than the determination of the thermodynamic data. The reason should mainly be sought in the great difficulty of separating various superimposed factors. It is also not always possible to transform the results of technological processes to a laboratory scale, which makes the determination of the various factors even more difficult. The investment for research in this field may therefore prove very high and this explains why intensive research in this direction has been undertaken only within the last ten years in connection with development of vacuum-metallurgical processes on a large industrial scale, together with the effort for their optimization. In some fields, for instance in the discontinuous degassing of homogeneous metal melts, useful models have been developed which make possible the mathematical treatment of the material transport, thus giving a concept of the reaction-kinetic course. As the results derived from these models show excellent agreement with those obtained in technology it may be assumed that the principles on which these models have been based are correct. However, these considerations cover only some experimental fields of vacuum metallurgy and even then it has not been sufficiently verified how far their range of validity extends. Further extension of this research seems therefore a matter of priority.

It is hoped that this survey, restricted to some sectors of thermodynamics important in vacuum metallurgy and to the basic mechanisms and relationships governing the reaction kinetics, may serve as an introduction and as a basis for a better understanding of the phenomena underlying the vacuum metallurgical processes described in the following chapters.

2. Thermodynamic Principles of Vacuum-metallurgical Processes

2.1 The law of mass action

The basis for any calculation of reaction equilibria is the law of mass action. This postulates how reactants and reaction products are partitioned into the gas phase and the condensed phase at equilibrium in the heterogeneous reactions usually met within vacuum-metallurgical processes.

The pure condensed phases do not need to be taken into account for such considerations; only those components of the reaction should be looked at which are either in solution or in the gas phase.

As an example we may take a reaction equilibrium of the following form:

$$\langle Al_2O_3 \rangle + 3[C]_{Fe} = 3(CO) + 2[Al]_{Fe}$$

which represents the reaction of an alumina crucible with the carbon of an iron melt in this crucible forming carbon monoxide. For a certain temperature an equilibrium constant of

$$K = \frac{p_{CO}^3 \cdot C_{Al}^2}{C_C^3}$$

is found, which relates the equilibrium partial pressure of the carbon monoxide above the melt with the aluminium and carbon concentrations in the melt*.

The number of moles of each substance which takes part in the reaction appears in this equation as a superscript either of the pressure or of the concentration. It is customary to place the reaction products in the numerator and the reactants in the denominator of the fraction. If this custom is followed, high K values indicate that the driving force of the reaction, which leads to the desired product, is relatively large. More favourable conditions for the reaction may be created by the removal of the reaction product from the system, for instance in the reaction quoted earlier by pumping off the carbon monoxide.

The pressure is usually expressed in atmospheres, the concentration as molar concentration, mole fraction or in atomic or weight percent. K is, however, in most cases only constant over a narrow range of concentrations within which no change of the interaction between dissolved substance and solvent occurs in relation to the concentration. This usually applies only to highly dilute solutions. In all other cases the activity of the dissolved substance should therefore be used in this equation instead of the concentration (*cf.* Subsection 2.5).

2.2 Fundamental thermodynamic relationships

On the basis of the fundamental relations which have their origin in the first and second laws of thermodynamics, two simple relations important to vacuum-metallurgical reactions can be derived. They are simple because the pressure dependence of the enthalpy is usually negligible in vacuum metallurgy as pressures in excess of more than one atmosphere are not, as a rule, employed. The enthalpy is therefore identical with the heat content. Another reason is that the gas phase usually obeys the ideal gas law

$$pV = RT. \tag{1}$$

Starting with the relation

$$G = H - T \cdot S, \tag{2}$$

which connects the three most important state variables

---

* It is common practice in English nomenclature to use ⟨angular brackets⟩ in reaction equilibria as an expression of solid phases, {braces} for liquid phases, (light parentheses) for gaseous phases and [square brackets] for solutions.

References pp. 92–93

G = free enthalpy (free energy)
H = enthalpy (in vacuum-metallurgical processes equal to the heat content)
S = entropy
of each individual starting and terminal product of a reaction or transformation, an equation is obtained for the reaction or transformation of the form:

$$\Delta G = \Delta H - T \cdot \Delta S. \tag{3}$$

This is therefore a relation between the changes in the values of the state variables occurring during the reaction or transformation, where

$$\Delta G = \sum G_n - \sum G_m$$
$$\Delta H = \sum H_n - \sum H_m$$
$$\Delta S = \sum S_n - \sum S_m,$$

and $m = 1, 2, 3\ldots$ is the number of reactants and $n = 1, 2, 3\ldots$ the number of reaction products. If, for instance, a compound is formed from its components at a reaction temperature T, then ΔG is the change of free enthalpy, ΔH the enthalpy of formation of the compound and ΔS is the change in entropy of the system caused by this reaction. For exothermic reactions, *i.e.* when heat is liberated, ΔH carries the negative sign.

Equation (3) is the first of the two relations which are so important in vacuum metallurgy. (The second equation will be derived in the paragraphs which follow.)

The quantity ΔG is often also called ΔF. The greater the decrease in free enthalpy the greater is the driving force of the reaction, or expressed in a different way, the more negative ΔG is, the more probable it is that a reaction will take a course in the desired direction.

The equilibrium state of a reaction is characterized by the fact that the reaction is reversible. This means that as soon as this state is attained, the change in free enthalpy becomes zero ($dG_{TP} = 0$) and that at equilibrium the sums of the free enthalpies of the reactants of the left-hand and of the right-hand sides of the reaction equation must be equal.

2.3 Equilibria between a pure condensed phase and an ideal gas

It should be pointed out that in computations involving thermodynamic state variables, the same standard state of the substances considered should always be used (related to 1 mole or g-atom). This is characterized, for instance, for pure phases by the substance being present in a state of high purity, at a pressure of 1 atm and in a state of aggregation which is typical of it in the range of temperatures involved.

As mentioned earlier, in most cases of vacuum metallurgy we are concerned with a heterogeneous equilibrium between a condensed phase and an ideal gas. If we consider such an equilibrium, for instance one between a liquid metal and its vapour, which may be expressed by an equation of the form

$$M_{cond} = M_{gas}, \qquad (4)$$

we obtain for this transformation

$$G_{cond} = G_{gas}, \qquad (5)$$

as the free enthalpy of the components of the reaction at the left and the right side must be equal at equilibrium.

The free enthalpy of the vapour phase is derived by calculation of its increase or decrease of the standard state G^0_{gas} by compression or expansion to the equilibrium pressure p_e for constant temperature T. This change in enthalpy for a certain temperature is obtained by integration of the equation

$$dG_T = V dp_T \qquad (6)$$

taking into account eqn. (1).

As a result of this integration the following equation is obtained

$$G_{gas} = G^0_{gas} + RT \cdot \ln(p_e/p^0_e). \qquad (7)$$

For expressing the pressure in atm and using as the standard state 1 atm

$$G_{gas} = G^0_{gas} + RT \cdot \ln p_e. \qquad (8)$$

A similar calculation for the condensed phase, the metal, is not necessary as the change in free enthalpy caused by the change in pressure is negligible. Therefore

$$G_{cond} = G^0_{cond} \qquad (9)$$

and using eqn. (5) we find

$$G_{gas} = G^0_{cond}.$$

We may now write eqn. (8) in the following form:

$$G^0_{cond} = G^0_{gas} + RT \cdot \ln p_e. \qquad (10)$$

The change in enthalpy is therefore

$$G^0_{gas} - G^0_{cond} = \Delta G^0 = -RT \cdot \ln p_e. \qquad (11)$$

If the temperature dependence of ΔG^0 is known, the pressure/temperature diagram of the equilibrium vapour pressure may be plotted.

If an analogous calculation is carried through for the dissociation pressure of metal–gas compounds, an equation of the form

$$G^0 = -RT \cdot \ln p^n_e \qquad (12)$$

is obtained, where n is the number of moles of gas which take part in the reaction with one mole of the starting material.

References pp. 92–93

Where a chemical reaction is involved, as, for instance, in the reduction of an oxide by carbon:

$$\langle MO \rangle + \langle C \rangle = \langle M \rangle + (CO) \tag{13}$$

we obtain according to eqn. (8) for the CO gas

$$G_{CO} = G_{CO}^0 + RT \cdot \ln p_{CO}. \tag{14}$$

Besides this

$$G_C = G_C^0, \; G_{MO} = G_{MO}^0, \; G_M = G_M^0.$$

As the sum of the free enthalpies of the reaction products must at equilibrium again be that of the reactants we find:

$$G_{MO}^0 + G_C^0 = G_M^0 + G_{CO}^0 + RT \cdot \ln p_{CO} \tag{15}$$

or

$$\Delta G^0 = -RT \cdot \ln p_{CO}. \tag{16}$$

The logarithmic quantity in eqn. (16) is also always the equilibrium constant K of the reaction, which results from the law of mass action at the chosen temperature. Therefore the following relation is also valid:

$$\Delta G^0 = -RT \cdot \ln K. \tag{17}$$

This is in fact the second important function for vacuum-metallurgical reactions mentioned at the beginning.

2.4 Calculation of the temperature function of the equilibrium constants

As according to eqn. (3)

$$\Delta G^0 = -RT \cdot \ln K = \Delta H^0 - T \cdot \Delta S^0, \tag{18}$$

the equilibrium constant may be calculated from the standard heat of formation and the entropy change of the reaction, if the temperature function of the free enthalpy is not known. If the temperature range covered is not very wide, ΔH^0 and ΔS^0 may be assumed to be constant, temperature-independent quantities. Therefore for $\ln K$ or $\log K$ an equation of the following form is obtained:

$$\log K = \frac{A}{T} + B. \tag{19}$$

This is the most simple form of the temperature function of K.

If ΔS^0 is also not known, the standard heat of formation and the free enthalpy may be used for a computation. ΔS^0 is then derived from the equation

$$\frac{\Delta H^0 - \Delta G^0}{T} = \Delta S^0. \tag{20}$$

This equation may also be used for the interpolation of the standard free enthalpy of formation for points between known quantities of other temperatures, since, as mentioned earlier, ΔH^0 and ΔS^0, in contrast with ΔG^0, are almost constant over a limited range of temperature.

If these thermodynamic data are not known and the calculation of $\Delta H°$ and ΔS^0 has to be based on the values of the heat capacities, the computation becomes much more difficult. The methods to be used in this case will be found in textbooks on thermodynamics[3].

Reactions in which more than three components are taking part and for which the thermodynamic data are not known may be divided into several partial reactions, for which the relevant quantities may be taken from published tables.

2.5 Thermodynamics of solutions

So far only equilibria between ideal gases and pure condensed phases have been considered here. However, in metallurgical processes metallic solutions of several components are usually involved. The thermodynamic treatment of these solutions is considerably more difficult. Starting points in this case are the so-called partial values of the state variables, which are derived for solutions from measurement of the partial molar free enthalpy of solution and the enthalpy and entropy of solution, if one mole of a substance is dissolved in a very large quantity of a solvent.

In Subsection 2.3, eqn. (7), it was shown how the free enthalpy of the gas or vapour phase increases or decreases on change of pressure from p^0 to p_e. It may also be shown[3] that if the partial pressure p_i above a dilute solution of a substance i in a solvent is used instead of p_e, and the vapour pressure p_i^0 of the pure substance i for the same temperature is substituted for p_e^0, the free enthalpy is identical with the partial free enthalpy of solution, which is liberated during the solution of one substance in another substance. For 1 mole of dissolved material the partial quantity is

$$\overline{\Delta G_i} = RT \cdot \ln (p_i/p_i^0) = RT \cdot \ln a_i^R. \tag{21}$$

The ratio $p_i : p_i^0$ is called the activity a_i^R of substance i in the solution*. If x_i is the molar fraction of a component i in the solution, the following relation exists between its activity a_i^R and the concentration x_i:

$$a_i^R = \gamma_i \cdot x_i, \tag{22}$$

where γ_i is the activity coefficient of element i.

---

* The activity a_i is related to the so-called chemical potential μ_i of the dissolved substance i by the equation

$$RT \cdot \ln a_i^R = \mu_i - \mu_i^0,$$

where μ_i^0 is the chemical potential in the standard state (pure substance).

References pp. 92–93

Equation (21) becomes therefore

$$\overline{\Delta G_i} = RT \cdot \ln \gamma_i + RT \cdot \ln x_i. \qquad (23)$$

For $\gamma_i = 1$, $a_i^R = x_i$, we have the case of an ideal solution without interaction of the components. Such a solution obeys Raoult's law.

If γ_i is not unity, but is constant over a certain range of concentration, as often happens for dilute solutions, a positive or negative deviation from Raoult's law is observed according to whether γ_i is greater or smaller than unity. If γ_i is greater, there will be a tendency for separation and if it is smaller, there will be a tendency for compound formation between solvent and dissolved substance.

If γ_i is greater than unity, a reduced solubility is observed for a given partial pressure of the component i above the melt, and if γ_i is smaller than unity a higher solubility is found than for an ideal solution.

The constant value of the activity coefficient in dilute solutions is called γ^0.

We therefore obtain from

$$p_i = p_i^0 a_i^R \qquad (24)$$

with

$$a_i^R = \gamma_i^0 x_i \qquad (25)$$

$$p_i = p_i^0 \gamma_i^0 x_i = k_H x_i \qquad (26)$$

which is Henry's law and where k_H is Henry's constant.

Figure 1 shows a diagram of the partial pressures of the components of a binary alloy system with positive deviation. If Raoult's law had been obeyed, the partial pressures would lie on the straight lines which characterize this law. Henry's

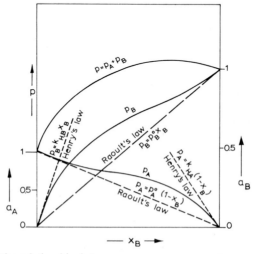

Fig. 1. Example of the relationship between vapour pressure and concentration in a binary system A–B with a positive deviation from Raoult's Law.

lines are the tangents to the activity curve for x_A or $x_B \to 0$, respectively. As may be seen from the diagram γ_A and γ_B are constant only over a very small range of concentration. The partial pressure curves also demonstrate—as typical for most partial pressure curves—that for high concentrations Raoult's law is obeyed. This is quite comprehensible, if it is kept in mind that a new atom of the other component on entering the system finds itself surrounded almost entirely by identical atoms, in contrast with the conditions at low concentrations. No change in the interaction is therefore to be expected.

As solubility is in many cases low, it is usually not opportune to refer to the pure substance as in a "reference and standard state". It is therefore more convenient to base the activity on Henry's line and, similarly to Raoult's activity a^R, to define a "Henry activity" a^H, where

$$a_i^H = f_i \cdot C_i, \tag{27}$$

where f_i becomes unity, when $C_i \to 0$. The concentration C_i is conveniently expressed in weight percent. Henry's activity a_i^H is therefore in ideal dilute solutions equal to the concentration of the dissolved substance C_i in weight percent.

Unfortunately, the activity coefficient is constant over the whole range of interest in only a very few cases, so that more or less complex functions must be employed for the numerical calculations and so-called interaction coefficients must be introduced.

The interaction coefficients which were introduced by Wagner[8] are also of importance for the consideration of the effect of addition elements on the activity of a dissolved substance.

The term $RT \cdot \ln \gamma_i$ in eqn. (23) is the contribution which represents the interaction portion of the free enthalpy or the chemical potential of the solution of the component i. If the function $\ln \gamma_i$ according to Wagner is expanded into a Taylor series, where the molar concentrations x_i of the components 2, 3... i are the independent variables, the following expression is obtained as a first approximation with the zero- and first-order terms for low values of $\ldots x_2, x_3, x_4 \ldots$:

$$\ln \gamma_2(x_2, x_3, x_4 \ldots) = \ln \gamma_2^0 + x_2 \varepsilon_2^{(2)} + x_2 \varepsilon_2^{(3)} + \ldots \tag{28}$$

or if changed to Henry's activity coefficients

$$\log f_2(C_2, C_3, C_4 \ldots) = C_2 e_2^{(2)} + C_3 e_2^{(3)} + C_4 e_2^{(4)} + \ldots, \tag{29}$$

where

$$\varepsilon_2^{(2)}, \varepsilon_2^{(3)}, \varepsilon_2^{(4)} \ldots \quad \text{and} \quad e_2^{(2)}, e_2^{(3)}, e_2^{(4)} \ldots$$

are Wagner's interaction coefficients, which are defined as:

$$\varepsilon_2^{(2)} = \partial \ln \gamma_2 / \partial x_2$$
$$\varepsilon_2^{(3)} = \partial \ln \gamma_2 / \partial x_3 \quad \text{and so on}$$

References pp. 92–93

$$e_2^{(2)} = \partial \log f_2/\partial C_2$$
$$e_2^{(3)} = \partial \log f_2/\partial C_3 \quad \text{and so on.}$$

Such interaction coefficients have been determined for many alloying elements, particularly for iron, as the solvent (i = 1). With their aid the effect of other additions (i = 3, 4...) on the activity coefficient of a dissolved substance (i = 2) can be calculated in a simple way using eqn. (29), at least for dilute solutions.

For wider solubility ranges the introduction of terms of a higher order of the Taylor series may become necessary. Relevant expressions have been derived, for instance by Lupis and Elliott (see ref. 9).

Table 1 shows as an example a compilation of interaction coefficients between nitrogen and other alloying elements in liquid iron according to Simkovich[10]*.

If the interaction coefficient $e_N^{(i)}$ of an alloying element i is known, the nitrogen solubility for the concentration C_i (wt.%) of this element in the ternary system Fe–N–i may be calculated for a given partial pressure p_{N_2} with $f_N^N = 1$ from

$$C_N = \frac{(p_{N_2})^{\frac{1}{2}}}{K \cdot f_N^i}$$

(*cf.* also Subsection 2.6), where

TABLE 1

SUMMARY OF ALLOYING ELEMENT–NITROGEN INTERACTION COEFFICIENTS IN LIQUID IRON AT 1600°C[10]

Alloying element	$e_N^{(i)} = \dfrac{\partial \log f_N}{\partial C_i}$
C	+0.13
Si	+0.047
Co	+0.010
Ni	+0.010
Sn	+0.006
Cu	+0.006
W	−0.002
Mo	−0.011
Mn	−0.023
Al	−0.028
Ta	−0.032
Cr	−0.045
Nb	−0.061
V	−0.093
Ti	−0.53
Zr	−0.63

---

* An extensive list of literature references will be found in H. SCHENCK AND E. STEINMETZ, *Wirkungsparameter von Begleitelementen flüssiger Eisenlösungen und ihre gegenseitigen Beziehungen*, (in German) [Interaction coefficients of accompanying elements of liquid iron solutions and their mutual relations], Verlag-Stahleisen, 1968, second revised edition.

$$\log f_N^i = C_i \cdot e_N^{(i)}.$$

As the interaction coefficients $e_N^{(i)}$ of Table 1 are constant for most alloying elements over a concentration range of several weight percent[11], we may therefore write

$$\frac{\partial \log f_N}{\partial C_i} = \frac{\log f_N}{C_i}.$$

2.6 Solubility of gases

The term "gas" is used here for matter which is gaseous in the standard state at room temperature. The most important substances in vacuum metallurgy are oxygen, hydrogen, nitrogen and carbon monoxide. As these gases are dissolved in metals almost solely in their atomic state, the following reaction equation applies for instance for nitrogen

$$\tfrac{1}{2}(N_2) = [N]_M \tag{30}$$

with the equilibrium constant

$$K_{N_2} = \frac{a_N}{(p_{N_2})^{\frac{1}{2}}} = \frac{f_N \cdot C_N}{(p_{N_2})^{\frac{1}{2}}} \tag{31}$$

which is, of course, different for each metal. It may also be affected by other alloying elements[12] (see also preceding subsection and Fig. 2).

For nitrogen, for instance, in the concentration range in which the activity

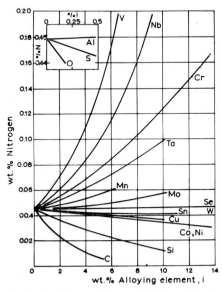

Fig. 2. Solubility of nitrogen in liquid iron alloys at 1600°C and 1 atm N_2-pressure (Pehlke and Elliott; Schenck, Frohberg and Graf), from ref. 9.

References pp. 92–93

coefficient is almost constant,

$$C_N, \text{wt. \%} \sim (p_{N_2})^{\frac{1}{2}} \tag{32}$$

This is the well-known relation between solubility and the partial pressure p_{N_2} above the melt, which was established by Sieverts for diatomic gases as far back as 1911.

For oxygen (as well as for sulphur) constancy of the activity coefficient can, as a rule, be expected only in low concentration ranges because of the more vigorous chemical interaction compared with that of hydrogen or nitrogen. In the

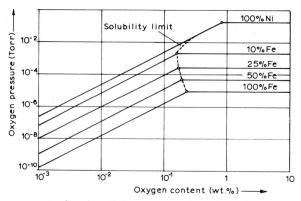

Fig. 3. Oxygen pressure as a function of the oxygen content of pure iron, pure nickel and several iron–nickel alloys at 1600°C[13].

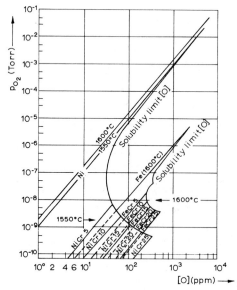

Fig. 4. Oxygen pressure as a function of the oxygen content of several iron–chromium and nickel–chromium alloys at 1600°C and 1550°C respectively, from ref. 14.

range of the limited solubility the deviation from Henry's line here too is usually only small, as may be seen from Fig. 3. Here the solubility of oxygen in iron, nickel and alloys of these metals[13] is plotted and the fact is demonstrated, that the maximum attainable equilibrium pressure for a given temperature corresponds to the decomposition pressure of the metal–gas compound which is in equilibrium with the melt at saturation. Figure 4 shows the solubility of oxygen in alloys of iron and nickel with chromium[14].

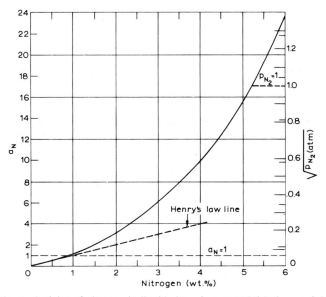

Fig. 5. Activity of nitrogen in liquid chromium at 1745°C, from ref. 9.

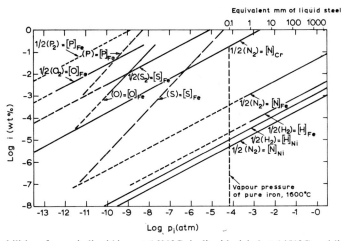

Fig. 6. Solubilities of gases in liquid iron at 1600°C, in liquid nickel at 1450°C and liquid chromium at 1745°C

References pp. 92–93

The variability of the activity coefficient can become quite noticeable in systems of markedly higher solubility ranges. For instance in the chromium–nitrogen system an appreciable deviation occurs from Henry's line at about 1 wt.% N_2, as may be seen from Fig. 5. The solubility of nitrogen is particularly high in this system and amounts at 1 atm to 5.6 wt.% or 18.7 at.%.

Figure 6 shows a survey of the pressure dependence of the solubility of various gases (also those of sulphur and phosphorus) in liquid iron, nickel and chromium according to Elliott[9] and to computations by the present author. It may be seen that only for hydrogen and nitrogen are the equilibrium pressures even at low concentrations still so high that a vacuum removal of the elementary gases is feasible solely by lowering the pressure. The reason for this phenomenon is not only that a technologically important material transport requires high gas densities, but also because only at a sufficiently high molar concentration of the gas in the vapour of the solvent is degassing possible without too high a loss in solvent (cf. Subsection 3.3). It should also be borne in mind that rapid degassing can only be expected if it is possible for bubbles of the gas to form in the melt. In this case the equilibrium pressure of the gas must be higher than the pressure which results from the surface tension and the depth of the liquid column at their place of origin (cf. Subsection 3.1.2e). As an indication, the vapour pressures of the solvent iron and the equivalent pressures of various depths of the bath are also given in Fig. 6 (upper right-hand corner). In this case the reaction kinetics must be taken into account, which will be dealt with in detail in Section 3 of this chapter.

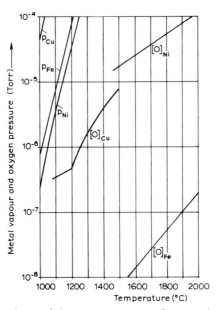

Fig. 7. Temperature dependence of the oxygen pressure of copper, iron and nickel containing 100 ppm oxygen and of the vapour pressure of the pure metals.

Figure 7 shows for iron, nickel and copper[13] the temperature dependence of the equilibrium pressure of oxygen at a given oxygen level of 100 ppm. The slope of these curves is markedly flatter than that of the vapour pressure curves of the solvents. It must therefore be concluded that no improved degassing conditions can be obtained for oxygen by raising the temperature. In fact, the contrary is true. The same applies for the elimination of nitrogen and hydrogen from metals of the iron group, because in some of these cases the solubility increases with increasing temperature and therefore the equilibrium pressure falls.

The temperature dependence of the solubility of hydrogen and nitrogen in iron is plotted in Fig. 8. As may be seen, more favourable degassing equilibria exist in the solid than in the liquid state. However, in technology, degassing will always be considered only in the liquid state within the given limits, as only in this case is a rapid material exchange possible.

The temperature dependence of the solubility of gases varies greatly from metal to metal, particularly in the case of hydrogen. This shows an increase with increasing temperature for Ni, Fe, Co, Cr, Cu, Al, Ag, Mo, W and Pt, and a decrease for Ti, V, Zr, Nb, La, Ce, Ta, Th, Pd and Hf. Detailed data on the solubility of gases in metals may be found in reference 15.

2.7 Deoxidation by a reduction process or by evaporation of suboxides

As the oxygen equilibrium pressures above common metals and alloys containing oxygen are so low that, with high-purity starting materials, a pick-up of oxygen from the residual gas atmosphere or the oxide-ceramic crucible materials rather than a removal should generally be expected in vacuum metallurgy, the only possibility for the deoxidation via the gas phase is by a reduction with carbon or hydrogen, or by volatilization via a suboxide.

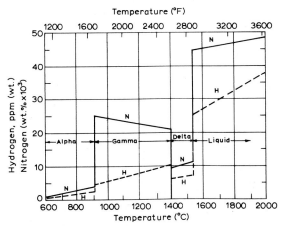

Fig. 8. Effect of temperature on the solubility of hydrogen and nitrogen at 1 atm pressure in iron, after J. Chipman.

References pp. 92–93

2.7.1 Reduction by carbon

The reaction

$$[O] + [C] = (CO) \tag{33}$$

having an equilibrium constant

$$K = \frac{p_{CO}}{a_C \cdot a_O} = \frac{p_{CO}}{f_O C_O \cdot f_C C_C} \tag{34}$$

has been extensively studied for liquid iron. As the activity coefficients at low carbon levels are almost unity, the equation for the equilibrium constant may be expressed as

$$K = \frac{p_{CO}}{C_O \cdot C_C}, \tag{35}$$

which is also known as the "Vacher–Hamilton equilibrium". At higher carbon contents the influence of the activity coefficient can no longer be disregarded and a deviation from the 45° line is observed for carbon contents of more than 0.1%, if $\log C_O$ is plotted against $\log C_C$ (see Fig. 9).

These theoretical values are only seldom obtained in industrial smelting practice for the reason that equilibrium cannot be attained in the time available for degassing and also because side reactions with the refractories lead to a supply of oxygen. Only a dynamic equilibrium can therefore exist between these competing reactions.

The effect of other addition elements on the elimination of oxygen is shown in Fig. 10. The quantity plotted on the ordinate (the inverse of the products of

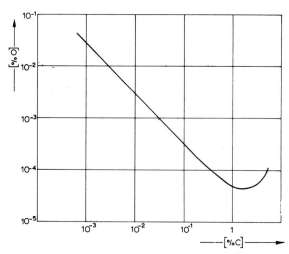

Fig. 9. Oxygen–carbon diagram of iron in equilibrium with 0.01 atm CO gas at 1600°C.

the activity coefficients which result from the interaction with the addition element i) indicates by what factor the oxygen content changes under the effect of these addition elements for a given carbon level.

It is very interesting to compare the reducing effect of carbon with that of other reducing agents. In Fig. 11 the oxygen contents of iron melts are depicted, which are in equilibrium with the oxides of various elements i in relation to the concentration of the corresponding elements, % i. The curve designated Al repre-

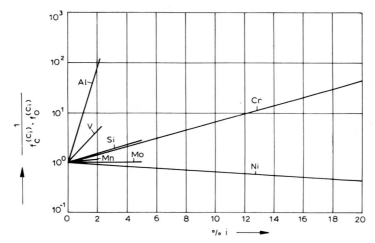

Fig. 10. Factors for consideration of the effect of an additional element from calculations by A. Gerhardt and G. Kaipl.

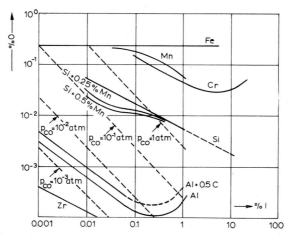

Fig. 11. Comparison of deoxidation by carbon and other deoxidizing elements. Oxygen content of iron melts which are in equilibrium with the oxides of various elements i in relation to the concentration of the elements concerned and influence of the CO partial pressure on the oxygen content of Fe–C melts.

References pp. 92–93

sents the reaction equilibrium

$$2[Al] + 3[O] = \langle Al_2O_3 \rangle,$$

where [Al] and [O] are the concentrations of these elements in the melt in wt. %. At the same time the [O] · [C] equilibrium of the reducing agent i = C has been plotted in relation to the CO partial pressure above the melt. These curves show that iron oxide is formed only at about 0.23% oxygen; Al_2O_3 on the other hand is already being produced at about 1 ppm oxygen, if $[Al]_{Fe} > 0.1\%$. For an equilibrium pressure p_{CO} below 10^{-3} atm, i.e. at a moderate vacuum of about 1 Torr, an even lower oxygen content could theoretically be obtained with as little as 0.05% C. This demonstrates the intense deoxidation effect of carbon at a reduced pressure.

2.7.2 Reduction by hydrogen

The reduction by hydrogen according to the reaction equation

$$[O]_M + (H_2) = (H_2O) \tag{36}$$

with

$$K = \frac{p_{H_2O}}{a_O \cdot p_{H_2}} \tag{37}$$

is then of particular interest, when the reduction by carbon cannot be carried out, for instance because of the formation of stable carbides. This method has also the advantage that the hydrogen dissolved in excess can be pumped off after completion of the reduction process. Equation (37) shows immediately that a reduction of the pressure above the melt does not offer any advantage. Not the partial pressures of hydrogen and water vapour *per se* but the partial pressure ratio $p_{H_2}:p_{H_2O}$ is decisive for the oxygen equilibrium obtainable. If a certain partial pressure or dew point of the water vapour is attainable in the drying of the hydrogen by one of the well known techniques, the reduction equilibrium becomes the more favourable the higher the hydrogen pressure. It is therefore more advantageous in this deoxidation method to operate at an increased pressure.

2.7.3 Deoxidation by evaporation of suboxides

The transport processes in the vapour or gas phase have become of steadily increasing importance in extractive metallurgy during the last few years. Side by side with methods for the distillative separation of metals, processes have been developed in which easily volatile metal compounds are formed. In this way possibilities for the extraction of difficult volatile metals have been accomplished. A "vaporization chemistry"[16] has been developed in which new, so far unknown, vapour species of oxides and other chemical compounds were found, *e.g.* with the aid of the mass spectrometer.

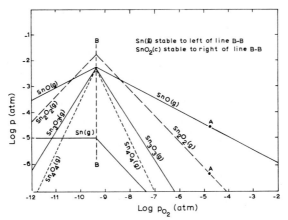

Fig. 12. Vapour pressures in the system Sn–O at 1500 K, from ref. 16.

In the tin–oxygen system, for instance, quite a number of monomeric and polymeric oxygen compounds exist in the vapour phase, which are markedly more volatile than tin itself. Their ranges of stability and their partial pressures are a function of the O_2 partial pressure. The partial pressures of the various oxides are shown in Fig. 12. It may be seen that a maximum for the volatilization of tin in the form of these oxygen compounds is obtained at a critical O_2 pressure around 10^{-9} atm.

One or more suboxides exist for many technologically important metals, which have a substantially higher vapour pressure than that of the basic metal. Some of these, for instance SiO, SnO, GeO, AlO, Al_2O or ZrO, and probably their polymers, are stable only in the vapour phase.

The thermodynamic data of many suboxides have been determined during the last few years (see mainly references 4, 17 and 18). The data usually refer to the reaction

$$\tfrac{1}{2}(O_2) + \langle M \rangle = (MO) \tag{38}$$

with

$$K = \frac{p_{MO}}{(p_{O_2})^{\frac{1}{2}}}. \tag{39}$$

A plot of the relation (39) between the oxygen and the suboxide partial pressures is shown for various metals in Fig. 13[9]. The suboxide partial pressure attains its peak when it is in equilibrium with the solid or liquid suboxide. Using the equilibrium data we are also able to calculate the equilibrium partial pressure of the suboxide above an oxygen-bearing metal in relation to the oxygen concentration, *i.e.* the reaction

$$[O]_M + \langle M \rangle = (MO) \tag{40}$$

with

References pp. 92–93

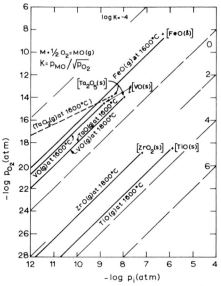

Fig. 13. Oxygen potential of metal in equilibrium with its volatile oxide(s). (+) Pressure below which a solid (or liquid) oxide of the composition indicated is no longer stable, from ref. 9.

$$K = \frac{p_{MO}}{f_O \cdot C_O}. \tag{41}$$

These data are obtained if the reaction expressed in eqn. (38) is combined with the reaction

$$[O]_M = \tfrac{1}{2}(O_2), \tag{42}$$

$$K = \frac{(p_{O_2})^{\frac{1}{2}}}{f_O \cdot C_O}. \tag{43}$$

Equation (38)+eqn. (42) gives the required eqn. (40).

Whether or not a deoxidation via the evaporation of suboxides will be successful does not depend only on their partial pressures, just as in the evaporation of other impurities. The vital point is whether or not the concentration of the component to be removed is higher in the vapour phase than in the starting material. Only in this case can a reduction of its concentration in the melt be expected.

The quantity ratio in the vapour corresponds—if kinetic factors be ignored—approximately to the ratio of the partial pressure of the suboxide to the vapour pressure of the solvent, which was neglected in eqns. (39) and (41). Therefore

$$x_{MO} = \frac{p_{MO}}{p_{MO} + p_M} = x_D$$

must be compared for a first approximation with the molar fraction x_C for MO in the condensed phase. (For a more accurate computation please see Subsection 3.3 of this chapter.)

The ratios between the vapour pressures of various suboxides and their basic metals, as shown in Table 7, Chapter V, are therefore not a sufficient criterion as to whether deoxidation via the suboxide is, in principle, possible or not. The only statement that can be made is that a deoxidation is certainly possible if the ratio exceeds unity, even in the case in which kinetic factors, which usually have a detrimental effect, play a part.

This fact and the detailed numerical calculation may be demonstrated by an example[9]:

The partial pressure ratio of FeO to pure iron at 1600°C is, as may be seen from Figs. 6 and 13, about 10^{-2} and according to the criteria based on Table 7, Chapter V, deoxidation should not be possible.

From the difference between

$$\tfrac{1}{2}(O_2)+\{Fe\} = (FeO), \text{ with } \log K_{44} = -10800/T+3.91 \tag{44}$$

and

$$\tfrac{1}{2}(O_2) = [O]_{Fe}, \text{ with } \log K_{45} = 6120/T+0.15 \tag{45}$$

we obtain

$$[O]_{Fe}+\{Fe\} = (FeO), \text{ with } \log K_{46} = -16920/T+3.76 \tag{46}$$

and at 1600°C

$$K_{46} = \frac{p_{FeO}}{f_O \cdot C_O} = \frac{p_{FeO}}{[\%O]} = 5.2 \times 10^{-6}.$$

For low oxygen levels, if $x_C \ll 1$,

$$p_{FeO} = 1.49 \times 10^{-4} x_C \text{ (atm)}$$

and the molar fraction in the vapour phase

$$x_D = 2.1 x_C, \text{ or } \frac{x_D}{x_C} \sim 2.$$

Therefore, theoretically, degassing should be possible for any concentration of oxygen, even the lowest, despite the unfavourable partial pressure ratio of FeO to Fe. In reality this process cannot be carried out owing to the too low partial pressure of the FeO and the high evaporation losses of the solvent. For an $x_D/x_O = 2$ even under ideal conditions, 50% of the melt would evaporate during the lowering of the oxygen content to one-half of the starting level.

In cases where the vapour pressure of the suboxide of the basis metal is too low, the possibility should be considered as to whether or not deoxidation is feasible by addition of other alloying elements with substantially higher suboxide vapour pressures. The vapour pressure of this suboxide must, however, in such a

References pp. 92–93

case be higher by several orders of magnitude than that of the basis metal. Furthermore, the alloying addition must not exercise any undesired side effects. It is also an advantage if the alloying addition has itself so high a vapour pressure that it can be distilled off later.

The best known and most important application of a deoxidation by evaporation of suboxides is that for tantalum and niobium, both in the solid and the liquid state. This application will be considered in detail in Chapter IV, Section 2 and in Chapter V, Section 4.

2.8 Equilibria of metal vapours

The thermodynamic equilibria so far discussed here represent, in fact, borderline cases, which can be realized in metallurgical practice in rare cases only. However, the knowledge of these equilibria is important as they form the starting point for the analysis and prediction of the actual processes which arise from the creation of distortions of such equilibria. The same applies to the equilibria of metal vapours which will now be discussed.

The relation (11) derived in Subsection 2.3

$$\Delta G^0 = -RT \cdot \ln p_e \tag{11}$$

between the heat of evaporation and the vapour pressure of a metal may also be written on the basis of eqn. (3) as

$$\ln p_e = \frac{\Delta H^0}{RT} + \frac{\Delta S^0}{R}, \tag{47}$$

which represents an expression of the form

$$\log p_e = -\frac{A}{T} + B, \tag{48}$$

if ΔH^0 and ΔS^0 are regarded as nearly constant. This is in fact the Clausius–Clapeyron equation representing approximately the temperature dependence of the vapour pressure. If the substance to be evaporated is dissolved in a metal M, we may write

$$n[\text{i}]_M = (\text{i}_n) \tag{49}$$

with

$$K_i = \frac{p_{i_n}}{(a_i)^n}, \tag{50}$$

and

$$\Delta G^0 = -RT \cdot \ln \frac{p_{i_n}}{(a_i)^n} = -4574 \log \frac{p_{i_n}}{(a_i)^n}. \tag{51}$$

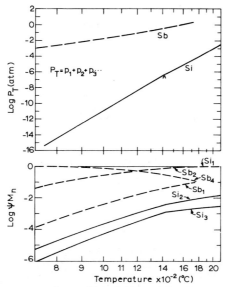

Fig. 14. Total vapour pressures of pure Si and Sb, and fractions ψM_n of species in vapour, from ref. 9.

In most cases n equals unity* and the proportions of the polymers ($n = 2, 3, \ldots$) in the vapour phase are negligibly small. On the other hand, there are also elements, such as silicon and antimony, in which the proportions of polymers are significant. These elements have polymers with up to 3 or 4 atoms, respectively, when equilibrium exists between vapour and condensed phase.

The total pressure of all species and their proportions of this total pressure in the vapour phase are plotted in Fig. 14 in relation to the temperature. As may be seen, silicon behaves just like the metals, whereas for antimony the dimer and at very low temperatures even a higher polymer outweigh the monomers. Despite this exception the monomers in dilute solutions will, here too, prevail in the vapour phase, as the partial pressure according to eqn. (50) decreases proportionally with $(a_i)^n$. This decrease is the faster, the larger n becomes.

Evaporation, particularly of the more volatile metals, plays an important role in both extractive metallurgy and in refining by vacuum-metallurgy. Just as in the evaporation of the suboxides (and other impurities) here too the ratio is decisive, to a first approximation, of the molar fractions of the evaporating metals in the vapour phase and in the solution, whether or not the possibility of a separation by this method exists.

---

* i.e. $K_i = \dfrac{p_i}{a_i}$, where $a_i = \dfrac{p_i}{p_i^0}$. The vapour pressure of the dissolved substance is therefore the product of the vapour pressure of the pure metal, p_i^0, and its activity a_i (Raoult's activity) in the solution.

References pp. 92–93

The curves of the vapour pressures of the pure metals in the solid and liquid ranges are shown in Figs. 4a and 4b of Chapter III.

2.9 Reactions with refractory oxides and slags

Reactions between the melt and the crucible or lining, in induction melting as well as in the degassing of steel, affect the quality of the product and the service life of the refractories. As has already been mentioned in Subsection 2.7.1, the constant supply of oxygen from the refractories is one of the causes of the fact that the theoretical equilibria with oxygen can only rarely be obtained in a vacuum during reduction by carbon. Equally important are the reactions with dispersed or floating oxides or slags. The possibility of a decomposition of these impurities in a vacuum is in many cases a prerequisite for obtaining the required degree of purity of the final product. The thermodynamics of these reactions are therefore of great technological importance.

The following compilation provides a survey of some important reactions which may occur between currently used refractory oxides and metal melts[9]:

$$\langle Al_2O_3 \rangle = 2[Al] + 3[O] \tag{52}$$
$$\langle Al_2O_3 \rangle = 2(AlO) + [O] \tag{53}$$
$$\langle Al_2O_3 \rangle = (Al_2O) + 2[O] \tag{54}$$
$$\langle MgO \rangle = (Mg) + [O] \tag{55}$$
$$\langle MgO \rangle = [Mg] + [O] \tag{56}$$
$$\langle CaO \rangle = (Ca) + [O] \tag{57}$$
$$\langle ZrO_2 \rangle = [Zr] + 2[O] \tag{58}$$
$$\langle ZrO_2 \rangle = (ZrO) + [O] \tag{59}$$
$$\langle SiO_2 \rangle = [Si] + 2[O] \tag{60}$$
$$\langle SiO_2 \rangle = (SiO) + [O] \tag{61}$$

Further reacting substances are spinels, silicates and mixed oxides, which are also used as crucible materials or which are formed at the surface of the crucible during the melting process as a result of reaction with the melt, for instance:

$$\langle Al_2O_3 \rangle + \{Fe\} + [O] = \langle FeO \cdot Al_2O_3 \rangle.$$

There are also slags of various origins. These may be dispersed in the melt or they may float on it.

In the reactions (55)–(57) magnesium and calcium are formed. As they are not soluble in liquid iron, they are continuously distilled off in a vacuum in this case either from the surface of the bath or through the crucible wall if this is porous[19]. Displacement of the reaction equilibria (55) and (57) to the right-hand side leads to continuous decomposition of the oxides and theoretically to an increase in the oxygen content of the melt up to the point at which a slag is formed with FeO. This is demonstrated in Fig. 15[9]. That this is never observed in practice is because of the reaction kinetics, *i.e.* owing to the very slow decomposition rate of the MgO and CaO (*cf.* Subsection 3.4). (However, as magnesium is soluble in nickel, a different reaction mechanism, one analogous to eqn. (56), should be expected.)

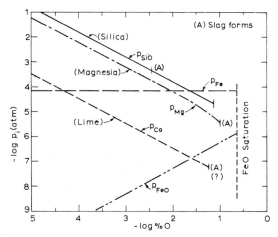

Fig. 15. Partial pressures of primary species in equilibrium with liquid iron and refractory crucibles at 1600°C, from ref. 9.

The equilibrium constant of the reaction (55) for, *e.g.*, 1600°C[3] is given by

$$K = p_{Mg} \cdot [\%O] = 2.9 \times 10^{-7} \text{ atm. wt. } \%, \tag{62}$$

which means that the oxygen concentration is the higher the lower the vapour pressure of the magnesium above the melt, in other words, the faster the magnesium formed is removed. A more rapid decomposition is therefore to be expected in a vacuum than at ambient pressure.

A crucible material, the reaction products of which are soluble in the melt, as for instance alumina, will behave in an entirely different manner.

The relation between the aluminium and oxygen contents of an iron melt, *i.e.* the reaction according to eqn. (52), has been shown earlier in Fig. 11*. The vapour pressure of the aluminium above the melt corresponding to the Al–O equilibrium, as shown in Fig. 11, is plotted in Fig. 16. The vapour pressures of AlO and Al_2O according to reaction eqns. (53) and (54) are also plotted in Fig. 16. (At a high oxygen content it might be possible for hercynite ($FeO \cdot Al_2O_3$) to form.)

As no substantial degree of evaporation of the aluminium or of its suboxides

---

* Using a high-purity iron melt and taking into account that the molar ratio of the aluminium to be dissolved to oxygen, O:Al, should be 1.5, the equilibrium is defined by a point at 27 ppm Al and 24 ppm O on this curve.

References pp. 92–93

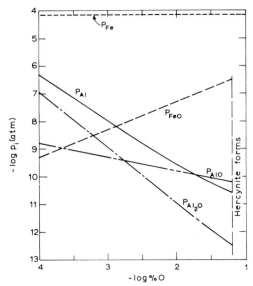

Fig. 16. Partial pressures of primary species over liquid iron in an alumina crucible at 1600°C, from ref. 9.

can be expected, the disintegration of the crucible material will progress at a much reduced rate as soon as a certain amount of aluminium has gone into solution and a corresponding oxygen content is established in the melt.

Kubaschewski et al.[3] studied, *inter alia*, the thermodynamic equilibrium which exists between a 10% Cr–Ni alloy and Al_2O_3. Assuming that aluminium and oxygen dissolve in the melt, similarly to their behaviour in iron, according to the reaction equation

$$\langle Al_2O_3 \rangle = 2[Al]_{CrNi} + 3[O]_{CrNi}, \tag{63}$$

equilibrium quantities for oxygen and aluminium are obtained which are only a little below those for iron. If, however, the possible equilibrium between dissolved chromium and Al_2O_3 is also to be taken into consideration, leading to the formation of Cr_2O_3 which is materially more stable than either Fe_3O_4 or NiO and which is precipitated at very low oxygen concentrations,

$$\langle Al_2O_3 \rangle + 2[Cr]_{CrNi} = 2[Al]_{CrNi} + \langle Cr_2O_3 \rangle, \tag{64}$$

it will be found that equilibrium exists at a substantially higher aluminium content and an oxygen level which is only about one-tenth of the value according to eqn. (63).

This example shows that all conceivable reaction possibilities, in particular those with components of high affinity for oxygen, must be examined before a final statement on the equilibrium that may eventually be attained can be made.

This situation is fundamentally changed at the moment at which the melt contains carbon which, according to eqn. (33),

$$[C] + [O] = (CO), \tag{33}$$

leads to a reaction which results in a steady removal of oxygen. The strong reducing effect of carbon at low pressures and the low oxygen concentration which is therefore attained, accelerate the disintegration of refractory oxides and of slags.

According to the $[\%O] \times [\%Al]$ product of the equilibrium constant of reaction (52), the lowering of the oxygen concentration leads to a corresponding increase of the aluminium content of the melt. Beside this the following reaction might also be expected

$$\langle Al_2O_3 \rangle + 3\langle C \rangle = 3(CO) + 2[Al] \tag{65}$$

with

$$K = \frac{p_{CO}^3 \cdot a_{Al}^2}{a_C^3}. \tag{66}$$

This reaction can also lead to a rapid enrichment of aluminium in the melt and, if the carbon monoxide is pumped off continuously, may finally result in a complete replacement of the carbon by aluminium.

Regardless of whether reaction (52)+(33) or (65) governs the decomposition of the Al_2O_3, the effect is the same.

For material containing silica, the reactions

$$\langle SiO_2 \rangle + 2[C] = [Si] + 2(CO) \tag{67}$$

and

$$\langle SiO_2 \rangle + [C] = (SiO) + (CO) \tag{68}$$

may take place along with reactions (60) and (61). From the two reactions (67) and (68) it is more likely that reaction (67) will dominate, since it was found[13] that the rise in silicon content of carbon-bearing steel melts is only about one-half that of the carbon burn-off.

Even in cases in which the liberated metal does not go into solution, an acceleration of the reaction is to be expected in the presence of carbon as the oxygen concentration gradient at the phase border increases.

Krone et al.[14] calculated for various oxides the carbon monoxide partial pressure at which the reaction between a carbon-containing iron melt and the relevant oxide starts. It was assumed in this calculation that the carbon level is 0.1 wt. % and that for volatile metals, such as Mg and Ca, $p_M = p_{CO}$ and that for non-volatile metals, such as Al and Zr, the concentration in the melt is also 0.1 wt. %. The results obtained in this calculation are compiled in Table 2. As may be seen, CaO has the highest stability.

The determination of the carbon loss from the melt in relation to time as a measure of the CO evolution is therefore the simplest method of testing the stability

References pp. 92–93

TABLE 2

REDUCTION PRESSURES OF VARIOUS CRUCIBLE MATERIALS WITH IRON MELTS AT 1600°C, ACCORDING TO KRONE et al.[14]

Crucible materials	p_{CO} (Torr)
CaO	0.3
ZrO_2	1
MgO	4
Al_2O_3	4
SiO_2	610

of the various refractory oxides to carbon-containing steel melts[13,19,20].

All investigations carried out on reactions between a melt and oxide-refractory materials in a vacuum showed that it is only very rarely possible to correlate the results for the stable oxides Al_2O_3, MgO, CaO and ZrO_2 in the presence of carbon with thermodynamic equilibrium calculations. (This is, however, partly due to the fact that the accuracy of the determination of the oxygen content of melts was inadequate[21].) Here too, the reaction kinetics are in most cases more decisive for the result than the position of the thermodynamic equilibrium.

The situation is different for oxides and oxide mixtures of lower chemical stability, for instance for refractory materials containing high proportions of SiO_2 or Fe_2O_3. As an example, measured data of the liberation of gas from various refractory materials used for linings in ladle degassing are shown in Table 3[22].

This gas consists of a portion liberated from the porous materials, where it was originally absorbed or adsorbed, and of a portion which is given off continuously as a result of the reaction with the carbon-containing steel melt. As may be seen, gas liberation from the "brick–metal reaction" for ($Al_2O_3 + SiO_2$)-bearing refractories increases with increasing SiO_2 content, as would be expected. It may further be seen that higher reaction rates are also obtained for refractories on a MgO basis, which are most stable in their pure form, by additions of less stable

TABLE 3

EXPERIMENTAL DATA ON GAS EVOLUTION FROM VARIOUS REFRACTORIES[22]

Approx. chemical composition of refractory		Porosity (%)	Amount of gas from brick at 1400°C (l/kg)	Gas evolution from brick–metal reaction (l/m² min)
$Al_2O_3 + SiO_2$	90–10	22	50	19
$Al_2O_3 + SiO_2$	72–28	23	240	20
$Al_2O_3 + SiO_2$	62–38	22	125	24
$Al_2O_3 + SiO_2$	40–60	9–12	180	67
$Al_2O_3 + SiO_2$	38–62	14–18	380	55
$Al_2O_3 + SiO_2$	30–70	16–20	490	58
$MgO + SiO_2$	98–2	15	240	14
$MgO + Al_2O_3$ $+ Fe_2O_3 + SiO_2$	78–12 –5–5	18	570	27

oxides. (It should be pointed out that these results still do not tell us anything of the usefulness of the refractory materials in technology. For such a decision, other properties, such as thermal shock resistance, abrasion resistance, etc., also play an important role (see also Chapter IV)).

2.10 Refining in a vacuum

The counterpart of a deoxidation with carbon, *viz.* the elimination of excess carbon by reaction with gaseous oxygen or by addition of oxides in a vacuum, has attained increasing importance in the production of high-alloy steels during the last few years[23]. This vacuum process, which is called "selective refining", is mainly employed for the decarburization of stainless and other chromium-bearing steels. The vacuum treatment, for instance in the form of ladle degassing, follows a steelmaking process which is conducted under ambient pressure. If low-carbon grades are to be produced by this two-stage process, it is no longer necessary to attain low carbon levels in the first stage under ambient pressure, as the carbon may be removed selectively in the second stage, the vacuum treatment, *i.e.* at such low oxygen contents that a slagging of chromium and other elements with a high affinity for oxygen is avoided. This type of working leads to a substantial simplification and a marked saving in time in the first stage of the process and makes possible a decarburization down to about 0.01 % C with hardly any loss in chromium. In this process, therefore, the affinity of carbon for oxygen is made use of. This affinity is very high at low pressures and low oxygen levels compared with other deoxidizing agents.

A schematic diagram of the phenomena occurring at the various stages of

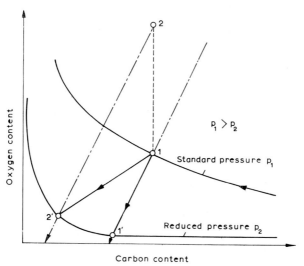

Fig. 17. Schematic representation of the change of carbon and oxygen contents of a steel melt during decarburization[23].

this technique when applied to a carbon-containing ferrous alloy is presented in Fig. 17. The curve "standard pressure p_1", which represents the function $C_O \cdot C_C = K \cdot p_{CO}$ for $p_{CO} = 1$ atm, shows the change in oxygen and carbon contents during the refining in an oxygen top-blown converter at about 1800°C. At point (1) the heat is transferred into the vacuum vessel, where it should be at a somewhat lower temperature. The equilibrium curve "reduced pressure p_2" corresponds, for instance, to a pressure $p_{CO} = 0.1$ atm and a temperature of 1700°C. Without any further additions of oxygen the composition of the steel would shift from (1) to (1'). In order to arrive at lower carbon contents, it is necessary to add oxidizing agents, for instance iron ore, up to the theoretical oxygen requirement indicated by point (2). In this way a renewed removal of carbon monoxide and therefore a change in concentration in the stoichiometric direction along the line (1)–(2') is obtained, which leads to the maximum possible decarburization.

Refining reactions in a vacuum are also employed for the decarburization of tantalum and niobium, in most cases, however, in the solid state, particularly so if an extremely high degree of purity is to be attained. The methods to be applied, which are dictated mainly by the reaction kinetics and by the requirement that the losses in metal by evaporation should not be too high, are described by Melchior[24].

3. Kinetics of Vacuum-metallurgical Processes

3.1 Kinetics of degassing

3.1.1 *The various stages*

The transition of a dissolved gas from a solid or liquid metal to the gas phase—or vice versa—comprises a number of stages:

(i) Transport from the interior to the interface of the metal. In the solid state this transport takes place only by diffusion of the gas atoms or ions through the crystal lattice of the metal and in the liquid state first by convective transport processes in the interior towards a zone near the surface of the metal and from there by diffusion to the interface.

(ii) Transition of the gas atoms from the dissolved state to the adsorbed state at the surface.

(iii) Reaction of the gas atoms in the adsorbed layer either with each other with formation of molecules by association, or with atoms of a different type, as, for instance, during the formation of carbon monoxide or of suboxides.

(iv) Desorption of the gas molecules from the surface.

(v) Diffusion of the gas molecules into the gas space and removal by the pumping system.

The postulates for continuity from one state to the next must, of course, be

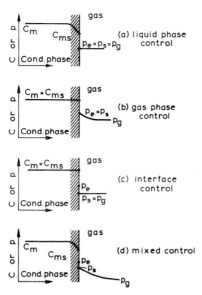

Fig. 18. Mass transfer across a gas–liquid interface. Time average concentration and partial pressure profiles for different types of control[25].

fulfilled, *i.e.* at least a quasi-stationary state must have been attained during the individual stages. The slowest of these stages is then rate-determining for the material transport.

Figure 18 shows schematically the concentration changes in the condensed and in the gas phases according to which stage determines the rate of transport[25]. C_m is the mean concentration of the dissolved substance in the condensed phase and C_{ms} that at the surface, p_e is the equilibrium pressure corresponding to this concentration, p_s is the pressure at the surface if one of the stages (ii)–(iv), *e.g.* the evaporation process, becomes rate-determining and p_g is the mean pressure in the gas space.

The following considerations are confined to the mass transport from a liquid metal into the gas phase. The kinetics of degassing in the solid state will be dealt with in detail in Chapter IV, Section 2.

Gas liberation may take place either through the free surface of the metal melt or through the crucible wall or lining provided that these are porous. If formation of gas bubbles is possible in the interior of the melt, additional exchange faces are created, which are in most cases very much larger at the beginning of the degassing process than that of the free surface of the melt, but their size decreases as degassing progresses.

3.1.2 Transport in the liquid phase

(a) *The mass transfer coefficient*

The mass transport in the interior of the liquid phase is, as mentioned above, effected by convection, which may have several causes. Convection may be generated by artificial means, for instance by induction stirring or by blowing gases through the melt. On the other hand, it may be induced spontaneously by differences in the density of the melt, for instance by heat radiation from the surface of the melt when the uppermost layers become cooler than those underneath.

This convection can effect equalization of the concentration only within the melt. In most cases this equalization takes place sufficiently rapidly for the concentration in the interior to correspond to the bulk concentration. The more important process for material exchange is in general the transport to the interface. This transport takes place by diffusion in a boundary layer.

Figure 19 shows how the stream lines of convective flow from the interior towards the surface of the melt change their direction at the surface and proceed parallel to it for a certain distance. After this they return to the interior of the melt. The diffusion boundary layer is embedded in this surface current. The transport velocity in this boundary layer usually determines the overall mass transport. The driving force for it is the concentration difference which builds up perpendicular to the direction of flow.

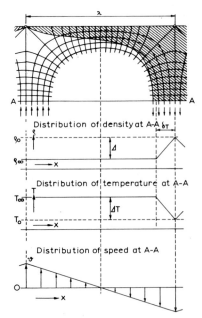

Fig. 19. Diagrammatic representation of convection at the free surface.

Figure 19 is not only a model for the formation of the diffusion boundary layer near the surface of a melt, for it applies equally to other types of phase boundaries, for instance, to the phase boundary between melt and an ascending gas bubble (*cf.* Fig. 30), even if the stream lines are different.

The mass transport through an element of the interface is proportional to the concentration difference between the interior and the surface of the melt. The quantity of gas given off per unit area of surface in unit time is therefore.

$$\frac{dn}{dt} = -\beta(C_m - C_{ms}), \tag{69}$$

where C_m is the concentration in the interior of the melt and C_{ms} that immediately at the surface which is in equilibrium with the partial pressure of the substance to be removed above the surface.

These two quantities are given. β is the so-called material or mass transfer coefficient or transport coefficient, which is defined by this equation. It has the dimensions cm/s, and expresses a fictitious velocity at which the substance transported traverses the phase boundary. If this quantity is expressed in relation to the free surface of the melt, as is the usual practice in vacuum metallurgy, it may be determined by measuring the quantity of gas given off from the melt.

As

$$\frac{dn}{dt} = \frac{V}{A} \cdot \frac{dC_{mt}}{dt}, \tag{70}$$

where V is the volume and A the surface area of the melt, combining this with eqn. (69) gives

$$\frac{dC_{mt}}{dt} = -\frac{A}{V} \beta(C_{mt} - C_{ms}). \tag{71}$$

Integration of eqn. (71) gives

$$\ln \frac{C_{mt} - C_{ms}}{C_{mo} - C_{ms}} = -\frac{A}{V} \beta t, \tag{72}$$

where C_{mo} and C_{mt} are the concentrations in the interior of the melt initially and at time t, respectively. C_{ms} is usually also not constant, but it may be ignored if $C_{ms} < C_{mt} < C_{mo}$, so that we finally obtain

$$\log \frac{C_{mt}}{C_{mo}} = -\frac{A\beta}{2.3V} \cdot t. \tag{73}$$

If C_{mo} and C_{mt} are determined from samples taken from the melt at the beginning and after time t, and if $-\log \frac{C_{mt}}{C_{mo}}$ is plotted as function of t, the material transfer coefficient β may be calculated from the slope of the curve.

Several theories have been developed for the calculation of the material transfer and its time dependence[26-28]; all of these are based on the concept that volume elements of the melt are transported from the interior towards the surface

References pp. 92–93

as a consequence of the convection current and that these volume elements move—as described above—as a current parallel to the surface for a certain contact time t_s before returning to the interior of the melt. The diffusion boundary layer is formed during this contact time; its thickness increases to a maximum which is determined by the diffusion coefficient D and the contact time t_s. If the flow processes can be expressed in mathematical terms and if therefore a calculation of this contact time is possible, a calculation of the mass transfer coefficient is also possible, at least to a first approximation. Measured and calculated results may then be compared and it can be considered how far the model of the flow phenomena used for the calculation corresponds to reality.

As has been mentioned earlier, the transport velocity in the diffusion boundary layer is in most cases rate-determining for the resulting material transport. For a computation of β the transport in this boundary layer must therefore be expressed in mathematical quantities. This procedure is based on Fick's first law which states that the flux density, *i.e.* the number of moles which pass through a plane perpendicular to the concentration gradient per unit of area and time, is proportional to the concentration gradient:

$$\frac{dn}{dt} = -D\frac{dc}{dx}, \qquad (74)$$

where D is the diffusion coefficient of the material transported.

Computation of the change in concentration in the diffusion boundary layer leads to a mathematical expression which corresponds to a Gaussian type error function, which is depicted in Fig. 20. The concentration difference $C-C_{ms}/C_m-C_{ms}$ is plotted on the ordinate and the relative distance from the interface x/δ on the abscissa[29]. The ordinate 0 corresponds therefore to the surface ($C=C_{ms}$)

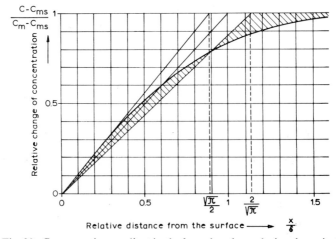

Fig. 20. Concentration gradient in the boundary layer during degassing.

and the ordinate 1 to the interior of the melt ($C = C_m$).

The best approximation to this function is given by the straight line of slope $2/\sqrt{\pi}$. The tangent at the origin gives for dc/dx and therefore also for the flux density a value which is too high by 27%. It is therefore the usual custom to use for this computation an approximation given by the straight line of slope 1, as then theoretically a quantity is obtained which is about only 12% too high. The equation of this line is

$$\frac{C - C_{ms}}{C_m - C_{ms}} \approx \frac{x}{\delta}. \tag{75}$$

By differentiating we obtain

$$\frac{dc}{dx} = \frac{C_m - C_{ms}}{\delta}. \tag{76}$$

Therefore δ may be regarded as the thickness of the virtual boundary layer, in which the concentration decreases linearly from C_m to C_{ms}. It depends on D and t_s. By substitution we get from eqn. (74):

$$\frac{dn}{dt} = -\frac{D}{\delta}(C_m - C_{ms}). \tag{77}$$

The change of δ with time which occurs during the dwell time of the current in the boundary phase may be derived to a first approximation from Fig. 21[29]. It shows how much the quantity of a substance contained in a volume element of thickness δ changes, if the thickness of the boundary layer increases by $d\delta$. This quantity must be equal to the quantity of material dn transferred, i.e.

$$\frac{dn}{dt} = -\tfrac{1}{2}(C_m - C_{ms})\frac{d\delta}{dt}. \tag{78}$$

From eqns. (78) and (77) it follows that

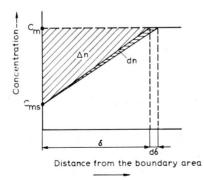

Fig. 21. Change of concentration in a volume element in the boundary layer.

References pp. 92–93

$$\frac{d\delta}{dt} = \frac{2D}{\delta} \tag{79}$$

and after integration we obtain

$$t_s = \frac{\delta^2}{4D} \tag{80}$$

or

$$\delta = 2\sqrt{Dt_s}, \tag{81}$$

with an average of

$$\delta_a \approx \sqrt{Dt_s}, \tag{82}$$

i.e. the thickness of the boundary layer increases as the square root of the time. Figure 22 shows, for instance, the change in the course of concentration between $t_s = 1$ and $t_s = 2$ in arbitrary units.

From eqns. (69) and (77) we obtain

$$\beta = D/\delta \tag{83}$$

with an average

$$\beta_a = \sqrt{\frac{D}{t_s}} \tag{84}$$

or

$$\beta_a = \sqrt{D\frac{v}{s}}, \tag{85}$$

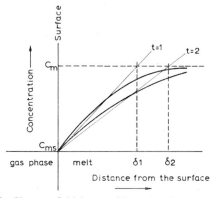

Fig. 22. Change of thickness of boundary layer with time.

where s is the distance covered by the volume element of the melt in the boundary layer and v is the flow velocity. The mass transport coefficient is expressed here in quantities which are calculable by the mathematical treatment of flow models as will be shown in the following examples.

(b) The model of Machlin

Formerly when no theory for the computation of the boundary layer existed the calculation was based on a fictitious stationary boundary layer, the so-called Nernst boundary layer. The thickness of this boundary layer was determined by regression and then used for the calculation of other transport problems. This procedure was, of course, rather arbitrary. Machlin[30] was the first to derive the thickness of this boundary layer from the parameters of the system in order to obtain a more reliable basis for the computation of the material transfer in refining processes and in reactions between melt and crucible. He developed the so-called "rigid-flow" model which is, however, applicable only to induction-stirred melts or to melts with similarly forced convection flow.

The forces created by the interaction between the medium frequency field of the induction coil and eddy currents induced in the melt produce a flow such as is depicted in Fig. 23. Machlin based his concept on the assumption that the melt moves like a rigid body along the melt–gas or melt–crucible interface and that the velocity gradient normal to the surface is almost zero. The melt ascending in the centre of the crucible flows radially outwards and then along the crucible wall downwards. The thickness of the diffusion boundary layer, which is shown schematically in the flow line network of Fig. 24, is assumed to be materially thinner than the depth of flow. Based on this model the time dependence of the concentration change dC_m in the melt is

$$-\frac{dC_m}{dt} = 2(C_m - C_{ms})(2Dv/\pi rh^2)^{\frac{1}{2}}, \qquad (86)$$

where v is the flow velocity near the crucible wall, r is the crucible radius, and h is the depth of the bath. In this equation D and v are the only quantities which carry some degree of uncertainty. The velocity v was determined by observation of the surface of the bath in induction furnaces of various sizes ranging from 100 g to 1 ton to about 10 cm/s, accurate to within one order of magnitude. Since only

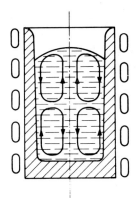

Fig. 23. Flow lines in an induction heated melt.

References pp. 92–93

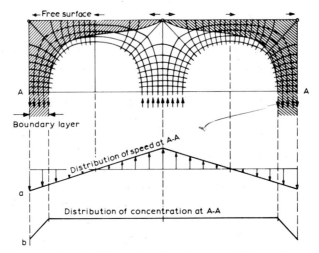

Fig. 24. Flow line network and diffusion boundary layer at the free surface of an inductively stirred melt (not to scale)[29].

the square root of a potential error enters into the results, the inaccuracy of the result caused by it will be materially reduced.

A comparison of the mass transport calculated from this theory with measured quantities in the refining of iron-base melts and for reactions between such melts and Al_2O_3 or MgO crucibles shows fair agreement within the limits of error of D and v. However, this result is not an adequate basis for exact quantitative calculations and Machlin himself drew attention to the limitations of this very simplified model. He pointed out, however, that his model would prove useful as a starting point for the study of kinetic problems, particularly for an estimate of the course of reactions. It should be noted that the stirring effect in the melt is a function of the heat input and that therefore v may fluctuate greatly during the degassing process. Furthermore, the stirring effect increases with decreasing frequency at the same power input. (In modern induction furnaces there is often a facility for adjusting the heating and stirring independently by employing two widely differing frequencies (cf. Chapter V, Section 2).) The range of variation of v may therefore be wider than Machlin assumed and an exact prediction of the material transport therefore becomes even more difficult.

As will be shown presently, it is probable that other transport phenomena will be dominant at low velocities and will lead to a higher rate of degassing than seems possible by Machlin's theory at weak agitation. On the other hand, there exists no theory at present for induction-stirred melts, at least where v is above 10 cm/s, that makes a more accurate calculation possible.

(c) The model of Kraus

Kraus[29] also tried to find a solution for the calculation of the material exchange which is free from arbitrary assumptions. His theory is more universally applicable and makes possible a rather more accurate computation of the mass transfer on the basis of parameters of the system at least in those cases in which, for instance, during the degassing of steel and for only slightly agitated induction-heated melts, natural convection currents are created by radiation from the surface of the melt and mass transfer via gas bubbles is not possible.

The causes setting up material transport in free convection will now be discussed in more detail.

If, as the result of free radiation from a melt in the upward direction, heat transport occurs in a direction opposite to the force of gravity, the temperature gradient which is caused by this heat transport will be in a labile state. This heat transport leads to the superposition on warmer layers of cooler, specifically heavier zones in the melt. This creates convection currents, similarly to those in a model described by Lord Rayleigh[31]. The colder parts of the boundary layer will therefore try to flow downwards, so that hotter parts will replace them at the surface.

Figure 25 may serve as a demonstration of this phenomenon. It shows a sector of the flow line network perpendicular to the surface and also indicates schematically the distribution of the sources and sinks as they are seen on looking at the surface of the melt. The stagnation points shown in the upper half of the diagram lie very probably on lines which in fact are usually curved. The diffusion boundary layer should be imagined to be embedded in this flow diagram by analogy with Fig. 24.

We now assume that the difference in weight between the cooler boundary layer as a result of the heat radiation and the hotter melt emerging from the interior of the crucible is the driving force for the circulation. This driving force is used for overcoming the flow resistance which exists as a consequence of the

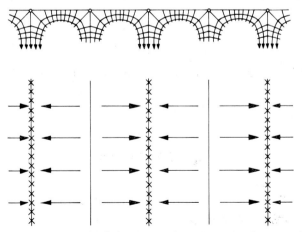

Fig. 25. Flow line network and distribution of the stagnation points in the surface (schematic).

References pp. 92–93

viscosity of the melt flowing in the opposite direction, *i.e.* between the ascending and descending parts of the melt.

As such convection currents cannot be made visible in liquid metals, Kraus[29] studied the water–carbon dioxide system as a model. He succeeded in demonstrating the generation of these convection currents by blowing carbon dioxide on to the surface of the water and in photographing the effect. The results of these experiments are an important support for his concept.

The calculation of this flow process leads to the expression for the mean dwell time of the melt at the surface of

$$t_s = \frac{s}{v} = \sqrt{\eta \cdot c_p/g\gamma q} \qquad (87)$$

(please see the Appendix to this chapter for the meaning of the symbols).

The heat flow in the upward direction, which is the only major variable of the system and which cannot be determined with accuracy amounts, for instance, for an iron melt with a surface free from slag at 1600°C and a spatial angle of radiation of 180°, to 4.72 cal.cm²/s. (The heat flow of course diminishes if the environment of the melt is also at a high temperature.) Using this quantity a mean contact time t_s of about 0.1 second is obtained for such a melt.

From the formulae derived only the ratio $s/v = t_s$ is obtained, and neither s nor v itself. These quantities are obtained by making certain assumptions, for instance, that the distance of the stagnation points corresponds to about twice the thickness of the friction boundary layer in the countercurrent in the interior of the melt. In this case we obtain a distance between the stagnation points of the hydrodynamic model of $s \approx 1$ mm and a laminar circulation velocity $v \approx 1$ cm/s. Choosing another distance between the stagnation points, *i.e.* longer or shorter, has no effect on the magnitude of the contact time t_s.

Based on eqn. (84) the mean value of the transport coefficient may be expressed as

$$\beta_a \approx \sqrt[4]{D^2 g \gamma q / \eta \cdot c_p} \qquad (88)$$

For an iron melt at 1600°C and a diffusion coefficient $D = 6 \times 10^{-5}$ cm²/s (as for C, O, N) we get

$$\beta_a \approx 2.5 \times 10^{-2} \ cm/s$$

and according to eqn. (83)

$$\delta_a \approx 2.5 \times 10^{-3} \ cm$$

also on the average.

These quantities are in fair agreement with the data obtained by experiments on bubble-free degassing ($\beta_a = (2-4) \times 10^{-2}$ cm/s).

The transport coefficient calculated by Machlin for his model is

$$\beta_a \approx 1.6\sqrt{Dv/r},$$

but with $D = 6 \times 10^{-5}$ cm^2/s, $v = 10$ cm/s and $r = 6$ cm

$$\beta_a \approx 1.6 \times 10^{-2} \text{ cm/s},$$

i.e., smaller than the quantity obtained with the Kraus model. This difference becomes even more pronounced with increasing crucible radius. It is therefore quite conceivable that at least for low velocities of a forced flow in induction-stirred melts the Kraus model gives a transport coefficient which comes nearer to reality. However, we do not know how long it takes for the free convection flow to become established, i.e., if, in a forced circulation, there is sufficient time for the natural convection current to become superimposed on the forced circulation. All we can say is that it is more likely that the Kraus model will represent the phenomena more truly for extremely low flow velocities, whereas the Machlin concept will apply to very high velocities. On the other hand, it is quite possible that the transport coefficient passes through a minimum between these two extremes.

(d) Effect of foreign layers on the exchange surface

So far it has been assumed that the flow is free from friction. This is of course not always true. As soon as film formation occurs by surface-active substances or by oxidation, the flow velocity approaches zero at the surface, i.e. a velocity gradient is created within the current. The mean velocity v_a of the liquid layer, in which the diffusion boundary layer develops, is approximately to the velocity v of the circulation as the thickness of the diffusion boundary layer δ is to the thickness of the friction boundary layer δ_v:

$$\frac{v_a}{v} \approx \frac{\delta}{\delta_v}. \tag{89}$$

Using eqn. (81) and the relations derived by Kraus[29] for δ_v, eqn. (89) may be written as

$$\frac{v_a}{v} \approx \left(\frac{D}{v}\right)^{\frac{1}{3}}, \tag{90}$$

where v is the kinematic viscosity. The contact time increases in the same ratio. Computation shows for the case of the CO degassing of steel melts that in the most unfavourable circumstances the material transfer coefficient will be reduced by a factor of 5 compared with that existing at the surface in friction-free flow.

Layers of foreign matter or slag at the surface of the melt may therefore reduce the material transfer in two ways:

(i) These layers may form a diffusion barrier which leads to an increase in the equilibrium concentration C_{ms} at the surface and therefore to a flattening of the concentration gradient.

References pp. 92–93

(ii) This already reduced gradient in the liquid boundary layer below these layers of foreign matter may further be diminished as a result of friction.

In liquid iron mainly oxygen and sulphur and to a lesser degree also carbon and phosphorus are substances of a surface-active nature. The first two elements reduce the interfacial tension to a high degree at rather low concentrations and therefore collect at all interfaces between melt and gas phase, not only at the free surface but also at the interfaces of ascending gas bubbles and consequently also affect the material transport of other elements. The inhomogeneity in the interface created by this adsorption affects the transport in the same manner as does a macroscopic oxide film or a layer of slag, in that the flow is no longer free from friction. This effect was long ago detected as a hindrance to mass transport of nitrogen in liquid iron[32,33]. For instance an increase in the oxygen level from 0.009 to 0.08 wt.% lowers the transport coefficient to as little as one-sixth. The effect of sulphur is less.

It is surprising that no such lowering of the transport coefficient could be found for hydrogen[34]. This may—at least to some extent—be explained by the more favourable value of the ratio v_a/v because the diffusion coefficient of hydrogen is about 20 times greater than that of nitrogen (cf. eqn. (90)).

(e) Spontaneous degassing by gas bubbles

I. *Formation of bubble nuclei.* Degassing by gas bubbles depends on the presence of bubble nuclei which must be able to grow. The equilibrium pressure p_e, which corresponds to the concentration of the dissolved substance in the melt and which exists in the interior of a stationary bubble in equilibrium with the melt, must be at least as high as the sum of the three types of pressure which act externally. They are: (i) the ambient pressure p_g resting on the liquid, (ii) the hydrostatic pressure which is the result of the column of liquid above the bubble $\rho g(h+r_B)$, where h is the distance of the apex of the bubble from the surface of the liquid, and (iii) the capillary pressure $2\sigma/r_B$ which is caused by the surface tension.

$$p_e > p_g + \rho g(h+r_B) + 2\sigma/r_B \tag{91}$$

If we designate the sum of all items which are or are almost independent of the bubble diameter as p_{ext} we may write

$$p_{ext} = p_g + \rho g(h+r_B) \tag{92}$$

and therefore express the bubble radius as

$$r_B > \frac{2\sigma}{p_e - p_{ext}}. \tag{93}$$

There is therefore a critical bubble radius r_B which must be exceeded if the bubble is to remain stable. This radius is the greater, the higher is the surface tension σ.

Owing to the high surface tension with the same equilibrium pressure the minimum size of bubble nuclei is much greater in metals than, for instance, in water or organic liquids.

A spontaneous, homogeneous nucleation of bubbles is therefore not conceivable under these conditions. For instance an internal pressure of about 20 atm would be required for the formation of a bubble nucleus of a radius of 10^{-4} cm in liquid steel, in order to remain stable. The rate of nucleation of hydrogen in steel as a function of the hydrogen partial pressure is depicted in Fig. 26[25]. As may be seen, equilibrium pressures of about 10^4 atm are necessary in order to attain a sufficiently high probability. Therefore, only heterogeneous nucleation at solid interfaces, such as slag particles or crucible walls, may be considered as a source of bubbles.

For a bubble nucleus on a flat substrate the contact angle θ is a function of the interfacial tensions of the three adjoining phases. For a given bubble volume the contact angle determines the curvature and therefore the minimum internal pressure which is required in order to keep this nucleus stable (see also Fig. 27). Bradshaw[25] showed that this contact angle must approach 180°, until the internal pressure becomes lower than the equilibrium pressure met with normally, *i.e.* the

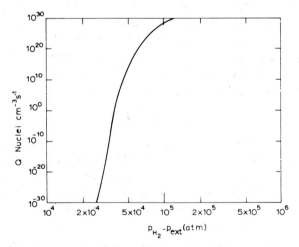

Fig. 26. Rate of homogeneous nucleation of hydrogen in steel *vs.* partial pressure, from ref. 25.

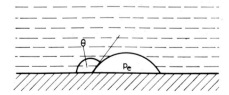

Fig. 27. Contact angle of a bubble on a flat substrate.

References pp. 92–93

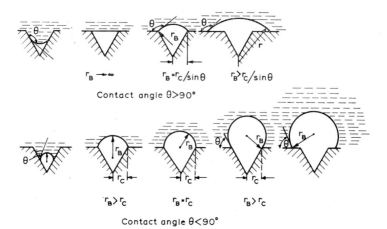

Fig. 28. Sketch of growth of bubble from a cavity in a solid surface, from ref. 25. Lower figure: contact angle <90°, upper figure: contact angle >90°.

melt may scarcely wet the surface. This means that only pores, crevices or other recesses of the solid phase boundary remain as nucleation sources. Here too the internal pressure depends on the contact angle θ as well as on the shape of the recess. Bradshaw assumes as a model a recess in the form of an acute cone with a radius of the lip of the cavity r_c. Figure 28 shows schematically the growth of bubbles for various contact angles below and above 90°. From this picture it is obvious that the bubble radius remains sufficiently large during increase in bubble volume for contact angles of more than 90°, so that further growth of the bubble is possible. The smallest bubble radius which exists is

$$r_B = r_c / \sin \theta. \tag{94}$$

For a contact angle of 160° as found by Kosakevitch[35] for liquid iron on aluminium oxide, the radius of the recess should have an r_c greater than 10^{-3} cm, so that the equilibrium pressure required for growth remains below 0.7 atm. It is also important that the recess be not rounded off at its root as otherwise complete wetting might occur. If these presuppositions are fulfilled, such a recess or cavity represents a continuous source of bubbles.

This or a similar mechanism of bubble formation is of importance not only for the degassing of melts in containers but also, very probably, in stream degassing.

In this method of degassing vigorous disintegration of the stream of liquid metal is observed as a result of the gas liberation on entry into the vacuum chamber. The question as to whence the bubble nuclei come which trigger this effect has not so far been answered with certainty.

Warner[36] carried out model tests for a clarification of this problem on steel and silver melts in which increasing quantities of oxygen were dissolved. He found that this disintegration of the stream does not continuously increase with the lowering of the pressure in the vacuum chamber, but that it passes through a maximum at a pressure of a few Torr. This quite surprising result must be connected with bubble nucleation, and Warner offered an explanation for a mechanism which might be typical for stream degassing processes.

According to Warner, only three of the many possibilities of a nucleation should be considered:

(i) The heterogeneous nucleation in pores and other cavities in the nozzle which was mentioned earlier. The necessary condition for nucleation is that the bubble should not be carried along during its formation by the stream of molten metal before it has attained a size which makes possible further growth, *i.e.* the entrained bubble must be sufficiently large.

(ii) The injection of gas bubbles into the stream of liquid metal which are sucked into this stream from the outside through pores or cracks in the refractory lining of the holding ladle or tundish.

(iii) The most probable of these, according to Warner, is "that gas entrainment could occur within the nozzle as a result of the metal stream trapping pockets of gas substantially at ambient vacuum chamber pressure against the wall of the nozzle".

These gas particles are pulled along by turbulence and other instabilities of the stream and so get into the interior of the metal stream where they either collapse or grow by diffusion of gas from the surrounding liquid.

A study of the stability of bubbles which have originated from such gas pockets, in relation to starting volume and internal pressure (equal to the pressure in the vacuum chamber), shows that unless a certain minimum pressure exists in the gas pocket at the beginning, the bubble forming from it cannot be stable. This explains why the disintegration of the stream is probably smaller in a high vacuum than at a pressure of, for instance, 10 Torr. Warner also demonstrates that for a certain volume of the gas pocket the growth rate of the bubble is greater at a pressure in the vacuum chamber of about 10 Torr than at lower pressures, for the reason that the original diameter of the entrained bubble is larger. This allows at least a qualitative explanation of the observed phenomena. The validity of this concept is further supported by the observation that the dispersion of the stream is more vigorous after flowing through a long parallel-sided orifice than through a sharp-edged orifice. Several high-speed photographs of silver streams with various oxygen levels and for various pressures of the vacuum chamber are shown in Fig. 29.

II. *Conditions for the growth of bubbles.* From eqn. (91) the minimum pressure prevailing in the inside of a bubble at which it is no longer capable of existence may be calculated. This minimum pressure may be derived from:

$$\frac{dp_e}{dr_B} = \rho g - \frac{2\sigma}{r_B^2} = 0, \tag{95}$$

References pp. 92–93

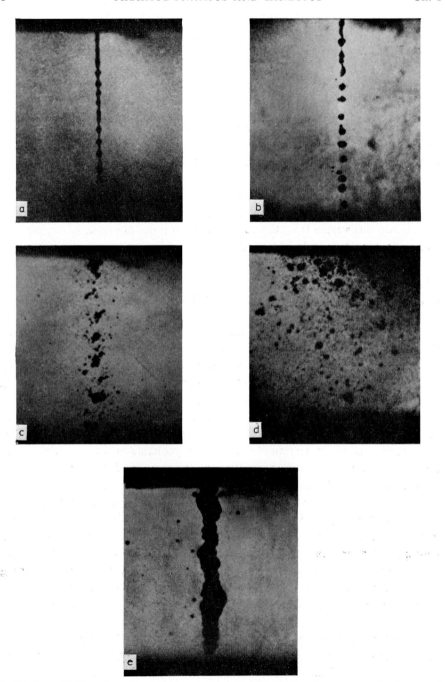

Fig. 29. Jets of falling silver through a low pressure (0.5 Torr). The silver was previously treated with gas mixtures of oxygen and nitrogen containing the following % by volume of oxygen (a) 0%, (b) 10%, (c) 20%, (d) 100%, (e) as for (d) but with the surroundings at 50 Torr, from ref. 25.

where

$$r_{B\,min} = \sqrt{2\sigma/\rho g}. \tag{96}$$

The minimum pressure $p_{e\,min}$ is found for bubbles in the upper part of the liquid column, which are formed immediately below the surface of the liquid at an ambient pressure of zero. We obtain[37]

$$p_{e\,min} = 2\sqrt{2\sigma\rho g} \tag{97}$$

and for a steel melt of $\sigma = 1400$ dyn/cm

$$p_{e\,min} \sim 0.01 \text{ atm.}$$

This pressure corresponds to an equilibrium concentration of

    2.5 ppm H
    40   ppm N
    10   ppm O at 0.02% C
    1   ppm O at 0.2% C.

It should, however, be noted that these theoretical minimum limits must be greatly exceeded in reality if the nucleation frequency is to be obtained which is required for vigorous bubble formation to get going also in greater depth. At this equilibrium pressure the diameter of the bubble must amount to at least 0.6 cm. The existence of much larger bubble nuclei is unlikely and also impossible because of the too high ferrostatic pressure underneath it. Therefore, in order that substantially smaller nuclei are able to grow, correspondingly higher equilibrium pressures must exist in the melt.

III. *Growth of ascending gas bubbles.* For the computation of the mass transport between a liquid and a gas bubble it is necessary to know the flow behaviour of the liquid at the interface with the bubble and the rate of rise of the bubble in the liquid.

The rate of ascent of the bubbles and the maximum pressure which is attained in their interior just before they emerge from the surface is given by the size of the bubbles which may be calculated from the growth rate on the basis of the theory of material exchange between liquid and gas phase for forced flow previously described. It is here also assumed that the flow around the bubble is free from friction.

The flow line network which develops in this case and which is derived from conformed mapping of parallel flow is shown in Fig. 30. The diffusion boundary layer is also shown which forms around the bubble during its ascent. Both phenomena were made visible by Kraus[29] with CO_2 bubbles in water with the aid of the colour change of a CO_2 indicator and of suspended particles and then photographed.

Figure 31 shows how the current approaches the upper stagnation point, how

References pp. 92–93

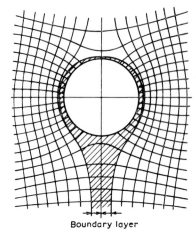

Fig. 30. Flow line network with diffusion boundary layer around an ascending gas bubble (not to scale).

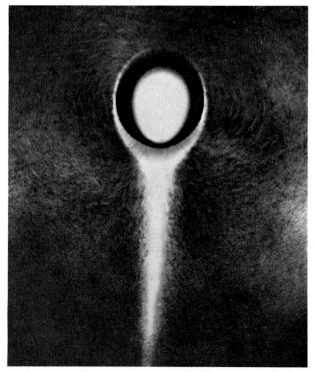

Fig. 31. Picture of an ascending gas bubble. The diffusion boundary layer and the absolute flow lines have been made visible.

it flows along the surface of the bubble, how the diffusion boundary layer is developed in the current and how the current with its diffusion zone leaves the bubble behind at the lower stagnation point. The change of the thickness of the diffusion boundary layer and therefore the material exchange in the bubble may be calculated on the basis of this model for any given flow or ascent rate. In this case too it is again assumed that mass transport takes place only in a direction perpendicular to that of the current and that the mass transport is the rate-determining step of this process.

Kraus calculates the rate of ascent of the bubbles under the assumption that they remain globular and that that force is decisive which is required for the acceleration of the quantity of liquid being displaced by the bubble per unit time.

Under this assumption v may be expressed as

$$v \approx \sqrt{\frac{4}{3} r_B g}. \tag{98}$$

This equation is regarded as valid for bubble radii r_B greater than 0.1 cm in steel melts.

The increase in the quantity of gas of an ascending bubble by diffusion may be calculated as

$$\frac{dn}{dt} \approx (C_m - C_{ms}) \sqrt{6\pi D v \eta / C_g}, \tag{99}$$

if C_g is the particle density of the gas in the interior of the bubble. The relation

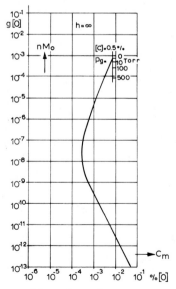

Fig. 32. Effect of oxygen content of the melt on the gas content of CO bubbles emerging from the surface of a steel bath containing 0.5% carbon.

References pp. 92–93

between gas take-up by the bubble and the particle density C_g in the bubble is found for bubbles in which the pressure caused by the surface tension is negligible compared with the hydrostatic pressure from the differential equation

$$\frac{dn}{dC_g} \approx -\frac{(C_m - C_{ms})RT}{\rho \cdot g} \sqrt{6\pi D\eta/vC_g}. \tag{100}$$

The gas content of the bubble on emerging from the surface could be calculated by integration. The results are plotted in Figs. 32, 33a and b, and 34a and b for CO bubbles in liquid steel.

The oxygen content derived from the CO content of the ascending bubbles is plotted on the ordinate of Fig. 32 and the oxygen concentration C_m of the melt on the abscissa. The curve shown represents, towards the left-hand side, the limit of existence of the bubbles which are capable of growing in the melt for a carbon concentration $[C] = 0.5\%$. The lower branch of this curve indicates the gas contents of the smallest bubbles which are still capable of growing at the relevant oxygen concentration. For instance, for $[O] = 10^{-3}\%$ the smallest bubble is defined by an oxygen content of 5×10^{-10} g, if both the external pressure above the melt and the hydrostatic pressures converge towards zero. As equilibrium with the saturation pressure of the melt exists here, the diameter of the bubbles may be calculated, which is about 2×10^{-2} cm. The capillary pressure of such a bubble amounts to about 0.3 atm, which must be quite close to the equilibrium pressure which corresponds to the carbon and oxygen concentrations. Besides these, bubble

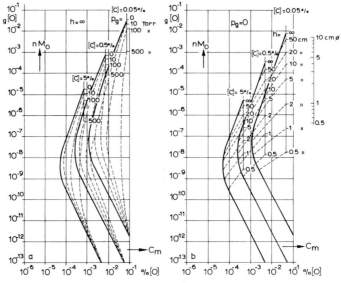

Fig. 33a, b. Effect of oxygen content of the melt on the gas content of CO bubbles emerging from the surface of a steel bath containing various amounts of excess carbon.

nuclei of a larger diameter and consequently greater oxygen content are capable of existence. Their location in the melt may be at a greater depth the larger their diameter is, because with decreasing capillary pressure the ferrostatic portion of the pressure may be higher.

As long as the bubbles are small and the surface to volume ratio large, the carbon monoxide partial pressure in the bubble is only slightly less than the saturation pressure of the melt. However, as soon as the diameter reaches the order of about 0.1 cm and the bubble attains its maximum ascent rate (see Fig. 35), the growth rate of the bubble becomes so large, as a result of its relatively larger surface area and of the reduction of the capillary pressure and the hydrostatic pressure, that the resistance to the mass transport within the diffusion boundary layer becomes noticeable. An increasing concentration gradient develops, and with it the molar concentration in the bubble deviates more and more from the saturation concentration.

The size of the bubble, of course, increases with the distance over which the bubble has travelled in the bath towards the surface. It is also the greater, the larger the original diameter of the bubble nucleus. The maxima are obtained at a depth of bath as small as 50 cm.

Figure 33a shows how the limiting curve is displaced more and more to the right-hand side if the ambient pressure above the melt increases. The maximum diameter of the bubble when it emerges at the surface of the melt becomes smaller and so therefore does the quantity of material transported, since the depth of the

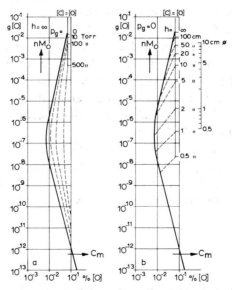

Fig. 34a, b. Effect of oxygen content of the melt on the gas content of CO bubbles emerging from the surface of a steel bath, calculated for $[C] = [O]^{29}$.

References pp. 92–93

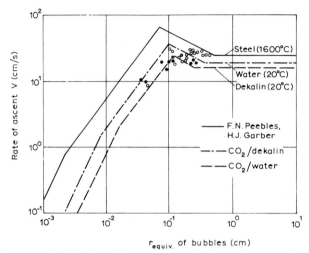

Fig. 35. Rate of ascent of bubbles in relation to their equivalent radius according to Peebles and Garber[38] and Bogdandy, Rutsch and Stranski[39].

bath at which the nuclei are capable of growing is the less the higher p_g is.

The scale at the right-hand side shows the diameter size of the bubble which corresponds to the value on the ordinate for the oxygen contents. (This scale is valid only for the upper branch of the limiting curves.)

Figure 33b depicts the relation between bubble size and the depth h of the bath for an ambient pressure of zero. The broken curves represent the upper limit of the oxygen content and of the bubble diameter, which is attainable in relation to the depth of the bath. This upper limit of the bubble diameter lies, for example, for 0.05% C and 10^{-2}% O at a depth of bath of 20 cm at about 3 cm. As would be expected, the smaller h is, the smaller is the maximum bubble size.

Figures 34a and b depict the same relation for melts in which the molar concentrations of C and O are equal, *i.e.* where no excess carbon content exists. These curves would, as a good approximation, apply also to nitrogen, if the nitrogen concentration or the corresponding saturation pressure were high enough to make possible the growth of bubbles.

The quantity of gas liberated from a melt could be calculated from these data if the nucleation frequency in relation to the pressure and the concentration were known. As this function is, however, not known, the computation is restricted to the limiting case of an excess of bubble nuclei. With this assumption the maximum quantity of gas liberated is

$$\frac{dn}{dt} \approx -2\psi A \cdot C_m \sqrt[3]{Dg/L}, \tag{101}$$

where L is Ostwald's solubility coefficient. $L = \dfrac{C_m}{C_g}$ is the ratio of the concen-

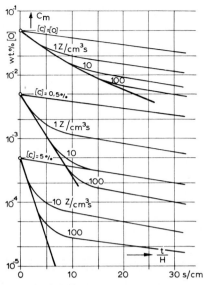

Fig. 36. Effect of time and nucleation frequency on the degassing of steel melts at various carbon levels.

trations, in mol/cm³, in the melt and in the gas space, which according to the law of mass action exists at the interface at equilibrium. ψ is the population density of bubbles near the surface, *i.e.* the ratio volume of gas: total volume in the surface zone of the melt and A is the escape area of the gas flow.

The rate of ascent of the bubbles was assumed to be constant as a first approximation and this seems to be true. The results of model tests are shown in Fig. 35[38,39]. As may be seen, the rate of ascent in liquid steel is almost constant if the radius of the sphere equivalent to the volume of the bubble exceeds 0.1 cm.

The mass transfer coefficient is

$$\beta \approx 2\psi \sqrt[3]{Dg/L}. \tag{102}$$

For $\psi \approx 0.5$ a mass transfer coefficient of $\beta = 0.1$–0.3 cm/s is obtained for the CO formation in steel melts of carbon levels between 0.05 and 0.5%. ($\psi \approx 0.5$ corresponds to an almost closest packing of the bubbles near the surface.) This quantity of β agrees well with those found in ladle degassing and in the DH process.

Finally, Fig. 36 shows the course of degassing as derived on the basis of this model for ladles of great depth. ψ was again assumed to be 0.5. In this case the time constant of degassing is proportional to the depth of the bath. For this reason, the degassing time was expressed as per cm of the depth of the bath. In this calculation it was in one case assumed that an excess of bubble nuclei was present, *i.e.* it was assumed that all bubble nuclei begin to grow in that depth in

References pp. 92–93

the bath at which they are capable of growing according to the concentration in melt (heavy curve). In another case it was estimated which course the curve would take, if the nucleation frequency $Z/s \cdot cm^3$ were lower.

The lowest zone in which bubbles capable of growing may exist is gradually raised during the degassing process until bubble formation ceases. In the end all curves must join that which expresses bubble-free degassing at natural convection owing to heat radiation. The transport coefficient decreases and finally equals that which is found from eqn. (88).

The model developed by Kraus[29] for degassing through ascending bubbles assumes that the bubbles have the shape of a sphere, but during the ascent of a bubble its shape changes with its size. It was found that the originally spherical bubbles, if their equivalent diameter amounts to 7–8 mm, assume, during subsequent growth, the shape of an oblate ellipsoid of rotation and that on further growth they take the form of a spherical cap[40]. This deformation of the bubble leads to an increase in mass transfer and to a slowing down of the rate of ascent compared with spherical bubbles or to a constant v value, as shown in Fig. 35. Kraus, however, postulated this constancy in his calculation.

In comparison, the parameter ψ contains a markedly higher uncertainty. In view of the indeterminacy of the nucleation frequency, and of the possible simultaneous reaction between melt and refractory lining for carbon-bearing steels, a check of the range of validity of this theory is extremely difficult for the degassing of steel and other metals. However, this theory indicates an upper limit for the mass transfer coefficient, which very likely cannot be exceeded.

(f) Degassing by scavenging gas

In the foregoing discussions of the degassing process, which involves the generation of bubbles, carbon monoxide formation in liquid steel has been chosen as an example. This is, in practice, the only case in which equilibrium pressures attained are so high that bubble nuclei are able to grow at great depths in the bath.

Bubble formation of hydrogen and nitrogen at the concentrations usually occurring is possible only in the uppermost layer of a melt (the lowest concentration, below which no bubbles can form, was given as 2.5 ppm H and 40 ppm N in Subsection 3.1.2e). No time would therefore be available for further growth.

The elimination of hydrogen and nitrogen in the form of gas bubbles is therefore possible only in combination with carbon monoxide, *i.e.* CO acts as a scavenging gas. In all cases in which spontaneous bubble formation is not feasible either because the equilibrium pressure of the dissolved gas is too low or because the probability of nucleation is too small, the application of a scavenging gas is the only means possible for promoting mass transfer by increasing the surface area available for exchange. The use of a scavenging gas results at the same time in agitation of the melt and so in an equalization of the concentration. This procedure is also useful in all those cases in which spontaneous bubble formation would be feasible, but would be restricted to the uppermost liquid layer. In this case complete degassing is obtained in a shorter period of time than would be possible by the application of a vacuum alone.

Argon is usually employed as the scavenging gas, and is injected in the finest possible state of dispersion through porous refractory inserts.

Bradshaw and Richardson[41] were probably the first to calculate the mass transfer between a steel melt and an ascending argon bubble with the aid of a computer. Later Lange, Ohji, Papamantellos and Schenck[42] also tackled this problem. They too used a computer and also calculated among other items the pressure dependence of the mass transfer during the elimination of hydrogen and nitrogen from iron melts. However, in their calculation a substantially smaller quantity of scavenging gas per bubble, i.e. a smaller initial volume, was assumed. The deduction was based on eqns. (98) and (100) derived by Kraus[29]. For this calculation it was also assumed that the bubbles retain their globular shape and that their radii are greater than 10^{-2} cm and that therefore the capillary pressure as compared with the ferrostatic pressure is negligible. Similarly to the calculation of Kraus for the case of spontaneous bubble formation, it was postulated that the transport within the boundary layer of the liquid is alone rate-determining for the total process, that equilibrium at the phase boundary exists with a concentration C_{ms} at the surface, and finally that the melt is otherwise homogeneous. This last-named assumption is unfortunately not always fulfilled, as a depletion of the dissolved substance may occur in those regions through which the ascending bubbles pass. This is very pronounced in the vicinity of the bath surface, where the mass transfer rate is particularly high. It may also occur when the convection currents are not strong enough to effect an equalization. This may easily happen if the quantity of scavenging gas used is small.

The two quantities C_{ms} and C_g of eqn. (99) may be expressed with the aid of the mass action constant K and the generalized gas equation as the partial pressure p_i and the total pressure p_g in the bubble[42]. In this case a differential equation is obtained which permits the calculation of the change of gas content of the bubble in relation to the distance x through which the bubble has risen.

$$\frac{dn}{dx} = \phi n^{5/12} \cdot p_g^{-5/12} \left(c_m - K \sqrt{\frac{n - n_{w,o}}{n}} \, p_g \right), \tag{103}$$

where

$$\phi = \left[\frac{6\pi \cdot D \cdot R \cdot T}{\frac{4}{3} g (3RT/4\pi)^{\frac{1}{6}}} \right]^{\frac{1}{2}},$$

$n_{w,0}$ is the quantity of scavenging gas, in moles, at the initiation of the bubble, n is the quantity of gas in the bubble at any time and C_m, in mol/cm^3, is the concentration of the substance scavenged from the melt.

Computer evaluation of this differential equation made it possible to cover

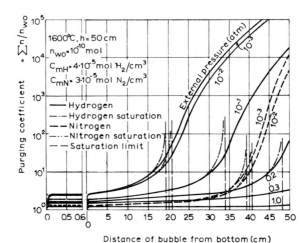

Fig. 37. Relationship between bubble location at various external pressures (constant quantities of scavenging gas, constant gas content of the melt) and the purging coefficient for hydrogen and nitrogen, after ref. 42.

the various quantities important for an analysis of the degassing process. Figure 37 shows the relation between the so-called purging coefficient and the bubble location for hydrogen and nitrogen in liquid iron at 1600°C for various external pressures. The initial content of scavenging gas in the bubble was always 10^{-10} mol. The molar concentration C_m was assumed to be 4×10^{-5} mol/cm^3 for hydrogen and 3×10^{-5} mol/cm^3 for nitrogen. The purging coefficient is the ratio of the total quantity of gas n to the quantity of scavenging gas $n_{w,0}$ in the bubble.

The vertical straight lines represent the saturation limits. They indicate the depth of bath at which the saturation pressure of the melt is equal to the total external pressure. Spontaneous bubble formation, provided that bubble nuclei are present, should be possible from these limiting lines onwards. This saturation limit is, of course, the lower, the more the external pressure is reduced. Above this saturation limit the gas intake of the bubble increases greatly and the quantity of gas taken up by the bubble becomes the larger, the longer the path is which the bubble traverses in rising to the surface. The quantity of gas taken up by the bubble should approach infinity—as indicated in the figure by the broken line near the saturation limit—if the internal pressure remains near equilibrium. But the rate of gas supply from the melt cannot match the rapid increase in volume. Therefore the partial pressure in the bubble decreases. This is confirmed by the curves in Fig. 38, which depicts the partial pressure of hydrogen and nitrogen in an ascending bubble in relation to the length of ascent under equal initial conditions. As soon as the inert gas bubble is formed and becomes detached, the saturation pressure of the gas dissolved in the melt is very quickly attained in the bubble. This equilibrium is maintained almost until the external pressure to which the bubble is exposed becomes lower than the saturation pressure. As a result of the

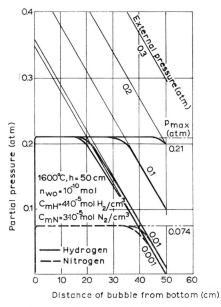

Fig. 38. Effect of bubble location at various external pressures (constant depth of bath, constant gas content in the melt, constant quantities of scavenging gas) on the partial pressure of hydrogen and nitrogen in the ascending bubble, after ref. 42.

rapidly increasing bubble volume the partial pressure will fall. This was also shown by previous calculations[41] for bubbles with a large initial volume. It should be pointed out that the effect of the capillary pressure has not been taken into account in these calculations. The quantity of gas taken up at low external pressures might therefore be slightly lower.

Calculations show that the degassing rates attained by employing a scavenging gas are several orders of magnitude higher at pressures below 0.1 atm than at ambient pressure and that therefore the use of a vacuum offers considerable advantages.

These investigations were restricted to hydrogen and nitrogen. Qualitatively similar results should be expected for CO degassing and almost the same values as for nitrogen for melts with a stoichiometric ratio of [C] to [O], as oxygen and nitrogen show almost the same diffusion coefficients. In both cases Sieverts' square root law applies for the relation between gas concentration in the melt and the equilibrium pressure.

The results of calculations by Bradshaw[25] are plotted in Figs. 39, 40 and 41. The initial volume of the bubbles is in this case greater by a factor of almost 10^4 and the bubbles have a size at nucleation which in the preceding example for nitrogen is reached only at about 10 cm below the surface. Bradshaw also assumes a much greater depth of bath.

It is interesting to compare the gas take-up of the bubbles where only one

References pp. 92–93

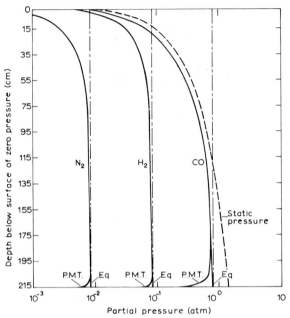

Fig. 39. Simultaneous transfer of N_2, H_2 and CO. Volume at breakaway: 0.1 cm³. Partial pressures of individual gases. The division of each curve close to the bottom refers to bubbles which break away prior to mass transfer (P.M.T.) and of those which leave in equilibrium (Eq.), from ref. 25. ———— Total ferrostatic pressure. —·—·—·— Equilibrium pressure of appropriate species.

type of gas is dissolved in the melt with that where all three gases, nitrogen, hydrogen and oxygen, are present. A plain carbon steel with 0.11 % carbon was adopted with the following molar concentrations:

hydrogen	2.8×10^{-5} mol/cm³
nitrogen	1.0×10^{-5} mol/cm³
carbon	6.5×10^{-5} mol/cm³.

Although the equilibrium pressures which correspond to the hydrogen and nitrogen concentrations are lower than those of Figs. 37 and 38, the gas take-up of the bubbles with simultaneous transfer starts at markedly lower depths of the bath owing to the fact that, as a result of the high equilibrium pressure of CO (about 0.8 atm), growth can start at lower depths. CO therefore acts here as an additional scavenging gas.

A comparison of the results of calculations for the take-up of nitrogen and hydrogen by small and large bubbles of scavenging gas according to Figs. 37 and 40 is made in Table 4 for the case of independent mass transfer. A direct comparison is not possible because of the different initial concentrations in the melt as these affect not only the rate of mass transfer from the liquid into the bubble but also the distance which the bubble travels to the surface after the saturation

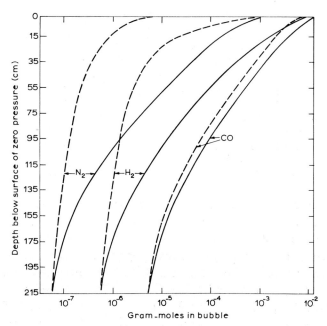

Fig. 40. Transfer of N_2, H_2 and CO to a bubble, initial volume 0.1 cm^3, leaving in equilibrium with metal, from ref. 25. ———— Simultaneous transfer of all three gases. ----- Separate transfer of each gas.

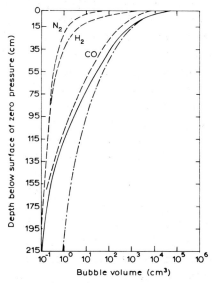

Fig. 41. Bubble volume vs. height for bubble leaving in equilibrium with metal, from ref. 25. Initial volume 0.1 cm^3. ———— Simultaneous transfer of N_2, H_2 and CO. ----- Separate transfer of each species. Initial volume 1.0 cm^3. —·—·— Simultaneous transfer of N_2, H_2 and CO.

References pp. 92–93

TABLE 4

COMPARISON OF THE UPTAKE OF GAS BY SCAVENGING GAS BUBBLES OF VARIOUS SCAVENGING GAS CONTENTS

	Case 1	Case 2
Content of scavenging gas of the ascending bubble, mol	10^{-10}	6.5×10^{-5}
N_2 content of the melt, mol/cm³	3×10^{-5}	1.1×10^{-5}
Depth of bath at saturation point, cm	11	1.5
N_2 uptake of the bubble, mol	10^{-6}	8×10^{-6}
H_2 content of the melt, mol/cm³	4×10^{-5}	2.8×10^{-5}
Depth of bath at saturation point, cm	27	15
H_2 uptake of the bubble, mol	4×10^{-5}	10^{-3}

point has been exceeded. As has been shown previously, the uptake of gas is greatest during this part of the path, particularly in the case of a low scavenging gas content in the bubble.

Considering that the quantity of scavenging gas is larger by a factor of almost 10^6 in Case 2 than it is in Case 1, the quantity of nitrogen or hydrogen taken up per mol of scavenging gas is relatively very low. This comparison confirms the hypothesis that the quantity of gas taken up from the melt per mol of scavenging gas is the larger, the lower the gas content of the bubble in scavenging gas. This is also true for the case where the saturation pressure of the melt is so low, that it always remains below the external pressure on the bubble.

For a number of reasons these calculations cannot accurately reflect the true conditions. At best it is possible to calculate to a first approximation the mass transport in relation to the quantity of scavenging gas for an external pressure of zero. The mass transport is very difficult to evaluate, particularly in the upper zones of the melt, as in this region in which the growth of the bubbles becomes very fast we have, besides the three pressure components (hydrostatic pressure, capillary pressure and dynamic compression), the momentum transfer to the melt as a result of expansion. An exact calculation of the bubble growth is therefore extremely difficult. It should also be taken into account that the bubbles are heavily oblated or that they assume the shape of spherical cups. It is very difficult to appraise how large the exchange area is in reality and how great the impoverishment in gas concentration of the upper zones is as a result of the extensive mass transfer.

The experimental investigations[41] have also hitherto shown too large deviations from the theoretical prediction, so that we are here not on sure ground. This can certainly also be attributed to the difficulty of establishing well defined experimental conditions, particularly for steel melts because of the many possible side reactions. Any calculation therefore can only indicate the order of importance of the various factors involved. The relations existing in scavenging gas treatment at ambient pressure are very much easier to appraise.

In summing up it may be said on the basis of these calculations that scavenging gas treatment under a vacuum offers material advantages only if either the saturation pressure is so low that no spontaneous bubble formation is possible or where no spontaneous bubble formation of another gas occurs, such as the formation of CO which may affect the scavenging of another gas, or if the saturation pressure of the gas to be eliminated is so low that bubble formation is restricted to the uppermost zone of the melt.

(g) Degassing of falling metal drops

In stream degassing two degassing processes act side by side: first, the gas liberation connected with bubble formation, provided that the saturation pressure of one of the dissolved gases is sufficiently high so that bubble nuclei can grow, and second, the gas liberation at the surface of falling bubbles and drops. In stream degassing a minimum run-out velocity must be maintained in order to prevent too high a loss of heat. For the same reason the depth of fall for prolonging the dwell time cannot be made too great. The concentration level of the dissolved gas in the metal at the end of the degassing process is therefore materially higher than that which would correspond to the pressure inside the vacuum chamber in equilibrium.

The various causes for bubble formation in stream degassing were discussed in Subsection 3.1.2e. These phenomena cannot, however, be calculated as yet. Determination of the quantity liberated by degassing through the surface of falling drops is also difficult, but at least an estimate may be made of which process will contribute most to the degassing effect.

In practice, about 75% of the hydrogen content during the degassing of steel can be eliminated (see also Chapter IV). If it is assumed that according to eqn. (84) gas liberation would be proportional to the square root of the diffusion coefficient, the corresponding reduction of the nitrogen or oxygen content should amount to about 22%. For degassing at 1600°C the following diffusion coefficients were used as a basis for this estimate:

hydrogen $\quad\quad\quad\quad D = 1.5 \times 10^{-3}$ cm^2/s
nitrogen and oxygen $D = 7 \times 10^{-5}$ cm^2/s.

In reality the nitrogen liberation and in particular the oxygen liberation are usually higher and it should therefore be assumed that a different mechanism is at work for the liberation of these gases than that for hydrogen[41].

In those cases in which spontaneous formation of bubbles occurs—this is always the case in the stream degassing of steel—the difference may simply be attributed to the fact that nitrogen and oxygen are eliminated almost exclusively through gas bubbles, where the CO acts as a scavenging gas for nitrogen, whereas hydrogen, besides this type of transport, is also liberated by diffusion from the surface of the bubbles and drops as a result of its high diffusion coefficient. This mechanism is evident from calculations of the gas liberation of falling drops by Bradshaw and Richardson[41].

References pp. 92–93

The time of fall t is

$$t = [(H+h_f)^{\frac{1}{2}} - H^{\frac{1}{2}}](2/g)^{\frac{1}{2}}, \tag{104}$$

where h_f is the height of fall and H is the pressure head, being the sum of the pressure of the liquid column in the holding ladle and the atmospheric pressure resting on this column, under which the metal enters the degassing chamber at the beginning of the degassing process. During the casting operation the height of fall decreases, but so also do the pressure H and consequently the inlet velocity, so that these effects partly cancel each other out. For $H = 330$ cm and an initial height of fall of 550 cm a time of fall of about 0.5 second is obtained, which remains the same if the height of fall is reduced to 370 cm at the end of the degassing process.

If a partial pressure of 0.1 Torr be assumed in the vacuum vessel for each of the gases H_2, N_2 and CO and if it be also assumed that the concentration at the surface of the drops is in equilibrium with this partial pressure and that further no circulation within the drop exists owing to lack of gaseous friction, the minimum quantities of gas liberated from drops of 1–3 mm diameter and for times of fall of 0.5–1 second are obtained which are shown in Table 5. These quantities are related to the initial content of the particular gas and the quantities N/N_{tot} shown therefore represent that fraction of the gas content originally contained in the drop which was liberated during the fall. The quantities shown in Table 5 refer to a steel melt with initial concentrations of 0.0005 wt.% H, 0.004 wt.% N, 0.015 wt.% O and 0.10 wt.% C.

For stream degassing with a typical drop radius of 1–5 mm, the theoretical liberation of hydrogen from the surface of the drops is almost as great as is observed in practice, whereas the liberation of nitrogen and oxygen is markedly lower and must therefore take place mainly in the form of gas bubbles. High-speed cinephotographs revealed that the major part of the sputtered jet consists of bubbles and not of drops[43]. Most of the drops are formed only if the bubbles burst. It is therefore reasonable to assume that the outward exchange surface area per g steel

TABLE 5

GAS LIBERATION FROM FALLING DROPS DURING STREAM DEGASSING

Gas	Drop radius (mm)	N/N_{tot} Time of fall	
		0.5 sec	1 sec
H_2	1	0.66	0.80
	3	0.26	0.37
N_2	1	0.17	0.23
	3	0.05	0.08
O_2	1	0.19	0.26
	3	0.06	0.09

is even larger for these bubbles than for drops and that it is therefore of greater importance for hydrogen, because of the higher diffusion coefficient, than for nitrogen and oxygen. The conclusion drawn above carries therefore even greater weight.

Besides the liberation of gas during the free fall the continued degassing in the receiving ladle or the casting mould through rising gas bubbles from the melt should not be overlooked. Since sufficient bubble nuclei are provided by the stream and as the melt is strongly agitated by the incident stream, the mass transfer through bubbles even at this stage should not be ignored; it might even be dominant for the elimination of oxygen and nitrogen.

In order to obtain effective degassing in the stream degassing process it is therefore imperative to promote bubble nucleation by suitable shaping of the nozzle. Bubbles not only provide large exchange areas but also ensure the formation of the desired fine droplets on bursting. Total pressures lower than 1 Torr are not necessary if a sufficiently vigorous CO formation occurs which by its scavenging effect in the bubbles and also in the vacuum vessel itself creates a partial pressure sufficiently low to obtain low hydrogen levels.

3.1.3 Mass transfer from the melt into the gas phase

It was assumed in the evaluation of the mass transport discussed in the preceding subsection that the transport in the melt is alone rate-determining. We shall now examine how far or under what conditions this assumption is correct.

A schematic presentation of the distribution of the molar concentrations and of the transport processes in the various stages which a gas particle traverses during the transition from one phase to another is shown in Fig. 42 as a complement to Fig. 18. There are three stages depicted:

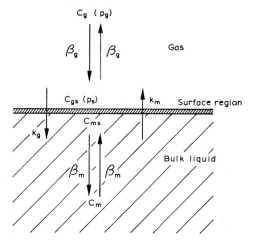

Fig. 42. Mass transfer across a gas–liquid interface.

References pp. 92–93

1. Transport within the diffusion boundary layer of the melt.
2. Phase transition.
3. Transport in the phase boundary of the gas phase.

It was shown in Subsection 3.1.1 that the transfer in the reaction zone is characterized by taking place in several stages: stages 2, 3 and 4. As the individual stages are not, or are only in exceptional cases, calculable, they are now combined to a single stage under the heading "phase transition". According to Kraus[37], there exists the following reaction scheme involving these three stages:

$$C_m \underset{-\beta_m \cdot C_{ms}}{\overset{\beta_m \cdot C_m}{\rightleftharpoons}} C_{ms} \underset{-k_g \cdot C_{gs}}{\overset{k_m \cdot C_{ms}}{\rightleftharpoons}} C_{gs} \underset{-\beta_g \cdot C_g}{\overset{\beta_g \cdot C_{gs}}{\rightleftharpoons}} C_g.$$

Transport within the diffusion boundary layer of the melt

Phase change in the reaction zone

Transport in the gas phase

The flux density in the melt dn_m/dt may be expressed as

$$\frac{dn_m}{dt} = -\beta_m(C_m - C_{ms}). \tag{69}$$

The flux density is the difference between mass transport toward the surface which is proportional to the concentration C_m in the interior of the melt and transport in the opposite direction from the inside surface of the melt at which the concentration C_{ms} exists. The rate of transport and consequently the transport coefficient β_m must be equal for the two fluxes.

The second stage, the phase transition, may also be called in general terms "degassing" or "engassing" depending on the direction of the transport. The proportionality constants k_m and k_g for the two directions of transport are in this case different from each other and strongly temperature-dependent. For this stage we obtain the relation:

$$\frac{dn_g}{dt} = -k_m \cdot C_{ms} + k_g \cdot C_{gs}, \tag{105}$$

where C_{gs} is the molar concentration of the material transported in the gas phase in the immediate vicinity of the surface.

Analogous to the diffusion boundary layer in the metal, the gas transport in the gas phase is the difference between two gas streams away from and towards the surface. These two streams may be regarded as independent, provided that the pressure is sufficiently low; in other words, if the mean free path is sufficiently long. Each stream is proportional to the particle density at the surface or in the gas space, respectively, and is also proportional to the molecular gas velocity β_g, which under isothermal conditions and at low pressure may be assumed to be equal for the two directions.

For this stage we therefore obtain

$$\frac{dn_g}{dt} = -\beta_g(C_{gs} - C_g) = -\frac{\beta_g}{R'T}(p_s - p_g). \tag{106}$$

Equation (105) must also be valid in the case of heterogeneous gas equilibrium, where $dn_g/dt = 0$. From eqn. (105) therefore we get

$$\frac{k_g}{k_m} = \frac{C_{ms}}{C_{gse}} = L. \tag{107}$$

L is Ostwald's solubility coefficient for the material. (It should be borne in mind that L is constant only if a linear relation exists between concentration in the melt and the partial pressure of the dissolved substance above the melt.) C_{gse} is the molar concentration in the gas phase at equilibrium. It is also true that

$$\frac{dn_m}{dt} = z\frac{dn_g}{dt}, \tag{108}$$

if z gas atoms of an element leaving the melt associate to form a molecule.

From eqns. (69) and (105)–(108) we get

$$\frac{dn_m}{dt} = -\frac{C_m(1 - L \cdot C_g/C_m)}{1/\beta_m + 1/zk_m + L/z \cdot \beta_g}. \tag{109}$$

If

$$1/\beta = 1/\beta_m + 1/z \cdot k_m + L/z \cdot \beta_g, \tag{110}$$

we obtain

$$\frac{dn_m}{dt} = -\beta C_m(1 - L \cdot C_g/C_m), \tag{111}$$

an equation which resembles eqn. (69). (If the resistance in the condensed phase dominates we get $L = C_{ms}/C_{gse} \ll C_m/C_g$ and $L \cdot C_g/C_m \ll 1$.)

The quantity β is the overall mass transfer coefficient and its reciprocal, the total resistance to the mass transport, is the sum of the individual resistances in the melt at the phase transition and in the gas space[37].

The conditions under which $1/z \cdot k_m$ and $L/z \cdot \beta_g$ may become dominant will now be examined.

For the calculation of k_m we must consider that, at equilibrium, the same number of molecules are transferred from the surface into the gas phase as are absorbed from the gas phase per unit of area and of time. The number of moles absorbed may be calculated on the basis of the kinetic theory of gases:

$$\frac{dn_g}{dt} = -\varepsilon \frac{p_e}{R' \cdot T_{ms}}\sqrt{\frac{RT_{ms}}{2\pi M}}, \tag{112a}$$

where p_e is the equilibrium pressure corresponding to the surface concentration C_{mse}. T_{ms} is the temperature of the gas at the surface of the melt and is assumed to be equal to the temperature of the liquid at the surface. ε is the condensation or sticking coefficient. ε represents that proportion of the molecules that is absorbed or condensed out of the total number of molecules impinging on the surface of the melt, whereas $\varepsilon-1$ is that proportion which is reflected from the surface. Only if $\varepsilon = 1$ are all molecules arriving at the surface condensed or absorbed.

The number of gas moles given off from the melt per unit of area and time is therefore

$$\frac{dn_g}{dt} = -\varepsilon \frac{p_e}{R'} \sqrt{\frac{R}{2\pi M T_{ms}}}. \tag{112b}$$

Applying the law of mass action the equilibrium pressure p_e of this equation may be expressed by the concentration in the liquid boundary layer C_{ms}:

$$p_e = \left(\frac{C_{ms} \cdot f_m}{K}\right)^z, \tag{113}$$

or, if f_m converges towards unity,

$$p_e = \left(\frac{C_{ms}}{K'}\right)^z. \tag{114}$$

Therefore

$$\frac{dn_g}{dt} = -\varepsilon \left(\frac{C_{ms}}{K'}\right)^z \frac{1}{R'} \sqrt{\frac{R}{2\pi M T_{ms}}}, \tag{115a}$$

or

$$\frac{dn_m}{dt} = -C_{ms} \frac{\varepsilon \cdot z \cdot C_{ms}^{z-1}}{K'^z \cdot R'} \sqrt{\frac{R}{2\pi M T_{ms}}}. \tag{115b}$$

k_m can now be expressed as

$$k_m = \frac{\varepsilon \cdot C_{ms}^{z-1}}{K'^z \cdot R'} \sqrt{\frac{R}{2\pi M T_{ms}}}, \tag{116a}$$

and the transport resistance

$$\frac{1}{z \cdot k_m} = \frac{K'^z R'}{z \cdot \varepsilon \cdot C_{ms}^{z-1}} \sqrt{\frac{2\pi M T_{ms}}{R}}. \tag{116b}$$

$k_m = k_{m1}$ becomes independent of C_{ms} for $z = 1$.

For the C–O reaction we get

$$p_e = \frac{C_{ms}^{(O)} \cdot C_{ms}^{(C)}}{K'}, \tag{117}$$

and therefore for the deoxidation

$$k_m^{(O)} = \frac{\varepsilon C_{ms}^{(C)}}{K'R'} \sqrt{\frac{R}{2\pi M T_{ms}}}, \qquad (118)$$

and for the decarburization

$$k_m^{(C)} = \frac{\varepsilon C_{ms}^{(O)}}{K'R'} \sqrt{\frac{R}{2\pi M T_{ms}}}. \qquad (119)$$

$k_m^{(O)}$ and $k_m^{(C)}$ can only be regarded as constant if a surplus of carbon or oxygen respectively exists.

Equation (116b) shows that the transport resistance may approach infinity: (a) for diatomic gases, such as hydrogen or nitrogen, if C_{ms} approaches zero; (b) for the C–O reaction, if in eqns. (118) and (119) $C_{ms}^{(C)}$ and $C_{ms}^{(O)}$ attain very low values. In both cases the phase transition may therefore become rate-determining. The same is true if ε becomes very small.

For nitrogen in liquid iron, e.g. with $\varepsilon = 1$, values comparable with the transport resistance in the melt ($1/k_m \approx 100$) are reached only at very low surface concentrations of about 2×10^{-7} wt.%. Even if ε should be smaller by two orders of magnitude, the transport resistance at the phase transition would still be relatively low at the concentrations obtained under normal conditions.

Since no reliable experimental data are available for the condensation coefficient of gases at high temperatures, accurate computation of the transport resistance at the phase transition is unfortunately not possible. Experimental investigations have been made with solid metal surfaces, e.g. thin wires or foils, where the transport resistance in the condensed phase may be neglected (see Chapter V, Section 2).

With liquid metals, but restricted to diatomic gases, there is also a possibility to find out by experiment whether the transport in the boundary layer of the melt or the phase transition is rate-determining, because in this case the two reactions are of different order.

For the transport in the condensed phase, eqn. (69) is valid, which corresponds to a first-order reaction.

For the phase transition we get from eqns. (105), (108) and (116a) with k_{m2} for $z = 2$ and $C_{gs} \to 0$

$$\frac{dn_m}{dt} = -2k_{m2}C_{ms} = -\frac{2\varepsilon \cdot C_{ms}^2}{K'^2 \cdot R'} \sqrt{\frac{R}{2\pi M T_{ms}}}, \qquad (120)$$

which corresponds to a second-order reaction. By following up the change of concentration during degassing we may therefore determine which type of reaction is prevailing, and thus which stage is rate-determining.

For instance in degassing of vanadium–nitrogen alloys it was found[44,45] that its course can only be explained by a second-order rate law. It has to be concluded

References pp. 92–93

therefore that here the phase transition is rate-determining.

In other degassing reactions, *e.g.* via CO with a surplus of carbon in the melt, or by evaporation of suboxides, the phase transition proceeds according to a first-order reaction. Therefore a differentiation between the two stages is not possible here.

Kraus[37] made an estimation of the maximum value of the solubility coefficient L below which the phase transition will not be rate-determining, which will be discussed in the following lines.

In eqn. (116a) we can substitute the equilibrium constant K' by L. From

$$p_e = \left(\frac{C_{ms}}{K'}\right)^z = R'T_{ms}C_{gse} = \frac{R'T_{ms}C_{ms}}{L}, \qquad (121a)$$

we get

$$(K')^z = \frac{L \cdot C_{ms}^{z-1}}{R'T_{ms}}. \qquad (121b)$$

By substitution we get from eqn. (116a)

$$k_m = \frac{\varepsilon}{L}\sqrt{\frac{RT_{ms}}{2\pi M}}. \qquad (122)$$

The square root expression is the mean velocity of the gas particles in the gas phase:

$$\frac{\bar{v}}{4} = \sqrt{\frac{RT_{ms}}{2\pi M}} = \beta_g, \qquad (123)$$

which is of the order of magnitude of 3×10^4 cm/s, or about of the order of the speed of sound. It is also equal to the transport coefficient β_g in the case of unrestricted molecular flow. We get therefore

$$k_m = \frac{\varepsilon}{L}\beta_g. \qquad (124)$$

If we neglect the transport resistance in the gas phase (see later) we obtain from eqn. (110)

$$1/\beta = 1/\beta_m + L/\varepsilon \cdot z \cdot \beta_g. \qquad (125)$$

The resistance of the phase transition may be ignored only if

$$L/\varepsilon\beta_g \ll 1/\beta_m$$

or if

$$L \ll \varepsilon\beta_g/\beta_m. \qquad (126)$$

The mass transfer coefficient in the liquid phase β_m of oxygen in liquid iron, for instance, has been determined earlier. The theoretical and experimental data

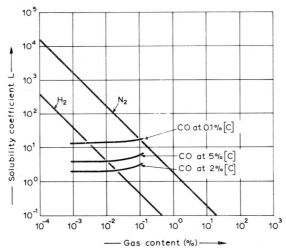

Fig. 43. Solubility coefficients of CO, N_2 and H_2 in relation to the oxygen, nitrogen and hydrogen concentrations in iron melts.

showed values between 0.02 and 0.3 cm/s, depending on whether the degassing process via the C–O reaction occurs with or without formation of bubbles.

With the values above for β_g and β_m we get from eqn. (126) $L \ll \varepsilon \cdot 10^6$, whereby $\varepsilon \lessgtr 1$.

Figure 43 shows that for diatomic gases like hydrogen and nitrogen high values of L are possible. But the concentrations where the critical values are surpassed are very low here. On the other hand, if we look at Fig. 6 we see that with other elements in liquid iron like oxygen, phosphorus and sulphur in equilibrium with the gas in atomic or molecular form, these values are already reached at higher concentrations. The resistance of the phase transition can therefore not be ignored here*.

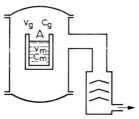

Fig. 44. Schematic diagram of a vacuum chamber containing a melt and attached pumping system.

---

* Because of the very low equilibrium pressures in comparison with the vapour pressure of iron (see Subsection 3.3), this is here also only of theoretical interest.

References pp. 92–93

Finally, if we consider the transport in the gas space, we find that the transport resistance here is of importance only in those cases where the gas density or the partial pressure of the residual gas atmosphere is so high that the flow of gases from the surface will be impeded. If we have molecular flow, i.e. a sufficiently large free mean path of the molecules leaving the surface, they will move into the gas space with the same mean velocity β_g with which they left the surface. Therefore, in this case, an additional flow resistance in the gas space does not exist.

3.1.4 Degassing in crucibles and ladles in relation to time and the parameters of the system

The following discussion is based mainly on the mathematical derivations by Kraus[46].

Figure 44 shows schematically a vacuum melting unit, the vacuum chamber with the pumping system and the melt contained in the melting unit. C_g and C_m are the concentrations of the substance to be removed in the gas space and in the melt in mol/cm^3; V_g and V_m are the volumes of the vacuum chamber and the melt, respectively; and A is the surface area of the melt. For a computation of the change of concentration with time for given initial conditions in relation to the system parameters, the differential equation (69) may be used

$$\frac{dn}{dt} = -\beta(C_m - C_{ms}). \tag{69}$$

The current density of the material transport through the surface of the melt dn/dt is equal to the change in concentration dc_m with time dt in the volume of the liquid below the surface. The total quantity of material liberated from the total surface area A in time dt is

$$A\,dn = V_m \cdot dC_m = -\beta A(C_m - C_{ms})\,dt \tag{127a}$$

or

$$V_m\,dC_m \approx -\beta A(C_m - L.C_g)\,dt. \tag{127b}$$

For a solution of this differential equation the concentration C_g in the gas space must be known. This concentration depends not only on the gas liberation from the melt, but also on the pumping speed S_p of the vacuum pump and on the volume V_g of the vacuum chamber. The quantity of gas transported by the pump in mol/s is equal to the product of the pumping speed S_p, and the concentration C_g. The quantity of gas taken up by the vacuum chamber in the time dt is $V_g \cdot dC_g$. We therefore obtain the relation

$$V_m\,dC_m = -S_p C_g\,dt - V_g\,dC_g. \tag{128}$$

The course of degassing may be calculated by solution of the two simultaneous differential equations (127b) and (128) if some simplifying assumptions are made. It will be assumed that the solubility coefficient L is constant, a condition which

KINETICS

is fulfilled to a large degree in the case of CO formation in liquid steel. It is further assumed that the pumping speed of the pumping system is constant over the pressure range considered. This condition, however, is fulfilled in the high vacuum range only if diffusion pumps or steam ejectors are used, and at higher pressures than 1 Torr if rotary oil pumps are employed. It is not fulfilled in the conventional working range of vacuum metallurgical melting processes, that is between 10^{-3} and 1 Torr. In this case the only possibility remaining is to estimate later the influence which a change in pumping speed may have had on the results calculated. The mass transfer coefficient β is also assumed to be constant. Further details of the computation will not be discussed here.

The results of the calculation of the course of degassing with the aid of a computer starting with the initial concentrations C_{mo} and C_{go} in the melt and in the gas space are plotted in Figs. 45 and 46. As parameters the dimensionless quantities

$$F_1 = \frac{S_p}{\beta \cdot A \cdot L} \quad \text{and} \quad F_2 = \frac{S_p \cdot V_m}{\beta \cdot A \cdot V_g}$$

were used. In the plot depicted in Fig. 45 it was assumed that the concentration C_g in the gas space is initially zero. The change in concentration of the melt related to the initial concentration is plotted on the ordinate, and the time related to the time constant t_k of the degassing process on the abscissa. According to eqn. (127b),

$$\frac{dC_m}{dt} = -\frac{\beta A}{V_m}(C_m - L \cdot C_g). \tag{129}$$

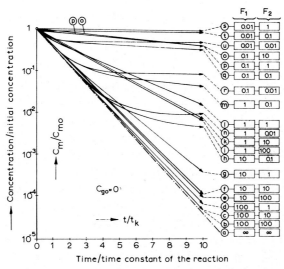

Fig. 45. Course of concentration change in the melt during degassing.

References pp. 92–93

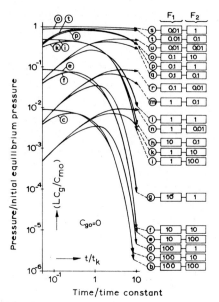

Fig. 46. Pressure change during degassing.

If C_g tends towards zero, and this is the case if F_1 and $F_2 \gg 1$, we obtain by integration

$$C_m = C_{mo} \cdot \exp[-t\beta A/V_m], \tag{130a}$$

or

$$2.3 \log \frac{C_m}{C_{mo}} = -t\beta A/V_m. \tag{130b}$$

The time constant t_k of the degassing process is derived from this equation as

$$t_k = \frac{V_m}{\beta A}. \tag{131}$$

This is the lowest value attainable at a given transport coefficient under optimum conditions. In this limiting case the degassing rate depends only on β and the ratio of $A:V_m$, and no longer on the solubility coefficient L. As may be seen from Fig. 45, degassing takes place at a particularly high rate if for a given β not only the ratio surface area to volume, A/V_m, but also F_1 and F_2, are always very much larger than unity. In this case, it does not matter if S_p and L are not constant.

This family of curves may be used for calculating the required pumping speed for given values of A, V_m, and V_g and the possible maximum values of β and L. The pumping speed will be determined so that the operation can be carried out near to the optimum degassing curve, in other words, so that the degassing process is not choked by the pumping system. In this derivation particular attention should be paid to the solubility coefficient as it usually fluctuates over far wider ranges than does the mass transfer coefficient. If the variation of L is too large, this calculation has possibly to be carried out individually for several concentration ranges.

On the other hand, these curves may also serve for the determination of the mass transfer coefficient on the basis of experimental results of the change in concentration with time, provided that L is known. In such a determination we may start by using an estimated β value and then select that curve which corresponds to the system parameters, so that finally the (t/t_k) value is obtained which matches the concentration change. From the relation

$$(t/t_k) = t_m \cdot \beta \cdot A/V_m, \tag{132}$$

where t_m is the experimentally determined time, a new β value may be determined, which is closer to the true value. This new value may be used as a starting point for a more accurate evaluation of β.

In vacuum induction melting it is common practice to start with $C_g = 0$. If we begin with the equilibrium value $C_{g0} = C_{mo}/L$, which corresponds to the initial concentration, approximately that which exists at the start of a ladle degassing operation, then a family of curves is obtained which deviates noticeably from that depicted in Fig. 45, if $F_2 < F_1$. The degassing process is then slightly retarded at the beginning until the pressure in the vacuum chamber has been sufficiently lowered.

The time dependence of the pressure in the vacuum chamber in relation to the initial equilibrium pressure is shown in Fig. 46. From this family of curves it can be seen which is the pressure course in relation to the values of F_1 and F_2. In this case again, $C_g = 0$ was supposed to be the initial pressure.

A compilation of some constants used in this computation is shown in Table 6.

TABLE 6

SOME CONSTANTS FOR STEEL MELTS

Mass transfer coefficient, β:	
of CO in a boiling steel melt	0.3 cm/s
of CO in a non-boiling steel melt	0.03 cm/s
Solubility coefficient, L:	
of CO in molten steel with 0.1% C	18
of CO in molten steel with 1% C	1.8
Pumping speed, S:	
1 m³/h	2.78×10^2 cm³/s
1 kg CO/h . Torr (at 20°C)	1.81×10^5 cm³/s
Molar concentration of the melt, C_m:	
1 wt.% O in molten steel	4.38×10^{-3} mol/cm³
1 wt.% C in molten steel	5.38×10^{-3} mol/cm³
Molecular density of the gas, C_g:	
1 Torr (at 20°C)	5.47×10^{-8} mol/cm³
1 Torr (at 1600°C)	8.56×10^{-9} mol/cm³

References pp. 92–93

3.2 Evaporation of metallic elements

The following discussions are mainly concerned with the difference existing between the evaporation of metallic elements and gas removal. The effect of various other factors on mass transfer, in particular such influences as concentration of the material to be evaporated, temperature, pressure conditions and foreign gas atmosphere during all stages from the initial phase to the condensing surface, will be dealt with in detail during the discussion on metal distillation in the fourth section of Chapter III.

The various individual stages of the transport process for metals are, in principle, almost the same as those for gases, although an association of atoms to molecules does not usually occur with metals. However, important differences exist in that the vaporization and the transport in the gas space are more likely to become rate-determining for the mass transfer, that spontaneous bubble formation occurs only in exceptional cases, and that the pumping speed of the system has no direct influence on the pressure of the metal vapour above the condensed phase. The reason for the last-named is that the inner surfaces of the vacuum chamber or, in the case of distillation processes, the condenser above the melt, act as a condensation pump. The pumping speed of the external pumping system is then of influence only if sources of permanent gases exist in the system at the same time.

Dissolved metals usually show a materially higher solubility coefficient than do gases. Therefore the phase transition may become rate-determining. Only metallic components with an extremely high vapour pressure and at high concentrations behave similarly to gases; *i.e.*, only in these cases may the transport in the boundary layer of the metal melt become rate-determining.

Many of the relations and all the various factors which promote or retard mass transfer derived for gases apply also to metals. They were described in the preceding sections for the transport in the liquid phase with and without a scavenging gas, for the phase transition, and for transport in the gas space. Vice versa, the relations described in the fourth section of Chapter III on the effect of a foreign gas pressure on the transport coefficient of metal vapours do also apply to the transport of gas molecules. Furthermore, the statement[47] is important, that for pressures below a partial residual gas pressure of about 10^{-2}–10^{-1} Torr the rate of evaporation becomes pressure-independent and that only above this pressure do range diffusion and convection in the gas space become rate-determining for the transport.

According to Langmuir[48] the rate of vaporization of a pure metal in g-atom/cm^2 is

$$\frac{dn_g}{dt} = -k_L \, \varepsilon \, p_e \sqrt{\frac{1}{M_i T_{ms}}}, \tag{133}$$

where k_L is a constant (see list of symbols) and ε the condensation coefficient. It is common practice to extend the validity of this equation also to alloy evaporation.

According to Langmuir's hypothesis[48] this equation is valid even when no state of equilibrium exists, i.e. the rate of vaporization does not depend on the pressure in the gas space (with the limitations mentioned above), but only on the concentration C_{ms} in the surface and on the temperature T_{ms} of the surface.

The rate of vaporization should be distinguished from the effective rate of evaporation, which is the result of the difference between the rate of vaporization as defined above and the material transport, which comes from the gas phase and is absorbed by the condensed phase. Calculation of this quantity is extremely difficult, firstly because the flux towards the surface is usually at a temperature different from that of the surface, and secondly because no isotropic conditions exist at low pressures. If we approach the region of molecular flow, pressure and temperature can no longer be defined. In this case we can only compute from particle densities and velocity vectors. Only the two limiting cases are possibly computable; i.e. if the material transport takes place near equilibrium conditions or if the conditions of molecular distillation are fulfilled and the mass transport towards the evaporating surface approaches zero. Even in these cases the condensation coefficient ε must be known.

If in eqn. (110) the quantity $1/\beta_m \ll 1/\beta$, i.e. if the phase transition and the transport in the gas phase become rate-determining, we obtain instead of eqn. (109) with $z = 1$:

$$\frac{dn_m}{dt} = \frac{dn_g}{dt} = -\frac{C_{ms}(1-C_g/C_{gs})}{1/k_m + L/\beta_g}. \tag{134}$$

As $C_{ms}/C_{gse} = L$ we may express eqn. (134) after substitution as

$$\frac{dn_g}{dt} = -\frac{C_{gse} - C_g/\varepsilon}{1/k_m L + 1/\beta_g}. \tag{135}$$

If we consider only the phase transition or evaporation we get

$$\frac{dn_g}{dt} = -k_m L(C_{gse} - C_{gs}). \tag{136}$$

If the molar concentrations C_{gse} and C_{gs} are replaced by the corresponding partial pressures and if we use also eqns. (123) and (124) for substituting k_m we obtain the expression

$$\frac{dn_g}{dt} = -\varepsilon \frac{(p_e - p_s)}{R'} \sqrt{\frac{R}{2\pi M T_{ms}}}, \tag{137}$$

which is the well known formula for the computation of the rate of evaporation, in which p_e is the saturation pressure above the melt, p_s is the partial pressure of the vapour in the gas space near the surface and ε is the condensation coefficient, which for metal vapours normally approaches unity.

References pp. 92–93

3.3 Evaporation losses of the basis metal during degassing and evaporation of alloying elements and impurities

It has been pointed out in Section 1 of this chapter that degassing and evaporation of impurities can be carried out economically only if the loss in solvent, *i.e.* of the basis metal, during this process is not too heavy. The criterion for a decrease of concentration or enrichment of the impurity or alloying element in the basis metal is the ratio of the relative concentrations in the vapour mixture leaving the surface and in the metal. A reduction of the concentration is to be expected only if this ratio is greater than unity.

For an evaluation the following prerequisites have to be fulfilled:

(1) Atoms or molecules evaporating from the surface of the melt do not return to the melt. Further, the mean free path above the melt is assumed to be so large, that no collisions between evaporating particles and between evaporating particles and residual gas molecules are possible, as these would lead to back-scattering. The relation between the vapour pressure of the evaporating substances and the largest diameter of the evaporating surface, for which this prerequisite is fulfilled, has been derived by Onillon[49].

This condition can be realized either by evaporation at sufficiently low temperature and correspondingly low vapour pressure or by keeping the distance between evaporation surface and condenser sufficiently small, provided that materials are involved the condensation coefficient of which converges towards unity. It is also important that no back-scattering should take place on hot surfaces, such as crucible walls, as the condensation coefficient of the evaporating species might be different at these surfaces[50].

(2) The equalization of the concentration in the metal must be sufficiently rapid so that the concentration gradient in the boundary layer remains sufficiently small. This is always the case when the phase transition or the evaporation process is rate-determining, and also when the rate of gas removal or evaporation is sufficiently small.

We have to distinguish between the following four cases:
1. Degassing via a metal–gas compound *e.g.* a suboxide.
2. Molecular degassing of diatomic gases.
3. Degassing via a nonmetal–gas compound; *e.g.* CO.
4. Evaporation of impurities or alloying elements in elementary form.

Degassing via a metal–gas compound and the second case have been treated theoretically and experimentally by Hörz[45]. The following discussion of the first two cases will be mainly based on his work.

If x_g and x_m are the mol fractions of the element to be removed in the vapour mixture leaving the surface and in the metal we get in the first case, where the degassing is governed by a first-order rate law (see also Subsection 3.1.3), the following equation:

$$x_g = \frac{x_m}{2x_m + k_1/k_2}, \tag{138}$$

and in the second case, which is governed by a second-order rate law,

$$x_g = \frac{x_m^2}{x_m^2 + (1-x_m)k_1/k_3}, \tag{139}$$

where k_1, k_2 and k_3 are rate constants according to the reaction equations

$$\frac{dn_{g1}}{dt} = k_1(1-x_m) \tag{140}$$

for the evaporation of the basis metal,

$$\frac{dn_{g2}}{dt} = k_2 x_m \tag{141}$$

for the degassing in the first case, and

$$\frac{dn_{g3}}{dt} = k_3 x_m^2 \tag{142}$$

for the degassing in the second case*.

In order to find out in which range of concentration a decrease of the gas concentration in the metal is possible, in the first case, when gas removal takes place by evaporation of a gas–metal compound, Hörz[45] comes to an equation,

$$x_{m\,\text{crit.}} = \frac{1 - k_1/k_2}{2}, \tag{143}$$

which is valid for $k_1/k_2 < 1$ and not too high x_m values ($x_m < 0.1$).

This equation defines the critical concentrations above which, depending on the k_1/k_2 ratio chosen, no reduction of the gas concentration in the metal occurs. This is because of the simultaneous loss of metal via the metal–gas compound. Figure 47, which represents eqn. (138), shows that for $k_1/k_2 < 1$ the critical concentration is the higher the lower k_1/k_2 was chosen, as one would expect.

The change of the gas concentration in the metal with time we obtain from

$$\ln \frac{x_m}{x_{mo}} = \left(\frac{k_2}{k_1} - 1\right) \ln(1 - Ak_1 t/n_1), \tag{144}$$

where $Ak_1 t < n_1$ is the amount of g-atoms of the basis metal evaporating in time t from the surface A and n_1 its total quantity at the beginning. $Ak_1 t/n_1$ represents

---

* If k_1, k_2, k_3 are expressed in the coefficients k_{m1} and k_{m2} related to C_{ms1} and C_{ms2} (see Subsection 3.1.3) we get $k_1 = k_{m1}(C_{ms1} + C_{ms2})$, $k_2 = k_{m1}(C_{ms1} + C_{ms2})$, $k_3 = k_{m2}(C_{ms1} + C_{ms})^2/C_{ms2}$.

References pp. 92–93

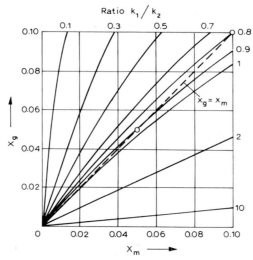

Fig. 47. Relationship between the mole fractions x_g and x_m in the vapour mixture and in the base metal of a gas leaving the surface as a gas–metal compound with simultaneous evaporation of the basis metal for different ratios k_1/k_2 of the rate constants, from ref. 45.

therefore the relative loss of metal.

For $k_1 \ll k_2$ we get from eqn. (144):

$$\frac{x_m}{x_{mo}} = \left(1 - \frac{Ak_1 t}{n_1}\right)^{k_2/k_1}; \qquad (145)$$

and if the metal loss is negligible ($Ak_1 t \ll n_1$)

$$\frac{x_m}{x_{mo}} = \exp\left(-\frac{Ak_2 t}{n_1}\right). \qquad (146)$$

By using this equation, k_2 can be determined for different temperatures within the temperature range in which the prerequisites above are fulfilled and indirectly also by extrapolation towards higher temperatures. With this extrapolation we have therefore the possibility of calculating approximately the course of degassing at any temperature chosen.

In the second case when the degassing is molecular, corresponding to a second-order rate law, we get from eqn. (139) for $k_1/k_3 < 1$ an analogy with eqn. (143) giving the relationship between a critical concentration and the ratio k_1/k_3:

$$x_{m\ crit.} = k_1/k_3. \qquad (147)$$

In contrast with the first case we get an enrichment in the metal if the concentration at the beginning is below this critical value for a certain ratio k_1/k_3. A decrease of concentration takes place only for higher concentrations and is restricted to the limit given by eqn. (147). This is also demonstrated by Fig. 48

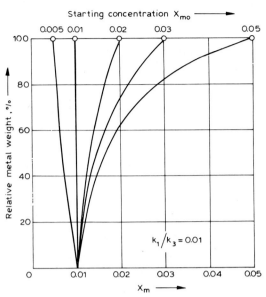

Fig. 48. Relationship between the concentration x_m and relative metal weight for different starting conditions with simultaneous basis metal evaporation and degassing according to a second-order rate law, from ref. 45.

showing the relationship between the concentration and the relative metal weight for different starting conditions for $k_1/k_3 = 0.01$. Concentrations below this limit can only be reached if the gas is simultaneously released as atomic gas[13,51].

As in the first case, the concentration in the metal always increases if $k_1/k_3 \geqq 1$.

Figure 49 shows the relationship between x_m, x_g and k_1/k_3 and indicates in which ranges of x_m and k_1/k_3 a lowering of the concentration can be expected. It is the area above the straight line representing $x_g = x_m$. On this line also the limiting concentrations are found which finally will be reached with different k_1/k_3 ratios.

The analogy with eqn. (146) describing the change of the concentration in the metal with time if the metal loss is negligible and the concentration at the beginning is high compared with the limiting concentration is

$$\frac{x_{mo}}{x_m} = \frac{Ak_3 x_{mo} t}{n_1} + 1. \tag{148}$$

With the C–O reaction a continuous change from the first-order to the second-order rate law can take place if we start with a surplus of carbon and the carbon content slowly decreases by reaction with a continuous supply of oxygen from an oxide crucible or from the surrounding atmosphere. If x_{mO} and x_{mC} are the mole fractions of oxygen and carbon in the metal we get the following equation:

References pp. 92–93

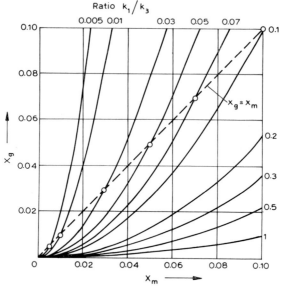

Fig. 49. Relationship between the mole fractions x_g and x_m in the vapour mixture and in the basis metal of a gas leaving the surface as a molecule with simultaneous evaporation of the basis metal for different ratios k_1/k_3 of the rate constants, from ref. 45.

$$x_{gO} = \frac{x_{mO} - x_{mC}}{2x_{mO}x_{mC} + (1 - x_{mO} - x_{mC})k_1/k_4}. \qquad (149)$$

Here k_4 is the rate constant of the degassing equation

$$\frac{dn_{gCO}}{dt} = k_4 x_{mO} \cdot x_{mC}. \qquad (150)$$

For $x_{mO} < x_{mC}$ and $x_{mC} \ll 1$ we get an expression quite similar to eqn. (143):

$$x_{mO\,crit.} = \frac{1 - k_1/k_4 x_{mC}}{2}. \qquad (151)$$

Figure 47 is therefore approximately valid also in this case if we replace k_1/k_2 by $k_1/k_4 x_{mC}$.

For $x_{mO} = x_{mC}$ we get an equation which equals eqn. (147):

$$x_{mO\,crit} = k_1/k_4. \qquad (152)$$

The limiting concentrations for the deoxidation are therefore the same as those which are shown in Fig. 49.

The third case, simultaneous evaporation of basis metal and metallic alloying elements and impurities, has been studied by Olette[52], Fischer[53] and Fischer and Derenbach[50].

Olette[52] defined an "evaporation coefficient" for binary alloys as follows:

$$\alpha = \frac{\gamma_B}{\gamma_A} \cdot \frac{p_B^0}{p_A^0} \sqrt{\frac{M_A}{M_B}}. \tag{153}$$

(For dilute solutions $\gamma_A = 1$.) A denotes the solvent and B the impurities and M_A and M_B are their molecular weights. The "evaporation coefficient", provided that the impurities are metals, is therefore the product of the ratio of the evaporation rates of the pure substances at the corresponding temperature and the ratio of their activity coefficients. With the aid of α, the weight fractions x and y (as per cent) of A and B which have evaporated may be evaluated from the following equation:

$$y = 100 - 100\left(1 - \frac{x}{100}\right)^\alpha. \tag{154}$$

According to whether α is greater or less than unity, the melt will be impoverished or enriched in B.

In order to clarify the relationship between the change in concentration of the substance to be removed and the corresponding decrease in the weight of the melt in relation to α, which demonstrates more clearly at what values of α refining by evaporation is still economical, x and y must be expressed in terms of the concentrations C_0 and C_t and the weights of the melts m_0 and m_t at the beginning, *i.e.* at the time t_0, and at any later time t.

We then obtain

$$x = 100 \frac{m_0 - C_0 - m_t + C_t m_t/m_0}{m_0 - C_0}, \tag{155}$$

and

$$y = 100 \frac{C_0 - C_t m_t/m_0}{C_0}. \tag{156}$$

By substitution in eqn. (154) we obtain

$$\frac{C_t}{C_0} = \left(\frac{m_t - C_t m_t/m_0}{m_0 - C_0}\right)^\alpha \cdot \frac{m_0}{m_t}, \tag{157}$$

or if $C_0 \ll m_0$ and $C_t \ll m_t$,

$$\frac{C_t}{C_0} = \left(\frac{m_t}{m_0}\right)^{\alpha-1}, \tag{158a}$$

or

$$\log \frac{C_t}{C_0} = (\alpha - 1) \log \frac{m_t}{m_0}. \tag{158b}$$

This equation is identical with the one that Fischer[53] derived on the basis of Nernst's partition coefficient. This is defined as the ratio of the mole fractions of the impurities in the melt and in the vapour phase

References pp. 92–93

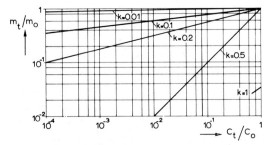

Fig. 50. Change of the weight of the melt with the concentration of the impurity for various partition coefficients.

$$k = \frac{[x_B]}{(x_B)}. \tag{159}$$

If $x_B \ll 1$, then according to Fischer and Derenbach[50]

$$k = 1/\alpha. \tag{160}$$

The relationship expressed in eqn. (158b) using k instead of α as a parameter is plotted in Fig. 50. As may be seen, values of k above 0.1 are of no interest, because more than 20% of the melt must be evaporated in order to lower the impurity level to one-tenth of the initial content.

A relation between the partition coefficient and the molar concentration of the impurity, valid also for higher concentrations and variable activity coefficients, has been derived by Fischer and Derenbach[50]:

$$k'_B = \frac{\gamma_A \cdot x_A \cdot p_A^0}{\gamma_B \cdot p_B^0} \sqrt{\frac{M_B}{M_A}} + x_B = \frac{x_A}{\alpha} + x_B = C_B(1-x_B)+x_B, \tag{161}$$

where x_A and x_B are the molar ratios of the solvent A and the dissolved substance B in the melt. C_B is a constant, if Henry's law is obeyed, i.e. γ_B is constant. The partition coefficient, however, becomes dependent on the concentration, if C_B is either not constant or small. The latter is the case if p_B^0 is relatively high, as, for instance, for manganese in an iron alloy.

The numerical computation of the evaporation process, taking into account the dependence of the partition coefficient on the content of alloying element, makes possible a prediction of the changes in the concentration in relation to the weight loss of the melts even for the more volatile alloying elements.

3.4 Reactions between melts, refractories and slags

The experiments carried out so far on crucible reactions, mostly on a laboratory scale in crucibles of 0.1–4 litre capacity[13,19,20,22,53–55], show that besides the composition the following items are important for the course of the reaction:
(a) The structure of the raw material used for the manufacture of the crucible.

(b) The structure of the sintered crucible.
(c) The level of impurities of lower chemical stability in the crucible material, the content of adsorbed and combined gases, and the history of the crucible.
(d) The pressure.
(e) The temperature of the melt.
(f) The wetting of the crucible by the melt.

The significance of the first four items, a–d, has been elucidated by the systematic investigations of Brotzmann[19] on the behaviour of iron melts in MgO crucibles. While in former research work[56–58] after consumption of the carbon a rapid increase was found in the oxygen content approaching the vicinity of the oxygen equilibrium values under high vacuum conditions, according to Chipman[59] and to Kubaschewski et al.[3] of about 0.1%, recent studies[19,60] have shown that the preparation of high-purity iron in MgO crucibles is quite possible provided that the crucible has been manufactured from high-purity electrically melted MgO and has been sintered to a high density at high temperatures. Although decomposition of the oxide still takes place, it is so slow that an increase in the oxygen level in the melt during the usual melting times is not detectable.

3.4.1 Effect of the structure of the raw material

If melted oxides, forming crystals with smooth cleavage faces during crushing, are used for the manufacture of the crucibles, a crucible surface is obtained after sintering (without binders) with a microroughness substantially lower than that of a sintered, fine-grain powder, such as magnesite. The reactive surface area is therefore considerably smaller.

3.4.2 Effect of the structure of the crucible

A mixture of suitable grain-size, the application of high sintering temperatures and the absence of any binders, particularly of those of low chemical stability, result in a relatively dense, homogeneous and mechanically strong surface, from which no spalling of oxide particles will occur during the gradual disintegration of the surface.

The porosity of the crucible can, as will be shown presently, exercise a strong influence on the position of the reaction equilibria and on the transport of the reaction products to the outer surface. A completely dense crucible with an extremely smooth surface is not advantageous for the degassing of carbon-containing melts because no sources of bubble nuclei are present and because no liberation of gases through the crucible wall can take place. The latter is of particular disadvantage in the case of bubble-free degassing. On the other hand, the area of the reactive surface is reduced. A dense crucible also should have a beneficial effect on the reaction equilibrium, the reason being that if the crucible is permeable to gases, the vacuum conditions at the melt/crucible interface will, with increasing porosity, approach more and more those which exist outside the crucible. As we have seen

References pp. 92–93

earlier, this affects the reaction equilibria in an adverse direction.

3.4.3 Reactive impurities in the crucible

It has already been pointed out in Subsection 2.9 how important is the effect of impurities and other contaminants in the crucible. These may originate from the raw material of the crucible, or may have been precipitated at the crucible surface during previous melting operations, or may have become embodied in the crucible walls by diffusion. A small portion of foreign oxides in the raw material for the crucible, such as Fe_2O_3 or SiO_2, can often not be avoided, if the costs for the refractory materials are to be kept within reasonable limits. These additional sources of oxygen may quite possibly have a greater effect on the refined metal or alloy than the crucible material itself[21].

Fischer and Hoffmann[58] have pointed out that a partition of the oxygen between melt and crucible material may occur. Such a partition was, for instance, observed in iron melts in Al_2O_3, ZrO_2 and MgO crucibles. Brotzmann[19] also confirmed this. He found that during the melting of iron with an initial oxygen content of 0.15% in MgO crucibles the oxygen level was reduced by one-half, whereas the FeO content in the crucible surface rose from the initial 0.4 to 5.2%. The diffusion rate must therefore have been relatively high. An estimate showed that within one hour the FeO diffused up to 5 mm into the MgO crucible material. If later a charge with a lower oxygen content is melted in this crucible, the iron oxide content will slowly be lowered again by the oxygen pick-up of the melt.

The SiO_2 content in the surface zone may also change, depending on the history of the crucible, and may therefore also act as an additional source of oxygen. A rapid decomposition of the SiO_2 occurs in the layers adjacent to the surface, particularly with carbon-containing iron melts, even if a further supply of oxygen by diffusion from deeper layers is possible[19]. If the oxygen level of the melt temporarily increases, for instance as a result of an increase in pressure or of an addition of oxygen-bearing alloying elements, the silicon which has been dissolved in the melt is slagged and precipitated on the crucible wall, where it might later be decomposed during periods of low pressure or by subsequent heat. Similar reactions between melt and crucible are also known to occur with other elements and alloys.

Beside the impurities discussed the moisture in the crucible is also an important source for an oxygen supply to the melt[21]. Every time the melting unit is opened to the atmosphere, atmospheric water vapour will be adsorbed. Kashin and Samarin[61] noticed a marked difference in the quantity of gas liberated when the second heat was charged into the hot crucible without the melting unit being opened to the atmosphere. They also found that the quantity of gas given off became lower with every new charge, because the gases sorbed at the colder parts of the crucible insulation or at the lip of the crucible were gradually liberated. It may also well be that the inner surface layer of the crucible changes, as a result of a decomposition of fine-grain material and of the formation of a less reactive structure, or that by taking up of reaction products a densening of the surface layer occurs and consequently a reduction in porosity.

3.4.4 Effect of pressure

The thermodynamic equilibrium between melt and crucible and the reaction rate depend only on the temperature and are independent of the pressure above the melt, if the crucible material consists of an oxide containing a metal component of low vapour pressure (Al_2O_3, ZrO_2) and if the melt is free from carbon (see also Subsection 2.9). In all other cases, *i.e.* when CO or a volatile metal is formed during the decomposition of the oxides (MgO, CaO) and the gases and vapours are continuously pumped off, the partial pressure which develops in the reaction zone at the surface of the crucible or on the surface of the melt is crucial for the equilibrium which will be reached and for the reaction rate depending on it.

As has been shown in Subsection 2.9, the carbon content of the melt is used for the decomposition of the oxides and when the reduced metal is not volatile and is soluble in the melt, it will replace the carbon. The loss in carbon is therefore a direct measure of this exchange provided that other oxygen sources can be excluded[22].

It should, however, be borne in mind that the pressure at the melt/crucible interface is not the same as at the free surface. If the crucible be absolutely gastight, the CO or metal vapour partial pressure in the reaction zone should be at least as high as the hydrostatic pressure. Further, if bubble nuclei exist in surface recesses with incomplete wetting, these bubble nuclei may grow by diffusion from the adjacent reaction region provided that the equilibrium pressure is sufficiently high. However, a direct reaction of the oxide with carbon, for instance according to reaction eqn. (65), will still not be possible. Only liberation of oxygen from an oxide, for instance by reaction eqn. (52), can take place which, after diffusion into the surface of the bubble nucleus, is transformed to CO. At the high CO partial pressure, which should exist in the lower parts of the melt at the crucible surface, the oxygen equilibrium concentration would, however, remain relatively high and the decomposition of the crucible would therefore proceed rather slowly. Mainly the removal of the oxygen by convection currents towards the surface and the reaction taking place there with carbon would control the oxygen exchange. It should also be borne in mind that the velocity vector of the liquid current in the boundary layer at the surface of the crucible converges towards zero. This means, as has been shown in Subsection 3.1.2, that the mass transfer is materially slower than for a friction-free current. It is therefore very difficult to predict which of the following three stages will become rate-determining: (1) decomposition of the oxide; (2) transport to the melt surface; (3) diffusion through the boundary layer.

The reaction is slowed down even more if the formation of an insoluble, volatile metal is involved, as for instance during the reaction of liquid iron with MgO.

Rather high metal vapour pressures are built up in the reaction zone, which, according to eqn. (62), lead to extremely low oxygen equilibrium concentrations. If the magnesium partial pressure exceeds 1 Torr, the oxygen equilibrium concentration falls below 2 ppm. In this case too a high reaction rate is conceivable only in the zone of the crucible near the surface of the bath.

These considerations show that in a vacuum the reaction equilibria over the major part of the melt-crucible reaction zone are not so unfavourable as one would imagine at first glance.

Rammed or isostatically pressed, sintered crucibles, as conventionally used these days in vacuum melting practice, are, however, always more or less porous, so that, also in the lower parts of the melt, removal of gaseous reaction products through the wall of the crucible is still possible. Brotzmann[19] was able to show that the Mg partial pressure in the reaction zone in rammed and sintered MgO crucibles probably amounts to several Torr and that the magnesium vapour liberated by reaction with an iron melt, as well as the carbon monoxide which was formed in carbon-containing melts, escaped through the crucible wall. Evidence for this is the fact that the quantity of gas liberated remained almost constant when the melt solidified at the surface and no noticeable quantity of gas could pass through the free surface.

We must therefore assume that the reaction products diffuse into gas pockets at the surface of the crucible. From here the reaction products escape through pores and crevices in the crucible. As the flow resistance is sufficiently high so that a pressure gradient of several Torr may build up, the quality of the high vacuum is no longer of great importance for the reaction. The pressure conditions in the reaction zone, in particular the magnesium partial pressure, are here solely responsible for the position of the reaction equilibrium and for the reaction rate. This behaviour is the explanation for the fact that, despite the not particularly high chemical stability of MgO, it is possible to produce high-purity iron melts of low oxygen content in MgO crucibles.

How far an influence of the external pressure on carbon-containing melts exists under such conditions is depicted in Fig. 51[19]. The curves show how the reaction rate increases with falling CO partial pressure. The reason for the flattening of the curves at about 5 Torr, instead of their getting steeper, is the fact that the above-mentioned pressure gradient is built up within the crucible wall and that therefore the internal pressure cannot be materially changed by lowering the external pressure. If the gas permeability of the crucible had been greater, this increase would surely have continued to lower pressures.

Figure 51 also shows that the reaction rate increases with increasing carbon content, the reason being that in the [C] · [O] product, which corresponds to the CO partial pressure in the reaction zone, the oxygen concentration becomes the smaller, the higher is the carbon content. The lower this concentration, the more rapidly will the decomposition of the MgO proceed. From the known values of

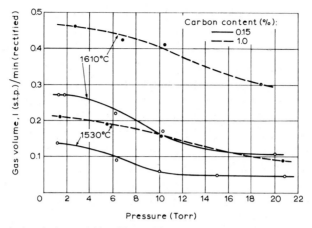

Fig. 51. Gas evolution during melting of iron with 0.15 or 1.0% C in a MgO crucible at different temperatures and pressures, from ref. 19.

gas liberation, the decomposition rate of the crucible was calculated as varying between 0.002 and 0.01 mm/min at a temperature of 1530°C; this may rise to 0.025 mm/min for a temperature of 1610°C.

The effect of the crucible reaction on the melt, *e.g.* with respect to the reduction of the carbon content with time, depends not only on the nature of the reaction participants but also on the ratio of the volume of the melt to the reactive surface area. The greater this ratio, the less will the crucible reaction interfere with the deoxidation of the melt. Furthermore, contamination by the decomposition products of the refractory materials will be lower.

3.4.5 Effect of melting temperature

The effect of the melting temperature on the position of the reaction equilibria is less than on the rate of exchange. This is also evident from Fig. 51. It is, for instance, known that the purity of high-temperature alloys on a nickel–chromium basis, which is attained during remelting in vacuum precision-casting units, depends greatly on the melting temperature. It was found *inter alia* that the number of inclusions is reduced after repeated remelting near the melting point, but that the number of inclusions again increases when the remelting is carried out at a temperature higher by about 100 °C. Superheating of the melt in the production of barstock material leads also to increased crucible reaction and slag formation[55] and should therefore be avoided.

3.4.6 Effect of wetting

The wetting properties of a melt also have a great effect on the reaction rate and the service life of the crucible. It is, for instance, known that oxygen, sulphur and to a lesser degree carbon are surface-active in liquid iron and that they lower the interfacial tension between melt and an oxide material, in other words, promote wetting.

Unfortunately, the investigations on the reaction behaviour and its effect on wettability by determining the change of the contact angle of a metal drop on an oxide-ceramic substrate with time[54] cannot directly be applied to industrial melts as the ratio of melt volume to reaction surface area is greater by several orders of magnitude. For instance, for carbon-bearing alloys it might be quite possible that the carbon content of the metal drop is consumed in a few seconds and that in a relatively short time a reaction equilibrium is attained which under technological conditions probably rarely occurs.

The wettability of the crucible plays an important role, in particular in the melting practice of nickel-rich vacuum casting alloys which are used for precision-casting of high-temperature components and which frequently contain titanium and aluminium as alloying additions[55]. Suppression of wetting is a vital presupposition for the production of high-quality castings and for keeping the percentage of rejects low. Experience shows that with complete wetting, *i.e.* when the melt can penetrate into the pores of the crucible, a marked increase in the crucible reaction occurs. This is in line with expectation as the reactive surface area is greatly increased. In addition, the layer of metal which remains firmly adherent to the crucible wall spalls off parts of the crucible surface during shrinkage so that the subsequent melt becomes contaminated with refractory particles. Wetting and erosion of the crucible are promoted if binders of lower chemical stability are used during the manufacture of the crucible, as these are preferentially decomposed.

However, even in crucibles of a high degree of purity wetting may be observed, if the melting practice is not successful in obtaining low oxygen levels in a short time. There are several causes for this, *e.g.* the leak rate being too high, too much water vapour being desorbed from the walls of the melting unit, or the pumping speed of the pumping system being too low particularly during the first phase of degassing when increased quantities of CO are given off by the melt. Especially if recirculation scrap has to be worked up, it may happen that it is not possible to lower the pressure beyond that critical point at which the liberation rate of oxygen from the melt in the form of CO becomes greater than the amount of oxygen available from the atmosphere above the melt. In addition, it is then not possible to arrive below the oxygen concentration at which the liberation of gas decreases quickly, probably as a result of the reduced wetting. Only if the O_2 and H_2O partial pressures are sufficiently low can the low oxygen concentrations necessary for decomposition of the oxide impurities (which consist mainly of Al_2O_3) be obtained in the melt, and that degree of purity be reached which in turn seems to be the prerequisite for non-wettability. If this cannot be attained, the carbon content of the melt is eliminated mainly through the reaction with the crucible and this to such an extent that a sufficient degree of deoxidation of the melt might become impossible.

3.4.7 Reactions with slags

What has been said about the reactions between metal melts and oxide crucibles applies almost equally to the reactions with slags. This is also evident from the observations described here earlier.

Reactions with oxide particles dispersed in the melt are subject to the same retardation as those occurring in a dense crucible below the surface as was described before. The probability for the formation of CO gas bubbles as a result of the C–O reaction is even lower, as the existence of bubble nuclei in surface cavities is very unlikely. This, however, is different with slags which float on the surface. Because of the absence of hydrostatic pressure and because the removal of gaseous decomposition products is not impeded, very favourable conditions for a decomposition of these oxide particles exist.

Slags may originate from the charge materials or from the crucible, or they may be the result of too high a partial pressure of oxygen or water vapour during the melting process. In the latter case a thin oxide skin is usually formed on the surface, which greatly impedes the mass transfer into the vacuum space, as has been shown in Subsection 3.1.2d.

Since the deoxidation of a carbon-containing melt (which for low oxygen contents takes place without the formation of bubbles) proceeds sufficiently fast only if the crucible reaction is materially slower than the oxygen transport through the free surface of the melt, to avoid or remove such layers of foreign matter might be decisive for the success of a vacuum treatment. The absence of such layers is also necessary because, as for instance in precision casting, it is not always possible to keep these oxide skins and slags away from the casting even by pouring through an intermediate ladle.

Table 2 shows that a CO partial pressure below about 1 Torr should be sufficient for the decomposition of even the most stable oxides in the presence of carbon. This applies even if substantial quantities of either or both of the more highly reactive metals aluminium and titanium are dissolved in the melt. In this case an equilibrium concentration of less than 1 ppm O_2 exists in the contact zone between melt and oxide directly below the surface for 0.1 % carbon. The same applies for carbon-bearing nickel–chromium alloys, the reaction equilibria of which with oxides are not materially more unfavourable than those for iron, as has been shown in Subsection 2.9.

Acknowledgement

The author wishes to thank Professor Dr. M. G. Frohberg and Dr. T. Kraus for helpful suggestions during the preparation of this paper.

References pp. 92–93

List of Symbols

A	surface area (cm²)
a_i	activity of component i (i = 1,2,3... or A,B,C..., or symbol of element)
C_i	concentration of component i (wt % or mol cm^{-3})
c_p	specific heat, with iron melts 0,19 (cal g^{-1} K^{-1})
D	diffusion coefficient (cm² s^{-1})
f_i	$= \dfrac{a_i^H}{C_i}$, activity coefficient (Henry) of element i, C_i in wt %
g	acceleration due to gravity (cm s^{-2})
G	free enthalpy or free energy (cal)
ΔG	free enthalpy of formation or reaction (cal)
$\overline{\Delta G}$	partial molar free enthalpy of solution (cal)
h	hydrostatic head (cm)
H	heat content, enthalpy (cal)
ΔH	heat of formation or reaction (cal)
k_g, k_m	rate constants (cm s^{-1})
k_H	Henry's constant
k, k'	distribution coefficient
k_L	constant of the Langmuir equation (when p in Torr, $k_L = 58.33 \times 10^{-3}$)
K	equilibrium constant expressed in terms of pressures, concentrations or activities
L	$= \dfrac{C_{ms}}{C_{gse}}$, Ostwald's solubility coefficient (C_{ms} and C_{gse} in mol cm^{-3})
ln	natural logarithm, log$_e$
log	log$_{10}$
M	symbol for metal
M_i	molecular or atomic weight of component i (g)
n_i	number of moles of component i (mol)
dn/dt	mass transport flow density (mol cm^{-2} s^{-1})
p	pressure (atm or Torr)
q	density of upward directed thermal flow of metal melts with radiating surface (cal cm^{-2} s^{-1})
r	radius (cm)
R	gas constant, 8.31×10^7 (erg mol^{-1} K^{-1})
R'	gas constant, $84.7 \rightarrow$ (atm cm³mol^{-1}K^{-1})
s	distance of stagnation points (see Figs. 24 and 25) (cm)
S	entropy (cal K^{-1})
ΔS	entropy of reaction (cal K^{-1})

LIST OF SYMBOLS

S_p	pumping capacity (cm³ s⁻¹)
t	time (s)
t_s	contact time (s)
T	absolute temperature (K)
v	velocity (cm s⁻¹)
V	volume (cm³)
x_i	mole fraction of component i
Z	number of bubble nuclei
z	number of atoms of an element which form a molecule at transfer from the condensed phase to the gas phase

Greek letters

α	evaporation coefficient
β	mass transfer coefficient (cm s⁻¹)
γ	cubic coefficient of thermal expansion, with iron melts approx. 2.2×10^{-4} (K⁻¹)
γ_i	$= \dfrac{a_i^R}{x_i}$, activity coefficient (Raoult) of element i
δ	thickness of boundary layer (cm)
δ_v	thickness of friction boundary layer (cm)
ε	condensation or sticking coefficient
η	dynamic viscosity, with iron melts 4.9×10^{-2} (g cm⁻¹ s⁻¹)
Θ	contact angle (deg)
μ_i	chemical potential of component i (cal)
ν	kinematic viscosity, with iron melts 7×10^{-3} (cm² s⁻¹)
ρ	density (g cm⁻³)
σ	surface tension (dyn cm⁻¹ = g s⁻²)
ψ	degree of filling

Subscripts

a	average value
B	bubble
c	cone
e	equilibrium
ext	external
g	gas
i	component i
m	melt
M	metal
o	starting value
s	surface
t	at time t

References pp. 92–93

Superscripts

H	Henry
i	component i
o	standard state
R	Raoult

References

1 L. S. Darken, in J. F. Elliott (Ed.), *The Physical Chemistry of Steelmaking*, Technology Press (M.I.T.) and Wiley, New York, 1958, p. 101.
2 R. Hultgren, R. L. Orr, P. D. Anderson and K. K. Kelley, *Selected Values of Thermodynamic Properties of Metals and Alloys*, Wiley, New York, 1963.
3 O. Kubaschewski, E. Ll. Evans and C. B. Alcock, *Metallurgical Thermochemistry*, Pergamon Press, London, 4th edn., 1967.
4 *Janaf Thermochemical Tables*, Thermal Laboratory, Dow Chemical Co., Midland, Michigan, March 1961.
5 J. F. Elliott and M. Gleiser, *Thermochemistry for Steelmaking*, Vol. I and Vol. II (with V. Ramakrishna), Addison-Wesley, Reading, Mass., 1960, 1963.
6 *Standard Reference Data System*, U.S. Nat. Bureau of Standards, Washington, D.C.
7 Gmelin-Durrer, *Metallurgie des Eisens*, Vol. 1, Verlag Chemie, Weinheim, 4th edn., 1964.
8 C. Wagner, *Thermodynamics of Alloys*, Addison-Wesley, Cambridge, Mass. 1952, p. 51.
9 J. F. Elliott, *Vide*, 135 (1968) 123.
10 A. Simkovich, *J. Metals*, 18 (1966) 504.
11 J. Chipman and D. A. Corrigan, *Trans. AIME*, 223 (1965) 1249.
12 L. S. Darken and R. W. Gurry, *Physical Chemistry of Metals*, McGraw-Hill, New York and London, 1953.
13 O. Winkler, *Met. Rev.*, 5 (1960).
14 K. Krone, J. Krüger and H. Winterhager, *Beitrag zum Schmelzen von NiCr-Basislegierungen im Hochvakuum, Forschungsber. des Landes Nordrhein-Westfalen, No. 1825*, Westdeutscher Verlag, Köln und Opladen, 1967.
15 S. Dushman, in J. M. Lafferty (Ed.), *Scientific Foundations of Vacuum Technique*, Wiley, New York, 2nd edn., 1962, Sect. 8.
16 H. H. Kellogg, *Trans. AIME*, 236 (1966) 603.
17 M. Gleiser, *Trans. AIME*, 211 (1958) 300.
18 L. Brewer and G. M. Rosenblatt, *Chem. Rev.*, 61 (1961) 247.
19 K. Brotzmann, *Arch. Eisenhüttenw.*, 31 (1960) 67.
20 A. Simkovich, *Trans. Vacuum Met. Conf.*, 1967, Am. Vacuum Soc., New York, 1968, p. 361.
21 C. W. Hunter and S. K. Tarby, *Trans. Vacuum Met. Conf.*, 1968, Am. Vacuum Soc., New York, 1968, p. 355.
22 C. B. Griffith and M. P. Fedock, *Trans. Vacuum Met. Conf.*, 1968, Am. Vacuum Soc., New York, 1968, p. 463.
23 M. Schmidt, O. Etterich, H. Bauer and H.-J. Fleischer, *Stahl Eisen*, 88 (1968) 153.
24 G. Melchior, *Vakuum-Tech.*, 18 (1969) 1.
25 A. V. Bradshaw, *Vide*, 138 (1968) 376.
26 R. Higbie, *Trans. Am. Inst. Chem. Engrs.*, 31 (1935) 365.
27 P. V. Danckwerts, *Ind. Eng. Chem.*, 43 (1951) 1460.
28 M. C. Kishinevsky, *Zh. Prikl. Khim.*, 24 (1951) 542 and *J. Appl. Chem. USSR*, 24 (1951) 593.
29 T. Kraus, *Schweiz. Arch. Angew. Wiss. Tech.*, 28 (1962) 452.
30 E. S. Machlin, *Trans. AIME*, 218 (1960) 314.
31 (Lord) Rayleigh, *Phil. Mag.*, [VI] 32 (1916) 529.
32 P. Kosakevitch and G. Urbain, *Rev. Met. (Paris)*, 60 (1963) 143.
33 R. D. Pehlke and J. F. Elliott, *Trans. AIME*, 227 (1963) 844.

REFERENCES

34 T. R. MEADOWCROFT AND J. F. ELLIOTT, *Trans. Vacuum Met. Conf.*, *1963*, Am. Vacuum Soc., Boston, Mass., 1964, p. 1.
35 P. KOSAKEVITCH AND G. URBAIN, *Mem. Sci. Rev. Met.*, *58* (1961) 401, 517, 931.
36 N. A. WARNER, *J. Iron Steel Inst. (London)*, *207* (1969) 44.
37 T. KRAUS, *Tagungsber. 2e und 3e Balzers Kundenkolloquium, 1966 und 1967*, Balzers, Liechtenstein. Rept. published by Balzers AG, FL 9496 Balzers, also T. KRAUS, *Balzers High Vacuum Rept. 10*, 1967.
38 F. N. PEEBLES AND H. J. GARBER, *Chem. Eng. Progr.*, *49* (1953) 88.
39 L. V. BOGDANDY, W. RUTSCH AND I. N. STRANSKI, *Chem. Ingr.-Tech.*, *31* (1959) 580.
40 P. H. CALDERBANK AND A. C. LOCHIEL, *Chem. Eng. Sci.*, *19* (1964) 485.
41 A. V. BRADSHAW AND F. D. RICHARDSON, *Spec. Rept. Iron Steel Inst. (London)*, No. 92, 1965, 24.
42 K. W. LANGE, M. OHJI, D. PAPAMANTELLOS AND H. SCHENCK, *Arch. Eisenhüttenw.*, *40* (1969) 99.
43 J. D. SHARP, *Spec. Rept. Iron Steel Inst. (London)*, No. 92, 1965, 50.
44 J. KRÜGER, *Thesis*, Aachen, 1966.
45 G. HÖRZ, *Z. Metallk.*, *60* (1969) 115, 121.
46 T. KRAUS AND O. WINKLER, *Spec. Rept. Iron Steel Inst. (London)*, No. 92 (1965) 45.
47 B. ILSCHNER AND I. HUMBERT, *Z. Metallk.*, *51* (1960) 626.
48 I. LANGMUIR, *Phys. Rev.*, *2* (1913) 329.
49 M. ONILLON, *Thesis*, Univ. Bordeaux, 1967.
50 W.-A. FISCHER AND M. DERENBACH, *Arch. Eisenhüttenw.*, *35* (1964) 307, 391.
51 T. KRAUS AND O. WINKLER, in R. BAKISH, *Introduction to Electron Beam Technology*, Wiley, New York, 1962.
52 M. OLETTE, *Mem. Sci. Rev. Met.*, *67* (1960) 467.
53 W.-A. FISCHER, *Arch. Eisenhüttenw.*, *31* (1960) 1.
54 E. SNAPE AND P. R. BEELEY, *J. Am. Ceram. Soc.*, *50* (1967) 349.
55 J. B. BARBER AND R. C. HAMBLETON, Investigation into cleanliness of nickel-rich vacuum-casting alloys, *9th Ann. Conf. British Investment Casters Tech. Assoc.*, *1968*, New Avenue Press, Feltham, Middlesex.
56 J. H. MOORE, *Metal Progr.*, *64* (1953) 103.
57 F. WEVER, W.-A. FISCHER AND H. ENGELBRECHT, *Stahl Eisen*, *74* (1954) 1515.
58 W.-A. FISCHER AND A. HOFFMANN, *Arch. Eisenhüttenw.*, *27* (1956) 343.
59 J. CHIPMAN, Physical chemistry of liquid steel in *Basic Open Hearth Steelmaking*, AIME, New York, 2nd edn., 1956, p. 676.
60 W.-A. FISCHER, H. TREPPSCHUH AND K. H. KÖTHEMANN, *Arch. Eisenhüttenw.*, *27* (1956) 563.
61 V. I. KASHIN AND A. M. SAMARIN, *The Uses of Vacuum in Metallurgy*, transl. by E. Bishop, Oliver and Boyd, London, 1964.

Chapter II

Vacuum Engineering

B. D. POWER AND M. E. HARPER

1. Sources of Gas in Metallurgical Process Plant

1.1 Processing pressure

Absolute pressures below atmospheric are most commonly measured in terms of the "Torr", which is almost equivalent to the "mm Hg" of a barometric column. Thus 760 Torr correspond to a standard "atmosphere"; 1,000 "millitorr" correspond to 1 Torr.

The pressure range of most interest for the various vacuum metallurgical processes extends from 100 Torr down to 10^{-6} Torr, a few processes being performed beyond these limits (see Fig. 1).

Selection of processing pressure can greatly influence both choice of vacuum equipment and design of processing plant from the vacuum point of view. If the process pressure can be increased the capital cost is likely to decrease sharply, particularly for processes undertaken in the high vacuum pressure region. To conduct a process at 10^{-4} Torr rather than at 10^{-5} Torr requires pumps of one-tenth of the original pumping speed, with correspondingly smaller high-vacuum valves and connecting pipes. It is important therefore properly to assess the vacuum requirements. For example, a relatively high process pressure is sometimes acceptable if the gases present mainly emerge from the product during processing, and

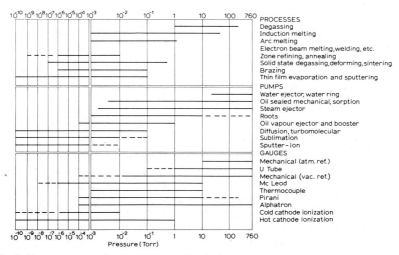

Fig. 1. Pressure ranges for vacuum metallurgical processes and for pumps and gauges.

References p. 143

atmospheric gases have previously been eliminated to a low level.

1.2 Effective pumping speed

There have been many cases of properly selected pumps being rendered ineffective by the use of unduly restrictive connecting pipes, etc. This danger is accentuated under high vacuum conditions.

The mean free path λ travelled by air molecules between collisions with other air molecules at room temperature is given approximately by

$$\lambda = \frac{5 \times 10^{-3}}{p} \text{ cm} \tag{1}$$

where p is the pressure in Torr.

Though turbulent flow may be encountered near the start of the pump-down, streamline or "Poiseuille" flow is commonly obtained in vacuum pump connecting pipes or ducts during much of the pump-down period until the mean free path approaches about one-hundredth of the duct diameter or cross dimension. For air at room temperature in a 50-mm-diam. pipe this corresponds to a pressure of about 0.1 Torr.

The pipe resistance (impedance) to streamline flow varies with pressure. Its influence on the effective volumetric pumping speed at pipe entry can be estimated in terms of the pump speed and inlet pressure and the pipe dimensions at a number of pump inlet pressures and a curve of effective pumping speed against pressure at pipe inlet can be derived. For this purpose the fundamental Poiseuille formula can be arranged:

$$\frac{p_E}{p_P} \approx \sqrt{1 + \left[\frac{L}{91D^4} \cdot \frac{S_P}{p_P}\right]} \tag{2}$$

also, since mass flow is continuous,

$$S_E p_E = S_P p_P \tag{3}$$

where

p_P, p_E are pump inlet pressure and effective pressure at pipe entry in Torr.

S_P, S_E are pump inlet speed and effective speed at pipe entry in l/s.

L, D are pipe length and diameter in cm.

The Poiseuille equation is arranged for air at room temperature and the continuity equation assumes uniform temperature.

Molecular or "Knudsen" flow is approximately obtained in pump connecting pipes, traps, baffles, etc., whenever the mean free path exceeds about one-third of the duct cross dimension. For air at room temperature this corresponds to pressures below about 3×10^{-3} Torr in a 50-mm-diam. pipe or below about 3×10^{-4} Torr in a 500-mm-diam. pipe.

The pipe or component resistance (impedance) to molecular flow does not vary with pressure. Pipes or components in the flow path can be allocated individual conductances (conductance = 1/impedance) which depend only on geometry and on gas type and temperature. The effective volumetric pumping speed at entry to a series of pipes and components connected between a process chamber and the pump inlet can be assessed in terms of their individual conductances and of the pump speed by the reciprocal formula:

$$\frac{1}{S_E} = \frac{1}{S_P} + \frac{1}{C_1} + \frac{1}{C_2} + \cdots \frac{1}{C_n} \qquad (4)$$

where $C_{1,2\ldots n}$ are the conductances,
 also $S_E p_E = S_P p_P$ as above.

The conductance for the molecular flow of air at room temperature of a round tube of diameter D and length L cm is given by

$$C = 12.1 \frac{D^3}{L} \text{ l/s} \qquad (5)$$

The conductance under similar conditions of an aperture or orifice of any shape and of area A cm^2, when the flow enters the aperture from a region of much larger area for flow, is given by

$$C = 11.6 \, A \text{ l/s} \qquad (6)$$

The effective speed at the entry to a pipe carrying molecular flow and connecting a vessel to a pump can be assessed by reciprocally adding the speed of the pump and the conductances of the pipe *and* of its entry aperture (and of any other components in the flow path) to find the reciprocal of the effective speed.

At pressures when the flow is transitional between Poiseuille and Knudsen, both calculating methods may be used, and a reasonable interpolation made.

The brief account above is intended to provide an initial appreciation of the influence of pumping duct dimensions. For a fuller treatment see Bibliography, references 1, 2 and 3.

1.3 Initial gas content

The initial content of atmospheric gas and vapour in the gaseous phase is readily assessed in terms of the total internal free volume of the plant. Over the pressure range where the initial content provides the main pumping load the rate of pumpdown is easy to estimate in terms of the effective volumetric pumping speed. This can be assessed from the speed of the pumps provided and the dimensions of the pumping lines (eqn. (4)). (In many cases the impedance of properly designed vacuum lines does not greatly reduce the pumping speed available over the pressure range in question.)

The curve or the formula in Fig. 2 can be applied to assess the progress of pump-down until other sources of gas load become important. The curve and formula are for constant volumetric pumping speed. If the effective pumping

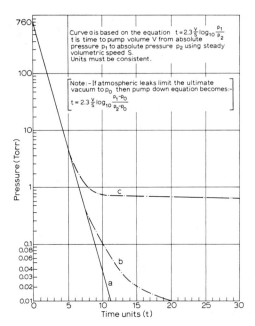

Fig. 2. Pump-down curves for "unit steady pumping speed" pumping "unit chamber volume".
(a) Theoretical curve; time T to pump between any two pressures in volume V using speed S is found by multiplying t from graph by V/S, using consistent units.
(b) Practical curve showing effect of outgassing from an empty steel vessel.
(c) Curve obtained in the presence of a leak which limits the ultimate vacuum to 0.7 Torr.

speed is not constant during pump-down, approximately constant speeds may be assumed over limited bands of pressure and the curve or formula can be applied successively to these. For a fuller treatment of pump-down calculations see Bibliography, references 1, 2 and 3.

1.4 Leakage

Methods for assessing leak rate are discussed in Subsection 4.1. Leak rate is commonly measured in "Torr litre per second" units (Torr l s^{-1}).

The Torr litre is a measure of quantity proportional to mass at constant temperature.

Leakage alone, of magnitude Q_L Torr l s^{-1}, will prevent a pressure lower than p_L Torr being reached such that $p_L S = Q_L$, where S l s^{-1} is the effective pumping speed available at pressure p_L. The manner in which a leakage of Q_L affects pump-down time to higher pressures is illustrated and indicated by formula in Fig. 2. It is seen that pump-down time to a pressure of 10 p_L is not significantly affected and for a process conducted at 10 p_L about 90% of the pumping speed would be available for gas loads other than leakage (assuming steady volumetric pumping speed).

However, other considerations affect the maximum tolerable leak rate.

In many metallurgical processes the in-leaking gas can contaminate the material being processed and an upper limit for tolerable leakage can be assessed on the assumption that all the leakage during the process time will be absorbed by the metal being processed.

If both above considerations suggest that relatively high leak rates are permissible, nevertheless it is unwise to relax standards which can be reasonably held by sound workmanship, since poor leak tightness may be related to faulty structure of a weld or seal so that rapid degeneration may occur in use. From the standpoint of sound structure the following minimum standards are reasonable:

Each component attached to the process vessel: $< 10^{-5}$ Torr l s^{-1}

Each "O" ring seal not above 8 cm diam.: $< 10^{-5}$ Torr l s^{-1}

Each "O" ring seal above 8 cm diam.: $< \dfrac{\text{diam. (cm)}}{8} \times 10^{-5}$ Torr l s^{-1}

Process vessel: A maximum leak of 10^{-4} Torr l s^{-1} is allowed for each 1,000 l of volume.

Leak testing equipment is available which indicates the magnitude of leakage present, see Section 4.

1.5 Virtual leakage

This may arise when a trapped volume has a leakage path to the system interior, for example, when a seam is welded inside and out and some space remains between the welds and the inner weld is leaky. Trapped volumes should be avoided at the design stage, and double welds, for example, should be executed with a full internal weld and an intermittent external weld sufficient to provide adequate strength while still providing adequate venting.

Under this heading may be mentioned the danger of employing materials such as lubricating greases in regions where temperature may be high during the vacuum process. Vapour pressures of such materials increase so rapidly with temperature that they may constitute undesirable sources of contaminating vapour.

1.6 Outgassing

The common structural materials of a high vacuum chamber (metals, sealing elastomers, glasses, etc.) absorb atmospheric gas and water vapour on to their surfaces at atmospheric pressure and this gas and vapour are released continuously though at progressively decreasing rate after high vacuum has been re-established.

When an empty chamber is pumped down from atmospheric pressure gas and (predominantly) water vapour from this source begin to contribute noticeably to the gas load at about 100 millitorr and have become much more important than the residue of the initial gas content when a pressure of a millitorr is approached.

Outgassing behaviour of materials can depend greatly on their nature, sur-

References p. 143

face finish, cleanliness, recent history and temperature. Data are available for most common structural materials of vacuum systems under defined conditions of use. Some data are presented in Table 1 by way of example. In general the manufacturing method and surface finish of a metal are more important than its type. Rolled sheet or forgings are preferred to castings and smooth or polished surfaces are superior to rough or rusty surfaces.

If materials are sensibly selected, outgassing from the empty chamber is usually much less important than outgassing from its contents during a process run. Metallurgical process plant may contain various materials of a porous or hygroscopic nature which can absorb (or may retain from their manufacture)

TABLE 1

OUTGASSING DATA FOR SYSTEM MATERIALS

After one hour under vacuum subsequent to atmospheric exposure:

10^{-4}–10^{-5} Torr l s^{-1}cm^{-2}
  Laminated fibre ("Tufnol" type)*
  Resin-coated fibre glass (coil insulation, etc.)*
  Asbestos-based board ("Syndanyo" type)*
  Ramming mix (fritted in air and after use for one melt)*
  Coarse refractories and refractory bricks*
  (The above materials tend to become deeply permeated with atmospheric water vapour on exposure and this is only slowly released unless the materials are heated during use.)

5×10^{-6} Torr l s^{-1}cm^{-2}
  Pure alumina or mixtures of carbon, silica, silicon carbide and alumina ("Super Salamander") for crucibles.

10^{-6}–10^{-7} Torr l s^{-1}cm^{-2}
  PVC
  Glazed porcelain
  Hot face insulating bricks (alumina/silica mixtures)*
  Araldite
  PTFE
  Vacuum grease ("Apiezon M")
  Slightly rusty mild steel.

10^{-8} Torr l s^{-1}cm^{-2}
  Borosilicate glass
  Polished non-porous metals (rolled or forged, not cast) such as mild steel, nickel or chromium plated mild steel, stainless steel, aluminium.
  (The outgassing rates of the above non-porous materials would drop by about an order after ten hours under vacuum).

After a full bake under high vacuum (400 °C for 16 hours) and subsequent cool-down:

5×10^{-14}–5×10^{-15} Torr l s^{-1}cm^{-2}
  Smooth or polished rolled or forged metals (stainless steel, aluminium, etc.) borosilicate glass.

* Data provided by B. H. Colwell.

large amounts of gas and water vapour through the bulk of the material and available for progressive release during the later stages of pump-down. Some examples of such materials are: resin-impregnated fibre glass tape or laminated fibre for electrical insulation; asbestos composition, carbon, alumina, etc. for crucibles; ramming mixes and coarse refractories; powdered, coarsely sintered or spongy metals. In particular water vapour from such sources begins to evolve copiously at around 10 Torr during pump-down and provides the predominant gas load to be pumped until the processing pressure is reached.

Some values of outgassing rates for some of these materials are indicated in Table 1. For very porous materials, shape and thickness are important; these are ignored in the table, which merely indicates likely orders of magnitude at room temperature. In many cases the desorption is greatly accelerated when heat is applied in the later stages of pump-down prior to processing. Sharp peak gas loads from refractories, radiation shields, etc. should be expected at temperatures between 150 °C and 400 °C. Hygroscopic structural or insulating media more remote from the process region may attain such temperatures after the processing temperature has been reached, with accompanying evolution of water vapour at a time when this can be harmful to the process.

A particular difficulty is the great variation of the magnitude of this sort of gas source with such factors as atmospheric humidity and time of exposure to atmosphere. Exposure to atmosphere should be as brief as possible and it is beneficial if the process vessel is warm (say above 50 °C) during the period of exposure.

The types of material discussed above can release such large amounts of water vapour that the effect on the operation of the pumps must be considered and these must be so selected and operated that they can perform the duty required.

1.7 Gases released during the process

These are often the main gas load to be considered. Vacuum metallurgical processes are, however, so various that only some brief generalities are presented here and details must be sought in the specialist chapters dealing with the various processes.

To design the vacuum system it is desirable to know the nature of the gases, the pressure to be maintained during their evolution, the amounts likely to be released and the time period during which the release takes place. (Sometimes the rate of processing can be controlled to be compatible with the pumping speed provided.) In most cases this information should be sufficiently available on the basis of theoretical treatments and previous practical experience.

Gases may arise from the process as follows:

(i) Direct removal of dissolved gases (*e.g.* nitrogen and hydrogen from molten steel, hydrogen from solid titanium). The way the metal is presented to the vacuum (stirring, pouring, etc.) affects to a great extent the rate of gas release.

(ii) Removal of impurities in the form of gaseous products after chemical

References p. 143

reaction. Typical is the formation of carbon monoxide from carbon and dissolved oxygen or as the result of the reduction of a metal oxide by carbon. Reactions take place, in general, at phase boundaries at a rate controlled by the arrival there of the reaction materials. The crucible wall itself is an important example of such a boundary and it can also provide nucleation sites for the formation of gas bubbles. Reactions may take place here which may also involve the crucible material itself (*e.g.* carbon in the melt reducing the refractory oxide wall to produce copious carbon monoxide).

The speed of such reactions is affected by stirring. Sometimes they are promoted by the deliberate addition of one of the reaction materials.

(iii) Removal of gases introduced to control the process. Examples are the introduction of helium, argon or other gases as a stirring means for molten metal or of argon up to about 30 Torr to suppress too violent agitation in a crucible and requiring removal when this has subsided.

Typical pressure ranges for operating some vacuum metallurgical processes are indicated in Fig. 1. As suggested above, sudden fluctuation in the pumping load may occur during a process run and widely different pressures may be required at different stages in a process. The pumping system must be controlled so that suitable pumps are employed at the various stages.

The most frequently encountered gases are water vapour, nitrogen, hydrogen, carbon monoxide, sulphur dioxide and argon where this is used for process control.

2. Pumps and Pumping Systems

2.1 Water ring pumps

Approximate operation pressure range:

Single-stage: atmospheric pressure to 100 Torr.

Two-stage: atmospheric pressure to 25 Torr depending on operating water temperature.

Capacities available: between about 3 m^3/h and 8000 m^3/h free air displacement.

Running speeds: 1450–700 r.p.m.

Example of service requirements: pump rated at 1500 m^3/h might require 110 l/min make-up water and 30 kW for driving motor.

The rotation of a multi-vane impeller in an eccentric circular or an oval stator compels water contained in the stator to travel in a peripheral path around it so that voids or pockets are formed each bounded by a pair of vanes, the impeller hub and the water ring (see Fig. 3).

As each pocket moves round with the impeller the water moves in and out in a manner controlled by the stator shape and provides a liquid piston in the

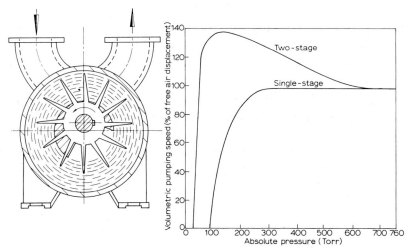

Fig. 3. Construction of water ring pump and examples of performance curves for single- and two-stage pumps.

pocket. Suitably placed inlet and outlet ports in the stator back-cover or in a static drum within the vanes coincide with the pockets at appropriate periods of the liquid piston movement so that a pumping action is produced. Pumps may be connected in series for two-stage work or may be manufactured as two-stage units.

Bearings are external and conventionally lubricated. There are no rubbing parts or close clearances inside the pump and no valves. Water from an exterior reservoir is continuously supplied, usually to points just inside the shaft seals, to cool the pump and maintain its liquid content. Excess water is ejected through the outlet port and returned to the reservoir. New water for cooling is continuously supplied to mix with the circulating water on its way to the pump and the reservoir continuously overflows to waste.

The water ring pump can pump water vapour and many other vapours over its whole working range and can accept moderate amounts of liquid in the gas load without difficulty. Dust and grit are likely to be flushed through the pump without causing damage. Hot gases are cooled and contracted on entering the pump and vapours are partially condensed so that the pumping capacity is increased.

The ultimate pressure of two-stage pumps always exceeds by a small amount the vapour pressure of the sealing water at outlet temperature. This depends on the flow rate of "make-up water" as well as on the make-up water temperature. The limited vacuum performance renders the water ring pump used alone unsuitable for many vacuum metallurgical applications, but it is useful as a primary pump in association with other types (*e.g.* Roots pumps, steam ejectors) to provide a pumping combination with a considerably extended operating pressure range.

References p. 143

2.2 Water-operated ejectors

Approximate operating pressure range: atmospheric pressure to 25 Torr (depending on temperature of operating water).

Capacities available: up to about 2000 m³/h free air displacement (excluding ejector jet condensers).

Example of service requirements: pump rated at 150 m³/h might require 40 l/min make-up water and approximately 20 kW for water pump.

Water supplied under pressure is accelerated through the nozzle to emerge as a high velocity stream into the space around the nozzle mouth (see Fig. 4). The

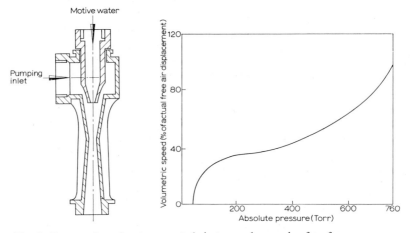

Fig. 4. Construction of water-operated ejector, and example of performance curve.

surface of the water jet becomes sufficiently broken and diffuse in this region to entrain gas and compress it through the throat of the mixing tube and into the diffuser along which the pressure rises as the water stream decelerates until a sufficient pressure to discharge to atmosphere is reached. In some types multiple nozzles are used which provide additional jet surfaces for entrainment. In some cases it is necessary to "drown" the outlet port by providing a tailpiece having a "U" bend or dipping into a well. Operating water may be supplied by a water pump from a storage reservoir into which the used water is returned. A small amount for fresh water is supplied to the reservoir to limit the temperature rise and the excess overflows to waste. Water ejectors can tolerate some liquid content in the gas load and tend to condense vapours and to cool hot gases. They can also tolerate some dust or grit content in the gas load. Although they are of low mechanical efficiency their use is sometimes justified because of these characteristics. They can be used as primary pumps in combination with other types to provide a greatly extended operating pressure range.

2.3 Oil-sealed rotary mechanical pumps

Approximate operating pressure ranges:

Single-stage: atmospheric pressure to 2×10^{-2} Torr (total pressure).

Two-stage: atmospheric pressure to 5×10^{-3} Torr (total pressure).

Capacities available: all sizes up to about 1,700 m³/h.

Running speeds: usually between 350 and 750 r.p.m. (faster for some small pumps).

Example of service requirements: pump rated at 200 m³/h might require 2–3 l/min cooling water and 6 kW (peak) for motor.

2.3.1 Operation

The oil-sealed rotary pump is a positive displacement mechanical pump able to deliver against atmospheric pressure and to pump at approximately uniform speed over a wide range of pressures. Figure 5 illustrates the two most commonly encountered mechanical arrangements and provides examples of typical speed curves. The volumetric pumping speeds are about the same for all types of gas.

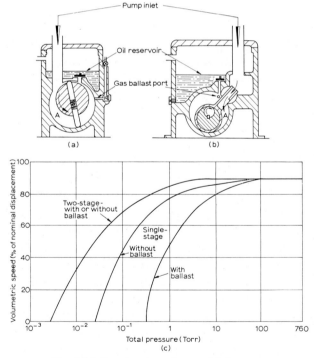

Fig. 5 (a) Construction of rotary blade pump.
(b) Construction of rotary piston pump.
(c) Performance curves for single- and two-stage pumps with and without gas ballast (based on total pressure).

References p. 143

Rotation of the eccentric cam of the rotary piston pump (Fig. 5b) and of the slotted rotor of the rotary blade pump (Fig. 5a) causes the pumping chamber A in communication with the inlet pipe to grow so that gas flows into it. A stage is reached where further rotation isolates this space from the inlet side and transfers it to the outlet side of the pump, where its volume diminishes so that its gas content is compressed and forced out through the outlet port and valve. The valve is submerged in oil to prevent back flow of gas. Oilways (not shown) allow oil from the reservoirs continuously to enter the pumps where it serves to seal against gas leakage the fine clearances between relatively moving parts. Oil is ejected back to the reservoirs through the outlet valves at the end of each compression stroke. The ultimate vacuum is controlled by the more volatile constituents of the oil and by any air, water, etc. which may be dissolved in the oil and which come out of solution when the oil enters the low pressure regions of the pump. Two-stage pumps contain two pumping mechanisms, usually mounted in a single casing and connected to pump in series. Oil is fed first to the outlet stage for "degassing" before entering the inlet stage. Two-stage pumps are not built in very large sizes, but a similar effect is obtained by connecting a small single-stage pump to the outlet of a large one which has been suitably prepared for series operation.

2.3.2 Pumping of vapours

Vapours (such as water vapour) entering the pump inlet may be compressed to condensation point in the pump and ejected as liquid into the oil reservoir. Thereafter they re-circulate into the pump with the lubricating oil and degrade the vacuum that the pump can produce. Such contamination is most commonly controlled by admitting a regulated amount of atmospheric air into the pumping chamber after this has become isolated from the inlet pipe. This "gas ballast" is compressed together with the vapour and contributes the greater part of the pressure needed to open the outlet valve so that the vapour can emerge through the outlet valve before its condensation pressure is reached. Inlet pressures of 15–30 Torr of water vapour can typically be tolerated without pump contamination. Full vapour-handling capacity does not develop until the pump has reached operating temperature with the gas ballast turned on. Some ballast gas leaks back to the inlet side so that the ultimate vacuum is affected as illustrated by the speed curves in Fig. 5c, but the ballast flow can be reduced or discontinued when the main vapour pumping duty is over. Alternatively a two-stage pump which uses gas ballast in the outlet stage only has a performance very little affected by the ballast flow.

A combination of a large and a small pump in series with the small pump using gas ballast can only pump water vapour without contamination with the same mass flow rate as could the small pump working alone. If a water-cooled condenser is included between the two to limit the vapour load on the small pump, the vapour-handling capacity is so enormously increased that it is unlikely ever to be exceeded in practical applications.

2.3.3 Ingress of liquids or dusts

In circumstances where process vapours may condense during their passage towards the mechanical pump and there is in consequence a danger of *liquid* entering the pump, catchpots can be provided on the inlet side where liquid can accumulate. These may require periodic draining.

Oil-sealed mechanical pumps should be protected from grit or dust arising from the process chamber and various types of filter have been used, including fine wire mesh and fabric types. Large area fabric filters of about 15 microns pore size give very adequate protection but impose some loss of pumping speed and require periodic cleaning. (For example, a clean filter of area 1000 cm^2 would reduce the speed of a 100 m^3/h pump by about 7% over the level part of the speed curve.)

2.3.4 Inhibited oils

For process plant use where pump contamination is likely, pump oils may be employed including low-volatility additives which improve resistance against corrosion or gum or lacquer formation inside the pump and provide some protection against seizure. Regular oil change is desirable, according to the nature of the application.

2.3.5 Traps to control backstreaming

These are only required when mechanical pumps are used for the "rough pumping" of process or investigational chambers where extremely clean or ultra-high vacuum conditions are required. The danger of organic vapours migrating from the rotary pump against the pumping direction and producing some contamination in the pumped system exists only during the later stages of rough pumping (at pressures below about 1 Torr).

Cold traps cooled to liquid nitrogen temperature or sorbent traps containing molecular sieve, activated alumina or other sorbent, or traps which cause the breakdown of backstreaming organic vapours in a cold cathode discharge, are available to control such back migration. The last named are not very effective when inhibited oils are used, but the use of such oils is, in any case, avoided for very clean system work.

2.3.6 Arrangements on the outlet side

Where considerable gas loads are pumped or where gas ballast is in continuous use the gas emerging from the outlet side is loaded with fine oil mist. It is normal, in all but the smallest installations, to connect the discharge side of the pump to a pipe leading outside the building. There is danger that any condensable vapours in the pumping load may condense in this pipe and run back into the pump; a catchpot should, therefore, be included directly connected to the outlet port of the pump to catch any liquid draining back from the pipe. This must occasionally be drained. In cases where a discharge pipe is inconvenient, a mist filter, which employs a filtering medium of very fine pore size, can be employed instead and will capture nearly all of the emerging oil mist.

2.3.7 Applications

The oil-sealed rotary pump is widely used on all sorts of process plant where the pumping duty is within its working range, but proper provision must always be made for vapours and dust in the pumping load (as indicated above). It is sometimes advantageous to employ a bank of two or three pumps, each connected to a manifold through an isolation valve and having a small valve to admit air when the pump is idle; this allows flexible operation and the servicing of a pump with less interruption to the operation of the plant.

At pressures below 10 Torr to 1 Torr (depending on system conditions) most process plants begin to evolve such large volumes of gas and vapour that it is no longer economical to employ oil-sealed rotary pumps alone to perform the pumping duty. High volumetric speeds at low pressures are more economically provided by combinations of oil-sealed rotary pumps in series with Roots pumps, or vapour pumps, or turbomolecular pumps (see Subsections 2.4, 2.6 and 2.9) according to the pressure range of interest. Such combinations continue to operate at pressures much lower than those that the oil-sealed rotary pump alone can achieve.

Oil-sealed rotary pumps are also used for the preliminary evacuation of plant, etc., to be pumped at high vacuum by getter or getter-ion pumps (see Subsection 2.7) which are unable to operate until a low pressure has been reached.

2.4 Roots pumps (mechanical boosters)

Approximate operating pressure range: 10 Torr to 10^{-3} Torr (depending on primary pump characteristics).

Approximate capacities available: 150–100,000 m^3/h.

Running speeds: 3,000–500 r.p.m.

Example of service requirements: pump rated at 3,600 m^3/h (1,000 l/s) might require 10 l/min cooling water (including oil-sealed backing pump) and 20 kW for motors (including backing pump).

2.4.1 Description

The Roots pump is a positive displacement pump incorporating twin-lobed rotors of approximate "figure of eight" section rotating in opposite directions in an oval stator (see Fig. 6). The rotors are synchronised by gears at the shaft ends and are so contoured that they rotate with very small clearances maintained between them and between each rotor and the stator walls. Bearings and gears are segregated in compartments at the stator ends and are usually water-cooled. The rotors can run at high rotational speeds giving large volumetric displacements in pumps of moderate size.

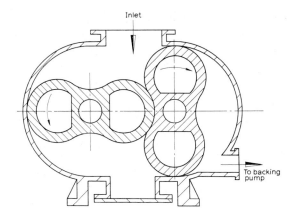

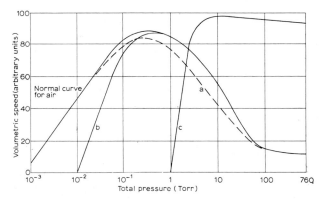

Fig. 6. Construction of Roots pump and example of performance curve for a pump driven through a fluid coupling and backed by a single-stage oil-sealed mechanical pump having about one-eighth of its nominal displacement. The curves show (a) the speed curve pumping hydrogen and (b) the effect on the air speed curve of using gas ballast on the backing pump. The curve (c) illustrates the performance when backed by a two-stage water-ring pump of about the same displacement as the Roots pump.

The rotors are intended to run dry (without lubricating or sealing oil) and so there is a continuous leak back of pumped gas. Roots pumps in high vacuum use are always used in series with a primary pump (usually an oil-sealed rotary pump) and are not engineered to pump a gas load corresponding to their full rated displacement at pressures near atmospheric. The primary pump may be used to evacuate the whole system to a pressure of about 10 Torr, either through a by-pass pipe or *via* the Roots pump (which is allowed to idle around) before the Roots pump is started. Alternatively, the Roots pump may be used to assist the primary pump during the whole period of pump-down from atmospheric pressure if special drive or by-pass arrangements are provided to limit the pressure rise between inlet and outlet.

References p. 143

Normal operation at full rotational speed and without by-pass is possible at pressures below about 10 Torr (depending on backing pump size) but some Roots pumps are not continuously rated until somewhat lower inlet pressures are reached, because of the difficulty of dissipating the heat of compression from the rotors.

2.4.2 Operational characteristics

Because the unsealed clearances exist, Roots pump performance depends sensitively on backing pump pumping speed and ultimate vacuum. A ratio of peak speeds of about 8:1 is usual between the Roots pump and an oil-sealed rotary backing pump, and a compression ratio of about 30:1 is typical at ultimate vacuum. An ultimate approaching 10^{-4} Torr is possible using a two-stage rotary backing pump or pump combination but this could rise to above 10^{-2} Torr if a single-stage backing pump using gas ballast were substituted. A large Roots pump may sometimes have a small Roots pump interposed between it and the backing pump to reduce the dependence on backing pump performance. Speeds with hydrogen are somewhat less than with air, owing to more rapid leak-back through the unsealed working clearances.

Because of their tolerance for vaporous or dusty gas loads, two-stage water ring pumps are sometimes used as primary pumps to back Roots pumps where such loads are encountered. The best ultimate vacuum such a combination can achieve is about 1 Torr, or about 5×10^{-2} Torr if two Roots pumps, in series, are backed by a water ring pump (depending on operating water temperature). See Fig. 6. Roots pumps should, nevertheless, be protected from any significant amounts of grit or dust in the load. Fabric or fine wire mesh filters are often appropriate, but filtration can present difficult problems (see the comments in Subsection 2.6.2 on filters for vapour pumps). If protection is inadequate, abrasion of the rotors may occur particularly where the dry condition of surfaces within the pump chamber is affected by system vapours or oil seepage from bearing housings. Such liquid contamination may cause dust to stick on internal surfaces. Roots pumps must be isolated from water ring pumps, condensers, etc. when idle to prevent rusting.

2.4.3 Applications

Combinations of Roots pumps and backing pumps may be used as the main pumping means for metallurgical process plant operating about in the 10 Torr to 10^{-2} Torr region or they can be used in association with other sorts of pump. If used in parallel with vapour pumps, high volumetric speeds can be obtained over a very extensive pressure range below about 10 Torr. If used for backing and rough-pumping duties in association with vapour pumps, fast pump-down to vapour pump operating pressures can be economically achieved in face of heavy gas loads.

2.5 Steam ejectors

Usual operating pressure ranges:
Single-stage: atmospheric pressure to 75 Torr.
Two-stage: 125 Torr to 20 Torr.
Three-stage: 30 Torr to 2.5 Torr.
Four-stage: 5 Torr to 3×10^{-1} Torr.
Five-stage: 5×10^{-1} Torr to 3×10^{-2} Torr.
Six-stage: 5×10^{-2} Torr to 3×10^{-3} Torr.
(Also seven-stage ejector groups have been built.)
Capacities available: all sizes from below 10 to above 100,000 l/s, but not below about 500 l/s for operation at 10^{-1} Torr and not below about 5,000 l/s for operation at 10^{-2} Torr.
Example of service requirements:
A 4-stage unit for 2000 m³/h (550 l/s) at 1 Torr might require 90 l/min condenser water and 120 kg/h steam.
A 6-stage unit for 100,000 m³/h (28,000 l/s) at 10^{-2} Torr and 50,000 m³/h at 10^{-1} Torr might require 1400 l/min condenser water and 1100 kg/h steam.

2.5.1 Description

A steam ejector stage (see Fig. 7) depends for its pumping action on a very high speed jet of steam produced by a convergent–divergent nozzle and directed down a carefully contoured tube (combining tube or diffuser) to mix with the gas load and carry it through in the pumping direction. The steam supply pressure to the nozzle may vary between about 14 kg/cm² and only about 0.2 kg/cm² de-

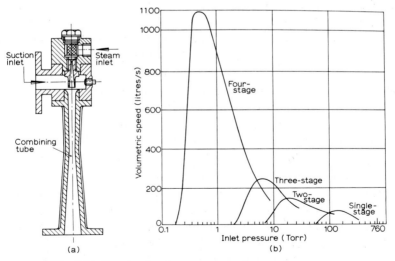

Fig. 7 (a) Constructional arrangement of steam ejector stage.
(b) Examples of performance curves of ejector groups.

References p. 143

pending on the operating pressure range of the stage. Its velocity at the nozzle exit may exceed 1,000 m/s.

Unlike oil and mercury vapour pumps, steam ejectors do not employ condensation on the pump casing walls and a particular stage tends to develop its optimum performance over a fairly narrow band of pressures determined by its geometry and steam supply pressure.

It is uneconomic to use a single ejector stage to produce a pressure difference much greater than 7 to 1, so that multi-stage units are used to operate at pressures below about 80–100 Torr, as indicated above. All but the simplest two- and three-stage units employ condensers between the stages to relieve the next stage in line from the need to pump the spent steam from the upstream stage. An after-condenser at the outlet from the last stage may be used to reduce noise and the annoyance of steam discharge or to recover heat. Units with four or more stages commonly employ condensers only at the outlets of the last three stages since the spent steam emerges from earlier stages at too low a pressure to be conveniently condensed.

The most efficient and economical type of condenser is the direct contact condenser in which the condenser water is made to fall in sprays or curtains through the condenser casing to mix directly with the steam and pumped gas discharged from the ejector outlet. Both the condenser water and the condensate must be extracted from the condenser, which is, in most cases, operating at vacuum. This is most economically done by mounting the condenser at a height greater than the barometric height of a column of water (10.5 m) and providing a drain pipe with its lower end submerged in a "hot well" from which the water can drain away (Fig. 8).

Surface condensers usually consist of a casing containing closely banked tubes through which the condensing water passes and on the outside of which condensation takes place. Initial cost is higher and condensing efficiency is lower than with direct contact types, but only the condensate needs to be drained, so that such condensers may be more convenient for low-level mounting and draining by extraction pump. They may also be used where the condensate may have some noxious content so that it must not be discharged with the condensing water.

Component geometries and steam and water supplies for a multi-stage unit are so selected that optimum performance is reached when the duty required by the particular application is being performed, and these factors would need to be modified for best results performing a different duty. Flexibility of performance is sometimes obtained, for example, by providing two or more ejectors in parallel for certain stages, or by providing separate ejectors of a high throughput type merely for initial pump-down to processing pressure. If multiple ejectors in parallel are fully valved it is possible, in the case of ejector failure, to isolate the faulty component by closing the inlet, outlet and steam valves. The component can then be changed while the rest of the unit remains in operation.

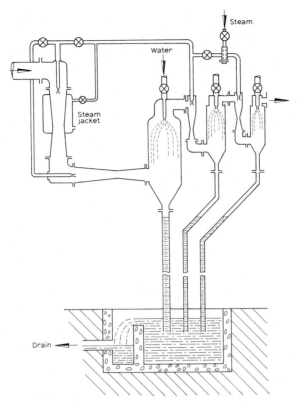

Fig. 8. Four-stage ejector group having direct-contact inter-condensers and after-condenser. Condensers drain barometrically into a common hot-well.

Economy of operation is sometimes obtained, particularly when steam supply or supply pressure is limited, by replacing the two stages at the discharge end of a multi-stage unit by a water ring pump or water operated ejector.

2.5.2 Operational characteristics

Figure 7 presents examples of speed curves for air. The volumetric speed of an ejector group for water vapour is a little greater than that for air. The volumetric speed for hydrogen may be over $2\frac{1}{2}$ times the air speed. Ejector performances are often quoted on a "mass per unit time" basis which would indicate a hydrogen pumping speed only one-fifth the air speed for a $2\frac{1}{2}:1$ volumetric speed ratio.

Compared with other pumping means working in the same pressure range, steam ejector units have exceptional tolerance for water vapour, chemical gases and vapours, dusts, etc. in the pumping load, so that protective filters, etc. are not usually required. Erosion due to liquids and dusts and corrosion due to chemical attack do occur, but periodic replacement of nozzles and diffusers is relatively cheap and special constructional materials can be chosen in difficult cases.

References p. 143

Low-pressure steam ejector units are particularly suitable for large scale operations. The components of a large multi-stage unit are bulky, but they are relatively light and can be positioned in so many alternative ways that roof space or other "dead" space can often be employed to house them.

Typical steam and water supplies for a large steam ejector unit together with performance data are listed at the beginning of this section. At ultimate vacuum, water vapour from the ejector group pervades the pumped system, but this is suppressed by the gas flow towards the pump when any significant gas load is present.

Steam ejector units have no moving parts (except where condenser feed pumps or extraction pumps are employed) but they may suffer loss of performance owing to various causes. Corrosion or erosion of ejector parts may change important dimensions and affect performance, and corrosion of these and of condensers may eventually cause air leaks. If dirt or scale accumulates in nozzles, diffusers or condensers, or if scale occurs inside the tubes of surface condensers, ejector performance or condenser efficiency may be progressively impaired. Strainers or throttles in the steam supply line may be blocked by scale, etc., or strainers or spray nozzles in condenser water circuits may become blocked by débris.

Degeneration of performance due to some of the above causes may render one of the ejector stages unable to sustain operation against the outlet pressure it is subjected to. "Break back" of the motive steam may then occur and so it emerges from the inlet of the ejector stage, possibly accompanied by some condenser water.

Liberal provision of valves between stages and in water and steam lines, and of pressure gauges or gauge points, can greatly facilitate location of faults and the isolation of parts which may require replacement. Many of the likely fault conditions are approached gradually so that routine periodic strip-down for service can greatly reduce incidence of failure in operation.

2.5.3 Applications

Steam ejector units are suitable for large-scale industrial vacuum metallurgical application where high gas throughputs at pressures within their working range are required. They are used, for example, to evacuate large consumable electrode arc furnaces, large induction furnaces and especially steel degassing plant. Particularly important is their application to the stream degassing or vacuum ingot casting of steel; for this duty their ability to pump very high gas loads at low pressures and their high volumetric speed for hydrogen particularly recommend them.

2.6 Vapour ejector and diffusion pumps

Approximate operating pressure range:
Vapour ejector pumps: below about 1 Torr to 10^{-4} Torr.

Diffusion pumps: $\sim 10^{-2}$ Torr to $< 10^{-10}$ Torr.
Capacities available:
Vapour ejector pumps up to 30,000 l/s.
Diffusion pumps up to 90,000 l/s.
Examples of service requirements: an ejector pump rated at 1000 l/s or a diffusion pump rated at 5000 l/s might require 10 l/min cooling water (including backing pump), 6–7 kW for heaters and 4 kW (peak) for backing pump motor.

2.6.1 Description

Oil and mercury pumps employ multiple jets of vapours of high molecular weight fluids, to entrain the gas to be pumped and to drive it in the pumping direction.

The maximum pressure against which such pumps can discharge the pumped gas is usually between a few Torr and a few tenths of a Torr so that a "backing pump" (usually an oil-sealed rotary pump) must be connected to their outlet. Very high pumping speeds can, however, be achieved over a wide band of pressures down to the lowest of vacuum metallurgical interest.

Early diffusion pumps employed mercury as a working fluid, but numerous organic fluids of different volatilities are now available for vapour pumps, and fluids are chosen and pumps are designed to have optimum pumping characteristics in different bands of pressure as required.

Figure 9b illustrates an oil-vapour diffusion pump intended to work in a very low pressure region and Fig. 9c illustrates an oil-vapour ejector pump which employs much more powerful jets to handle greater mass flows of gas at considerably higher pressures. Figure 9a presents typical speed curves for the types of pump illustrated.

Referring to Figs. 9b and 9c, when a sufficiently low starting pressure has been produced by a primary pump, the heaters (1) cause the special oil in the boiler (2) to boil so that a vapour pressure is generated in the space (3) and high velocity vapours emerge from the nozzles (4). Gas above the pump inlet migrates into the pump and becomes mixed with and entrained by the first vapour jet and is driven towards the outlet through the pump by subsequent jets in the same way. The vapour jets impinge on the water-cooled pump walls where they condense to liquid which runs back to the boiler.

Some operating fluid vapour from the pump mouth can migrate back into the pumped system where it contributes to the residual pressure and may contaminate surfaces, gauge filaments, etc. These effects are adequately controlled in almost all cases by the provision of an array of water-cooled surfaces above the pump mouth (baffles) arranged to be optically opaque so that most migrating molecules must strike a cool surface. Such baffles control system contamination, but the vapour pressure of the pumping fluid still pervades the system; this is quite acceptable in most cases. Where it is necessary to reduce or eliminate this vapour pressure, the baffle surfaces must be more strongly cooled by mechanical refrigeration, or a trap which provides condensing surfaces cooled by liquid nitrogen must be used. Such a trap is almost always required with a mercury pump. When an isolation valve is fitted to the mouth of a vapour pump, the valve is often designed to serve as a water-cooled baffle when it is open (baffle valve). The diffusion pump in Fig. 9 is illustrated with a baffle fitted above its inlet.

Some examples of fluids commonly employed in vapour pumps are included in Table 2 together with some of their properties. Vapour ejector pumps most commonly employ chlorinated diphenyl fluids* which are resistant to chemical breakdown when exposed to air or water vapour at operating temperature. Among the numerous diffusion pump fluids available, the silicones are outstanding from this point of view. Many of the other fluids suffer degradation if exposed to atmospheric air at operating temperature. Even with stable fluids it is undesirable, in most applications, to expose to atmospheric pressure vapour pumps at operating temperature. For processes involving regular pump down from atmospheric pressure it is, therefore, usual to employ a system of valving arranged to permit the vapour

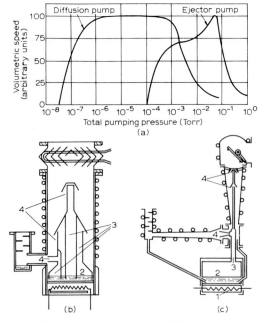

Fig. 9 (a) Examples of performance curves for a diffusion pump and an ejector pump.
(b) Diffusion pump with water-cooled baffle.
(c) Ejector pump with baffle valve.

---

* These have recently become suspect as persistent pollutants. Alternatives include stabilized hydrocarbon oils.

TABLE 2

MAIN TYPES OF OPERATING FLUIDS FOR VAPOUR PUMPS

(A variety of fluids of different vapour pressures, etc. is normally available for each type. These fluids are marketed under numerous different brand names. The appropriate spread of characteristics is indicated below.)

	Molecular wt.	Boiling point at 1 Torr (°C)	Estimated true vapour pressure at 20 °C (Torr)	Comments
Mercury	200.6	127	1.1×10^{-3}	Chemically robust; nearly always requires liq. nitrogen trap. Toxic characteristics must be understood
Paraffinic hydrocarbons	250–580	150–260	5×10^{-5} to 5×10^{-9}	Avoid atmospheric exposure of fluid at working temperature
Esters of phthalic and sebacic acid	278–419	102–215	1.5×10^{-5} to 1×10^{-8}	Avoid atmospheric exposure of fluid at working temperature. Avoid use when pumping load includes much water vapour
Mixed chlorinated diphenyls	290–330	137–150	1.5×10^{-4} to 8×10^{-5}	Chemically robust. Most used for ejector pumps. Slight toxic characteristics must be understood
Silicones (Methyl- and phenyl-methylpolysiloxanes)	480–570	173–254	10^{-6} to 3×10^{-10}	Chemically very robust
Mixed 5-ring polyphenyl ethers	447	285	Below 10^{-9}	Chemically very robust

Note: Ionization gauges calibrated for air normally indicate the vapour pressure of a pump oil about an order greater than the true vapour pressure present, owing to their high sensitivity for such vapour.

pump to remain under vacuum when the process chamber is opened to the atmosphere, and to be isolated from its backing pump when the latter is used to reduce the pressure in the process chamber from atmospheric to a safe operating pressure for the vapour pump. Sometimes a small mechanical pump is provided merely to hold a large vapour pump at low pressure during the period when the main mechanical pump is "rough pumping" a process chamber from atmospheric pressure (holding pump).

Figure 10 illustrates a typical fully valved system. The main isolation valve (1) in this case is water-cooled and serves as a baffle when it is opened. When the

References p. 143

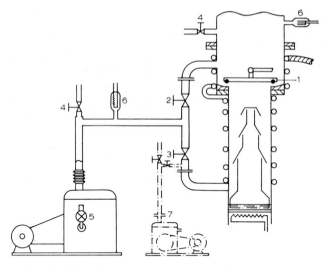

Fig. 10. Fully valved vapour pump system for process plant.

chamber is at atmospheric pressure this valve and the roughing valve (2) are closed and the backing valve (3) is opened. During the rough pumping period the backing valve (3) is closed, and the roughing valve (2) is opened, and at the end of rough pumping, valve (3) is opened, valve (2) is closed and the main isolation valve (1) is opened. If a holding pump is fitted it is brought into use during the rough pumping period.

2.6.2 Operational characteristics

Some examples of performance characteristics are presented in Fig. 9a. All gases and vapours likely to be encountered can be effectively pumped and the low-pressure performance is well maintained in routine use. Of particular value in some metallurgical applications is the very high pumping speed of the vapour ejector pumps for hydrogen. This may be about twice the speed for air.

When working in the millitorr region or at lower pressures, it is particularly important that pumping ducts, baffles, valves or other accessories interposed between the vapour pump and the pumped system provide wide passages for gas flow. Even so, their impedance to flow considerably reduces the effective pumping speed. Manufacturers' data are usually available for the impedance of accessories and methods for estimating impedance and their effect on pumping speeds have been widely published (see Bibliography, references 1, 2 and 3, also Subsection 1.2).

Vapour pumps employing robust fluids can generally pass on to the backing pump volatile vapours, like water, without condensation or chemical interaction with the pumping fluid, but dirt and grit should be prevented from entering the vapour pump if possible. Filtration can present a difficult problem in many metallurgical processes. Fabric or even fine gauze filters may be rapidly impregnated with fine dust so that the effective pumping speed is very greatly reduced. It is often more practicable to use a coarse gauze filter to capture particles of diameter greater than 1.5 mm which would interfere with the sealing of isolation valves and to allow the fine dust to enter the vapour pump. If the rate of entry of fine dust is too great to permit prolonged vapour pump operation between cleanout operations, then a large settling chamber can be interposed between the process chamber and the coarse gauze filter and can be provided with an internal system of baffle plates to impose multiple changes of direction on the flowing gas. Coating of the baffle plates with a low vapour pressure liquid can assist in capturing the fine dust. Such settling chambers should be regularly emptied and any air admission valves should be so positioned that the admitted air does not blow dust towards the pumps. Dusts, etc. which do enter the vapour pump usually accumulate in the boiler, where they may combine with decomposition products of the oil to produce tarry residues which must be cleaned out occasionally.

In operations where large throughputs of gas are regularly pumped some carryover of the pump fluid into the backing system is likely, necessitating occasional replenishment. Large vapour pumps have convenient arrangements for measuring boiler fluid charge and for replenishing or draining the boiler.

2.6.3 *Mercury diffusion pumps with high backing pressure*

Small sizes of mercury vapour pumps are available, capable of delivering the pumped gas at outlet pressures as high as 30 to 40 Torr. Such pumps may be used for gas handling in apparatus designed to determine the amounts of different gases contained in metallurgical specimens. The mercury fluid has the extra advantage that it is inert to most gases and does not contribute any decomposition products, etc. to the gases which are to be measured.

2.6.4 *Applications*

Vapour diffusion and ejector pumps are used either alone or in combination with each other or with other types of pumps on a great variety of vacuum metallurgical process or experimental plant, the operating pressure range and the scale of the operation being important criteria in making a selection.

Vapour pumps are employed to evacuate all types of vacuum furnaces for melting or casting, or for sintering, brazing, heat treatment or degassing. They are employed on some steel degassing plant and on plant for purification of metals, for crystal pulling, for evaporative deposition of thin films and for electron beam welding. They are employed in pumping systems for gas analysis plant.

A large single installation may employ multiple pumping systems of identical pumps, but vapour ejectors are sometimes used in parallel with diffusion pumps to provide high pumping speed over a broad band of pressures, or interposed between diffusion pumps and mechanical backing pumps to increase the degree of compression before the gas reaches the mechanical pumps. Similar mixed systems incorporating vapour pumps and Roots pumps are in use.

2.7 Pumps depending on gettering and ionization

In the last ten years numerous pumps have been described which depend for their operation on the intermittent or continuous deposition of a film of a highly reactive metal on internal surfaces in the pump. The film is almost always of titanium (although other metals are sometimes used, usually in association with titanium) and can "pump" active gases in the system at high speed by chemical combination to form stable compounds. Some pumps which employ the mechanism described above include also an electrode system to provide a continuous ionizing discharge arranged so that a proportion of the ions formed from residual gas is accelerated to impinge on surfaces where the reactive metal film is being deposited. These ions have such high velocity on impact that many remain on the surface for long enough to be buried under the depositing film (some unionized gas molecules seem to be buried in the same way). A positive pumping speed also for inert gases is thus provided. Pumps depending on gettering and ionization are most useful at very low pressure (below about 10^{-6} Torr) and at ultra-high vacuum. They can provide clean and relatively organic-free vacua without the need for traps and baffles. In general, they can operate without fore pumps after the "start-up" procedure has been completed.

2.7.1 Sublimation pumps

Usual operating pressure range: 10^{-2} Torr → $<10^{-10}$ Torr (10^{-1} Torr to assist preliminary pump-down).

Sizes available: up to several thousand l/s.

The most common form of sublimation pump (Fig. 11) employs a central

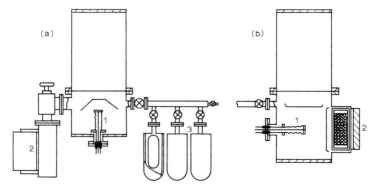

Fig. 11 (a) Example of a plant evacuated by a sublimation pump (1) situated in a well beneath the process chamber in conjunction with a sputter-ion pump (2) provided with an isolation valve. Preliminary evacuation is by sorption pumps (3) used in sequence.
(b) Possible integrated construction if isolation valve is omitted.

group of titanium-bearing filaments, each capable of subliming titanium of the order of a gramme and requiring a few hundred watts of low tension power for its operation. The titanium is deposited on the water- or air-cooled wall of the pump casing, or sometimes on an internal condensing surface which may be water-cooled or liquid-nitrogen-cooled. This last method of cooling modifies the structure of the deposited film and greatly increases the pumping capacity.

Normal operation consists in a preliminary pump-down to about 10^{-2} Torr (using trapped mechanical pumps or sorption pumps), and at this pressure the filaments are heated sufficiently to release sorbed gases, etc. After this, one or more filaments may be operated to deposit titanium to a schedule depending on the requirements of the particular system being evacuated. Automatic intermittent operation is usual, the frequency of operation reducing as lower pressures are reached.

Sublimation pumps are frequently designed to comprise the lower part of an integral structure, the upper part of which is the process or investigational chamber being evacuated (see Fig. 11). The high pumping speeds are, therefore, available immediately adjacent to the source of gas load. Some shielding must be interposed to prevent deposition of titanium within the chamber.

Sublimation pumps do not pump inert gases, and the active films may be "poisoned" by organic vapours; they are therefore always used in association with some other form of pump (*e.g.* sputter-ion pump or trapped diffusion pump), which only needs to have a relatively small speed in most systems. Such combinations may be more economical than the use of a single large pump of a type which can pump both active and inert gases (see Subsection 2.10.6).

2.7.2 *Sputter-ion pumps*

Usual operating pressure range: 10^{-2} Torr $\rightarrow$ $<10^{-10}$ Torr (below 10^{-4} Torr for continuous operation).

Sizes available: up to several thousand l/s.

These pumps employ a continuously sustained cold cathode discharge within an electrode system having titanium negative electrodes. The discharge produces copious ionization of residual gas. The ions bombard the titanium electrodes with sufficient energy to sputter the electrode material on to all internal surfaces accessible to it. Continuously renewed reactive films are thus provided to pump active gases. Those inert gas ions which impinge on surfaces where more titanium is arriving than is being sputtered away are "pumped" by burial. Light organic vapours are decomposed in the discharge and gaseous decomposition products are pumped as above.

The most common electrode structures are: a diode structure with a multi-cell anode (usually of stainless steel) supported between twin titanium cathode plates, and a triode structure where the cathode plates of the diode are replaced by robust titanium grids and an extra pair of electrode plates usually of stainless steel at

References p. 143

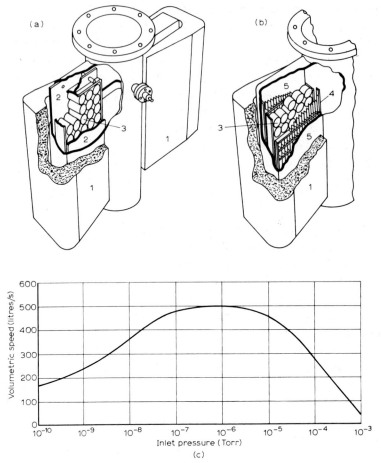

Fig. 12 (a) Constructional arrangement of diode sputter-ion pump.
(b) Constructional arrangement of triode sputter-ion pump.
1 Magnets; 2 Titanium cathodes (earthed); 3 Cellular anode; 4 Titanium grid cathode; 5 Casing acting as collector anode.
(c) Example of performance curve for sputter-ion pumps.

anode potential is provided outside them. These are illustrated in Figs. 12a and b. The electrode potential difference is usually 4–7 kV and a magnetic field of 1–2 kilogauss is provided normal to the plane of the electrodes by permanent magnets. In combination with the symmetrical electrode arrangement, this ensures very long electron trajectories and high ionizing efficiency so that the discharge is sustained to indefinitely low pressures. At high vacuum the discharge current is approximately proportional to pressure so that the sputtering rate is regulated according to the gas load and the current has some usefulness for pressure indication. Typical performance characteristics for a 500 l/s pump are illustrated in Fig. 12 c.

Normal operation consists in preliminary pump-down, preferably to well below 10^{-2} Torr, using trapped mechanical pumps or sorption pumps. The voltage supply to the sputter-ion pump is then progressively increased, care being taken that gas evolution on warm-up does not cause too great a pressure rise or power consumption. As soon as a positive pumping action has been established the preliminary pumping system is isolated.

Sputter-ion pumps can be baked to promote the degassing of their internal surfaces, and many types can continue to operate while subjected to baking at about 300 °C, so that the pumping surfaces themselves consume gases released from elsewhere in the pump. Thorough baking is essential if good performance at ultra-high vacuum (as in Fig. 12 c) is to be achieved.

Sputter-ion pumps are best employed in clean systems operating at below 10^{-6} Torr, and they should not be operated for long at above 10^{-4} Torr. Cathode life at 10^{-6} Torr is 20,000 to 40,000 hours, and is proportionately shorter at higher pressures and longer at lower pressures.

Light organic gases are effectively pumped, but organic vapours may seriously interfere with good operation. Possible operational troubles are electrical short circuits or arcing due to "whiskers" on electrodes or insulator contamination or damage.

Figure 11a illustrates a vacuum chamber evacuated by a sublimation pump together with a sputter-ion pump provided with an isolation valve. Preliminary pumping is accomplished by a battery of three sorption pumps. The sputter-ion pump may be isolated and kept in operation when the chamber is opened to atmosphere. In a baked ultra-high vacuum system the main isolation valve is usually omitted and in this case sublimation and sputter-ion pumps may be integral in a single casing (Fig. 11b).

2.7.3 Applications

The use of sublimation and sputter-ion pumps in the field of vacuum metallurgy is limited, at present, to the evacuation of laboratory apparatus where exceptionally clean conditions are desired, with freedom from all risk of contamination by organic vapours. An example of such apparatus is a plant for the deposition of thin films which must be of exceptional purity. This arises particularly with films of semi-conducting materials or of materials with special magnetic or superconducting properties. Another example is the ultra-high vacuum furnace used in the investigation of gas–metal reactions or of high-temperature properties of reactive materials.

2.8 Cryosorption pumps

Usual operating pressure range: atmospheric pressure to 5×10^{-3} Torr.
Capacities available: suitable, in general, for systems not exceeding 200 l volume.
Certain highly porous materials have the ability to absorb and "condense"

gases and vapours on to their internal surfaces when they are strongly cooled, for example, by liquid nitrogen. The sorptive properties of activated charcoal were used by early vacuum workers seeking the lowest possible pressures, and today numerous materials are being studied for diverse applications over a wide range of pressures.

The present discussion is limited to the use of liquid-nitrogen-cooled cryosorption pumps for the pre-evacuation of clean or ultra-high vacuum systems prior to the employment of some other type of pump to complete the process of evacuation.

Almost invariably the sorbent employed in such pumps is one of the synthetic dehydrated crystalline aluminosilicates known as "molecular sieves", the calcium aluminosilicate with approximately 5 Å pore size being most commonly used. At liquid nitrogen temperature a gramme of this material can sorb between 1 and 10 Torr l of nitrogen to an equilibrium pressure of 10^{-3} Torr and 100 Torr l to an equilibrium of 10^{-2} Torr. Most other atmospheric constituents are strongly sorbed, but the inerts (neon, helium, etc.) are very poorly sorbed.

A cryosorption pump may consist merely of a cylindrical metal canister (stainless steel or aluminium) for containing the sorbent in pellet form, and having a tube extending from its top end cover as a pumping inlet and for filling and emptying the sorbent. The canister is immersed in liquid nitrogen during operation, and may contain radial metal fins to help conduct heat from the sorbent bed to the cooled outer walls, and a perforated duct or partition to allow some of the pumped gas to flow deeply into the sorbent bed before permeating through the tight packet pellets. Other constructional geometries are used. Pumps normally employ between about 100 and 2,000 g of sorbent, although larger sizes have been built.

Pumps absorb system gases when they are cooled. After use, they are valved off from the system and allowed to warm up. The sorbed gases are released and the pump should be vented. A blow-out bung is usually provided in case this is forgotten at any time. After some use, performance may deteriorate on account of accumulation of water in the sorbent; performance can be restored by baking the pump. This is usually carried out for several hours at about 300 °C with the pump vented to atmosphere. Periodically, the sorbent may deteriorate owing to dust formation, etc. and must be renewed.

On all but the smallest systems, cryosorption pumps are used in batteries of two or three pumps, each separately valved, mounted on a manifold with a main valve to isolate it from the system being evacuated (see Fig. 11). The pumps are used in sequence, so that the first pump sorbs the major part of the gas load. The last pump can therefore reach a much lower final pressure without becoming saturated.

Cryosorption pumps are less convenient to use than mechanical pumps and are only used where complete freedom from all risk of organic vapours in the system must be assured. This usually arises in ultra-high vacuum systems used for certain surface investigations or thin-film processes. The cryosorption pumps are employed to provide preliminary vacuum for high vacuum pumps of types which do not require a continuously operating backing pump (such as sublimation pumps, sputter-ion pumps and cryopumps).

2.9 Turbomolecular pumps

Usual operating pressure ranges: 10^{-1} to 5×10^{-10} Torr.
Sizes available: 70–10,000 l/s.
Rotational speeds: 6,000–32,000 r.p.m.

This is a mechanical high vacuum pump akin to a multi-stage axial flow turbo pump but capable of producing organic free ultra-high vacuum.

A pump consists of numerous stages each comprising a fixed ring of blades supported from the cylindrical pump casing and a rotating ring of blades carried on the single axial shaft (see Fig. 13). Drive and bearings are arranged remote from the pump inlet so that organic vapours from lubricants would have to traverse the pump against the pumping direction to enter the pumped system. The pump mostly operates in a pressure region where molecular mean free paths are long

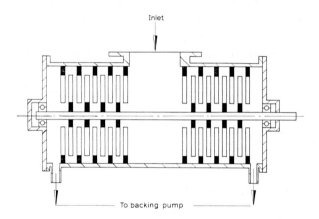

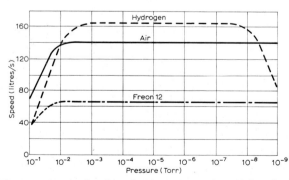

Fig. 13. Schematic arrangement of turbomolecular pump and examples of performance curves.

References p. 143

compared with the distances between adjacent blades and blade rings. Gases are driven in the pumping direction by a series of favourable molecular impacts with fixed and moving blades. This mode of operation results in greatly increased pressure ratios across each stage and across the pump as compared with the performance of turbo compressors operating in the continuous flow region.

The turbomolecular pump must operate in series with a "backing pump", which is responsible for evacuating the turbomolecular pump and the pumped system to a suitable operating pressure for the former and thereafter for maintaining a low pressure at the turbomolecular pump outlet. A two-stage oil-sealed rotary pump is commonly employed as a backing pump. This would typically have about one-twentieth of the rated pumping speed of the turbomolecular pump. The two pumps can be switched on together and the turbomolecular pump reaches its full rotational speed as its normal operating pressure region is approached.

The compression ratio the pump can sustain at ultimate vacuum increases with the molecular weight of the pumped gas. With nitrogen a pressure ratio of above 10^7 may be possible, but with hydrogen the ratio is only 200 or 300. Atmospheric water vapour absorbed on the blades, pump casing, etc. near the inlet end of the pump tends to desorb slowly under vacuum and to limit the pressure which can be reached. To reach the lowest pressures this can be removed by a long moderate bake while the pump is operating and a system of operation can be used which minimizes exposure to humid gas thereafter. Hydrogen diffusing from the backing side may then provide the main constituent of the ultimate vacuum.

Turbomolecular pumps are mostly applied to provide organic free high or ultra-high vacuum for research purposes when pumping cycles are relatively long. They have also been applied for thin-film deposition and other processing purposes where clean operating conditions are mandatory. They require no traps or baffles and can pump a wide variety of gases at roughly comparable speeds over a wide range of pressures.

2.10 Systems for ultra-high vacuum (U.H.V.)

2.10.1 The pressure region

A possible definition of the pressure below which the *high-vacuum* region begins is the pressure below which collisions between molecules start to become less important than molecular collisions with internal system surfaces. A figure of 10^{-3} Torr might then be taken as the threshold pressure (depending on the dimensions of the particular system). In the same way, the threshold pressure for the *ultra-high-vacuum* region may be defined as the pressure below which a surface which has been rendered perfectly clean stays at least substantially free from absorbed gas for a usefully long period for experimental or process purposes. At 10^{-9} Torr a surface rendered gas-free by strong heating or cleavage or film deposition can sustain the collisions of residual gas molecules for about half-an-hour before it becomes fully covered with a single layer of absorbed gas (and proportionately longer at lower pressures). By this definition 10^{-8} Torr might be taken as the threshold pressure for ultra-high vacuum. In ultra-high-vacuum systems a high degree of freedom from organic vapours is frequently sought in addition to a low rate of surface bombardment by residual gas.

2.10.2 Applications in metallurgy

These are not numerous at present, but may become increasingly important. They include the deposition of high-quality thin films of semi-conducting, magnetic or super-conducting materials, zone melting or annealing to produce very pure metals, and investigating the high-temperature properties of reactive materials. In each case a low flux of residual gas to the important surfaces is desired to limit contamination or surface–gas reactions.

2.10.3 Gas load

If the gas load from a conventional system normally operating at 10^{-6} Torr were to be handled at 10^{-9} Torr the volumetric speed (and size) of the pumps would need to be increased a thousandfold. This is normally impracticable. It is necessary instead to design and operate the system in such a way that the gas load is reduced so that it can be handled by pumps of practical size at the required low pressure. Table 1 gives examples of degassing rates of system materials. For ultra-high-vacuum systems the materials with the high degassing rates must be avoided and the materials used must be freed from most of their surface layers of gas and vapour by baking under vacuum during the early part of the pumping cycle. An appropriately improved standard of leak tightness must also be stipulated.

Baking is usually carried out at between 200 °C and 400 °C for between 4 and 16 hours with the system at high vacuum. In addition to the vacuum chamber, it may be applied to part of the pumping system or even to the complete high-vacuum pumps (according to type). The degassing procedure includes the strong heating of filaments and electrode structures. The various events, including the cooling of traps and baffles after bake-out and the start-up of the main pumps, must be performed in their proper sequence. The low pressure desired should then be obtained when the chamber and its contents have cooled down after baking.

Table 1 includes examples of the spectacular improvement in outgassing rate achieved by the baking process.

2.10.4 Pumps for U.H.V.

These include oil or mercury diffusion pumps with suitable traps and baffles, sputter-ion pumps, turbomolecular pumps, sublimation pumps (in co-operation with other types), cryo and cryosorption pumps (usually in co-operation with other types). Most of these have been described in Section 2. A diffusion pump for U.H.V. is usually backed by a second, smaller diffusion pump to help keep the

References p. 143

main pump free from gas and volatiles and (with mercury pumps) to provide vacuum during the bake so that the mouth of the main pump may be baked. Types of sputter-ion pump used for U.H.V. can operate while undergoing moderate baking in with the rest of the system.

2.10.5 System components for U.H.V.

These employ materials selected for their low gas-evolution rates and their ability to withstand baking temperatures. Main constructional materials are usually glass, quartz or stainless steel. Stainless steel flange joints usually employ gold, copper or aluminium gaskets. Valves employ stainless steel bellows for shaft seals and copper, gold, nickel, etc. for seat gaskets; the mechanism is usually on the atmospheric side of the seal and uses a high-temperature lubricant. Electrical feedthroughs employ glass or ceramic insulators.

Plain and modulated Bayard–Alpert gauges (Subsection 3.9) and partial pressure gauges (Subsection 3.10) are suitable for ultra-high vacuum use. Several other special types of ionization gauge employing the cold cathode or hot cathode principles are also available.

2.10.6 Systems and plant for U.H.V.

Many varieties of system exist. Figure 14 illustrates some examples which are described below. It is impossible in so brief an account properly to discuss factors of choice.

(a) The main pump is a mercury diffusion pump surmounted by a refrigerated baffle and a two-hit liquid nitrogen cooled trap. This is backed by a second mercury diffusion pump backed, in turn, by a mechanical pump provided with an oil vapour trap. Separate ovens are provided to bake the vacuum chamber and the top of the main pump with its traps. The chamber can therefore be kept warm while the main pumping system is started up. Characteristics include excellent low-pressure performance, exceptional freedom from organics and the ability to pump all gases and vapours stably and well. Mercury vapour pervades the system during the bake, but not after.

(b) An example of a differentially pumped system where the U.H.V. chamber (with bake out heaters attached) is enclosed in a conventional chamber held at ordinary high vacuum. A lighter construction for the U.H.V. chamber and a much less exacting standard of leak tightness for its seals, etc. are possible. Oil diffusion pumps are used in this example; the pump connected to the U.H.V. chamber is provided with a pair of baffles, one water-cooled and the other refrigerated. The baking does not extend below the refrigerated baffle to reduce the risk of migration of organic vapours into the system. For oil-pumped U.H.V. systems the ultimate pressure is not usually below about 10^{-10} Torr (pump mouth and second baffle are unbaked). The pumping groups are simple and economical but are less desirable where complete freedom from organics is paramount.

(c) An integrated system employing sputter-ion pump and sublimation pump built into the lower part of the main vacuum chamber. Preliminary pumping is by a battery of sorption pumps used in sequence. Two ovens are provided so that the pumps and the upper part of the chamber can be baked independently and at

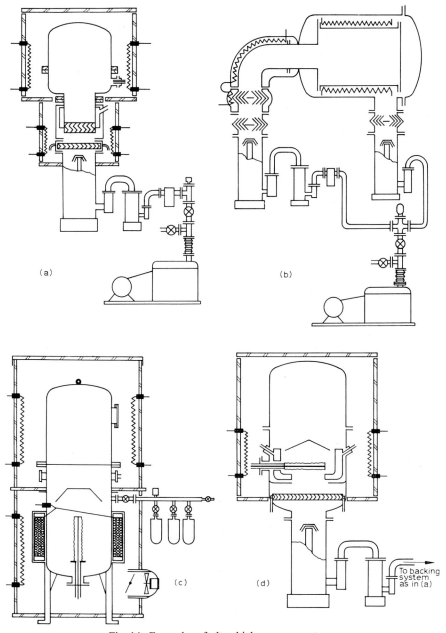

Fig. 14. Examples of ultra-high vacuum systems.

different temperatures if desired. A fan is provided to reduce the cool-down time of the pumps after baking. The sputter-ion pump normally operates during the bake. Low-pressure performance is good and freedom from organic vapours is assured (though light organic gases may be present). The vacuum system is entirely self-contained after pre-pumping and is thus inherently safe in the event of power failure.

(d) A system employing a combination of oil diffusion pumps and a sublimation pump provided with a cooled condenser for the titanium film. The two diffusion pumps and the mechanical backing pump are connected in series so that the main diffusion pump is continuously purged of any volatiles which might tend to accumulate there. The housing above the main diffusion pump contains a water-cooled baffle (immediately above the pump mouth) and an annular reservoir liquid nitrogen trap which also acts as a condensing surface for the titanium sublimation source mounted within it. (This trap could also be operated with water cooling, since the titanium film can effectively trap small traces of organic vapour.)

For lowest pressure operation the vacuum chamber is mounted immediately above the trap housing without any intervening valve (though a flap is sometimes used to suppress pump oil vapour migrating during pump warm-up). The system is baked down to the level of the water-cooled baffle with both the diffusion pumps in operation. The combination provides high pumping speed for all gases and a rapid pump-down to ultra-high vacua comparable with those obtained with systems (a) and (c).

3. Pressure-measuring Devices

Numerous gauges employing various operating principles have been developed to measure a vast range of pressures below atmospheric and down to about 10^{-13} Torr. Most of these gauges can be arranged to operate switch contacts at one or more pre-set pressures and in some cases gauges incorporating pressure switches (or pressure switches without pressure indication) are commercially available. Such switches can be used to operate indicators or safety locks or to initiate an event when a suitable pressure has been reached.

The response of some gauges depends on types of gas present. Some gauges are sensitive to contamination by vapours or solid particles.

The gauge types of most interest for metallurgical applications are briefly reviewed below and the useful pressure ranges of their application are indicated in Fig. 1. Mention is also made of instruments capable of indicating the partial pressures of the various constituents of the residual atmosphere present in a vacuum system since such instruments are beginning to be used in association with certain sophisticated metallurgical plant and their use for process control is likely to increase.

3.1 Mechanical gauges with atmospheric pressure reference

Pressure range: atmospheric pressure → ~ 10 Torr (lower limit set by likely variations in atmospheric reference pressure).

Indication: continuous and true total pressure of gases and vapours present.

Reliability: robust and little affected by contamination in metallurgical systems. Reading varies directly with atmospheric pressure variations which can produce an increasing percentage error as pressure is reduced (unless a correction is applied).

Applications:

Monitoring pump-down and for general system use within working range.

Interstage pressures of steam ejector and other pumping groups within working range.

Leak testing by pressure rise method.

Working principle: The elastic deflection of a Bourdon tube, diaphragm, capsule or capsule stack when a sub-atmospheric pressure is applied inside it is used to operate a linkage situated on the atmospheric side of the deflecting element to transmit movement to a pointer.

3.2 Mechanical gauges with high vacuum reference

Pressure range: 760 Torr → ~ 1 Torr. Accuracy is typically $2\frac{1}{2}\%$ of full scale deflection; types exist with limited ranges down to about 20 Torr full scale deflection. (See also under "Working principle (iii)".)

Indication: continuous. True total pressure of gases and vapours present.

Reliability: Types with mechanical linkage and pointer in atmospheric pressure are little affected by contamination in metallurgical systems. Types with mechanical linkage and pointer in the vacuum space are sensitive to sticking owing to condensation or dust. (In troublesome cases an associated lamp may supply warmth sufficient to prevent condensation and a filter may protect against dust.) Senstitive mechanism should be protected from mechanical shock.

Application: as for Subsection 3.1, but where lower pressure readings are required.

Working principles:

(i) A capsule or capsule stack is evacuated and sealed. Its elastic deflection when a varying sub-atmospheric pressure is applied outside it is used to operate a linkage and an associate pointer, both of which are situated in the varying sub-atmospheric pressure region.

(ii) A capsule or capsule stack has the varying sub-atmospheric pressure applied inside it and operates a linkage and pointer situated in the atmospheric pressure region. A second exactly similar capsule or capsule stack is evacuated and sealed and is so connected to the first that deflecting forces due to atmospheric pressure variation are exactly cancelled out.

References p. 143

(iii) Sophisticated diaphragm gauges exist where a diaphragm has a continuously pumped high vacuum maintained on one side of it and the pressure to be measured on the other side. The diaphragm is sustained in the null position (or its small movements are sensed) electronically. Such instruments can measure down to about 10^{-4} Torr.

3.3 Liquid "U" tube manometer

Pressure range: mercury manometers—atmospheric pressure (according to type) $\rightarrow$ 1 Torr (with accuracy $\sim$ 1 Torr). (Range extended to a few tenths of a Torr for oil manometers or tilted mercury types.)

Indication: continuous, true total pressure of gases and vapours present.

Reliability: Contamination may affect sensitivity. Gas entering the sealed and evacuated arm can introduce error. Such deviations are usually apparent. Oil-filled types may give trouble owing to release of dissolved gas.

Applications: as for Subsection 3.1, but more fitted for laboratory and test use than for industrial use. Calibration standard for mechanical and other gauges covering same range.

Working principle: Transparent "U" tube half-filled with mercury (or oil) having one arm sealed or continuously pumped at high vacuum so that gas pressure in other arm produces an equivalent difference in liquid level in the two arms.

3.4 McLeod gauge

Pressure range: 10 Torr $\rightarrow$ 10^{-5} Torr ($\rightarrow$ 10^{-8} Torr for special laboratory instruments).

Indication: intermittent, several minutes between successive readings (according to type). True total pressure of permanent gases. Response to vapours often grossly erroneous owing to sorption and condensation effects. Vapours can sometimes be accurately measured by a "gas barrier" technique.

Reliability: Contamination of mercury and of glass capillaries can introduce errors, particularly at pressures near low-pressure limit—so also can gas leakage into "lift" mechanism. Such deviations are usually apparent.

Applications: as for Subsection 3.3. Encased types for industrial use are, however, made and a miniature quick-reading type has found considerable industrial application.

Working principle: A glass bulb carrying a closed-ended capillary tube is connected from its underside to the system where the pressure is to be measured. Mercury is made to ascend from a reservoir so as first to seal the bulb from the system, and then to fill the bulb so that its gas content is compressed into the top of the closed capillary. Mercury also ascends a reference tube so that the pressure difference between the gas compressed in the capillary and that in the system under vacuum can be observed. System pressure can then

3.5 Thermocouple gauge

Normal pressure range: 10 Torr → 10^{-3} Torr.

Indication: continuous. Responds to all gases and vapours present, but sensitivities for different gases differ by factors which vary with pressure. Sensitivity is highest for hydrogen. Usually calibrated in terms of dry air or nitrogen.

Reliability: simple, rugged and compact. Contamination can seriously affect calibration.

Applications:

Monitoring pump-down and interstage pressures of pumping groups. General system and process use within working range.

Leak testing by pressure rise method. (Limited use for leak-finding (see Subsection 4.3.3).)

Economical general-purpose industrial and laboratory gauge. Can operate recording and switching devices.

Working principle: The hot junction of a thermocouple (*e.g.* copper constantan) is supported on a filament inside the gauge envelope and is heated at approximately constant power by the application of a constant voltage across the filament. Thermocouple output is displayed on a millivolt meter. Over the operating pressure range, thermocouple temperature varies with pressure because heat loss by gas conduction varies with pressure. At higher pressures this is not so, and at lower pressures heat loss by conduction becomes negligible compared with heat loss by radiation.

3.6 Pirani gauge

Normal pressure range:

~ 10 Torr → 10^{-4} Torr for the more conventional "constant voltage" instrument. (A single such instrument is normally limited to four decades covered in two ranges.)

~ 200 Torr → 10^{-3} Torr for "constant temperature" instruments. (Typically this pressure span is covered in three ranges with automatic range switching.)

Indication: as for thermocouple gauge above.

Reliability: simple and capable of rugged construction. Contamination can seriously affect calibration. Zero drift may require occasional correction with some types. Pirani gauges are available with pre-blackened filaments so that contamination is less likely to affect calibration (see below).

Applications: as for thermocouple gauge. Its sensitivity as a leak finding device (using hydrogen as a search gas) can be improved by use of a special purpose control circuit (Subsection 4.3.3). The wide pressure span of the

References p. 143

"constant temperature" type renders it useful for monitoring almost the whole of pump-down from atmospheric pressure and for monitoring processes including operations at pressures above 10 Torr.

Working principles:

Constant voltage type: A filament supported in the gauge envelope is externally connected to form one arm of a bridge circuit across which a constant voltage is applied so that the filament is electrically heated. In the operating pressure range, filament temperature varies with pressure as for the thermocouple gauge, and so the filament resistance varies. The bridge circuit is balanced at atmospheric pressure for best accuracy at the higher operating pressures or at high vacuum for best accuracy at the lowest pressures in the operating range. Within the operating pressure range the "out of balance" bridge current is used to indicate pressure.

Constant temperature type: A very fine filament supported in the gauge envelope is supplied with a heating current which is made to vary in such a way that the filament temperature remains constant however the pressure varies. The driving voltage required provides an indication of pressure. The method permits the design of instruments of wide pressure range.

Blackened filaments: Contamination may increase filament emissivity and radiant heat loss and thus affect calibration. Pre-blackened filaments initially possess emissivity near unity and are unlikely to be changed by contamination, so that calibration is better maintained (at the expense of some sensitivity).

3.7 "Alphatron" ionization gauge

Pressure range: 760 Torr → 10^{-4} Torr.

Indication: continuous. Linear response. Responds to all gases and vapours present. Sensitivities for different gases differ by approximately constant factors (*e.g.* response for hydrogen about one-quarter of response for same pressure of air). Usually calibrated in terms of dry air.

Reliability: The pressure-sensing head itself is simple and its characteristics are stable. The signal current produced is very small (*e.g.* 10^{-13} A at 10^{-3} Torr) and sensitive amplification is employed.

Applications: as for thermocouple gauge. The continuous indication from atmospheric pressure extends its usefulness.

Working principles: A few hundred micrograms of radium stored within the gauge envelope are used as a stable source of alpha particles to ionize gas and vapour molecules. The rate of production of ions is directly proportional to the residual gas pressure. Only about 30–40 V are required to attract the ions to a collector electrode for measurement. A pre-amplifier is carried on the gauge head. Two ionization chambers (small and large) are used to cover the full pressure range. The radioactive hazard is very small, but its presence should be recognized.

3.8 Cold cathode ionization gauge (Penning type)

Pressure range: 10^{-2} Torr → 10^{-6} Torr (can be extended to 10^{-10} Torr).

Indication: continuous. (Accuracy is not high.) Responds to all gases and vapours present. Response is non-linear with pressure and sensitivity depends on gas type. "Pumping action" of gauge necessitates short wide connection to main system to ensure correspondence of pressures.

Reliability: Rugged internal structure and simple control circuit. Sustains atmospheric exposure and is reliable on clean systems. Organic vapours are broken down by cold cathode discharge. Contamination of electrodes or insulators by heavy vapours, breakdown products or system dust can seriously affect operation. Some protection is obtained by a baffle plate overlapping entry to gauge head and spaced off. Heads for industrial use are best made demountable for cleaning with detergents, solvents, acids or abrasives.

Applications: Monitoring the later stages of pump-down. General system and process use within working range, can operate recording and switching devices. One of the most economical and rugged instruments covering an important industrial pressure range and widely used on process plant working in the 10^{-3}–10^{-5} Torr region.

Working principle: Twin flat-plate cathodes, usually at earth potential, are arranged face-to-face within the gauge envelope with a ring-shaped or cylindrical anode between them. Electrode potential difference is 2–3 kV and a permanent magnet external to the gauge envelope provides a field of 300–400 oersteds perpendicular to the cathode surfaces. At low pressures a cold cathode discharge is obtained. The electrons emitted from the cathodes by ion impact are constrained by the magnetic and electrostatic field arrangements to travel in such very long helical and oscillatory paths that the ion yield is high and the discharge is well sustained. The ion current varies with pressure and is sufficient, over the main working range, to operate a microammeter directly without amplification.

3.9 Hot cathode ionization gauge

Normal pressure range:
10^{-2} → about 10^{-7} Torr.
Bayard–Alpert type 10^{-3} → about 10^{-10} Torr.
Above with modulating electrode 10^{-3} → about 10^{-12} Torr.
Special "high pressure gauge" 1 → about 10^{-6} Torr.

Indication: continuous. Linear response. Responds to all gases and vapours present. Sensitivities for different gases differ by approximately constant factors (*e.g.* response for hydrogen about one-third of response for same pressure of air). Usually calibrated in terms of dry air. Gauge electrode structure may project unshrouded into main vacuum system for quick response to true system pressure.

References p. 143

Reliability: Hot tungsten filaments must eventually fail quickly if exposed to high pressures. For special "high pressure" gauges, therefore, iridium filaments coated with thoria are used which may even be operated at atmospheric pressure. Contamination of electrodes can affect calibration and contamination of insulators can cause leakage currents affecting particularly low pressure indications. Control circuitry is elaborate. Gauges for such low pressures are refined instruments and must be operated with understanding. Accuracy is superior to that of cold cathode type.

Applications: Monitoring the later stages of pump-down. Experimental and test use. General systems and process use where superior accuracy or low-pressure performance justify extra cost and reduced ruggedness (*cf.* cold cathode type). Can operate recording and switching devices. Limited use for leak-finding; this can be improved by special circuitry (Subsection 4.3.2).

Working principle: The gauge electrodes comprise a filament directly heated by the passage of a current, a grid having a positive potential with respect to the filament of about 150 V and (beyond the grid) a collector electrode having a negative potential (with respect to the filament) of about 20 V. The filament temperature is so controlled that a constant emission current passes to the grid, some of the electrons having trajectories which take them beyond the grid before falling back into it. The electron stream produces ionization of residual gas, and favourably situated ions travel to the ion collector to provide an ion current which is directly proportional to the pressure for any particular gas. The calibration depends on gauge geometry and operating conditions and on gas type. Typically, the ratio of ion to electron current is about 10 per Torr for nitrogen.

Electrons arriving at the grid generate soft X-rays and some of these strike the ion collector, producing some electron emission which has the effect of a spurious ion current in the pressure-measuring circuit. Gauges with plate-type collectors exhibit spurious current equivalent to around 10^{-8} Torr of nitrogen, and so become untrustworthy at pressures below 10^{-7} Torr.

Bayard–Alpert gauges have a single fine wire as an ion collector. The target for X-rays is thus made of such reduced area that the spurious indication is limited to about 10^{-11} Torr nitrogen pressure and the gauge can be used with reasonable accuracy down to about 10^{-10} Torr. Such gauges are less resistant to contamination effects at the higher pressures of possible operation.

For use down to about 10^{-12} Torr, Bayard–Alpert gauges can be provided with an extra electrode (modulating electrode) near the collector electrode. By varying the potential of this electrode the gauge can be made to operate in alternative modes, which permits spurious currents to be identified and allowed for.

3.10 A note on partial pressure gauges (mass spectrometers)

It is increasingly desirable to know the types and individual amounts of the gases which comprise the residual atmosphere of high vacuum apparatus. There are cases where a very low total pressure is less important than a low partial pressure of a particular gas so that, if this can be measured, the plant operation can begin safely at a higher total pressure and after a shorter pumping period. Information about the nature of the residual gases in a vacuum system can often help in identifying faults or in interpreting experimental observations. By monitoring the changing gas composition during a process the progress towards completion of the process can sometimes be assessed.

An extensive family of high vacuum instruments is available for analysing the composition of gases by ionizing some of the gas (as in an ionization gauge) and measuring separately the currents of ionized molecules of different molecular masses. Some of these instruments are elaborate and costly, having their own pumping systems and capable of sampling a complex gas mixture from a region at any pressure and analysing it comprehensively.

Instruments for direct mounting on vacuum systems are relatively simple and the sensing head may be no bigger than a conventional gauge head. There are several basic types which employ different methods to separate streams of ions of different masses for collection and measurement. They are sometimes referred to as "partial pressure gauges".

A frequently encountered instrument suitable for mounting on vacuum systems is the magnetic deflection mass spectrometer. A controlled electron stream ionizes all species of residual gas molecules in a small ionization chamber. A beam of ions is directed out of the ionization chamber by accelerating electrodes and travels through a uniform transverse magnetic field so that the various ionized species are deflected to travel circular paths of different radii depending on their masses and degrees of ionization. The radii of the paths of the ions can also be varied by changing the accelerating voltage so that the ions of each species in turn can be made to pass through a slit in a shield to reach a collector electrode for the ion current to be measured.

It is apparent that the instrument can only measure the ion current corresponding to one species of ion at a time so that the mass range of interest must be "scanned" by progressively varying the voltage and observing the peaks of ion current on a recorder trace (or alternative indicator). (This is the same with almost all mass analysers.) The calibration factor for converting ion currents to partial pressures differs for different gases. Some gases and vapours are partially "cracked" into components which produce a group of peaks on the recorder trace; these "cracking patterns" are so characteristic of the original gas or vapour that they come to be recognized as a means for confirming its presence.

Different designs of such instruments may have widely different performance characteristics; the following provide a typical example.

References p. 143

Max. operating pressure: 10^{-4} Torr.
Lowest partial pressure detectable: 10^{-11} Torr.
Capable of operation as *total* pressure gauge.
Mass ranges: 1–10 and 10–100 atomic mass units (a.m.u.).
Scanning time per range: 0.5–5 min (according to operating conditions).
Capable of being "tuned" to monitor a single peak.
Resolving power: Can distinguish equal peaks at about mass 40–50 only 1 a.m.u. apart without the "valley" between the peaks exceeding 1 per cent of peak height.

4. Leak-proving and Leak-finding

Leak rates are measured in Torr l s^{-1}. The Torr litre (product of pressure and volume) is proportional to mass of gas if temperature is stable. Values for tolerable leak rates are discussed in Subsection 1.4.

Initial leak-proving and leak-finding for components, chambers, etc. should be undertaken before these are painted.

4.1 Leak-proving methods

4.1.1 Pump-down

Evacuate with pump of known speed S at pressure p such that $pS = Q_L$ max., the maximum tolerable leak rate. If pressure falls below p then leak rate is below tolerable limit.

System outgassing contributes fictitious leakage (see Subsection 1.5) and so system is best tested clean and empty of gassy materials; a liquid nitrogen trap in the gauge line helps to suppress errors due to system outgassing. The method is capable of great sensitivity with clean systems; the convenience and cost of its use should be weighed against alternative methods in each particular case.

4.1.2 Leak-up (or pressure rise)

Assess system volume V. Evacuate system and gauge to low pressure and isolate from pump. Record pressure rise against time, possibly for many hours.

Leak rate $Q_L = \dfrac{\Delta p}{\Delta t} \cdot V$

System outgassing disturbs results as in Subsection 4.1.1 above and the same comments on system cleanliness and the use of a cold trap apply. Effects of outgassing are usually revealed by initial curvature in the plot of p against t. When they subside, a straight-line relationship should persist at least up to about 75 Torr and the slope of this should be used to derive Q_L. Sensitivity depends greatly on apparatus used and system conditions.

4.1.3 Gas detection methods with internal pressurization

The inside of the component or system to be tested is pressurized with a test gas and some form of detector is provided externally to detect the emergence of the test gas through any leak. This is primarily a leak-finding rather than a leak-proving method and does not readily yield a quantitative measure of total leak rate unless the component to be tested is placed in an enclosure from which the whole of the effluent test gas can be pumped to the detector. It is, however, often used for both purposes together in simple cases.

Methods of detection include: immersion in water or painting with soap solution and watching for bubbles (a leak of 10^{-4} Torr l s^{-1} should be just detectable); sucking the gas from an enclosure around the object being tested (or through a probe tube held near a suspected area) and delivering it to a detector sensitive to the particular gas used.

Sensible precautions against pressurizing unsuitable systems and against accidentally applying excessive internal pressure should always be taken (pressure-regulator, blow-off valve, rupture disc, etc.).

4.1.4 Gas detection methods with internal evacuation

The component or system to be tested is evacuated and connected to a sensing device sensitive to the test gas to be used. The component or system is entirely covered with a hood and the air inside this is displaced by test gas. Desirably this is a type of gas unlikely to be already present inside the evacuated system. Ingress of the test gas through leaks is indicated by the sensing device. The sensing device can be calibrated by measuring its response to the entry of test gas through high impedance ducts of measured impedance ("standard leaks"). The method is, therefore, quantitative.

If the response of the sensing device is specific to the test gas, degassing effects are eliminated and the leak rate is indicated rapidly. If this is unacceptably high, then the hood can be removed and individual leaks can be sought by probing with a fine jet of test gas. The method is widely used for leak-proving and leak-finding for high vacuum components and for leak-finding for high vacuum plant.

A note on response time: When search gas is applied to a leak, it will begin to enter the system and its partial pressure within the system will rise until its removal rate by the pump balances its entry rate through the leak. The rate of pressure rise to this value is exponential, depending on system volume and geometry and pumping speed. Sensitivity depends on partial pressure of test gas reached in the system and can often be usefully increased by throttling the pumps; but time to approach maximum test gas pressure in the system is correspondingly increased. (A ratio of system volume to gas volume pumped per second of about 10 is best if system conditions permit.) Similar considerations control the optimum arrangement when the sensing device is used in the backing line of a vapour pump. A normal rate of traverse for the search gas probe is about 5 mm per second.

References p. 143

4.2 Systematic leak-finding

Leak-finding facilities should be considered at the design stage of complex plant. These may include:
(i) Grooves for test gas around main flange seals (with entry and exit holes).
(ii) Valves for isolating different regions of a system.
(iii) Valved couplings for connecting extra gauge heads or leak-finding apparatus (with associated electrical socket outlets).

As an example of a leak-finding routine, consider a furnace pumped with an oil vapour pump and mechanical backing pump, and provided with valved couplings for gauges, etc. on both sides of the backing pump isolation valve and on the pump side of the vapour pump isolation valve (see Fig. 15).

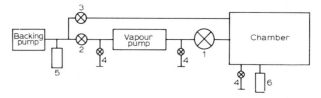

Fig. 15. Process plant and pumping system with valves and couplings to facilitate location of leaks.
    1 Vapour pump isolation valve   4 Valved couplings
    2 Backing valve   5 Backing pressure gauge
    3 By-pass valve   6 High vacuum gauge

(i) With backing pump isolation valve closed, confirm backing pump ultimate vacuum. (Watch for pressure build-up at vapour pump outlet.) If satisfactory, proceed to (ii).

(ii) With vapour pump isolation valve closed, confirm vapour pump ultimate vacuum. (Watch for rapid pressure build-up in main chamber.) If satisfactory, proceed to (iii).

(iii) Connect leak-finding apparatus sensing head up-stream of backing pump isolation valve (which can sometimes be part-closed with advantage to sensitivity), or to main chamber, according to type.

(iv) Probe suspect regions of chamber in systematic order; *e.g.* any regularly troublesome seal—seals where motion is transmitted—seals of regularly opened or recently opened doors or ports—other demountable seals—welded joints. (It is sometimes possible to hood parts of the system for more speedy examination, as in Subsection 4.1.4.)

4.3 Some leak-sensing devices and search gases

4.3.1 Sputter-ion pump (where this is already in use on system, see Subsection 2.7.2)

Pressure range of use: below 10^{-4} Torr.

Usual search gases:
Helium (pump current increases).
Oxygen (pump current decreases).
Minimum detectable leak: approx. 10^{-10} Torr l s^{-1} depending on system conditions. For maximum sensitivity a special circuit is used which measures only the change in pump current on applying the search gas to the leak, and can therefore employ high amplification.

4.3.2 Hot cathode ionization gauge (usually where this is already in use on system)
Pressure range of use: below 10^{-4} Torr.
Usual search gas:
Helium or hydrogen (gauge current decreases).
Butane or acetone (gauge current increases).
Minimum detectable leak: approx. 10^{-8}–10^{-10} Torr l s^{-1} depending on system conditions. For maximum sensitivity a special circuit is used as for above (Subsection 3.9).

4.3.3 Thermal conductivity gauges (Pirani or thermocouple gauge)
Pressure range of use: below 10 Torr but best 10^{-1} to 10^{-2} Torr.
(Very suitable for use on backing side of vapour diffusion pumps.)
Usual search gases: Hydrogen (gauge current increases).
Minimum detectable leak: 10^{-3}–10^{-5} Torr l s^{-1}. (For maximum sensitivity a special circuit is used which measures only the change in gauge current when applying the search gas to the leak, so that high amplification can be used.)

4.3.4 Halogen detector
Description: A special-purpose leak-finding instrument. The emission of a hot-cathode diode with platinum electrodes increases sharply when subjected to traces of halogen vapour. Sensitive to organic contamination and overdoses of search gas. May respond to halogen vapours arising from vapour pump boilers where chlorinated diphenyl pumping fluid is used.
Pressure range of use: atmospheric pressure to 10^{-3} Torr.
Search gas: Halogen vapour such as Freon 12.
Minimum detectable leak: 10^{-5}–10^{-6} Torr l s^{-1}.

4.3.5 "Snifters"
Special versions of both the thermal conductivity and the halogen detector may be used in conjunction with a search probe which samples the atmosphere by suction. Such devices are used where the component or plant under test has been internally pressurized with search gas. Leaks are located where the search gas escapes to atmosphere. Halogen snifter can detect down to 10^{-4} Torr l s^{-1}. See also Subsection 4.3.8.

References p. 143

4.3.6 *Analytical mass spectrometer*

Description: The use is increasing of small mass spectrometers (see Subsection 3.10) as part of the normal instrumentation of certain refined plant. Such instruments are readily used for leak-finding by tuning to any convenient and stable peak where high sensitivity can be used (*e.g.* helium, argon), or even by tuning to nitrogen and observing a reduction in the nitrogen ion current when a search gas of different character finds the leak.

Pressure range of use: usually below 10^{-3} Torr.

Minimum detectable leak: $\sim 10^{-9}$ Torr l s^{-1} (depending on class of instrument.

4.3.7 *Leak-test plant*

Mass spectrometers and thermal conductivity and halogen detectors (and other types not included here) are suitable for incorporating in integrated leak-testing plant incorporating facilities for their optimum use, *e.g.* see below.

4.3.8 *Leak-test mass spectrometer*

Description: This is the most sophisticated leak proving and finding apparatus available. It consists of a mass spectrometer designed for tuning to the search gas (almost invariably helium) and mounted together with its own special-purpose traps, pumping system, instrumentation and reference leaks in the form of a mobile plant with versatile operational possibilities. Facilities are usually provided for the direct connection of small and medium sized components to the mass spectrometer pumping system which can evacuate them to a suitable pressure for leak-proving (by hood) or leak-finding (by gas probe) while the sensing head remains ready for use under vacuum. Very large components or large plant may need to be evacuated by other pumps, the leak-test apparatus being connected to a suitable port and valved-in to sample the gas content of the plant after this has been evacuated. To find leaks on a plant using vapour pumps, the leak-test apparatus is conveniently connected to a valved coupling in the backing line upstream of the backing pump isolation valve. Calibrated reference leaks are usually provided to permit the magnitude of any leak found to be determined from the response of the spectrometer.

Minimum detectable leak: $\sim 10^{-11}$ Torr l s^{-1}.

"Snifter" probes (see Subsection 4.3.5) are available for use with leak test mass spectrometers, the minimum detectable leak being increased by this mode of operation.

BIBLIOGRAPHY

BOOKS ON HIGH VACUUM TECHNOLOGY

General:
1. S. DUSHMAN AND J. LAFFERTY, *Scientific Foundations of Vacuum Technique*, Wiley, New York, 1962, p. 806.
2. M. PIRANI AND J. YARWOOD, *Principles of Vacuum Engineering*, Chapman & Hall, London, 1961, p. 578.
3. H. A. STEINHERZ, *Handbook on High Vacuum Engineering*, Reinhold, New York, 1963, p. 358.
4. N. T. DENNIS AND T. A. HEPPELL, *Vacuum System Design*, Chapman & Hall, London, 1968, p. 223.

Pumps and Pumping Systems:
5. B. D. POWER, *High Vacuum Pumping Equipment*, Chapman & Hall, London, 1966, p. 412.

Pressure Gauges and Vacuum Instrumentation:
6. J. H. LECK. *Pressure Measurement in Vacuum Systems*, 2nd edn., Chapman & Hall, London 1957, p. 221.

Leak-proving and Leak-finding:
7. *Methods for Proving the Gas Tightness of Vacuum or Pressurized Plant*, British Standard 3636; 1963, British Standards Institution, London, 1963, p. 118.

Ultra-high Vacuum:
8. R. W. ROBERTS AND T. A. VANDERSLICE, *Ultra High Vacuum and its Applications*, Prentice-Hall, New Jersey, 1963, p. 199.
9. N. W. ROBINSON, *The Physical Principles of Ultra-high Vacuum Systems and Equipment*, Chapman & Hall, London, 1968, p. 292.
10. E. A. TRENDELENBURG, *Ultrahochvakuum*, G. Braun, Karlsruhe, 1963.
11. P. A. REDHEAD, J. P. HOBSON AND E. V. KORNELSEN, *The Physical Basis of Ultrahigh Vacuum*, Chapman & Hall, London, 1968, p. 498.

Chapter III

Use of Vacuum Techniques in Extractive Metallurgy and Refining of Metals

J. KRÜGER

1. Introduction

The number of metals in technological use is so large that it could not be expected that a standard process for their production would exist. Dependent on the starting materials and their properties there are, therefore, a great variety of technological processes in which, among many others, vacuum technology also has its place. There are several guide lines which in principle indicate the potential use of vacuum technology in metallurgy.

A simplified scheme for the production of high-purity metals is shown in Fig. 1. In this arrangement the metals are grouped in the order of their affinity for oxygen. Process stages in which a vacuum might be employed are marked off in this diagram. However, it should be borne in mind that even in these individual stages, vacuum-metallurgical reactions may, in many cases, be replaced by entirely different processes.

The advantages of vacuum-metallurgical reactions can be characterized in a few words:

1. By lowering the pressure of the gaseous reaction products the concentration gradient is increased, so that in many instances the reaction rate may be raised.

2. By pumping off gaseous reaction products the equilibria are displaced in the direction of the desired reactions, so that either the yield is increased or the working conditions are improved.

3. High vacua reduce the extent of reactions between the corresponding metal vapours and the residual gases in the gas atmosphere as well as at the phase boundary between base material and gas.

4. A reduction of the residual gas pressures to such low values that the mean free path of vapour or gas molecules becomes large compared with the dimensions of the reactor results in maximum evaporation rates almost independent of any further reactions taking place.

5. The absence of an oxygen or nitrogen atmosphere increases the stability of a large series of compounds, which, in many cases, show higher vapour pressures and greater differences in vapour pressure than do the corresponding metals. These compounds are, therefore, better suited to distillation than the metals.

For the reactions named under (1) only so-called rough vacua are required,

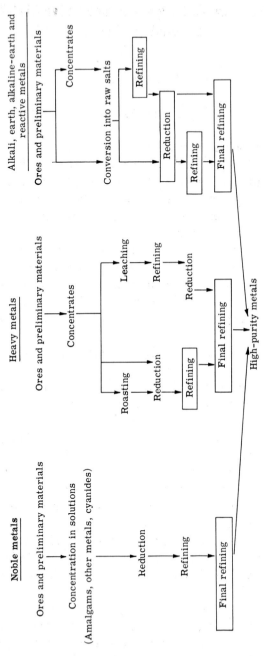

Fig. 1. Scheme of potential vacuum-metallurgical process stages in the production of high-purity metals.

as the concentration gradient is only insignificantly increased when the pressure, originally 760 Torr, is reduced to 10^{-6} Torr instead of 1 Torr. However, this is different for the other items. In item (2) the required vacuum depends on the equilibrium condition and a rough, medium or high vacuum is necessary according to the reaction. The processes named under the other items normally require operation in a high vacuum.

The quantity of gas that is to be pumped off may decide the choice of a vacuum process. If, for instance, Nb_2O_5 is reduced with either C or NbC according to the following overall equations:

$$\langle Nb_2O_5 \rangle + 5\langle C \rangle = 2\langle Nb \rangle + 5(CO) \tag{1}$$

$$\langle Nb_2O_5 \rangle + 5\langle NbC \rangle = 7\langle Nb \rangle + 5(CO) \tag{2}$$

and if a temperature of the gases of 200°C at the pumps is assumed, 798 or 228 m^3 CO must be pumped off per kg Nb produced if a pressure of 1 Torr prevails in the reaction vessel, but at a pressure of 10^{-3} Torr 798000 or 228000 m^3 CO must be handled.

This means that such processes must be divided into several individual steps in order for the large quantities of gas to be pumped off at the highest possible pressure. In practice the major portion of the gases formed is pumped at about 100 Torr or even at atmospheric pressure, so that under high vacuum conditions only a relatively small quantity of gas has to be removed.

Working conditions in the distillation of condensable metals or their compounds are much more favourable. In this case the pumps are used only for the initial pump-down of the plant and for maintaining the predetermined vacua.

2. Vacuum Technology in the Preliminary Stages of Metallurgical Reduction Processes

2.1 Physical processes

Vacuum-technological processes in metallurgy are not only—as shown in Fig. 1—employed for the reduction and refining of metals. In many cases further process stages, such as degassing or de-airing, sublimation, filtration, pumping off of liquid metals, crystallization, concentration and drying may be carried out with advantage in a vacuum. It is characteristic of all these processes that they are conducted in the rough or medium vacuum ranges and that only in rare cases are chemical reactions involved. All these processes have been established in the chemical industry for a long time. At present, vacuum milling and vacuum magnetic separation are in the experimental stages. Some of these processes take place without thermal effects (degassing, de-airing, filtration, milling, magnetic separation), some others are accompanied by thermal effects and some of these are

References pp. 321–335

endothermic (sublimation, concentration by evaporation, drying) and others exothermic (crystallization).

In ore mining considerable quantities of fines are often present which must be agglomerated at as low a cost as possible. This also applies to flue dust from smelting plants and the red mud in alumina production. These materials may be de-aired to a high degree in a vacuum press; in many cases this is a requisite for the compacting of such fine-grain substances. The compacts may develop sufficient strength for use in the shaft furnace if compacted at pressures in excess of 1000 atm. A schematic presentation of a two-stage press as used in experiments in the United Kingdom is shown in Fig. 2[1].

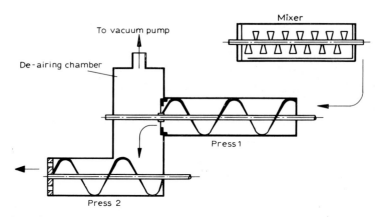

Fig. 2. Schematic presentation of a two-stage vacuum press incorporating a de-airing chamber[1].

The pressures employed vary between 50 and 150 Torr. In the same way the red mud from alumina production may be compacted, but owing to the very high moisture content (50%) additional drying or addition of, for instance, dry sinter, is necessary.[2]

Vacuum filtration with drum filters or disc filters offers a number of advantages. Filter efficiency is greatly increased by a vacuum; on the other hand there is a limit to the vacuum owing to the partial vapour pressure of the solution to be filtered so that hot slurries can be filtered only within limits. Operation is continuous; throughputs of 5–15 ton/h m^2 filter area have been attained. The pressure obtainable depends on the operational conditions and varies between 150 and 500 Torr. The potential applications in hydrometallurgy are numerous: in the preparation of high-purity salts or oxides of high-melting-point metals, in hydrometallurgical processes for the winning of the common heavy metals and in the production of alumina.

In the production of alumina the red mud has at first to be separated from the aluminate solution. The hot leaching solution is then cooled from about 80° to 55°C at a pressure of 300–400 Torr which is maintained by a low-pressure steam

Fig. 3. Partial view of a vacuum drum filter plant[3].

ejector pump. The use of higher vacua would result in an unwanted higher temperature loss. The filter cake would then be more compacted and consequently the filter throughput could not then be raised. Such a filter plant is shown in Fig. 3.

A fabric of stainless chromium–nickel steel has proved most suitable as the filter cloth. By employing drum filters, the labour cost and the rinse water consumption could be lowered and the yield increased[3]. In a similar way, filtration of the aluminium hydroxide obtained is feasible.

The size of such plants may be judged from the following data:

	Drum filter (red mud)	Disc filter (aluminium hydroxide)
Filter area	45 m²	37.5 m²
Throughout dry substance/m²h	100 kg	1500 kg
Pressure	380 Torr	500 Torr
Water ring pumps	800 m³/h	1500 m³/h

Vacuum grinding of dry materials is in the experimental stage. Ocepek and Eberl[4] report an increase of 20% in the specific surface area of the milled product at a pressure of 20 Torr.

According to Planiol[5], ores may be ground to a much narrower grain-size range in a vacuum centrifugal breaker-plate mill than in air, as may be seen from

References pp. 321–335

the following data:

	Largest grain diameter	50% <	
Air grinding	70 μm	10 μm	12% < 2 μm
Vacuum grinding	29 μm	9 μm	5% < 2 μm

The quantity of residual gas present at a pressure of between 10^{-2} and 1 Torr is negligible, consequently the entire kinetic energy is available at the moment of impact and therefore creates the best conditions for grinding.

Similar vacuum conditions also give better results in magnetic separation, as the viscosity of the air usually limits the separation rates. A combination of the two types of apparatus gives rise to the expectation of a separation of even intimately intergrown minerals[5].

2.2 Chemical processes

Although the vapour pressures of the metals cover a very wide range (see Fig. 4a, b), the vapour pressures of many metals differ too little for complete separation. In many cases the temperatures are already so high at the pressures necessary for economical operation that a separation by distillation is out of the question.

In many cases halides, oxides and suboxides as well as sulphides may be separated by distillation from one another or from any gangue under much more favourable conditions. The vapour pressures of the pure compounds are shown in Figs. 5–10.

In this connection a critical review by Kellog[9] should be mentioned, according to which many of these data require revision.

Mass-spectroscopic studies showed that the vapours above the oxides and sulphides of zinc, cadmium and mercury do not consist of molecules of the simple composition alone, but consist mainly of metal and metalloid vapours as well as of a negligibly small quantity of the compound. For other elements a number of gaseous polymers would be expected, for instance, in the oxides and sulphides of lead and tin.

From the diagrams it is evident that metals of comparable low vapour pressures may quite possibly form compounds of high vapour pressures at any predetermined, equal temperatures, or vice versa.

A number of technological processes have been developed on the basis of this knowledge. However, as the thermochemical data of many compounds have still not been determined, it should be pointed out that we are only at the beginning of a so-called "distillation or volatilization metallurgy".

It is not always possible in an ore preparation process to obtain rich concentrates. The high vapour pressures of some sulphides, however, make possible

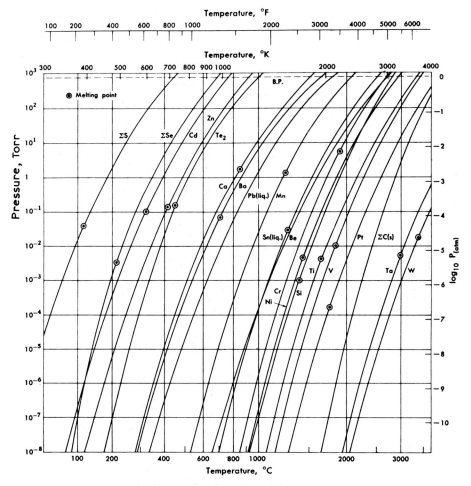

Fig. 4 (a). Vapour pressure curves of metals[6].

their volatilization from solid, finely ground raw materials in a vacuum. Preliminary results of the volatilization of tin in a vacuum via the sulphides were reported in the Soviet Union as far back as 1957[10]. By addition of pyrites the volatilization of tin yields 97–98% at 900–1050°C according to the equation:

$$3\langle SnO_2\rangle + 3\langle FeS_2\rangle = 3(SnS) + 2\langle FeO\rangle + 2(SO_2) + \langle FeS\rangle. \tag{3}$$

An isothermal phase diagram for this reaction already exists, as shown in Fig. 11. From this diagram it may be seen that even at low S_2 and SO_2 pressures, volatile tin sulphides, particularly SnS, are formed also from SnO_2 as base material. The other tin compounds, such as Sn_2S_2, SnO, Sn_2O_2, etc., play a minor role only. According to recent investigations intensive sublimation was already occurring at between 750° and 1000°C[11]. As the vapour pressure of SnS at 1000°C does not

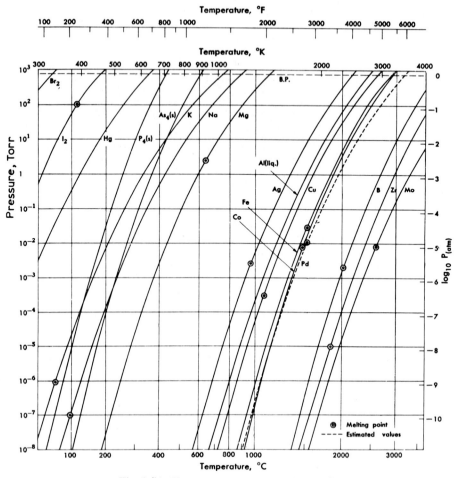

Fig. 4 (b). Vapour pressure curves of metals[6].

quite attain 1 Torr, a pressure of at least 0.5–1 Torr should be maintained so that this low rate of evaporation is not further lowered by collisions in the gas space. The major portion sublimes within the first 30 min. Attempts to sulphidize SnO_2 with elementary sulphur were not very successful; the sublimation yield was not more than 50%.

In a similar way bismuth may be volatilized as Bi_2S_3 from a sulphide concentrate at 1000°C and 0.1 Torr with a yield of 93% in 30 min. In addition PbS (90%) and ZnS (73%) will sublime[12].

The high vapour pressure of lead and zinc sulphides suggested the carrying out of preliminary experiments to volatilize these two substances from lead–copper mattes. From the data of Table 1 and from Fig. 12 it may be seen that lead

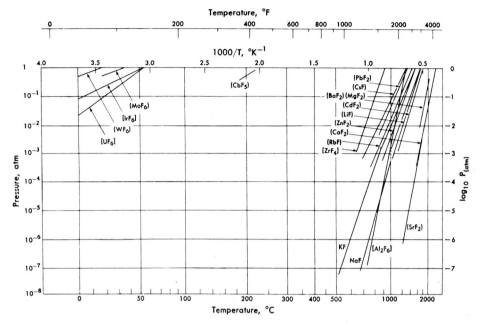

Fig. 5. Vapour pressure curves of fluorides[6-8].

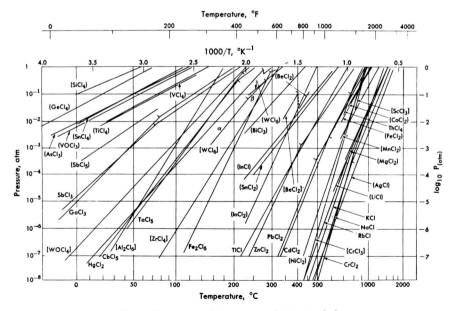

Fig. 6. Vapour pressure curves of chlorides[6-8].

References pp. 321–335

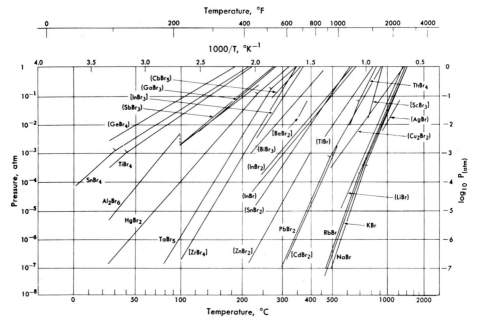

Fig. 7. Vapour pressure curves of bromides[6-8].

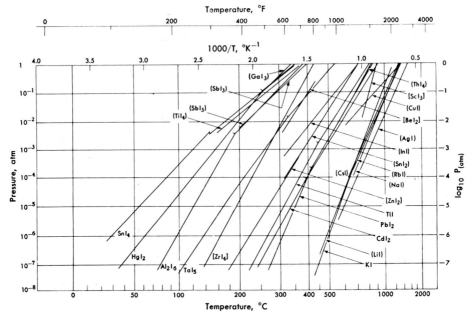

Fig. 8. Vapour pressure curves of iodides[6-8].

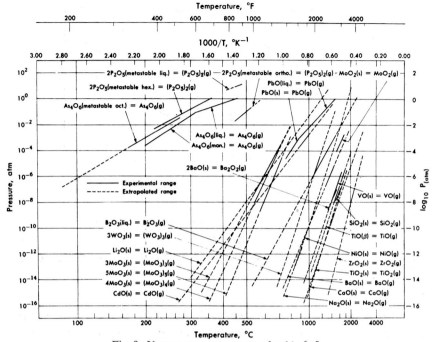

Fig. 9. Vapour pressure curves of oxides[6-8].

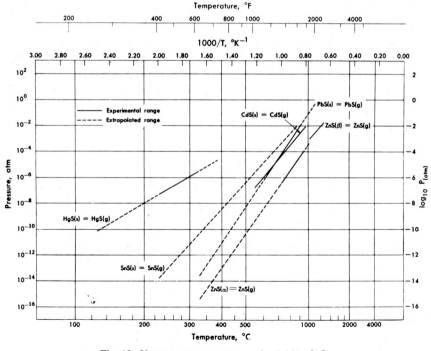

Fig. 10. Vapour pressure curves of sulphides[6-8].

References pp. 321–335

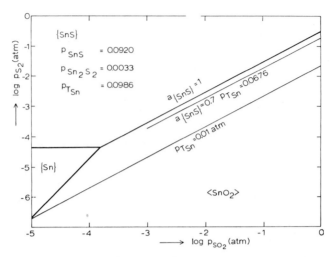

Fig. 11. Thermodynamic equilibrium diagram of the Sn–O–S system at 1300 K according to Kellog[9].

may be volatilized to more than 99% and zinc up to 97%. The fine dust obtained is used in the production of lead, whereas the copper-enriched matte may be directly blown. From Fig. 12 it may further be noted that evaporation is markedly reduced at pressures as high as 6 Torr. Mainly PbS, PbO, Pb and ZnS evaporate as their vapour pressures are between 10^{-1} and 100 Torr in the temperature range 900–1200°C. At these temperatures it is also possible to volatilize As, Sb and Cd[13]. Higher ratios of iron to copper in the matte result in a heavier volatilization of PbS[16]. In general, PbS in Cu_2S–FeS solutions shows a negative deviation from Raoult's law; only at very high PbS levels are the conditions reversed[17]. Also Au–As concentrates may be almost completely de-arsenized by a vacuum sublimation of the As_2O_3 in 15 min at 900°C and 1–10 Torr. The residue may be used

TABLE 1

VOLATILIZATION OF LEAD AND ZINC FROM LEAD–COPPER MATTE IN A VACUUM

Time (min.)	Pressure (Torr)	Temp. (°C)	Composition of the matte (wt. %)					Yield (%)		Composition of the distillate (wt. %)					Ref.
			Pb	Cu	Fe	Zn	S	Pb	Zn	Pb	Cu	Fe	Zn	S	
	Starting material		33.7	21	15.4	5.3	20.3								
10	1	900	5.7	39.5	27.0	2.3	26.0	97	51						
20	1	900	3.2	38.0	26.5	1.1	28.5	98	75	66.1	–	1.7	7.4	11	13
30	1	900	2.2	38.8	28.9	0.5	28.4	>99	90						
	Starting material		10	13–20			8–16								
30	0,5	1150	0.1				1	99	90						14
30	0.1–1	1200						99	97						15

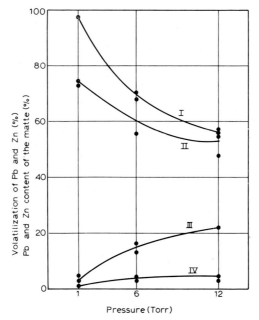

Fig. 12. Volatilization of lead and zinc from lead–copper mattes in relation to the residual gas pressure at 900°C after 20 min[13].
I = Volatilization of lead   III = Lead content of the matte
II = Volatilization of zinc IV = Zinc content of the matte

in the production of noble metals. The sublimate contains about 70% As[18].

As an extension of their experiments on the volatilization of zinc, Decroly and Ghodsi[19] made attempts for the partial reduction and volatilization of SnO_2 with Fe, FeO and Fe_3O_4 at reduced pressures. Certain quantities of SnO_2 are reduced to gaseous SnO by Fe, FeO and Fe_3O_4 above 900°C at low pressures and then volatilized. Mixtures of oxides of the rare-earth metals are only partly reducible between 1100° and 1400°C by a great excess in carbon. Despite thermogravimetric measurements, it could not be clarified whether in this process the metals or the suboxides of Sm, Eu, Lu, Tm and Yb are volatilized. However, this makes possible a separation of these rare-earth metals from the others[20].

Decroly and Ghodsi[21] reduced GeO_2 with Fe and FeO at a pressure of about 0.1 Torr above 800°C and volatilized GeO. The corresponding reactions take place according to:

$$\langle GeO_2 \rangle + \langle Fe \rangle = \langle FeO \rangle + (GeO) \tag{4}$$

$$\langle GeO_2 \rangle + 3\langle FeO \rangle = \langle Fe_3O_4 \rangle + (GeO) \tag{5}$$

During the subsequent condensation GeO decomposes according to:

$$2(GeO) = \langle Ge \rangle + \langle GeO_2 \rangle \tag{6}$$

References pp. 321–335

While at a temperature of 800°C only 15% has reacted in 5 h, at 950°C 95% are volatilized during the same period of time.

While the reduction of ZnO with Fe at a pressure of about 0.1 Torr at about 1000°C does not meet the slightest difficulty[22], in fact an almost complete exchange is obtained within 3 h according to:

$$\langle ZnO \rangle + \langle Fe \rangle = \langle FeO \rangle + (Zn) \tag{7}$$

the reduction of $ZnSiO_4$ by Fe takes place at a much lower rate. In this case, however, an addition of CaO or FeO for increased silicate formation may accelerate the reaction, but the rate of the pure ZnO reduction will not be reached[23].

Extraction and separation of zinc from zinc-containing calcined pyrites[22, 24] and slags from non-ferrous metallurgy[25, 26] have become of great interest. Although these zinc contents may be reduced without difficulty by carbon or by metallothermic reduction (iron)[22, 26], and then volatilized and finally used, there is also possible an exchange reaction of the zinc carrier with FeO, but only in a vacuum. The various zinc compounds react according to:

$$\langle ZnO \rangle + 3 \langle FeO \rangle = \langle Fe_3O_4 \rangle + (Zn) \tag{8}$$

$$\langle ZnS \rangle + 4 \langle FeO \rangle = \langle Fe_3O_4 \rangle + \langle FeS \rangle + (Zn). \tag{9}$$

The lead contents of such slags may also be utilized to a large extent in this way[25, 26]. During the dezincification of calcined pyrites or other oxides the required FeO is first produced from the Fe_2O_3 content of the calcined pyrites by partial reduction at a maximum temperature of 850°C[24]. Shaft furnace slags from lead and copper metallurgy already contain an adequate FeO content (about 30%) besides lower contents of Fe_2O_3 (maximum 3%), and slags from reverberatory smelting show somewhat higher Fe_2O_3 contents (maximum 5%), but their FeO content is usually sufficient for reduction of the ZnO portion. Higher Fe_2O_3 contents may be present only in the slags from converter and other special processes. In this case a reduction of the ZnO portion would not take place, or, if it took place, only at a strongly retarded rate, as may be derived from data by Decroly and Ghodsi[26]. But even under these conditions partial evaporation of zinc is possible, because with increasing temperature additional dissociation, and therefore evaporation of zinc, and a sublimation of ZnO take place. The experimental results obtained by Decroly and Ghodsi[22, 26] are plotted in Fig. 13 and from the diagram it is evident that sublimation and dissociation of ZnO assume considerable importance at temperatures above 1250°C, whereby about one-half of the yield is by sublimation, the other half by dissociation. Medium or high vacuum conditions have no noticeable effect on the volatilization of zinc. Amounts in excess of 100% are caused by an oxidation of the condensate or by sublimation of oxides. Zinc reduction and volatilization at atmospheric pressure by FeO dis-

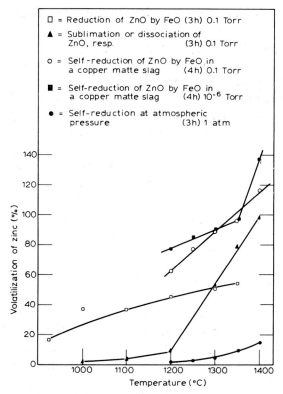

Fig. 13. Volatilization of zinc from ZnO and zinc-containing slags under the conditions stated[22,26]. Composition of the slag: ZnO 16.9%; FeO 28.9%; PbO 1.3%; SiO_2 32.3%; CaO 6.8%.

solved in the slag result only in a minimum yield as may also be seen from Fig. 13[26]. Kim and Miller also examined end slags in which an increase in the Fe_3O_4 equivalent to a decrease in the zinc and lead contents was found. This increase is in accord with eqn. (8). The distillates consisted of the metals and small amounts of their sulphides[25].

It is interesting to note that the reduction of pure zinc oxide by FeO takes place only to about 50%. It appears that further reaction between the two oxides is blocked by association above 1100°C[22].

A special case is the Merrill–Crowe process for the winning of gold. This may be regarded as an example of the use of vacuum technique in hydrometallurgy. Gold is dissolved from the minerals by a sodium cyanide solution in the presence of oxygen. The gold in the solution is then precipitated by means of zinc dust. This reaction is, however, only possible if the oxygen content is below 1 mg/l. It is for this reason that the gold-containing solutions are de-aerated in a vacuum of 50 Torr in order to reduce the oxygen level from 6.5 to 0.5 mg/l[27].

References pp. 321–335

3. Vacuum-metallurgical Reduction Processes

In metallurgical technology only relatively few vacuum-metallurgical reduction processes exist which are established in practice and which are used exclusively for less-common metals. Comparison of the Richardson diagram for the more important oxides shows that carbon at about 1750°C and at a CO pressure below 10^{-3} atmosphere surpasses all other reducing agents with regard to efficiency. For most metals even higher CO pressures and lower temperatures seem to be feasible.

In practice, however, the conditions for a reduction are not so ideal, as side by side with the reduction, the formation of carbides may take place according to the following total reaction:

$$\langle MeO_2 \rangle + 2\langle C \rangle = \langle Me \rangle + 2(CO) \tag{10}$$

$$\langle Me \rangle + \langle C \rangle = \langle MeC \rangle \tag{11}$$

On the basis of the metallurgical phase diagrams a possible reduction or carbide formation of a metal may be anticipated. In this case a distinction should be made between the so-called Pourbaix–Ellingham and the thermodynamic constitution diagrams according to Knacke[28] (see also Figs. 14, 15, 20, 21 and 23).

From these diagrams it may be seen that the formation of carbides can be prevented in most cases only by application of high temperatures, certain $CO:CO_2$ ratios, and high vacua.

For lowering the quantities of gas liberated during the reaction, the reduction of the oxide by the carbides of the same metal may be used according to:

$$\langle MeO_2 \rangle + 2\langle MeC \rangle = 3\langle Me \rangle + 2(CO). \tag{12}$$

This reaction plays a considerable role in the reduction of high-melting-point metals.

At high temperatures, however, either the vapour pressures of the metals are often of the same order of magnitude as those of the CO pressure, so that intensive evaporation of the metal occurs during reduction, or the formation of carbides is almost inevitable. With simultaneous evaporation of metal and CO it is almost impossible to prevent a reaction in the undesired opposite direction during condensation at low temperatures according to:

$$(Me) + (CO) = \langle MeO \rangle + \langle C \rangle. \tag{13}$$

For this reason metallic compounds or metals, such as FeSi, Ca, Al, CaC_2, etc., are employed as reducing agents.

During the metallothermic reduction of high-melting-point metals in a vacuum mainly volatile suboxides (or halides, etc.) of the reducing agents are formed, but in the reduction of easily volatile metals the formation of such suboxides

(SiO, AlO, Al$_2$O) must be avoided. By use of lower reduction temperatures (below 1200°C) their formation may be almost completely suppressed.

Another method is the metallothermic reduction (with Al, Ca, Na, Mg) of oxides and halides of high-melting-point metals under protective gases with subsequent evaporation of the oxidation products and excess reducing agents in a vacuum. In this way, too, relatively high-purity metals are produced.

All the reactions considered here so far have the shortcoming that the oxygen and the reducing agents (carbon or metals) cannot be completely removed from the reduced metal.

Better degrees of separation are obtained by reduction of oxygen-free, gaseous halides by hydrogen according to:

$$(MeCl_2)+(H_2) = \langle Me \rangle + 2(HCl) \tag{14}$$

Also dissociation at high temperatures in a vacuum may play an important role (van Arkel–de Boer process):

$$(MeCl_2) \rightarrow \langle Me \rangle + (Cl_2) \tag{15}$$

3.1 Carbothermic reduction processes

3.1.1 *Carbothermic reduction processes without volatilization of the metals to be reduced*

(a) *Carbothermic reduction of vanadium, niobium and tantalum*

Darling[34], in his comprehensive review of carbothermic reduction in a vacuum, could only report in 1961 on the successful reduction of the niobium oxides. In the meantime such a reduction has been made possible for all metals of the Fifth Group of the Periodic Table (V, Nb, Ta). The possibilities and conditions for a reduction may be expected on the basis of the Pourbaix–Ellingham or Knacke[28] constitution diagrams. Diagrams for the system V–O–C constructed on the basis of identical thermodynamic data[29] are shown in Fig. 14 a, b. Both types of diagram show a degree of correspondence, but the constitutional diagram according to Knacke appears to be the more instructive. It should, however, be noted that a number of phase boundaries are pressure-dependent, so that for this isobaric presentation, displacements of certain phase fields will result from falling total pressure. Triple points change their temperature with a change in pressure (and vice versa) and correspond to the lines which characterize the four phases of the Pourbaix–Ellingham diagrams.

If it is assumed that no further phases exist, then the carbothermic reduction to vanadium should be interpreted on the basis of these diagrams and Ostwald's rule of successive reactions as follows:

$$\langle V_2O_5 \rangle + (CO) = 2\langle VO_2 \rangle + (CO_2) \tag{16}$$

$$2\langle VO_2 \rangle + (CO) = \langle V_2O_3 \rangle + (CO_2) \tag{17}$$

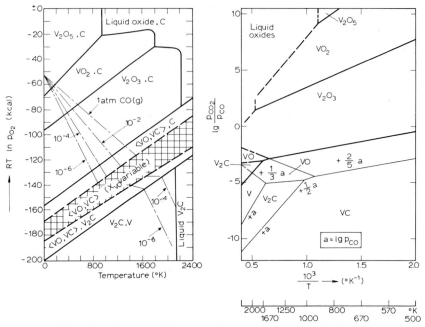

Fig. 14(a). Pourbaix–Ellingham diagram for the system V–O–C[29].
Fig. 14(b). Phase diagram for the system V–O–C– according to Knacke for $p_{CO} = 1$ atm (heavy line) and $p_{CO} = 10^{-6}$ atm (thin line).

Instead of CO, C or H_2 may used here. Then in the reactions (16) and (17) either CO_2 or H_2O is almost exclusively formed.

$$\langle V_2O_3 \rangle + 5 \langle C \rangle = 2 \langle VC \rangle + 3(CO) \tag{18}$$

$$2 \langle V_2O_3 \rangle + \langle VC \rangle = 5 \langle VO \rangle + (CO) \tag{19}$$

$$\langle VO \rangle + 3 \langle VC \rangle = 2 \langle V_2C \rangle + (CO) \tag{20}$$

$$\langle VO \rangle + \langle V_2C \rangle = 3 \langle V \rangle + (CO) \tag{21}$$

From these reactions it may be ascertained that neither VO_2 nor V_2O_3 or VC can exist in equilibrium with V. These compounds must be reduced to VO or V_2C before V can form. These rather complex reactions indicate that partial steps may govern the reaction rate. From Fig. 14b it is evident that for $p_{CO} = 1$ atm no reduction to the metal can be expected even at 2500 K, but that the range of existence of V metal is extended by a lowering of the pressure and that in this way technologically feasible reduction conditions may be created. The required minimum reduction temperatures for V, Nb and Ta are shown in Table 2. Kieffer et al.[35] found in reduction experiments at 10^{-7} atm at temperatures increasing from 800°C in steps of 200°C a reduction in oxygen and carbon content in agreement with the above-mentioned reactions.

In this presentation (Fig. 14 a, b) the mutual solubilities of carbon and oxygen

TABLE 2

REDUCTION TEMPERATURES (K) OF THE METALS OF THE FIFTH GROUP OF THE PERIODIC TABLE

CO Pressure (atm)	V	Nb	Ta*	Ta**
1	2700	2780	2860	3170
10^{-3}	2020	2095	2130	2320
10^{-6}	1620	1695	1705	1850

* According to Fig. 20a.
** According to Fig. 20b.

in vanadium were not taken into account, because the levels of the concentration products $\%\,[O]_V \times \%\,[C]_V$ in both solid and liquid vanadium are relatively low at sufficiently low pressures. The question, however, remains, whether or not states of equilibria were reached in these experiments (see also Table 3). While Joly[37] pre-reduced V_2O_5 with carbon to VO_2 according to Fig. 14b and eqn. (16), Karasaev et al.[36,38] carried out a pre-reduction of V_2O_5 with H_2 at 500–600°C via VO_2 to V_2O_3 (see also Table 4). Joly continued the reduction process at increasing temperature (up to 1350°C) at a final pressure of 10^{-1} Torr and obtained an intermediate product which Kieffer and Braun[43] called "raw carbide". This may be a mixture of V_2C and VO according to the thermodynamic conditions, as may be seen from the following comparison:

	%V	$\%\,[O]_V$	$\%\,[C]_V$
Actual product	86–87	7–8	5–6
Theoretical composition ($VO+V_2C$)	84	10.3	6

The production of a relatively pure metal is feasible in two further process stages.

Karasaev et al.[36,38] pre-reduced V_2O_3 with lampblack at 900–1000°C for two hours. After the conventional intermediate hydrogenation, final reduction takes place at an increasing temperature up to 1700°C and decreasing pressures down to 10^{-4} Torr in five process stages which cover an overall period of 34 hours. Data of Karasaev et al.[36] show that according to Fig. 14b only VO is obtained in a reduction of V_2O_3 by lampblack at 1450°C and a CO pressure above 4 Torr, whereas reduction to the metal takes place at 1575°C and 10^{-2} Torr, although again

TABLE 3

$\%\,[O]_V \cdot \%\,[C]_V$-PRODUCTS FROM THE ANALYSIS OF VANADIUM

$\%\,[O]_V \cdot \%\,[C]_V$	Temperature (°C)	Pressure (Torr)	Reference
11.6×10^{-3}	1625–1650	10^{-4}	36
7.2×10^{-3}	1700	5×10^{-5}	37
3.8×10^{-3}	1680–1750	$10^{-3} – 5 \times 10^{-4}$	38
1.7×10^{-3}	1650–1675	$< 5 \times 10^{-5}$	39
5.6×10^{-3}	2000–2100	$\leq 5 \times 10^{-5}$	40

References pp. 321–335

TABLE 4

WORKING CONDITIONS FOR THE CARBOTHERMIC REDUCTION OF VANADIUM IN A VACUUM

	O/C Ratio	Temperature (°C)	Pressure towards the end (Torr)	Time (h)	Analysis (wt. %)			Ref.
					O	C	N	
$2\langle V_2O_5\rangle + \langle C\rangle = 4\langle VO_2\rangle + (CO_2)$		530–540						
(see eqns. (17)–(20))		750–1350	10^{-1}	3	7–8	5–6	a	
$\langle VO\rangle + \langle V_2C\rangle = 3\langle V\rangle + (CO)$	1.0	1500	10^{-3}	10	2–3	1–1.5	a	37
		1500–1700	5×10^{-5}	12	0.06	0.12	a	
$\langle V_2O_5\rangle + (H_2) = 2\langle VO_2\rangle + (H_2O)$		500–600			a	a	a	
$2\langle VO_2\rangle + (H_2) = \langle V_2O_3\rangle + (H_2O)$								
Pre-reduction (see above)	0.95	900–1000		2	a	4–5	a	36
$\langle VO\rangle + \langle V_2C\rangle = 3\langle V\rangle + (CO)$		1100–1700	10^{-4}	34	0.045	0.2	0.02	
Starting mixture	1.0	–	–	–	13.5	9.8	a	
1. Indirect sintering (see eqns. (19) and (20))		1450	5×10^{-4}	8	7–9	5–6	a	
2. Indirect sintering (see eqn. (21))		1500	1×10^{-4}	9	3.5	2.6	a	39
After pre-sintering direct sintering (see eqn. (21))		1650	$<2\times 10^{-5}$	3	0.17	0.01	0.01	
After correction post-sintering		1675	$<5\times 10^{-5}$	3	0.014	0.12	0.01	
Starting mixture		–	–	–	13.5	10	a	
1. Indirect sintering		1450	–	12	5.6	4.3	a	
2. Direct sintering		1500	–	3	0.85	0.3	a	35
3. Direct sintering		1650	max 10^{-5}	3	0.115	0.07	a	
Starting mixture 50:50 V–Cr		–	–	–	13.1	9.8	a	
1. Indirect sintering		1400		10	0.9	1.3	a	35
2. Direct sintering		1600	–	4	0.13	0.04	a	42

a Not known.

in several process stages. The corresponding technological data are shown in Table 4.

Kieffer et al.[39] also obtained a cold-workable vanadium in a multi-stage process (see also Table 4). They used as starting materials V_2O_3 and VC in order to avoid the low-melting-point oxide V_2O_5. Reduction experiments in the arc furnace were not successful. A combination of two indirect sinter processes (in the vacuum induction furnace) and two direct sinter processes (direct resistance heating in a vacuum) gave the best results. In a similar way vanadium alloys with either 40% Nb or 40% Ta were reduced at temperatures 50°C higher. In these reactions Nb and Ta were used in the form of oxides and V as carbide. The alloys obtained were cold-workable because the oxygen level was below 0.1% and the carbon content below 0.08%. If, on the other hand, the temperature was 50°C lower V–Cr-alloys of composition 50:50 could be produced, whose oxygen content was finally only 0.1% and that of the carbon only 0.04%[35,42] (see also Table 4). Although the yield is higher than 95%, the time required is so considerable that so far these processes have not found an industrial application.

However, the situation is different for niobium. Carbothermic reduction has been carried out on an industrial scale for a number of years[44]. The thermodynamic constitutional diagrams which illustrate the following reduction reactions are shown in Fig. 15a, b:

$$\langle Nb_2O_5 \rangle + \langle C \rangle = 2\langle NbO_2 \rangle + (CO) \tag{22}$$

$$\langle Nb_2O_5 \rangle + 7\langle C \rangle = 2\langle NbC \rangle + 5(CO) \tag{23}$$

$$3\langle Nb_2O_5 \rangle + \langle NbC \rangle = 7\langle NbO_2 \rangle + (CO) \tag{24}$$

$$2\langle NbO_2 \rangle + 10\langle NbC \rangle = 6\langle Nb_2C \rangle + 4(CO) \tag{25}$$

$$3\langle NbO_2 \rangle + \langle Nb_2C \rangle = 5\langle NbO \rangle + (CO) \tag{26}$$

$$\langle NbO \rangle + \langle Nb_2C \rangle = 3\langle Nb \rangle + (CO) \tag{27}$$

From these equations it may be inferred that neither NbO_2 nor NbC but only NbO and Nb_2C are stable in the presence of Nb. Equations (22), (23), (26) and (27) were confirmed by Sazhin et al.[45,46]. For eqns. (24) and (25) formulas on the basis of a subcarbide $NbC_{0.8}$ have been quoted, which cannot be brought into agreement with Gibb's phase rule, but which, according to the authors, are used only for an overall explanation of a complex reaction. Other suboxide phases, on which Niebuhr[47] reports, should not play a role at the high reduction temperatures used. In this case, too, no reduction by carbon is possible at atmospheric pressure up to 2500 K (see also Table 2), but by applying ultimate pressures of 10^{-3} Torr technologically suitable working conditions are reached. In Fig. 15a, b the solubility data for carbon and oxygen are not quoted.

From preliminary equilibrium determinations by Sibert and Steinberg[48] the

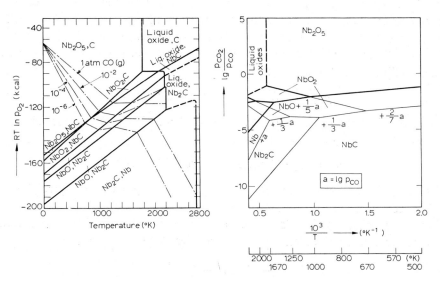

Fig. 15(a). Pourbaix–Ellingham diagram for the system Nb–O–C[29].
Fig. 15(b). Phase diagram for the system Nb–O–C according to Knacke for $p_{CO} = 1$ atm (heavy line) and $p_{CO} = 10^{-6}$ atm (thin line).

References pp. 321–335

TABLE 5
% $[O]_{Nb}$ · %$[C]_{Nb}$-PRODUCTS FROM THE ANALYSIS OF NIOBIUM

% $[O]_{Nb}$ · % $[C]_{Nb}$	Temperature (°C)	Pressure (Torr)	References
0.16	1600	6.5×10^{-2}	
0.63	1700	8×10^{-1}	
0.12	1800	7.2×10^{-1}	48
0.11	1900	1.65	
0.015–0.03	1750–1800	5×10^{-3}	49, 50
0.01–0.04	2150	10^{-4}	51
0.005	2100	10^{-4}	37
0.08	1700	10^{-1}	
0.005	2000	10^{-1}	52

data which are shown in Table 5 were obtained; these are supplemented by a few data from reduction experiments and further equilibrium determinations by Fromm[52,53].

These data are in most cases slightly higher than those for vanadium, but refining to still lower values may be carried out by a vacuum treatment at temperatures between 2000° and 2500°C without any difficulty (see also Subsection 4.2.2). At higher temperatures (above 1900°C) a study of the C–O equilibria is very difficult, as evaporation of suboxides (NbO and NbO_2) is superimposed, which proceeds to a steadily increasing extent with increasing temperatures.

Results of studies by Fromm[52,53] on the Nb–C–O system are also available, but incontestable measurements were only possible in the temperature range between 1700° and 2000°C. At lower temperatures the pressure of formation of CO was too low, at higher temperatures oxygen evaporation via suboxides was so intensive that only a carburization could be detected. For pure engassing with CO the equilibrium relation exists*:

$$p_{CO} = C_{CO}^{1.6} \times 10^6 \times \exp[-67000/RT],$$

where p is in Torr, and C in mol %. If the C:O ratio in the niobium is not unity, the equilibrium may also be expressed by:

$$p_{CO} = C_C \cdot C_O \times 1.6 \times 10^6 \exp[-67000/RT].$$

Saturation concentration is reached at about 0.3 mol % CO, when precipitation of a carbide, or even possibly of a carboxide, can be observed. On the basis of further engassing experiments with O_2 and CO a formulation of the equilibrium constants was derived which deviates from Sievert's $p^{\frac{1}{2}}$ law:

$$K = p_{CO}/C_C \cdot C_O^{0.6}$$

Gel'd and Shveikin[50] were able to confirm the breakdown of Nb_2O_5 via NbO_2

---
* List of symbols, p. 320.

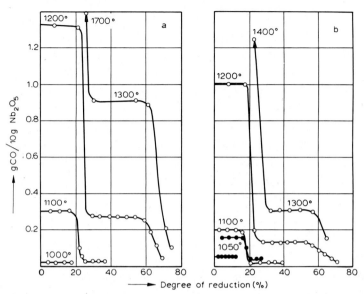

Fig. 16. The effect of temperature and reduction agent on the step-wise decomposition of Nb_2O_5 is clearly recognizable from the change in reaction rate at the steps[50].
a Pitch; *b* Lampblack. Temperatures in °C.

(Nb_2O_4) to NbO which should be expected from Fig. 15b and eqns. (22) and (26). This breakdown is shown in Fig. 16, in which a degree of reduction of 20% corresponds to NbO_2 and of 60% to NbO.

When using NbC as a reducing agent for Nb_2O_5 corresponding stages in the reduction are noticeable, although these are not quite so pronounced (see also Fig. 17). X-Ray analysis showed NbO_2 and NbC for the first stage (about 15% reduction, see also eqn. (24)), NbO_2, NbO and NbC for the second stage (42% reduction, see also eqns. (25) and (26)), and NbO, Nb_2C and a small amount of Nb in the final third stage (70% reduction, see also eqn. (27)). A final pressure of 10^{-3} Torr was attained in the experimental apparatus, thus roughly confirming the thermodynamic constitutional diagram.

The reduction to the metal was studied by Lyubimov et al.[54] according to the following overall equations:

$$\langle NbO_2 \rangle + \langle C \rangle = \langle NbO \rangle + (CO) \tag{28}$$

$$\langle NbO \rangle + \langle C \rangle = \langle Nb \rangle + (CO) \tag{29}$$

From X-ray analysis it was found that NbO could be identified in the reaction according to eqn. (28) only after a temporary appearance and disappearance of NbC. In the reaction according to eqn. (29) NbC was formed as primary product, and was then converted to Nb_2C. Only after the disappearance of the NbC was it possible to identify Nb besides Nb_2C and NbO. For final pressures of 10^{-3}

References pp. 321–335

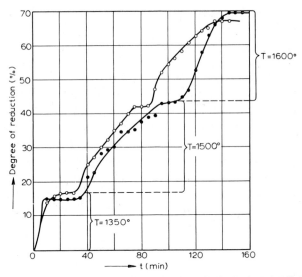

Fig. 17. Effect of temperature and time on the reduction of Nb_2O_5 by NbC[49, 50]. Temperatures in °C.

Torr agreement with the thermodynamic considerations is excellent. The same knowledge may be obtained from a Table by Sazhin et al.[45], but it should be pointed out that the reduction has already begun at 1050°C to a marked extent if lampblack instead of NbC is used, and that it is almost complete at 1730°C (see also Table 6).

Investigations on the reaction kinetics carried out by Gel'd and Shveikin[49, 50] as well as by Sazhin et al.[45] revealed the following picture: The reaction rate of the lower oxides is only slightly increased by increasing the compacting pressure (1.25–20 ton/cm^2) in the preparation of the oxide–NbC briquettes[54]. In the V_2O_5 reduction of the V_2O_5–VC briquettes the compacting pressure (0.5–8 ton/cm^2) has no effect whatsoever[36]. The reactivity and surface area of the carbon used for the reduction play a role in both vanadium[36] and niobium[50, 54]. But these effects decrease with lower oxides and increasing temperatures. Additions of alkali metal carbonate to the briquettes increase the reaction rate to a marked extent. A high vacuum or a medium vacuum do not affect the reduction of the niobium oxides[49, 50, 54]; pressures above 10 Torr lead, for Nb_2O_5 at 1050°C, only to the NbO_2 stage[49, 50], and for V_2O_5 only to the VO stage[36]. The low-temperature modification of Nb_2O_5 shows a somewhat better reducibility[49, 50].

From these results, the authors conclude that Nb_2O_5 is reduced predominantly by CO. It is thought that the reaction:

$$(CO_2) + \langle C \rangle = 2(CO) \tag{30}$$

should be rate-determining particularly at the start of the reduction of the higher oxides. The activation energy at the beginning of the NbO_2 reduction varied around

TABLE 6
REDUCTION OF Nb_2O_5 BY NbC[45]

Temperature (°C)	Reaction time (h)	NbO_2	NbO	NbC	Nb_2C	Nb	Degree of reduction (%)
1200	24	++	++	++	—	—	29.3
1300	12	++	++	++	—	—	29.1
1400	3	+	++	++	++	—	47.5
1400	6	—	++	——	++	+	62
1400	12	—	++	——	++	++	84.3
1700	3	—	+	——	+	++	97.4
2300	3	—	—	——	—	++	100.0

++ Strong X-ray lines.
+ Weak X-ray lines.

50 kcal/mol. For higher compacting pressure the diffusion lengths for CO and CO_2 are considerably shortened.

It is thought that oxy-carbides which are difficult to reduce are formed at higher CO pressures. Such intermediate phases break up with falling pressure into NbO or Nb_2C, respectively. The reduction of these compounds may be expressed by parabolic time laws. For this reason diffusion phenomena in the solid bodies may characterize the final stage of the process. All these considerations concern reactions taking place at between 1000° and 1600°C. For higher temperatures, reactions between volatile NbO, NbO_2 and solid Nb_2C may play a role, in which case the formation and desorption of the oxides (CO, NbO or NbO_2[55,56], respectively) may be rate-determining (see also Chapter IV, Section 2).

The technological process of the production of Nb is usually carried out as a three-stage process:

1. After production of NbC, compacts of Nb_2O_5 and NbC are prepared, in which an O:C ratio of about 1.15 should be maintained.

2. In the following vacuum reduction several working stages are distinguished as is shown in Table 7 and Fig. 18. While the loss in weight in this second stage is attributable to a marked degree to liberation of CO, the loss in weight in the third

TABLE 7
REDUCTION OF Nb_2O_5 BY NbC TO Nb IN A VACUUM[51,57]

Working range	Temp. (°C)	Final pressure (Torr)	Time (min)	Analysis (wt. %)			Weight loss (%)
				O	C	N	
Starting mixture	—	—	—	10	7	a	
1	1500	10^{-3}	80	6	2	0.39	
2	1800	10^{-2}	100	2.5	1.1	0.41	12
3	2100	10^{-4}	140	<0.1	<0.2	0.02	16

a Not known.

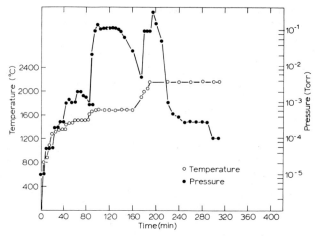

Fig. 18. Pressure and temperature during the carbothermic reduction of Nb_2O_5[51].

stage should be attributed mainly to the evaporation of the suboxides. The reduction is usually carried out in resistance- or induction-heated vacuum furnaces with high pumping speeds.

3. After an intermediate hydrogenation, crushing and adjustment of composition, a post refining by sintering or melting in a high vacuum is carried out.

For reducing the relatively long reaction times Kimura and Sasaki[58] carried out reduction in the liquid state in an arc furnace above 2500°C, which, however, involved a number of technological difficulties (see also Table 8). According to Shveikin and Gel'd[59], if a preliminary reduction of the Nb_2O_5 with C to NbO_2 and NbC is carried out at atmospheric pressure and 1400°C, there are lower quantities of gas to be handled in the subsequent reduction operation in a vacuum. Here too three working stages are distinguished (1400°, 1500° and 1750°C at a final pressure of 2×10^{-3} Torr). The final product contains C 0.1–0.5% and O less than 0.1%.

After a preliminary reduction it should be possible to reduce and to refine such a material directly in the electron beam furnace[43].

TABLE 8

REDUCTION OF Nb_2O_5 BY NbC IN THE ARC FURNACE

Number of melts	Time (min)	Pressure (atm)	Analysis (wt. %)		
			Nb	O	C
Starting mixture	–	–	82.0	10.53	7.44
1	2	1.0	90.2	4.5	3.6
2	2	0.5	94.1	2.7	1.8
4	2	0.5	98.3	0.7	0.6
5	2	0.5	98.7	0.3	0.3
6	2	0.5	99.7	a	0.2

a Not known.

Similar to the carbothermic production of high-purity niobium metal in a vacuum, direct alloy reduction is also possible. Besides the production of Nb–Ta alloys by reduction at 1800°C and 10^{-2} Torr[60] the reduction of a number of other alloys is feasible. In these cases the maximum contents of oxygen, nitrogen and carbon are found, for reaction temperatures of 1900–2100°C and pressures of 10^{-3} Torr, to be 0.1%, but in most cases they are between 0.02 and 0.05%. For tungsten, molybdenum and tantalum additions the starting alloy compositions are maintained after the reduction, whereas Al is almost completely lost and Ti also shows large losses, while the losses of V and Zr depend mainly on the reaction temperature. Final melting in the electron beam furnace only lowers the gas contents (see also Table 9).

The multi-stage (four main and two intermediate stages) direct reduction of alloys with 70% Nb and 30% Zr seems also possible without difficulty at temperatures between 1500° and 1950°C and ultimate pressures of 10^{-4}–3×10^{-5} Torr. The carbon and oxygen contents vary around 0.1%[35].

In a similar way the production of a Nb-containing ferro-chromium addition for steelmaking is possible at 1500°C[62]. In this process high-carbon, crushed (0.07–0.037 mm in diameter) ferro-chromium is mixed with Nb_2O_5 to an O:C ratio of 1–1.2, compacted and treated below 1500°C in a vacuum (below 3 Torr). Oxygen and carbon are almost completely removed by this process.

Although tantalum was reduced by v. Bolton[43], using a carbothermic process in a vacuum, as far back as 1905, this method seems only recently to have been introduced into technology, despite its simplicity. Whereas some low-melting-point components exist of niobium (Nb_2O_5 at about 1500°C) and of vanadium (V_2O_5 at 675°C, VO_2 at 1360–1545°C) which therefore do not permit the use of high reduction temperatures, the melting point of Ta_2O_5 in the Ta–C–O system is as high as 1870°C, so that the whole reduction process may be carried out at high temperatures.

TABLE 9

CARBOTHERMIC PRODUCTION OF NIOBIUM ALLOYS IN A VACUUM[61]

Expt. No.	Starting composition (wt.%)			Temp. (°C)	Composition after the reaction (wt.%)					
	W	Mo	Zr		W	Mo	Zr	C	N	O
1	—	4.5	0.8	1900	—	4.25	0.7	0.14	0.17	0.08
2	—	4.5	2.5	1900	—	4.9	1.2	0.03	—	0.10
3	—	5	3.0	1900	—	4.9	1.85	0.04	0.09	0.04
4	—	4.5	2.5	2000	—	4.4	1.35	0.04	0.06	0.04
5	—	4.5	2.5	2100	—	4.6	0.87	0.03	0.05	0.02
6	15	5	1	1900	14.1	5.0	1.1	0.02	0.04	0.03
2	—	4.9	1.2	Remelting in the	—	4.9	1.1	0.03	0.02	0.06
3	—	4.9	1.85	electron-beam furnace	—	4.6	1.2	0.02	0.01	0.03

Certain difficulties occur, however, in the thermodynamic constitutional diagrams. According to Worrell and Chipman[29] only the oxide phase Ta_2O_5 exists, whereas Niebuhr[63] reports the existence of TaO and TaO_2. The two systems are compared in Fig. 20a, b. The thermodynamic data for TaO and TaO_2 are partly taken from other workers[63,64], partly estimated.

While, according to Worrell and Chipman[29], the following equilibrium equations apply:

$$\langle Ta_2O_5 \rangle + 7\langle C \rangle = 2\langle TaC \rangle + 5(CO), \tag{31}$$

$$\langle Ta_2O_5 \rangle + 12\langle TaC \rangle = 7\langle Ta_2C \rangle + 5(CO), \tag{32}$$

$$\langle Ta_2O_5 \rangle + 5\langle Ta_2C \rangle = 12\langle Ta \rangle + 5(CO), \tag{33}$$

from Fig. 20b equations corresponding to the niobium reduction are derived (see also eqns. (22)–(27)), in which the very small range for the existence of TaO_2 is striking.

Fromm and Heinkel[52,65] measured the CO pressures above Ta–O–C solid solutions. As in the case of niobium the studies were restricted to temperatures between 1700° and 2000°C as too low a CO pressure and suboxide evaporation determine the limits. For pure CO engassing the following equation applies:

$$p_{CO} = C_{CO}^{1.6} \times 7.3 \times 10^4 \exp[-65000/RT],$$

where p is in Torr and C in mol %.

If the C:O ratio in tantalum is not unity, equilibrium may be expressed as:

$$p_{CO} = C_O \times C_C \times 1.2 \times 10^5 \exp[-65000/RT].$$

In this case, similar to the other metals of the Fifth Group of the Periodic Table, very small % $[C]_{Ta} \times \%[O]_{Ta}$ products result. Some data obtained by the above-mentioned authors are shown in Table 10.

In the reduction of Ta_2O_5 by C or TaC at 1200°C in a vacuum only Ta_2O_5 and carbide phases (Ta_2C, TaC_{x+1}) were found in accordance with the phase diagram. Although TaO and TaO_2 should be found, their existence does not seem to be certain at low temperatures. The rate of reaction with carbon is determined by the history of the oxide, the gas pressure and the contact between particles, the rate of reaction with TaC by the shape of the particles, the density of the charge and the pressure of the gas phase[66].

TABLE 10

$\%[O]_{Ta} \cdot \%[C]_{Ta}$-PRODUCTS FROM THE ANALYSIS OF TANTALUM[52,65]

$\%[O]_{Ta} \cdot \%[C]_{Ta}$	Temperature (°C)	Pressure (Torr)
6.5×10^{-4}	1900	7×10^{-3}
2.1×10^{-4}	1900	2.7×10^{-3}
1.5×10^{-3}	2000	10^{-2}
5.6×10^{-5}	2000	10^{-3}

From the experiments it emerged that it is advantageous to maintain an O:C ratio of 1.03, but this ratio (and the time of mixing and also the compacting pressure in the production of the briquettes) had no marked effect on the quality of the reduced metal[67]. This indicates that during reduction of Ta_2O_5 above 1700°C neither the reduction by CO with temporary formation of CO_2 (see also Fig. 20b) nor a solid-state reaction would be expected to play a role. Diffusion of C and O in the metal as well as the desorption of TaO or TaO_2[55,56,67] respectively, and subsequent reactions with Ta_2C, are more likely to determine the rate of reduction. The technological practice of the reduction is carried out by analogy with niobium in the same apparatus as for that metal (see Fig. 19). Corresponding data are shown in Table 11. Here too several temperature stages are distinguished in the reduction. The final product is directly cold-workable, just as is niobium.

Whereas the pure metal may be reduced without any difficulty at temperatures between 2000° and 2700°C, reduction temperatures between 1600° and 1900°C should be maintained in the direct reduction of alloys of 80% Ta, 10% Hf and 10% W. The working conditions and analytical results are also shown in Table 11[35].

(b) Carbothermic reduction of titanium, zirconium and hafnium

During the last few years a number of attempts have been made to reduce Ti, Zr and Hf by carbon or by carbides in a vacuum. However, the equilibrium

Fig. 19. Induction furnaces for the niobium and tantalum reduction[43].

References pp. 321–335

TABLE 11

WORKING CONDITIONS FOR THE REDUCTION OF TANTALUM IN A VACUUM

Reaction step	Temp. (°C)	Pressure (Torr)	Time (h)	Analysis (wt.%)			Ref.
				O	C	N	
1	2000			0.2	0.05	0.05	68
2	2500–2700			0.005	<0.01	0.003	
1	1700–1900	1×10^{-2}	2		0.04		
2	2400	5×10^{-3}–5×10^{-4}	2				67
3	2700–2750	5×10^{-3}–5×10^{-4}	2–3	0.001	0.007	0.006	
Starting mixture (80 Ta, 10 Hf, 10 W)	—	—	—	10.2	7.6	a	
1. Indirect sintering	1500	10^{-4}	8	6.3	4.6	a	
2. Indirect sintering	1600	10^{-4}	8	3.4	2.0	a	35
3. Direct sintering	1700	5×10^{-5}	3	0.34	0.1	a	
4. Direct sintering	1950	3×10^{-5}	3	0.14	0.09	a	

a Not known.

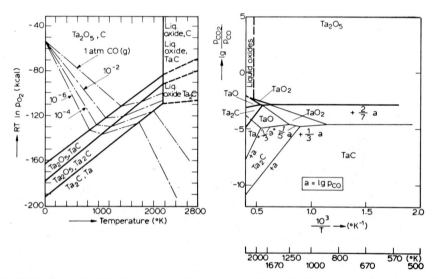

Fig. 20(a). Pourbaix–Ellingham diagram for the system Ta–O–C with Ta_2O_5 as the sole oxide[29].
Fig. 20(b). Thermodynamic equilibrium diagram of the system Ta–O–C according to Knacke for $p_{CO} = 1$ atm (heavy line) and $p_{CO} = 10^{-6}$ atm (thin line) with the oxide phases Ta_2O_5, TaO_2 and TaO.

phase fields of the pure metals, without taking into account the O and C solubilities, lie, even in a vacuum, at such high temperatures that an economically feasible reduction by carbon does not seem promising.

Accordingly, attempts to reduce TiO_2 by carbon between 1200° and 1800°C and at a CO pressure of 10^{-2} Torr in an arc furnace resulted in a reaction product containing 7.7% $[O+C]_{Ti}$. The O level could, however, be lowered by addition of Si or CaSi, respectively[69,70]. Also, reduction experiments by Meerson et al.[71] only produced mixed phases between Ti, TiO and TiC, the oxygen levels of which were dependent to a large extent on the CO partial pressure. Kieffer and his co-workers[35] found only traces of metal in experiments carried out at about 1750°C. On the other hand, carbothermic alloy reduction of Mo–Ti alloys seems to be feasible, as Kieffer et al.[35] report. An alloy of 40% Ti and 60% Mo was produced in a multi-stage reduction process at temperatures between 1600° and 2000°C. This alloy finally showed only 0.11% oxygen and 0.14% carbon.

In addition, experiments for the reduction of Zr by carbon in an electron beam furnace resulted in either a zirconium with 0.7% oxygen and 0.01% carbon, or, if higher carbon additions were used, in the primary precipitation of carbides. These were even enriched in the melt by partial evaporation of the zirconium[72,73]. Kosolapova et al.[70] found only oxycarbides in the reaction of ZrO_2 and ZrC between 1800° and 2000°C after 1 h. No reaction took place with HfO_2–HfC.

(c) Preparation of carbides in a vacuum

In contrast with the reduction of the pure metal a straightforward preparation of TiC is successful at 1600°C and a pressure of 1 Torr. If the pressure is lowered, a carbide with a deficiency in carbon is obtained. At higher CO pressures, of the order of 100 Torr, higher oxide levels were found in the carbide. Even with continuously increasing temperatures during carbide preparation, the stepwise reduction of TiO_2 can be identified[74].

Vanadium pentoxide may be converted by a multi-stage process with gaseous hydrocarbons first at 600°C and subsequently at 1050°C into vanadium oxycarbide. If then an adequate quantity of carbon is added, a high-purity carbide may be obtained[75] at 1400–1500°C and a pressure of 0.05 Torr.

Makarenko and Kvas[76] report on methods for the preparation of rare-earth dicarbides such as GdC_2, TbC_2, DyC_2, ErC_2, TmC_2 and LuC_2, using for all materials a temperature of 1800°C and a vacuum.

In a similar way Achard[20] produced the dicarbides of La, Ce, Pr, Nd, Gd, Dy, Er, Lu and Y working at a temperature of 1500°C and a vacuum of 10^{-4} Torr. Usually less than 1% of the rare-earth metals is volatilized during this process. The reactions start for

La and Nd at 1250°C
Gd and Dy at 1300°C
Er and Y at 1330°C
Lu at 1400°C

For Ce and Pr the reaction began at a temperature as low as 1000°C, but this should be attributed to a pre-reduction of oxides. Reaction rate constants were also determined.

References pp. 321–335

Kieffer and Rassaerts[77] report on attempts to produce carbides of the metals of the Fourth, Fifth and Sixth Groups of the Periodic Table. They melted the relevant metals together with a carrier metal, such as Co, Ni, Fe, Ni–Fe, in a graphite crucible. After evacuation to 10^{-5} Torr the melt was heated to 1800–2000°C for 2–3 h. After the reaction was completed the crucible was broken up and the fractured metal–carbide mixture leached with a mineral acid. In this process the carbides and the precipitated elementary graphite are not dissolved; they may subsequently be separated mechanically from each other. The level of free carbon is below 0.1 % in the relevant carbides, that of oxygen below 100 ppm and that of nitrogen below 10 ppm.

(d) Carbothermic reduction of chromium, molybdenum, tungsten and uranium

The carbothermic reduction of chromium and tungsten seems to be nowadays of great technical interest, although Sully[78] was still denying this possibility in 1954. In the meantime a number of positive experimental results have been reported. Figure 21 shows the thermodynamic equilibrium diagram according to Pourbaix–Ellingham[30] and that according to Knacke for the system Cr–O–C.

From these diagrams the following course of reduction should be assumed:

$$3\langle Cr_2O_3\rangle + 13\langle C\rangle = 2\langle Cr_3C_2\rangle + 9(CO) \tag{34}$$

$$5\langle Cr_2O_3\rangle + 27\langle Cr_3C_2\rangle = 13\langle Cr_7C_3\rangle + 15(CO) \tag{35}$$

$$\langle Cr_2O_3\rangle + 3\langle Cr_7C_3\rangle = \langle Cr_{23}C_6\rangle + 3(CO) \tag{36}$$

$$2\langle Cr_2O_3\rangle + \langle Cr_{23}C_6\rangle = 27\langle Cr\rangle + 6(CO) \tag{37}$$

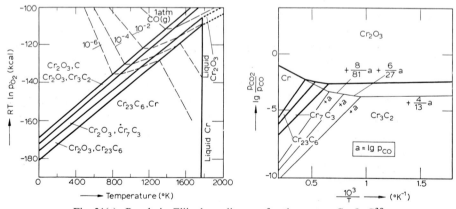

Fig. 21(a). Pourbaix–Ellingham diagram for the system Cr–O–C[30].
Fig. 21(b). Thermodynamic equilibrium diagram of the system Cr–O–C according to Knacke for $p_{CO} = 1$ atm (heavy line) $p_{CO} = 10^{-4}$ atm (thin line).

TABLE 12

REDUCTION TEMPERATURES OF THE METALS OF THE SIXTH GROUP OF THE PERIODIC TABLE

CO-Pressure (atm)	Cr	Mo (°C)	W
1	2000	1070	900
10^{-2}	1550	825	685
10^{-4}	1260	655	540
10^{-6}	1050	535	430

From these diagrams it is evident that a reduction to the metal is not possible at atmospheric pressure. But the field of existence of metallic chromium is extended by application of a vacuum.

Table 12 shows the minimum temperatures for a reduction of Cr, Mo and W. According to these temperatures favourable conditions for a reduction of all three metals in a vacuum should be expected. Of particular interest in this connection is the simultaneous solubility of C and O in chromium; a number of data are compiled and shown in Table 13.

These %C · %O products show that it is possible to obtain relatively low values, with the C content lying below 0.1%. Samsonov and Kosolapova[81] were able to identify with certainty two, possibly even three, reduction stages in reduction experiments of Cr_2O_3 by Cr_3C_2 at a pressure of 10^{-2} Torr and temperatures increasing in steps of 50°C (see Fig. 22). These reduction stages correspond to eqns. (35), (36) and (37). Whereas the formation of the lower carbides occurred only at temperatures higher than those corresponding to the equilibrium diagram (1200° and 1250°C instead of 800° and 1000°C), a reduction to the metal was observed at temperatures from 1300°C upwards, as should be expected. The reduction was completed at 1600°C.

Two reduction stages (1200° and 1500°C) could be inferred in the reduction of Cr_2O_3 by Cr_7C_3 and one reduction stage (1500°C) in the reduction of Cr_2O_3 by $Cr_{23}C_6$ from the data by Naeser et al.[82], who used a heating rate of 300°C/h. These reduction stages indicate the occurrence of reactions according to eqns. (36) and (37), particularly as the weight losses in the reduction by Cr_7C_3 are 5 and 15%, respectively, and in the reduction by $Cr_{23}C_6$ 10%. The same authors also assume the temporary formation of an unstable CrO. The reaction rate is

TABLE 13

$\%[O]_{Cr} \cdot \%[C]_{Cr}$-PRODUCTS FROM THE ANALYSIS OF CHROMIUM

$\%[O]_{Cr} \cdot \%[C]_{Cr}$	Temperature (°C)	Pressure (Torr)	Ref.
2.8×10^{-3}			
16×10^{-3}	Melted in the vacuum		79
5×10^{-3}	induction furnace		
24×10^{-3}	1400	$1-5 \times 10^{-2}$	80

References pp. 321–335

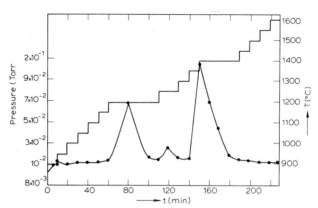

Fig. 22. Pressure/time/temperature curves of the reduction of Cr_2O_3 by Cr_3C_2[81].
—·— Pressure; ——— Temperature.

said to increase by addition of alkali metal salts. Diffusion of carbon and oxygen in the solid state is thought to be decisive.

Tonomura et al.[83], in the reduction of Cr_2O_3 by graphite between 850° and 1300°C, found only formation of carbide, but from 1300°C upwards and at a pressure of 10^{-2} Torr metallic chromium was also formed. Kosolapova et al.[70] confirmed the presence of metallic chromium above 1200°C. At 1400°C marked evaporation of chromium was noticeable.

In the reduction experiments two reaction periods were distinguished, the first of which takes place in two steps.

Investigations[84] showed that a metallic phase is formed even in the first reaction period and that the transport of the carbon into the metal for the formation of the carbide involves a further process. In both processes the gas phase ($CO-CO_2$) takes part. The activation energy amounts to 34.4 kcal/mol[84], whereas Karsanov et al.[80] report 30.5 kcal/mol [for coke] and 27.9 kcal/mol [for charcoal]. The quality of the reduced metal depends mainly on the rate and the extent of the second reaction period, in which diffusion plays an important role. According to Karsanov et al.[80] the transition to the second reaction period was observed to take place at a carbon content of about 5–6% when the oxygen content may vary between 4% and 10%, which indicates an irregularity in the Boudouard reaction. If the carbon levels of the starting mixture are lower, the diffusion phase starts at a lower degree of reduction. Carbon contents different from these affect the onset of the second reaction period only if the Cr_2O_3 used is sufficiently fine-grained. Increasing particle size of the Cr_2O_3 influences the beginning of the second reaction period, which starts the earlier the larger the Cr_2O_3 particles are. In this case the Boudouard reaction should not determine the reaction rate of the first period. Carbon monoxide pressure differences between 0.05 and 1 Torr have no effect on the reaction. For pressures of 0.1 Torr the losses in chromium owing to evaporation vary between 3% and 6% over 6–8 hours, but these losses may increase considerably at

TABLE 14

WORKING CONDITIONS FOR THE CARBOTHERMIC REDUCTION
OF CHROMIUM IN A VACUUM

	O/C Ratio	Temperature (°C)	Pressure (Torr)	Time (h)	Analysis (wt.%)			Ref.
					O	C	N	
		1275–1400	3×10^{-1}		0.04	0.015	0.001	68
	0.95	1200–1600	10^{-2}	6.8	low	0.09	a	81
		1400	10^{-2}	7	0.8	0.02	a	80
Eqns. (34) and (35)		1300–1400	ambient pressure	3.5	7–8.2	5.2–6.8	a	
Eqns. (36) and (37)		1400	10^{-2}	4	0.5	0.02	a	80
Starting mixture		—	—	—	11.3	8.7	a	
Indirect sintering		1250		12	0.6	0.2	a	35
Direct sintering		1400		4	0.04	0.08	a	

a Not known.

higher vacua. If the pressure is maintained at about 0.1 Torr during the whole reduction operation the losses may be kept within reasonable limits.

The economic importance of the Cr_2O_3 reduction by C or Cr_3C_2 may be judged from the number of patents granted[85]. The quality of the metal obtained is good; this may also be seen from the data shown in Table 14. It should also be pointed out that Si, Fe and Al levels in particular in the reduced metal are very low.

Worrell[30] reports on the possibility of a reduction of Mo and W as well as on the production of their carbides in a vacuum on the basis of the established thermodynamic equilibrium diagrams of Mo–O–C and W–O–C. According to Mettler[86] the carbothermic reduction of WO_3 proceeds to completion at 1100°C and at reduced pressure.

Although the reduction of uranium may be carried out with advantage by magnesium from uranium fluoride, the reduction of the oxide is also possible by carbon in a vacuum. In the same way the production of uranium carbide may be carried out by reaction of uranium with carbon or by a reaction of uranium oxide with carbon. The strong affinity of oxygen for uranium requires in both cases working in a vacuum. The working conditions may be derived from a preliminary thermodynamic equilibrium diagram (Fig. 23), which clearly shows that, for atmospheric pressure, temperatures around 2000°C are required for the formation of carbides.

The temperatures may be lowered by working in a vacuum. In this presentation the ternary phase U(C, O) is not shown, as no adequate thermodynamic data exist, but only preliminary phase field outlines[33, 87]. Because of this mixed phase a further vagueness exists. While good agreement between calculated[31] and experimental data (dark points)[32, 33] is found for the equilibrium UO_2–C–UC_2, no such agreement exists for the remainder of the experimental data. The experimental results, which are shown as crosses in Fig. 23, should correspond to the

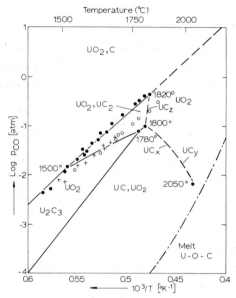

Fig. 23. Thermodynamic equilibrium diagram of the system U–O–C[31], supplemented by additional data[32,33].

equilibrium UO_2–UC–UC_2[32], and the empty circles to the equilibrium UO_2–UC_2–$U(C, O)$[33]. In no case was U_2C_3 detected by X-ray analysis.

A temperature of 800–1000°C is sufficient for the formation of carbide from the metal and carbon, while reaction times of 15–16 h seem to be necessary. For the reaction between the oxide and carbon, the onset of the reaction was found to occur at about 1100–1300°C, in agreement with the constitution diagram, with the main reaction taking place at 1250–1600°C[88,89]. The large differences in the temperatures quoted should be attributed to the pronounced pressure dependence of the reactions (see also Fig. 23). At 1900°C the processes for the formation of carbides are complete.

The reduction of uranium by carbon in a vacuum seems to take place in three stages. Below 1000°C all higher uranium oxides are converted into UO_2. At 1300–1550°C the reduction of UO_2 to carbide occurs and up to 1800°C possibly the formation of a lower oxide. The last stage in the reduction, the reaction between carbide and the remainder of the oxide, requires temperatures between 1900° and 2100°C[90]. In a further publication by the same authors[91] more details are made known.

For the two last stages the following reactions are quoted: for 1400–1600°C:

$$\langle UO_2 \rangle + 2\langle C \rangle \rightarrow \tfrac{1}{3}\langle UO_2 \rangle + \tfrac{2}{3}\langle UC \rangle + \tfrac{4}{3}(CO); \qquad (38)$$

for 1800–1900°C:

$$\tfrac{2}{3}\langle UC \rangle + \tfrac{1}{3}\langle UO_2 \rangle \rightarrow \langle UO_{0.5}C_{0.5} \rangle + \tfrac{1}{6}(CO) \qquad (39)$$

and

$$\langle UO_{0.5}C_{0.5}\rangle \to \{U\} + \tfrac{1}{2}(CO). \tag{40}$$

Reaction (38) should, according to the constitutional diagram, take place in intermediate stages.

The metal yield attains 92%, the uranium contains about 2500 ppm C and 500 ppm O. By a vacuum remelting treatment the carbon level is lowered to less than 600 ppm and the oxygen level to less than 100 ppm[91].

Bazin[92] also carried out investigations on the carbothermic reduction of UO_2 with special attention to the formation of the mixed phase $U(C_{1-x}, O_x)$. At the same time the occurrence of the reactions according to eqns. (39) and (40) was confirmed. In addition, $U(C_{1-x}, O_x)$ is expected to react with excess UO_2 to U and CO.

3.1.2 Carbothermic reduction processes involving volatilization of the reduced metal

The main difficulties in the carbothermic reduction of metals of the Fourth, Fifth and Sixth Groups of the Periodic Table in a vacuum are: coping with the large quantities of gas to be pumped off and the lowering of the quantities of oxygen and carbon which remain in the dissolved state after the reduction has been completed. The losses in reduced metal owing to evaporation may be kept within limits even with chromium.

Further difficulties are met in the carbothermic reduction of the alkali and

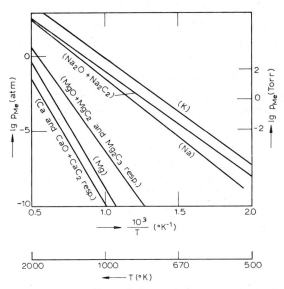

Fig. 24. Effect of temperature on the formation pressures of alkali and alkaline earth metals calculated from eqns. (41)–(47).

the alkaline-earth metals. The reduced metals evaporate and have to be condensed from the gas phase at low temperatures. This involves the risk of a retro-reaction or of the formation of carbides. The possibility of a reduction of the oxides is shown in Fig. 24. The corresponding CO pressures follow from the reduction equations:

$$\langle Na_2O\rangle + \langle C\rangle = 2(Na) + (CO) \tag{41}$$

$$\langle K_2O\rangle + \langle C\rangle = 2(K) + (CO) \tag{42}$$

$$\langle MgO\rangle + \langle C\rangle = (Mg) + (CO) \tag{43}$$

$$\langle CaO\rangle + \langle C\rangle = (Ca) + (CO) \tag{44}$$

or from the corresponding carbide–oxide reactions:

$$2\langle Na_2O\rangle + \langle Na_2C_2\rangle = 6(Na) + 2(CO) \tag{45}$$

$$2\langle MgO\rangle + \langle MgC_2\rangle = 3(Mg) + 2(CO) \tag{46}$$

$$2\langle CaO\rangle + \langle CaC_2\rangle = 3(Ca) + 2(CO) \tag{47}$$

In using Mg_2C_3, only very small differences in comparison with the data of eqn. (46) were found. The calculated lowest required reduction temperatures according to eqns. (41)–(44) are contained in Table 15.

In the reduction of the alkali metal oxides with the carbides even more favourable reduction conditions exist, but according to the results obtained by Kozhevnikov[93] no alkali carbides are stable at temperatures above 800°C or in a vacuum below 10 Torr.

During condensation the metal vapour must not come into contact with graphite below 800°C, as carbide formation is likely[94]. In the condensation of the alkali metals it was found appropriate to introduce the gaseous mixture of metal vapour and carbon monoxide at a temperature of 1000–1200°C and a pressure of 0.5–10 Torr into the water-cooled condenser. At lower temperatures (900° to 700°C) yield and activity of the metal fall to 60–40%.

On the admission of the gas mixture into the condenser, supersaturation in the gas phase may occur at rapid cooling rates. In order to keep the dwell time of the

TABLE 15

EFFECT OF REDUCTION TEMPERATURE ON THE VAPOUR PRESSURES OF ALKALI AND ALKALINE-EARTH METALS DURING CARBOTHERMIC REDUCTION

Pressure (Torr)	Temperature (°C)			
	Na	K	Mg	Ca
1	575	450	1270	1450
10	675	530	1440	1650
100	820	640	1630	1900
760	960	750	1840	2170
Melting point	97.8	63.5	650	843
Boiling point	882	779	1105	1483

supersaturated vapour in the condenser as short as possible, the volume of the condenser should be kept to a minimum; at the same time its surface area should be as large as possible. At temperatures below 200°C no reaction with CO would be expected[93].

It is not only the pure oxides that may successfully be reduced in a vacuum. A reduction of the alkali silicates[95] and aluminates[96,97] and of other intermediate products[98] also is feasible. Good conditions for a reduction of these compounds exist, however, only at temperatures higher than those for the reduction of the pure oxides.

In the silicate reduction according to:

$$\langle Na_2SiO_3 \rangle + \langle C \rangle = 2(Na) + \langle SiO_2 \rangle + (CO) \qquad (48)$$

the yield is increased materially if CaO is added to the starting mixture up to a CaO:SiO$_2$ ratio of two. The overall reaction (48) is then changed to:

$$\langle Na_2SiO_3 \rangle + \langle C \rangle + 2\langle CaO \rangle = 2(Na) + \langle Ca_2SiO_4 \rangle + (CO) \qquad (49)$$

The addition of CaO affects the reaction in a positive sense only at temperatures of 1100–1200°C or more, as the formation of orthosilicates is greatly retarded at temperatures lower than those quoted (see also Fig. 25). The rate of the Na reduction is according to the authors[95] governed by the rate of the formation of Ca$_2$SiO$_4$. The optimum conditions for a reduction are compiled in Table 16.

The reduction of aluminates takes place under similar conditions at 1100–1200°C, so that the reduction of Al$_2$O$_3$ is prevented just as was previously that of SiO$_2$; only at temperatures above 1600°C was the formation of volatile aluminium suboxides observed. The reduction begins already for the aluminates at 276–557°C (K) and 610–630°C (Na)[96]. The residues consist of Al$_2$O$_3$ up to 95%. If the starting materials also contain iron oxides, a reduction of the Fe, Al and Si to an Al–Fe–Si alloy[97] takes place (see also Fig. 26).

The carbothermic reduction of magnesium by the Hansgir process[99,100]

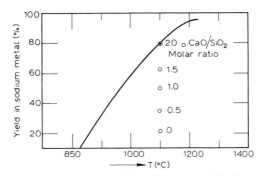

Fig. 25. Effect of temperature and the CaO:SiO$_2$ ratio on the yield of metal in the carbothermic reduction of Na$_2$SiO$_3$[95].

TABLE 16

EXPERIMENTAL CONDITIONS DURING THE REDUCTION OF ALKALI SILICATES AND ALKALI ALUMINATES WITH DOUBLE THE STOICHIOMETRIC QUANTITY OF CARBON ADDED

Starting material	Temp. (°C)	Pressure (Torr)	Time (h)	Yield (%)	Ref.
Na_2SiO_3	1100	0.5–2.5	3	95 with CaO-addition	95
$KAlO_2$	1100	0.8	2–3	90–93	96
$NaAlO_2$	1200	0.8	2–3	80–82	96
$KAlO_2 + NaAlO_2$ 10–50% 90–50%	1100	1	4	85 Na, 100 K	97

was of economic importance for a period of time. The reaction according to:

$$\langle MgO \rangle + \langle C \rangle = (Mg) + (CO) \qquad (43)$$

takes place at 2000°C; according to Table 15 a vapour pressure of Mg above 1 atm is therefore to be expected, but difficulties may arise in the condensation, as the temperature of condensation is considerably below that for reduction. According to Guljanitzki[101] in this case intensive retro-reaction should be expected. By quenching with H_2 or natural gas metallic Mg could anyway have been produced on that occasion, but this had to be subjected to a subsequent vacuum distillation.

More favourable conditions for reduction are obtained by reduction in a vacuum[102]. From the calculated data of Table 15 it may be seen that a noticeable reduction is possible even at 1400°C. This considerably reduces the difficulties met with in the condensation. It appears that the reduction takes place according to the equation:

$$\langle MgO \rangle + (CO) = (Mg) + (CO_2) \qquad (50)$$

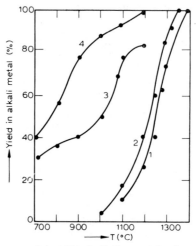

Fig. 26. Effect of temperature on the yield of alkali metal in the carbothermic reduction of $NaAlO_2$ (1, 3) and $KAlO_2$ (2, 4) at ambient pressure (1, 2) and in vacuum (3, 4)[97].

in which the reduction of CO_2 on excess carbon seems to be the rate-determining step for the whole reaction. The optimum CO pressure for 1400°C is 0.1 Torr. The reaction rate decreases with pressures both higher and lower than this. On the other hand, the rate of the pure solid-state reaction (eqn. (43)) rises with rising temperature. The activation energies for the reduction in a high vacuum between 1200° and 1700°C are 93 kcal/mol, for the reduction at a pressure of 0.1–1 Torr Ar or CO 55 kcal/mol, for the reduction at a pressure of 10 Torr in Ar 88 kcal/mol, for a reduction at even higher pressures 90–107 kcal/mol. From these data it must be inferred that different rate-determining steps apply in the various pressure ranges.

The carbothermic reduction of CaO or the treatment of CaC_2 contaminated with CaO is also of technological interest. Calcium oxide can be reduced by Al_4C_3 at 10^{-1} Torr and 1150–1200°C. The addition of 2.5% CaF_2 substantially increases the Ca yield. Besides carbon, aluminium too acts as a reducing agent; the slag formed in this reaction consists mainly of $5CaO \cdot 3Al_2O_3$[103]. In the treatment of impure CaC_2 carbon monoxide is also formed, although in small quantities only, which can be taken up by Ti or Zr[104]. The carbon monoxide may also be removed at a temperature of 1500°C and at a pressure below 5 Torr before the onset of the actual carbide decomposition. This decomposition takes place subsequently at higher temperatures[105,106].

More recently, the first results on the dissociation of alkali metal carbides have been reported. According to these reports alkali metal carbonates are mixed with carbon and heated in an electric furnace to 1100–1200°C. At this temperature metal vapours and CO are formed. While in the processes described so far carbide formation was to be avoided, in this process the evaporated metals are completely converted into carbides as the gas passes a carbon filter which is heated to a temperature between 550° and 600°C. The carbides formed are afterwards decomposed preferably in a vacuum at the temperatures indicated below and the alkali metal vapours are precipitated in a water-cooled condenser[107,108]. The following dissociation temperatures are quoted for the various alkali metal carbides:

	Li	Na	K
In a protective atmosphere	1400–1500°C	900–950°C	700–900°C
In a vacuum (0.5–2 Torr)	1000–1200°C	600–700°C	500–600°C

3.2 Metallothermic reduction processes

3.2.1 Metallothermic reduction processes without volatilization of the metals to be reduced

  (a) *Metallothermic reduction of the halides of titanium, zirconium and hafnium*

The reduction of the halides of the high-melting-point metals (Ti, Zr, Hf, V or Ta) by Mg or Na is usually carried out in a protective atmosphere, using either helium or argon, at a pressure slightly above atmospheric and between 750° and

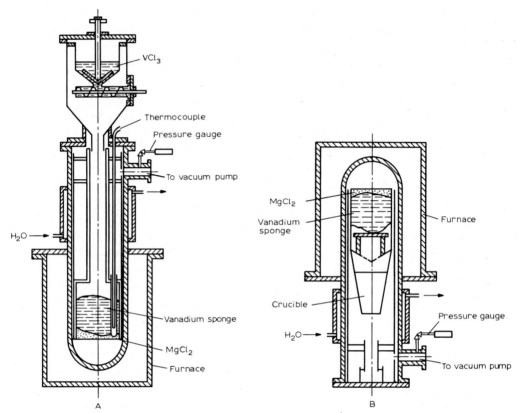

Fig. 27(a). Apparatus for the production of high-melting point metals from their chlorides[43].
Fig. 27(b). The same apparatus after modification for the vacuum distillation of the chlorides formed and of the excess reducing metal[43].

920°C over a period of several hours. The reaction products (metal sponge, $MgCl_2$ or NaCl, excess Mg or Na, respectively) are intimately mixed, but may be separated by leaching or vacuum distillation. This distillation proceeds for all metals almost to completion at a temperature of about 900°C and a vacuum of 10^{-3}–10^{-1} Torr, requiring distillation times of 8–35 h. The Mg levels achieved are below 0.1% and the Cl levels below 0.02%. A combined reduction and distillation apparatus is shown in Fig. 27. In Japan charges of 1.4 ton of titanium are already being vacuum-treated at 1000–1050°C at a pressure of 5×10^{-5} Torr for 26 h, thus obtaining final levels of 0.02% of both chlorine and magnesium[109].

The reduction of PuO_2, ThO_2 and U_3O_8 is possible under similar conditions using a liquid Mg–Zn alloy and an addition of alkali halides or alkaline-earth halides as a flux. Magnesium and zinc may subsequently be completely removed by vacuum distillation[110].

It would be more advantageous if the reduction could be carried out in a vacuum, because the yield in metal sponge decreases with increasing pressure and the quantity of fines in the sponge increases[111]. But these reduction processes are exothermic and control of the temperature is very difficult to attain in a vacuum.

Proft *et al.* tried to carry out the Kroll processes of the Ti reduction[112] and Zr reduction [113] completely in a vacuum. Here reduction to Ti metal is carried out in a retort which is provided with several heating groups and which is pumped down before the reduction. The retort is then maintained under a static vacuum. After being heated to 700°C, $TiCl_4$ vapour is admitted into the retort. The quantity control of the vapour is effected by a separate heating unit of the $TiCl_4$ storage vessel. The vapour pressure of the $TiCl_4$ in the reactor should be higher than that of the Mg, so that the main reaction takes place in the Mg reservoir. Owing to the liberated heat of reaction, the heating unit may be switched off; in spite of this a temperature of 800–900°C is maintained.

The end of the reduction may be noted by the fall in temperature and pressure. External heating is then again required for the distillation, the end of which is identified by a further fall in pressure and a simultaneous rise in the temperature. During the distillation excess Mg and the $MgCl_2$ formed evaporate. They condense in a cooling trap and are collected in a drip catcher. The reduction is carried out at an argon partial pressure of 50 Torr, in order to obtain an improved heat exchange and to prevent boiling of the Mg (see also Fig. 28a).

The zirconium reduction is carried out under similar conditions. Magnesium contained in the reaction vessel is fused; the pressure of the $ZrCl_4$ is again higher than that of the Mg and the main reaction therefore takes place in the reaction vessel. After distillation the main portion of the zirconium is obtained as a solid

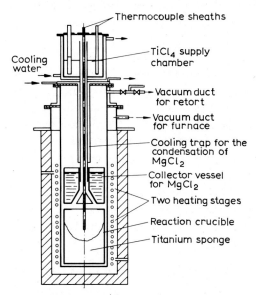

Fig. 28(a). Vacuum apparatus for the reduction of $TiCl_4$ by magnesium and subsequent vacuum distillation for the purification of the titanium sponge formed[112].

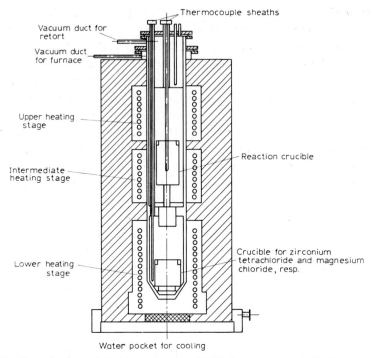

Fig. 28(b). Vacuum apparatus for the reduction of $ZrCl_4$ by magnesium and subsequent vacuum distillation for the purification of the zirconium formed[113].

plate at the bottom of the reaction vessel. This plate can be handled with lower hazards than can sponge and can be further refined. The somewhat higher Mg content of 0.5% in the as-distilled metal is of no great disadvantage, as it can be removed in the subsequent refining stage (see Fig. 28b).

Based on similar considerations, experiments for the magnesiothermic reduction of $NbCl_5$ and $TaCl_5$ were carried out in a vacuum[114]. These processes too are controlled by the rate of evaporation of the pentachlorides. The main reaction in these processes takes place at 550–600°C. Distillation of the $MgCl_2$ and of the excess Mg is carried out at 800–900°C. Unfortunately, a highly pyrophoric metal sponge is formed so that it can be removed from the retort only under a protective atmosphere. The impurities for both niobium and tantalum are about 0.2% Mg, 0.3–0.7% O, 0.03–0.08% N and 0.01–0.02% H, which can be almost completely removed by remelting in the electron beam furnace.

The reduction of $TiCl_4$ by Mg in the gas phase at 900–950°C and a vacuum of 10^{-3} Torr according to:

$$(TiCl_4) + 2(Mg) = 2(MgCl_2) + \langle Ti \rangle \tag{51}$$

may be regarded as another major technological advance[115]. The production of titanium of very high purity is effected in one step. The titanium corresponds in

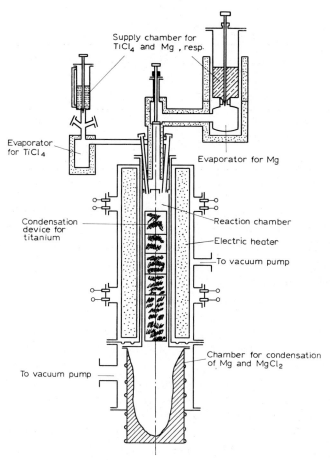

Fig. 29. Vacuum furnace for the reduction of $TiCl_4$ by magnesium in the gas phase[115].

purity and mechanical properties almost to iodide titanium. The yield of $TiCl_4$ is around 98%, that of Mg about 80%.

In this technology Mg and $TiCl_4$ are heated directly and indirectly, respectively, in separate chambers an devaporated. The rate of evaporation is governed by the heat input. The vapours are introduced separately into the reactor, where they react very rapidly and almost completely at about 950°C. The Ti is condensed at a Ti frame which is wound with flattened Ti wires, where it grows to large crystals. Magnesium chloride and magnesium are collected separately in a water-cooled vessel below the reactor (see Fig. 29). Owing to the high-vacuum conditions in the reactor the Fe, N, H, C, Mg and Cl levels are all below 0.01%.

(b) *Metallothermic reduction of the oxides of titanium, zirconium, and hafnium*

Attempts have been made at the metallothermic reduction of the oxides of Ti^{116}, $Zr^{117,118}$ and $Hf^{117-119}$ in a vacuum, using Si, $CaSi_2$ and Al as reducing

References pp. 321–335

TABLE 17
REDUCTION OF HfO_2 BY SILICON IN A VACUUM[117-119]

Reaction step	Composition						Reducing agent
	of the evaporated oxides (%)			of the metal button (%)			
	O	Si	Hf	O	Si	Hf	
1	28.4	38.7	32.9	a	a	a	Si
2	9.4	5.2	85.4	a	a	a	Si
3	4.7	2.0	93.3	0.153	0.146	99.7	Si
–	a	a	a	0.036	0.089	a	$CaSi_2$

a = Not known.

agents. During the reduction in a high vacuum (use of an electron-beam furnace) mainly gaseous suboxides like SiO and Al_2O are formed. Besides these the volatile suboxides of Zr (ZrO) and Hf(HfO) are also formed.

The reduction of hafnium as well as zirconium by Si proceeds at temperatures between 2200° and 3200°C according to:

$$\langle HfO_2 \rangle + 2\{Si\} = \{Hf\} + 2(SiO) \qquad (52)$$

A slight deficiency in silicon should be maintained in the starting mixture in order to avoid a residual silicon content in the final product which is difficult to remove. According to the authors, excess oxygen is mainly removed as gaseous HfO. The composition of the evaporated oxides changes with increasing reaction time, and also the losses in metal owing to evaporation increase, as the composition of the analysed condensates shows (see Table 17).

The metal obtained is workable only within limits. Reduction by $CaSi_2$, which appears to be more favourable, results in a more thorough deoxidation, but the nitrogen content of the $CaSi_2$ (0.04%) is entirely transferred into the hafnium, which will contain 0.15% N[119].

The reduction of ZrO_2 by $CaSi_2$ may be expressed theoretically by:

$$3\langle ZrO_2 \rangle + 2\{CaSi_2\} = 3\{Zr\} + 2\langle CaO \rangle + 4(SiO) \qquad (53)$$

At a substoichiometric composition the reaction will not be complete. Besides ZrO_2, $CaSi_2$ will also be present. At stoichiometric or hyperstoichiometric compositions, zirconium and an unidentified zirconium silicide will be formed. It was not possible to remove all the silicon by vacuum remelting[117].

As a result of the strong affinity between Si and Zr the silicon contents of the distillate are only slightly higher than those of the melt[120].

The reduction of TiO_2 or TiO by aluminium in a vacuum of 10^{-3}–10^{-4} Torr takes place above 1200–1300°C with great intensity. The reaction may be expressed below 1300°C by:

$$\langle TiO_2 \rangle + 7\{Al\} = \{TiAl_3\} + 2(Al_2O), \qquad (54)$$

up to 1500°C by:

$$\langle TiO_2\rangle + 5\{Al\} = \{TiAl\} + 2(Al_2O) \qquad (55)$$

and up to 1900°C by:

$$\langle TiO_2\rangle + 3\{Al\} = \{TiAl\} + 2(AlO), \qquad (56)$$

when at the higher temperatures the aluminium content is reduced owing to increasing evaporation and reaction products richer in titanium are formed[116].

The reduction of ZrO_2 by aluminium is carried out in two stages. At a temperature of about 900–1150°C the following reaction takes place within an hour under a protective atmosphere:

$$3\langle ZrO_2\rangle + 13\{Al\} = 2\langle Al_2O_3\rangle + 3\langle ZrAl_3\rangle. \qquad (57)$$

The reduction of the sponge-like intermediate product is then continued in a high vacuum at temperatures above 2500°C

$$2\{Al_2O_3\} + 3\{ZrAl_3\} = 3\{Zr\} + 6(Al_2O) + (Al). \qquad (58)$$

The zirconium obtained may be cold-rolled up to 95%. A reduction in one stage would result in extremely high Al losses owing to evaporation[117,118].

Since the reduction of HfO_2 by Al takes place only at very high temperatures, very high Al losses owing to evaporation should be expected, if complete reduction is attempted[117–119].

3.2.2 Metallothermic reduction processes with volatilization of the reduced metals
(a) Metallothermic reduction of rare-earth metals

The rare-earth metals, with the exception of samarium, europium and ytterbium, in the form of their fluorides, may be reduced by calcium in tantalum or tungsten crucibles either in an inert gas atmosphere or in a vacuum[121]. In this reaction CaF_2 and the respective rare-earth metals are formed. The reduction starts at 800–1000°C. For rare-earth metals of higher melting points these temperatures have to be raised by about 50°C in order to obtain complete reaction and satisfactory separation of metal and slag. The yield in metal is about 97–99%; the main impurity is Ca at 0.1–2%[122–124]. Subsequent remelting in a vacuum further increases the purity so that finally 0.01% Ca, 0.005% N, 0.03–0.1% O and 0.07% C remain in solution.

TABLE 18

LANTHANOTHERMIC AND CEROTHERMIC REDUCTION OF RARE-EARTH METALS

	La	Ce	Sm	Eu	Tm	Yb	Ref.
Melting point (°C)	920	795	1072	826	1545	824	122
Boiling point (°C)	3470	3470	1900	1440	1725	1425	122
Reduction temperature (°C)			1200	875–900	1400	1200	see below
References	—	—	123, 126, 131–133	123, 127, 128	129	123, 130	

If the calcium is distilled immediately before the reduction in order to reduce its oxygen content, rare-earth metals of a higher purity may be produced. A further improvement is obtained in the reduction of YF_3 by Ca vapour in a vacuum at 1600°C, as the final oxygen levels of the yttrium are substantially lower than those in the more conventional metallothermic process[125].

This method fails to work for the three metals Sm, Eu and Yb mentioned above, as the reaction leads only to highly stable difluorides[121-123]. These three rare-earth metals, together with thulium, have considerably lower boiling points and therefore higher vapour pressures than the rest of the rare-earth metals, particularly Ce and La, as is shown in Table 18. Consequently, a reduction of the oxides of these rare-earth metals is feasible by La or Ce. In this process Sm, Eu, Tm or Yb is volatilized as for instance:

$$2\{La\} + \langle Sm_2O_3 \rangle = \langle La_2O_3 \rangle + 2(Sm) \tag{63}$$

The reaction is carried out also in a tantalum or molybdenum crucible, which is inserted into a ceramic retort, if the charge is small, and into retorts of high temperature steels, if the charge is large. The retorts together with their condensation devices are similar to those used in the reduction of magnesium[121,131,132] (see also Fig. 34).

As a preliminary, briquettes are produced at a compacting pressure of 2.5–5 ton/cm^2; up to a 2.75-fold quantity of lanthanum and up to a three-fold quantity of cerium are used for reduction. In this way yields of up to 99% are obtained. During the reduction a vacuum of about 10^{-3} Torr is maintained.

Table 19 contains a number of results. From Fig. 30 the pressures of formation of Yb, Tm and Sm in the lanthanothermic and cerothermic reductions in relation to the temperature are shown. Attempts to use aluminium for the reduction were unsatisfactory, as heavy evaporation of Al was observed at temperatures as low as 1050°C. Zirconium appears to be usable as a reducing agent[132] (see also Table 19).

In a similar way, Dy, Ho, Er and Y may be prepared. But the distilled metals still contain certain quantities of oxides, which co-distil as volatile suboxides

TABLE 19

REDUCTION OF SAMARIUM AND EUROPIUM

Starting material	Excess of reducing agent (%)	Temp. (°C)	Time (h)	Yield (%)	Reference
Sm_2O_3–La	25	1000	1/6	68	
Sm_2O_3–Ce	0	1200	1	75	
Sm_2O_3–La	35	1030	1	67	
Sm_2O_3–Zr	100	1450	1	81	132
Eu_2O_3–La	50	900	16	93.5	
Eu_2O_3–La	30	1050	4	91	
Eu_2O_3–La	25	1200	1.5	99	
Eu_2O_3–La	~5	1200	8	95	128
Sm_2O_3–Ce	175	1200	1	90	133

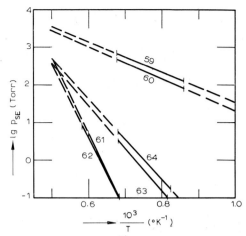

Fig. 30. Effect of temperature on the formation pressures of ytterbium[130], thulium[129] and samarium[126] calculated from the relevant equations.

Equation	No. of equation	Reference
$\langle Yb_2O_3 \rangle + 2\{La\} = 2(Yb) + \langle La_2O_3 \rangle$	59	130
$\langle Yb_2O_3 \rangle + 2\{Ce\} = 2(Yb) + \langle Ce_2O_3 \rangle$	60	130
$\langle Tm_2O_3 \rangle + 2\{La\} = 2(Tm) + \langle La_2O_3 \rangle$	61	129
$\langle Tm_2O_3 \rangle + 2\{Ce\} = 2(Tm) + \langle Ce_2O_3 \rangle$	62	129
$\langle Sm_2O_3 \rangle + 2\{La\} = 2(Sm) + \langle La_2O_3 \rangle$	63	126
$\langle Sm_2O_3 \rangle + 2\{Ce\} = 2(Sm) + \langle Ce_2O_3 \rangle$	64	133

during the reduction. The oxide level may be lowered by a subsequent additional distillation.

Similarly to the oxides, the carbides of Dy, Ho, Er and Tm may be reduced by lanthanum at 1600–1800°C in a high vacuum and distilled. Only very small quantities of carbon are found in the distilled metal.

During the reduction of the above-mentioned rare-earth metals, alloys are temporarily formed with the metal used for reduction. The reaction rate is governed at the beginning by the evaporation rate of the reduced metal from this temporary alloy; but later on it is governed by diffusion of the reactants through the oxides formed during the process[126,127,129]. The grain size of the materials used for briquetting (0.1–1 mm in diam.) does not exercise a noticeable effect on the reaction rate. The C, N, O and H contents were all below 100 ppm[126,127,129].

(b) Metallothermic reduction of alkali metals

Alkali metals may be prepared by electrolysis of their halides, but the yield may not be high because of the high melting points of the halides and the high vapour pressures of the metals (Rb, Cs)[134,135]. In some cases the production of water-free halides meets with difficulties (Li)[136].

Although it is said that oxides, hydroxides or carbonates of the alkali metals

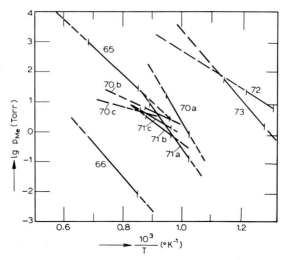

Fig. 31. Effect of temperature on the formation pressure of lithium, sodium and potassium in the metallothermic reduction of their oxides and halides calculated from the relevant equations.

Equation	No. of equation	Reference
$3\ Li_2O \cdot Al_2O_3 + 2\ Al = 6\ Li + 4\ Al_2O_3$	65	136
$3\ Li_2O + Si = 4\ Li + Li_2O \cdot SiO_2$	66	136
$2\ KCl + CaSi_2 = CaCl_2 + 2\ Si + 2\ K$	70	138
a : $\dfrac{KCl}{CaCl_2} = \dfrac{100\ mol\ \%}{0\ mol\ \%}$		
b : $\dfrac{KCl}{CaCl_2} = \dfrac{67\ mol\ \%}{33\ mol\ \%}$		
c : $\dfrac{KCl}{CaCl_2} = \dfrac{50\ mol\ \%}{50\ mol\ \%}$		
$2\ NaCl + CaSi_2 = CaCl_2 + 2\ Si + 2\ Na$	71	139
a : $\dfrac{NaCl}{CaCl_2} = \dfrac{100\ mol\ \%}{0\ mol\ \%}$		
b : $\dfrac{NaCl}{CaCl_2} = \dfrac{67\ mol\ \%}{33\ mol\ \%}$		
c : $\dfrac{NaCl}{CaCl_2} = \dfrac{50\ mol\ \%}{50\ mol\ \%}$		
$2\ KF + CaSi_2 = CaF_2 + 2\ Si + 2\ K$	72	140
$2\ NaF + CaSi_2 = CaF_2 + 2\ Si + 2\ Na$	73	141

can be reduced by iron in a vacuum at about 1200°C[137], aluminium or the alkaline-earth metals are more frequently employed as reducing metals. The results of the reduction experiments of spodumene [$3Li_2O \cdot Al_2O_3$] by Al (eqn. (65)) and Li_2O by Si (eqn. (66)) are plotted in Fig. 31. The experiments were carried out at temperatures between 900° and 1200°C and a vacuum of 10^{-5} Torr for up to 2 hours. The purity of the lithium is higher when produced by reduction by Al than by electrolysis. For the temperature range quoted an activation energy of 33.1 kcal/mol was found for the main period of reduction. This indicates that the reduction is governed by the desorption rate of the lithium[136]. However, it is thought that with increasing temperatures and increasing reaction times, diffusion of the reaction partners determines the reaction rate. For this reduction the reactors employed for the metallothermic Mg production may be used[136]. Experiments for the aluminothermic reduction of Li_2O, which were carried out by Szarowicz and Orman[142], gave the best results at a temperature of 1100°C and pressure of 10^{-3} Torr with a yield of 84%. Decroly[143] reports a similar success in the reduction of Li from amblygonite under similar conditions.

Experiments for the reduction of sodium aluminate by aluminium in a two-stage process at a temperature of 800–1000°C in a vacuum of 0.2–0.3 Torr over a period of 2 hours led to an extensive reduction (60–70% of the original sodium content was reduced) according to:

$$12\langle Na_2O \cdot Al_2O_3\rangle + \tfrac{22}{3}\{Al\} = 22(Na) + \tfrac{11}{3}\langle Al_2O_3\rangle + \langle Na_2O \cdot 12Al_2O_3\rangle. \quad (67)$$

At 1100–1200°C the final reduction takes place (up to 99.7% of the sodium was reduced) according to:

$$\langle Na_2O \cdot 12Al_2O_3\rangle + \tfrac{2}{3}\{Al\} = 2(Na) + \tfrac{37}{3}\langle Al_2O_3\rangle. \quad (68)$$

The Al_2O_3 formed has purity of about 99.5%. The optimum working conditions are: temperature 1200°C, pressure 0.2–0.3 Torr, reduction time 2 h, Al excess 50%, a relatively low compacting pressure of 210 kgf/cm² and a grain size of the Al of less than 0.2 mm. If a higher compacting pressure is employed the porosity of the compacts decreases and the evaporation of the Na is made difficult. On the other hand, for lower compacting pressures the mechanical strength of the compacts was too low[144].

With the aid of CaC_2 the sodium of zeolites may be reduced and volatilized. In this process zeolite and CaC_2 are mixed and briquetted. An intensive reduction takes place at about 1200°C in steel retorts. This reduction leads within 15–30 min to a yield of more than 80%. If the briquettes after the reaction are melted at 1400–1500°C, an alloy of 32% Si, 12% Al and 42% Fe is obtained[145]. The reduction of spodumene[146] or alkali chlorides[147] is said to be improved by use of calcium carbide with an addition of lime. The reaction of the chlorides takes place at about 900–1000°C according to:

$$2\{NaCl\} + \langle CaC_2\rangle = 2(Na) + \{CaCl_2\} + 2\langle C\rangle. \quad (69)$$

The mixture of $CaCl_2$ and C formed is drawn off as a liquid.

The optimum composition of the charge for potassium is 50% KCl, 34% CaC_2 and 16% CaO. The main reaction starts in the pressure range between 20

References pp. 321–335

and 1 Torr. In this process the evaporation of the potassium is usually slower than the reduction process. The optimum yields after a reaction time of one hour are about 24%, but on the average only 8.5% is attained. After 6 h the yield reaches 51%. In this process too it is evident that the vacuum conditions and, in the second place, the heat input exercise a dominant effect on the course of the reduction[148].

Results of experiments on the reduction of KCl[138] and NaCl by $CaSi_2$[139] are plotted in Fig. 31. According to these studies the reaction is substantially affected by the formation of a mixed phase $MeCl-CaCl_2$ and in order to obtain the same efficiency higher temperatures need to be employed. This effect is more pronounced for KCl than for NaCl. In the subsequent condensation of the potassium, falling residual gas pressures (from 20 to 0.1 Torr) displace the limit of condensation from 320–200°C to 160–130°C. This displacement was greatest at 5–0.1 Torr[149]. In Fig. 31 the reduction of NaF[141] and KF[140] by $CaSi_2$ is also shown. The substantially more favourable conditions for reduction are brought about by the greater stability of the CaF_2 formed. Optimum working conditions are for Na, 900°C and for K, 800°C in a vacuum. The overall reaction rate is governed by the evaporation rate of the Na and K, respectively[140,141].

The difficulties involved in the electrochemical winning of caesium led to the development of reduction processes, which are conducted either in a protective atmosphere or in a vacuum[135,150]. One of these processes is the direct reduction of pollucite

$$2Cs_2O \cdot 2Al_2O_3 \cdot 9SiO_2 \cdot H_2O$$

by a twelve-fold excess of Ca or a four-fold excess of Mg. The results are compiled in Table 20.

The evaporation yield could be even further increased by vacuum reduction, although the condensation in the pilot plant was still unsatisfactory. Further distillation of the raw metal is necessary in any case, as pollucite contains impurities, such as Rb, K, Na and Li, which are also reduced and are co-distilled in part. In the reduction of high-purity compounds by lithium a high-purity caesium should immediately be expected[134]. The yield also is high (more than 90%). In the re-

TABLE 20

EXPERIMENTS ON THE REDUCTION OF POLLUCITE[151]

Reducing agent	Temp. (°C)	Time (h)	Atmosphere	Yield (%)		Composition of the condensed metal (%)					
				Evaporation	Condensation	Cs	Na	K	Rb	Ca	Mg
Ca	>700	3	Argon	92	72	90.1	3.0	3.0	3.0	0.4	–
Mg	>700	3	Argon	96	88	87	3	3	3	–	3
Ca			Vacuum	96	55	91	2	3	3	0.4	–
Mg			Vacuum	99.5	79	86	a	a	a	a	8

a Not known.

duction of CsCl or CsBr by lithium at 350–800°C a caesium metal of a purity of more than 99% should be distilled if the pressure is below 10 Torr[152].

The results of studies of the systems CsCl–CaSi$_2$[153] and RbCl–CaSi$_2$[154] are plotted in Fig. 32. In this case the formation of a mixed phase MeCl–CaCl$_2$ is even more pronounced than with NaCl or KCl. An increase in temperature for raising the yield does not bring about the desired result.

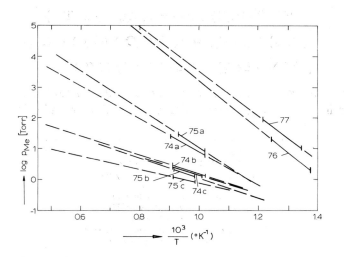

Fig. 32. Effect of temperature on the formation pressure of caesium and rubidium in the metallothermic reduction of their halides calculated from the relevant equations.

Equation	No. of equation	Reference
2 CsCl + CaSi$_2$ = CaCl$_2$ + 2 Si + 2 Cs	74	153
a : $\dfrac{\text{CsCl}}{\text{CaCl}_2} = \dfrac{100 \text{ mol \%}}{0 \text{ mol \%}}$		
b : $\dfrac{\text{CsCl}}{\text{CaCl}_2} = \dfrac{67 \text{ mol \%}}{33 \text{ mol \%}}$		
c : $\dfrac{\text{CsCl}}{\text{CaCl}_2} = \dfrac{50 \text{ mol \%}}{50 \text{ mol \%}}$		
2 RbCl + CaSi$_2$ = CaCl$_2$ + 2 Si + 2 Rb	75	154
a : $\dfrac{\text{RbCl}}{\text{CaCl}_2} = \dfrac{100 \text{ mol \%}}{0 \text{ mol \%}}$		
b : $\dfrac{\text{RbCl}}{\text{CaCl}_2} = \dfrac{67 \text{ mol \%}}{33 \text{ mol \%}}$		
c : $\dfrac{\text{RbCl}}{\text{CaCl}_2} = \dfrac{50 \text{ mol \%}}{50 \text{ mol \%}}$		
2 CsF + CaSi$_2$ = CaF$_2$ + 2 Si + 2 Cs	76	155
2 RbF + CaSi$_2$ = CaF$_2$ + 2 Si + 2 Rb	77	156

References pp. 321–335

(c) Metallothermic reduction of alkaline-earth metals

A number of investigations on the equilibria in the metallothermic reduction of magnesium are available, the results of which are plotted in Fig. 33.

Economically interesting vapour pressures of magnesium between 1 and 100 Torr in the temperature range 1000–1400°C should theoretically be obtained for the reactions:

$$2\langle CaO \cdot MgO\rangle + \langle Si\rangle = 2(Mg) + \langle Ca_2SiO_4\rangle \tag{78}$$

$$8\langle MgO\rangle + \langle Al_4C_3\rangle = 2\langle MgAl_2O_4\rangle + 3\langle C\rangle + 6(Mg) \tag{79}$$

$$\langle MgO\rangle + \langle CaC_2\rangle = (Mg) + \langle CaO\rangle + 2\langle C\rangle \tag{80}$$

In practice, however, the vapour pressures of magnesium are highest in the reaction according to eqn. (79), but theoretically 25% of the MgO added to the starting

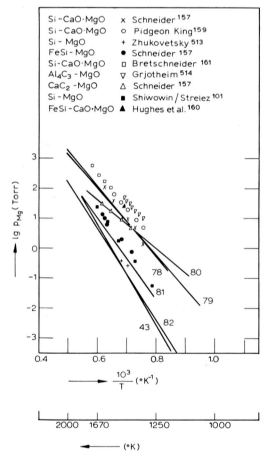

Fig. 33. Vapour pressure of magnesium above certain reaction mixtures measured by various authors or calculated from thermodynamic data[8] based on eqns. (43) and (79)-(82).

charge remains unused. While the Mg vapour pressures at about 1000°C in reactions (78)–(80) are rather similar, the reduction of dolomite at 1400°C in particular offers certain advantages.

The measured data of the reaction (80) agree with the calculated values, but the calculated vapour pressures are somewhat lower for the reactions (78) and (79). This should mainly be attributed to inaccurate thermodynamic data for carbides and complex oxides, as has already been pointed out by Kubaschewski[8].

Less favourable conditions of reduction exist in the silicothermic reduction of MgO:

$$4\langle MgO \rangle + \langle Si \rangle = \langle Mg_2SiO_4 \rangle + 2(Mg) \tag{81}$$

as measured data and calculated vapour pressures of Mg in Fig. 33 show. For the sake of comparison the corresponding Mg vapour pressures as existing in the carbothermic reduction of MgO (eqn. 43) are also plotted in this diagram. In the reduction by carbides the carbon does not appear to be the reducing agent, as was established during investigations of the reaction according to eqn. (80)[157]. This is also evident from a comparison of the vapour pressures of Mg obtained from thermodynamic calculations based on eqn. (82):

$$3\langle MgO \rangle + \langle CaC_2 \rangle = 3(Mg) + \langle CaO \rangle + 2(CO) \tag{82}$$

The values calculated with the aid of eqn. (82) are substantially lower than the experimental data.

Very exhaustive kinetic investigations, particularly on the reduction of dolomite by ferro-silicon, were carried out by Schneider et al.[157,158], Pidgeon and King[159] and Hughes et al.[160]. According to these investigations, the formation of calcium silicide occurs during heating of the reaction mixture. This silicide formation is explained by Schneider[158] as follows:

$$4\langle CaO \rangle + 5\langle Si \rangle = 2\{CaSi_2\} + \langle Ca_2SiO_4 \rangle. \tag{83}$$

The reduction of the MgO subsequently takes place by a reaction with $CaSi_2$ formed in the meantime:

$$10\langle MgO \rangle + 2\{CaSi_2\} + 6\langle CaO \rangle = 10(Mg) + 4\langle Ca_2SiO_4 \rangle \tag{84}$$

which yields in total eqn. (78).

According to Schneider et al. this intermediate formation of silicide is found above 600°C[157], whereas Hughes et al.[160] observed metallic particles ($CaSi_2$) only above 950°C and extensively above 980°C. Since the briquettes react first in the interior, Hughes et al. assume that this primary reaction (eqns. (83) and (85), respectively) is slightly exothermic; this is confirmed by the thermodynamic data. The formation of $CaSi_2$ should indeed take place according to:

$$5\langle CaO \rangle + 5\langle Si \rangle = 2\{CaSi_2\} + \langle Ca_3SiO_5 \rangle \tag{85}$$

as during the formation of calcium silicide, instead of 40% of the CaO (according to eqns. (83) and (84)), 60% of this quantity is always used for calcium silicate formation.

It is assumed that the reduction of MgO involves the formation of another intermediate product $(Ca_3Mg(SiO_4)_2)$ which has been found by X-ray analysis in partly reacted compacts. The principal stage of the reduction therefore takes place according to:

$$6\langle MgO\rangle + 2\langle CaO\rangle + \{CaSi_2\} = \langle Ca_3Mg(SiO_4)_2\rangle + 5(Mg) \qquad (86)$$

and towards the end of the reduction the formation of Ca_2SiO_4 occurs according to:

$$5\langle Ca_3Mg(SiO_4)_2\rangle + \{CaSi_2\} + 8\langle CaO\rangle = 12\langle Ca_2SiO_4\rangle + 5(Mg) \qquad (87)$$

Contrary to the statements of the authors, these equations with the known thermodynamic data do not lead to higher calculated reaction pressures of Mg, so that a small difference between measured and calculated vapour pressures of Mg remains:

$$p_{Mg} \text{ calc.}/p_{Mg} \text{ meas.} = 0.5\text{--}1.$$

The Mg reduction is greatly accelerated by the temporary formation of liquid phases; for instance by use of ferro-silicon as reducing agent, where a low-melting-point ternary eutectic (55:25:20 Si–Fe–Ca) is formed at 890°C. If the temperature rises, $CaSi_2$ too becomes liquid[158,161].

Additions of CaF_2 and $AlNa_3F_6$, respectively, lead to even more accelerated reactions[101,161]. It is probable that at higher temperatures MgO and CaO can also form with CaF_2 a liquid phase, which on contact with liquid $CaSi_2$ results in a further increase in reaction rate. Pidgeon and King[159] showed that the equilibria are not displaced by an addition of CaF_2. This CaF_2 effect is most pronounced when neither CaO nor Ca is present in the mixture for reaction.

According to Schneider et al.[158] the rate of reduction of dolomite by ferro-silicon is determined by the evaporation of calcium from the silicide. Pidgeon and King were able to show that neither silicon nor silicon monoxide vapour affects the reaction rate[159]. Strelez et al. reported the results of similar experiments; according to their findings Si, Al, SiAl, FeSi and Ca vapour each has no noticeable effect on the rate of the reduction. In experiments in a current of inert gases a yield of 4.1% was attained by employment of Ca; yields between 1.2 and 3.2%[101] were obtained for the other reducing agents. These authors prefer to regard diffusion phenomena as rate-determining.

The vapour pressure of Ca over $CaSi_2$ is so low that the high degrees of material exchange observed are not explicable[158]. Calcium vapour appears not to play a decisive role in the main phase of the reaction.

On the other hand, according to Hughes et al.[160] the vapour pressure of the Mg very quickly attains equilibrium conditions at the reduction front. Further reduction is then determined by the rate at which the magnesium is escaping, so that the reduction front is moved into the interior of the compacts. The porosity of the compacts is therefore decisive for the reaction rate; this means that the reaction rate in this phase of the reaction decreases with increasing compacting pressure. In the partly reacted zones the reduction proceeds at a rate which is proportional to the concentration of the remaining MgO and which therefore formally corresponds to a reaction of the first order. The reaction rate in this phase is independent of the compacting pressure. It is, however, affected by the degree of milling and the degree of mixing of the reaction components.

In practice a time law of the first order applies in the retort process in which, according to Hughes et al.[160], the heat transfer determines the rate of magnesium winning.

According to Schneider[157] diffusion processes are assumed to determine the reaction rate of the reduction of MgO by CaC_2. While changes in the grain size of the components have hardly any effect in the reduction of dolomite by ferro-silicon, they are noticeable in the reduction by carbide.

Of the potential metallic reducing agents only ferro-silicon and calcium carbide are of importance for economic reasons. Owing to the great demand for magnesium during World War II, intensive technological development of metallothermic reduction processes took place during that time, although its share of the total production of magnesium never exceeded 25%. Production plants for the metallothermic reduction of magnesium exist in the United States, Canada, England, Japan, France, Italy, China and Germany. Some data of these plants are shown in Table 21. With the exception of the plants in the U.S.A., they are used to full capacity[162].

The different plants work mainly by the following processes: Retort process,

TABLE 21

COUNTRIES IN PRODUCTION WITH THERMAL MAGNESIUM PROCESSES[162–164]

Country	Process	Starting material	Production capacity (ton/year)
USA	Retort	Dolomite–FeSi	6250
USA	Retort	Dolomite–FeSi	5000
Canada	Retort	Dolomite–FeSi	10000
Italy	Retort	Dolomite–FeSi	6600
Great Britain	Retort	Dolomite–FeSi	5000
France	Magnetherm	Dolomite–FeSi	3500
Japan	Retort	Dolomite–FeSi	6000
Japan	Retort	MgO from seawater	5000
Federal Republic of Germany	Thermic	Various	500
China	Thermic	a	1000
Hungary	Rotary furnace	Dolomite–FeSi	a

a Not known.

References pp. 321–335

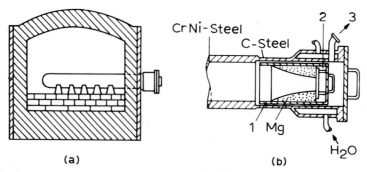

Fig. 34(a). Arrangement of retorts in the furnace. (b) Diagram of the condenser. 1 Condenser insert; 2 Alkali collector; 3 Vacuum pump.

Rotary furnace process, Magnetherm process, Knapsack process, which operate with calcined dolomite as starting material and ferro-silicon as the reducing agent. There is also the VAW process*, in which calcined magnesite is reduced by calcium vapour obtained from the decomposition of calcium carbide[100].

The greatest economic importance attaches to the retort process, in which briquettes of calcined dolomite, ferro-silicon containing 75% Si, and an addition of fluorspar are reduced in chromium–nickel steel retorts. Figure 34a shows in principle the arrangement of the retorts in the furnace, which are usually gas-fired. Each furnace contains about 40 retorts (Fig. 34c); the daily output of a retort varies between 40 and 60 kg. One working cycle takes about 8 hours. The restricted high-temperature strength of the retorts limits the reduction temperature to a maximum of 1170°C. The vacuum is maintained around 0.1 Torr. The yield attains only

Fig. 34(c). Two furnaces with the retorts loaded for the thermic production of magnesium, during actual operation[164].

---

* VAW = Vereinigte Aluminium Werke AG.

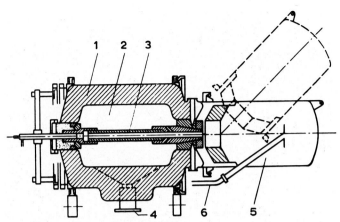

Fig. 35. Original vacuum rotary kiln furnace of IG-Farbenindustrie Bitterfeld. 1 Masonry structure; 2 Reaction chamber; 3 Carbon electrode; 4 Opening for charging and tapping; 5 Condensation chamber; 6 Vacuum duct.

65–70% relative to the silicon. The part protruding from the furnace consists of the water-cooled condenser, the principle of which is shown in Fig. 34b. Since the heat transfer is probably rate-determining in this process, an increase in the size of the retorts (length 3 m, diameter 250 mm, wall thickness 30 mm) results in an increase of the reduction time[99,100,161]. By using electric internal heating, a production of 220 kg magnesium is said to have been achieved per retort and per charge in the U.S.A., and 450 kg magnesium in Italy[163].

In order to overcome these difficulties (low reaction temperature, poor heat transfer, low throughput) IG-Farbenindustrie, Bitterfeld, developed a vacuum rotating drum furnace during the time of World War II. A schematic presentation of the process is shown in Fig. 35. Heating is effected by heating elements of carbon, which are arranged in axial fashion in the furnace. The charge again consists of briquettes (calcined dolomite, ferro-silicon and fluorspar), which have been hardened under hydrogen at about 1000°C so that they do not disintegrate during rotation of the furnace. Reduction takes place at 1300–1400°C, the working cycle requires about $3\frac{1}{2}$–4 hours. The daily output varies around 1000 kg magnesium with a yield of 80%, again related to the starting quantity of silicon. The evaporating magnesium is collected in a detachable, cooled condensation chamber, which consists, in contrast with the ceramic-lined reduction chamber, of high-temperature cast steel. The residuals consist mainly of calcium orthosilicate. They decompose during cooling and may be removed without any difficulty from the reduction vessel[100,161,165]. More recent investigations at VAW led to a modification of the process to complete perfection in industrial working. In this modification the hardening process before the reduction is omitted. The mechanical strength of the briquettes is greatly improved if the calcination of the dolomite is carried out in the presence of small quantities of chlorine[100]. Figure 36 shows

References pp. 321–335

Fig. 36. Vacuum rotary drum furnace for the thermic winning of magnesium[166].

the pilot plant of VAW for the thermal production by the rotary drum furnace process. In these furnaces almost 5000 tons of magnesium were produced from dolomite and ferro-silicon between 1955 and 1965 despite interruptions[166].

According to experiments carried out in a rotary drum furnace in Hungary, the yield in magnesium could be increased to 83% even with very finely ground ferro-silicon (below 0.06 mm). The yield increases with increasing compacting pressure in the production of the compacts, but decreases at very high compacting pressures (see also above) owing to a decrease in porosity[167,168]. In this category also belong attempts for the realization of a semi-continuous process with liquid condensation of magnesium[167,169,170]. Despite a low power input not all the technological difficulties have yet been overcome[169].

Although the retort and, to an even greater extent, the rotary drum furnace were modified and improved, both processes have the inherent disadvantage that working is done in batches. Knapsack–Griesheim AG therefore developed a continuous vacuum shaft furnace process, the principle of which is depicted in Fig. 37. A granulated mixture of dolomite and ferro-silicon is fed continuously into the furnace on top of a column of reaction material that has completed the reaction. The mixture is heated by electric radiation heating to about 1600°C. In order to prevent a sintering of the charge as a result of these high temperatures, a 100% excess of dolomite is used. The operating pressure may vary between 1 and 50 Torr. The residuals of the reduction are slowly lowered by a water-cooled grate. As they consist of calcium orthosilicate, they disintegrate with falling temperature and may be transferred to a bunker for residuals. The magnesium vapour produced is passed through a filter and is subsequently condensed to a liquid. The liquid magnesium is continuously tapped via a barometric seal. Above the liquid Mg condenser, two condensers for solid Mg are provided, through

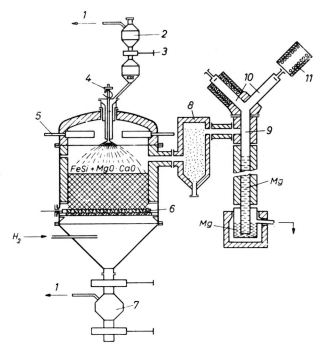

Fig. 37. Furnace plant for the continuous silicothermic magnesium winning by the Knapsack–Griesheim process.
1 To vacuum pump; 2 Hopper; 3 Lock system; 4 Distributor; 5 Electric heating elements; 6 Grate; 7 Residue collector [Ca_2SiO_4+CaO · MgO+FeSi contg. 30% Si]; 8 Filter; 9 Condenser for liquid Mg; 10 Condenser for solid Mg; 11 Electric resistance furnace.

which the whole system is evacuated by turns. In certain intervals of time, the magnesium dust condensed in these condensers is melted out and the liquid magnesium added to the bulk of it[100,161,171].

A comparable process, the so-called Magnetherm process, was developed in France in which the residuals of the reaction are withdrawn in the liquid state. In order to lower their melting point, alumina is added at a ratio of 1 ton alumina per ton of magnesium to the reaction mixture consisting of calcined dolomite and ferro-silicon containing 75% silicon. The mixture of dolomite and alumina is fed in portions every three minutes and the ferro-silicon continuously on to the liquid slag in the furnace (see Fig. 38). The furnace is heated electrically; the graphite lining of the furnace is used as one electrode, and a graphite block forms the other electrode. The reaction mixture slowly dissolves in the slag. The magnesium vapour produced is condensed in liquid form at about 650°C in a condenser. When the slag has completed its reaction, the reduction process is interrupted, the furnace is opened to the atmosphere, and part of the slag and the ferro-silicon are tapped and separated. The slag may be granulated; the condensed magnesium is teemed. Part of the slag remains in the furnace in order to maintain the

References pp. 321–335

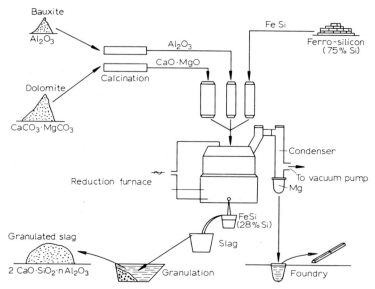

Fig. 38. Diagram depicting the magnetherm process for the reduction of MgO · CaO by FeSi in the liquid state.

electrical connection between the electrodes in the subsequent cycle. The working temperature is at about 1500–1550°C, and one working cycle takes about 12 hours. The pressure is maintained at about 25 Torr. The yield in relation to silicon is about 70% at a production rate of 1400 kg per working cycle[100,161,172,173].

According to Japanese sources this process may be operated with a slag without Al_2O_3 addition (44–64% SiO_2, 30–53% CaO, 3–14% MgO) at 1600°C, but if Al_2O_3 is added (final slag 62% SiO_2, 15% Al_2O_3, 23% CaO) a temperature of 1300–1400°C only should be required[174]. In the Magnetherm process the slag contains about 33% SiO_2, 59% CaO and 8% MgO[173]. In Table 22 once again the most important data are compiled.

The economic production of the reduction metals is of great importance in the metallothermic magnesium winning processes[176]. Owing to the high price of the reduction agent, ferro-silicon (which amounts to about 30% of the total production costs), only smaller production plants (less than 10000 tons Mg/year) are more profitable than electrolysis plants. For this reason it was proposed to replace ferro-silicon by other reducing agents. Besides the very cheap carbon (cf. Subsection 3.1.2) and the expensive ferro-silicon, calcium carbide and AlSi should in particular be considered, as may be seen from the following comparison[101]:

	Power consumption (kWh) for the production of the reducing agent used for the production of 1 ton of Mg
FeSi 75%	7000
CaC_2 85%	9000
Al 99%	13500
AlSi 45–50% Al	6700

TABLE 22

DATA ON THERMIC REDUCTION PROCESSES FOR MAGNESIUM[100,161,165,172,175]

Process	Temp. (°C)	Time for one cycle (h)	Daily output (kg/d)	Yield related to Si (%)	Pressure (Torr)	Specific power consumption (kWh/kg Mg)
Retort	1170	8	40–60*	65–70	1	
Rotary furnace	1300–1400	4	1000	80	1–2	8–10
Magnetherm	1500–1550	Semi-continuous	2800	70	25	10.8
Knapsack	1600	Continuous	2000**		1–50	

* Possibly more (cf. ref. 163).
** Estimated.

The reduction of MgO by CaC_2, previously carried out at 1200°C in the retort, corresponded otherwise to the retort process for silicothermic reduction, but a considerably greater wear of the retorts took place[175]. According to a recent suggestion MgO may be reduced by calcium vapour obtained by decomposition of CaC_2. The method appears to be substantially more expensive, although the yield is said to exceed the 70–80% previously mentioned. Calcium vapour reacts with MgO very rapidly, completely and with liberation of heat, so that this process takes place without a large input of heat in contrast with the silicothermic reduction of dolomite. A schematic presentation of the technological solution is shown in Fig. 39. Calcined magnesite is introduced into the reactor via a lock system without cooling. The rate of lowering is controlled by the tapping mechanism. The reaction takes place in the lower third of the reactor according to:

$$\langle MgO \rangle + (Ca) = (Mg) + \langle CaO \rangle. \tag{88}$$

The Mg vapour is precipitated in the adjacent removable condenser as a solid. The calcium vapour is produced from commercial CaC_2 in a vacuum arc furnace. The residual graphite in the arc furnace as well as the CaO in the reactor are removed via a vacuum lock system. The process is still in the development stage[100].

From the compilation of potential reducing agents the employment of AlSi appears to be very promising. In the reduction in the retort the yield could be substantially increased at the same working temperatures of 1170°C. Even at lower temperatures and acceptable working times good yields were still obtained (see Table 23). These data confirm results obtained by Tajc[177] who also reported that higher yields were found with AlSi at 1200°C, a reduction time of 2h and lower power consumption than in the reduction by FeSi. Aluminium-silicon is also suggested as a reducing agent in a large number of patent applications for the thermal winning of Mg.

In the metallothermic reduction of the other alkaline earth metals, Al granules have so far been almost exclusively used as reducing agent, but according to more recent investigations AlSi appears to gain in importance. The reduction temper-

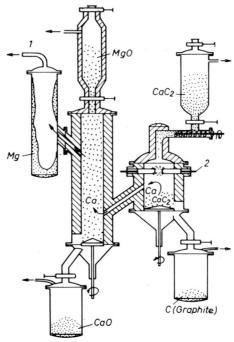

Fig. 39. Furnace plant for the thermic winning of magnesium by the calcium vapour process of Vereinigte Aluminiumwerke AG (VAW). 1 To vacuum pump; 2 Electrodes.

atures of Ca, Sr and Ba are so low that the formation of gaseous suboxides of aluminium can almost be prevented.

Only beryllium, besides its difficult reducibility, possesses such a low vapour pressure that temperatures around 1500°C and pressures of about 10^{-4} Torr must be maintained for obtaining technologically satisfactory distillation rates. As reducing agents only Zr, Y and La need consideration, as they have sufficiently low vapour pressures as well as high oxygen potential. The results of reduction experiments carried out in crucibles of tantalum are shown in Table 24. Yields and purities could, according to the authors, still be materially improved[178].

TABLE 23

MAGNESIUM YIELD IN % FOR VARIOUS REDUCING AGENTS[176]

	Mole 75% FeSi/Mole CaO · MgO			Mole AlSi(50%)/Mole CaO · MgO		
	1.25	1.0	0.8	1.25	1.0	1.0
After 2h reduction	a	a	a	a	76	a
After 4h reduction	82	69	61	92	87	a
After 8h reduction	88	76	64	95	a	85
Temperature (°C)		1170			1170	1100

a Not known.

TABLE 24
METALLOTHERMIC REDUCTION OF BeO[178]

Starting material	Excess of reducing agent (%)	Temp. (°C)	Time (h)	Yield (%)	Impurities (%)	
					Reducing agent	O
BeO–Zr	50	1650	40	85	0.08	0.31
BeO–Y	25	1600	30	97	2.8	a
BeO–Y	50	1450	60	89	0.07	a
BeO–Y	50	1500	30	86	3.0	0.17
BeO–La	0	1425		70		
		1425	64.8	20	<1.0	0.15

a Not known.

Attempts at the reduction of BeO by aluminium at 1350–1450°C and 10^{-2}–10^{-3} Torr were shown to take place incompletely with formation of an Al–Be alloy containing only 2% Be[179].

The methods of reduction for Ca, Sr and Ba differ very little. The reaction with Al granules takes place according to the composition of the mixture as:

$$6\langle CaO\rangle + 2\{Al\} = \langle 3CaO \cdot Al_2O_3\rangle + 3(Ca) \tag{89}$$

$$5\langle CaO\rangle + 2\{Al\} = \langle 2CaO \cdot Al_2O_3\rangle + 3(Ca) \tag{90}$$

$$4\langle CaO\rangle + 2\{Al\} = \langle CaO \cdot Al_2O_3\rangle + 3(Ca). \tag{91}$$

Fair agreement was found between these ratios of the mixture and experimental data. Only the reaction:

$$3\langle CaO\rangle + 2\{Al\} = \langle Al_2O_3\rangle + 3(Ca) \tag{92}$$

proceeds incompletely with formation of $CaO \cdot Al_2O_3$ or $CaO \cdot 2Al_2O_3$[180,181] as may be seen from the data shown in Table 25. The reduction takes place with considerable evolution of heat, and in it, temporary formation of an alloy, for instance Al_3Ca or Al_4Ca, cannot be excluded. With increasing temperatures increased formation of aluminate was found for constant CaO:Al ratios, similar to increasing Al quantities at constant temperatures.

TABLE 25
REDUCTION OF CaO BY Al

Ratio of mixing (Mole CaO/ Mole Al)	Temp. (°C)	Time (h)	Pressure (Torr)	Yield in %			Reference
				Theor. Ca	Actual Ca	Al	
5:2	1190	3	10^{-3}	60	61	97	180
4:2	1195	3	10^{-3}	75	72	93	180
3:2	1180	10	10^{-3}	100	33	a	181

a Not known.

Fig. 40. External view of vacuum furnaces for the aluminothermic reduction of calcium.

The reduction is carried out in retorts which correspond to those used in the winning of Mg. A number of vacuum furnaces for the aluminothermic reduction of calcium is shown in Fig. 40. Briquettes of CaO and Al granules are reduced at about 1170–1200°C in a vacuum. The distillates are caught in three condensation stages (for a first refining) at 710°C, 440°C and 300°C[182,183] (see Figs. 41 and 42).

Aluminium–silicon may also be used as reducing agent. The reactions (89), (90) and (91) take place primarily starting at 950°C and 0.01–0.02 Torr, but intensively only at 1200°C. Above 1300°C, Si too begins to act as a reactant, although it is considerably more sluggish:

$$4\langle CaO\rangle + \langle Si\rangle = 2(Ca) + \langle Ca_2SiO_4\rangle \qquad (93)$$

but only if all the aluminium has been consumed[184].

According to Seliger[185], Ca_2Si is formed temporarily during the reduction

Fig. 41. Upper part of a retort of the vacuum furnaces shown in Fig. 40, without condenser.

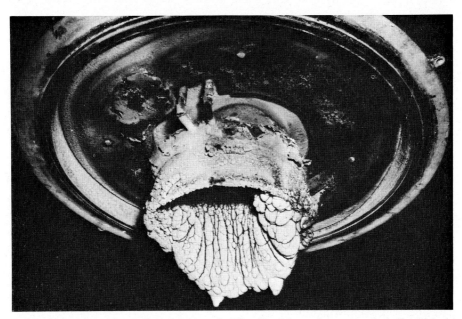

Fig. 42. Lid of the vacuum furnace shown in Fig. 40, with condenser attached, in which the calcium metal is precipitated.

by silicon:

$$4\langle CaO\rangle + 2\langle Si\rangle = \{Ca_2Si\} + \langle Ca_2SiO_4\rangle \tag{94}$$

$$4\langle CaO\rangle + \{Ca_2Si\} = \langle Ca_2SiO_4\rangle + 4(Ca) \tag{95}$$

References pp. 321–335

Calcium carbide can therefore only be reduced by Si if a certain quantity of CaO is added, so that Ca_2Si may be formed according to eqn. (94). According to Kaess et al.[186] CaC_2 is therefore reduced by Ca_2Si directly in a retort at 1100–1200°C and 10^{-3} Torr. In all cases mentioned here an addition of CaF_2 accelerates the reaction.

Reaction temperatures of about 1200°C and vacua of between 10^{-3} and 10^{-2} Torr always cause the formation of gaseous suboxides of the reducing agents, although to a small extent only. However, the calcium produced is contaminated with them.

The reduction of strontium and barium[165] is feasible in the same way and takes place under almost similar conditions (1100–1200°C, pressure lower than 10^{-2} Torr) with the formation of monoaluminates in the reduction by aluminium[187,188], and the additional formation of orthosilicates in the reduction by AlSi[189].

The apparatus used is similar to the retorts of the Mg reduction. In the case of barium too the temporary formation of an alloy $BaAl_4$ occurs. Although the reduction should take place at about 700°C, temperatures of over 1100°C are required in order to distil off the barium[68].

4. Refining of Metals in a Vacuum

The refining of metals at high temperatures and at atmospheric pressure involves in most cases an electrochemical exchange reaction, such as a reaction with slag, or a partition of one or more components between two phases, for instance, the desilvering of lead bullion by zinc. The equilibrium condition of the reaction indicates which refining effect is obtainable. The application of counterflow processes increases the reaction rate, as for instance in the desulphurization of pig iron in the electromagnetic channel or in the continuous desilvering of lead bullion by zinc.

Gaseous reaction products are formed in several important refining processes, for instance, in the decarburization of steel and the blowing of copper matte. The refining effects obtainable are again determined by the equilibrium conditions, for instance, by the pressure of the reaction gases above the melt. In this connection distillation processes should also be mentioned, which, if not carried out in vacuum, are usually carried out with exclusion of air, for instance the separation of cadmium and zinc or in the distillation of $GeCl_4$ or $SiCl_4$.

Electrochemical exchange reactions with slags do not occur in refining processes which are carried out in a vacuum. On the other hand, partition of one or more components between two phases is of great importance for the special case of zone melting, in which the vacuum in most applications plays the role of a protective gas. Neither before nor during or after the passage of the molten zone should the metal have any opportunity to be contaminated by the residual gas atmosphere. On the contrary, a number of volatile components may evaporate in the vacuum.

If gaseous reaction products are formed during a refining process, the reaction rate increases by application of a vacuum and the equilibrium is displaced to lower levels of the impurities. Degassing, deoxidation and decarburization are examples of this phenomenon.

Vacuum distillation plays an important role in the refining of metals. By application of the vacuum the distillation temperatures may be lowered, the evaporation rate increased, and the chances for a separation of the components improved. Furthermore, unwanted side reactions can be almost avoided.

A special case of distillation is the van Arkel–de Boer process which is a combination of distillation and chemical reaction for the processing of halides of metals.

4.1 Separation of metals by distillation in a vacuum
4.1.1 Discussion of the rate-determining steps

In the distillation of metals in a vacuum four processes may be distinguished:
(i) heat and materials transport in the metal to be distilled,
(ii) evaporation,
(iii) materials transport through the gas space,
(iv) condensation.

The pressure, concentration and temperature distribution during distillation are shown in the form of a diagram in Fig. 43. It is depicted in a somewhat simplified manner and with the assumption that diffusion of the metal vapour particles takes place through a residual gas atmosphere. The more the mean free path of the metal vapour particles is extended—effected by lowering the residual gas pressure —or the distance between evaporator and condenser is shortened, the more the

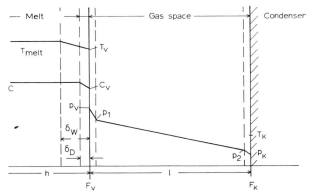

Fig. 43. Diagram indicating vapour pressures, concentration and temperatures between evaporator and condenser surfaces, for an explanation of the various equations.

References pp. 321–335

diffusion process changes into a directional molecular flow process so that isotropic conditions no longer exist. Because the vapour pressure is not definable under such conditions it is necessary to use for calculations the concentration in the gas space rather than the vapour pressure.

Independent of these considerations the rate of distillation is governed by the slowest process. As will be discussed later, each of the individual steps may be rate-determining. But a stationary state may develop during distillation after a very short time, in which the rate-determining process can no longer be co-ordinated to an individual step, as the necessary data required for the calculation cannot be measured.

(a) Diffusion in the melt

In the transport of a substance dissolved in a melt to the surface, diffusion through a fictive boundary layer δ_D plays an important role in many cases. The quantity of the substance which is transported may be expressed for metals as a first approximation by Fick's First Law:

$$i = -D\Delta C/\delta_D = -D \cdot \frac{C - C_V}{\delta_D} \tag{96a}$$

where D is the diffusion coefficient depending on composition, concentration and the temperature of the metal, and $C - C_V$ the concentration gradient of the metal to be distilled between the mass of the melt (C) and its surface (C_V).

According to Kraus[190,191], Machlin[192] and Bradshaw[193] this diffusion boundary layer δ_D is not stationary, but flows parallel to the surface, at a speed u. Including Fick's Second Law it follows for a melt with induction stirring[192]:

$$i = -2(C - C_V)(2Du/\pi r h^2)^{\frac{1}{2}} \tag{96b}$$

where h is the depth of the pool of molten metal and r is the radius of the crucible containing the melt (see also Fig. 43 and Chapter I, Section 3).

With increasing pressure of an inert gas, or if the crucible dimensions are considerably larger than the mean free path, diffusion or convection of the metal molecules evaporated through the gas space may become rate-determining.

(b) Convection in the melt or in the gas space

Convection in the gas space as well as in the melt will, in most cases, be sufficiently rapid. Besides the diffusion of the metal molecules in the gas space, convection flow processes may develop if the residual gas pressure is above about 1 Torr. In such flow processes the residual gas takes on the role of a carrier gas.

In some cases the vapour pressure of the metal may attain such high values, however, that a viscous flow is created and maintained solely by the metal vapour. In such a process it may be assumed that most of the residual gas particles will be excluded from the space in which that flow takes place by unidirectional collisions with metal atoms, provided that the gas particles do not emerge from the melt[194].

In the most simplified form the quantity of material transported may be written as:

$$i = vC, \qquad (97)$$

where v is the rate of the gas flow and C is the concentration of the metal vapour.

(c) Diffusion in the gas space

If a diffusion is predominant the quantity of a substance diffusing through an inert gas may be expressed, according to various authors[195-198], for residual gas pressures above about 1 Torr as:

$$i = \frac{MDP}{RTl} \ln\left[\frac{P-p_1}{P-p_2}\right], \qquad (98)$$

where P is the total pressure in the distillation space (i.e. $P = p_{\text{inert gas}} + p_{\text{metal vapour}}$), the vapour pressure p_1 is only slightly lower than p_V (the equilibrium vapour pressure at the evaporation surface) and p_2 is slightly higher than p_K (the equilibrium vapour pressure at the condenser surface). As the term DP depends almost solely on the temperature, the diffusion in the gas space depends in turn almost entirely on the diffusion length and the pressure difference (see also Fig. 43).

(d) Heat transport

The heat transport in the melt may also assume great importance, as the rate of distillation of metals depends mainly on the heat input owing to their large heat of evaporation. In many cases this quantity of heat is made available at the evaporation surface by heat conduction and convection from the interior of the melt. Immediately below the surface of the melt a temperature boundary layer, δ_W, is created which may be defined by the following equation:

$$\frac{\Delta T}{\delta_W} = \frac{\Delta H_V}{M\lambda} \cdot i \qquad (99)$$

where ΔT is the temperature difference which forms within this temperature boundary layer. (For the other coefficients see List of Symbols used (p. 320) and Fig. 43.)

In contrast with these conditions, the zone of highest temperature in electron beam melting is the surface of the melt, as most of the kinetic energy of the electrons is converted into heat in a very thin surface layer.

(e) Evaporation

Evaporation of metals may take place via bubble formation or directly from the free surface. Bubble formation within the melt is very rarely observed in the evaporation of metals unless gases are evolved at the same time, as the pressure of the liquid column is very high, the thermal conductivity and the heat of evaporation of metals attain high values and the surface tension is also relatively high.

Evaporation from the free surface changes to molecular distillation if the residual gas pressure is so low that the mean free paths of the evaporated particles become at least as large as the distance between evaporator and condenser.

In this case the rate of distillation can attain the maximum evaporation rate:

$$i_{max} = 0.0583 \cdot \varepsilon_V p_V \left(\frac{M}{T_V}\right)^{\frac{1}{2}}, \tag{100a}$$

where p_V is the vapour pressure in Torr at the evaporation temperature T_V, ε_V is the condensation coefficient, which is the ratio of atoms condensed to atoms impinging on the evaporating surface under equilibrium conditions (see also Chapter I, Section 3). In most cases ε_V attains unity for liquid metals. However, especially with metals of high vapour pressure, lower values also occur.

(f) Superposition of several transport phenomena

Sometimes the condenser has such a high temperature that a marked degree of re-evaporation takes place. Under conditions of molecular distillation a transport of material in the opposite direction will therefore be superimposed to the material transport from the evaporating surface F_V to the condenser, and the resulting total material flow may be derived from eqn. (100b), where p_K is the vapour pressure in Torr of the condensate at the temperature of the condenser T_K:

$$i = 0.0583 \, \varepsilon_V (p_V - p_K) \left(\frac{M}{T_V}\right)^{\frac{1}{2}}, \tag{100b}$$

These equations for molecular distillation are valid without reservation only at total pressures below 0.1 Torr. The same equations apply accordingly to the condensation process.

At higher pressures of the residual gases frequent collisions between particles of the residual gases and of the evaporating metals occur. In this process a proportion of the metal particles is back-scattered in the direction of the surface of the melt, *i.e.* with increasing residual partial gas pressure again an increasing material transport in the opposite direction is created. The distillation rate therefore becomes pressure-dependent; it falls, and diffusion and convection in the gas space must be taken into account. In this case the materials transport may be expressed by:

$$i = 0.0583 \, \varepsilon_V (p_V - p_1) \left(\frac{M}{T_V}\right)^{\frac{1}{2}}, \tag{100c}$$

(see also Fig. 43).

This fact makes itself felt during the distillation of metals in the pressure range 10^{-2}–1 Torr and appears, according to Ilschner and Humbert[200], to depend

TABLE 26

SO-CALLED "CRITICAL PRESSURES" AND CORRESPONDING TEMPERATURES OF A NUMBER OF METALS AND ALLOYS

Evaporating metal	Basis metal	Working temperature (°C)	Corresponding vapour pressure (Torr)	Critical residual gas pressure (Torr)	Reference
Zn	Zn	465*	0.55*	0.3	
Zn	Zn	440**	0.3*	0.06	201
Zn	Zn	421*	0.14*	0.1	
Ag	Ag	1050	0.11	0.15	
Ag	Ag	1150	0.6	0.31	200
Al	Al	1050	0.0019	0.01	
Al	Al	1150	0.011	0.04	200
Al	Al	1250	0.05	0.035	
Mn(0.75–10%)	Ni	1550	~0.3–3	~0.1	202
Mn(0.1–5%)	Fe	1530	0.02–1	~0.2	
Mn(0.1–5%)	Fe	1600	0.04–2	~1	203
Mn(0.7–0.1%)	Fe	1580	0.04–0.3	0.07	204
Cu(1–5%)	Fe	1530	0.01–0.2	~0.03	203
Cu(0.2–4.5%)	Fe	1600	0.01–0.2	~0.03	
Al(63.8%)	Si 32.7%	1035–1245	<0.018	0.05	205

* Calculated. ** Experimental 497°C.

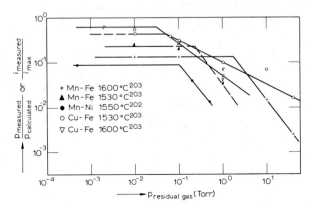

Fig. 44. Effect of residual gas pressure on the relative evaporation rate of metals.

on such atomic properties as effective scattering cross-section and mass. In Table 26 a number of established and converted data, the so-called critical pressures, are listed for several pure metals and alloys. For pressures higher than these, a pressure-dependent evaporation pattern should be expected. Figure 44 shows data computed from experiments for several alloys.

Friedrichs et al.[206] also emphasize that several individual steps may determine the rate of distillation in the above-mentioned pressure range of 10^{-2}–1 Torr, whereas at higher pressures the materials transport through the gas space becomes rate-determining.

But even in the pressure-independent range, i_{max} will not be attained in many cases. This is attributed by Ilschner and Humbert[200] and by Spendlove[201] to back-scattering effects originating from the prevailing experimental conditions.

Ward[204] distinguishes three stages in his studies of the pressure dependence of the evaporation of manganese from iron (0.7–0.1% Mn)$_{Fe}$ at 1580°C:

(1) Between 2×10^{-3} and 7×10^{-2} Torr the distillation rate of the manganese is independent of the residual gas pressure. The transfer may depend on the diffusion rate through the boundary layer in the melt as well as on the evaporation rate at the surface itself.

(2) Between 7×10^{-2} and 7×10^{-1} Torr the distillation rate decreases and is governed by the diffusion rate in one of the adjacent boundary layers in the metal or in the gas space.

(3) Between 7×10^{-1} and 760 Torr the distillation rate of the manganese is mainly governed by the transport through the boundary layer in the gas space.

Ward and Aurini[207] also determined the temperature dependence of the distillation rate of the manganese between 1168° and 1810°C in the low pressure range ($<7 \times 10^{-2}$ Torr). For temperatures above 1430°C the diffusion process in the melt is rate-determining; at temperatures lower than this it is the evaporation from the surface. These authors particularly point out that for large $\gamma P°$ products diffusion, and for small $\gamma P°$ products evaporation determine the distillation rate. From this it follows that the distillation rate of zinc and lead from iron is diffusion-controlled, but that the distillation rate of chromium, copper or aluminium from iron is governed by the evaporation process over a wide range of temperatures.

From Fig. 45 it may be seen that many metals already show a vapour pressure of the order of magnitude of these critical pressures at the melting point or attain it with slight superheating. This relatively high metal vapour pressure ($p_{metal} \geq 0.1$ Torr) may greatly suppress the evaporation of a number of impurities, particularly when these have a lower mass. These particles may be reflected into the melt by collisions with metal atoms, as was considered by Bunshah[195].

It should, however, be pointed out that surface films of oxides or any contamination of the surface also may strongly impair the distillation of metals or the evolution of dissolved gases[208-210]. Results of investigations on this effect were presented by Walsh and Burnett[211] for Sn–Zn, Cd–Bi and In–Zn.

In the vacuum distillation of almost pure metals heat transport may determine the distillation rate. This is expressed by Friedrichs, Jauer and Knacke[210] by:

$$i = \frac{\lambda \Delta T}{\Delta H_V \delta_w} = 0.0583 \, \varepsilon_V p_V \left(\frac{M}{T_V}\right)^{\frac{1}{2}} \qquad (101)$$

Only at low temperatures and therefore at low evaporation rates does the temper-

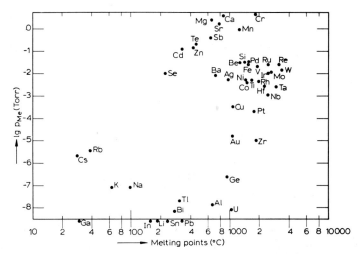

Fig. 45. Vapour pressures of metals at the melting point.

ature gradient at the surface of the melt become negligibly small, and only in this case can the evaporation process by itself determine the rate of distillation. Conversely, the effect of heat transport increases with increasing temperatures.

If the pressure of the residual gas atmosphere attains or surpasses the critical values mentioned earlier, the distillation rate may be controlled solely by the flow velocity of the gas or by the cooling of the surface of the melt, either by a high evaporation rate, or by a cooler gas flowing over it. According to the same authors[210], if the distillation rate is determined by evaporation as well as by convection in the gas space it may be expressed by:

$$i = \frac{\varepsilon_V}{1-\varepsilon_V} 0.0583 \, (p_V - p_1) \left(\frac{M}{T_V}\right)^{\frac{1}{2}} \qquad (102)$$

under the condition that $\varepsilon_V < 1$. The various limiting cases of distillation are discussed by these authors in detail.

If evaporation and diffusion within the melt are rate-determining for the distillation, the transport equation may be written in a simplified form as:

$$i' = \frac{C \cdot F_V}{\dfrac{\delta_D}{D} + \dfrac{C}{p_V \, 0.0583 \, \varepsilon_V} \left(\dfrac{T_V}{M}\right)^{\frac{1}{2}}} \qquad (103)$$

This equation expresses the fact that diffusion through the liquid phase becomes rate-determining for low solubility or low concentration, C, of the element to be distilled from the melt in combination with a relatively high vapour pressure, p_V, whereas the evaporation process becomes the rate-determining step for high solubility or high concentration in combination with a low vapour pressure.

References pp. 321–335

By consideration of the effects of diffusion in the melt, evaporation and condensation as well as the differences in the evaporation surface, F_V, and the condensation surface, F_K, and of the difference in evaporation temperature, T_V, and condensation temperature T_K, a simplified equation[194] may be written as:

$$i' = \frac{\varepsilon_K F_K F_V (C \cdot K) \left(\frac{M}{2\pi R T_V}\right)^{\frac{1}{2}} - p_K \left(\frac{M}{2\pi R T_K}\right)^{\frac{1}{2}}}{(\varepsilon_K F_K + F_V) + \frac{\delta_D}{D} K \varepsilon_K F_K \left(\frac{M}{2\pi R T_V}\right)^{\frac{1}{2}}}. \tag{104}$$

Correction factors for pressure changes at the condenser or evaporation surfaces caused by the flowing vapours may be added to this expression according to Davey[198] (see also ref. 194).

Müller[196] expresses the relationship between the materials transport by evaporation and diffusion in the gas space for the case that

$$\frac{i \cdot l \cdot R \cdot T}{8289{,}4 \cdot D \cdot P \cdot M} \gg 0.5$$

as:

$$i = 0.0583 \, \varepsilon_V p_V \left(1 - \frac{p_{ig}}{p_V}\right) \left(\frac{M}{T_V}\right)^{\frac{1}{2}}. \tag{105}$$

where p_{ig} the partial pressure of the residual inert gas atmosphere. Transport of material can only be expected if $p_{ig} < p_V$, provided that convective flow processes are absent.

If the evaporation process, diffusion in the gas space and the condensation process become rate-determining at the same time[212], this may be expressed by:

$$i = 0.0583 \left[(p_V - p_{ig}) \left(\frac{M}{T_V}\right)^{\frac{1}{2}} - p_K \left(\frac{M}{T_K}\right)^{\frac{1}{2}}\right] \tag{106}$$

The distillation, or more precisely the evaporation, of several components has been expressed by Olette[213] with:

$$y = -\frac{B}{A^\alpha}(A-x)^\alpha + B \tag{107a}$$

where

$$\alpha = \frac{\gamma_B}{\gamma_A} \cdot \frac{p_B^0}{p_A^0} \left(\frac{M_A}{M_B}\right)^{\frac{1}{2}} \tag{108}$$

Langeron[214,215] formulates this slightly differently although he too starts in the same way from the Langmuir equation:

$$\lg \frac{C_B}{C_B^0} = (\alpha - 1) \lg \frac{A-x}{A} \tag{107b}$$

(For explanation of the symbols please see Appendix, p. 320. See also Chapter I, Subsection 3.3.)

Good agreement between theory and experiment was found for a considerable number of iron melts which were alloyed with various elements. Deviations (Al, Si) could be explained by the presence of side reactions, such as the formation of gaseous suboxides. In a more recent paper by Onillon and Olette[216] the necessity of avoiding the occurrence of back-scattering effects is emphasized, thus resulting in an even better agreement between calculated and measured data.

Friedrichs, Jauer and Knacke in later work[206] dealt with stationary states during the separation by distillation of mixed binary phases. They took into account the transient change in concentration of the melt, the diffusion within the melt through a border zone, the evaporation, the convection or diffusion in the gas space, the heat transport and, finally, the activity conditions. The authors succeeded in formulating easily solvable equations for a number of limiting cases.

For evaporation in a high vacuum, heat transport is disregarded in one case, in another case the activity constants are regarded as unchanging and, finally, both limiting cases are combined. Diffusion in the gas space during distillation is especially considered with special reference to the case of a non-negligible pressure of the residual gas atmosphere.

In a similar way, Bradshaw[193] describes stationary states in the distillation and evaporation in a vacuum with special consideration of the transport in the liquid phase and in the gas phase and of reactions at the melt/gas phase boundary.

4.1.2 Methods of metal distillation

The application of the results of laboratory tests on vacuum distillation to industrial practice has in most cases been characterized by the fact that the industrial enterprises have developed, designed and constructed their own prototypes of the vacuum apparatus. In the beginning of vacuum distillation it was therefore the usual practice simply to attach cooled condensers to the top of crucibles, if liquid charges were used, and to the top of retorts, if solid charges were employed (see also Figs. 34 and 52).

After it had been realized that large surfaces and small volumes are advantageous to distillation, the path for the development of the thin film distillation technique was shown.

A thin-film distillation[217] with evaporator and condenser surfaces facing one another can be found in the dezincing of lead (see Fig. 53) and probably also in the distillation of cadmium from Cd–Pb–Bi alloys[218]. Plants for centrifugal molecular distillation[217,219] have been up to now employed only in the organic chemical industry. In such an arrangement an inverted hollow cone is rotated around its axis so that the liquid to be distilled is forced to climb up the wall forming a thin continuously changing film. From this thin film rapid distillation is achieved. Condensation takes place at a cooled condenser facing the cone. The residue after separation of the

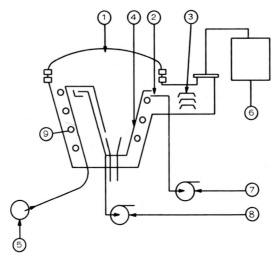

Fig. 46. Diagram showing the principle of a centrifugal distillation plant[217]. 1 Vacuum vessel; 2 Residue overflow; 3 Cooling trap; 4 Rotor; 5 Inlet; 6 Vacuum pumps; 7 Pump for residues; 8 Pump for distillate; 9 Heating.

distillate is discharged over the edge of the cone into a catcher channel, from whence it is pumped off and brought into the atmosphere via a barometric seal. Such plants should be suitable in principle also for liquid metals for large throughputs in at least semicontinuous operation. In all thin-film distllation plants pre-degassing of the liquid is necessary in order to prevent spattering on entry into the vacuum chamber and—in most cases—contamination of the distillate. A diagrammatic presentation of a centrifugal distillation plant is shown in Fig. 46.

There are also suggestions for the vacuum rectification or fractionating distillation of metals in columns in which trays are provided and in which, in addition, the vapour stream is diverted several times in its flow direction by baffles, so that partial condensation may take place at these baffles.

4.1.3 Distillation processes in the metallurgy of lead

Vacuum technique has attained very great, although not much publicized, importance in the refining of lead. Almost the total annual world production of lead of about 3.3×10^6 ton is subjected to the so-called vacuum de-zincing process. An intermediate product obtained in the refining of lead bullion—the so-called Reichschaum (silver crust)—may be vacuum-distilled in a further process stage.

(a) Distillation of silver crust

Lead is a collector for precious metals, such as gold and in particular silver. Both precious metals are, therefore, almost entirely contained in the reduced lead.

TABLE 27

DATA ON VACUUM DISTILLATION OF SILVER CRUST, COMPOSITION IN wt.-%

Temperature (°C)	Pressure (Torr)	Time (h)	Starting material			Distillate			Residue			Ref.
			Ag	Pb	Zn	Ag	Pb	Zn	Ag	Pb	Zn	
	0.005		30.3	8.1	56.0				95			220
1000	0.021	3	9.5	33.8	54.5	0.8	34.1	63.5	74.2	0.6	0	221
950	0.017	7	26.9	8.6	63.7				95.7	1.7	0.2	
900	1–2	8	20	5	68				63	11	3	212
750–800	10	12	29.2	19.2	50	0.15	1.8	98	38	57	1.8	222
920	0.5–5	7–10	10	60.3	26.3	1.7	12.4	85.9	12.1	87	0.9	223

During the refining of lead, the precious metals, silver and gold, are separated from the lead. The precious metals are converted with the aid of zinc into intermetallic compounds, which precipitate from the pool of molten lead, float up and may be skimmed off, while the lead after de-silvering contains less than 10 ppm silver. The skimmed off silver crust which carries the precious metals still consists of more than 50% zinc, as may be seen from Table 27. The zinc may be distilled off without any difficulty in a vacuum and may again be used for the separation of the precious metals.

In most cases retorts with external heating are used, similar to those used in the winning of magnesium (see also Fig. 34), but in Tschimkent a retort was equipped with internal electric heating[223]. About 2.7 tons of silver crust may be processed in one furnace unit per day.

The Société Minière et Métallurgique de Pennarroya[222] designed a special furnace with internal heating for this purpose (see Figs. 47 and 48) in which two liquid alloys (see Table 27) are obtained. The temperature of the condenser is about 420–450°C[222, 224]. The throughput of the furnace is 2 tons, *i.e.* the daily production of silver amounts to about 500 kg.

In these vacuum processes the yields of zinc and silver exceed 95%, whereas in the conventional Faber du Faur process (distillation at atmospheric pressure) yields of only 75% for zinc and 80–90% for silver are obtained. Another advantage

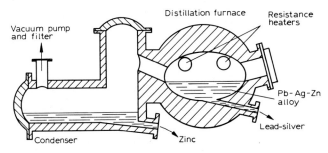

Fig. 47. Diagram of a furnace for the vacuum distillation of silver crust[222].

References pp. 321–335

Fig. 48. Overall view of the plant for the distillation of silver crust at Noyelles–Godault[222].

is that the working temperatures of the conventional process of 1100–1350°C are lowered to below 1000°C by the use of a vacuum.

Depending on the temperature and pressure, lead is codistilled ($T > 950°C$, $p < 0.1$ Torr) or kept in the liquid residue ($T < 900°C$, $p > 1$ Torr), in the first instance the evaporation of silver is also increased. A slight evaporation of silver is of no consequence, as the zinc which contains the silver is recirculated in the de-silvering process.

(b) *Vacuum de-zincing of lead bullion*

The refined lead still contains about 0.6% zinc in solution after completion of the de-silvering process. While in former times this zinc was oxidized (involving the simultaneous oxidation of a rather large quantity of lead) which was therefore not directly available for re-use, nowadays most of the dissolved zinc is distilled off—optimum final zinc levels of 0.01% are obtained—and may therefore immediately be re-used for the separation of the precious metals.

A miscibility gap also exists in the liquid field of the lead–zinc system, thus maintaining a high activity of the zinc at low zinc levels in the lead and therefore a relatively high vapour pressure of the zinc. The vapour pressures of zinc over lead–zinc melts are shown for various temperatures in Fig. 49. To this diagram a number of experimental as well as works data from the de-zincing of lead bullion are added; from these it may be seen that the distillation is stopped at zinc pressures of about 0.05 Torr. These pressures correspond in almost every case to the vacua quoted. It should, however, be pointed out that the pressure measurements were

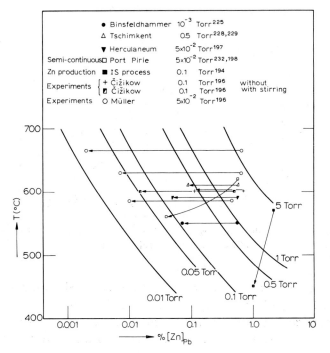

Fig. 49. Effect of temperature on the vapour pressure of zinc over lead–zinc melts during vacuum dezincing showing the initial and final zinc contents; data taken from the literature.

in very few cases carried out directly at the vacuum vessel and that therefore the true conditions may not necessarily be presented. The starting and final zinc levels obtained by the Imperial Smelting process (IS) are also plotted in Fig. 49. This process is, in fact, a method for the production of zinc. It will be discussed later.

In the dezincing of lead bullion the distillation process is conducted in such a way that as much zinc as possible is recovered and that a lead of the highest possible purity is obtained. In contrast with this process the distillation in the Imperial Smelting process is carried out only in that range which gives a zinc of a refined grade at a high distillation rate. In batch processes heat is introduced by external heating of the vessel, but in continuous or semi-continuous processes the heat of evaporation of the zinc is mainly extracted from the pool of molten metal without adding large quantities of heat. Consequently the temperature of the pool falls quite considerably.

Only a few results are available on the kinetics of the distillation of zinc; these are also contradictory among themselves and refer to different processes. A number of measurements were carried out in a discontinuous plant at Binsfeld-hammer (Germany)[225] (see also Fig. 50). The results obtained when plotted on semilogarithmic paper are represented by straight lines down to zinc levels of 0.08%. This formally corresponds to a time law of the first order.

References pp. 321–335

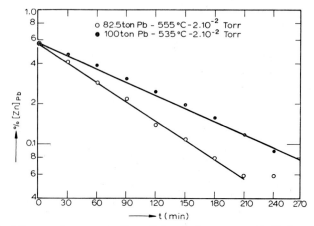

Fig. 50. Effect of time on the concentration of zinc during vacuum distillation after desilvering of lead bullion, according to ref. 225.

Heat transport in the melt may in this case be excluded as the rate-determining step despite the high heat of evaporation of the zinc. Owing to the unusual activities the evaporation process should also be eliminated as a rate-determining step because of the too high values found if the quantities of zinc which could be transported by evaporation are estimated. Now diffusion in the gas phase and in the melt, and condensation, should be considered.

At a residual gas pressure of 10^{-3} Torr the mean free path in the vessel becomes so large that diffusion in the gas space may be discounted as the rate-determining step. For an estimate of the quantity of zinc transportable by diffusion in the melt in unit time, a formula based on Fick's law may suffice. Some assumptions and simplifications should first be made for this diffusion process:

As a first approximation[194] D_{Zn} should be about 8×10^5 cm²/sec, δ_D with effective stirring about 3×10^{-3} cm. At a surface area of 6.027 m² and 100 or 82.5 tons lead[225] in the kettle the following distillation times for final levels of 0.08% $(Zn)_{Pb}$ are obtained if the change of D with T is taken into account:

Vessel content (ton)	82.5	100
Temperature (°C)	555	535
Time, calculated (min)	180	260
Time, measured (min)	155	210

This relatively good agreement between calculated and measured data indicates that diffusion in the melt may not be excluded as the rate-determining step.

Davey[197] carried out theoretical studies of the rate-determining step for the batch and semi-continuous process practised at Port Pirie. He concluded that the evaporation, diffusion in the gas space, and the condensation process together determine the rate of zinc distillation. Large deviations between measured and calculated results in the batch process are attributed to premature condensation of the zinc in the gas space and fall-back into the melt. In a more recent publication[198] by Davey the conditions in the continuous and semi-continuous vacuum de-zincing processes are discussed in a more refined treatment. It was found that isothermal or adiabatic diffusion in the gas space bring about only insignificant differences. It is also assumed that evaporating zinc carries impulses in a preferred direction and consequently carries away most of the residual gas from the evaporating surface. On the other hand, a higher pressure is generated at the condenser surface by the impinging zinc particles. This formulation proposed by Davey has been simplified by Kidyarov[226].

Agreement is said to exist between calculated and theoretical data for the semi-continuous distillation of zinc at Port Pirie. On the other hand, Fick's First Law alone makes it possible to bring about agreement between calculated and measured data; this again indicates that in this case also diffusion in the melt plays a decisive role in determining the rate of reaction. On the assumption that diffusion in the melt is rate-determining, a final zinc level of $0.045\% (Zn)_{Pb}$ is calculated. The formulation due to Davey[198] leads to almost the same quantity, with $0.04\% (Zn)_{Pb}$.

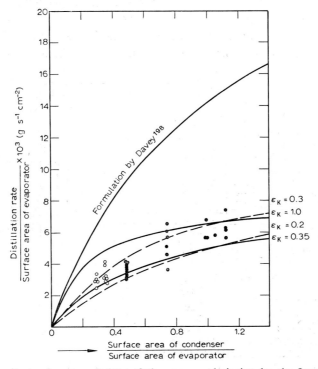

Fig. 51. Comparison of measured data of the vacuum dezincing by the Imperial Smelting process, according to Warner's theory[194], at 530°C and an initial zinc content of 2.1%.
——— Equation (104);   • Pilot plant;   – – – – Supplemented eqn. (104);   ○ Works process.

References pp. 321–335

Warner[194], on the other hand, cannot obtain agreement between practice with the Imperial Smelting process and values calculated with the equation proposed by Davey. However, he finds agreement between calculated and measured data if he uses an equation in which diffusion in the melt, evaporation and condensation are taken into consideration. It was found that only values between 0.2 and 0.3 should be used for the condensation coefficient ε_K in the Imperial Smelting process. This is attributed mainly to the fact that the condenser surface is only partly (assumed to be 50%) covered with liquid zinc. The results obtained on pilot and industrial plants are plotted in Fig. 51. Warner also attributes great importance to the material transfer coefficient in the melt, which could only be increased by more vigorous bath agitation. This, however, involves the risk that lead could be splashed on to the zinc condenser, thus impairing its quality.

Friedrichs, Jauer and Knacke[206] concluded, on the basis of the formulations quoted earlier, that during conventional de-zincing in a kettle, evaporation and diffusion in the melt may be rate-determining only in the temperature range 320–475°C, *i.e.* not under conditions usually met with in technology, and that heat transport may also be excluded as the rate-determining step. Only at temperatures above 900°C do heat transport and evaporation attain any equal order of magnitude. Diffusion in the melt, which is normally the rate-controlling step here, exercises at these high temperatures of course an even greater effect on the distillation rate than it does under the usual technological conditions. Besides this the interdependent transport processes of evaporation and diffusion in the gas space

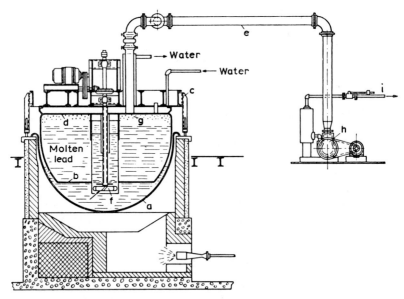

Fig. 52. Vacuum dezincing plant according to Isbell using solid zinc condensation.
a Lead kettle;   b Bell-shaped shell;   c Condenser frame;   d Water jacket;   e Vacuum duct;   f Stirrer;   g Condensed zinc;   h Vacuum pump;   i Exhaust.

may at the temperatures of conventional operation exercise a certain influence on the distillation rate.

The first technological solution for the distillation of zinc was devised by Isbell[227] in 1947, after Kroll in 1935 had already suggested such a process. Today the major part of the world's lead production is de-zinced by the batch process developed by Isbell. The plant for this process is depicted in Fig.52. A bell-shaped shell which is water-cooled at the top and which carries a stirring device is inserted into the conventional lead kettle of 100–150 ton capacity. When vacuum is applied to the shell, the level of the lead bath in the shell rises a little, while it falls outside the shell. The final difference in height between inside and outside the shell is about 90 cm. The lead bath seals the vacuum against the atmosphere. The working temperature varies between 550° and 650°C, and the pressure is lower than 1 Torr. The heat required for evaporation is supplied by heating the kettle. The bath is stirred at about 150 rev/min for promoting distillation. The evaporated zinc condenses as a solid on the top of the shell, from which it is detached when the distillation has been completed. The working conditions of such a plant are listed in Table 28. By this process 80–90% of the zinc is recovered in metallic form. Further distillation would lower the zinc level down to 0.001%, but is economically not acceptable[234].

The semi-continuous plant[232] at Port Pirie, which is depicted in Fig. 53, is a further development of the process just described. De-silvered zinc-containing

TABLE 28

DATA ON VACUUM DE-ZINCING OF LEAD

Name of works	Capacity of the plant (ton)	Temperature (°C)	Pressure (Torr)	Time (h)	Throughput (ton Pb/h)	Starting content (wt.% Zn)	Final content (wt.% Zn)	Refs.
Binsfeldhammer (Germany)	100	550	10^{-3}	4–5	23	0.57	0.07	225
Tschimkent (Soviet Union)	150	600–620	0.5	4–5	32	0.55–0.57	0.08–0.1	228, 229
	50	580–620		7–8	6.5	0.56–0.63	0.11	228, 229
Herkulaneum (USA)	115	590	5×10^{-2}	5	23	0.5–0.6	0.05	196
Příbram (CSSR)		600	0.2	5–6				230
Rönnskär (Sweden)	100	600	0.2					231
Port Pirie-semi-continuous (Australia)	–	560 Average surface temperature	0.05–0.1	–	27	0.58	0.04	198, 232
ISP-vacuum-de-zincing Avonmouth	36 ton Zn/day	570	0.1	–	Several hundred	2.2	~ 1.0–1.5	194, 233
Swansea (Great Britain)	110 ton Zn/day	550		–	1200–1700			

References pp. 321–335

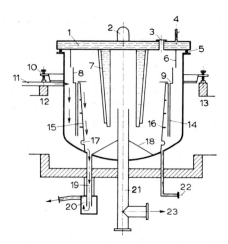

Fig. 53. Semicontinuous plant for vacuum dezincing using solid zinc condensation.
1 Water-cooled condenser; 2 Hook; 3 Sight glass; 4 Cooling-water inlet; 5 Rubber seal; 6 Baffle; 7 Cooling fins; 8 Baffle; 9 Channel for lead; 10 Level adjustment; 11 Lead inlet; 12 Heating channel; 13 Exhaust channel; 14 Inlet; 15 Running-down surface; 16 Reinforcing rings; 17 Bellow; 18 False bottom; 19 Off-take of lead; 20 Siphon for lead; 21 Vacuum duct; 22 Drain-off system; 23 To vacuum pumps.

lead is fed through an inlet (11) into the vacuum vessel into a pool of molten metal. Through a number of risers (14) the metal enters a distributor channel (9) from which it runs down the wall as a thin film. During this descent, intensive de-zincing takes place. The heat of evaporation is extracted from the metal which has previously been heated to 620°C, and which cools during the de-zincing process to 560°C. According to Davey[197] a theoretical loss in temperature of 82°C should be expected, which is, however, not attained through additional heating (12, 13). The evaporated zinc condenses as a solid at the water-cooled condenser (7). The de-zinced lead leaves the vessel through a barometric seal (19, 20). The vacuum is maintained by continuous pumping (21, 23). As soon as a predetermined quantity of zinc (2–3 tons) is condensed, the process is interrupted for changing the condenser. The zinc may be removed from the condenser by melting in a lead bath at 450°C. Technological data are shown in Table 28.

A continuous plant using liquid zinc condensation is employed in zinc production by the Imperial Smelting process[194, 198, 233, 235, 236]. In this process the zinc is reduced in a shaft furnace and volatilized and then condensed in lead up to a maximum quantity of 2.5% zinc. The temperature of the lead carrier bath rises from 450° to 570°C during condensation. Previously, in the subsequent separation of the zinc from the lead, the miscibility gap in the lead–zinc system was utilized and zinc containing about 1.2% lead was precipitated by cooling to 450°C. The lead-bearing zinc was finally refined. For obtaining 1 ton of zinc, 400 tons of lead had to be circulated.

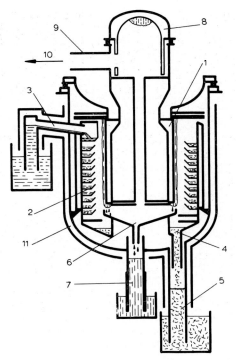

Fig. 54. Continuous plant for vacuum dezincing using liquid zinc condensation.
1 Water-cooled condenser; 2 Helical channel; 3 Lead inlet with siphon; 4 Lead take-off; 5 Lead siphon; 6 Zinc take-off; 7 Zinc siphon; 8 Cooled dome for final condensation; 9 Vacuum duct; 10 To vacuum pump; 11 Auxiliary heater.

For distilling off the zinc from the lead, the semi-continuous distillation process was used as model after extensive modification, as may be seen from Fig. 54. With the very large quantities of lead recirculating, no coherent metal film could be maintained on the original running-down surface in the vessel. The lead splashed and contaminated the zinc. In order to prevent this happening, the zinc-containing lead is made to flow in a helical channel from the top to the bottom of the vessel while the zinc evaporates. In the centre of the vessel the water-cooled cylindrical condenser is set again in the vertical direction, on which the zinc is deposited as a solid layer about 5 cm thick. The temperature in this layer rises gradually to 420°C, so that the zinc finally condenses as a liquid, running down the surface of the condenser. The zinc-containing lead is fed into the vacuum vessel through a barometric seal and the liquid zinc and the liquid lead are drawn off separately, also through barometric seals.

The zinc which has not condensed in the vacuum vessel proper is precipitated in a cooled bell which is mounted on top of the plant.

The quality of the zinc depends mainly on the temperature at which the lead is fed into the de-zincing plant. In a pilot plant at Avonmouth, with a daily pro-

References pp. 321–335

Fig. 55. Upper part of the vacuum dezincing plant at Swansea[236] with the dome for final zinc condensation in position.

duction of 35 tons of zinc, there was no difficulty in obtaining zinc with a lead level of 0.05% or even 0.03%, while the iron content could be kept below 0.005%. In the preliminary experiments in the production plant at Swansea, with a daily capacity of 110 tons of zinc, the lead content of the distilled zinc fell from 0.25% to 0.07% when the entry temperature of the lead was reduced from 550° to 510°C. In order to prevent condensation of the zinc at the walls of the distillation plant, additional heat must be supplied. Photographs of the production plant at Swansea are shown in Figs. 55 and 56. The output figures are listed in Table 28. These show that the output of the two plants at Avonmouth and Swansea is 5 tons of zinc per hour and the quantities of lead recirculated are several hundred and 1500 ton/h, respectively, and far exceed that of the plants previously discussed. In the first-named, batchwise plants, the heat of evaporation for the zinc must be supplied externally, whereas at Port Pirie, and particularly in the Imperial Smelting process, the heat for the evaporation of the zinc is extracted from the liquid metal, whereby the lead is cooled to about 450°C. In the Imperial Smelting process there is possible primarily only separations of Pb (less than 0.06%) and of As (less than 0.001%), whereas cadmium is collected in the distillate. According to an American patent[237], cadmium separation from the zinc should be feasible by multi-stage distillation and fractional condensation in a vacuum.

It has further been proposed to carry out the de-zincing of lead bullion in a continuous vacuum distillation column. In order to obtain sufficiently high ex-

Fig. 56. Lower part of the vacuum dezincing plant at Swansea[236] showing the barometric seal for the zinc take-off and the zinc collector furnace.

change rates and to collect the zinc in the liquid state, the distillation temperature should be about 680°C at a residual gas pressure of about 0.3 Torr. In a very short time the zinc content of the lead may thus be lowered to levels below 0.05%[238].

At the relatively high distillation temperature of 650°C, the distillate contains about 1% Pb, and, if the temperature is further increased to 750°C, it contains as much as 15% Pb. Single distillation at 750°C and a zinc level of less than 2% in the lead bullion results in a zinc content of about 0.1% in the refined lead. A second subsequent distillation under the same temperature conditions approaches the limiting level of 0.01% zinc[229,239].

(c) Proposals for the distillative separation of the components of further lead products

Based on the distillation of zinc from lead bullion, attempts have also been made to separate As, Sb, Bi, Te, Mg and Cd from the lead by distillation. It is possible to evaporate As and Cd to extents of 70% and 95%, respectively, at 700°C and 10^{-2} Torr, with lead losses of 0.5%. Antimony and bismuth may be reduced by volatilization only to 8–10% of the initial levels. Raising the temperature to 800–1000°C only leads to evaporation of all components[201,240]. Müller, Spendlove and others[196,201,234] found, for antimony under similar conditions, that even an enrichment of the melt in antimony up to 15% could occur, this quantity being increased to higher levels with increasing temperature. Tellurium may be almost eliminated (0.01%) from lead at 900–1100°C and 10^{-2} Torr, but the distillate also consists mainly of lead (more than 70%). For magnesium a content of less than 0.6% Mg in the lead could not be obtained by distillation at 900°C and 10^{-2} Torr. Here too the simultaneous evaporation of lead is considerable[201].

An attempt was also made to work up the dross precipitated on cooling of raw lead by vacuum volatilization[241]. With a composition of 70.4% Pb, 16.5% Cu, 3.4% As and 0.65% Sb, it was possible to volatilize 92.2% of the lead at 1000°C. In the following addition of sulphur or pyrite, respectively, Pb could be volatilized at 800°C to 95.6%, As to 94%, Sb to 87.5%, within one hour by addition of sulphur, at 1000°C and by addition of pyrite, the Pb to 99.6% and the As to 98.3%. The residue could be worked up directly for copper.

4.1.4 Distillation processes in the metallurgy of the remaining low-melting-point heavy metals, in particular cadmium, zinc and tin

Zinc and cadmium already possess at their melting points of 421° and 327°C respectively, vapour pressures above 10^{-1} Torr. The rectification of zinc is carried out on large industrial scales above 1000°C at a vapour pressure of the zinc of 1 atm (in closed vessels) by the New Jersey process. The use of vacuum processes should be possible at considerably lower temperatures, giving also improved separation effects. The high heat demand for the zinc and cadmium distillation may be cut down by employing multi-stage distillation and then using the heat obtained by the condensation in the distillation at atmospheric pressure for a further refining distillation in a vacuum[242].

So far two technological methods have been used on an experimental basis, vacuum rectification[243] and vacuum distillation[244]. In the latter method condensation is carried out along a temperature gradient. In addition, baffles are provided

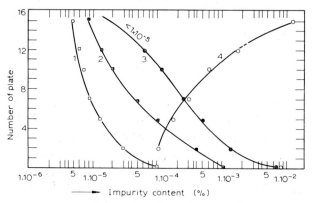

Fig. 57. Impurity distribution at various plates during the vacuum rectification of zinc using a graphite column[243]. 1 = As; 2 = Cu; 3 = Pb; 4 = Cd.

in the lower part of the condenser, these to prevent evaporated particles from getting into the upper part of the condenser without deflection or collision. The rectification was carried out in graphite columns (15 plates) and in quartz columns (7 plates). The distribution of several impurities in these experiments are plotted in Fig. 57. From this diagram it may be seen that the more volatile elements collect in the coolest zones, the less volatile elements in the hotter zones of the condenser. Distillation and rectification gave the same results. Measurements of the electrical resistance ratio showed that the best values were obtained on material from the middle plates. Relevant data and the temperature distribution are listed in Table 29a.

The distillation of cadmium may be carried out in a similar way with the evaporation temperatures a little lower than those for zinc. After experiments in small pieces of apparatus made of quartz[247, 249] proved successful, a continuous pilot plant was constructed of steel which gave equally good results (see Fig. 58).

In this apparatus cadmium is melted at atmospheric pressure (1), the molten metal rises through a tube (3) by the atmospheric pressure into the evaporator vessel (2). In the evaporator cadmium is preferentially evaporated and is condensed in liquid form (4). The distillate is drawn off through a discharge tube (6) and a barometric seal (5). In the evaporator the cadmium is enriched by substances of low volatility. This metal mixture may be drained off through a second discharge tube and another barometric seal (7, 8). Further technological data are contained in Table 29a.

The effect of additions on the rate of evaporation of zinc and cadmium is still completely inexplicable. In the distillation of zinc and cadmium from silver–zinc

TABLE 29a

DATA ON VACUUM DISTILLATION OF ZINC AND CADMIUM

Metal	Process	Temperature of evaporator (°C)	Temperature of condenser (°C)	Residual gas pressure (Torr)	Purity Before (wt. %)	Purity After (wt. %)	R_{298K} / $R_{4.2K}$	Refs.
Zn	Distillation	460	300–100	10^{-4}	99.98	99.99997	26000	244
Zn	Rectification Graphite	700	700–300	10^{-4}			12500	243, 245, 246
Zn	Rectification Quartz			10^{-2}			7500	243, 245, 246
Cd	Rectification						12000	245
Cd	Distillation Quartz	400–500		10^{-4}	99.998	>99.999	14000	247
Cd	Distillation	400–520		0.05–0.15		99.999		248
Cd	Distillation Quartz	440–460	340–350	10^{-1}	~95	99.98		249
Cd	Distillation Steel	430–460	335–350	0.5–1	99.95	99.992		249
Cd	Distillation Steel	430–460	335–350	0.5–1	~95	99.98		249

References pp. 321–335

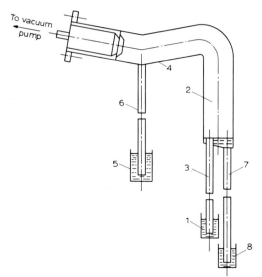

Fig. 58. Continuous plant for the vacuum distillation of Crude cadmium[249]. 1 Cd storage vessel; 2 Evaporator; 3 Inlet siphon; 4 Condenser; 5 Cd collector vessel; 6 Cd take-off siphon; 7 Residue siphon; 8 Residue collector vessel.

and silver–cadmium solid solutions in the solid state at 650°C and 10^{-4} Torr, aluminium and tin additions (of about 0.2%) act as an accelerator for cadmium, and as a retarder for zinc. Additions of lead initially promote the evaporation of both metals, while later the distillation rate is greatly reduced. If the distillation rate is raised the porosity will increase and the contraction in volume will decrease[250].

Mercury, cadmium and zinc may be almost totally separated from such low-melting-point heavy metals as Sn, Ga, In and Tl, which all have low vapour pressures (see Fig. 45) by vacuum distillation. In this process the distillates may also be obtained in the high-purity form[201,251].

In a special process for the production of cadmium and thallium[252] these two metals are cemented with a zinc amalgam up to total replacement of the zinc (see Fig. 59). From this Cd–Tl amalgam, cadmium and mercury are volatilized simultaneously in a vacuum (d), whereas thallium is retained and forms a pool (g). Cadmium accumulates in the lower part of a fractionating column (n) and is drained off through a barometric seal (r). The mercury is condensed after leaving the fractionating column in a mercury cooling trap (k) and then recirculated into the process via another barometric seal (l). The cadmium obtained has a purity of 99.95%. Heitmann and Knacke[253] carried out investigations on the distillation rate of cadmium amalgam, which showed that the evaporation process determines the distillation rate.

Distillation experiments on Cd–Bi–Pb alloys were carried out at about 9×10^{-5} Torr over the temperature range 450–550°C with liquid-state condensation

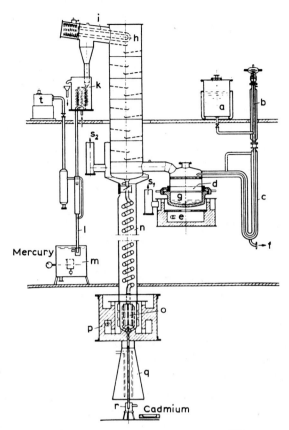

Fig. 59. Vacuum distillation plant for the separation of Cd–Tl–Hg[252]. a Amalgam storage vessel; b Control valve; c Inlet system; d Boiling pan; e Heating furnace; f Condensed water outlet; g Thallium pool; h Return condenser; i Cooler; k Hg cooler; l Hg take-off tube; m Hg collector vessel; n Cd distillation coil; o Cd collector; p Heating furnace; q Heating furnace; r Cd take-off siphon; s Exhaust ducts; t Vacuum pump.

at 340°C in a short-path distillation plant[218]. The molten alloy to be distilled flows down an inclined baffled distribution plate, where the liquid metal stream frequently changes direction. This procedure facilitates a thorough and rapid homogenization of the concentration of the melt. On the other hand, the molten alloy to be distilled could also be continuously recirculated. The distillation rate attains, on the average, 65 to 75% of the theoretical values, and in single individual cases even up to 90%. The highest purity of the distilled cadmium is about 99%, but this decreases with lower initial content, with increasing distillation temperatures and with higher flow rates in the distillation channels, as may be seen from Fig. 60.

The distillation of bismuth may also be carried out in a plant which is almost identical with that shown in Fig. 58 for the distillation of cadmium. Bismuth of a

References pp. 321–335

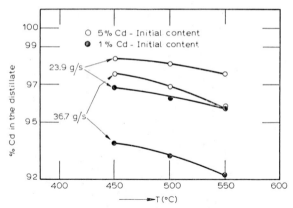

Fig. 60. Effect of temperature, initial cadmium content in Cd–Pb–Bi alloys and flow rate on the purity of distilled cadmium[218].

purity of 99.999–99.9999% may be obtained from a 99.98% metal by distillation at a temperature of 950–1000°C and a pressure of 10^{-2} Torr, when only 20% of the starting quantity remains as a residue[254].

Crude tin frequently contains relatively large quantities of lead and bismuth. These two impurity elements may be separated by distillation in multi-plate columns at a temperature of about 1100°C and at pressures not higher than 0.2 Torr. The operation, which is continuous, reduces the lead content from 0.2 to 0.03% and the bismuth level from 0.037 to 0.008%. The distillate consists of a Sn–Bi–Pb alloy; the lead and bismuth content varies according to the starting material between 20 and 60% lead and 3 and 20% bismuth. Using fractional distillation high-purity metals may also be obtained. Pressures higher than 0.08 Torr lead to a marked reduction in the distillation output[238].

Using a semi-industrial plant with a thermal rating of 100 kVA, both lead and bismuth could be reduced in crude tin from 1.5 to 0.02% Pb and from 0.3 to 0.005% Bi by a single vacuum refining. Even alloys containing up to 27% lead and 10% bismuth showed, after a three-fold distillation, only 0.008% Pb and 0.001% Bi. If the distillation temperature is raised to 1200–1300°C, distillation may even be satisfactorily carried out at 0.2–0.3 Torr and a lowering of the evaporation output occurs only at residual gas pressures higher than 2 Torr. The throughput of the plant varies, according to the quality of the starting material, between 240 and 820 kg/m²h[255].

The low-melting-point metals arsenic and antimony belong to the group of metals which have acquired great importance in semiconductor technology, but they must be subjected to special methods of refining to be used for this purpose. Both sublimation and distillation are made use of in this refining operation.

The sublimation of arsenic is carried out at 600°[256], 500°[257] or 400°C[258] and the condensation in a temperature gradient at temperatures which are at least 150°C lower. The less volatile elements remain in the residue and the most volatile elements or compounds (As_2O_3, As_2S_3) condense at the coolest zones of the condenser, but the separation, particularly that of the sulphur, is inadequate. Markedly better results should be expected from the distillation of arsenic dissolved in lead[259]. In the dissolution of arsenic in lead, the sulphur dissolved in the arsenic reacts with the lead forming PbS, which has a substantially lower vapour pressure. If the temperature of the bath is maintained at 550–600°C, practically only arsenic evaporates. This is condensed at lower temperatures. Sulphur as well as selenium and tellurium co-distil in only minute quantities. On the other hand, an increase in distillation temperature to only 700°C substantially impairs the refining effect, as may be seen from the following data:

	Sulphur (ppm)	Selenium (ppm)	Tellurium (ppm)
Starting material	1.1–3.9	100–170	0.13–0.35
Distillate,			
535°C	$< 1 \times 10^{-3}$	a	a
600°C	1.1×10^{-3}	2.7×10^{-2}	2×10^{-2}
700°C	6.6×10^{-2}	7.4×10^{-1}	2×10^{-2}

a Not known

These results have also been confirmed by various other authors[259,260]. In the distillation of arsenic some lead is carried over, but this lead content may be lowered to about 1 ppm by a two-fold back sublimation. Also lead does not exercise the same detrimental effect as does sulphur, selenium or tellurium[256].

The main impurities of antimony, such as Pb, Ag, Sn, Cu, Fe, Ni and Si, show a lower vapour pressure than that of antimony, but the vapour pressure of arsenic is higher than that of antimony. One distillation of antimony may therefore be sufficient to produce a metal of very high purity[259], since any arsenic present can very effectively be removed by subsequent zone-melting in the presence of small amounts of aluminium. An addition of 0.1% Al brought about the best results. But the complete removal of carbon met with difficulties. Starting levels of about 100 ppm carbon could only be reduced to one-half that level by zone melting. A substantially improved lowering was possible only by subsequent vacuum distillation, in which carbon enrichment was not found in the residue, but in the coolest zones of the condenser. This occurrence leads to the conclusion that these are residues of an organic compound[259].

Bonnier and Charveriat[261] found, during their experiments on the sublimation of antimony at a temperature of 500–600°C and a pressure of 5×10^{-5} Torr, that Ag, Al, Cr, Cu, Fe, Mg, Mn and Ni remain in the residue, and that As, Pb, Sn and S are to a large extent carried over into the distillate, as are Si and C. An addition of 2% manganese to the starting material keeps almost all the Sn and S and part of the As and Pb in the residue. By addition of 2% manganese and 3%

TABLE 29b

SUBLIMATION OF ANTIMONY (ppm)[261]

Impurity	Laboratory tests			Pilot-plant tests				Recirculation of antimony	
	Starting material	Straight sublimation	After addition of 2% Mn	After addition of 2% Mn + 3% Al	Starting material	Straight sublimation	After addition of 4% Mn	After addition of 1% Mn + 2% Al	Before sublimation	After sublimation
Al	1	3			10	<5	<5	<5		
Ag	<2	<2								
As	340	440	250	170	500–1000	1500	1400	<400	880	200
B	<1	<1							<5	<5
Cu	<5	<5			500–1000	<5	<5	<5	700	1
Cr	<2	<2								
Fe	30	6	6	<3	2000–5000	<5	5	<5	80	4
Mg	9	3	5	1.5					35	2
Mn	5	<2	<3	<2	5	3	3	<2	40	1
Ni	11	<8								
Pb	280	180	120	20	4000–6000	900	210	<30	600	90
S	100	70	<40	<40	100		40	<40	120	<40
Si	21	25	40	34						
Sn	450	>400	<40	<40	4000–6000	60	<20	<10	900	<20
C	100	>200							40	40
Residual gas pressure (Torr)	5×10⁻⁵	5×10⁻⁵	5×10⁻⁵	5×10⁻⁵		2×10⁻⁴	3×10⁻⁴	2×10⁻⁴	10⁻⁴	10⁻⁴
Sublimation temperature (°C)	600	580	620			600	600	620	600	600
Condenser temperature (°C)	100–300	a	a			100–400	100–400	100–500		

a Not known

aluminium to the starting material the greater proportions of the As, Pb, Sn and S remain in the residue. The intense evaporation of tin during the sublimation of antimony in the absence of any additions is attributed to the formation and subsequent evaporation of SnS. No explanation was offered by these authors[261] for the reason for the presence of silicon and carbon in the distillate.

Pilot-plant tests with starting material of considerably lower purity also gave excellent results. All data (analyses as well as experimental conditions) are compiled in Table 29b.

For the production of high-purity silicon, according to a suggestion by Bonnier and Charveriat[261], antimony is used as bath metal in re-circulation. It is, however, necessary to sublime it for purification at intervals. The results obtained are also shown in Table 29b.

4.1.5 Distillative separation and refining of rare-earth metals

All rare-earth metals, with the exception of lanthanum and cerium, are to be distilled in a high vacuum, since their purity level hardly ever exceeds 99% after reduction. Distillation is carried out in induction-heated high-vacuum furnaces, in which the conventional oil-diffusion pumps have recently been replaced by ion-getter pumps in order to exclude any possibility of a reaction of the metals with oil vapours during the course of distillation. Usually the pressure attained was around 10^{-5} Torr or lower, but in plants equipped with ion-getter pumps vacua of 5×10^{-8} Torr are now obtained. The distillation temperatures applied depend on the vapour pressures of the individual metals. The condensation temperatures used are so high that the more volatile elements do not condense. Heavy metals of about the same vapour pressure, such as Fe, Ni and Ti, may be retained in the melt by addition of tungsten to the starting material. A two-stage distillation for high-purity refining may also be carried out successfully in the electron-beam furnace[262]. Table 30 lists the technological data for the distillation of the rare-earth metals.

TABLE 30

DATA ON DISTILLATION OF RARE-EARTH METALS

Metal	Temperature of evaporator (°C)	Temperature of condensor (°C)	Distillation rate (g/h)	Yield (%)	Pressure (Torr)	References
Y	2000–2200	1300–1400	65	80	5×10^{-4}	122, 263
Tb	1900–2100	1200–1300	50			122
Dy	1600–1700	900–1000	700	>98		122
Ho	1600–1700	900–1000	400			122
Er	1600–1700	900–1000	150			122
Tm	1400–1500	800–900	1100	>98		122
Lu	2000–2200	1300–1400	25			122
Sc	1550	1050		99.5	$<10^{-3}$	264
Sc	1650–1700				10^{-5}	265

The complete removal of oxygen from the distillate meets with great difficulties. It is assumed that the oxygen is transported in the form of gaseous suboxides of the crucible materials (usually tantalum) or even of the rare-earth metals themselves into the distillate. Moreover the oxygen content of the distillate slightly increases with increasing condenser temperature[263]. The purity levels obtained are shown in Table 31. The main impurity of the distillates is tantalum from the crucibles. In addition to tantalum, silicon and oxygen are also found in considerable quantities if a quartz apparatus is employed for the distillation[267].

According to Vorob'ev[268] the amount of dissolved gases and carbon in vacuum-distilled yttrium is mainly determined by the residual gas pressures during distillation. Distillation under a static vacuum only leads to better results if the metal to be distilled is most carefully degassed and the distillation apparatus thoroughly outgassed.

A number of the rare-earth metals are reduced by magnesium under a protective atmosphere, thus forming magnesium alloys containing up to 80% of the rare-earth metals. These alloys are subsequently distilled in a vacuum for the

TABLE 31

IMPURITY LEVEL OF DISTILLED RARE-EARTH METALS (ppm)

	Ca	Ta	Si	Fe	Cu	Mg	O	C	N	F	H	Al	Ni	Ti	Cr	Zr	Refs.
Y	10	60	10	100	50	5	120	40	25	50	4	60	9	150	50	60	263
Y	10	400	30	150	–	5	300	150	10	100	10						123, 266
Ce	10	100	25	250	–	–	345	200	150	50	10						266
Nd	10	≤500	25	80	–	15	155	65	55	110	–						266
Sm	20	≤500	<50	5	–	–	–	95	20	35	–						266
Gd	50	≤500	≤250	20	–	–	90	15	9	28	–						266
Yb	–	≤500	25	5	–	–	–	35	50	50	–						266
Lu	20	40	25	50	–	–	210	20	12	64	–						266
Sc	<100	<500	b	1100	100	–	2900	280	<100	–	630	1000	b	b	100	b	264

b Not found.

TABLE 32

REDUCTION OF HALIDES OF RARE-EARTH METALS BY MAGNESIUM, FOLLOWED BY VACUUM DISTILLATION

Metal	Reduction			Vacuum distillation				Ref.
	Composition		Yield (%)	Temperature (°C)	Pressure (Torr)	Time (h)	Impurity level of magnesium (%)	
	% reduced metal	% Mg						
Sc			90	1200			<0.01	266
Y	76	24	99	900 950	3×10^{-5}	6 25	0.01	269
Lu			>99	1200			<0.01	266
Th	80	20	94	750 920		2 24	0.01	270

removal of the magnesium. The distillation temperatures applied are chosen so that only the magnesium is distilled off and the rare-earth metal remains as the residue. Relevant data are compiled in Table 32. A second distillation is required for obtaining these rare-earth metals in the high-purity forms.

4.1.6 Distillative separation and refining of the alkali and alkaline-earth metals

Alkali and alkaline-earth metals are required to an ever increasing extent and in extremely high purity for the reduction of metals of high oxygen affinity and for uses in reactor technology. The purity of conventional, commercial metals is in almost all cases inadequate. In many instances only alloys are obtained in the reduction. High-purity metals may then be obtained by distillation or rectification; in these processes working in a vacuum is preferred in order to exclude any reaction with oxygen, nitrogen, or other residual gases present in conventional protective atmospheres. The distillation vessels resemble the retorts used for the magnesium reduction with the modification that a baffle is inserted between raw material and condenser, in order to condense the less volatile substances.

(a) Distillation of the alkali metals

In the distillation of the alkali metals, the distillation of lithium has been the object of a number of investigations. Some of the main impurities, such as K, Na and Mg, show higher vapour pressures than lithium, and others, such as Ca, Al, Si and Fe, show lower vapour pressures, so that a single distillation cannot attain the required separation. Only fractional distillation[271,272] or rectification[271,273] brings about the desired success. Starting with lithium of 99% purity, single distillation results in 99.9% lithium[271], fractional distillation gives a lithium of 99.99%[271]; if lithium of 99.99% purity is used as starting material a metal of 99.9999% purity may be obtained, if the distillation is carried out under most carefully selected conditions[272]. A still higher purity may, however, be achieved by rectification[271,273].

During the straight lithium distillation at about 800°C and at residual gas pressures of between 5×10^{-5} and 10^{-3} Torr, the less volatile elements such as Fe, Si, Cr and Ni remain in the residue and only less than 1 ppm is found in the distillate. The readily volatile elements, such as Na and K, are separated by maintaining the condenser temperature at 200°C or even higher. This results in a lowering of the sodium levels from 1.44 to 0.02–0.06%, and those of potassium by one-half, from 0.03 to 0.015%. In this distillation process the condensation of the lithium takes place in the liquid form[274].

Results of the simple distillation of lithium using the distribution of sodium as a criterion are shown in Fig. 61, in which the residual sodium content is plotted on the ordinate. The starting level of the sodium was about 0.6%[271]. Although the more volatile impurities are easier to separate at lower distillation temperatures (a), the total quantity distilled in unit time falls markedly. Schmidt[272] therefore

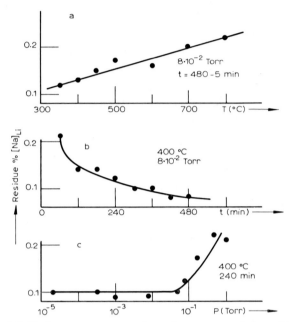

Fig. 61(a). Effect of distillation temperature on the sodium content of the residue. Starting level $C_{Na} = 0.6\%$.
61(b) Effect of distillation time on the sodium content of the residue.
61(c). Effect of residual gas pressure on the sodium content of the residue.
a, b, c = Quantities evaporated constant; a, c = Quantity of residue constant.

suggests 450°C as the minimum temperature. The preferred distillation of the sodium becomes evident also with a prolonged distillation time (b); the sodium level of the residue approaches a limiting value, at which the sodium and the lithium show such evaporation rates that the composition is not further changed. Residual gas pressures above 8×10^{-2} Torr also reduce the evaporation rate in the distillation of lithium, whereas evaporation is unaffected at pressures lower than that (c) (see also Fig. 44). Since during distillation in glass apparatus reactions with the walls are unavoidable[275], the best results are still obtained in stainless-steel apparatus[271-273]. In the production of high-purity lithium[272] the addition of small amounts of B_2O_3 to the starting material resulted in distillates of markedly higher purity.

Sodium of the highest purity containing 0.6–5 ppm oxygen can also only be obtained by fractional distillation. In this process a number of impurities are removed by gettering with a foil of zirconium heated to 650°C[275a].

(b) Distillation of alkaline earth metals

The distillation of calcium is carried out preferably at 10^{-3} Torr and 900°C with the condenser at a temperature of 300–400°C[276]. It was found that magnesium evaporates in the initial stages of the distillation at a high rate, *i.e.*, it enriches

TABLE 33

DATA ON VARIOUS PROCESSES FOR CALCIUM PRODUCTION (IMPURITY LEVELS IN ppm)

Impurity	Electrolysis after Rathenau	Subsequent vacuum distillation	Ca–Cu-Electrolysis and vacuum distillation	Al-reduced calcium	Subsequent distillation	Double distillation
Mg	<200	<100	<100	5000	300–3000	<3
Si	<500	<100	<10	200	<1	<2
Cu	<50	<10	<10		<3	<1
Fe	<3000	<40	<20	100	5	<3
Al	<3000	<100	<100	1000	<3	<2
Mn	<70	<20	<30	500	45	50
N	<300	<50	<30	600	175	<10
Cl	<16000	<8000	<2000	Max. traces	Max. traces	Max. traces
O				3000		<500
Power consumption kWh/kg Ca	70–80	12	55			
Yield	50%		80%			35–40%
Reference	279	279	279	183, 276	183, 277	183, 277

the first fractions[277]. On the other hand, magnesium may be sublimed to a large extent also from solid calcium at 600°C[276]. The effect of nitrogen in the residual gas atmosphere and in the starting material on the nitrogen level in the distilled metal was studied by Decroly et al.[278]. They obtained by multistage distillation at 5×10^{-4} Torr minimum levels of nitrogen of about 40 ppm, but by sublimation about 10 ppm was attained.

Multistage distillation is carried out preferably at 850–950°C in a helium atmosphere of 2 Torr, this procedure leading to calcium of 99.95% purity[183,277]. Nitrogen and magnesium levels in particular could be considerably reduced, as may be seen from Table 33. Akerman and Orman[181] were able to show that the impurity levels could be kept particularly low if high-purity starting materials were used for the reduction.

The electrochemical "Kombinat" at Bitterfeld perfected, after the last world war, the so-called Rathenau process for the electrolysis of $CaCl_2$ for the winning of calcium by precipitating calcium in a liquid copper cathode[279,280]. According to the constitutional diagram of the calcium–copper system alloys of between 30 and 60% calcium are still liquid at 700°C, i.e. at a temperature at which the calcium vapour pressure is still very low. Using a Ca–Cu alloy enriched to about 60% Ca, this metal is distilled off at 1050°C and 0.1–0.2 Torr until the calcium level of the alloy has been reduced to about 30%. This low-calcium alloy is then used as new cathode material in the electrolysis. In the initial stages of the distillation evaporated calcium getters any remaining oxygen or nitrogen. A diagram of the distillation retort is shown in Fig. 62. In the centre of the retort Raschig rings of iron are placed on a perforated plate for the condensation of entrained or evaporated particles ($T = 950°C$). In the condenser compartment a temperature of about 850°C is maintained, but the condenser itself is water-cooled. The precipitation of the metal takes the shape of a so-called "calcium regulus". Not only the retort, but also the furnace chamber, is kept under vacuum in order to prevent destruction of the retort by the ambient atmosphere. An industrial plant consisting of 28 furnaces for an annual production of 480 tons calcium is shown in Fig. 63.

In this combined electrolysis–distillation process a metal of increased purity is obtained, and at the same time the power requirements are lower, as may be seen from Table 33. More recently suggestions for the continuous distillation of the calcium–copper alloy have been put forward[281,282].

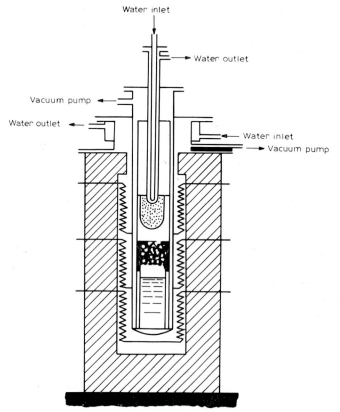

Fig. 62. Vacuum retort made of steel for the distillation of calcium from a calcium–copper alloy[279].

The large-scale, industrial production of silicon–calcium and aluminium–calcium alloys is feasible by electrothermic or aluminothermic processes. In both cases strong binding forces between the metals (formation of intermetallic compounds) exist, which make the vacuum distillation of these alloys probable only

Fig. 63. Production plant for the distillation of calcium from calcium–copper alloys[279].

within limits. According to suggestions made by Bonnier and Andrieux[283,284], a 64:36 mixture of aluminium–17% calcium and silicon–30.5% calcium alloys is distilled at a temperature of 1300–1400°C and a pressure of 10^{-3}–10^{-4} Torr. Calcium is distilled off, while a pool of molten residue consisting of a silicon–aluminium alloy is formed. As may be seen from Fig. 64, the calcium distillate yield amounts at 1300°C to 96.5%; the condensed distillate contains 1.6% aluminium and 0.17% silicon. The yield of silicon–calcium [T_s] and aluminium–calcium [T_a] alloys distilled on their own as well as the impurity levels of the distillate are also plotted in this diagram, and as may be seen, the results are markedly poorer[284].

The distillation of calcium, manganese or aluminium from silicides is facilitated if the vacuum distillation is carried out after an addition of carbon, which promotes the formation of silicon carbide[285]. In the publication of Bonnier and Andrieux[284] already mentioned a process is described for the production of calcium at a temperature of 1150–1250°C and a pressure of 10^{-4} Torr from silicon–calcium or aluminium–calcium alloys with nickel or cobalt additions. In this process the residue consists of alloys which may be used for the preparation of nickel or cobalt catalysis by the Raney or Fischer–Meyer processes.

The distillation of the other alkaline earth metals, strontium and barium, is possible in the same way, but it is of no technological importance because of the low demand for these metals. Here too the difficulty met with is the elimination of oxygen and nitrogen. Both impurity levels may be lowered in barium by melting the metal in a titanium crucible in a vacuum[286].

References pp. 321–335

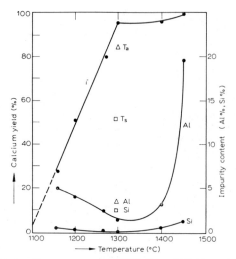

Fig. 64. Data of a simultaneous vacuum distillation of Si–Ca and Al–Ca alloys[284].
 T Yield of calcium in simultaneous distillation
 T_a Yield of calcium in the distillation of the Al–Ca alloy
 T_s Yield of calcium in the distillation of the Si–Ca alloy
 Al, Si respective contents in the distillate.

(c) Distillation and refining of beryllium

The preparation of high-purity beryllium has become of great importance in reactor technology, particularly on account of its low capture cross-section for neutrons, its high strength and low density. Impurities particularly impair the first two properties named. Although distillation of beryllium is possible, the vapour pressures of the most detrimental impurities, such as manganese, aluminium and silicon, are very similar to that of beryllium. In order to obtain technologically valuable distillation rates (250 mg/cm²h) temperatures as high as about 1350°C are required. The condensation of the beryllium takes place over the temperature range 900–1300°C; beryllium of the highest purity condenses at temperatures between 1050° and 1100°C[287–292]. At these temperatures only relatively small quantities of Mn, Na, Mg, Zn, Ca and Al are condensed. The less volatile elements, such as Fe, Cu, Ni, W, Nb and Ta, are held back in the residual melt, which may be distilled up to as much as 85% before higher levels of these impurities are detected in the distillate[287,288]. Despite the use of very high vacua (better than 10^{-6} Torr), relatively high oxygen and carbon contents are found in the beryllium. These are attributed to a codistillation of beryllium suboxide from the crucible and to reactions with residual hydrocarbon gases[288]. Finally, the leak rate of the distillation furnace may have an effect as was shown by Hordon and Hess[293]. The oxygen contents of the distillate are usually much lowered if high distillation rates are employed. Sinelnikov et al.[294] were able to show how the oxygen level of the distillate depends on the residual gas pressure under otherwise constant conditions. At 10^{-3} Torr the oxygen content was 0.125%, at 10^{-4} Torr 0.05%,

TABLE 34
REFINING EFFECT BY VACUUM DISTILLATION OF BERYLLIUM (QUANTITIES IN PPM)

	Starting material	After One distillation	After Two distillation	Starting material	Single distillation Without baffles in the column	Single distillation With baffles in the column	Starting material A	Starting material B	Zone-refined A	Zone-refined B	After distillation A	After distillation B	Distillation under static vacuum
Na							<40	<40	4	1	≤0.2	≤1	
K									0.06	0.3	0.3	0.3	
Mg							<4	5.2					40
Ca							<7	<7	0.6	2.0	—	0.03	
Al	1000	20	10	10	7	5	<10	10	0.6	1	—	<0.2	60
Si	3000	100	30	150	10	10	<5	26	0.6	0.3	≤0.3	<0.3	
Sn				1000	40	<10							
Pb	50	10	10	300	100	20							
				30	2	<1							
Ti	13	10	10	50	3	b							
Cr	200	40	—	10	5	b	<2	<2	0.025	0.05	0.04	0.01–0.1	20
Mn	500	20	10	200	40	10	<0.8	<0.8	0.2	<0.07	0.1	0.1	
Fe	6500	60	12	500	20	5	2.4	2.4	3.5	1.0	0.04	0.02	100
Co	10	0.5	0.5	6000	60	<10							
Ni	300	10	10	10	3	b	1.5	<4	5	1.5	0.03–0.5	0.01–0.06	10
Cu	250	5	5	300	10	5	<0.7	1.7	0.7	0.5	0.2	0.6	20
Zn				250	5	5	<11	<11	0.02	0.08	0.01	0.005	<10
C				10	<10	<10			5	10	2	2	
N									<0.02	≤0.02	0.3	<0.2	
O	100	10	10	100	10	200	1020	1730	10	8	3.0	5.0	10
F						10			≤0.6	0.3	0.3	<0.2	
S						400			0.2	0.2	—	<0.2	
Cl				10	<10	<10			1.5	2.0	0.4	0.1	
Total									32.9	28.6	7.8	10.3	300
Reference	287				288					295			296

b Not found.

TABLE 34 (continued)

	Starting material	Double sublimed	Starting material	Distilled and sublimed	Starting material	Sublimed	Starting material	Sublimed
Na	<50		0.6	0.2			5–10	0.04–1
K							0.4–0.25	0.05–1
Mg							3–5	0.05–1
Ca					30	0.4		
Al	<120	5–20	15	0.14	165	3	10–20	0.3–3
Si	20	2	22	<15	100	1	4	0.2–3
Sn								
Pb								
Ti								
Cr	3	1	3	0.2	25	2	0.5–0.8	0.1–0.4
Mn	3	2	5	1.5	15	10	1.2 2	0.2–1
Fe	49	3	25	4	210	0.7	2.5–4	0.03–0.3
Co								
Ni	29	2.3	35	0.6	15	0.15	1.2–1.5	0.05–0.3
Cu	<5	1.5	3.2	0.4	2	0.1	0.4–0.15	0.07–0.27
Zn								
C	58		36	57				
N								
O								
F								
S								
Cl								
Reference	297		297		298		298	

TABLE 34 (continued)

	Starting material	6× Electron beam melted	After distillation Residue	After distillation Distillate	Be-Powder compact starting material	Distillate	Starting material	4× Induction-melted	After Distillation distillate
Na		0.5	<0.1	0.3		1.0	59	8	1
K									
Mg	<3	0.055	0.25	0.06	30	0.4	4	0.03	0.3
Ca	<5	0.2	0.06	0.05					
Al		5	0.8	0.07	165	3	10	4	2.2
Si		5	12.6	0.5	100	1	5.0	5.0	1.2
Sn									
Pb									
Ti	<1	0.3	2.6	0.005	25		0.3		
Cr	<1	1.8	13.3	0.05	15	2	0.3	2.0	0.6
Mn	2	3	3.6	3.5		10			
Fe	3.2	3.4	16	0.045	210	0.7	1.6	4.5	0.15
Co									
Ni	4.5	6.5	32.5	<0.005	15	0.15	5.0	7.5	0.5
Cu	1	9.0	14	0.28	20	0.1	1.5	1.6	0.25
Zn	2	0.17	0.03	0.15		0.08	4.1	0.2	0.05
C	95	30	40	17.5	675	13	73	9	4
N		1.25	3.3	0.1	65	3	a	2	0.5
O	108	25	50	35	3300	100	216	10	8
F									
S									
Cl	135	2.1	0.4	0.25	12	50	150	3	0.3
Total					4600	235	523	~50	~20
Reference		209			208			208	

a Not known.

at 10^{-5} Torr and better 0.03% was obtained as limiting value. This high limiting value must be attributed to reactions with the BeO crucible. In agreement with other authors, they too consider the presence of residual hydrocarbon gases to be the cause of the relatively high carbon contents.

The results of the distillation process are compiled in Table 34. The hardness could be lowered from 450 kgf/mm^2 to a minimum of 130 kgf/mm^2.

Considerably improved results of distillations may be obtained if ion-getter pumps[295] or sublimation getter pumps combined with cryo-pumps[297] or turbomolecular pumps are employed instead of the conventional oil diffusion pumps. The results obtained under static vacuum conditions were markedly poorer[296] (see also Table 34). Furthermore, previous zone refining has a beneficial effect on the reduction of the oxygen content, as BeO is soluble in solid beryllium only in traces (see also Table 34). In this way, the evaporation of beryllium suboxide is almost excluded in the subsequent distillation from a water-cooled copper crucible[295].

Bunshah and Juntz[208,209] have also shown the decisive effect both of a treatment of the beryllium before the distillation and of the working conditions during the distillation on the purity of the distillate. The levels of Na, K, Cl, Mg, Mn, F and Al are greatly reduced by evaporation during repeated melting in water-cooled copper crucibles (induction-heated[208] or by electron beams[209]) in a high vacuum of about 10^{-6}–10^{-7} Torr. Carbon and oxygen are precipitated as Be_2C or BeO during solidification and are removed mechanically. The less volatile impurities are enriched in the melt during this operation, as the evaporation losses of beryllium may be quite considerable.

In the subsequent distillation, condensation was carried out on a tantalum condenser. The deposition always took place in such a manner that the distilled metal could be fabricated directly into sheet.

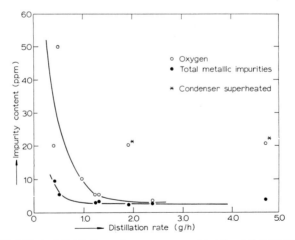

Fig. 65. Effect of distillation rate on oxygen and total metal contents[295].

While during operation in an induction furnace the beryllium could, for technological reasons, be only sublimed (1200°C) from the solid material, high distillation rates (up to 13.5 g/min) could be obtained in the electron-beam furnace at distillation temperatures of 1400–1500°C. In order to keep the oxygen and carbon levels low, the distillation chamber was made very small and was specially outgassed before the distillation started.

The impurity level seems also to depend on the rate of distillation, as may be derived from Fig. 65. A minimum distillation rate of about 2 g/h should be maintained for obtaining optimum quality of the distillate. According to more recent investigations, the metallic impurities may be reduced to a total of less than 10 ppm[295, 297, 298]. However, with regard to the levels of the non-metals there still exist many uncertainties[297]; in many cases no data at all are available[287, 296–298].

4.1.7 Distillative separation and refining of aluminium, gallium and indium
(a) aluminium-containing alloys

The industrial, large-scale production of aluminium is carried out almost exclusively by a two-stage process (alumina production–molten salt electrolysis) which has only been modified in technological details during the last seven decades. Although the initial very high power consumption of about 30 kWh/kg of aluminium in the electrolysis could be reduced to less than one-half that value, numerous attempts for a carbothermic reduction of aluminium-containing raw materials in the electric furnace were made. There is, however, a risk of the formation of aluminium carbide in this process, which may almost be suppressed by simultaneous reduction of silicon. In most cases, therefore, Al–Si or Al–Si–Fe alloys containing 30–60% Al and 70–40% Si were produced[299]. The capacity of these pilot plants was not greater than 10000 ton/year. In no instance was it possible to obtain commercially valuable alloys directly, or even aluminium of commercial purity.

From the electrothermically produced Si–Al–Fe alloys, aluminium may be

TABLE 35

DATA ON VACUUM-DISTILLATION OF ALUMINIUM–SILICON ALLOYS

Starting alloys (wt. %)				Residue (wt. %)				Distillate (wt. %)			Remarks	Ref.
Al	Si	Fe	Balance	Al	Si	Fe	Balance	Al	Si	Fe		
63.8	32.7	3.5	–	10	82.4	7.6	–	95.5	3.5	1.0	1250°C Dist. temp.	205
45	35	a	a	a	a	a	a	94	5.4	0.8	580°C Extraction temp. / 620°C Dist. temp.	201
60	33	3	4	a	a	a	a	99.7	0.19	0.08	450°C Extraction temp.	300
60	33	3	4	a	a	a	a	86.6	13	0.4	650°C Extraction temp.	300

a Not known.

separated by distillation, for which a vacuum of at least 5×10^{-2} Torr at a temperature of about 1250°C is necessary. In this process aluminium is distilled off to the extent of 98 %, iron is distilled off only to 15 % and silicon to 7 %, as is shown by the analytical data of Table 35[205].

Spendlove[201] suggests extraction of the aluminium by zinc in a vacuum at 580°C from the Si–Al–Fe alloys produced electrothermically. From the Zn–Al alloy the zinc is distilled off at 620°C and condensed at 425°C. It is then recirculated in the process. The aluminium so produced contains about 94 % Al, 0.8 % Fe and 5.4 % Si, and corresponds almost exactly to the above quoted distillate. However, substantially lower working temperatures are required.

Pechiney and Ugine carried out similar experiments on a pilot-plant scale. They melted these Si–Al–Fe alloys together with zinc. Depending on the composition and melting temperature a solid Si–Fe-rich and a liquid Zn–Al-rich phase are formed. The zinc content of the latter varies between 65 and 85 % according to the extraction temperature, 650° and 450°C, respectively. After the Zn–Al phase has been separated from the other phase, it is distilled in a vacuum. The residue consists of aluminium of commercial purity (99.7 % Al, 0.19 % Si, 0.08 % Fe) if the extraction takes place at 450°C; and of an Al–Si alloy (86.6 % Al, 13 % Si, 0.4 % Fe) if the extraction temperature is raised to 650°C.

The zinc must also be distilled off from the other, solid Fe–Si-rich phase before this can be used, for instance, as reduction metal. This process has not attained technological importance, probably bacause of economic considerations. The aluminium yield is given as 83 %[300]. The results of these experiments are once again summarized in Table 35.

The greatest prospects for technical realization were shown by refining of the electrothermically produced aluminium alloys through subchloride distillation. Alcan started operation for the refining of Al–Fe–Si alloys by this process in 1961, with an annual capacity of 7000 tons of aluminium, but the plant ceased operation in 1967 for economic reasons[301]. Also, Ginsberg and Sparwald[302] showed that there is no chance for lowering the power consumption of that process.

In the subchloride process, $AlCl_3$ vapour acts on the impure Al alloy and according to:

$$2\{Al\} + (AlCl_3) \underset{600°C}{\overset{1200°C}{\rightleftharpoons}} 3(AlCl) \tag{109}$$

forms volatile AlCl, which decomposes on cooling. With increasing temperatures and lowered pressures the equilibrium is shifted to the right, and so the process for the formation of AlCl is carried out at higher temperatures (1100–1200°C) and lower pressures than the decomposition (about 600°C) (see also Fig. 68). Manganese, iron, silicon and titanium distil off only to a small extent and may even—at least partly—be separated by an intermediate rectification column.

A diagram depicting the plant of the Alcan process is shown in Fig. 66.

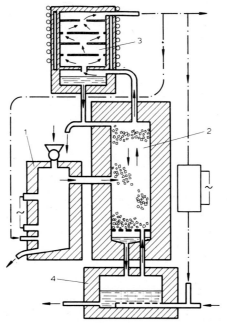

Fig. 66. Diagram of the subchloride plant of Alcan[300].
1 Reactor; 2 Rectification column, filled with alumina rings and cubes; 3 AlCl decomposer; 4 Furnace for collecting impurities and for the generation of additional heat.

The equilibria of the reaction according to eqn. (109) have frequently been studied, in many cases the continuous flow method having been employed. The results of these studies are plotted in Fig. 67.

The equilibrium values given by Weiss et al.[303] are regarded as too high by Gross[304] and by Ginsberg and Sparwald[302], as the gaseous diffusion of aluminium chlorides can be superimposed on the flow at slow flow rates.

The theoretical degree of reaction, α (quantity of $AlCl_3$ reacted), may be calculated from

$$K = \frac{27\alpha^3 p^2}{(1+2\alpha)^2(1-\alpha)};$$

α is, as may be seen from Fig. 68, dependent on both pressure and temperature. In this diagram are also plotted the optimum working conditions according to Stroup[299].

For pressures below 0.1 Torr the reaction is governed, according to Hirschwald and Knacke[305], only by the rate of the chemical reaction; at higher pressures the reaction rate decreases owing to diffusion in the gaseous phase exercising a retarding effect. Ginsberg and Sparwald[302] also interpret the results of their experiments with argon as the carrier gas ($P_{Total} = 1$ atm) with the assumption that the diffusion of the $AlCl_3$ to Al is rate-determining. However, their reaction rate

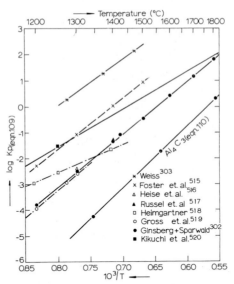

Fig. 67. Plots of the equilibrium constant of the reaction
$$2\{Al\}+(AlCl_3) = 3(AlCl)$$
in relation to the temperature according to various authors (Ginsberg and Sparwald, ref. 302, and others).

is substantially lower than that of the first-named authors.

While relatively favourable conditions exist for aluminium of commercial purity (solid lines), the equilibria are shifted to higher operating temperatures when a 50:50 Fe–Al alloy (broken line) or Al_4C_3 (dot–dash line) is used as the starting material (Fig. 68). Despite this shortcoming, the aluminium production from alumina via Al_4C_3 has gained a certain interest.

In this process corundum is obtained by selective reduction of Fe, Ti and Si oxides by carbon from Al-containing raw materials. The corundum is subsequently further reduced and evaporating Al, Al_2O, and CO react with excess carbon to form aluminium carbide. In a further step this carbide is decomposed at about 2000°C and a residual gas pressure of about 0.5 Torr. These experiments have recently advanced to a stage that enabled the Pechiney–Ugine Combine to put into operation an industrial pilot plant of an annual capacity of 3000–5000 tons of aluminium[299, 300, 306].

Ginsberg and Sparwald[302] carried out studies on the equilibria involved in the reaction:

$$\tfrac{1}{2}\langle Al_4C_3\rangle +(AlCl_3) = \tfrac{3}{2}\langle C\rangle +3(AlCl) \tag{110}$$

The results show that this reaction is shifted to temperatures about 200°C higher than those in the reaction with aluminium (see Fig. 68, dot–dash line).

A combination of carbothermic reduction and subchloride extraction in one stage is considered by Gross[304] impossible owing to difficulties in the condensation.

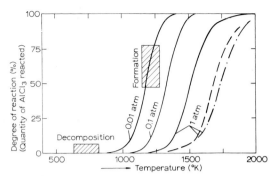

Fig. 68. Effect of pressure and temperature on the extent of the reaction (quantity of $AlCl_3$ reacted) during the reaction with: Al (solid line); FeAl (broken line); Al_4C_3 (dot–dash line).

(b) Distillation of gallium and indium

High-purity gallium and indium are required for the manufacture of semiconductors[259]. As the vapour pressures of both metals are still very low far above their melting points, distillative separation from various volatile elements is feasible.

Rosi et al.[307] refined gallium in a quartz crucible at a temperature of 650°C and a pressure of 10^{-7} Torr. The evaporating impurities are condensed or frozen out immediately above the melt. Such a vacuum treatment may take seven days or even longer. Comparison tests showed that distillation at this low temperature and prolonged periods of time is more effective than at higher temperatures (800–1000°C) and a somewhat higher pressure. In such a vacuum treatment particularly Hg and Zn, then Mg and Ca, and finally Cu, Pb, Ag, Fe and Si, could be removed. The choice of the crucible is still a problem as, according to Foster and Kramer[308], SiO_2 and Ga react in a vacuum, forming volatile Ga_2O and Si, the latter dissolving in the gallium. It appears so far that graphite crucibles are more suitable for this purpose.

In a similar manner, Hulme and Mullin[309] distilled off, in particular, cadmium and zinc from indium at a temperature of 700°C and a pressure lower than 10^{-5} Torr over a period of 24 hours. Sulphur, selenium and tellurium could not be separated, probably because of their affinity to indium. Effer[310] kept indium in the liquid state at a temperature of 800°C and a pressure of 10^{-6} Torr for up to 48 hours in order to distil off impurities. Optimum values were obtained from an indium prerefined by amalgam electrolysis.

Sergeeva and Shtrum[311] distilled indium directly. The distillation was carried out also in quartz vessels. A first fraction, about 4% of the total quantity, was distilled off at 825–925°C. It contained the most volatile elements, such as Zn, Cd, Pb, etc. The main quantity of the indium was distilled at 1025°C and represented the fraction of the highest purity, but even this was still contaminated by Si and O, so that a refining electrolysis had to follow. The residue from the distillation—about 20% of the starting material—contained the less volatile elements, such as copper, iron, etc.

References pp. 321–335

Weisberg et al.[260] report on obtaining improved results by fractional distillation at a temperature of 900°C and a pressure of 10^{-7} Torr. About 75% of the original charge was distilled into three fractions about equal in quantity. The total impurity level of the last two fractions was about 9 ppm for an impurity content of 123 ppm of the starting charge.

4.1.8 Distillative separation of the iron-group metals

Owing to their high vapour pressures, from the iron-group metals manganese and chromium in particular may be refined by vacuum distillation.

The distillation of ferro-manganese is carried out at about 1300–1400°C, whereas condensation takes place between 700° and 800°C. Analytical data are compiled in Table 36. As may be seen from the data, a manganese of 99.96% purity is obtainable in a single distillation. The distillate yield is about 80%.

The distillation of chromium is carried out preferably at pressures lower than 10^{-5} Torr and at a temperature of 1200–1500°C with condenser temperatures of 950–1200°C. The metals of the iron group co-distil almost in the same proportion; a marked reduction of the C, Si and Al contents can only be attained by addition of tungsten to the starting metal. By distillation at 10^{-7} Torr, the microhardness of the distilled chromium is lowered to about 120 kgf/mm² compared with 180 kgf/mm² for a metal distilled at 10^{-5} Torr. If chromium is stored in air, it takes up large quantities of gases during the first three days, as may be seen from Table 36.

The possibilities of a manganese reduction, volatilization and condensation were also considered, and relevant experiments were carried out. In a volatilization in a vacuum or at ambient pressure, a yield of over 95% was obtained, but despite the high yield the process was economically not promising[314].

TABLE 36
DISTILLATION OF IRON-GROUP METALS

	Starting material	Mn	Fe	C	Si	P	Al	Ni	Ca					Ref.
Manganese	Starting material	73	18	7	2	0.22	0.1	0.001	0.5				wt. %	312
	Distillate	Balance	40	a	100	60	50	20	16				ppm	
Chromium	Starting material		0.75		0.17		0.6						wt. %	313
	Distillate		6500		10		30						ppm	
	(W-Addition)									H	N	O		
Chromium	Electrolytic	5	160	300	300		100	70		5	130	210	ppm	
	Chromium													
	Distillate	5	120	50	10		20	60		14	5	3	ppm	313
	(W-Addition)	After three days								31	70	10	ppm	

a Not known.

4.1.9 Distillation of selenium and tellurium

Refining of selenium by distillation involves certain difficulties, as the impurities present have higher (Hg, S), about the same (As, HgSe, Cd), or else lower vapour pressures. Most selenides (PbSe, CdSe, ZnSe) have relatively low vapour pressures.

Mercury may evaporate in the elementary form or as selenide, and condenses at two temperature levels or in two zones, *i.e.* at 210–220°C and 290–300°C. A relatively mercury-free selenium can therefore be condensed only between 240° and 270°C.

These considerations were confirmed by the results of experiments with fractional condensation. In the distillation of selenium at between 370° and 430°C and a residual gas pressure of about 0.2 Torr, components of very low vapour pressures are deposited on inserted baffles. The phases of the highest purity condense between 240° and 270°C. If the temperatures are lower, enrichment in mercury is observed[315]. An exhaustive review of the production of high-

TABLE 37

MULTIPLE-DISTILLATION OF SELENIUM

Distillation apparatus	Distillation temperature (°C)	Condenser temperature (°C)	Residual gas pressure (Torr)	Reference
Glass	450		10	317
Quartz	260	20–60	10^{-3}	318
Glass	330			319
Quartz	400		5	316
	400–420		5–7	320
	300–320		5×10^{-2}	321
Quartz-steel	370–430	240–270	0.2	315
Glass	350	~240	2×10^{-2}	254

purity selenium by distillation in a vacuum and at ambient pressure is given by Pleteneva[316]. The conditions for distillation of selenium are compiled in Table 37.

In a more recent publication by Richter and Müller[254], selenium distillation is effected in a plant similar to that shown in Fig. 58 for the distillation of cadmium with the modification that two condensing devices are provided in series. As expected, Ag, Cu, Ni, Pb and Sn were enriched in the residue. The results of the purification may be seen from the following comparison of the impurity levels, in ppm:

	Ag	As	Bi	Cd	Cl	Cu	Ni	Pb	S	Sb	Sn	Tl
before	0.2–1.0	<1	<0.9	<1	7–14	0.7–1.9	0.5–1.0	0.7–2.7	2–20	<1.6	0.4–0.6	<1.6
after	0.02–0.08	<1	0.05–0.08	<0.6	<2	0.2–0.4	<0.4	0.1–0.7	a	<0.4	0.2–0.3	<0.2

a Not known

References pp. 321–335

TABLE 38
RESULTS OF DISTILLATION EXPERIMENTS ON TELLURIUM[322]

Remarks	Se	Cu	Sb	Bi	Pb	Fe	Si	Al	Mg	Au	Na	Vacuum (Torr)
Starting material Approx. 20 g at	10	50	3	2	20	2	30	3	2	1	50	—
500° C distilled 23 min. 95% yield	1	2	b	<1	1	<1	2	2	<1	b	b	1×10^{-3}
Starting material 1 kg at 500° C distilled	60	70	>50	<1	>100	>100	>100	>100	100	–	50	
1–2 h, 35–57% yield	<1	<1	b	b	5	1	1	1	<1	b	b	3×10^{-2}
at 550° C distilled 2.5h, 80% yield	<1	2	b	b	30	1	1	1	<1	b	b	3×10^{-2}

b Not found.

Smirnov and Bibenina[322] carried out experiments on the vacuum distillation of tellurium on a laboratory and pilot-plant scale with 20–30 g and 1 kg starting charges, respectively. The working temperature in the laboratory experiments varied between 400° and 600°C. During a reaction time of up to 60 min between 30% and more than 99% of the starting charge was distilled. The experiments in the pilot plant were carried out at 440–550°C and 35–80% of the starting material was distilled over a period of 1–3 hours. The optimum results are shown in Table 38.

4.1.10 Distillation of heavy non-ferrous metal scrap

The quantities available of ungraded and contaminated heavy non-ferrous alloy scrap, in particular brass, are extremely large. This scrap is used for the recovery of the pure metals. So far the industrial process employed has almost exclusively been melting in hot-blast converters or shaft furnaces. Highly impure copper and, in the filters, a mixture of tin, zinc and lead oxides are obtained by this process, while the various losses of metal are very high.

Attempts, particularly in the Soviet Union, have been made to separate the zinc from brass scrap by vacuum distillation or vacuum sublimation and to recover it in the metallic form. For a sublimation at 800–900°C very fine chips must be used if the zinc content is to be reduced to a residual level below 1% in acceptable times. If the chips have a thickness of 2 mm sublimation no longer gives the desired effect[323,324]. The distillation is carried out at between 1100° and 1200°C. In this operation the pressure should at the beginning be not lower than about 200 Torr in order to prevent too heavy evaporation of the zinc. When the zinc content is reduced to less than 10%, the pressure is greatly lowered and the temperature raised further until the ultimate temperature of 1200°C is reached. Under these conditions the evaporation of copper is still relatively small, whereas lead co-distils up to about 50%. The high energy requirement for the zinc distillation makes internal heating in ceramic-lined vacuum furnaces imperative. After sublimation at about 800°C the sublimed zinc contains less than 0.2% lead, less than 0.024% tin and less than 0.01% copper. During a distillation at 1200°C, on the other hand, up to 5% copper and about 50% of the original lead are carried over, while tin is found only in traces in the distillate. The purity of the zinc in this case varies between 90% and 95%[325]. Table 39 shows a number of experimental data.

TABLE 39

DATA ON DISTILLATION OF HEAVY-METAL, NON-FERROUS ALLOY SCRAP

Composition of scrap (%)				Temperature (°C)	Pressure (Torr)	Time (min)	Residue (%)				Refs.
Cu	Zn	Pb	Sn				Cu	Zn	Pb	Sn	
69.2	30.8	—	—	1200	0.3	180	99.4	0.4	—	—	323, 325
79.5	16.2	0.5	0.2	1200	0.3	60	95	0.2	0.3		
82.9	14.0	2.0	0.2	1150	0.3	180	95.5	0.3	0.8	0.3	326
~60	~40			1200	5			0.25–0.5			324
				1150	1.5			0.7			
				1100	0.1	300		1.17			
	23.6	2.1	1.24	1200	10	240		0.3–0.5	1.2		327

Frade et al.[328,329] studied the rate of sublimation of zinc from alpha-brass at between 450° and 900°C and a pressure of 10^{-6} Torr. At the lower end of this temperature range mainly grain-boundary diffusion was found, which, however, becomes negligibly small compared with volume diffusion at temperatures above 650°C. If, on the other hand, such alloying elements as tin or arsenic are contained in the brass, grain-boundary diffusion exercises a marked influence even at higher temperatures. Such effects are not so pronounced in aluminium–zinc alloys.

In the electrochemical "Kombinat" at Bitterfeld, a vacuum distillation process for the recovery of light-metal scrap (Al–Mg–Zn alloys) was developed after the last world war. In this process zinc and magnesium are simultaneously distilled. For further distillation, the amount of silicon necessary for the formation of Mg_2Si is added to the distillate in the form of aluminium–silicon. First the zinc is evaporated at 650°C and a residual gas pressure of 0.1 Torr. The distilled zinc still contains about 2% magnesium. Then the magnesium is distilled off at 950–1000°C, and it contains less than 5% zinc after this distillation. The Al–Si is recirculated[330].

This work was extended on an experimental basis to Mg–Al, Zn–Al and Mg–Zn–Al alloys. Distillation at 850°C and a residual gas pressure of 0.2 Torr results in zinc and magnesium yields of 98–99% in the condensate, while the aluminium remains in the residue as high-purity metal with less than 0.05% Zn, Mg. In pure Zn–Al alloys addition of magnesium increases the volatility of the zinc[331].

References pp. 321–335

4.2 Deoxidation and decarburization of non-ferrous metals in a vacuum

4.2.1 Decarburization of ferro-chromium in a vacuum

Carbothermic reduction of chromium–iron ores to high-percentage ferro-chromium leads only to a ferro-chromium with high carbon content owing to the high affinity of chromium for carbon. In most cases the decarburization of this ferro-chromium is carried out by a silicothermic exchange process, or by the so-called Perrin process in several stages at ambient pressure, although the decarburization of ferro-chromium in a vacuum, besides the carbothermic production of tantalum by v. Bolton, is one of the oldest vacuum-metallurgical processes in use on an industrial scale.

Rohn more than 40 years ago decarburized liquid ferro-chromium in a vacuum by addition of iron oxide or chromium ore in what was then the Heraeus Vakuumschmelze AG at Hanau. This procedure was successful in lowering the carbon levels from 1–3% to 0.04%. Figure 69 shows two of the vacuum induction furnaces used at that time, which had a capacity of 4 tons. These furnaces even at that time were suitable for pouring in a vacuum.

Extensive studies on the decarburization of ferro-chromium in the liquid state were started also in the Soviet Union about 30 years ago. These led to the establishment of a technologically very interesting process, in which about 400 kg ferro-chromium is decarburized in about 1 hour from 0.06 to 0.015% carbon in vacuum induction furnaces at a temperature of almost 1700°C and a residual

Fig. 69. Two vacuum induction furnaces of 4 ton capacity each for the decarburization of ferro-chromium, which were installed by Rohn more than 40 years ago[332].

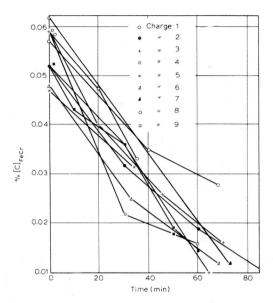

Fig. 70. Decarburization of ferro-chromium in the vacuum induction furnace at almost 1700°C and a residual gas pressure of 1–2 Torr[333, 334].

gas pressure of 1–2 Torr. The change in carbon contents of such industrial melts with time is plotted in Fig. 70. Diffusion of the dissolved carbon and oxygen through the phase boundary of the melt should be regarded as the rate-determining step in agreement with theoretical considerations. The working conditions were selected so that the losses in chromium owing to evaporation do not exceed about 3%.

Recently, on the basis of experiments carried out in a 200-kg vacuum induction furnace, decarburization of ferro-chromium was studied in a 1.2-ton furnace[335]. At 1750–1800°C in 1½–2 hours by continuous addition of Fe_2O_3, the carbon level was lowered from less than 0.5% carbon to 0.01–0.02%. The lowest limit of the carbon concentration achievable is probably at about 0.004% carbon. Owing to the relatively high vapour pressure of chromium, a vacuum of only about 0.1 Torr is applied; with this arrangement the losses owing to evaporation vary between 4 and 6%. The vacuum induction furnace plant employed for this purpose is shown in Fig. 71.

During this vacuum treatment the phosphorus, nitrogen and tin contents are not changed, but lead, arsenic, zinc, bismuth and copper are completely evaporated; cadmium remains in traces only, so that a high-quality ferro-chromium is produced[333–335]. Silicon may be removed to the extent of 30–50% by formation of gaseous SiO. The power requirement is about 500 kWh per ton of ferro-chromium[333, 334], and 1750 kWh per ton of ferro-chromium, including melting down[335].

References pp. 321–335

Fig. 71. 1.2 ton vacuum induction furnace for the experimental decarburization of ferro-chromium shown in the open position. The plant has a nominal rating of 900 kW. (Courtesy Balzers–Elphiac)

An attempt was also made to add oxygen, not in the form of chromium oxide or iron oxide, but by blowing gaseous oxygen in a vacuum on to or into the ferrochromium at 1540–1800°C. In this process ferro-chromium with an initial carbon content of 4–6.5% was blown with oxygen at atmospheric pressure until the carbon level was reduced to 1%. Blowing was then continued under vacuum (usually 200–350 Torr). Although ultimate carbon contents of down to 0.03% may be obtained, in most cases there remained about 0.4% carbon. The yield in metal amounted only to 35–70%; the yield in chromium metal was even lower, at 30–65%. In this process the silicon content is also lowered, from over 1% to an average of 0.3%. Chromium and silicon are slagged and evaporated[336].

In addition, the sulphur content of liquid ferro-chromium may also be reduced by a vacuum treatment. At a residual gas pressure of 2–9 Torr, the sulphur level of the metal is lowered by 28–85% in a period of time of 5–25 min[337]. Desulphurization is higher in graphite crucibles than in alumina crucibles, but the carbon level is then raised again by carbon pick-up. According to the equilibrium conditions, the formation of atomic gaseous sulphur should be expected to take place according to:

$$[S]_{FeCr} \rightarrow (S) \qquad (111)$$

and of COS according to:

$$[C]_{FeCr} + [O]_{FeCr} + [S]_{FeCr} \rightarrow (COS) \tag{112}$$

If stream degassing is applied during pouring, the sulphur content is lowered by 14–31% in less than 1 min at a residual gas pressure of 18–48 Torr[337].

It is advisable to decarburize ferro-chromium in the liquid state only if the initial carbon level is lower than 1% in order to avoid splashing from the bath owing to vigorous boiling.

High-carbon ferro-chromium may be decarburized with advantage in the solid state in a vacuum. In this case the same conditions apply as for the carbothermic reduction of Cr_2O_3 (see Subsection 3.1.1). In this process oxygen carriers, in the form of oxidized ferro-chromium or SiO_2, are added to the high-carbon ferro-chromium for oxidation of the carbon. In the former case the following are the reactions taking place according to Chadwick[338]:

$$5\langle Cr_2O_3\rangle + 27\langle Cr_3C_2\rangle = 13\langle Cr_7C_3\rangle + 15(CO) \tag{35}$$

$$\langle Cr_2O_3\rangle + 3\langle Cr_7C_3\rangle = \langle Cr_{23}C_6\rangle + 3(CO) \tag{36}$$

$$2\langle Cr_2O_3\rangle + \langle Cr_{23}C_6\rangle = 27\langle Cr\rangle + 6(CO) \tag{37}$$

These reactions are in agreement with the curves shown in Fig. 21. According to these curves the reaction occurs for ambient pressure theoretically at 1240 K for reaction (35), for reaction (36) at 1600 K, and for reaction (37) at 2000 K, but for about 10^{-4} atm it starts at 780, 990 and 1260 K, respectively.

The reaction temperatures may be even further lowered if Fe_2O_3 is added instead of Cr_2O_3. Theoretically the temperature decrease obtained should even be of the order of 500°C[339]. Data of several processes for the decarburization of ferro-chromium in the solid state are shown in Table 40; the Simplex process of the Union Carbide Metals Co. is probably the most important of them[338].

In the last-named process high-carbon ferro-chromium is ground. The coarse grains are returned, and the fine-grained ferro-chromium is decarburized. One of the two above-named oxidizing agents is then added. Part of the pulverized ferrochromium can be oxidized in the presence of an excess of oxygen while suspended

TABLE 40

DATA ON VACUUM DECARBURIZATION OF SOLID FERRO-CHROMIUM

Temperature (°C)	Time (h)	Pressure (Torr)	Carbon levels (wt.%)		Reference
			Initial	Final	
1000–1450	6	10^{-2}	5.6		340
1270		10^{-1}	3–8	0.03	339
Up to 1350		<3	5–6	0.01–0.02	338
1250		0.1	<8	<0.01	82

Fig. 72. Plant consisting of four vacuum furnaces for the decarburization of ferro-chromium by the simplex process in operation. The furnace doors are water-cooled[338]. View of discharge head of furnace.

Fig. 73. View into an opened furnace being charged with small briquettes on a hearth car[338].

Fig. 74. Withdrawal of a hearth car loaded with large brick-shaped products after vacuum treatment[338].

in a gas stream. Alternatively, very finely ground quartz may be employed as oxidizing agent.

High-carbon, pulverized ferro-chromium is then mixed with oxidized ferro-chromium or quartz sand, moistened and compacted with the aid of a binder, for instance 3% molasses[340], to briquettes of about 150 g (see Fig. 73) or 10 kg (see Fig. 74) in weight. The briquettes are dried in gas-fired ovens before being charged into the vacuum furnaces. The briquettes are stacked on 36-m-long hearth carriages (see Fig. 74), which are pushed directly into the vacuum furnaces (see Fig. 73). These furnaces are 42.5 m long and have a diameter of 4.5 m; the required vacuum is generated by a five-stage steam-ejector pump, the furnace is heated electrically by means of graphite resistors and the nominal rating is 4000 kW per furnace. The plant at Marietta has eight of these furnaces. Figure 72 shows four of them.

These furnaces are heated to a predetermined programme, as it should be remembered that a eutectic with a melting point of only 1260°C is formed at a carbon content of 2.75%. Once the carbon content has fallen below this level the furnace may be heated to 1350°C, but if higher temperatures are used the evaporation of chromium substantially increases[340]. After the decarburization is completed the furnace is flooded with argon and is then cooled. The furnaces are finally opened and the hearth carriages withdrawn (see Fig. 74).

The products show a slightly different composition according to the oxidizing agent used, as may be seen from the following table:

References pp. 321–335

Oxidizing agent	Composition		
	%C	%Cr	%Si
SiO_2	0.008	67.0	6.5
Oxidized ferro-chromium	0.008	69.5	1.2

The grinding, oxidation and mixing process before the vacuum treatment may be replaced by atomization of liquid, high-carbon ferro-chromium with air or steam to an equivalent oxygen and carbon content. After cooling the oxide–carbide–metal mixture is then processed as described earlier[82,341]. It is probable that diffusion processes in the reaction mixture are rate-determining towards the end of the reaction. By introducing reaction gases into the furnace space it should be possible to increase the reaction rate[342].

4.2.2 Deoxidation via suboxides

In the course of the last decade numerous publications on gaseous oxides of metals have appeared, particularly those by Brewer, Chupka, Gleiser, Inghram and their coworkers[343,344]. From these publications two features emerge:
(i) the valency of the metals is lower than usual
(ii) most of these oxides are stable at high temperatures only.

The studies also showed that oxides of numerous high-melting-point metals have a higher vapour pressure than the metals and that therefore separation of metal and oxide by distillation becomes possible.

Smith[345] in 1958 pointed out the potential importance of deoxidation of a metal via its suboxides. On the basis of the thermodynamic data known at that time he regarded deoxidation as possible for Y, Th, Zr, Hf, Nb, Ta, Mo, W and B, but not for Be, Ti, V, Cr, Mn, Fe and Ni.

Brewer and Rosenblatt[343], in a comprehensive review, showed which of the metals might be separated from their residual oxides by distillation, which metals might be deoxidized by evaporation of their suboxides or of the oxygen, and for which metals no refining effect or only a slight one would be expected. The results of these studies are shown in Table 41. With the exception of beryllium, the refining effects calculated by Brewer and Rosenblatt show excellent agreement with technological experience. The possible degree of purification is expressed by an evaporation ratio R, which is defined as the ratio of the molar concentrations of oxygen in the evaporating mixture and in the melt

$$R = \left(\frac{O}{Me}\right)_{gas} \bigg/ \left(\frac{O}{Me}\right)_{metal}$$

This ratio is calculated from thermodynamic data alone, ignoring kinetic factors. A favourable purification effect can only be expected if R becomes materially greater than 10, as for $R = 10$ about 10% of the metal must be evaporated in order

TABLE 41

DATA ON THE SEPARATION OF METALS FROM DISSOLVED OXYGEN BY DISTILLATION

Metal	Mainly evaporating oxide	Evaporation ratio R	Metal	Mainly evaporating oxide	Evaporation ratio R
Preferred metal evaporation, oxide enrichment in residue					
Li	Li_2O	10^{-8}	Be	Be_3O_3	
Na	Na_2O		Mg	MgO	
K	K_2O		Ca	CaO	
Rb	Rb_2O		Sr	SrO	
Cs	Cs_2O		Zn	ZnO	$<10^{-6}$
Mn	MnO	10^{-4}	Cd	CdO	
Metals and oxides evaporate simultaneously at almost identical rates					
Ba	Ba_2O	1	Cr	CrO	$10^{0.5}$
Sc		1	Fe	FeO	$10^{0.4}$
Ti	TiO	1	Co	CoO	10^1
			Ni	NiO, O	10^1
Preferred oxide or oxygen evaporation					
B	BO and polymers		Zr	ZrO	10^1
Al	Al_2O	10^3	Hf		
Ga	Ga_2O	10^6	V	VO	10^1
In	In_2O	$>10^3$	Nb	NbO	10^3
Tl	Tl_2O	10^4	Ta	TaO	10^5
Si	SiO	10^8	Mo	Mo_3O_9, MoO_3, MoO_2, O	10^6
Ge	GeO and polymers	$10^{9.5}$	W	W_3O_9, WO_3, WO_2, O	10^6
Sn	SnO and polymers	$10^{6.3}$	Tc, Re, Ru		
Pb	PbO and polymers	$10^{2.6}$	Os, Rh, Ir	O_2, O	$\geqslant 10^1$
Bi	BiO	$10^{1.5}$	Pd, Pt, Ag, Au		

to reduce the oxygen content to one-half of its initial level.

The refining of alkali, alkaline-earth and the low-melting-point heavy metals has already been dealt with in Subsections 4.1.6, 4.1.3 and 4.1.4. The possible suboxide distillation of high-melting-point metals has also been mentioned earlier, *viz.* in Subsections 3.1.1 and 3.2.1, and will therefore not be described here in detail.

(a) Purification of titanium, zirconium and hafnium

Brewer and Rosenblatt[343] showed in their calculations that there is no possibility of deoxidizing titanium via the suboxides, a limited possibility for zirconium, and good possibilities for hafnium.

As a result of their calculations Trouve and Accary[346] also considered that titanium could not be deoxidized via its suboxides, but they confirmed the possibility, discovered experimentally by Lehr, Albert and others[72,73,347-349], of deoxidizing zirconium by preferential evaporation of ZrO. These authors gave as the limiting value for 2500 K a quantity of 0.7 wt.% or 3.5–4 at.% oxygen. For

References pp. 321–335

lower oxygen contents an enrichment of the melt in oxygen will occur, whereas for greater oxygen contents the oxygen level will be lowered by evaporation of ZrO.

While Albert et al.[348] deoxidized oxygen-rich zirconium–oxygen alloys by evaporation of ZrO in a dynamic vacuum of 10^{-5} Torr and high temperatures, Ailloud et al.[214,347,349,350] carried out experiments on zirconium which contained only 0.1275 wt.% or 0.725 at.% oxygen. In these experiments samples 2–4 g in weight were kept in the liquid state on a water-cooled copper support for 20 hours at a temperature that was only slightly higher (about 50°C) than the melting point. The samples were previously degassed in a static vacuum of 2×10^{-7} Torr. During the 20-hour treatment the oxygen level of the samples was lowered to 0.0382 wt.%. It is obviously easier to attain the required low oxygen partial pressures if a closed system with a low outgassing rate of its components is employed and if use is made of the getter effect of the evaporated zirconium. The loss in metal during the evaporation amounted to only 3.8%. The reaction rate followed the Langmuir equation for evaporation (see page 216 and following pages). The relevant data are compiled in Table 42.

Besides oxygen, other non-metallic and metallic impurities may also be separated from zirconium by distillation. During the vacuum melting of high-purity zirconium at 2700°C and 10^{-5} Torr, Cr, Al, P and As could be reduced to less than one-tenth of the initial level in 15 min, whereas Fe, Cu, Au, Mn and Sb were lowered to one-fifth. Sulphur and molybdenum have a similar evaporation rate to that of zirconium; their concentration, therefore, does not change, while tungsten and nitrogen accumulate in zirconium[349,350].

TABLE 42

DATA ON THE REFINING OF ZIRCONIUM[349–351] AND HAFNIUM[352–354] IN A VACUUM

	Melting temp. (°C)	Vacuum (Torr)	Time (h)	Analysis (wt. ppm)			Evaporation loss (%)
				O	N	Al	
Starting material Zr	—	—	—	300	20	26	—
After remelting		5×10^{-5}	—	70	15	7	20
After remelting		5×10^{-5}	—	117	15	2	20
Starting material Zr	—	—	—	30000	320	a	—
After 1st remelting	~2500	~10^{-5}		7000	a	a	
After 2nd remelting	~2500	~10^{-5}		350	130	a	50
Starting material Zr	—	—	—	1275	a	a	—
After remelting	~1900	2×10^{-7}	20	382	a	a	3.8
Starting material Hf	—	—	—	1500	50	66	—
After remelting		3×10^{-5}		400	27	<25	a
Starting material Hf	—	—	—	1870	95	a	—
After 1st remelting		8×10^{-5}		310	55	a	
After 2nd remelting		3×10^{-5}		170	25	a	5

a Not known.

TABLE 43
REFINING OF HAFNIUM IN THE ELECTRON BEAM FURNACE[355]

Melting processes	Analysis (ppm)							Brinell hardness	Electrical resistance ratio $R_{298K}/R_{4.2K}$	Weight loss per melting operation (%)
	Al	Cr	Cu	Fe	Mg	N	O			
Vacuum arc premelting	36	33	70	290	<10	28	800	205	4	–
Electron beam remelting 40 kg/h	22	5	56	70	<10	15	790	207	4.5	2.5
Electron beam remelting 30 kg/h	22	2	17	26	4	21	570	180	5	2.5
Electron beam remelting 20 kg/h	19	2	28	19	1	26	390	180	5.5	3
Electron beam remelting 10 kg/h	22	2	18	9	1	23	210	173	9	7
Electron beam remelting 5 kg/h						35	100	158	12	18
Vacuum arc premelting	34	26	94	170	<10	8	780	205	4	–
Electron beam remelting 40 kg/h	19	4	28	42	<1	12	570	195	5	1.5
Electron beam remelting 30 kg/h	19	2	18	10	<1	15	530	185	6	2.5
Electron beam remelting 20 kg/h	23	<2	12	8	2	17	300	175	7.5	4
Electron beam remelting 10 kg/h	19	<2	7	<8	2	20	290	170	9.5	8
Electron beam remelting 5 kg/h						25	80	150	14	11

The refining of hafnium may be carried out in a similar way, and results obtained on hafnium are shown in Tables 42 and 43. For hafnium, high superheating of the melt results in a particularly thorough purification and elimination of most of the impurities[354,356].

Armand, Givord and Herold[355] refined hafnium sponge to a high degree of purity by repeated remelting in the electron-beam furnace. The sponge, which was obtained by reduction of $HfCl_4$ with magnesium, was previously remelted in the vacuum arc furnace in order to remove chlorine and magnesium. It was found in these experiments that Cr, Cu, Fe and Mg could be eliminated to a high degree and even oxygen and aluminium to a marked degree, whereas for nitrogen, on the contrary, an increase was observed after remelting. The results obtained are shown in Table 43.

Remelting for refining and purification of the metals zirconium and hafnium and also of other high-melting-point metals is almost exclusively carried out in electron-beam furnaces, as in these furnaces any crucible reaction can be excluded. An electron-beam furnace for the refining, and also for remelting, of such high-melting-point metals as Zr, Hf, V, Nb, Ta and their alloys is shown in Fig. 75.

(b) Refining of vanadium, niobium and tantalum

The degassing of niobium and tantalum in the solid state was fundamentally investigated by Fromm, Hörz and co-workers. Besides the equilibria with the gaseous states of the systems Nb–O[358], Nb–N[359], Ta–O[358] and Ta–N[360], the stationary states were studied which are attained, *e.g.*, under non-isothermal conditions where no equilibrium exists. In the degassing of Nb–O[55,56,361–363], Nb–N[56,363,364], Ta–O[55,56,361–363] and Ta–N[363,364] the temperatures were:

References pp. 321–335

Fig. 75. Electron-beam furnace, incorporating four deflecting, separately evacuated Pierce electron guns, for an operating pressure of 10^{-3}–10^{-5} Torr for the melting of Nb, Ta, W and Th at hourly outputs of 100–150 kg[357]. (Courtesy Leybold–Heraeus)

Nb 1600–2200°C, Ta 1600–2500°C and the pressures between 5×10^{-5} and 2×10^{-3} Torr (see also Chapter IV, Part 2: Vacuum Degassing in the Solid State). The results show that, for both metals, the oxygen is given off in the form of suboxides and the nitrogen in the molecular form. As would be expected, the solubility of oxygen and nitrogen decreases with increasing temperatures at constant pressure, so that for temperatures sufficiently high and pressures sufficiently low a degree of degassing may be obtained which can be regarded as almost complete. In this process the degassing rate depends greatly on the temperature. In the elimination of oxygen a mixture of MeO and MeO_2 is evaporated; the MeO portion increases with increasing temperatures and decreasing pressures. For oxygen levels below 1 at. % in either metal it may be assumed that the formation and evaporation of these oxides are the rate-determining step[55,56]. For nitrogen the recombination of the atoms to form N_2 molecules and the desorption of the molecules are thought to be rate-determining[364,365], because the reactions seem to obey a second-order rate law.

TABLE 44

ANALYTICAL DATA ON THE REFINING OF NIOBIUM AND TANTALUM[367] (ppm)

Impurity	Starting powder Nb	Vacuum sintered 1 h, 1200 °C	Electron beam remelted	Starting powder Ta	Vacuum sintered 1h, 1200 °C	Electron beam remelted
Al	30	7	<1	10	10	<1
Ba	3	—	—	6	2	1
Bi	—	—	—	20	3	—
Ca	20	2	<1	20	10	<1
Co	1	<1	<1	40	<1	<1
Cr	10	10	<1	10	3	1
Cu	15	<1	<1	4	<1	<1
Fe	1300	400	3	600	20	<1
Hf	10	10	4	—	—	—
K	1000	280	2	90	80	<1
Mg	30	30	<1	10	4	<1
Mo	—	3	5	240	210	40
Na	2200	80	1	10	10	1
Nb	Main metal			3600	3100	1100
Ni	6	<1	<1	50	2	<1
Pb	1300	1	—	10	10	—
Re	20	4	1	—	—	—
Si	800	280	10	130	10	10
Ta	6200	6200	19000	Main metal		
Ti	500	1400	<1	300	90	5
V	<1	<1	<1	30	6	—
W	30	50	20	1100	900	1400
Zn	15	<1	—	2	1	<1
Zr	250	250	40	—	—	—
C	1000	25	15	100	60	10
H	100–1000	5	1	—	—	—
N	2000–5000	20	1	10–10000	20	1
O	5000–12000	600	20	500–10000	750	15
Loss			30%			45%

The reaction equations observed on solid niobium and tantalum may also be applied, as a first approximation, to the conditions prevailing in liquid niobium and tantalum, as has been shown by Fromm and his co-workers[55,365,366]. They correspond to a first-order rate law.

Zedler et al.[367] found in the refining of niobium and tantalum during an annealing treatment at as low as 1200°C a high degree of purification. They were able to reduce the level of most impurities below the 5 ppm limit by remelting in an electron-beam furnace, as may be seen from Table 44. Despite considerable losses owing to evaporation, only the tantalum level increased in niobium, and that of tungsten in tantalum.

It is worth mentioning that an increase in the residual gas pressure in the melting chamber did not have the expected negative effect. If the residual gas pressure was raised from 5×10^{-6} to 5×10^{-3} Torr the Vickers hardness of niobium increased from 30–50 HV to only 50–80 HV, and that of tantalum from

60–70 HV to 65–115 HV. Both metals retained their cold-working properties. If the pressure was raised to levels above 5×10^{-5} Torr for niobium a thin, hard surface layer was found, which showed a high oxygen content and rendered cold rolling difficult. The presence of oxygen in this thin layer indicates that the oxygen is introduced into the metal only after solidification of the metal. The high rate of evaporation of these two metals during the melting operation, owing to their high vapour pressures, very probably prevents the oxygen or nitrogen from reaching the surface of the pool of molten metal at all.

Kimura et al.[368] found during the refining of carbon-reduced niobium in the electron-beam furnace that in this operation the oxygen may easily be eliminated as NbO or CO, and carbon as well as CO. Of course the carbon is so eliminated to a sufficient degree only if an excess in oxygen is present. The decarburization rate as CO is in any case, even initially, about one-quarter the rate of the deoxidation as NbO. For an adequate elimination of the carbon the excess in oxygen should be four- to fivefold, as is evident from Fig. 76 (see also Table 45).

The conditions prevailing in vanadium closely resemble those in the two other metals. While it is also possible to remove nearly all the nitrogen from niobium and tantalum, this appears not to be easy for vanadium. Preliminary results of investigations on solid[369,370] as well as on liquid[370,371] vanadium indicate that this metal, containing 1 at. % nitrogen in solution, does not give off any at a temperature of 1550°C and a pressure of 7×10^{-10} Torr. For liquid vanadium a limiting value of about 0.3 wt. % $[N]_V$ was found at a temperature of 2000–2100°C and a residual gas pressure of about 2×10^{-5} Torr. For higher initial nitrogen levels degassing occurs during melting, but for lower levels an enrichment in nitrogen is observed owing to the loss of vanadium by evaporation which cannot be prevented.

The oxygen level of vanadium, on the other hand, is lowered by heating in

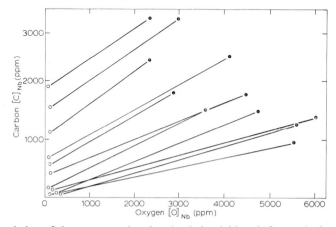

Fig. 76. Correlation of the oxygen and carbon levels in niobium before and after remelting in the electron-beam furnace for 30 min[368]. ● = Before remelting; ○ = After remelting.

TABLE 45

DATA ON THE REFINING OF VANADIUM, NIOBIUM AND TANTALUM

Metal	Melting process	Vacuum (Torr)	Analysis (ppm) O	N	H	C	Hardness (kg/mm²)	Reference
V	Starting material	–	1045	210	16			
	After 1st remelting	2×10^{-4}	325	140	12		90–110	354
	After 2nd remelting	8×10^{-5}	235	95	12		90–110	
V	Starting material	–	415	123	9			
	After 2nd remelting	1×10^{-4}	545	169	10		120	366
V	Arc melting		820	500	12	360	158	373
	Electron-beam melting		330	440	6	210	127	
V	Starting material	–	30000	460				371, 372
	After remelting	2×10^{-5}	2100	1400				
V	Starting material	–	4230	900	9			
	After remelting	2×10^{-5}	1235	1180	7			371, 372
V	Starting material	–	840	1030	107	450		
	Arc melting		920	1055	18	470		370
	Electron-beam melting	$<1 \times 10^{-4}$	95	1480	<1	156		
Nb	Starting material	–	3600	470		1460	>200	368
	20 min remelted	5×10^{-5}	80	70		250	70	
Nb	Starting material	–	2350	570		2410	>250	368
	20 min remelted	5×10^{-5}	100	100		1200	145	
Nb	Starting material	–	6010	470		1440	>200	368
	30 min remelted	5×10^{-5}	180	<50		150	45	
Nb	Starting material	–	2100	105	4	259		
	After 1st remelting	1×10^{-4}	80	40	1	50	55	354
	After 2nd remelting	5×10^{-5}	20	15	1	35	50	
Nb	Starting material	–	1460	56	40			
	After 2nd remelting 1 min	1×10^{-4}	100	40	5		71	366
	After 2nd remelting 2.5 min		55	18	4		55	
	After 2nd remelting 5 min		24	<3	3		45	
Ta	Starting material	–	2080	39	214			
	After 2nd remelting 1 min	1×10^{-4}	64	<1.2	4		76	366
	After 2nd remelting 2.5 min	1×10^{-4}	24	<1.2	4		66	
	After 2nd remelting 5 min	1×10^{-4}	30	<1.2	5		66	
Ta	Starting material	–	8900	100	4	500		
	After 1st remelting	2×10^{-4}	1000	60	1	95	89	354
	After 2nd remelting	5×10^{-5}	30	26	1	44	53	

the solid state[369,370], as well as by remelting in the electron-beam furnace[370,371]. Here too the elimination of oxygen seems to take place via a suboxide and the rate of elimination may be described by a first-order law with time[40,372]. However, the conditions in this case are not so favourable as with niobium or tantalum, as high evaporation losses occur owing to the relatively high vapour pressure of the vanadium (evaporation ratio $R = 10^1$)[370,371]. Table 45 shows a collection of data on the refining and purification of all three metals by remelting in the electron-beam furnace.

The reduction of the metals vanadium, niobium and tantalum by aluminium is feasible without any difficulty on an industrial scale in quantities of several hundred kg per charge, although considerable amounts of oxygen and aluminium remain in the metals in solution. Aluminium, oxygen and other impurities may, however, be removed to a high degree by distillation[374]. It is advisable to carry out this distillation process in the electron-beam furnace independent of the preceding method of reduction (carbothermic, metallothermic using aluminium, calcium or magnesium). The refining effects for all metals are high.

The results obtainable for vanadium may be seen from Fig. 77. Aluminium and oxygen levels may be lowered considerably by a distillation process first in the solid and then in the liquid state[376], or in two stages in the liquid state[40, 372]. In the first stage preferential evaporation of Al_2O may be assumed besides that of aluminium and VO; however, at lower initial aluminium levels mainly Al and VO should evaporate. The evaporation of VO may be formulated as a reaction of the first order, in which it is probable that the desorption of the VO is rate-determining.

The content of nitrogen in aluminium-reduced vanadium is of particular significance as the nitrogen level cannot be lowered in contrast with that of oxygen, as mentioned above. As a result of the inherent evaporation of the vanadium during melting in a vacuum, the nitrogen level is even raised, as may be seen from the data presented in Table 46. For this reason the reduction of V_2O_5 by aluminium should be carried out under a protective atmosphere of argon or in a vacuum[378] in order to exclude the pick-up of nitrogen as far as possible during this reaction.

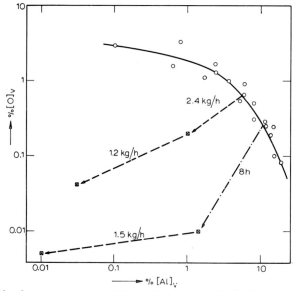

Fig. 77. Correlation between oxygen and aluminium contents in aluminium-reduced vanadium.
———— O–Al contents immediately after reduction; analytical data obtained on industrial charges of various origins[372, 375]
– – – – Change in the O–Al contents after double remelting in the electron-beam furnace[40]
– · – · – Change in the O–Al contents after distillation in the solid state and subsequent
– – – remelting in the electron-beam furnace[376].

TABLE 46

DATA ON THE REFINING REMELTING OF ALUMINIUM-REDUCED VANADIUM AND NIOBIUM

Melting processes	Temperature (°C)	Time	Pressure (Torr)	Impurity content (wt. %)			Refs.
				Al	O	N	
Starting material Al-reduced V	–	–	–	11.1	0.29	0.006	
Sublimation	1700	8 h	5×10^{-5}	1.4	0.01	0.008	376
Electron-beam remelting	~2000	1.5 kg/h	1×10^{-6}	0.01	0.005	0.008	
Starting material Al-reduced V	–	–	–	~6	0.6	0.075	
Electron-beam remelting	~2000	2.4 kg/h	5×10^{-5}	~1	~0.2	0.08	40, 372
Electron-beam remelting	~2000	1.2 kg/h	5×10^{-5}	0.04	0.04	0.10	
Starting material Al-reduced Nb	–	–	–	2.2	0.78	0.035	
Sublimation	2000	8 h	5×10^{-5}	0.12	<0.01	0.004	377
Arc remelting				0.12	<0.01	0.005	
Electron-beam remelting		1.5 kg/h		<0.002	<0.01	0.005	
Starting material Al-reduced Nb	–	–	–	2.2	0.78	0.035	
Arc remelting				0.92	0.44	0.033	377
Electron-beam remelting				0.0015	0.003	0.0025	
Starting material Al-reduced Nb				~4.8	1.5	0.20	
First electron-beam remelting		2 kg/h	$<1 \times 10^{-3}$	a	a	a	
Second electron-beam remelting		2 kg/h	$<2 \times 10^{-4}$	0.005	0.05	0.02	
Third electron-beam remelting		5.5 kg/h	$<5 \times 10^{-5}$	0.005	0.03	0.01	

a Not known

An initial sublimation from the solid state at about 1700°C with a subsequent electron-beam melting has been stated to produce a metal of increased purity[376]. In this case, however, the initial oxygen content was not high and the time consumed by the process is very considerable. In the two-stage remelting in the electron-beam furnace, on the other hand, additional melting losses of considerable magnitude occur, about 20%. An aluminothermic vacuum reduction process for V_2O_5 has been developed in which vanadium containing 10–16% Al could be produced. If this material is remelted in an electron-beam furnace the metallic impurities are reduced to 350 ppm and the interstitial elements also to a total of 350 ppm[378].

Much more favourable are the conditions for aluminium-reduced niobium, which— similarly to vanadium—may be subjected to a distillation first in the solid and then in the liquid state[377] (see also Table 46). Here too the time required is considerable; Nb–Al may therefore be remelted first in the vacuum arc furnace and then in the electron-beam furnace. If electron-beam furnaces are used which have electron guns that can be evacuated separately (see Chapter V, Section 4, Electron-Beam Melting) (Fig. 75), the first melting and the following remelting may be done in the same furnace and result in a purification of a high degree.

Niobium produced by carbothermic reduction at about 1800°C is hydrogenated at 450°C and then crushed to an average grain size of 0.08 mm. After dehydrogenation at 750°C, and after adjustment of the required oxygen and carbon levels,

References pp. 321–335

compacts are produced from the metal powder; these compacts are then remelted twice in the electron-beam furnace. In the first remelting operation granules are produced in order to have a homogeneous starting material for the second remelting process[379].

Tantalum obtained by aluminothermic reduction may also be purified by repeated remelting in the electron-beam furnace.

Further data on the refining of niobium and tantalum, as well as of other metals, by electron-beam remelting are reported by Lawley[380].

(c) Refining of molybdenum, tungsten and thorium

Molybdenum and tungsten, which are produced mainly by powder-metallurgical techniques, may also be subjected to electron-beam remelting for a lowering of the contents of metallic and non-metallic impurities, although it is not possible to obtain a metal which is cold workable. As little as 2 ppm oxygen and 8 ppm nitrogen substantially lowers the ductility[354]. The melting conditions and results of analyses are compiled in Table 47. From the analyses it may be seen that the oxygen and nitrogen levels were considerably lowered; the oxygen is removed mainly as the suboxide and only to a small extent as CO.

Rexer[382], too, did not obtain a cold-workable metal by electron-beam remelting of molybdenum and tungsten despite greatly reduced gas contents.

In contrast with vacuum arc melting, thorium may be purified to a high degree by electron-beam remelting. It seems that here too the oxygen escapes in

TABLE 47

DATA ON THE REFINING REMELTING OF MOLYBDENUM, TUNGSTEN AND THORIUM

Metal	Melting process	Vacuum (Torr)	Analysis (ppm)				Vickers hardness (kgf/mm^2)	Ref.
			O	N	H	C		
Mo	Starting metal	—	650	9	3	89		381
	Electron-beam remelting		5	9	0,2	<5		
Mo	Starting metal	—	810	51		170		354
	First electron-beam remelting	2×10^{-4}	105	15		64		
	Second electron-beam remelting	2×10^{-5}	6	3		25	145	
Mo	Electron-beam remelting	5×10^{-6}	~10	~1	~1	~5	153	382
Mo	Repeated electron-beam remelting		<40		<1	~40	175–190	383
W	Starting metal	—	4100	30	1	70		
	First electron-beam remelting	2×10^{-4}	115	11	1	45	235	354
	Second electron-beam remelting	2×10^{-5}	5	2	1	30	200	
W	Electron-beam remelting	5×10^{-6}	~10	~1	~1	~5	313	382
Th	Starting metal	—	4700	30		100		
	First electron-beam remelting	5×10^{-6}	80	50		60		270
	Third electron-beam remelting	5×10^{-6}	100	35	<1	25		
Th	Starting metal	—	1075	35		40		
	First electron-beam remelting	5×10^{-6}	200	40		55		270
	Third electron-beam remelting	5×10^{-6}	80	35	<1	50		

the form of suboxides. At the same time a number of metallic impurities evaporate. On the other hand, the nitrogen appears to be enriched to a small extent in the melt, as is evident from the data shown in Table 47, nevertheless the purity of the metal is comparable with that of iodide thorium[270].

4.3 Chemical transport reactions in a vacuum

A refining and purification of metals is also possible by reaction of the metal in the raw metallic state with a reactive gas forming a gaseous compound. After a transport of this compound to another location the reverse reaction is induced with formation of the reactive gas and the metal in a condition of high purity. A purification, of course, is possible only if the impurities do not react in the same manner as the base metal. Reaction, transport and the reverse reaction can take place only if the reactions are reversible and if a difference in the chemical potential exists in the two locations of reaction. This difference may be generated by differences in temperature or pressure or by a concentration gradient. A diagram illustrating these conditions is shown in Fig. 78. A metal, A_{impure}, reacts with the gas B at the temperature T_1 forming the gas $A_m B_n$. This gas is then transported to a location at temperature T_2, where it decomposes into the metal A_{pure} and the gas B.

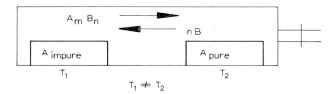

Fig. 78. Diagram indicating a chemical transport reaction in a vacuum.

In principle such reactions are conceivable for almost all metals in the form of numerous compounds. For metallurgical processes only the halides (see also Subsection 4.1.7, pages 254–256), in particular the iodides, have become of real importance. These processes are therefore generally called "iodide processes", or after their discoverers, "van Arkel–de Boer processes". They are also known as "hot-wire processes" with reference to the type of technique used. The iodides are extremely suitable for this purpose, as their thermal stability is relatively low. Moreover, iodine does not react with oxides, apart from the alkali and alkaline-earth metal oxides, and with nitrides and carbides only in very rare cases at the working temperatures usually applied.

According to van Arkel[384] favourable conditions exist for this process where:
(a) the metals form volatile iodides
(b) the melting point of the metals is higher than the dissociation temperature of the iodides
(c) the volatile iodide is formed at manageable temperatures

References pp. 321–335

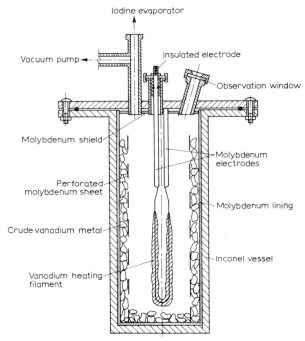

Fig. 79. Diagram depicting an installation for the refining of metals by the iodide process, exemplified by vanadium[43, 387].

(d) the iodide is easily decomposed at elevated temperatures
(e) the vapour pressure of the metal is still very low at the decomposition temperature of the iodide.

These conditions exist for Ti, Zr, Hf, V, Nb, Ta, Cr, Cu, Ag, Fe, B, Ge, Th, Y, Ni, U, W and Si, but only Ti, Zr, Hf, Th and Cr have been produced in small quantities during a certain period of time[385]. Only those metals offer a promise of economical refining by the iodide process which show, according to Table 41, an evaporation ratio R of between 10^{-1} and 10^1, i.e. in cases in which the separation of metal and oxygen (and usually nitrogen also) is difficult.

It is outside the range of this book to discuss in detail the great variety of metals produced on a laboratory scale by this method. An indication of the literature on this special subject may therefore suffice[384-386]. Even those metals which will be mentioned in the following examples are usually produced only in experimental plants of various sizes.

A plant for the refining of metals by the iodide process is depicted in Fig. 79. The raw material is placed at the outside wall of the reaction vessel. The centre of the vessel carries the heating filament made of the metal to be produced. The decomposition takes place at this filament, and the metal is precipitated on it. The filament therefore grows in thickness during the deposition and the heat input must constantly be adjusted for maintaining its temperature. The maximum attainable weights in most cases lie around several kg. After the reaction vessel has been loaded with the raw metal, the vessel is evacuated to pressures between 10^{-3} and 10^{-5} Torr and baked. The iodine is then introduced and the reaction temperatures adjusted so that the transport process may start. If the deposition on the filament is completed, the vessel is cooled and opened to the air. The metal crystals grown on the filament are removed; some are shown in Fig. 82. The reaction vessel is now ready for the start of a new process cycle.

The reaction proceeds in several stages all of which may be rate-determining:
(a) the heterogeneous reaction at the impure starting material (formation of the iodide)
(b) the transport through the gas phase
(c) the heterogeneous reaction at the high-purity material (decomposition of the iodide).

The transport velocity through the gas phase also depends on a variety of phenomena. In many cases diffusion and convection of the reacting gases as well as of the gaseous reaction products determine the reaction rate. Molecular movement becomes decisive if the pressures are very low and the mean free paths very long, and both the geometry of the apparatus as well as the temperatures of the materials affect the overall reaction rate.

Mathematical expressions for the quantitative prediction of the material exchange have been derived, in particular by Schäfer[386], Loonam[388] and Kesler[389].

4.3.1 Refining of vanadium, niobium, tantalum, uranium, chromium and thorium

Besides the above-mentioned metals which have been produced by the iodide process, vanadium has also been obtained by this method. The main advantage of employing the iodide process is the lowering of the nitrogen and oxygen contents of the raw metal, although no comparable success can be obtained in the removal of some metallic impurities, for instance iron and copper. The working conditions are listed in Table 48, while the degree of refining and purification is shown in Table 49[387]. The process has not succeeded in industrial practice.

In principle the refining of niobium and tantalum by the iodide process is also possible, but Rolsten[403,404] pointed out that the iodide process offers no advantages over a refining remelting in the electron-beam furnace.

The production of high-purity uranium[68,395] is also possible by the iodide refining process. Such a method is also described for chromium. This process, however, does not seem to be an economic proposition[405], although the oxygen levels may be lowered to 1/30 of the initial values, and those of nitrogen and hydrogen to 1/10[401].

The chances for the refining of thorium seem to be somewhat more promising. Some data on it are listed in Tables 48 and 49. Laboratory tests were carried out with temperatures between 400° and 600°C for the starting material and 1200° and 1500°C for the heating filament. A maximum in the reaction rate was observed

TABLE 48
DATA FOR THE PREPARATION OF METALS BY THE IODIDE PROCESS

Metal	Temperature of the raw metal (°C)	Dissociation temperature (°C)	Weight of the deposited metal per charge (kg)	References
Ti	150	1300		68
Ti	250	1200–1300		390
Ti	100		24	391
Zr		1200–1300		68
Zr	300	1200–1300		390
Zr	340		2.7	392
Zr			Max. 53	392
Hf	800	1500	25	355, 393
Hf	600	1600		68
Th	350–380	1200–1300	0.06	394
Th	455–485	900–1700		395–397
Th	450	1300		68
ThC$_2$	480	1300	0.2–0.25	398
U	300–400	1030–1100		395
V	850	1350	2	387
Cr	900	1000–1300		68, 399
Cr	800–850	1050–1100		400
Cr	800–850	1100–1150		401

at 6×10^{-2} Torr ThI$_4$. For very low pressures the dissociation is rate-determining and with rising pressures the transport through the gas space[406].

Scaife and Wylie[398] suggested the production of iodide thorium from thorium carbide, ThC$_2$, which is prepared at 2150°C, as thorium carbide reacts completely at a temperature as low as 480°C. Since the preparation of thorium carbide requires very high temperatures, many impurities are volatilized during this stage. As most of the other carbides do not react with iodine, or if they do, the reaction is very incomplete, a thorium of particularly high purity may be obtained in this way (see also Table 49).

4.3.2 The refining and purification of titanium, zirconium and hafnium

The iodide process has become of some importance in the refining of Ti, Zr and Hf, and certain amounts of these metals are refined by its use, although costs are very high. It seems that the iodide process for the preparation of high-purity titanium[391] and zirconium[396] is of great importance in the Soviet Union and other East European countries[407]. In this connection the numerous publications by Yemel'yanov and Yevstyukhin in the various volumes of *Metallurgiya i Metallovedenie chystich Metallov* should be mentioned.

The refining of zirconium by the iodide process has probably gained the greatest importance. Figure 80 shows the former pilot plant of Westinghouse. In the foreground the insert with the wires for deposition is shown, which has been

TABLE 49

ANALYSES OF SEVERAL IODIDE METALS (ppm)

Metal Impurity	Ti After	Ti Before	Ti After	Zr After	Hf After	Th After	Th Before	Th After	Th from ThC$_2$ Before	Th from ThC$_2$ After	V Before	V After	Cr After	Cr Before	Cr After
O	50–110			200	10–50	<300	1100	<100	500	20	600	40	4–10	3800	44
N	10–40			10	2.5	20	290	<100	8000	17	1800	<5	<5	60	13
H				20									1–40	56	0.8
C	100–300			100		200	1150	200	Carbide	38	50	10	10–40	50	20
S												150	3–80	190	110
Al	130–500	>1000	<300	30	<50	<50						b	<2	50	b
Ca		500	100	<50	<30						30	<20	<2	10	50
Cr				30	<10				1	<0.6	50	70			
Cu	15–20	10	10	<0.5	<10						100	30	1–5	20	<10
Fe	35–250	100	100	200	<100	<130			48	<48	<300	150	<2	b	30
Mg	15–20	300	50	<10	<10						50	<20	<2	20	20
Mn	50–130			<10	<10							b	<2		
Mo	15			<10	<10							b	1	100	50
Ni	30	50	<10	30	<10							20	<1	30	b
Si	<300	<10	20	30	<10	<100					<500	<50	<20	100	<50
Sn	<100	>1000	200	<10								b			
		30	30												
Reference	68	390		385	355	68	396, 397		398		387		402	399	

b Not found.

Fig. 80. Partial view of the former pilot plant of Westinghouse for the production of iodide zirconium at an annual output of 32 tons[392].

Fig. 81. Several reaction vessels for the refining of hafnium by the iodide process[68].

removed from the reaction vessel. In the left-hand corner are the installations for the evacuation and baking of the freshly filled reaction vessels, while in the background similar vessels can be seen during the deposition process. Heating is effected by means of salt-baths.

Extensive studies on the formation and dissociation of ZrI_4 were carried out by Shapiro[392]. He found that the reaction rate is depressed by additions of an inert gas, and the pressure should not exceed 7×10^{-1} Torr if no convection currents develop. Optimum temperatures of the raw material are given as 245° and 265°C, respectively, for filament temperatures between 1400° and 1500°C. Nickel, chromium, nitrogen and carbon can easily be separated from zirconium by the iodide process, but Fe and Al only with difficulty, and Ti and Hf not at all.

The formation of hafnium iodide requires particularly high temperatures, as do chromium and vanadium iodides, as may be seen from Table 48. A number of reaction vessels for the refining of hafnium by the iodide process are shown in Fig. 81.

Armand et al.[355] as well as Thien Chi et al.[393] report in detail on the refining and purification of hafnium by the van Arkel–de Boer process on an industrial scale.

In this process HfI_4 is formed at 800°C and decomposed on a hafnium wire at 1500°C. The deposit grows to an average weight of 25 kg during a period of one week. Figure 82 shows the hafnium refined by this method. The analyses of this product are listed in Table 49. Titanium and zirconium are transferred completely to the refined metal, and about 10% of the iron.

The temperature of the raw material in laboratory tests varied between 308° and 720°C, while that of the filament was maintained at between 1100° and 1500°C. Kinetic studies showed that for total pressures below 3.5×10^{-1} Torr the decomposition at the filament is rate-determining; for higher pressures it is the diffusion in the gas space. The maximum in reaction rate was found at a total pressure of HfI_4 of 7×10^{-1} Torr[408].

4.3.3 Refining of molybdenum

An example of the preparation of high-purity metals by the decomposition of compounds other than halides, and also in the solid state instead of from the gas phase, is that of the dissociation of MoS_2. On the basis of pilot plant experiments a plant for 40-kg charges was built. In a vacuum of $1–5 \times 10^{-3}$ Torr, MoS_2 is decomposed at 1650°C in the following steps:

$$4\langle MoS_2\rangle \rightarrow 2\langle Mo_2S_3\rangle + (S_2) \qquad (113)$$

$$2\langle Mo_2S_3\rangle \rightarrow 4\langle Mo\rangle + 3(S_2). \qquad (114)$$

The heating is effected by molybdenum heating elements inside the reaction vessel and the sulphur liberated is condensed outside the reaction vessel. In this process

Fig. 82. Iodide hafnium deposits of an average weight of 25 kg per bar[355].

it is possible to lower the sulphur content to about 0.02 wt. %.

Ingots melted from such molybdenum powder showed total impurity levels of 200 and 250 ppm and therefore a purity comparable with that of commercial molybdenum, although the aluminium, chromium and cobalt contents are slightly higher than usual[409].

The reaction of raw molybdenum and chlorine via gaseous $MoCl_5$ for obtaining high-purity molybdenum is also possible. Raw molybdenum and chlorine react at 300°C forming volatile $MoCl_5$ which may be decomposed on a deposition filament at 1300°C. The microhardness of the deposited metal is still high and varies between 175 and 225 kgf/mm^2, and so this metal also cannot be cold worked[410,411].

4.4 Zone melting in a vacuum

4.4.1 Basic principles in zone melting
(a) Solidification of melts

Although the techniques of recrystallization or of crystallization from solution or from the melt for refining have been known for many years and have been applied for certain metallurgical purposes, the zone melting process in the strict sense of the term was developed by Pfann[412] only about 18 years ago. Since that time a number of reviews on this field have been published, in which the whole field of zone melting[412-414] or in some cases only that of the refining of metals[415-417] has been covered. This process is based on the phenomenon that the compositions of the crystal and the melt of one and the same substance in equilibrium are different, unless the substance is absolutely pure, or, if several substances are involved, these behave as pure substances (maximum and minimum in Figs. 83d and 83e). Figure 83 shows a number of melting diagrams of binary systems, in which component B represents the contamination or impurity. From these diagrams it may be seen that a crystal solidified from a melt and containing predominantly component A may either be of higher purity (a, b, c, e) or of lower purity (d) than that of the melt from which it originated. It may also be seen that the freezing point of the residual melt may either be lowered by the solute B (a, b, c, e) or raised (d). The impurities collect either in the melt (a, b, c, e) or in the crystallized phase (d). Case (a) is of theoretical interest only, since a slight solubility in the solid state always exists for thermodynamic reasons. If an intermetallic compound between the two elements A and B is formed systems of types (a) and (b) do also exist if B is replaced by the compound A_xB_y. There are also peritectic reactions to be considered (Fig. 84), which, however, may be broken down to simpler types of diagrams.

Under the usual conditions solidification of a melt takes place so quickly that no equilibrium is reached at the liquid/solid phase boundary. In addition, freshly formed crystals may occlude some melt. If, however, normal freezing takes place, in many cases a preliminary separation of the substance can occur. If equilibrium exists during the solidification, an equilibrium distribution coefficient may be expressed for each component of the solution as:

$$K_0 = \frac{C_{0\ \text{solid}}}{C_{0\ \text{liquid}}}, \qquad (115a)$$

which, for lower concentration of substance B, is less than unity for the systems (a, b, c, e) and greater than unity for system (d). The coefficient K_0 can become constant only for very small concentrations of B, *viz.* in the range in which the solidus and liquidus lines are nearly straight lines. For this reason, the concentration range is usually quoted for which the K_0 value is valid. K_0 is also valid only under equilibrium conditions. In reality, K_0 is greatly affected by diffusion near the phase boundary in the melt, particularly if the solidification rate is greater than the diffusion rate. Kuchar[418] derived the value of K_0 for a large number of cases with the aid of thermodynamic data, obtaining the depression of the freezing point for different compositions. Using aluminium as the base metal, he found a dependence of K_0 on the atomic number of the elements.

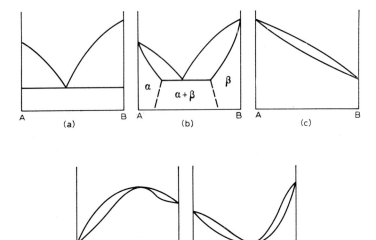

Fig. 83(a). Melting diagram of binary mixtures consisting of A and B, which form a eutectic but not a solid solution.
(b) Melting diagram of binary mixtures consisting of A and B showing limited solid solubility.
(c) Melting diagram of binary mixtures, which form liquid and solid solutions over the entire range of compositions.
(d) Melting diagram of binary mixtures, which form liquid and solid solutions with a melting point maximum.
e) Melting diagram of binary mixtures, which form liquid and solid solutions with a melting point minimum.

For the case where K_0 is less than unity, a concentration gradient of the form depicted in Fig. 85a exists for equilibrium at the phase boundary. Under practical conditions complete concentration equalization can no longer occur in the melt, i.e. a concentration gradient towards the bulk of the melt is being created (see Fig. 85b). As a result of the change in concentration at the phase boundary a new concentration ratio is created, which is called the effective distribution coefficient and is expressed by:

$$k = K_{\text{eff}} = \frac{C_{\text{solid}}}{C_{\text{liquid}}} \qquad (115b)$$

Burton, Prim and Slichter[412,413] related K_0 and K_{eff} by the expression

$$k = K_{\text{eff}} = \frac{K_o}{K_o + (1 - K_o)\exp[-f\delta/D]}. \qquad (116)$$

In other words, the effective partition coefficient depends on the rate of crystallization f, Nernst's boundary layer thickness δ, and the diffusion coefficient D. However, since the diffusion coefficient D is nearly constant over a wide range of

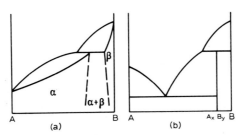

Fig. 84(a). Melting diagram of binary mixtures consisting of A and B with two types of crystal α and β and a peritectic point.
(b) Melting diagram of binary mixtures consisting of A and B forming a compound A_xB_y as well as a eutectic and a peritectic point.

concentration for temperatures only just above the melting point (about 10^{-4}–10^{-5} cm^2 s^{-1}), a possibility of influencing K_{eff} is essentially restricted to f and δ (10^{-1}–10^{-3} cm). An optimum k value should therefore be expected for a very low solidification rate and vigorous agitation of the melt. For $f = 0$, K_{eff} becomes K_0.

The concentration at a distance x of the solidified melt which originally extended from $x = 0$ to $x = L$, may be expressed by:

$$C_x = k \cdot C_0 \left(1 - \frac{x}{L}\right)^{k-1}, \tag{117}$$

where C_0 is the initial concentration and L the total extension of the melt, in the case under consideration a rod. However, k must be constant, which cannot always be assumed. The change in concentration for normal freezing of a melt is plotted in Fig. 86 for various k values. It is clearly evident that k must be either very much larger than unity or very much smaller in order to obtain satisfactory refining effects. Elements with values of k greater than unity collect in that part that solidified first, whereas those with a k smaller than unity in that part that solidified last.

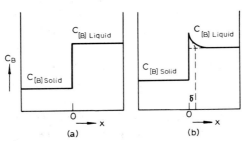

Fig. 85. Principle of the concentration distribution of an impurity with K_0 less than unity in the liquid and solid phase.
(a) In equilibrium (solidification rate < diffusion rate)
(b) In the stationary state (solidification rate > diffusion rate)

References pp. 321–335

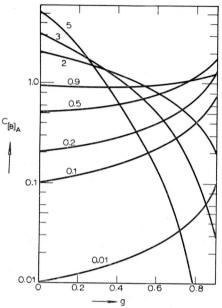

Fig. 86. Concentration distribution of $[B]_A$ in the solid for normal freezing for various values of k^{413}. g Fraction of the solidified melt.

(b) Theory of zone melting

Pfann's[412,413] considerations on zone melting resulted in the following conclusions:
(a) the impure materials should be used in the form of rods
(b) several travelling melting zones should be set up in a rod, which are separated by solidified zones
(c) migration of the melting zones should be brought about either by movement of the rods or by movement of the heating units.

The principle of zone melting and the change in concentration after one passage of the molten zone are depicted in Fig. 87a, b. In the actual operation a narrow zone near one end of the rod is melted, and this zone is slowly moved to the other end of the rod. Behind the molten zone l the metal solidifies again. In the zone that solidified first a concentration kC_0 is created, which slowly increases with increasing distance from the starting point to C_0, since the substance B in the molten zone is enriched above its initial concentration C_0. Adjacent to this zone there is a region of unchanged concentration C_0, since solidified and newly melting material have the same concentration C_0. At the other end of the rod there is in addition a zone of normal freezing which has the length of the melting zone l. In this zone the concentration of B increases with decreasing volume of the melt.

With the exception of this last zone the change in concentration may be expressed as:

$$C_x/C_0 = 1-(1-k)\exp[-kx/l]. \tag{118}$$

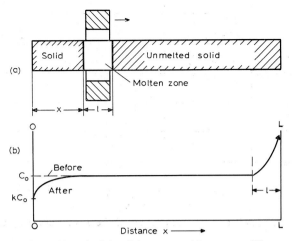

Fig. 87(a). Principle of the zone-melting process[412].
(b) Approximate concentration distribution of $[B]_A$ after one zone passage[412].

For further passages of the zone the concentrations change in the same directions. The concentration displacements obtainable after several passages of the zone may be calculated from equations derived by Hamming, Reiss, Lord[412], by Barthel[419] or by Helfand and Kornegay[420]. In these calculations a number of simplifications were introduced, which are not always in agreement with practice. Buhrig[421] points out that in some cases different results are obtained by application of the various methods for the determination of the distribution coefficient. On the other hand, he stresses that the experimental conditions must be kept constant with particular accuracy.

For an infinite number of zone passages a simple approximation may be given:

$$C_x = a \exp [bx], \tag{119}$$

where a and b are constants, which may be derived from:

$$k = \frac{b \cdot l}{\exp [bl] - 1}, \tag{120}$$

and

$$a = \frac{C_o \cdot L \cdot b}{\exp [Lb] - 1}; \tag{121}$$

again eqn. (119) is not valid for the last zone of normal freezing[412,413].

The change in concentration which follows from this equation is plotted in Fig. 87c.

If a high efficiency in zone melting is to be secured, the following points should be borne in mind:

(1) According to eqn. (118) the length of the molten zone for a single pass

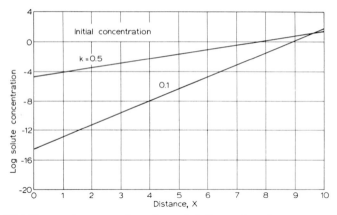

Fig. 87(c) Change of concentration after an infinite number of zone passages[312].

run should be as long as possible, as in this case C_x/C_0 has a minimum value. This requirement is, however, in contradiction with the insufficient stability of a large molten zone for crucible-free zone melting and, for multipass runs, with the occurrence of re-contamination of already refined sections of the rod. In general, a longer zone length provides a greater yield of pure material if the number of passes does not very much exceed the value L/l. For a greater number of passes the zone length should be shortened. More details have been described by Pfann[412].

(2) In order to obtain as nearly as possible the degree of partition indicated by eqn. (119) with as few passages and with as high a rate of travel as possible, an even distribution of the impurities in the molten zone is required. Since diffusion is a very slow equalization process, it might be necessary to employ mechanical means for achieving an even distribution of the impurities. Such an equalization might be effected by convection currents within the melt, which are generated by differences in the specific gravity of parts of the melt, which are at different temperatures, or, in induction heating, by a reaction of the field with the induced eddy currents. A reciprocal turning of the two solid parts of the rod around their mutual axis may also contribute to mixing or equalization.

(3) Any supercooling of the melt, which might lead to the formation of dendrites at the solidification front, must in any event be avoided as the dendrites may entrap parts of the melt which are enriched in solute. Dendrite formation may be suppressed by a sufficiently low cooling rate as well as by the above described means for a mixing of the melt.

(c) *Evaporation during zone melting in a vacuum*

Zone melting in a vacuum is possible only if the vapour pressure of the metals to be refined is so low that the losses from evaporation remain within acceptable limits. The vapour pressure at the melting point may serve as a yardstick. Vapour pressure data for the important metals are shown in Figs. 45 (Section 4). Assuming

TABLE 50

COMPUTED EVAPORATION LOSSES DURING ZONE MELTING

Pressure p_{Me} (Torr)	Quantities (mg)		
	1 min	10 min	60 min
10^{-4}	0.06	0.6	3.6
10^{-3}	0.6	6	36
10^{-2}	6	60	360
10^{-1}	60	600	3600

that the evaporation is not obstructed by outside means, the evaporation losses in g cm^{-2} s^{-1} for a mean atomic weight of 80 and 1 cm^2 evaporation surface area and an average temperature of the melt of 2400 K may be computed from:

$$i = 0.0583 \cdot p_{Me} \left(\frac{M}{T}\right)^{\frac{1}{2}}$$

The results are shown in Table 50.

According to these results, metals with a vapour pressure of 10^{-1} Torr and higher at the melting point cannot be subjected to vacuum melting as the losses arising from evaporation are much too high; Mg, Ca, Sr, Cr, Mn, Zn, Cd, As, Sb and Te belong to this group.

The melting conditions are still unfavourable if the vapour pressure is not lower than 10^{-2} Torr. Even then the evaporation rate is still relatively high if melting is carried out in a vacuum, although working in a vacuum may offer advantages for other reasons. In such a case the melting or travel rate may be maintained at a higher speed than that desired. Belonging to this group are Be, Ba, V, Re, Fe, Ru, Pd, Si, Se as well as Ir, Mo, and W. The last three metals, in contrast with the others, are almost exclusively zone-melted in a vacuum, as the major proportion of the impurities evaporates during this treatment. Excellent melting conditions exist for the high-melting-point metals, Ti, Zr, Hf, Nb, Ta, Rh, Co, Ni, Pt, Ge and U, as their vapour pressures are sufficiently low. The same applies for the low-melting-point metals Ga, In, Tl, Sn, Pb and Bi.

Owing to their high thermal conductivity, Ag, Cu, Au and Al are rarely zone-melted in a vacuum, but usually in a protective gas atmosphere for the improved extraction of heat from the solid parts.

The evaporation of dissolved substances during the zone-melting in a vacuum may result in additional refining effects. This phenomenon is utilized in many instances. In this operation the formation of gas bubbles in the molten zone must be prevented, as an escape of gas bubbles would destroy the stability of the melting zone. For this reason, a preliminary vacuum degassing of the starting materials is necessary. Pfann[412] and Klofáč[422] have derived mathematical solutions for the concentration partition of a number of these volatile elements.

On the other hand, the residual gas atmosphere may react with the rod to be zone-melted. In many cases reactions with the hot, although solidified, zones of the rod are possible so that the purified rod is re-contaminated. As is evident from a great number of experimental results, such contamination of zone-refined rods still occurs even at partial pressures of about 10^{-6} Torr of reactive residual gases owing to the long reaction time. In these cases, it may even be more advantageous to carry out the zone-melting in a closed system under a static vacuum, analogous to the process to be described later in Subsection 4.4.4. In this case, a piece of one end of the rod is evaporated before the zone-melting proper begins in order to getter any active gases that might be present[423].

4.4.2 Design of vacuum zone-melting apparatus

Melting for the zone-refining of metals is effected mainly either by induction heating or by electron bombardment. Either type of heating has its pros and cons[415,424]. In both cases melting may be carried out in crucibles or crucible-free. In crucible-free melting difficulties may arise in the processing of metals of high thermal conductivity. For instance, the molten zones can be kept small only with difficulty in the refining of Ag, Cu, Au and Al. If crucibles are employed, water-cooled copper crucibles may be used in electron-beam melting. In induction heating water-cooled copper pipes may be provided parallel to the rod so that they can support the rod. Also graphite or oxide ceramic boats may be used if there is no risk of a reaction taking place.

For the horizontal arrangement in the crucible several melting zones may be generated which travel along the rod at short distances from one another. In crucible-free zone refining the rod may be fixed in the horizontal or the vertical position. Naturally, only one melting zone is possible in this arrangement. The stability of the melting zone depends greatly on the surface tension of the liquid metal, the diameter of the rod and the specific gravity of the metal.

In order to increase the stability of the melting zone and the diameter of the rod in crucible-free induction melting it is advisable to operate the refining process with two superimposed frequencies, one for heating, employing a frequency of between 0.4 and 50 MHz, and the other for the electrodynamic support of the melting zone using a frequency of 1–50 kHz[425,426]. Warping and deformation of the molten zone are effectively restricted by a decrease—from bottom to top—in the number of turns per unit length of the induction coil and by an increasing coil diameter[426]. The arrangement of the coil is therefore similar to that used in levitation melting.

Induction heating may be carried out without any difficulty over a wide range of pressures, but the molten zone is in this case relatively wide. In electron bombardment heating it is possible to obtain very narrow heating zones by the use of magnetic or electric focusing, but until recently the melting operation had to be carried out in a high vacuum. Under certain circumstances high losses owing to evaporation occurred.

Recently means have been developed to make zone melting in this case

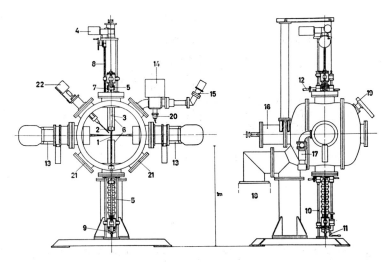

Fig. 88. Zone-melting equipment with two Pierce Guns[428]. 1 Sample; 2 Sample clamp; 3 Water cooled holder tube; 4 Driving motors; 5 Screening; 6 Screening; 7 Coupling system; 8 Driving screw for translation; 9 magnetic coupling; 10 rotation drive; 11 Fine adjustment; 12 Differential pumping; 13 Electron beam systems; 14 Trap for reflected electrons; 15 Quadrupol mass spectrometer; 16 Kryo pump; 17 Auxiliary pump for electron gun; 18 Main diffusion pump (1500 l/s); 19 Viewport; 20 Ionization Gauge; 21 Flanges for UHV-pumps.

possible at higher pressures also. One possibility is that of using a hollow cathode which surrounds the rod to be melted. A gas discharge is then produced between rod and hollow cathode. This method is described in Chapter V, Section 5 (Crystal growing). Another possibility is the employment of electron guns (Pierce guns) which incorporate a pressure stage. Their application to zone melting was described by Bas et al.[427,428]. The equipment used is depicted in Fig. 88. By providing another additional pressure stage[429] the working pressure in this method has lately been raised to 15 Torr.

The application of tungsten ring cathodes in electron-beam melting originally led to contamination of the zone-refined metals owing to evaporation of the tungsten, unless, of course, tungsten was zone-melted. The remedy is, therefore, to use a ring cathode made of the same metal as that to be refined if the melting point is high enough for adequate electron emission[430]. This mutual contamination can, however, be suppressed even if ring cathodes are employed by deflection of the electron beam by means of an electric field. The melting zone is then screened from the cathode (see Fig. 89). According to Barthel and Scharfenberg[431] the evaporation of tungsten from the cathode has not the detrimental effect which is often attributed to it, if the cathode is annealed and degassed before using it for the first time.

Gas eruptions and metal evaporation during electron-beam zone-refining may lead to arcing and even to short circuits between the ring cathode and the rod, which serves as the anticathode. Both arcing and short circuits limit the high

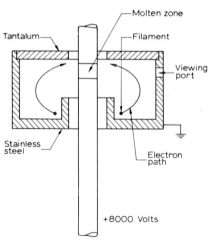

Fig. 89. Principle of an electron-beam heated zone-melting unit with ring cathode and electron-beam deflection[415].

voltage. Higher melting power can then only be obtained by employing higher emission current intensities, but this practice causes a broadening of the melting zone. In any case the power supply must be laid out so that any flash-overs lead to instantaneous interruption of the current; on the other hand, the restarting after de-ionization of the discharge space should take place so quickly that no noticeable cooling of the melting zone can occur.

If Pierce electron guns are used in electron-beam melting instead of ring cathodes, no flash-overs can occur between electron source and melting rod owing to the excellent screening. The voltage is therefore no longer limited and tungsten rods of a diameter of up to 15 mm have been zone-melted in such a set-up[429]. For this particular case, two electron guns arranged symmetrically to the tungsten rod are employed; they have a combined total input of 10 kW. The particular advantage of using the Pierce guns is the absence of any mutual contamination by deposition. Heating by this electron-beam system offers the additional advantage that, in principle, any mechanical movement of the rod or the heating device could be omitted, as the electron beam can be deflected continuously by magnetic means[424].

In crucible-free induction zone-melting, flash-overs may be prevented by superimposing a potential difference between melting rod and high-frequency coil, which counteracts the rectifier effect of the hot cathode formed by the melting zone[432].

An agitation of the molten zone is brought about in induction heating as a result of the presence of the axial magnetic field and the induced tangential eddy currents. In electron-beam melting, on the other hand, it is the usual practice to have the two ends of the rod rotated around their common axis in order to bring about better mixing in the melting zone. However, this method involves a certain

Fig. 90(a). Conventional arrangement of a number of high-frequency coils connected in series. This method is applied to the zone melting of germanium[414].

shortcoming in that the cross-section of the rod will no longer be so even and uniform along its length.

Most types of electron-beam heated zone-melting apparatus are based, with only slight modifications, on that developed by Calverley et al.[433,434]. The principle of such apparatus is shown in Fig. 1 of Chapter V, Section 5. The rod is fixed, and the electron-emitting ring electrode moves upwards and downwards. If the evaporation is not too intensive, the melting zone may easily be observed.

A similar arrangement is used in crucible-free zone melting by induction heating, but in many cases the inductor is placed outside the vacuum chamber.

The set-up for horizontal zone-melting is shown in Fig. 90a. In this arrangement melting can be carried out only in crucibles or sealed ampoules (see Fig. 90b). In many cases a quartz tube is employed into which ceramic or graphite boats are inserted. By the use of several induction coils a number of simultaneous melting zones can be set up. Such apparatus is common for the production of semiconductor materials.

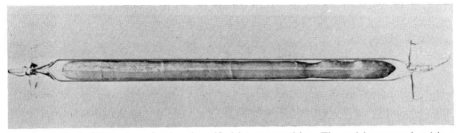

Fig. 90(b). Rod of tin to be refined and purified by zone-melting. The rod is encapsulated in a quartz ampoule[414].

References pp. 321–335

4.4.3 Appraisal of the effects of zone-melting

Any appraisal of the effects of zone melting meets with considerable difficulties. Although there are analytical methods in existence which extend into the ppb-range, either they require rather large quantities of the sample, or the accuracy is not at all satisfactory. Some of the methods also involve considerable technical expenditure. Therefore, if purity levels are quoted, the analytical method used for the determination should be named.

A relatively simple method of appraisal is the measurement of the electrical resistance at 4.2 K, the temperature of liquid helium, as it is rapid and highly sensitive, and a particular advantage of this method is that the resistance of the individual sections of the rod may be determined.

While the electrical resistance of dilute solid solutions at room temperature is mainly governed by the thermal vibrations of the particles, at 4.2 K physical defects (dislocations, vacancies, etc.) and chemical impurities affect the resistance. In order to eliminate geometrical effects, it is the usual practice to give the resistance ratio $R_{273K}:R_{4.2K}$. The greater the numerical value of this ratio is, the higher should the purity of the material be, but no discrimination between physical and chemical defects can be made[435].

4.4.4 Examples for vacuum zone-melting

(a) Zone-melting of zirconium

Zirconium is easily zone-melted in a high vacuum owing to its low vapour pressure at the melting point so that noticeable losses by evaporation do not occur. In previous experiments using induction heating difficulties were met in the stabilization of the melting zone[436,437], but these difficulties seem to have been overcome by electron-beam heating provided that the power input for melting is kept constant within $\pm 5\%$[438]. In this case the melting zone remains constant independently of the direction of passage, whereas in induction heating only melting in the downwards direction gave satisfactory results[437]. The working conditions for the zone-melting of zirconium are shown in Table 51.

When measuring the resistance ratio the highest values were obtained at the end of the rod where the zone-melting terminated, which means that the main impurities responsible for the change in resistance have k values above unity[437-439]. These impurities are mainly Hf, Hg and N as may be seen from Table 52. For a number of other impurities, such as Cu, Au, Fe, Mn, Cr, Al and O, the last-named as ZrO, additional evaporation plays an important role[439,440]. Higher iron levels are also greatly reduced by distillation[437].

Difficulties exist with the carbon, as no clear direction in distribution is found[438]. It seems, however, that the carbon originates from decomposition products of the diffusion-pump oils[438,439]. For this reason experiments were conducted in a static vacuum[437,439].

Mills and Craig[440] showed that no tungsten was detectable in zirconium after

TABLE 51

CONDITIONS FOR THE ZONE-MELTING OF ZIRCONIUM

Dimensions of rod		Rate of travel (cm/h)	Number of passages	Pressure (Torr)	Method of melting	Resistance ratio $R_{273\,K}/R_{4\cdot 2\,K}$	References
Diam.	Length (mm)						
3–6.5	100	15	max. 10	10^{-6}	Induction 220 kHz	b	436
7	200	3	4	$<5\times 10^{-7}$	Electron beam ring cathode	450	438
5	150	6	12–20	10^{-5}	Induction 600 kHz	20*	437
9.5	185	3–6	4–10	$3-6\times 10^{-7}$**	Induction 600 kHz	280	437, 439
3	130	0.4	5	10^{-8}	Electron beam ring cathode	b	440
11	–	12	3	10^{-3}	Induction	b	441

b Not known.
* $R_{293\cdot 6\,K}/R_{20\cdot 4\,K}$.
** Static vacuum.

zone melting using a screened tungsten ring cathode. They determined an effective k value for oxygen between 1.5 and 2. This determination was successful only if the travel speed was 0.4 cm/h; at higher speeds (11.4 and 3.8 cm/h, respectively) no such distribution was found. Good agreement existed between calculated and experimental values for oxygen related to the initial concentration.

(b) Zone-melting of vanadium, niobium and tantalum

Of the metals of the Fifth Group of the Periodic Table, niobium and tantalum have sufficiently low vapour pressures for the application of the zone-melting process in a vacuum. For vanadium, the vapour pressure at the melting point is of the same order of magnitude as that of molybdenum and tungsten, and the losses owing to evaporation are therefore correspondingly high at 30%[442]. For all three metals the evaporation of the impurities plays an important role during zone melting. Uniform resistance ratios over the whole length of the rod were observed for a high zone-melting speed of 12–30 cm/h in niobium[443]. For a slow melting speed of 1.25 cm/h gradients in the hardness[442] as well as in the resistance ratio[443] were found over the length of the rod which indicates that the first sections have a higher impurity level than the end parts. Whilst Tedmon and Rose[443] attribute this phenomenon to zone-melting effects ($K_{\text{eff}_W} > 1$, $K_{\text{eff}_{Ta}} > 1$), Calverley[442] assumes that reactions between the residual gas atmosphere and the freshly solidified, but still hot, rod are responsible for the renewed contamination. Tedmon and Rose[443] also pointed out the great influence of the residual gas atmosphere on the resistance ratio, even if the travel speeds are high. This is shown in Fig. 91. The

References pp. 321–335

TABLE 52

ANALYTICAL RESULTS AND DISTRIBUTION COEFFICIENTS FOR ZIRCONIUM (DATA IN ppm)

Element	Before	After	k	Before	After Beginning	After End	Before	After Beginning	After End	Before	After Beginning	After End	k	Before	After
Hf	125			50–200	95	80				65	65	60	>1	1000	b
Ti															
Cr	30	<2	0.35	<20	<10					0.86	0.06	0.3	0.25		
Fe				50–200	<10		880	8	92	5.1	1	2.7	0.34	3	b
Mn															
Ni	6	0.7	0.3–0.4	<100	<10		19	0.6	34	0.35	0.3	2.9	0.11	70	b
Hg										0.002	0.04	0.002	>1		
Cu				<100	<10					0.41	0.1	0.02	0.19	100	b
Ag										0.68	0.13	2.5	0.62		
Mg														20	b
C				5–100	20–60		180	140	130						
O				5–100	32	10	12	50	170	280	170	300			
N				<1	<1	<1				32	35	18	1.14		
Reference	436			438			437			437, 439				441	

b Not found.

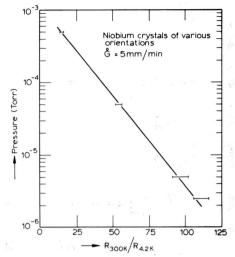

Fig. 91. Residual resistance ratios $R_{300K}/R_{4.2K}$ of niobium rods zone-melted at various residual gas pressures[443].

low resistance ratios observed by Ta and Nb are very likely caused by reaction of the residual gases with these metals. A compilation of results together with the experimental conditions is shown in Table 53.

The purity degree after one passage of the zone is only slightly changed after

TABLE 53

WORKING CONDITIONS FOR THE ZONE-MELTING OF V, Nb AND Ta

Metal	Dimensions of the rod		Travel rate (cm/h)	No. of passages	Pressure (Torr)	Method of melting	Resistance ratio R_{295K}/R_{10K}	Refs.
	Diam. (mm)	Length						
V	2–5	180			10^{-4}	Ring gun	27; 55	442
V	9		5.6	6	10^{-5}	Inductive 5MHz		444
V	11		12	4	10^{-3}	Inductive		441
V						3 Pierce guns	20*	473
Nb	5.5	125	7	1	1.6×10^{-6}	Ring gun		446
Nb	5.5	125	2.3	6	1.6×10^{-6}	Ring gun	63	446
Nb	2–5	180			10^{-4}	Ring gun	205	442
Nb	7	200	12.5	1	1.6×10^{-6}	Ring gun		447
Nb	7	200	3.75	1	1.6×10^{-6}	Ring gun		447
Nb	3		1.2; 12; 18; 24; 30	1	5×10^{-6}	Ring gun	80–90*	443
Nb	3		30	1	See Fig. 91	Ring gun	See Fig. 91	443
Ta	5.5	125	2.3	6	1.6×10^{-6}	Ring gun	250	446
Ta	2–5	180			10^{-4}	Ring gun	72	442
Ta			24	3		Ring gun		448
Ta	6	300			10^{-6}–10^{-7}	Ring gun		449

* 295 K/4.2 K
** 300 K/4.2 K

References pp. 321–335

TABLE 54

ANALYTICAL RESULTS OF ZONE MELTING EXPERIMENTS ON V, Nb AND Ta

Metal	V			V			Nb		Nb			Ta		Ta		Ta		Ta	
Reference	444			441			443		450			448		442		449		447	
	Before	After		Before	After		Before	After	Before	After		Before	After	Before	After	Before	After	Before	After
		Beginning	End		1	4				1	6								
Impurities (ppm)					Passages					Passages									
O	300	300	300				100	120	460	84	7	100	<30	10	3		4	60	30
N	b	<50	b				50	7	82	38	25	35	<10	7.5	b		8		
H	<300	100–300	<300				1	b						4.5	2.5		0.3		
C	<300	<10	b				20	5	51	35	33	30	10–30				10		
W					b	b				a	<5		1000						
Ti	300	300	b	300	100	30						10	b						
Si	<300	<50	b				50	a		a	<50	50	b						
Mn	<300	<10	b		3	7						50	1						
Mg	<10	<50	b																
Fe	<300	<50	<10	30	1000	800	50	a	10	<10	<10	20	b			5			
Cu	<50	<10	<10	3000	b	b						50	b						
Cr	<300	<10	<50	20	b	b			10	<10	<10	10							
Co	<300	<10																	
Ca	<10	<300	b																
Al	<10	<300	b		30	b	100	a				50	b						
Ta									1000	1000	1000								
Zr									5000	a	1000								
B										a	5								
Mo												25	b				15		
Nb												45	10				50		
Ni												40	b						

a = Not known   b = Not found

further passages. There are also almost no differences between beginning and end regions. It should, however, be mentioned here again that the accuracy of the analytical methods leaves much to be desired. Published results are shown in Table 54.

It is interesting to note from these results that it is very difficult to lower the oxygen and nitrogen levels below the 10 ppm limit; the best values[449] were obtained at pressures below 10^{-6} Torr. The high tungsten contents of tantalum in the zone-melting experiments of Ferriss et al.[448] can probably be attributed to the use of a tungsten cathode in the electron gun. Most authors employ a cathode made of the same metal as that to be refined in order to prevent contamination by the evaporating cathode metal.

(c) Zone-melting of molybdenum and tungsten

Although molybdenum and tungsten show a relatively high vapour pressure at the melting point (see Fig. 45 of Subsection 4.1), they can be zone-melted successfully in a vacuum and many experimental results have been published. Theoretically all elements may be eliminated to trace levels by evaporation from tungsten during conventional melting, and almost all elements, with the exception of Re, Ta and W, from molybdenum. This phenomenon is also present, of course, in zone-melting and intensifies its refining and purification effects. Analytical data for zone-melted tungsten and molybdenum as well as for the starting materials are compiled in Tables 55 and 56.

It should be remembered here that besides tungsten, silicon and carbon too are difficult to eliminate from molybdenum. Owing to the evaporation of most of the impurities, a distribution effect can be established with certainty only for carbon, although low values may be obtained only with great difficulty. Most authors attribute this to reactions taking place between the rods and the dissociation products of diffusion-pump oils. Ivanovsky and Zagorskaya[464] were able to demonstrate, during zone-melting experiments on molybdenum at a residual gas pressure of 10^{-7} Torr, that the residual resistance ratio $R_{298K}/R_{4.2K}$ was 1500 after refining in an oil-free vacuum vessel, whereas it was only 900 after refining in a vessel evacuated in the conventional way. Barthel and Petri[458] found a partition also for molybdenum in tungsten.

Hay et al.[465] showed by means of mass-spectrometric investigations that the majority of the impurities evaporate during the electron-beam melting of tungsten, though the analyses show an inhomogeneous distribution of the impurities, for which at present no plausible explanation can be offered. A distribution effect was established only for C, Ta and Re; all three elements accumulate at the end of the bar. On the other hand, molybdenum showed extensive evaporation and a uniform distribution over the length of the rod. Analytical data are compiled in Table 57; the experimental conditions are shown in Table 59. In the data of Table 57 it should be noted that part A was not melted, that position B marks the approximate

304   EXTRACTIVE METALLURGY   Ch. III

TABLE 55

ANALYTICAL RESULTS OF ZONE-MELTED MOLYBDENUM IN ppm

Element	Starting material	After one passage	After six passages	After	After	After one	After five	Arc-melted	After Beginning	After End	Before	After	Before	After three passages	After three passages*
						Passages	Passages								
W	<5	<5	<5	170											
Al	<5	<5	<5			1–5	2–4				50	50	135	162	0.2
Cr	10	<2	<2					10	<20	20	10	<0.2	9	0.18	0.1
Cu	10	<2	<2			1	1	40	11	1	5	<0.1	120	0.36	0.07
Fe	40	5	5			1–8	2–8	40	<1	<1			270	0.8	0.1
Mg	1	<1	<1			13–50	16–28				100	<0.1	800	3.2	
Mn	<2	<2	<2					—	<2	7					
Ni	<3	<3	<3			4	4	70	<1	<1	5	<0.2	6	0.12	0.02
Si	<15	<5	<5			5–13	5–10	20	25	20	50	50	80	4	0.1
Sn	20	<5	<5												
Cd				2											
Co								100	<2	<3					
O	60	<1	<1		4.5	3–20	3–24				10	<0.1		0.04	
N	20	<1	<1		1.0	0.3–1	0.3–3				P 5	<0.2			
H	<1	<1	<1		0.7	1–6	2–5				B 0.1	<0.1			
C	40	13	12		15	19	7–23	80	35	85	Be 5	5			
											V 1	1	0.4	0.08	
Reference		451	452	453	449	454		455			456, 457			458	

* Starting material of higher purity.

TABLE 56

ANALYTICAL RESULTS OF ZONE-MELTED TUNGSTEN (ppm)

Element	Before	After	Before	After	After	After	After	Before	After	After
Mg								<10	<1	0.2
Ca	10	b					1.5	10–100	<1	0.5
K	40	b						10–100	1–10	0.4
Na	20	b								
Al							0.1	<1	<1	0.5
Fe	10	b	40	<10			0.2	50	<1	1
Mo	40	1	80	<30			0.4	10–100	1–10	3
Si	20	b						10–100	1–10	1
In							0.05			
As							0.1			
Ge								10–100	1–10	
Zn							2			
Cu							0.2			0.5
Co							0.03	1–10	1–10	
Ni							0.3	<50	<1	0.5
Cr							0.5	1–10	<1	0.5
O	3	1	140	4	2.6	1				0.8
N	0.3	b			7.0	0.5				
H	0.1	–			0.4	0.3				
C	70	20				40				<1
Reference	459		460		449	461	462, 463	456		458

b Not found.

starting point of the zone-melting and that from here measurements were made every 2.5 cm over the whole length of the bar.

The data of Table 55 for Mo show that obviously after only one passage a degree of purity is obtained which is very similar to that found after several further passages. It may also be seen that only very slight concentration differences between beginning and end sections exist. This behaviour indicates that an evaporation of the impurities plays a decisive role in the refining. While Carlson[466] and Pugh[467] assume that tungsten is refined mainly by the evaporation of impurities, Schadler[459], Koo[454] and others showed, with the aid of resistance measurements, that true zone-melting effects exist in addition to the evaporation. Kovtun et al.[468] express a similar opinion with regard to molybdenum. Barthel and Petri[458] showed that for tungsten an increase in the refining effect exists after several zone passages for a number of impurity elements, for instance Mo, V, Ni, Mn, Cr, K, Al and C, while no such effect was observed for others, e.g. Cu, Fe, Ca and O.

The change in electrical resistance in a zone-refined tungsten rod after one zone passage and after five zone passages is plotted in Fig. 92, thus showing the difference in purity of the individual zones. As a result of the difficulties existing in the analytical determination, with this method differences are not recognisable. Lawley et al.[451] could not detect any concentration differences in the residual impurities after the first passage in zone-refining experiments on molybdenum

TABLE 57

MASS-SPECTROMETRIC ANALYTICAL RESULTS OF A
ZONE-MELTED TUNGSTEN ROD AFTER ONE PASSAGE[465]

Element	Impurity level (ppm) at position								
	A	B	C	D	E	F	G	H	I
B	1.3	0.08							
C	25	8	8	4	4	7	10	20	16
N	39	22	14	8	11	58	3	7	15
O	4	3	3	6	4	2	2	2	3
Mg	0.01	0.008	0.002	0.0009	0.0006	0.0003	0.003	0.001	0.008
Al	0.002	0.002	0.001	0.0006	0.0003	—	0.001	0.001	0.003
Si	0.07	0.03	0.02	0.02	0.01	0.07	0.03	0.02	0.07
S	0.04	0.08	0.05	0.03	0.01	0.01	0.03	0.02	0.1
K	0.13	0.05	0.02	0.01	0.02	0.03	0.02	0.01	0.03
Ca	0.01	0.01	0.003	0.002	0.001	0.0003	0.002	0.001	0.006
Tl	0.03	0.02	0.003	0.003	0.003	0.005	0.02	—	0.03
Ti	40	41	30	30	6				
V	0.2	0.1							
Ge	0.1	0.5	0.1	0.1	0.03	0.01	0.1	0.02	0.1
Fe	0.05	0.07	0.03	0.02	0.01	0.004	0.02	0.01	0.03
Ni	0.01	0.04	0.01	0.01	0.006	0.001	0.01	0.004	0.013
Co		0.06	0.01	0.002	0.003		0.03		
Cu	0.01	0.002		0.0002			0.001		
Mo	2.2	1.6	0.3	0.04	0.05	0.3	0.3		0.4
Ta	2.4	0.5	1.1	1.0	1.5	0.7	2.4	2.3	3.3
Re	17	5	3	3	3	4	10	7	16

compared with further passages, but they found a decrease in the hardness up to the fifth passage, after which the hardness remained constant. Barthel and Petri[458] showed during their systematic studies on molybdenum that the purity obtained after three passages is to a certain extent determined by the purity of the starting material.

The relatively high vapour pressures of molybdenum and tungsten have led to certain difficulties in a number of zone-refining experiments[455-457]. In most cases relatively high travel rates were employed for lowering the evaporation losses, as may be seen from Tables 58 and 59, in which the data on the experimental conditions are collected. Buehler[456,457], therefore, was only able to melt for 20 min at a time at his relatively slow travel rates owing to heavy evaporation of the tungsten and of the molybdenum. The refining process had to be interrupted as the windings of the induction coil became coated, and as the electrical power was absorbed by the deposit.

Rexer[462] and his co-workers[463] also carried out extensive investigations on the zone-melting of tungsten. The usual high travel rates (see Table 59) of about 20 cm/h were reduced by Rexer to relatively low travel rates of 1.5–6 cm/h in order to demonstrate the relationship between evaporation loss and resistance ratio. The evaporation losses or the evaporation rate, respectively, depend on the power input. For constant travel rate the evaporation losses vary as the residual

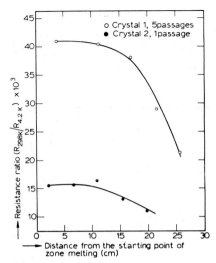

Fig. 92. Change of the residual resistance ratio $R_{298\,K}/R_{4.2\,K}$ over the length of zone-melted tungsten rods[454].

resistance ratio, as may be seen from Fig. 93. This effect is more pronounced for thinner than for thicker bars, *i.e.* a higher purification is obtained by the greater evaporation. On the other hand, it also becomes evident that evaporation alone does not determine the degree of refining. That transport processes in the melt also have a role to play is particularly evident from the data of Table 60, as the purity increases to a considerable degree with the number of passages for the same evaporation loss.

No improvements with regard to purity were obtained after more than three zone passages at these travel rates, and a rise of the pressure in the vacuum vessel to 7×10^{-5} Torr resulted in only slightly poorer resistance values. The many

TABLE 58

WORKING CONDITIONS FOR THE ZONE-MELTING OF MOLYBDENUM

Dimensions of the rod		Travel rate	No. of passages	Pressure (Torr)	Method of melting	Resistance ratio	References
Diam. (mm)	Length	(cm/h)				$R_{273\,K}/R_{4.2\,K}$	
4.2	80–110	18	3	5×10^{-6}	Ring gun	5270	452, 453
3	200	12	6	10^{-5}	Ring gun		451
3.5–6		22	3	10^{-6}	Ring gun	cf. Table 62	469
2		30	5–8	10^{-4}– 10^{-5}	Ring gun	900	470, 471
4.5	200–250	10	4	10^{-5}– 5×10^{-6}	Inductive 2–6 MHz	3300	456, 457
6–9.5		9 and 18			Ring gun		455
10					3 Pierce guns	15000	468, 472, 473
5–7		18	3	$<10^{-5}$	Ring gun		458

References pp. 321–335

TABLE 59

WORKING CONDITIONS FOR THE ZONE-MELTING OF TUNGSTEN

Dimensions of the rod		Travel rate (cm/h)	No. of passages	Pressure (Torr)	Method of melting	Resistance ratio $R_{298K}/R_{4.2K}$	Refs.
Diam. (mm)	Length						
3.5–6		22	5	10^{-6}	Ring gun	cf. Table 61	469
3.5–6		10	1–3	$10^{-4}, 10^{-6}, 10^{-8}$	Ring gun	cf. Table 61	469
1.6	350	15	1–3	5×10^{-6}–4×10^{-7}	Ring gun	cf. Figs. 94 and 95	474
3		18	2	5×10^{-5}	Ring gun	5400	459
3–4.5		15	1	2×10^{-6}	Ring gun		460
2–5	70–110	1.5–6	1–3	5×10^{-6}	Ring gun	cf. Table 60 and Fig. 93	462, 463
6	300			10^{-6}–10^{-7}	Ring gun	2365	449
3		18	5–7	10^{-4}–10^{-5}	Ring gun		461
3–4.5		4.5		1×10^{-5}–5×10^{-6}	Inductive 2–6 MHz		456
0.5	250	30	1	10^{-5}	Ring gun	15000	454
0.5	250	30	5	10^{-5}	Ring gun	40000	454
2			5–8	10^{-4}–10^{-5}	Ring gun	1400	471
5–7		18	3	$<10^{-5}$	Ring gun		458
1.5	250				Ring gun	21000–25000	465

analyses which were also carried out revealed the refining effect to a lesser extent than did the changes in the resistance ratio.

Belk[455] conducted zone-melting experiments on molybdenum and found evaporation losses of 27% at travel rates of 2.4 cm/h; these quantities are comparable with those obtained by Rexer for tungsten.

Murray et al.[469] pointed out, as a result of their studies on molybdenum and

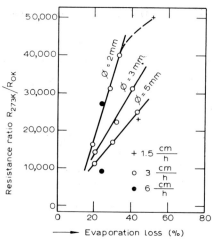

Fig. 93. Effect of evaporation loss—and therefore the melting temperature—on the residual resistance ratio of zone-melted tungsten rods in relation to the rod diameter[462, 463].

TABLE 60
EFFECT OF NUMBER OF PASSAGES ON THE RESISTANCE RATIO OF TUNGSTEN[462,463]

Diameter of the rod (mm)	Pressure (Torr)	No. of passages	Travel rate (cm/h)	Evaporation losses (%)	Resistance ratio R_{273K}/R_{0K}
2	5×10^{-6}	1	1.5	43	31 000
2	5×10^{-6}	2	3	43	55 500
5	5×10^{-6}	1	1.5	44	23 200
5	5×10^{-6}	2	3	44	83 000
5	5×10^{-6}	3	6	57	125 000

tungsten, the particular importance of a good vacuum. Their results are contained in Table 61, and the working conditions are collected in Table 59. These authors conclude, from the pronounced pressure dependence of the resistance ratio, that considerable evaporation of the impurities takes place at slow travel rates, and that, therefore, the transport in the gas phase may become the rate-determining step. They too consider, however, a reaction of the residual gas atmosphere with the hot, although not necessarily molten, zones of the bar.

These authors were able to demonstrate in further experiments that the diameter of the tungsten and molybdenum rods and the agitation of the molten zone exercise a decisive influence, thus indicating that the transport in the liquid phase should be rate-determining. The refining effect decreases with increasing diameter of the bars. If, however, the lower part of the bars is rotated at 240 rev/min, a considerably higher resistance ratio was obtained (Table 62).

Hay and Scala[474] studied the effect of an electrical diffusion by a superimposed electric field on the refining of tungsten during zone-melting in an electron-beam furnace. They used currents of 10 and 20 A, corresponding to current densities of 500 and 1000 A/cm², Potentials of 0.125 and 0.25 V/cm, respectively, were applied. The experimental data are contained in Table 59, whereas the results are shown in Fig. 94. The results, obtained without a superimposed direct current, are in agreement with those reported by Murray et al.[469]. If an electric field (+0.125 V/cm) is superimposed parallel to the direction of travel of the molten zone, poorer results were observed. However, reversion of the direction of the field and an increase in potential led to a further increase in the resistance ratio values.

TABLE 61
EFFECT OF PRESSURE AND NUMBER OF PASSAGES ON THE RESIDUAL RESISTANCE RATIO $R_{298K}/R_{4.2K}$ DURING THE ZONE-MELTING OF TUNGSTEN[469]

No. of passages	Pressure (Torr)		
	10^{-4}	10^{-6}	10^{-8}
1	3300	10 000	33 000
3	a	25 000	50 000
5	a	68 000	a

a Not known.

TABLE 62

RESISTANCE RATIOS $R_{298\,K}/R_{4\cdot 2\,K}$
OBTAINED IN ZONE-MELTING EXPERIMENTS ON MOLYBDENUM
AND TUNGSTEN IN RELATION TO THE DIAMETER OF THE BARS[469]

Diameter of the bar, mm	Molybdenum at 10^{-6} Torr and three passages	Tungsten at 10^{-8} Torr and a single passage
3.5	8 000	>40 000
6	2 000	4 000–10 000
6 with additional rotation of 240 rev/min	12 000	

The progress of refining can be described to a first approximation by a first-order law with time, as is evident from Fig. 95. From the results obtained in the experiments with a superimposed electric field, the authors[474] concluded that the diffusion at the solid/liquid phase boundary is mainly rate-determining. It is also assumed that the major part of the impurities migrates towards the anode, in other words, that their valencies should be regarded as negative.

(d) Zone-melting of the platinum-group metals

Of the platinum-group metals only platinum and rhodium have a sufficiently low vapour pressure at the melting point to be zone-melted in a vacuum without too great evaporation losses. On the other hand, the melting points of ruthenium, osmium and iridium are so high that any impurities present evaporate almost completely during zone-melting. Therefore, with a travel rate sufficiently high, the evaporation losses of the platinum-group metals can be kept within limits. The still higher vapour pressure of palladium results in considerable difficulties in the

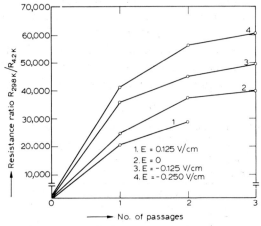

Fig. 94. Effect of a superimposed electric field on the purity of zone-melted tungsten rods in relation to the number of zone passages[474].

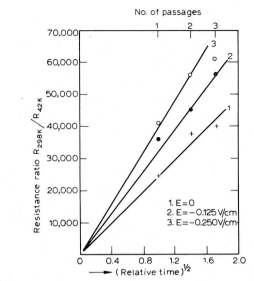

Fig. 95. Effect of melting time on the residual resistance ratio of zone-melted tungsten rods[474].

zone-melting using a ring cathode, but in principle zone-melting even of palladium is feasible[475]. In the zone-melting of ruthenium, employing a tungsten ring cathode, the cathode had to be exchanged after each passage owing to the heavy deposit of ruthenium metal on it[476].

Whereas Rhys[477] obtained stable melting zones even during the first zone

TABLE 63

WORKING CONDITIONS FOR THE ZONE-MELTING OF THE PLATINUM GROUP METALS

Metal	Dimension of the rods (mm)		Travel rate (cm/h)	No. of passages	Pressure (Torr)	Method of melting	Resistance ratio $R_{300K}/R_{4.2K}$	References
	Diam.	Length						
Ru	5		15	14		Ring gun		450, 477
Ru			7.5	2	10^{-5}	Ring gun		475
Ru			1.25	1	10^{-5}	Ring gun		475
Ru	5	100		3	10^{-6}	Ring gun	440	476
Ru	5	100		4	10^{-6}	Ring gun	710	476
Ru	5	100		6	10^{-6}	Ring gun	900	476
Ru	5	140		2	10^{-6}	Ring gun	750	476
Ru	5	140		5	10^{-6}	Ring gun	1000	476
Ru	5	140		6	10^{-6}	Ring gun	1100	476
Rh			15			Ring gun		478
Rh	~7	~120	7.5	2	10^{-5}	Ring gun		475
Rh	4					Ring gun		477
Ir				15		Ring gun		450, 477
Pt	5		60	12		Ring gun		477
Pt	5		15	15		Ring gun		450, 477
Pt	~7	~100	7.5	2	10^{-5}	Ring gun		475

References pp. 321–335

TABLE 64

ANALYTICAL RESULTS OF ZONE-MELTED PLATINUM GROUP METALS
Data in ppm with the exception of the basis metal

Element	Before	After beginning of the rod	Before	After		Before	After		Before	After	
				Beginning of the rod	End		Beginning of the rod	End		Beginning of the rod	End
Ru	Basis metal		c	c	c	200	200	c	c	c	c
Ir	100–1000	200–2000	99.856	99.997	99.959	c	c	c	c	c	c
Pt	10–100	10–<100	210	10	30	99.94	99.97	99.996	99.998	99.998	99.998
Pd	20–100	c–<5	370	c	140	80	10	30	9	4	5
Rh	10–100	c–<10	330	10	90	20	20	10	c	c	c
Au	2–20	2–20	Traces	c	c	1	c	c	1	c	c
Ag			Traces	c	c	100	1	1	2	c	c
Ca						3	1	2	1	2	2
Cu	2–20	c–<10				3	1	1			
Fe	20–200	5–50	520	10	20	150	10	10	10	10	10
Ni	20	20	10	c	130	20	10	10			
Mg	2–20	1									
Pb	5–50	c–<5	1	c	c	2	c	c			
References	450, 475		450, 477			477			450, 477		

c Not found

passage in the melting of Pt, Ru and Ir despite considerable evolution of gases, Schriempf[476] found it very difficult in his experiments on ruthenium to keep the melt in a stable condition owing to the low surface tension, the high vapour pressure, and the high thermal conductivity of this metal. The melting conditions are contained in Table 63.

Schriempf[476] obtained maximum values of 900 for the resistance ratio $R_{300K}/R_{4.2K}$ of ruthenium of a starting purity of 99.9% having a resistance ratio of 100; if 99.999% ruthenium was used as the starting material, the resistance ratio was increased from the original 500 to a maximum of 1100. On the other hand, Rhys[477] found a slight deterioration of the resistance ratio during the zone melting of platinum.

All authors attribute the refining effect mainly to the evaporation of the impurities. This is corroborated by the results of analyses, as about equally low impurity levels were found for the beginning and the end of the bars. Allred and others[475,478] found no marked changes in the concentration of the impurities, except for Cu in Pt and Rh and for Ca in Rh, in the zone-melting of Rh and Pt at the relatively high travel rates of 7.5 cm/h. Ruthenium also showed considerable changes in impurity concentration only after application of a much slower travel rate (1.25 cm/h), as may be seen from the data of Table 64. The working conditions are shown in Table 63.

In the zone-melting of Ru alloys (0.1% Ni or 0.1% Rh or 0.1% Ni+0.1% Rh) mainly an evaporation of the alloying elements (to 1–2 ppm Ni, 90–190 ppm Rh; 1–10 ppm Ni+10–250 ppm Rh) was observed besides a slight partition effect[477]. In all cases effective degassing of the molten zone took place.

(e) *Zone-melting of tin, lead and bismuth*

Of the low-melting-point metals, tin, bismuth and lead can easily be zone-melted with relatively poor vacua. In all cases melting is carried out in crucibles as no reaction with the crucible is likely owing to the low temperatures involved. The necessary working conditions are listed in Table 65. Geil and Ziegenbalg[480] found, in their experiments on lead, k values between 0.63 and 0.48 for silver. Aleksandrov and D'jakov[481] quote a purity of more than 99.999% for their zone-melted lead; the most difficult elements to eliminate were Sn ($k = 0.7$) and Sb, Bi, Mg and Na ($k = 0.4$–0.6).

In the zone-melting of bismuth, Wernick et al.[483] obtained for Ag, Cu, Pb, Sn, Ni, Mg and Ca k values below unity; for iron it is said to be above unity. Geil and Ziegenbalg[480] found, in their zone-melting experiments on bismuth, k values for Ag of 0.23–0.26 and for In of 0.19–0.18. With increasing travel rate the effective partition coefficients quickly approach unity, as may be seen from Fig. 96.

More recently, Geil[482] reported the concentration dependence of k_{Pb}. With low concentrations the k_{Pb} value is 0.5. Above 0.12 at.% Pb k quickly rises to unity; in other words, a separation of lead and bismuth by zone-melting is no longer possible.

Geil[485] intends to produce bismuth of a purity of 99.9999% by a multistage zone-melting process. Aleksandrov[484], by a combined distillation and zone-melting process, succeeded in producing bismuth of a higher purity than 99.999% from commercial bismuth containing Sb 0.7, Fe 0.01 and Cu 0.003%.

Tanenbaum et al.[479] reported for tin that Pb, Cu and Fe were the only impurities which were detected after zone-melting at the end of the bar.

TABLE 65

WORKING CONDITIONS FOR THE ZONE-MELTING OF TIN, LEAD AND BISMUTH

Metal	Dimensions of the rod		Travel rate (cm/h)	No. of passages	Pressure (Torr)	Method of melting	Resistance ratio $R_{273K}/R_{4.2K}$	Refs.
	Diam. (mm)	Length (mm)						
Sn		250		40	Vacuum			479
Pb		220	6	10–20	Vacuum	Induction 900 kHz C-Crucible		480
Pb	12 × 10	550	2.5	65	10^{-1}	Resistance Ta-Crucible	16000	481
Bi		220	6	10–20	$<10^{-4}$	Induction 900 kHz C-crucible		480, 482
Bi			4,5	45	2.5×10^{-2}	Induction Pyrex	600	483
Bi Analytically pure	11 × 14		2.5	60	10^{-1}	Resistance Ta-crucible	400	484
Bi Commercial purity	11 × 14		2.5	60	10^{-1}	Resistance Ta-crucible	240	484
Bi Hilger	4 × 4		2.5	8–30	10^{-1}	Resistance Ta-crucible	110	484

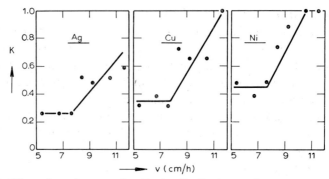

Fig. 96. Effect of travel rate on the effective distribution coefficients k_{Ag}, k_{Cu} and k_{Ni} in bismuth[485].

(f) Zone-melting of beryllium

Beryllium is usually zone-melted under a protective gas, such as helium or argon, because of its relatively high vapour pressure at the melting point. Only in a few cases was zone-melting carried out in a high vacuum[486] or at a reduced pressure of a few hundred Torr[487,488].

In this process the extensive evaporation of such easily volatile elements as Mg, Ca and Zn is so successful that zone-refining effects are no longer noticeable and that, therefore, these impurities are quite uniformly distributed over the length of the bar, whereas Al and Si accumulate at the end[487]. Edwards and Martin[488] found a reduction of the Fe, Al and Si contents. No marked changes were found in the contents of dissolved oxygen and hydrogen.

The refining effect could be markedly increased by means of a superimposed electric field, although according to Grinyuk et al.[486] the evaporation of volatile components also plays an important role. The conditions for zone-melting are shown in Table 66.

(g) Zone-melting of rare-earth metals

Refining of rare-earth metals is preferably carried out by distillation.

Experiments on the zone-melting of rare-earth metals have been carried out occasionally, but the results obtained were not very satisfactory[501-504]. While the fluorine levels could be effectively lowered from 0.06% to 0.002% by evaporation of oxyfluoride[501,502], Fe, Cr, Cu and Zr were only removed to a small degree, and O, N, Ni and Ti not at all[501,503]. According to Necker[502] metallic impurities could be reduced from 4300 to 1800 ppm after six passages. The travel rates varied between 6.5 and 15 cm/h. It is said that better effects are obtainable by use of a superimposed electric field[505].

(h) Zone-melting of uranium and plutonium

Uranium is zone-melted in a vacuum, observing special precautions. In this

TABLE 66

WORKING CONDITIONS FOR THE ZONE MELTING OF SEVERAL OTHER METALS AND INTERMETALLIC COMPOUNDS

Metal	Dimensions of the rod		Travel rate (cm/h)	No. of passages	Pressure (Torr)	Method of melting	Resistance ratio $R_{273\,K}/R_{4.2\,K}$	Refs.
	Diam. (mm)	Length (mm)						
Si			30	2	6×10^{-5}			489
Si	18	120	10	Up to 30	$<3 \times 10^{-5}$	High-frequency induction		490
Co	5		30	10	10^{-4}–10^{-5}	Electron beam		491
Co	3	200	25	1	10^{-5}	Electron beam		492
Ni	4.5		5.5	4	10^{-5}	High-frequency induction	835	444
Ni	6		4	5	5×10^{-5}	High-frequency induction	1300	493 494
Ni			2.5	3	5×10^{-6}	Electron beam	2780	494
Ti	6		3.1	6	10^{-5}	High-frequency induction		444
Al	4.5	250	3	10	10^{-5}	High-frequency induction		495
Al		600	4.2	Up to 15	10^{-4}–10^{-5}	High-frequency induction		496
B	6	200	0.6	Approx. 5		High-frequency induction		497
Be	16	100–150	3.8–7.5	12	125	High-frequency induction		487
Be	5–19	150–250	0.45–4.5	2	300	High-frequency induction		488
U		300	0.55	9	8.10^{-8}–3.10^{-7}	High-frequency induction	[298 K/ 20.4 K] 28	498
In		380		30	Low hydrogen pressure	Resistance heating elements		499
In		300	1 and 7.5 resp.	5 each	Low hydrogen pressure			500
InSb		200	1	10–15	Low hydrogen pressure			500
InAs			4	3	0.3 atm As_n pressure			499

process electrolytic uranium containing a total impurity level of about 200 ppm was refined to an impurity content of about 100 ppm after nine zone passages at a pressure of 10^{-7} Torr. The corresponding data are shown in Tables 66 and 67[498]. In addition to the impurities quoted in Table 67, the contents of Ta, Zr, As, Cl and S remain unchanged; those of Hf, Se, Zn, Co, Cr, V, Ti, K, F and B were reduced to a third or less. Resistance ratio measurements showed that the first part of the bar was considerably more purified than the end.

Bieber et al.[506] even succeeded in the preparation of uranium with a total impurity content of 30 ppm.

During normal freezing of plutonium, B, Cr and Fe could be accumulated in the residual melt, while no change could be detected for Mg, Mn and Si; the

TABLE 67

ANALYTICAL RESULTS FOR THE ZONE MELTING OF NICKEL, TITANIUM AND URANIUM (ppm)

Main metal impurity	Ni			Ti			Si			U	
	Before	High frequency 5×10^{-5} Torr 4 cm/h	Electron beam 5×10^{-6} Torr 2.5 cm/h	Before	After Beginning	End	Beginning	After Middle	End	Before	After
Ni		Main metal		620	180	51				20	1
Mn	a	a	a	150	b	b	0.1	0.2	1	1	3
Mg	3	1	2	100	43	43					
Fe	12	3	2	420	87	170	<5	<10	25	50	20
Si	15	12	13	2400	620	1100				20	7
Cu	3	1	1	60	8	7	0.5	1	5	1	0.2
Ca	1	—	—	330	140	140				0.3	0.3
Al	1	—	—	88	—	—	<10	50	150	15	15
Ag	1	—	—				<0.1	1	1	0.01	0.01
H	11	<3	<3				Pb <1	<1	8		
N	10	<3	<3				Sn 4.9	12	100		
O	30	12	<3				Sb 3.8	4.2	200		
Reference		494			444			489			498

a Not known.   b Not found.

quantity of inclusions was much lower at the beginning than at the end of the bar.

In horizontal zone-melting only a slight partition was found after six passages at a travel rate of 1.5 cm/h, but at 0.8 cm/h a marked separation was detected for Ag, B, Na, Si and Sn. In diffusion studies in the solid state at about 500°C with a superimposed direct current of a current density of 250 A/cm², Al, Cr, Fe, Mg, Mn, Ni and Ti migrated in the direction of the anode, and Ce and B towards the cathode[507].

(i) Zone-melting of cobalt, nickel and titanium

Nickel, cobalt and titanium, owing to their not too high vapour pressures at the melting point, still belong to the group of metals which may be zone-melted in a vacuum with some prospect of success. The melting conditions for these metals are compiled in Table 66.

The main refining effect is attributed in almost all cases[444,493] to the evaporation of the impurities, since the values of the residual resistance ratio at the end of the bar exceeded those of the beginning only very slightly.

Analyses carried out on titanium[444] and nickel[494] before and after zone-refining show the same values. If in vertical zone-melting the lower end of the bar is rotated[493], purification is improved.

In the zone-melting of nickel, a lowering of the travel rate, as well as improving the vacuum, results in an increase in the refining effect, as is evident from the rise in residual resistance ratio (Table 66) and from the analytical data (Table 67)[494].

(k) *Zone-melting of copper*

Although the vapour pressure of copper at the melting point is relatively low, copper requires a high degree of superheating because of its high thermal conductivity, so that both vapour pressure and evaporation losses increase accordingly. Superpurity copper may be prepared by zone-melting in a vacuum in a graphite crucible, using as starting material electrolytic copper with 500–900 ppm oxygen. If the very high travel rate of 8 cm/h is applied, the evaporation losses may be kept within limits even in a vacuum of 10^{-5} Torr. After 10 passages the oxygen level is lowered to 5–12 ppm and after 20 passages to 2–7 ppm[508].

(l) *Zone-melting of aluminium, boron and indium*

Zone-melting of aluminium is in most cases carried out under a protective atmosphere. Akerman et al.[496] made attempts to conduct the zone-melting process in a vacuum and obtained very satisfactory results. The operating conditions are contained in Table 66, and the analytical results are compiled in Table 68. Conventional zone-melting of up to 15 passages did not show the same good results which were obtained in experiments with nine passages in which a quarter of the rod at the most impure end was cut off after every three passages and was replaced by a piece of 99.999% purity.

The highest values of the residual resistance ratio which have been achieved so far in the zone-melting of aluminium in a graphite boat lie around 10000. Although this value deteriorates somewhat during fabrication as a result of the creation of lattice defects, such an increase in the low-temperature conductivity of aluminium is of immense technological importance as it may perhaps offer the possibility of greatly reducing the conduction losses in electric power transmission even without the application of superconductors.

Boron may be zone-melted in a crucible made of boron nitride. In this process the rod is degassed at 300° and at 600°C before the zone-melting proper, in order

TABLE 68

ANALYTICAL RESULTS FOR THE ZONE MELTING OF ALUMINIUM (ppm)[496]

	15 Passages		9 Passages with exchange of impure end					
	Before	After	Before	After	Before	After	Before	After
Cu	8	4.5	4.9	0.8	18.9	2.0	12.1	1.5
Cr	12.9	2.0	3.2	0.4	7.4	0.65	7.2	2.5
Mn	0.34	0.25	0.2	0.25	0.25	0.2	0.26	0.2
Sb	<0.1	<0.1	<0.1	<0.1	<0.1	<0.1	<0.1	<0.1
As	6.7	<0.5	<0.5	<0.5	<0.5	<0.5	<0.5	<0.5
Sc	0.52	0.48	0.18	0.18	0.28	0.20	0.18	0.20
Fe			26.1	<10	24.7	<10	83.7	<10
Zn			1.1	0.32	3.4	0.72	1.94	0.4
Co			0.046	<0.04	0.3	<0.04	3.2	<0.04
Ga			<0.5	<0.5	1.8	<0.5	1.6	<0.5

to reduce the content of H_3BO_3. During zone-melting at temperatures above 2000°C quantities of B_2O_3 are still liberated. The operating data are listed in Table 66. During melting Mg is evaporated. Aluminium and carbon show k values greater than unity, the k of Ti is about unity and it is somewhat lower for Si and Cu; but Fe, Ni and Cr have k values very much lower than unity. The highest degree of purity is therefore obtained in the middle zone of the bar[497].

Indium may be zone-melted in a quartz crucible[499], which has been covered with a layer of pyrolytic graphite, or in a graphite crucible[500]. In this process, before the zone-melting proper, the whole bar is melted and kept in the liquid state at 800°C for prolonged periods of time[499,509], especially in order to remove indium oxide. The melting conditions are also contained in Table 66. While Hulme and Mullin[445] did not find any difference in the purity after zone-melting, Harman[500] showed that tellurium migrates to the end of the bar. The zinc content is also lowered, but probably mainly by evaporation.

(m) Zone-melting of silicon

As silicon has a relatively high vapour pressure at the melting point, zone-melting is therefore in most cases not carried out in a vacuum. It is the usual practice to melt in a moist hydrogen atmosphere in order to oxidize and to volatilize boron, which is rather difficult to eliminate otherwise ($k = 0.9$)[510].

Similar difficulties exist in the elimination of the carbon. Zone-melting experiments with up to 30 passages in a high vacuum of 3×10^{-5} Torr at travel rates of 10 cm/h resulted in a partition coefficient of $k = 0.92$. This value does not make possible a separation by zone-melting[490].

Bonnier et al.[489] carried out zone-melting experiments by electron bombardment in a water-cooled copper crucible. Marked concentration displacements were found after two passages at a travel rate of 30 cm/h. The k-values determined for Cu, Fe and Mn were lower than 10^{-3}, for Sn and Sb below 2×10^{-2} and for Al below 2×10^{-3}. Experimental conditions and analytical data are shown in Tables 66 and 67, respectively.

(n) Zone-melting of Group III–V compounds

Group III–V compounds play an important role in semiconductor technology. Owing to their binding conditions they should be regarded not so much as metallic alloys but as true compounds of stoichiometric composition. For this reason the Group III–V compounds may be refined and purified by zone-melting to a much higher degree than can pure metals or other alloys.

The compounds used in semiconductor technology are nitrides, phosphides, arsenides and antimonides of boron, aluminium, gallium and indium. Melting points and dissociation pressures of compounds of the $A^{III}B^{V}$ type are shown in Table 69 and are compared with the data for the corresponding elements. From these data it may be seen that the melting points of these compounds are very

TABLE 69

MELTING POINTS AND DISSOCIATION PRESSURES OF GROUP $A^{III}B^{V}$ COMPOUNDS [497]

Element	Melting point (°C)				Dissociation pressure (atm) at the melting point of the compound with		
	Of the element	Of the compound with					
		Al	Ga	In	Al	Ga	In
P	44	>2000	1470	1060	a	15–35	15–60
As	~810	~1700	1238	942	~2	0.89	0.25
Sb	630.5	1065	706	530	~10^{-4}	~10^{-5}	Negligibly small
Al	660						
Ga	30						
In	157						

a Not known

much higher than those of the elements, and that the arsenides, and particularly the phosphides, show vapour pressures of up to several atmospheres, even at the melting point. It is therefore the usual practice to melt in closed pressure vessels and in order to suppress any dissociation, a reservoir of the component which evaporates more readily is provided in a cooler zone of the container. A pressure corresponding at least to the dissociation pressure is maintained.

For preventing any oxidation during the melting operation the apparatus is always first evacuated. Since oxygen is one of the most detrimental impurities, the evacuation is carried out down to 10^{-6} Torr. Baking of the apparatus is usually

TABLE 70

EFFECTIVE DISTRIBUTION COEFFICIENT k_{eff} OF THE VARIOUS ANTIMONIDES AND ARSENIDES

	GaAs	InAs	AlSb	GaSb	InSb
Cu	<2×10^{-3}		1×10^{-2}		6.6×10^{-4}
Ag	1×10^{-1}		1×10^{-1}		4.9×10^{-5}
Zn	1×10^{-1}	7.7×10^{-1}		2×10^{-1}	2.3–3.5
Cd	<2×10^{-2}	1.3×10^{-1}		<2×10^{-2}	2.6×10^{-1}
Ga	—			—	2.4
Tl					5.2×10^{-4}
Si	1×10^{-1}	4×10^{-1}	1×10^{-1}		1×10^{-1}
Ge	3×10^{-2}	7×10^{-2}		8×10^{-2}	1.2×10^{-1}
Sn	3×10^{-2}	9×10^{-2}	$\leq 1 \times 10^{-2}$	<2×10^{-2}	5.7×10^{-2}
P	2				1.6×10^{-1}
As	—	—		2–4	5.4
S	3×10^{-1}	~1		6×10^{-2}	1×10^{-1}
Se	5×10^{-1}	9.3×10^{-1}	4×10^{-1}	1.8×10^{-1}	1.7×10^{-1}–1.9
Te	3×10^{-1}	4.4×10^{-1}	3×10^{-1}	4×10^{-1}	4.7×10^{-1}–4.2
Fe	<2×10^{-2}		2×10^{-2}		4×10^{-2}
Ni	<2×10^{-2}		1×10^{-2}		6×10^{-5}
In	1×10^{-1}	—		~1	—
References	497	497	497, 512	497	497, 511

necessary as well as an intensive purification of the protective gases, such as argon, helium, hydrogen or nitrogen.

Only for the antimonides and arsenides may this process be considered as a vacuum zone-melting and even then to a limited extent only. Even at vapour pressures of 10^{-2} Torr high evaporation losses must be expected for the usual travel rates of a few cm/h. The antimonides are therefore also usually melted under high-purity protective gases.

Only for InSb are the vapour pressures of the elements at 530°C with 10^{-4} Torr (Sb_4) and 2×10^{-7} Torr (In) so low, that melting may be carried out in a vacuum without any difficulty[511]. In this operation zinc and cadmium in particular are evaporated; both of these show unfavourable distribution coefficients, besides silicon, selenium and tellurium, almost independently of the initial $A^{III}B^{V}$ compound, as is evident from Table 70.

The melting conditions are also contained in Table 66. In many cases a vacuum treatment of the starting material is carried out before zone melting in order to eliminate zinc and cadmium in particular.

LIST OF SYMBOLS

A quantity of substance A (g)
B quantity of substance B (g)
C concentration (g cm^{-3})
D diffusion coefficient (cm^2 s^{-1})
f rate of crystallization in zone melting (cm s^{-1})
F surface area (cm^2)
$\overset{\circ}{G}$ travel rate in zone melting (cm s^{-1})
H_v heat of evaporation (cal mol^{-1})
h depth of pool of molten metal (cm)
i material transport (g cm^{-2} s^{-1} or g s^{-1})
k effective distribution coefficient
K proportionality constant
l distance from evaporator to condenser or length of molten zone in zone melting (cm)
L total length of zone-melted rod (cm)
M molecular weight (g mol^{-1})
P total pressure (Torr)
p_0 vapour pressure of pure substance (Torr)
p partial pressure (Torr)
R gas constant
r radius of crucible (cm)

T temperature (K or °C)
u, v flow velocity (cm s^{-1})
x quantity of substance A evaporated (g)
y quantity of substance B evaporated (g)
α evaporation coefficient (after Olette[213])
δ thickness of boundary layer (cm)
ε condensation coefficient
γ activity coefficient
λ thermal conductivity (cal cm^{-1} s^{-1} K^{-1})

Indices

1, 2 near the evaporator or the condenser respectively
D diffusion
ig inert gas
K condenser
Me metal
SE rare-earth metal
V evaporator
W heat transfer
A, B elements A, B

Acknowledgement

I wish to express my thanks to the companies and their members who assisted me in supplying me with pictures and information during the preparation of this contribution.

I am especially grateful to:
Balzers AG, Balzers (Dr. O. Winkler),
Imperial Smelting Processes Ltd., London (Dr. A. Temple),
Leybold-Heraeus GmbH & Co., Hanau (Dr. F. Sperner and Dr. H. Stephan),
Magnetherm S.A., Paris (C. Faure),
Société Minière et Métallurgique de Penarroya, Paris (J. Iche),
Süddeutsche Kalkstickstoff-Werke AG, Trostberg (Dr. Kaess and Dr. Knahl),
Union Carbide Corporation, Niagara Falls (C.G. Chadwick),
Vacuumschmelze GmbH, Hanau (Dr. I. Pfeiffer),
Vereinigte Aluminiumwerke AG, Bonn and Lünen (Dr. Bergmann).

REFERENCES

1 A. STIRLING, *J. Iron Steel Inst. (London)*, 177 (1954) 25.
2 M. HÄUSER, *Erzmetall*, 10 (1957) 443.
3 F. KÄMPF AND J. TUSCHE, *Erzmetall*, 20 (1967) 402.
4 D. OCEPEK AND E. EBERL, *Rudarsko-Met. Zbornik*, 4 (1963) 49.
5 R. PLANIOL, *ATB Met.*, 5 (1964/65) 239; cf. also *Proc. Intern. Conf. on Vacuum Met.*, Brussels, 1965, p. 235.
6 J. F. ELLIOTT AND M. GLEISER, *Thermochemistry for Steelmaking*, Vol. 1, Addison–Wesley, Reading, Mass., 1960.
7 R. HÖRBE AND O. KNACKE, *Erzmetall*, 8 (1955) 556.
8 O. KUBASCHEWSKI, E. LL. EVANS AND C. B. ALCOCK, *Metallurgical Thermochemistry*, Pergamon, Oxford, 4th edn., 1967.
9 H. H. KELLOG, *Trans. AIME*, 236 (1966) 602.
10 D. N. KLUSHIN, O. V. NADINSKAYA AND K. G. BOGATINA, *Obogashch. i Met. Tsvetnych Metal.*, 13 (1957) 211.
11 T. N. KOMAROVA, YU. V. RUMYANTSEV AND E. YA. OGNEVA, *Izv. Fiz.-khim. Nauchn.-issled. Inst. pri Irkutsk Univ.*, 7 (1) (1966) 95.
12 V. N. NESTEROV, A. L. TSEFT, R. A. ISAKOVA AND S. NAIMANOV, *Tr. Inst. Met. i Obogashch., Akad. Nauk Kaz. SSR*, 5 (1962) 77.
13 P. I. MECHENOV, *Tsvetn. Metal.*, 31 (1958) 48.
14 V. S. ESYUTIN AND I. A. ONAEV, *Tr. Inst. Met. i Obogashch., Tsvetn. Met.*, 4 (1962) 43.
15 R. A. ISAKOVA, V. N. NESTEROV AND A. L. TSEFT, *Tr. Inst. Met. i Obogashch., Tsvetn. Met.*, 8 (1963) 13.
16 R. A. ISAKOVA, V. N. NESTEROV AND L. E. UGRYUMOVA, *Tr. Inst. Met. i Obogashch., Akad. Nauk Kaz. SSR*, 13 (1965) 16.
17 A. S. SHENDYAPIN, R. A. ISAKOVA AND V. N. NESTEROV, *Tr. Inst. Met. i Obogashch. Akad. Nauk Kaz. SSR*, 13 (1965) 25.
18 R. A. ISAKOVA, I. S. CHELOKHSAEV, L. E. UGRYUMOVA AND A. L. TSEFT, *Tr. Inst. Met. Obogashch. Akad. Nauk Kaz. SSR*, 13 (1965) 38.
19 C. DECROLY AND M. GHODSI, *Mem. Sci. Rev. Met.*, 63 (1966) 109.
20 J. C. ACHARD, *Rev. Hautes Temp. Réfract.*, 3 (1966) 281.
21 C. DECROLY AND M. GHODSI, *J. Less-Common Metals*, 6 (1964) 375.

22 C. Decroly and M. Ghodsi, *Mem. Sci. Rev. Met.*, 59 (1962) 829.
23 C. Decroly and M. Ghodsi, *Mem. Sci. Rev. Met.*, 58 (1961) 138.
24 H. Clasen, Vakuumentzinkung von zinkhaltigen Kiesabbränden, paper presented at *10. Hüttenausschussitzung für Zink der GDMB*; also *Erzmetall*, 12 (1959) 367; also *DAS 1,035,909* (19.1.1957), *DAS 1,010,740*.
25 G. W. Kim and O. G. Miller, *Izv. Akad. Nauk SSSR, Tekhn. Nauk*, (1962) 93.
26 C. Decroly and M. Ghodsi, *Mem. Sci. Rev. Met.*, 61 (1964) 299.
27 J. Krüger, A. Melin and H. Winterhager, Applications de la technique du vide en métallurgie extractive, *Proc. Intern. Conf. on Vacuum Met., Brussels, 1965*, p. 13.
28 W. Hirschwald, O. Knacke and P. Reinitzer, *Erzmetall*, 10 (1957) 123.
29 W. L. Worrell and J. Chipman, *Trans. AIME*, 230 (1964) 1682.
30 W. L. Worrell, *Trans. AIME*, 233 (1965) 1173.
31 O. Knacke, J. Krahe and F. Müller, *Z. Metallk.*, 58 (1967) 814.
32 J. R. Piazza and M. J. Sinnot, *J. Chem. Eng. Data*, 7 (1962) 451.
33 A. Heiss and M. Dode, *Rev. Hautes Temp. Refract.*, 3 (1966) 245.
34 A. S. Darling, *Metallurgia*, 64 (1961) 7; 64 (1961) 71.
35 R. Kieffer, F. Lihl and E. Effenberger, Über die karbidothermische Herstellung von Metallen der IVa-, Va- und VIa-Gruppe und deren Legierungen, *Plansee Proc. 6th Seminar, Reutte-Tyrol, 1968*.
36 R. A. Karasaev, V. I. Kashin, M. S. Makunin, A. Yu. Polyakov and A. M. Samarin, *Izv. Akad. Nauk SSSR, Otd. Tekhn. Nauk*, 4 (1956) 94.
37 M. F. Joly, Processes for the production of ductile vanadium and niobium from their oxides, *Proc. 2nd Intern. Conf. on Peaceful Uses of Atomic Energy*, Paper A/CONF. 15/P/1274, p. 309.
38 M. S. Makunin, A. Yu. Polyakov and A. M. Samarin, *Izv. Akad. Nauk SSSR, Otd. Tekhn. Nauk*, 2 (1959) 35.
39 R. Kieffer, H. Bach and H. Lutz, *Metall*, 21 (1967) 19.
40 J. Krüger and H. Winterhager, *ATB Met.*, 5 (1964/65) 397; *Metall*, 20 (1966) 430.
41 J. Krüger and H. Winterhager, *Proc. Intern. Conf. on Vacuum Met. Brussels, 1965*, 1966, p. 108.
42 E. Effenberger, *Thesis*, Technische Hochschule, Wien, 1967.
43 R. Kieffer and H. Braun, *Vanadium, Niob, Tantal*, Springer-Verlag, Berlin, 1963.
44 D. J. Soisson, J. J. McLafferty and J. A. Pierret, *Ind. Eng. Chem.*, 53 (1961) 861.
45 N. P. Sazhin, P. O. Kolchin and N. V. Sumarokov, *Izv. Akad. Nauk SSSR, Otd. Tekhn. Nauk Met. i Toplivo*, (1961) 6, 8.
46 P. O. Kolchin, *J. Nucl. Eng.*, (1957) 179.
47 J. Niebuhr, *J. Less-Common Metals*, 11 (1966) 191.
48 M. E. Sibert and M. A. Steinberg, Preliminary study of the equilibrium of carbon and oxygen in niobium with carbon monoxide above 1600°C. In W. R. Clough (Ed.), *Reactive Metals*, Interscience, New York, 1959, p. 171.
49 P. V. Gel'd and G. P. Shveikin, *Izv. Akad. Nauk SSSR, Otd. Tekhn. Nauk Met. i Toplivo*, 1, (1959) 44.
50 P. V. Gel'd and G. P. Shveikin, The kinetics of the vacuum reduction of niobium pentoxide by carbon. In E. Bishop (Ed.), *The Uses of Vacuum in Metallurgy*, Oliver & Boyd, Edinburgh, 1964, p. 91.
51 W. D. Klopp, C. T. Sims and R. I. Jaffee, Vacuum reactions of niobium during sintering. In B. W. Gonser and E. M. Sherwood (Eds.), *Technology of Columbium (Niobium)*, Wiley, New York, 1958, p. 106.
52 E. Fromm, Reaktionen beim Glühen von Niob und Tantal in Kohlenmonoxid bei hohen Temperaturen und niedrigen Drucken, *Compt. Rend. Congr. Intern. sur les Applications des Techniques du Vide à la Métallurgie, Strasbourg, 1967*, p. 108.
53 E. Fromm and G. Spaeth, *Z. Metallk.*, 59 (1968) 65.
54 V. D. Lyubimov, P. V. Gel'd, G. P. Shveikin and C. I. Alyamovsky, *J. Appl. Chem.*, 38 (1965) 2136; *Zh. Prikl. Khim.*, 38 (1965) 2174.
55 E. Fromm and H. Jehn, *Metall*, 19 (1965) 747.
56 E. Gebhardt, E. Fromm and D. Jakob, *Z. Metallk.*, 55 (1964) 432.

REFERENCES

57 H. KIMURA AND Y. SASAKI, *Trans. Nat. Res. Inst. Metals (Tokyo)*, 5 (1963) 213.
58 H. KIMURA AND Y. SASAKI, *Trans. Nat. Res. Inst. Metals (Tokyo)*, 5 (1963) 274.
59 G. P. SHVEIKIN AND P. V. GEL'D, *Tsvetn. Metal.*, 34 (1961) 39.
60 O. P. KOLCHIN AND N. P. CHUVELEVA, *Tsvetn. Metal.*, 32 (2) (1959) 60.
61 O. P. KOLCHIN, N. P. CHUVELEVA AND N. V. SUMAROKOVA, *Tsvetn. Metal.*, 37 (7) (1964) 73.
62 *Brit. Pat. 943,364* (8.1.1962).
   French Pat. 1,316,001 (9.1.1962).
63 J. NIEBUHR, *J. Less-Common Metals*, 10 (1966) 312.
64 B. R. ACKERMANN AND R. J. THORN, High-temperature thermodynamic properties of reactor materials, *Proc. 2nd Intern. Conf. on Peaceful Uses of Atomic Energy*, Paper C/CONF 28/P/715, p. 180–183.
65 E. FROMM AND O. HEINKEL, *Z. Metallk.*, 58 (1967) 805.
66 G. P. SHVEIKIN AND V. A. PERELYAEV, *Tr. Inst. Khim., Akad. Nauk SSSR, Ural'sk Filial*, 9 (1966) 33.
67 O. P. KOLCHIN AND I. K. BERLIN, *Tsvetn. Metal.*, 37 (8) (1964) 66 (English edition); also *At. Energ. (USSR)*, 17 (1964) 400.
68 C. A. HAMPEL, *Rare Metals Handbook*, Reinhold, New York, 2nd edn., 1961.
69 O. WATANABE, M. TEZUKA AND Y. HASHIMOTO, *Nippon Kinzoku Gakkaishi*, 28 (1964) 214.
70 T. YA. KOSOLAPOVA et al., *Izv. Akad. Nauk SSR, Neorgan. Mat.*, 2 (1966) 1516.
71 YA. I. GERASIMOV, A. N. KRESTOVNIKOV AND A. S. SHAKHOV, *Chemical Thermodynamics in Non-Ferrous Metallurgy*, Israel Program for Scientific Translations, Jerusalem, 1965.
72 G. GOSSE, L. RENUCCI, PH. ALBERT AND P. LEHR, *Mem. Sci. Rev. Met.*, 61 (1964) 716.
73 G. GOSSE, P. LEHR AND PH. ALBERT, *Proc. Intern. Conf. on Vacuum Metallurgy Brussels, 1965*, A. I. Ms., Mons, 1965, p. 48; also *ATB Met.*, 5 (1964/65) 273.
74 G. A. MEERSON, The high-temperature reduction of refractory metals by carbon. In E. BISHOP (Ed.), *The Uses of Vacuum in Metallurgy*, Oliver & Boyd, Edinburgh, 1964, p. 105.
75 D. O. BUKER AND T. W. MERRILL, *U.S. Pat. 3,342,553* (19.9.1967).
76 G. N. MAKARENKO AND O. F. KVAS, *Poroshkovaya Met.*, (8) (1967) 34.
77 R. KIEFFER AND H. RASSAERTS, *Intern. J. Powder Met.*, 2 (1966) 15.
78 A. H. SULLY, *Chromium*, Butterworth, London, 1954.
79 A. M. AKSOY, Thermodynamics and kinetics in vacuum induction melting. In R. F. BUNSHAH (Ed.) *Vacuum Metallurgy*, Reinhold, New York, 1958, p. 59.
80 G. V. KARSANOV, A. N. TIRKINA AND L. S. ODOEVSKY, *Stal'*, 20 (1960) 321.
81 G. W. SAMSONOV AND T. YA. KOSOLAPOVA, *Zh. Prikl. Khim.*, 34 (1961) 2780.
82 G. NAESER, W. SCHOLZ AND N. DAUTZENBERG, *Arch. Eisenhüttenw.*, 34 (1963) 27.
83 K. TONOMURA, S. NISHIMURA AND Y. KONDO, *Suiyokaishi*, 15 (1966) 502.
84 A. M. SAMARIN AND A. A. VERTMAN, *Tr. Inst. Met. im. A. A. Baikova, Akad. Nauk SSSR*, (1) (1957) 60.
85 *U.S. Pat. 2,833,645*, *U.S. Pat. 2,850,378*, *French Pat. 1,019,752*, *Ger. Pat. 830,839*, *Ger. Pat. 1,130,603*.
86 *U.S. Pat. 2,763,541*.
87 P. L. BLUM AND J. P. MORLEVAT, *Rev. Hautes Temp. Refract.*, 3 (1966) 253.
88 R. KIEFFER AND F. BENESOVSKY, *Hartstoffe*, Springer-Verlag, Vienna, 1963.
89 F. BENESOVSKY, *Pulvermetallurgie in der Atomkerntechnik*, Springer-Verlag, Vienna, 1963.
90 H. A. WILHELM, R. V. STRAIN AND E. P. NEUBAUER, Uranium metal by carbon reduction of uranium oxide in vacuum, *U.S. At. Energy Comm. Rept. IS-1379*, 1966.
91 R. V. STRAIN, Preparation of uranium metal by carbon reduction, *U.S. At. Energy Comm. Rept. IS-T-129*, 1966.
92 J. BAZIN, Thesis, Paris, 1968.
93 G. N. KOZHEVNIKOV, *Zh. Prikl. Khim.*, 38 (1965) 465.
94 G. N. KOZHEVNIKOV, *Min. Chern. Met. SSSR*, 11 (1966) 64.
95 G. N. KOZHEVNIKOV, A. S. MIKULINSKY AND L. D. BAKHIREVA, *Zh. Prikl. Khim.*, 38 (1965) 713.
96 V. S. MAL'TSEV, V. T. PANUSHKIN, S. M. ISABAEV AND V. D. PONOMAREV, *Izv. Vyssikh Uchebn. Zavedenii, Tsvetn. Met.*, 7 (6) (1964) 70; *Tr. Inst. Met. i Obogashch. Akad. Nauk Kaz. SSR*, 12 (1965) 125.

97 V. S. Mal'tsev, A. E. Buketov, V. T. Panushkin, S. M. Isabaev, V. D. Ponomarev and D. N. Abshev, *Izv. Vysshikh Uchebn. Zavedenii, Tsvetn. Met.*, 9 (2) (1966) 43.
98 G. N. Kozhevnikov and A. S. Mikulinsky, *Min. Chern. Met. SSSR*, 11 (1966) 77.
99 T. A. Tangen, *Erzmetall*, 14 (1961) 216.
100 P. Weiss, *Erzmetall*, 20 (1967) 249.
101 C. L. Strelez, A. J. Taiz and B. S. Guljanitzki, *Metallurgie des Magnesiums*, VEB Verlag Technik, Berlin, 1953.
102 B. S. Gulyanitsky and D. M. Chizhikov, *Izv. Akad. Nauk SSSR, Otd. Tekhn. Nauk*, 11 (1955) 13.
103 A. S. Burnazyan and M. V. Darbinyan, *Izv. Vysshikh Uchebn. Zavedenii, Tsvetn. Met.*, 4 (3) (1961) 81.
104 *U.S. Pat. 2,839,380* (13.6.1955).
U.S. Pat. 2,965,475 (13.1.1960).
105 *DAS 1,137,560* (15.5.1959).
106 L. Hackspill and N. Platzer, *Compt. Rend.*, 246 (1958) 2969.
107 G. N. Kozhevnikov and A. S. Mikulinsky, *Min. Chern. Met. SSSR*, 11 (1966) 84.
108 G. N. Kozhevnikov and A. S. Mikulinsky, *Tr. Inst. Met. (Sverdlovsk)*, (1966) 84.
109 T. Noda, *J. Metals*, 17 (1965) 25.
110 J. B. Knighton and R. K. Stennenberg, Reduction of U_3O_8, ThO_2 and PuO_2 by MgZn, *U.S. At. Energy Comm. Rept. ANL-7057, -7058, -7059*, 1965.
111 G. V. Forsblom and R. A. Sandler, *Tsvetn. Metal.*, 33 (10) (1960) 63.
112 R. Holst and R. Proft, *Neue Hütte*, 4 (1959) 106.
113 R. Proft, M. Suchi and R. Hollwitz, Die Reduktion von Zirkoniumtetrachlorid mit Magnesium in einer Vakuumretorte. In *Metallurgie der seltenen Metalle und der Spurenmetalle*, VEB Deutscher Verlag für Grundstoffindustrie, Leipzig, 1964, p. 130.
114 W. Thiel and S. Ziegenbalg, Die magnesiothermische Herstellung von Niob und Tantal. In *Metallurgie der seltenen Metalle und der Spurenmetalle*, VEB Deutscher Verlag für Grundstoffindustrie, Leipzig, 1964, pp. 204, 246.
115 S. Takeuchi, M. Tezuka, T. Kurosawa and S. Eda, A new development in the production of titanium by reaction in the gaseous phase. In G. R. St. Pierre (Ed.), *Physical Chemistry of Process Metallurgy*, Vol. 2, Interscience, New York, 1961, p. 745.
116 G. V. Samsonov and V. S. Sinel'nikova, *Izv. Vysshikh Uchebn. Zavedenii, Tsvetn. Met.*, 9 (4) (1966) 65.
117 G. Gosse, L. Renucci, Ph. Albert and P. Lehr, *Mem. Sci. Rev. Met.*, 61 (1964) 717.
118 G. Gosse, P. Lehr and Ph. Albert, *Compt. Rend. Coll. Intern. Met. sous Vide, Bruxelles, 1965*, p. 48.
119 G. Gosse, Ph. Albert and P. Lehr, *Mem. Sci. Rev. Met.*, 62 (1965) 407; also *Chem. Ing. Tech.*, 37 (1965) 343.
120 G. Gosse, P. Lehr and Ph. Albert, Contribution à l'étude des cinétiques de purification sous vide du zirconium au cours de sa fusion par bombardment d'électrons, *Congr. Intern. sur les Applications des Techniques du Vide à la Métallurgie, Strasbourg, 1967*, p. 87.
121 F. M. Lever and J. B. Payne, Separation of the rare earths and production of the metals, *Proc. Symp. on Advances in Extractive Metallurgy, London, 1967*, The Institution of Mining and Metallurgy, 1968, p. 789.
122 A. H. Daane, Metallothermic preparation of rare-earth metals. In F. H. Spedding and A. H. Daane (Eds.), *The Rare Earths*, Wiley, New York, 1961, p. 102.
123 F. H. Spedding and A. H. Daane, *Met. Rev.*, 5 (1960) 297.
124 L. A. Izhvanov and N. P. Vershinin, *Tsvetn. Metal.*, 32 (1) (1959) 44.
125 C. E. Habermann, A. H. Daane and P. E. Palmer, *Trans. AIME*, 233 (1965) 1038.
126 G. G. Gvelesiani and D. I. Bagdavadze, *Soobshch. Akad. Nauk Gruz. SSR*, 41 (1966) 657.
127 G. G. Gvelesiani and S. M. Bezarashvili, *Soobshch. Akad. Nauk Gruz. SSR*, 39 (1965) 669.
128 F. H. Spedding, J. J. Hanak and A. H. Daane, *Trans. AIME*, 212 (1958) 379.
129 G. G. Gvelesiani and D. I. Bagdavadze, *Soobshch. Akad. Nauk Gruz. SSR*, 42 (1966) 427.
130 A. A. Nadiradze and G. G. Gvelesiani *Soobshch. Akad. Nauk Gruz. SSR*, 40 (1965) 407.
131 E. Meckelburg, *Metall*, 15 (1961) 778.

132 T. T. CAMPBELL AND F. E. BLOCK, *J. Metals, 11* (1959) 744.
133 G. G. GVELESIANI AND D. I. BAGDAVADZE, *Soobshch. Akad. Nauk Gruz. SSR, 42* (1966) 151.
134 R. J. MOOLENAAR, *J. Metals, 16* (1964) 21.
135 C. E. BERTHOLD, *J. Metals, 14* (1962) 355.
136 T. F. FEDOROV AND F. I. SHAMRAI, The physicochemical principles of the vacuum thermite reduction of lithium. In E. BISHOP (Ed.) *The Uses of Vacuum in Metallurgy*, Oliver & Boyd, Edinburgh, 1964, p. 126.
137 *U.S. Pat. 2,710,798* (14.6.1955).
   U.S. Pat. 2,983,599 (2.7.1954).
138 G. N. ZVIADADZE AND O. V. SHENGELIYA, *Soobshch. Akad. Nauk Gruz. SSR, 42* (1966) 423.
139 G. N. ZVIADADZE AND O. V. SHENGELIYA, *Soobshch. Akad. Nauk Gruz. SSR, 42* (1966) 159.
140 G. N. ZVIADADZE, O. V. SHENGELIYA AND D. SH. OZIASHVILI, *Tr. Inst. Met. Akad. Nauk Gruz. SSR, 15* (1966) 27.
141 G. N. ZVIADADZE, O. V. SHENGELIYA AND D. SH. OZIASHVILI, *Tr. Inst. Met. Akad. Nauk Gruz. SSR, 15* (1966) 36.
142 T. SZAROWICZ AND M. ORMAN, *Prace Inst. Min. Hutnictwa, 7* (1955) 270.
143 C. DECROLY, *ATB Met., 5* (1964–65) 141; also *Proc. Intern. Conf. on Vacuum Metallurgy*, Brussels, 1965, p. 30.
144 S. M. ISABAEV, V. D. PONOMAREV, V. S. MAL'CEV AND R. U. ABISEVA, *Izv. Vysshikh Uchebn. Zavedenii, Tsvetn. Met., 9* (4) (1966) 39.
145 G. N. ZVIADADZE AND D. SH. OZIASHVILI, *Tr. Inst. Met. Akad. Nauk Gruz. SSR, 14* (1965) 279.
146 *UdSSR Pat. 142,430* (29.12.1955).
147 *Brit. Pat. 773,940* (1.5.1957).
148 A. S. MIKULINSKY AND I. E. SIPEIKO, *Tr. Inst. Met. (Sverdlovsk.)*, (1966) 94.
149 I. E. SIPEIKO AND A. S. MIKULINSKY, *Min. Chern. Met. SSSR, 11* (1966) 100.
150 T. D. BROTHERTON, O. N. COLE AND R. E. DAVIS, *Trans. AIME, 224* (1962) 287.
151 K. C. DEAN, I. L. NICHOLS AND B. H. CLEMMONS, *J. Metals, 18* (1966) 1198.
152 *U.S. Pat. 3,164,461* (15.1.1962).
153 G. N. ZVIADADZE AND O. V. SHENGELIYA, *Tr. Inst. Met. Akad. Nauk Gruz. SSR, 15* (1966) 50.
154 G. N. ZVIADADZE AND O. V. SHENGELIYA, *Tr. Inst. Met. Akad. Nauk Gruz. SSR, 15* (1966) 44.
155 G. N. ZVIADADZE, O. V. SHENGELIYA, L. S. TABATADZE AND D. V. VATSADZE, *Tr. Inst. Met. Akad. Nauk Gruz. SSR, 15* (1966) 3.
156 G. N. ZVIADADZE, O. V. SHENGELIYA, L. S. TABATADZE AND D. V. VATSADZE, *Tr. Inst. Met. Akad. Nauk Gruz. SSR, 15* (1966) 15.
157 A. SCHNEIDER, *Z. Metallk., 41* (1950) 205.
158 A. SCHNEIDER, J. F. CORDES, H. KRIBBE and H. RUNGE, *Erzmetall, 12* (1959) 103, 164, 224.
159 L. M. PIDGEON AND J. A. KING, *Discussions Faraday Soc., 4* (1948) 197.
160 W. T. HUGHES, C. E. RANSLEY AND E. F. EMLEY, Reaction kinetics in the production of magnesium by the dolomite-ferrosilicon process, *Proc. Symp. on Advances in Extractive Metallurgy, London, 1967*, The Institution of Mining and Metallurgy, 1968, p. 429.
161 F. RITTER and G. JAEKEL, Magnesium. In K. WINNACKER and L. KÜCHLER (Eds.), *Chemische Technologie*, Vol. 5, *Metallurgie/Allgemeines*, Hanser Verlag, Munich, 1961, p. 106.
162 N. HÖY-PETERSEN, Übersicht über elektrolytische und thermische Reduktionsverfahren, die heute industriell zur Herstellung und Anwendung von Magnesium in Gebrauch sind, *Freiberger Forschungsh. B120* (1967) 123.
163 J. D. HANAWALT, *J. Metals, 16* (1964) 559.
164 C. HAYASHI AND S. KASHU, Vacuum metallurgy in Japan. In E. L. FOSTER, *Trans. Intern. Conf. on Vacuum Metallurgy, New York, 1967*, Am. Vac. Soc., New York, 1968, p. 37.
165 H. SELIGER, Die thermische Herstellung von Erdalkalimetallen, insbesondere von Magnesium, *Freiberger Forschungsh. B34* (1959) 80.
166 Anon., *VAW Werkzeitschrift*, (2) (1967) 33.
167 L. JAKOBY, Die Bedeutung der Aufbereitung von Dolomit bei der silikothermischen Magnesiumgewinnung, *Freiberger Forschungsh. B28* (1958) 42.
168 T. G. GEDEON, A. SZULYOVSZKY AND K. A. CZAKÓ, Dolomit-redukction. In *A fémipari kutato intézet közleményéi*, Budapest, 1956, p. 293.

169 A. Szulyovszky and S. Baumann, *Kohasz. Lapok*, 97 (1964) 88.
170 V. M. Chel'tsov and I. D. Tsaregorodtsev, *Tsvetn. Metal.*, 34 (8) (1961) 49 (English edition).
171 *DAS 1,142,064* (18.10.1961).
 U.S. Pat. 3,151,977 (13.10.1961).
172 C. Faure and J. Marchal, *J. Metals*, 16 (1964) 721; also *Metallurgia Ital.* 58 (1966) 82.
173 *French Pat. 987,046* (8.8.1951).
 DAS 1,053,791 (4.4.1952).
 U.S. Pat. 2,971,833 (7.4.1959).
174 *U.S. Pat. 3,114,627* (24.5.1960).
 U.S. Pat. 3,129,094 (21.4.1960).
175 W. Moschel and O. R. Bretschneider, Magnesium. In *Ullmanns Encyklopädie der technischen Chemie*, Vol. 12, Urban und Schwarzenberg, Munich, 1960.
176 K. C. Dean, D. A. Elkins and B. H. Clemmons, *J. Metals*, 16 (1964) 564.
177 A. J. Tajc, *Tsvetn. Metal.*, 30 (1) (1957) 56.
178 T. T. Campbell, R. E. Mussler and F. E. Block, *Trans. AIME*, 236 (1966) 1456.
179 G. N. Kozhevnikov, V. V. Efremkin and A. S. Mikulinsky, *Tsvetn. Metal.*, 37 (1) (1964) 77.
180 P. Vignial and J. L. Andrieux, *Compt. Rend.*, 242 (1956) 709.
181 K. Akerman and M. Orman, *Acta Tech. Acad. Sci. Hung.*, 15 (1956) 179.
182 F. Ott, Die Alkali- und Erdalkalimetalle. In K. Winnacker and L. Küchler (Eds.), *Chemische Technologie*, Vol. 5, *Metallurgie/Allgemeines*, Hanser Verlag, Munich, 1961, p. 71.
183 H. A. Wilhelm and O. N. Carlson, *J. Metals*, 16 (1964) 170.
184 V. V. Zhukovetsky, *Izv. Vysshikh Uchebn. Zavedenii, Tsvetn. Met.*, 9 (1966) 49.
185 G. Schaufler, Calcium und Verbindungen. In *Ullmanns Encyklopädie der technischen Chemie*, Vol. 4, Urban und Schwarzenberg, Munich, 1953, p. 830.
186 *DAS 1,110,428* (10.1.1958).
187 V. V. Zhukovetsky, *Izv. Vysshikh Uchebn. Zavedenii, Tsvetn. Met.*, 3 (6) (1960) 119.
188 G. N. Kozhevnikov, *Tsvetn. Metal.*, 36 (2) (1963) 53.
189 G. G. Gvelesiani and N. P. Mgaloblishvili, *Tr. Inst. Met. Akad. Nauk Gruz. SSR*, 14 (1965) 205.
190 T. Kraus and O. Winkler, in R. Bakish (Ed.), *Introduction to Electron-Beam Technology*, Wiley, New York, 1962, p. 145.
191 T. Kraus, The mechanism of material exchange. In R. F. Bunshah, *Trans. Vacuum Met. Conf., New York 1963*, Am. Vac. Soc., Boston, Mass., 1964, p. 50.
192 E. S. Machlin, *Trans. AIME*, 218 (1960) 314.
193 A. V. Bradshaw, *Vide*, (1968) 376.
194 N. A. Warner, Kinetics of continuous vacuum dezincing of lead. In *Proc. Symp. on Advances in Extractive Metallurgy, London, 1967*, The Institution of Mining and Metallurgy, London, 1968, pp. 317, 403, 417.
195 R. F. Bunshah, Superpurification of metals by vacuum distillation: A theoretical study. In R. F. Bunshah, *Trans. Vacuum Met. Conf., New York, 1963*, Am. Vac. Soc., Boston, Mass., 1964, p. 121.
196 L. Müller, *Freiberger Forschungsh.*, B33 (1958).
197 T. R. A. Davey, *Trans. AIME*, 197 (1953) 991.
198 T. R. A. Davey, *Vacuum*, 12 (1962) 83; V. A. Pazukhin and E. E. Loukashenko, *Vacuum*, 14 (1964) 227.
199 O. Knacke and I. N. Stranski, The mechanism of evaporation. In B. Chalmers and R. King (Eds.), *Progress in Metal Physics*, Pergamon, London, 1956, p. 181.
200 B. Ilschner and J. Humbert, *Z. Metallk.*, 51 (1960) 626.
201 M. J. Spendlove, Experiments on vacuum distillation of non-ferrous metals and alloys in vacuum metallurgy, *J. Electrochem. Soc.*, (1955) 192; also *Vakuum-Tech.*, 6 (1957) 15, 36.
202 L. Schumann-Horn, A. Mager and W. Deisinger, *Z. Metallk.*, 47 (1956) 145.
203 H. Schenck and H. H. Domalski, *Arch. Eisenhüttenw.*, 32 (1961) 753; also H. H. Domalski, Thesis, Tech. Hochschule, Aachen, 1960.
204 R. G. Ward, *J. Iron Steel. Inst. (London)*, (1963) 11.
205 V. D. Burlakov, V. E. Ivanov and Yu. P. Kurilo, *Tsvetn. Metal.*, 37 (6) (1964) 71.

206 H. A. Friedrichs, H. Jauer and O. Knacke, personal communication.
207 R. G. Ward and T. D. Aurini, *J. Iron Steel Inst. (London)*, (1966) 920.
208 R. F. Bunshah and R. S. Juntz, Purification of beryllium by crucible-free vacuum melting and distillation processes. In *Beryllium Technology*, Vol. 1, Gordon & Breach, New York, 1966, p. 1.
209 R. F. Bunshah and R. S. Juntz, The purification of beryllium by vacuum melting followed by distillation and simultaneous deposition to sheet in an electron-beam furnace, *Trans. Vacuum Met. Conf., New York, 1966*, Am. Vac. Soc., Boston, Mass., 1967, p. 207.
210 H. A. Friedrichs, H. Jauer and O. Knacke, *Z. Metallk.*, 60 (1969) 635.
211 W. J. Walsh and G. Burnet, *Nucl. Sci. Eng.*, 25 (1966) 227.
212 L. Müller, *Freiberger Forschungsh.*, B6 (1954) 81.
213 M. Olette, Vacuum distillation of minor elements from liquid ferrous alloys. In G. R. St. Pierre, *Physical Chemistry of Process Metallurgy*, Vol. 2, Interscience, New York, 1961, p. 1065.
214 J. P. Langeron, *Ann. Mines*, 4 (1968) 76.
215 J. P. Langeron, *Vide*, (1968) 220.
216 M. Onillon and M. Olette, Thermodynamique et cinétique de l'évaporation sous vide des alliages ferreux binaires liquides dilués, *Congr. Intern. sur les Applications des Techniques du Vide à la Métallurgie, Strasbourg, 1967*, p. 55.
217 K. Lohwater and H. Eckstein, *Vide,16* (91) (1961) 44.
218 D. E. Westerheide and G. Burnett, *Ind. Eng. Chem., Process Design Develop.*, 4 (1965) 43; also *U.S. At. Energy Comm. Rept. IS-1329*, 1966.
219 *DAS 1,080,972* (22.10.1955).
220 A. W. Schlechten and C. H. Shih, *Eng. Mining J.*, 150 (1949) 80.
221 A. W. Schlechten and R. F. Doeling, *Trans. AIME*, 191 (1951) 327.
222 V. F. Leferrer, *Trans. AIME*, 209 (1957) 1459.
223 M. P. Smirnov, Ya. Z. Malkin, N. G. Tarkhov and V. Ya. Sergienko, *Tsvetn. Metal.*, 33 (5) (1960) 32.
224 Brit. Pat. 788,525 (2.1.1956).
225 H. Auler, *Diplomarbeit*, Tech. Hochschule Aachen, 1954, unpublished.
226 B. I. Kidyarov, *Zh. Fiz. Khim.*, 40 (1966) 1131.
227 W. T. Isbell, *Metal Ind. (London)*, 71 (1947) 47; and *Trans. AIME 182* (1949) 186.
228 M. P. Smirnov, N. G. Tarkhov and V. Ya. Sergienko, *Tsvetn. Metal*, 29 (5) (1956) 19.
229 A. Ya. Fisher and P. S. Shesternin, *Tsvetn. Metal.*, 29 (3) (1956) 15.
230 K. Vurm, *Hutnik*, 11 (1961) 383.
231 B. Lindvall and S. Walldén, *Erzmetall*, 11 (1958) 264.
232 R. Davey and K. C. Williams, *Proc. Australasian Inst. Mining Met.*, (1956) 207.
233 S. W. K. Morgan and D. A. Temple, *J. Metals*, 19 (1967) 23.
234 *Trennung von Metallen durch Vakuumdestillation*, Abschlussbericht des Metallhütteninstituts der Bergakademie Freiberg, 1959.
235 Anon., *Metallurgia*, 76 (1967) 249.
236 D. A. Temple, personal communication.
237 *U.S. Pat. 3,080,227* (14.10.1959).
238 V. S. Esyutin, *Tr. Inst. Met. i. Obogashch.*, 3 (1960) 184.
239 V. S. Esyutin, G. V. Kim and Zh. Sh. Taziev, *Tr. Inst. Met. i Obogashch., Tsvetn. Met.*, (1963) 3,8.
240 M. P. Smirnov, *Obogashch. i Met. Tsvetn. Met.*, 13 (1957) 235.
241 V. N. Nesterov and R. A. Isakova, *Tr. Inst. Met. i Obogashch.*, 2 (1960) 86.
242 *DAS 1,174,079* (31.8.1963).
243 V. N. Chernyaev and S. A. Ershova, *Zh. Prikl. Khim.*, 37 (1964) 2407.
244 V. M. Amonenko, A. A. Kruglykh and L. I. Papirov, *Fiz. Metal i Metalloved.*, 11 (1961) 633.
245 V. N. Chernyaev, V. B. Zernov, L. G. Povedskaya, S. A. Ershova and I. I. Klofach, *Zh. Prikl. Khim.*, 39 (1966) 1259.
246 V. N. Chernyaev, L. G. Povedskaya and Yu. T. Kovalev, *Zh. Prikl. Khim.*, 36 (1963) 56.
247 B. N. Aleksandrov and B. I. Verkin, *Fiz. Metal i Metalloved.*, 9 (1960) 362.

248 Zh. Sh. Taziev, V. S. Esyutin and A. L. Tseft, *Tr. Inst. Met. i Obogashch. Akad. Nauk Kaz. SSR*, *13* (1965) 11.
249 A. V. Volkovich, G. A. Komlev, A. A. Vasynkova and S. A. Kopytov, *Tsvetn. Metal.*, *37* (5) (1964) 86.
250 F. A. Santalov, *Fiz. Metal i Metalloved.*, *3* (1956) 247.
251 A. A. Shokol and L. F. Kozin, *Ukr. Khim. Zh.*, *28* (1962) 699.
252 W. Teworte, Cadmium und Cadmium-verbindungen, in *Ullmanns Encyklopädie der technischen Chemie*, Vol. 4, Urban und Schwarzenberg, Munich and Berlin, 1953.
253 G. Heitmann and O. Knacke, *Z. Physik. Chem. (Frankfurt)*, *7* (1956) 225.
254 S. Richter and L. Müller, Die Herstellung von Reinstmetallen mach dem Verfahren der halbkontinuierlichen Vakuumdestillation. In E. Rexer (Ed.,) *Reinststoffprobleme*, Vol. 1, *Reinststoffdarstellung*, Akademie-Verlag, Berlin, 1966, p. 453; and *Freiberger Forschungsh.*, *B112* (1966) 127.
255 Yu. S. Arzamastsev, L. D. Bakhireva, V. V. Efremkin, A. S. Mikulinsky et al., *Ministervo Chern. Met. SSSR*, *11* (1966) 30, 35, 69; also *Tr. Inst. Met.*, (11) (1966) 35.
256 J. M. Whelan, J. D. Struthers and J. A. Ditzenberger, *J. Electrochem. Soc.*, *107* (1960) 982.
257 D. P. Enright, *Navord Report 6024*, U.S. Naval Ordnance Lab., 1958.
258 J. T. Edmond, R. F. Broom and F. A. Cunnell, *Services Electronics Res. Lab. Technical J.*, *6* (1956) 123.
259 R. K. Willardson and H. L. Goering, *Compound Semiconductors*, Vol. 1, *Preparation of III–V-Compounds*, Reinhold, New York, 1962.
260 L. R. Weisberg, F. D. Rosi and P. G. Herkart, in H. C. Gatos, *Properties of Elemental and Compound Semiconductors*, Interscience, New York, 1960.
261 E. Bonnier and M. Charveriat, Purification de l'antimoine par sublimation, *Proc. Intern. Conf. on Vacuum Metallurgy, Brussels, 1965*, p. 94.
262 F. Trombe and G. Malé, *Compt. Rend.*, *264* (1967) 199.
263 C. E. Habermann and A. H. Daane, *J. Less-Common Metals*, *5* (1963) 134.
264 D. Geiselman, *J. Less-Common Metals*, *4* (1962) 362.
265 F. H. Spedding, A. H. Daane, G. Wakefield and D. H. Dennison, *Trans. AIME*, *218* (1960) 608.
266 C. E. Habermann, A. H. Daane and P. E. Palmer, *Trans. AIME*, *233* (1965) 1038.
267 A. A. Menkov, *Dokl. Akad. Nauk SSSR*, *144* (1962) 122.
268 V. V. Vorob'ev, *Ukr. Fiz. Zh.*, *12* (1967) 1110.
269 O. N. Carlson, J. A. Haefling, F. A. Schmidt and F. H. Spedding, *J. Electrochem. Soc.*, *107* (1960) 540; also F. H. Spedding and A. H. Daane, *The Rare Earths*, Wiley, New York, 1961.
270 D. T. Peterson, W. E. Krupp and F. A. Schmidt, *J. Less-Common Metals*, *7* (1964) 288.
271 T. F. Fedorov and F. I. Shamray, *Izv. Akad. Nauk SSSR, Otd. Tekhn. Nauk, Met. i Toplivo*, (6) (1960) 56.
272 P. H. Schmidt, *J. Electrochem. Soc.*, *113* (1966) 201.
273 *SUP 155,934* (14.1.1960).
274 Y. F. Bychkov, A. N. Rozanov, B. I. Gromov and V. I. Cheburkov, Laboratory apparatus for vacuum distillation of lithium and filling of crucibles. In *Metallurgy and Metallography of Pure Metals*, Gordon and Breach, New York, 1962, p. 171.
275 C. Kunze, *Vakuum-Tech.*, *8* (1959) 168.
275a R. L. McKisson, R. L. Eichberger and J. B. Ott, The construction and operation of a sodium purification apparatus, *U.S. At. Energy Comm. Rept. NAA-SR-12444*, 1967.
276 J. F. Smith, O. N. Carlson and R. W. Vest, *J. Electrochem. Soc.*, *103* (1956) 409.
277 W. J. McCreary, *J. Metals*, *10* (1958) 615.
278 C. Decroly and C. Den Tandt, *Mem. Sci. Rev. Met.*, *60* (1963) 609.
279 G. Wehner and H. Seliger, *Freiberger Forschungsh.*, *B17* (1956) 18.
280 *DWP 9540* (24.5.1952).
    DWP 9648 (24.5.1952).
281 *DAS 1,176,877* (17.10.1962).
282 J. Veprek-Siska, *Chem. Listy*, *56* (1962) 481.

283 *DAS 1,079,843* (4.10.1956).
    U.S. Pat. 2,867,527 (20.9.1956).
284 E. BONNIER AND J. L. ANDRIEUX, *Rev. Met.*, 54 (1957) 720.
285 *U.S. Pat. 3,136,628* (19.1.1961).
286 E. MILLER, K. KOMAREK AND I. CADOFF, *Trans. AIME*, 218 (1960) 978.
287 K. D. SINELNIKOV, V. E. IVANOV, V. M. AMONENKO AND V. D. BURLAKOV, Refining beryllium and other metals by condensation on heated surfaces, *Proc. 2nd Intern. Conf. on Peaceful Uses of Atomic Energy, 1958*, Vol. 4, P 2051, p. 295.
288 V. E. IVANOV, V. M. AMONENKO, F. G. TIKHINSKY AND A. A. KRUGLYKH, *Fiz. Metal. i Metalloved.*, 10 (1960) 581.
289 R. F. BUNSHAH, A status report on purification of beryllium and its properties, *U.S. At. Energy Comm. Rept. Conf-170*, 1965, p. 214.
290 E. W. HOOPER AND N. J. KEEN, Purification of beryllium metal by a distillation process, *U.S. At. Energy Comm. Rept. Conf-170*, 1965, p. 1; also *Trans. Inst. Mining Met.*, 75 (1966) (C) C262; also *The Metallurgy of Beryllium*, Chapman and Hall, London, 1963, p. 579; also *Bull. Inst. Mining Metallurgy, Trans.*, 75 (1966) C246.
291 J. P. PEMSLER, S. H. GELLES, E. D. LEVINE AND A. R. KAUFMANN, The purification of beryllium by distillation. In *The Metallurgy of Beryllium*, Chapman and Hall, London, 1963, p. 570.
292 K. L. EDWARDS AND A. J. MARTIN, The purification of beryllium by distillation and zone melting. In *The Metallurgy of Beryllium*, Chapman and Hall, London, 1963, p. 557.
293 M. J. HORDON AND W. T. HESS, Electron-beam vaporization of beryllium, *Beryllium Technology*, Gordon and Breach, New York, 1966, p. 67.
294 K. D. SINELNIKOV, V. E. IVANOV, V. M. AMONENKO AND G. F. TIKHINSKY, Some properties of high-purity distilled beryllium, *The Metallurgy of Beryllium*, Chapman and Hall, London, 1963, pp. 264.
295 G. J. LONDON AND M. HERMAN, Purification of beryllium by high-vacuum distillation, *Conf. Intern. sur la Métallurgie du Béryllium, 1965*, Presses Universitaires de France, Paris, 1966, p. 21.
296 V. E. IVANOV, V. M. AMONENKO, G. F. TIKHINSKY, I. I. PAPIROV, L. N. RYABCHIKOV AND V. N. GRINYUK, Preparation of high-purity beryllium, *Conf. Intern. sur la Métallurgie du Béryllium, 1965*, Presses Universitaire de France, Paris, 1966, p. 33.
297 B. L. BLANC, Contribution à l'étude de la purification du béryllium par sublimation et distillation, *Conf. Intern. sur la Métallurgie du Béryllium, 1965*, Presses Universitaires de France, Paris, 1966, p. 39.
298 R. F. BUNSHAH, Impurity removal by distillation of beryllium from the solid state, *Conf. Intern. sur la Métallurgie du Béryllium, 1965*, Presses Universitaires de France, Paris, 1966, p. 63.
299 P. T. STROUP, *Trans. AIME*, 230 (1964) 356.
300 E. HERRMANN, *Z. Metallk.*, 53 (1962) 617.
301 Anon., *Aluminium*, 43 (1967) 672.
302 H. GINSBERG AND V. SPARWALD, *Aluminium*, 41 (1965) 181, 219.
303 P. WEISS, *Erzmetall*, 3 (1950) 241.
304 P. GROSS, The subhalide distillation of aluminium, *Congr. Intern. de l'Aluminium, Paris, 1954*, p. 167.
305 W. HIRSCHWALD AND O. KNACKE, *Erzmetall*, 11 (1958) 99.
306 E. HERRMANN, *Aluminium*, 37 (1961) 143, 215.
307 F. D. ROSI, D. MEYERHOFER AND R. V. JENSEN, *J. Appl. Phys.*, 31 (1960) 1105.
308 L. M. FOSTER AND R. A. KRAMER, *J. Electrochem. Soc.*, 107 (1960) 189 C.
309 K. F. HULME AND J. B. MULLIN, *J. Electron. Control*, 3 (1957) 160.
310 D. EFFER, *J. Electrochem. Soc.*, 108 (1961) 357.
311 V. M. SERGEEVA AND E. L. SHTRUM, *Zh. Tekhn. Fiz.*, 27 (1957) 2698.
312 V. M. AMONENKO, B. M. VASYUTINSKY, V. V. LEBEDEV AND B. I. SHAPOVAL, *Fiz. Metal. i Metalloved.*, 7 (1959) 862.
313 V. M. AMONENKO, A. A. KRUGHLIKH AND G. F. TIKHINSKY, *Fiz. Metal. i Metalloved.*, 7 (1959) 868.

314 W. L. FALKE AND A. A. COCHRAN, Reduction volatilization processes for recovery of manganese from ores, *U.S. Bur. Mines Rept. Invest.*, 1966, p. 6738.
315 R. A. ISAKOVA, V. S. ESYUTIN, V. N. NESTEROV et al., *Tsvetn. Metal.*, 37 (4) (1964) 61 (English edition).
316 N. B. PLETENEVA, *Tsvetn. Metal.*, 34 (1) (1961) 57.
317 F. ECKART, *Ann. Physik*, 14 (1954) 6.
318 L. M. NIJLAND, *Philips Research Rept.*, 9 (1954) 267.
319 F. ECKART, *Naturwissenschaften*, 45 (1958) 14.
320 E. I. GULYAYEVA AND L. A. SOSHUIKOVA, *Byul. Tsentr. Inst. Inform. Tsvetnoi Met.*, (1957) 23.
321 N. B. PLETENEVA, M. P. SMIRNOV AND D. M. YUKHTANOV, *Byul. Tsentr. Inst. Inform. Tsvetnoi Met.*, (1960) 11.
322 M. P. SMIRNOV AND G. A. BIBENINA, *Tsvetn. Metal.*, 30 (12) (1957) 17.
323 D. N. KLUSHIN, I. A. BIBINA AND K. G. BOGATINA, *Obogashch. Met., Tsvetn. Met. i Metody Analiza*, Vol. 10, Moscow, 1955, p. 243.
324 A. SZULIOVSKY, *Kohász. Lapok*. 92 (1959) 316.
325 D. N. KLUSHIN, I. A. BIBINA AND K. G. BOGATINA, *Zh. Prikl. Khim.*, 28 (1955) 1242.
326 D. N. KLUSHIN, P. S. SHESTERNIN AND A. YA. FISHER, *Obogashch. Met., Tsvetn. Met.*, Vol. 13, Moscow, 1957, p. 192.
327 S. M. EGOROV, D. N. KLUSHIN AND A. YA. FISHER, *Tsvetn. Metal.*, 28 (6) (1955) 32.
328 M. G. FRADE, F. GUERIN AND M. P. LACOMBE, Etude comparée de l'évaporation sous vide élevé du zinc dans les laitons et les alliages Al–Zn à 11% de zinc, *Congr. Intern. sur les Applications des Techniques du Vide à la Métallurgie*, Strasbourg, 1967, p. 67.
329 G. FRADE AND P. LACOMBE, Study of the dezincification of alpha-brass of various compositions. In E. L. FOSTER (Ed.), *Trans. Intern. Conf. Vacuum Metallurgy, New York, 1967*, Am. Vac. Soc., New York, 1968, p. 321.
330 H. SELIGER, *Freiberger Forschungsh.*, B23 (1957) 69.
331 E. E. LUKASHENKO, G. G. ZYRYANOV AND V. P. KUZNETSOVA, *Izv. Vysshikh. Uchebn. Zavedenii, Tsvetn. Met.*, (3) (1967) 44.
332 *Die Heraeus-Vakuumschmelze 1923–1933*, G. M. Albertis-Verlag, Hanau, 1933.
333 Y. S. SHCHEDROVITSKY AND S. V. BESOBRAZOV, *Stal'* (in German), 6 (1966) 870.
334 S. V. BESOBRAZOV, K. N. KADARMETOV, G. V. CHARUSHNIKOVA, R. B. KRICHEVETS, YU. G. PONOMARENKO, N. A. TULIN, N. P. POZDEEV AND A. B. SERGEEV, *Stal'* (in English), (1965) 722.
335 H. VOLLMER, Entkohlung von flüssigem Ferrochrom im Induktions-Vakuumofen, *Congr. Intern. sur les Applications des Techniques du Vide à la Métallurgie*, Strasbourg, 1967, p. 165.
336 P. YA. SOROKIN, The production of low-carbon ferro-chromium by blowing *in vacuo*. In E. BISHOP (Ed.) *The Uses of Vacuum in Metallurgy*, Oliver & Boyd, Edinburgh, 1964, p. 84.
337 V. L. KOLOYARTSEV AND S. V. BEZOBRAZOV, *Izv. Akad. Nauk SSSR, Gorn. Delo.*, (5) (1963) 38.
338 C. G. CHADWICK, *J. Metals*, 13 (1961) 806; also *Trans. Vacuum Met. Conf., New York, 1963*, Am. Vac. Soc., Boston, 1964, p. 221.
339 G. ZHAK, The vacuum decarburization of ferro-chromium. In E. BISHOP (Ed.), *The Uses of Vacuum in Metallurgy*, Oliver & Boyd, Edinburgh, 1964, p. 114.
340 B. A. BAUM et al., *Stal'*, 27 (1967) 45.
341 *DAS 1,076,378* (30.10.1953).
342 *Belg. Pat. 635,292* (23.7.1963).
343 L. BREWER AND G. M. ROSENBLATT, *Trans. AIME*, 224 (1962) 1268.
344 M. GLEISER, *Trans. AIME*, 221 (1962) 300.
345 H. R. SMITH, Electron bombardment melting techniques. In R. F. BUNSHAH, *Vacuum Metallurgy*, Reinhold, New York, 1958.
346 J. TROUVE AND A. ACCARY, Obtention des métaux de transition réactifs par carboréduction, *Plansee Proc. 6th Seminar*, Reutte/Tyrol, 1968.
347 P. AILLOUD, L. RENUCCI AND J. P. LANGERON, *Compt. Rend.*, 266 (1968) (c) 366.
348 P. ALBERT, L. RENUCCI AND P. LEHR, *Bull. Soc. Chim. France*, (1962) 2091.
349 L. RENUCCI, Thesis, Paris, 1968.
350 P. AILLOUD, L. RENUCCI AND J. P. LANGERON, Zirconium pollution and purification mecha-

nism under vacuum. In E. L. FOSTER, *Trans. Intern. Conf. Vacuum Met.*, *1967*, Am. Vac. Soc., New York, 1968, p. 915.
351 P. LEHR AND P. ALBERT, *Rev. Hautes Temp. Refract.*, *2* (1965) 31.
352 H. SPERNER, *Metall*, *16* (1962) 679.
353 D. S. FAIRGRIEVE AND J. W. FORTNER, *J. Metals*, *12* (1960) 25.
354 F. SPERNER AND E. ERBEN, *Z. Metallk.*, *55* (1964) 674.
355 M. ARMAND, J. P. GIVORD AND R. HEROLD, Comparaison entre différents modes de purification du hafnium, *Plansee Proc. 6th Seminar, Reutte/Tyrol, 1968*.
356 H. R. SMITH, Electron beam melting. In R. BAKISH (Ed.), *Introduction to Electron Beam Technology*, Wiley, New York, 1962.
357 H. BOURGEOIS, *J. Four Elec.*, (1966) 193.
358 E. FROMM, *Z. Metallk.*, *57* (1966) 540.
359 E. GEBHARDT, E. FROMM AND D. JAKOB, *Z. Metallk.*, *55* (1964) 423.
360 E. GEBHARDT, H.-D. SEGHEZZI AND E. FROMM, *Z. Metallk.*, *52* (1961) 464.
361 E. FROMM AND H. JEHN, *Z. Metallk.*, *58* (1967) 61.
362 G. HÖRZ, *Z. Metallk.*, *59* (1968) 141, 180, 283.
363 G. HÖRZ, *Z. Metallk.*, *60* (1969) 115.
364 G. HÖRZ and E. GEBHARDT, *Z. Metallk.*, *57* (1966) 703, 737, 812.
365 E. GEBHARDT AND E. FROMM, Entgasungsreaktionen beim Schmelzen und Glühen von Niob und Tantal im Hochvakuum. In E. REXER (Ed.), *Reinststoffprobleme*, Vol. 1, *Reinststoffdarstellung*, Akademie-Verlag, Berlin, 1966, p. 373.
366 E. FROMM, *Z. Metallk.*, *56* (1965) 493.
367 E. ZEDLER, D. MÜLLER AND U. WIESNER, *Z. Metallk.*, *56* (1965) 316.
368 H. KIMURA, Y. SASAKI AND S. UEHARA, *Trans. Nat. Res. Inst. Metals (Tokyo)*, *8* (1966) 15.
369 G. HÖRZ, *Z. Metallk.*, *59* (1968) 832; *60* (1969) 50.
370 W. E. ANABLE, Electron-beam purification of vanadium, *U.S. Bureau of Mines, Rept. Invest. 7014*.
371 H.-J. GOTTWALD, K. KRONE AND J. KRÜGER, *Metall*, *23* (1969) 1284.
372 J. KRÜGER, *Thesis*, Tech. Hochschule Aachen, 1966.
373 S. R. SEAGLE, R. L. MARTIN AND O. BERTEA, *J. Metals*, *14* (1962) 812.
374 Brit. Pat. *985,253* (15.2.1963).
French Pat. *1,382,671* (14.2.1964).
375 H. J. GOTTWALD, *Diplomarbeit*, Tech. Hochschule, Aachen, 1967, unpublished.
376 O. N. CARLSON, F. A. SCHMIDT AND W. E. KRUPP, *J. Metals*, *18* (1966) 320.
377 H. A. WILHELM, F. A. SCHMIDT AND T. G. ELLIS, *J. Metals*, *18* (1966) 1303.
378 C. T. WANG AND E. F. BAROCH, *J. Metals*, (1969) 108A.
379 M. F. JOLY, *J. Four Elec.*, *72* (1967) 81.
380 A. LAWLEY, Properties of electron beam purified materials. In R. BAKISH (Ed.), *Proc. 4th Symp. on Electron Beam Technology*, AEC, Cambridge, Mass., *1962*, p. 133.
381 M. A. BADIALI, N. W. KIRSHENBAUM AND R. BAKISH, *Trans. AIME*, *227* (1963) 32.
382 E. REXER, On some properties of electron beam melted metals. In R. Bakish (Ed.), *Proc. 4th Symp. on Electron Beam Technology*, AEC, Cambridge, Mass., 1962, p. 245.
383 M. SEMCHYSHEN AND J. J. HARWOOD, *Refractory Metals and Alloys*, Interscience, New York, 1961.
384 A. E. VAN ARKEL, *Reine Metalle*, Springer-Verlag, Berlin, 1939.
385 R. F. ROLSTEN, *Iodide Metals and Metal Iodides*, Wiley, New York, 1961.
386 H. SCHÄFER, *Chemische Transportreaktionen*, Verlag Chemie, Weinheim/Bergstr., 1962.
387 O. N. CARLSON AND C. V. OWEN, *J. Electrochem. Soc.*, *108* (1961) 88.
388 A. C. LOONAM, *J. Electrochem. Soc.*, *106* (1959) 238.
389 G. H. KESLER, *Trans. AIME*, *218* (1960) 197.
390 N. D. VEIGEL AND J. M. BLOCHER, *J. Electrochem. Soc.*, *109* (1962) 647.
391 M. B. REIFMAN, A. I. GRIBOV, V. N. DMITRIEV AND M. A. LOSIKOVA, *Tsvetn. Metal.*, *34* (5) (1961) 50 (English edition).
392 Z. M. SHAPIRO, Iodide-decomposition process for production of zirconium. In B. LUSTMAN AND F. KERZE (Eds.), *The Metallurgy of Zirconium*, McGraw-Hill, New York, 1955, p. 135.
393 N. THIEN CHI, J. VERGNOLLE AND A. REMY, *J. Nucl. Mater.*, *12* (1964) 310.

394 V. S. Yemelyanov, A. I. Yevstyukhin and D. D. Abanin, Iodide method of refining thorium. In *Metallurgy and Metallography of Pure Metals*, Gordon and Breach, New York, 1962, p. 1.
395 R. G. Bellamy and N. A. Hill, *Extraction and Metallurgy of Uranium, Thorium and Beryllium*, Pergamon, London, 1963.
396 H. Spindler et al., *Werkstoffe der Kerntechnik*, Teil II, VEB Deutscher Verlag der Wissenschaften, Berlin, 1964.
397 N. D. Veigel, E. M. Sherwood and I. E. Campbell, *J. Electrochem. Soc.*, 102 (1955) 687.
398 D. E. Scaife and A. W. Wylie, A carbide–iodide process for high-purity thorium, *Proc. 2nd Intern. Conf. Peaceful Uses of At. Energy, 1958*, Vol. 4, P/1098, p. 215.
399 D. J. Maykuth, W. D. Klopp, R. I. Jaffee and H. B. Goodwin, *J. Electrochem. Soc.*, 102 (1955) 316.
400 N. V. Ageev and V. A. Trapeznikov, Methoden zur Herstellung reinen Chroms, in *Issledovanija po zapoprocnym splavam*, Moscow, 1959, p. 237.
401 V. S. Yemelyanov, A. Yevstyukhin, D. D. Abanin and V. I. Statsenko, Iodide method of refining chromium. In *Metallurgy and Metallography of Pure Metals*, Gordon and Breach, York, 1962, p. 10.
402 A. R. Edwards, J. I. Nish and H. L. Wain, *Met. Rev.*, 4 (1959) 403.
403 R. F. Rolsten, *Trans. AIME*, 215 (1959) 472, 478.
404 R. F. Rolsten, *J. Electrochem. Soc.*, 106 (1959) 975.
405 H. Winterhager and J. Holtkamp, *Metall*, 22 (1968) 33.
406 J. Gerlach, F. Pawlek and H. Probst, *J. Less-Common Metals*, 14 (1968) 355.
407 Das "van Arkel–de Boer-Jodidverfahren" zur Herstellung von reinen Titan, Zirkon, Hafnium, Silizium, Thorium, *Bibliographie No. 111 (1925–1956) des Forschungsinstitutes für NE-Metalle, Freiberg*.
408 J. Gerlach, J.-P. Krumme and F. Pawlek, *J. Less-Common Metals*, 15 (1968) 303.
409 W. G. Scholz, D. V. Doane and G. A. Timmons, *Trans. AIME*, 221 (1961) 356.
410 G. A. Leont'yev, Gewinnung von plastisch verformbarem Molybdän durch thermische Dissoziation des Pentachlorides. *Met. i Metalloved. Chistykh Metal., Sb. Nauchn. Rabot.*, (1960) 70.
411 V. S. Yemelyanov, G. A. Leont'yev and A. I. Yevstyukhin, Untersuchung der thermischen Dissoziation der Molybdänchloride, *Met. i Metalloved. Chistykh Metal., Sb. Nauchn. Rabot*, (1961) 137.
412 W. G. Pfann, *Zone Melting*, Wiley, New York, 2nd edn., 1966.
413 H. Schildknecht, *Zonenschmelzen*, Verlag Chemie, Weinheim, 1964.
414 N. L. Parr, *Zone Refining and Allied Techniques*, Newnes, London, 1960.
415 A. Lawley, Electron beam refining. In R. Bunshah (Ed.), *Introduction to Electron Beam Technology*, Wiley, New York, 1962, p. 184.
416 J. H. Wernick, *Purification of Metals by Zone Melting and Very-High-Temperature Melting Techniques in Ultra-High-Purity Metals*, ASM, Metals Park, Ohio, 1962, p. 55.
417 H. Hillmann, Das Verfahren des Zonenschmelzens und seine Anwendung auf Metalle, *Metall*, 15 (1961) 102.
418 L. Kuchar, Über eine neue Methode zur Ermittlung theoretischer Verteilungskoeffizienten. In E. Rexer (Ed.), *Reinststoffprobleme*, Vol. 1, *Reinststoffdarstellung*, Akademie-Verlag, Berlin, 1966, p. 523.
419 J. Barthel, Zur Theorie des Zonenschmelzens. In *Reinstoffe in Wissenschaft und Technik*, Akademie-Verlag, Berlin, 1963, p. 223.
420 E. Helfand and R. L. Kornegay, *J. Appl. Phys.*, 37 (1966) 2484.
421 E. Buhrig, Beitrag zur Ermittlung von Verteilungskoeffizienten aus Zonenschmelzkonzentrationsprofilen. In E. Rexer (Ed.), *Reinststoffprobleme*, Vol. 1, *Reinststoffdarstellung*, Akademie-Verlag, Berlin, 1966, p. 539.
422 J. Klofáč, *Hutnicke Listy*, 20 (1965) 730.
423 *U.S. Pat. 3,210,165* (29.12.1961).
424 N. F. Eaton, *J. Less-Common Metals*, 2 (1960) 104.
425 *DD 33861* (13.12.1963).
426 *DAS 1,212,495* (28.7.1955).

427 E. B. Bas and H. Stevens, *Z. Angew. Math. Phys.*, 18 (1967) 747.
428 A. Abdellatif, E. B. Bas and G. Wulff, *Z. Angew. Math. Phys.*, 18 (1967) 605.
429 E. B. Bas, personal communication.
430 E. Vanderscheuren and E. Votava, *Rev. Sci. Instr.*, 33 (1962) 389.
431 J. Barthel and R. Scharfenberg, Apparative Entwicklungen für die Einkristallzüchtung hochschmelzender Metalle. In E. Rexer (Ed.), *Reinststoffprobleme*, Vol. 1, *Reinststoffdarstellung*, Akademie-Verlag, Berlin, 1966, p. 495.
432 *DAS 1,126,245* (1.7.1964).
433 A. Calverley, M. Davis and R. F. Lever, *J. Sci. Instr.*, 34 (1957) 142.
434 F. E. Birbeck and A. Calverley, *J. Sci. Instr.*, 36 (1959) 460.
435 J. E. Kunzler and J. H. Wernick, *Trans. AIME*, 212 (1958) 856.
436 G. D. Kneip, Jr. and J. O. Betterton, Jr., *J. Electrochem. Soc.*, 103 (1956) 684.
437 J. P. Langeron, *Rev. Hautes Temp. Refract.*, 2 (1965) 137.
438 J. C. Wilson and M. L. Picklesimer, Variable-gradient electron-beam heating methods for growing single crystals of zirconium. In R. Bakish (Ed.), *Electron and Ion Beam Science and Technology*, Wiley, New York, 1965, p. 502.
439 J. P. Langeron, *Mem Sci. Rev. Met.*, 61 (1964) 637; also J. P. Langeron and M. Billion, *A.T.B. Met.*, 5 (1965–66) 259.
440 D. Mills and G. B. Craig, *Trans. AIME*, 236 (1966) 1228.
441 O. P. Kolchin, I. K. Berlin and N. V. Presnetsova, *Tsvetn. Metal.*, 36 (9) (1963) 59.
442 A. Calverley, The application of electron bombardment at S.E.R.L., *Services Electronics Res. Lab. Tech. Rept. No. 80*, p. 1.
443 C. S. Tedmon, Jr. and R. M. Rose, Substructure and impurities in niobium single crystals grown by the electron-beam, floating-zone technique. In R. Bakish (Ed.), *Electron and Ion Beam Science and Technology*, Wiley, New York, 1965, p. 521.
444 J. H. Wernick, D. Dorsi and J. J. Byrnes, *J. Electrochem. Soc.*, 106 (1959) 245.
445 K. F. Hulme and J. B. Mullin, *J. Electron. Control.* 3 (1957) 160.
446 B. B. Argent and G. J. C. Milne, The physical properties of niobium, tantalum, molybdenum and tungsten, *J. Less-Common Metals*, 2 (1960) 154; see also B. B. Argent, Zone refining of niobium, *Services Electronics Res. Lab. Tech. Rept. No. 80*, p. 32.
447 M. J. Leadbetter and B. B. Argent, *J. Less-Common Metals*, 3 (1961) 19.
448 D. P. Ferriss, R. M. Rose and J. Wulff, *Trans. AIME*, 224 (1962) 975.
449 S. Hurwitt, Diverse applications of electron floating zone apparatus, *Proc. 4th Symp. on Electron Beam Technology, AEC, Cambridge, Mass., 1962*, p. 232.
450 A. Lawley, Properties of electron-beam purified material. *Proc. 4th Symp. on Electron Beam Technology, AEC, Cambridge, Mass., 1962*, p. 133.
451 A. Lawley, J. van den Sype and R. Maddin, *J. Inst. Metals*, 91 (1962-63) 23.
452 E. Rexer, On the deformation, recrystallization and welding of molybdenum single crystals. In R. Bakish (Ed.), *Electron and Ion Beam Science and Technology*, Wiley, New York, 1965, p. 541.
453 E. Zedler, D. Müller, K. Schlaubitz and E. Rexer, *Z. Metallk.*, 55 (1964) 484.
454 R. C. Koo, *Acta Met.*, 11 (1963) 1083.
455 J. A. Belk, Zone melting of molybdenum, *Services Electronics Res. Lab. Rept. No. 80*, p. 9; also *J. Less-Common Metals*, 1 (1959) 50.
456 E. Buehler, *Trans. AIME*, 212 (1958) 694.
457 E. Buehler and J. E. Kunzler, *Trans. AIME*, 221 (1961) 957.
458 J. Barthel and R. Petri, Elektronenzonenschmelzen von Wolfram und Molybdän. In E. Rexer (Ed.), *Reinststoffprobleme*, Vol. 1, *Reinststoffdarstellung*, Akademie-Verlag, Berlin, 1966, p. 479.
459 H. W. Schadler, Electron bombardment melting of tungsten, *Proc. 1st Symp. on Electron Beam Melting, 1959*, p. 51; also *Trans. AIME* 218 (1960) 649.
460 W. R. Witzke, The zone melting of tungsten by electron bombardment, *Proc. 1st Symp. on Electron Beam Melting, 1959*, p. 73; see also W. R. Witzke, The purification of tungsten by electron-bombardment floating-zone melting. In R. F. Bunshah (Ed.), *Trans. Vacuum Metallurgy Conf., 1959*, New York University Press, 1960, p. 140.
461 B. C. Allen, D. J. Maykuth and R. I. Jaffee, *J. Inst. Metals*, 90 (1961–62) 120.

462 E. REXER, On some properties of electron beam melted metals, *Proc. 4th Symp. on Electron Beam Technology, AEC, Cambridge, Mass., 1962*, p. 245.
463 K. H. BERTHEL, H.-E. LONGO AND K. SCHLAUBITZ, Restwiderstandsmessungen an elektronenzonengeschmolzenem Wolfram, *Reinststoffe in Wissenschaft und Technik*, Akademie-Verlag, Berlin, 1963, p. 293.
464 G. F. IVANOVSKY AND T. N. ZAGORSKAYA, *Izv. Akad. Nauk SSSR, Metal.*, (3) (1965) 65.
465 D. R. HAY, R. K. SKOGERBOE AND E. SCALA, *J. Less-Common Metals*, 15 (1968) 121.
466 R. G. CARLSON, *J. Electrochem. Soc.*, 106 (1959) 49.
467 J. W. PUGH, On the purification of tungsten by electron beam refining, *Proc. 1st Symp. on Electron Beam Melting, 1959*, p. 89.
468 G. P. KOVTUN, A. A. KRUGLYKH AND I. G. D'YAKOV, *Izv. Akad. Nauk SSSR, Metal.*, (1) (1966) 71.
469 G. T. MURRAY, S. WEINIG AND I. DRANGEL, The relation between vacuum and purity in float-zone melting of refractory metals. In M. A. COCCA (Ed.), *Trans. Vacuum Met. Conf., 1964*, Am. Vac. Soc., Boston, Mass., 1965, p. 282.
470 E. M. SAVITSKY AND TAO TSZU-TSUN, *Izv. Akad. Nauk SSSR, Otd. Tekhn. Nauk, Met. i Toplivo*, (4) (1962) 133.
471 E. M. SAVITSKY, CH. V. KOPETSKY AND A. I. PEKAREV, *Izv. Akad. Nauk SSSR, Otd. Tekhn. Nauk, Met. i Toplivo*, (6) (1961) 74.
472 G. P. KOVTUN, A. A. KRUGLYKH AND V. A. PAVLOV, *Pribory i Tekhn. Eksperim.* (1966) 211.
473 G. P. KOVTUN, A. A. KRUGLYKH AND V. FINKEL', *Ukr. Fiz. Zh.*, 12 (1967) 1005.
474 D. R. HAY AND E. SCALA, Field-aided electron beam zone refining of tungsten. In M. A. COCCA (Ed.), *Trans. Vacuum Metallurgy Conf., 1964*, Am. Vac. Soc., Boston, Mass., 1965, p. 550; also *Trans. AIME*, 233 (1965) 1153.
475 W. P. ALLRED, R. C. HIMES AND H. L. GOERING, Electron-beam floating-zone melting of platinum-group metals, *Services Electronics Res. Lab. Tech. Rept. No. 80*, 1959, p. 27; also in J. S. HETHERINGTON, *Proc. 1st Symp. Electron Beam Melting, 1959*, Alloyd Electrochem Corp., 1959, p. 38.
476 J. T. SCHRIEMPF, *J. Less-Common Metals*, 9 (1965) 35.
477 D. W. RHYS, Electron-bombardment melting of some platinum metals, *Services Electronics Res. Lab. Tech. Rept. No. 80*, 1959, p. 22.
478 F. C. HOLDEN, R. W. DOUGLASS AND R. I. JAFFEE, *Symp. on Newer Metals*, 1959, ASTM Tech. Publ. 272, ASTM, Philadelphia, 1960, p. 68.
479 M. TANENBAUM, A. J. GOSS AND W. G. PFANN, *Trans. AIME*, 200 (1954) 762.
480 W. GEIL AND S. ZIEGENBALG, Über die Reinigung von Wismut, Antimon und Blei durch Zonenschmelzen, *Reinststoffe in Wissenschaft und Technik*, Akademie-Verlag, Berlin, 1963, p. 245.
481 B. N. ALEKSANDROV AND I. G. D'JAKOV, *Fiz. Metal. i Metalloved.*, 14 (1962) 267.
482 W. GEIL, Über die Beeinflussung der elektrischen Kenngrössen des Wismuts durch Verunreinigungen. In E. REXER (Ed.), *Reinststoffprobleme*, Vol. 1, *Reinststoffdarstellung*, Akademie-Verlag, Berlin, 1966, p. 559.
483 J. H. WERNICK, K. E. BENSON AND D. DORSI, *Trans. AIME*, 209 (1957) 996.
484 B. N. ALEKSANDROV, *Fiz. Metal. i Metalloved.*, 14 (1962) 733.
485 W. GEIL, Über die Herstellung von Reinwismut aus Rohwismut durch Zonenschmelzen, *Freiberger Forschungsh.*, B99, VEB Deutscher Verlag für Grundstoffindustrie, Leizpig, 1964, p. 93.
486 V. N. GRINYUK, I. I. PAPIROV, G. F. TIKHINSKY AND I. G. D'YAKOV, *Izv. Akad. Nauk SSSR, Metal.*, (4) (1967) 77.
487 W. R. MITCHELL, J. A. MULLENDORE AND S. R. MALOOF, *Trans. AIME*, 221 (1961) 824.
488 K. L. EDWARDS AND A. J. MARTIN, The purification of beryllium by distillation and zone melting. In *The Metallurgy of Beryllium*, Chapman and Hall, London, 1963, p. 557.
489 E. BONNIER, H. PASTOR AND J. DRIOLE, Sur une préparation de silicium de haute pureté, *Proc. Intern. Conf. on Vacuum Metallurgy, Brussels, 1965*, p. 75.
490 W. SCHRÖDER, Untersuchungen zur Möglichkeit der Abreicherung des Kohlenstoffs im Silicium durch tiegelfreies Zonenschmelzen im Hochvakuum. In E. REXER (Ed.), *Reinststoffprobleme*, Vol. 1, *Reinststoffdarstellung*, Akademie-Verlag, Berlin, 1966, p. 577.

491 H. Uwents, Formation of cobalt single crystals by electron bombardment, *Services Electronics Res. Lab. Tech. Rept. No. 80*, p. 36.
492 K. G. Davis and E. Teghtsoonian, *Trans. AIME*, *221* (1961) 1263; *227* (1963) 762.
493 G. Dressler, *Thesis*, Tech. Universität, Berlin, 1965.
494 G. Dressler, Die Herstellung von hochreinem Nickel. In E. Rexer (Ed.), *Reinststoffprobleme*, Vol. 1, *Reinststoffdarstellung*, Akademie-Verlag, Berlin, 1966, p. 467.
495 A. B. Berezin and Yu. D. Stepanov, Winning of ultra pure aluminium by zone melting, *Eksperim. Techn. i Metody Issled. pri Vysokikh Temperaturakh, Tr. Soveshch.*, (1959) 484.
496 K. Akerman et al., in E. Rexer (Ed.), *Reinststoffprobleme*, Vol. 1, *Reinststoffdarstellung*, Akademie-Verlag, Berlin, 1966, p. 509.
497 R. K. Willardson and H. L. Goering, *Compound Semiconductors*, Vol. 1, *Preparation of III–V Compounds*, Reinhold, New York, 1962.
498 G. Jayne, G. Cizeron and P. Lacombe, *Mem. Sci. Rev. Met.*, *62* (1965) 607.
499 D. Effer, *J. Electrochem. Soc.*, *108* (1961) 357.
500 T. C. Harman, *J. Electrochem. Soc.*, *103* (1956) 128.
501 O. N. Carlson, J. A. Haefling, F. A. Schmidt and F. H. Spedding, *J. Electrochem. Soc.*, *107* (1960) 540.
502 W. C. Necker, The zone purification of yttrium. In R. F. Bunshah (Ed.), *Trans. Vacuum Met. Conf., New York, 1960*, Interscience, New York, 1961, p. 289.
503 C. L. Huffine and J. M. Williams, Refining and purification of rare earth metals. In F. H. Spedding and A. H. Daane (Eds.), *The Rare Earths*, Wiley, New York, 1961, p. 145.
504 F. Hutchinson, Zone melting of gadolinium, *Services Electronics Res. Lab. Tech. Rept. No. 80*, p. 33.
505 V. M. Amonenko, A. A. Kruglykh and V. A. Pavlov, *Ukr. Fiz. Zh.*, *11* (1966) 1023.
506 J. L. Bieber, J. M. Schreyer and E. L. Williams, Purification of uranium by zone-refining techniques, *U.S. At. Energy Comm. Rept. Y-1564*, 1967.
507 B. Spriet, Plutonium purification by zone-melting and electro-diffusion, *J. Nucl. Mater.*, *15* (1965) 220.
508 K. G. Günther and H. Schreiner, Darstellung und Analyse hochentgaster Kupferlegierungen, *Congr. Intern. sur les Applications des Techniques du Vide à la Métallurgie*, Strasbourg, 1967.
509 R. H. Harada and A. J. Strauss, *J. Appl. Phys.*, *30* (1959) 121.
510 H. C. Theuerer, *Trans. AIME*, *206* (1956) 1316.
511 C. A. Hogarth, *Materials Used in Semiconductor Devices*, Interscience, New York, 1965.
512 D. Hazelby and J. L. Parmee, *J. Electrochem. Soc.*, *107* (1960) 144.
513 V. V. Zhukovetsky, *Izv. Vysshikh Uchebn. Zavedenü, Tsvetn. Met.*, *3* (1960) 115.
514 K. Grjotheim, O. Herstad and K. Stahl Johannessen, *Z. Anorg. Allgem. Chem.*, *328* (1964) 267.
515 L. M. Foster, A. S. Russel and C. N. Cochran, *J. Am. Chem. Soc.*, *72* (1950) 2580.
516 M. Heise and K. Wieland, *Helv. Chim. Acta*, *34* (1951) 2182.
517 A. S. Russel, E. Martin and C. N. Cochran, *J. Am. Chem. Soc.*, *73* (1951) 1466.
518 R. Heimgartner, *Schweiz. Arch.*, *18* (1952) 241.
519 P. Gross, C. S. Campbell, P. J. C. Kent and D. L. Levi, *Discussions Faraday Soc.*, *4* (1948) 206.
520 T. Kikuchi, T. Kurosawa and T. Yagihashi, *Trans. Nat. Res. Inst. Metals (Tokyo)*, *6* (1964) 214.

Chapter IV

Vacuum Degassing

Section 1

K. RÜTTIGER

Section 2

H. D. SEGHEZZI AND K. KÖSTLIN

1. Vacuum Degassing in the Liquid State

1.1 Introduction

Theory, experiments and processes which are concerned with the vacuum treatment of metal melts are here reviewed, the vacuum treatment being applied to molten metals produced in conventional melting units before the metal solidifies in casting moulds or ingot moulds. In contrast with the true vacuum melting processes which are very widely employed in metallurgical technology, vacuum treatment restricted to metals in the liquid state is used on an industrial scale almost only in steelmaking. The reason is purely one of economics as only very large melting furnaces or converters are economical in steelmaking. Even for high-alloy steel production rates of 20 ton/h are still at the lower limit of the economic efficiency, and for the melting rates required for conventional steelmaking present-day vacuum melting methods cannot be considered. In order to subject the very large quantities of molten metal usual in steelmaking to a vacuum treatment the only possible way was therefore to produce the molten metal conventionally in converters and furnaces and to apply the low-pressure treatment afterwards in a separate unit.

1.1.1 Historical survey

Every technological innovation is preceded by an abundance of concepts and ideas. For a long time technological suggestions were being made for the vacuum treatment of liquid metals which have been recorded in the patent literature[1-42]. For instance, Henry Bessemer suggested more than 100 years ago the application of an evacuated ingot mould which was to be filled with liquid steel by means of the atmospheric pressure[24] (Fig. 1). An American, Roman H. Gordon[25], obtained a patent in 1883 for a process for casting ingots in a vacuum, and in 1897 May[10] described a device for the pouring of several castings in an evacuated space (Fig. 2). According to his ideas a large number of moulds were to be arranged in a circle within a vacuum vessel. Pouring of the individual moulds was achieved by turning the launder. For another casting system which would exclude the ambient

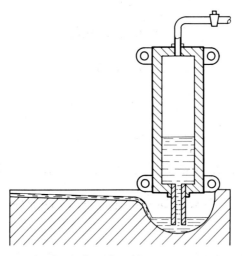

Fig. 1. Proposal of Bessemer for the casting of steel into an evacuated mould into which the steel is sucked by the vacuum.

atmosphere and which could have been employed for the production of mild steel a patent was granted in 1885[3]. This arrangement is shown in Fig. 3.

Towards the end of the last century ideas emerged for the arrangement of both the melting unit and the moulds in a vacuum chamber. As an example the apparatus suggested by Taussig[4] in 1889 may be mentioned, which consisted of a single-phase electric furnace and a mould in a vacuum vessel (Fig. 4). On the other hand, Simpson[9] proposed to enclose the ladle, instead of the melting furnace, together with the moulds in a common vacuum vessel (Fig. 5). A tiltable ladle is placed in this arrangement in a vacuum chamber which is separated by a low-melting-point diaphragm from a second vacuum chamber which contains the mould.

The alternative of carrying out degassing during the tapping of the melting unit into the ladle was also the object of several patents. For example, a proposal by Tholander[1] for the production of sound, dense converter-steel ingots may be mentioned here (Fig. 6). He proposed, *inter alia*, to connect a ladle, which could be evacuated, to the mouth of a converter in a vacuum-tight manner and to carry out the vacuum degassing of the steel in the ladle after tapping. Finally, the ladle and the converter were to be separated and the steel cast in the conventional way.

The technological difficulties, which prevented the realization of these ideas, led at the end of the last century to proposals to interpose a special device for vacuum degassing between the melting unit and the casting pit. The simplest technique, of moving the casting ladle into a vacuum chamber after tapping and then degassing, was regarded as having technological standing as early as 1882. Another patent[11] contains the idea of placing a preheated container filled with liquid steel in a preheated vacuum vessel (Fig. 7). It was pointed out that a particular advantage of this invention would be the fact that the vacuum treatment

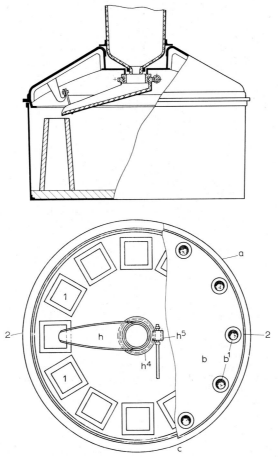

Fig. 2. Proposal of May[10] for the casting of ingots in a vacuum. a = Vacuum vessel; b = Lid; b^1 = Observation windows; h = Turntable tundish; h^4 = Inlet; 1 = Moulds.

could be extended for several hours because of the low heat losses, thus obtaining a complete degassing. Aitken[2] suggested, as shown in Fig. 8, that a special vessel, having a refractory lining and a suction pipe at the bottom, should be dipped into the ladle and by periodic evacuation and admittance of air, part of the melt should be sucked up into this special vessel so as to degas the melt in separate portions. Wainright[38] invented a similar vessel but with two intake pipes (Fig. 9). In this arrangement the steel flows through the vacuum chamber in a continuous stream. Each intake pipe carries at its lower part a siphon-type seal. The steel is poured into one of them, and leaves the degassing chamber through the other. The transport of the steel through the vacuum vessel is actuated by the hydrostatic pressure existing as a result of the difference in level between the two bath levels in the siphon-type seals. In order to keep heat losses low the suggestion was made to preheat the vacuum vessel when empty by gas-firing. Later, Williams[28] proposed

References pp. 507–515

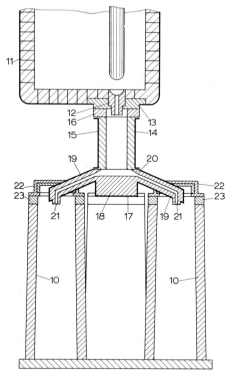

Fig. 3. Suggestion for casting with exclusion of air. For details, see ref. 3.

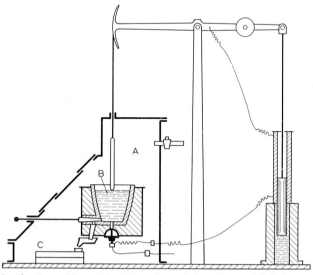

Fig. 4. Suggestion for the melting and casting of metals in a vacuum[4,7,8]. A = Vacuum vessel; B = Arc furnace; C = Mould.

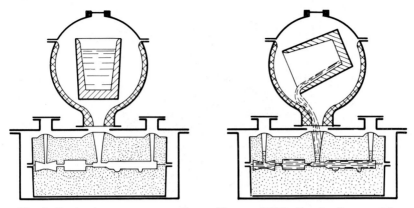

Fig. 5. Casting in a vacuum according to Simpson (1892). For details, see ref. 9

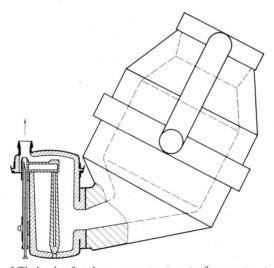

Fig. 6. Suggestion of Tholander for the vacuum treatment of converter steel during tapping[1].

a process based on the same principle but in which the rate of flow of steel is increased by injecting a gas into the intake pipe (Fig. 10). In this process it is also possible, just as it is in the technique suggested by Aitken, to discharge the degassed steel back into the original ladle. A modification of this design, also suggested by Williams, is shown in Fig. 11, where the intake and discharge pipes are arranged concentrically.

In the twenties a number of suggestions were made which were variants of the through-flow principle according to Wainright. Figure 12 depicts a modification after Betterton[27] in which the vacuum vessel is directly and permanently attached to the melting furnace. Similar proposals were made in a patent by

References pp. 507–515

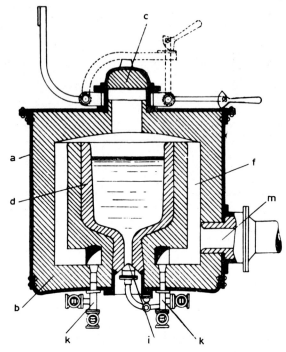

Fig. 7. Arrangement and installation for the vacuum treatment of high-grade steel in a heatable vessel containing the ladle. Heating is effected by burners (k) when the lid is removed.

Waldron[33], who also put forward some suggestions for an induction heating in the vacuum vessel (Fig. 13).

Entirely novel ideas on the degassing of liquid steel were developed in 1938 by Maré[31], who suggested subjecting the stream of molten metal to a vacuum degassing treatment during pouring or teeming. The stream of molten steel is introduced into an intermediate ladle, the sprue opening of which leads into a vacuum vessel. Another ladle is placed in this vacuum vessel for the reception of the stream-degassed steel (Fig. 14). After several trials had been made with the various potential processes on the laboratory and pilot-plant scale[43-49], this principle was eventually chosen for starting large-scale industrial steel degassing. The Bochumer Verein steelworks of Bochum must be credited with the development of the stream-degassing process to perfection for large-scale industrial use[50-55]; in fact this was the first vacuum process used industrially in the steel industry, and its success encouraged other steelworks to construct and test pilot plants for steel degassing.

Nowadays there are about twenty different processes known which have been tested for their industrial suitability, although only a few of them are actually being used in the steel industry. The individual processes may be classified into four principal groups:

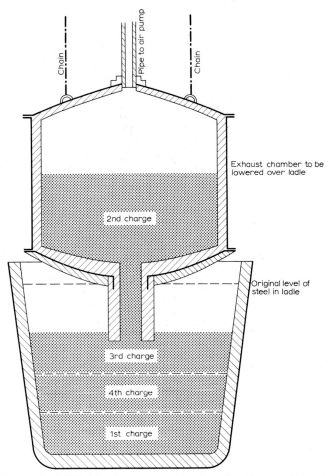

Fig. 8. Proposal of Aitken for the vacuum degassing of melts in partial quantities[2].

(a) *Ladle degassing processes.* In these the ladle filled with steel is closed with a vacuum-tight lid, or the ladle is placed in a vacuum vessel.

(b) *Stream degassing processes.* In these the stream of steel is collected in a vacuum vessel where it runs either into another ladle or into an ingot mould. Modifications are the so-called tap degassing and the Vac-I-Therm process.

(c) *Cycling and circulation degassing processes.* In these a separate vacuum chamber is placed on top of the ladle. The steel is sucked from the ladle into the vacuum chamber and after the degassing is discharged into the same ladle. Processes of this type are the so-called DH process, the RH process and the Thermo Flow process. Another modification is the so-called Transfer-Degassing, developed by BISRA, in which the steel enters the vacuum vessel through an entry sleeve and is discharged by an exit nozzle into a second ladle or into the tundish of a continuous casting machine.

References pp. 507–515

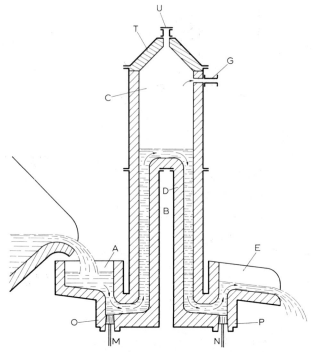

Fig. 9. Proposal of Wainright for a continuous degassing process of steel melts[38].

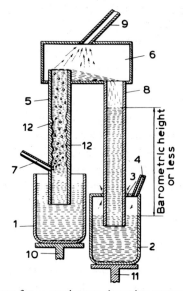

Fig. 10. Proposal of Williams for a continuous degassing process of metals with application of a carrier gas[28].

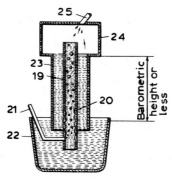

Fig. 11. Modification of the carrier gas process according to Williams[28]. Inlet and outlet tubes are arranged concentrically.

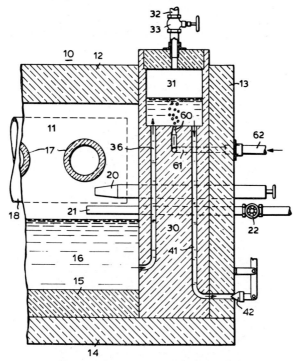

Fig. 12. Suggestion for the continuous vacuum degassing of metal melts during tapping[27].

(d) *Mould degassing processes.* In these the steel is degassed either during the teeming operation or by subjecting the mould to a vacuum after pouring.

The description of the individual processes will follow in subsequent sections, arranged, for the sake of clarity, according to this classification. It has not been established to what extent the development of the various plants now in existence were based knowingly or unknowingly on principles which had already been laid down at the turn of the century. Nevertheless, in many cases the similarity with

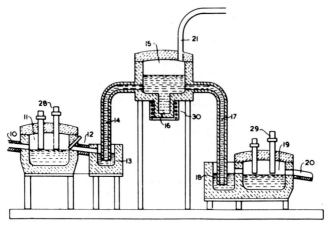

Fig. 13. Proposal of Waldron for the production of vacuum-degassed steel. During the dwell time in the vacuum vessel (15) the steel is heated by the induction coil (30).

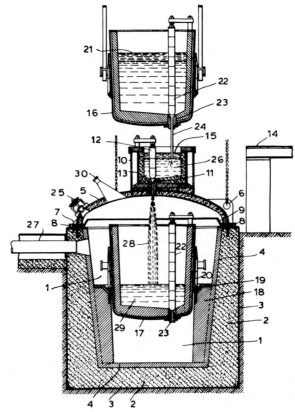

Fig. 14. Proposal of Maré for the vacuum degassing of steel melts. Figure taken from the patent application[31].

earlier proposals is rather surprising. However, before the individual processes are described in detail, some general points of view as well as the difficulties which are faced by any development of vacuum processes for the treatment of liquid steel will be discussed.

1.2 General points of view

1.2.1 Aims of a vacuum treatment of liquid steel

The incentive for large-scale technological experiments on the vacuum treatment of liquid steel originated in the difficulties met with in the production of large forgings[50–54, 56–59]. If the hydrogen content of the steel is high, a peculiar crack formation is observed; these defects in the steel are called flakes because of their shape. The cause of their appearance is the fact that hydrogen shows a very much lower solubility in steel at ambient temperature than at high temperature. During cooling hydrogen dissolved in the liquid steel is precipitated within cavities and other structural defects, creating such high pressures that the crystal grains are ruptured. In former times attempts were made to prevent the formation of flakes by annealing the forgings for days or even weeks at temperatures of 1200–1400°C, so as to lower the hydrogen level via a diffusion process. Much easier and very much more rapid is the elimination of the hydrogen from the steel in the liquid state by a vacuum treatment according to the equation:

$$2\,[H] \rightarrow H_{2\,\text{gas}}. \tag{1}$$

According to the law of mass action the hydrogen content is related to the hydrogen partial pressure[44, 50, 60] by Sieverts' law, *i.e.*

$$[H] = k_H (P_H)^{\frac{1}{2}}. \tag{2}$$

where [H] is the quantity of hydrogen dissolved in the steel, P_H is the hydrogen partial pressure above the melt, and k_H is a constant de pending only on the temperature of the steel. The same law applies to nitrogen dissolved in steel:

$$[N] = k_N (P_N)^{\frac{1}{2}}. \tag{3}$$

These two cases are plotted in Figs. 15 and 16.

Besides the reduction of the hydrogen content, the lowering of the oxygen content is also of great importance in steelmaking, as the oxygen dissolved in the melt reacts on addition of alloying elements with higher affinity for this element, forming oxides which precipitate in the steel as non-metallic inclusions, thus affecting the cleanness and the service properties of the steel. The fact that a lowering of the pressure above the melt also reduces the oxygen content of the steels, and therefore improves the purity with respect to oxide inclusions, contributed greatly to the expansion of the application of vacuum processes for the treatment of steel. In contrast with the elimination of hydrogen, the removal of oxygen from steel takes place only by reaction of the dissolved oxygen with the

References pp. 507–515

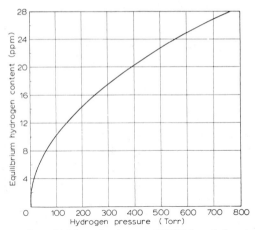

Fig. 15. Hydrogen saturation of high-purity iron at 1600°C in relation to the hydrogen partial pressure above the melt.

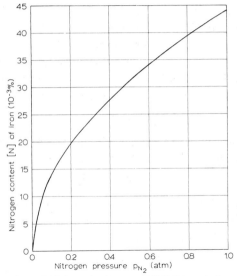

Fig. 16. Nitrogen saturation of high-purity iron at 1600°C in relation to the nitrogen partial pressure above the melt.

carbon present in the steel with formation of gaseous carbon monoxide according to the equation:

$$[C] + [O] = CO_{gas} \qquad (4)$$

and according to the law of mass action:

$$[C] \cdot [O] = k_{CO} \cdot P_{CO}, \qquad (5)$$

which means that the product of the carbon and oxygen contents of the steel is directly proportional to the carbon monoxide partial pressure above the melt[61,62].

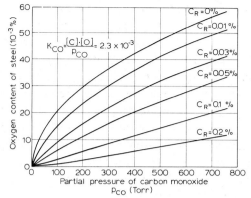

Fig. 17. Oxygen saturation of carbon-containing iron at 1600°C in relation to the carbon monoxide partial pressure above the melt.
Parameter $C_R = [C] - \frac{3}{4}[O]$; [C] = Carbon content of the melt; [O] = Oxygen content of the melt.

This linear pressure dependence of the carbon monoxide reaction makes it possible to obtain low oxygen contents and at the same time low carbon contents by reducing the pressure to a partial vacuum. The oxygen content theoretically attainable in relation to the carbon monoxide partial pressure is plotted in Fig. 17 for various carbon levels[61]. The constant k_{CO} was assumed to be 23×10^{-4} in accordance with experimental results obtained by Vacher and Hamilton. The quantities [C] and [O] are in wt.% and P_{CO} in atm in eqn. (5). The process for the removal of the oxygen is different for different carbon levels of the steel. As a simultaneous lowering of the carbon content takes place during the elimination of the oxygen, the carbon content [C] was not used as parameter for plotting these curves but instead the quantity $C_R = [C] - \frac{12}{16}[O]$, which is invariant during the degassing process provided that no further addition of oxygen or carbon takes place. The quantity C_R represents the residual carbon content which would remain in the melt if the total oxygen content were eliminated by formation of carbon monoxide.

Besides this reaction for the removal of dissolved oxygen from the melt, there also exist reactions in a vacuum which bring about the dissolution of previously formed oxide inclusions, or cause a reduction in their sizes[60,63,64]. All oxide inclusions in a steel show an oxygen solution pressure which is in equilibrium with the oxygen content of the melt. If during a vacuum degassing treatment the quantity of oxygen dissolved in the molten steel is lowered by the above carbon monoxide reaction, a concentration gradient of oxygen is created from the surface of the oxide particle to the melt which causes a migration of oxygen from the surface of the oxide particle by diffusion. The result is a gradual dissolution of the oxide particles. By vacuum treatment, therefore, particularly clean and sound steels may be produced which are superior in all their properties to conventionally melted steels.

For some qualities of steel it is desirable to eliminate the carbon as completely as possible. The conventional refining technique, *i.e.* addition of ore or the injection of air or oxygen at ambient pressure, attains low carbon levels only for very high oxygen contents which lead to a high iron oxide content of the slag and large iron losses. The reason for this behaviour may be seen from eqn. (5), as in this case the carbon monoxide partial pressure above the melt is about 1 atm. In some steelmaking processes the reduction of the carbon level is limited even by too heavy a slag formation. By employing a vacuum treatment it is, however, possible to lower the carbon content even further than by conventional methods because of the substantially lower partial pressure of the carbon monoxide above the melt. Transformer steels and austenitic stainless steels, as well as steels used for direct-on enamelling, are produced in a better quality by a vacuum treatment than by conventional production methods.

1.2.2 Choice of pumping system, and dust problems

It is a surprising fact that the vacuum processes, despite their promising advantages, have found application in the iron and steel industry only at a fairly recent date. This is mainly because, in addition to other technical difficulties, vacuum pumping systems which possess the required high pumping speeds and which would withstand the rugged conditions existing in steel works[65] were not available until about 10 years ago.

In the first plants started up the large quantities of dust created gave rise to serious difficulties. A number of elements which are present in the steel as alloying elements or as impurities show, at the temperature of liquid steel (about 1600°C), an appreciable vapour pressure, so that they evaporate during the vacuum treatment and, on cooling of the exhaust gases, precipitate as dust[60,66]. The vapour

TABLE 1

VAPOUR PRESSURES OF SELECTED METALS AT 1600°C

Metal	*Symbol*	*Vapour pressure (atm)*
Aluminium	Al	0.040
Barium	Ba	0.84
Calcium	Ca	1.88
Chromium	Cr	1.3×10^{-3}
Iron	Fe	2×10^{-4}
Lithium	Li	3.3
Lead	Pb	0.45
Magnesium	Mg	23.0
Manganese	Mn	0.035
Molybdenum	Mo	3×10^{-10}
Sodium	Na	58.0
Silver	Ag	0.018
Silicon	Si	8×10^{-4}
Tungsten	W	8×10^{-13}
Vanadium	V	5×10^{-7}
Tin	Sn	0.45
Zinc	Zn	76

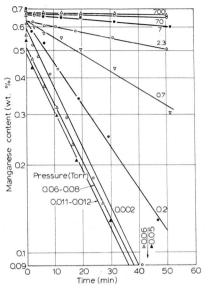

Fig. 18. Change in the manganese content of 12-kg melts with a carbon content of 0.25% in a vacuum at 1580±10°C. Parameter: Total pressure of the permanent gases in the space of a vacuum induction furnace. Results obtained by Ward[68].

pressure data of the pure elements P°_{Me} which are usually present in steel are compiled in Table 1. According to the concentration of these elements a partial pressure exists in the gas phase which may be expressed by the equation:

$$P_{Me} = a_{Me} \cdot P^\circ_{Me}, \qquad (6)$$

where a_{Me} is the activity of the element in the melt. Comprehensive studies of the evaporation rates were carried out by Schenck and Domalski[67] and by Ward[68,69]. As an example, the loss of manganese is plotted in Fig. 18 for a steel melt containing 0.5–0.7% Mn and 0.25% C for various pressure conditions during treatment at 1580°C. The sharp increase of the evaporation of manganese with decreasing pressure is obvious. For instance, dust quantities of 1 kg/ton of steel may have to be dealt with during the vacuum treatment of a steel containing about 1% Mn at pressures of around 1 Torr.

The dust produced during industrial degassing of steel consists mainly of iron and manganese. The dust precipitates on the cooler parts of the vacuum vessel as well as in exhaust ducts and in the pumps. In the intervals between the vacuum treatments the dust is oxidized and has then the appearance of a brown powder made up of a mixture of the evaporated metals and their oxide phases. The composition of precipitates which were found in laboratory experiments is shown in Table 2, while a compilation of dust precipitates found in industrial vacuum plants is presented in Table 3.

In the first years of the development of steel degassing on an industrial scale only mechanical pumps, such as oil sealed rotary vane pumps, rotary piston

TABLE 2

COMPOSITION OF THE PRECIPITATED METAL VAPOUR FROM AN IRON–MANGANESE MELT DEPOSITED ON A SUBSTRATE ABOVE THE SURFACE OF THE MELT AND ON THE FURNACE WALL[67]

Melt No.	Gas pressure (Torr)	melt % Mn	Composition of the precipitate			
			% Mn	% Fe	% C	Total (%)
			Substrate			
173	0.1	9.5	97	2.9		99.9
173	0.1	4.5	93.6	6.2		99.8
171	0.1	2.4	90.5	9.0	<0.01	99.5
171	0.1	1.4	78	20.8		98.8
			Furnace wall			
173	0.1		96	3.1	0.09	99.2
171	0.1		98.8	2.0	—	100.8

pumps, or Roots pumps in combination with water-ring pumps, were available[65]. These types of pumps are particularly susceptible to damage by dust. The hot waste gases had also to be cooled to the operating temperature of the pumps by special coolers or by a special arrangement of the exhaust ducts. Cyclones were found not to be sufficiently effective for the removal of the dust owing to its fineness. It is, however, not known whether electro-filters were employed for the precipitation of the dust. Electrostatic filters of conventional design are not suitable

TABLE 3

COMPOSITION OF DUST DEPOSITS IN VACUUM DUCTS OF VACUUM DEGASSING PLANTS

Sample No.	1	2	3	4	5	6	7	8	9	10
MgO	4.2	3.8	6.2	6.3	8.2	4.5	4.7	4.4	7.1	7.3
Al_2O_3	8.0	9.1	1.7	1.6	2.5	0.8	0.9	2.30	1.6	1.4
SiO_2	13.8	15.8	4.6	4.5	3.7	6.8	7.6	9.1	2.7	2.7
CaO	1.8	2.5	1.3	1.4	13.1	3.8	2.8	1.5	1.2	1.6
Cr_2O_3	1.2	1.2	1.1	1.1		<0.3	<0.3	<0.3	0.5	0.5
TiO_2	0.28	0.31	<0.1	<0.1	0.1	0.10	<0.10	0.14	0.10	0.10
MnO	31.0	24.8	36.0	35.0	8.5	47.5	52.0	43.0	25.5	23.6
$Fe_{met.}$	30.8	36.8								
FeO	0.00	0.13	10.98	10.21	11.38	5.73	5.34	0.26	12.68	8.46
Fe_2O_3	31.65	34.26	14.69	14.97	5.38	18.93	17.04	21.05	20.00	23.82
Na_2O	2.57	2.65	2.81	2.92	1.62	0.90	0.70	3.53	1.98	1.52
K_2O	1.31	1.52	1.81	1.78		2.02	1.13	0.45	2.26	2.23
ZnO			3.1	3.1		1.56	1.00	3.70		
PbO			3.5	3.5				1.14		
CuO						1.26	1.74	0.64		
P_2O_5								0.20		
SO_3								5.15	2.00	2.02
S total					0.5	0.37	0.18		0.87	0.82
CO_2								0.81	2.79	7.49
C total					21.0	0.48	0.30	0.24	0.90	0.75
H_2O									9.32	10.92
Loss on heating			8.2	8.2	36.5					

under vacuum conditions as gas discharges occur between the plates. Special mechanical filters were therefore developed with inserts consisting of metal or textile sieves. In order to prevent a clogging up of the pores these filters had to be supplied with shaking devices. Despite these precautions, fine dust particles still entered the pumps, causing contamination of the oil in oil-sealed rotary pumps and deposits on the rotors and the casing of Roots pumps. If these deposits grew to a certain thickness they were rolled to a hard layer. In some cases these hard layers were so thick that the casings of the Roots pumps burst.

The problem of dust precipitation was solved to a great extent by the employment of steam ejector pumps. It is sufficient to interpose relatively small cyclones or dust traps in the ducts in front of the steam ejector pumps, as these traps catch the main portion of the dust. If care is then taken to avoid any dead zones in the gas flow in the exhaust ducts between the vacuum vessel and the manifold of the first stage of the steam ejector pumps, the remaining dust will be carried into the first stage, where it is mixed with the steam and finally drained off with the cooling water from the condensers. Even substantial quantities of dust do not interfere with the working of steam ejector pumps. The exhaust gases of steam ejector pumps are free from dust and a special dust removal unit is therefore not necessary (see also Chapter II and ref. 70).

1.2.3 Heat losses

Particular attention should be paid in large-scale industrial plants to the question of the heat loss from steel melts[71]. As the fall in temperature is, as a rule, compensated for by superheating the melt in the steelmaking units, particularly expensive energy is used for this purpose. The heat loss may be retarded by reduction of the thermal emission and conduction losses in the vacuum treatment plants, by shortening the time required for the vacuum treatment, and also by additional heating during the vacuum treatment.

In order to give an idea of the magnitude of the possible heat losses, it will be assumed as an example that a fall in temperature of 50°C occurs during the vacuum treatment of a 100-ton steel melt in 20 min. The energy loss amounts to about 10^6 kcal or 1150 kWh (4×10^6 Btu). In order to compensate for this loss, an effective power of at least 3500 kW must be introduced into the melt during the period of the vacuum treatment. According to the efficiency of the heating system, effective power inputs of 5–7 MW would be required. These figures clearly show that it will in most cases be economically more advantageous to retard the heat losses from thermal radiation and thermal conductivity than to equip the vacuum treatment plants with highly rated heating units and the necessary auxiliary installations.

During the vacuum treatment no stationary thermal conductivity conditions are usually existent. In order to estimate the thermal losses under the prevailing conditions, it is necessary to study the non-stationary heat transport. A compila-

TABLE 4

FORMULAE FOR NON-STATIONARY HEAT TRANSPORT IN A MEDIUM WITH A ONE-SIDED BOUNDARY

$x[\text{m}]$ = coordinates $\quad t[\text{h}]$ = time $\quad k = \dfrac{\lambda}{c\rho}$ $\quad \Phi(y) = \dfrac{2}{\sqrt{\pi}} \int_0^y e^{-\eta^2}\, d\eta$

Title	Temperature distribution $u(x,t)$	Heat flow $j(x,t)$
1. Initial temperature u_0. Surface brought into contact with a body of temperature u_1	$u(x,t) = u_1 - (u_1 - u_0)\Phi\left(\dfrac{x}{\sqrt{4kt}}\right)$	$j(x,t) = \dfrac{(u_1 - u_0)}{\sqrt{\pi kt}} \lambda \cdot e^{-\frac{x^2}{4kt}}$
2. Initial temperature $= u_0 - \alpha x$ otherwise as (1)	$u(x,t) = u_1 - (u_1 - u_0)\Phi\left(\dfrac{x}{\sqrt{4kt}}\right) - \alpha x$	$j(x,t) = \left[\dfrac{u_1 - u_0}{\sqrt{\pi kt}} e^{-\frac{x^2}{4kt}} + \alpha\right] \cdot \lambda$
3. Surface preheated to temperature u_0 at the starting time t_0, then heated to temperature u_1 during time t	$u(x,t) = u_1 + (u_0 - u_1)\Phi\left(\dfrac{x}{\sqrt{4kt}}\right) - u_0 \Phi\left(\dfrac{x}{\sqrt{4k(t+t_0)}}\right)$	$j(x,t) = \lambda\left[\dfrac{u_1 - u_0}{\sqrt{4kt}} e^{-\frac{x^2}{4kt}} + \dfrac{u_0}{\sqrt{\pi k(t+t_0)}} e^{-\frac{x^2}{4k(t+t_0)}}\right]$
4. Surface heated with constant power input/m² = q[kcal/h m²]	$u(x,t) = \dfrac{q}{\lambda}\left[\sqrt{\dfrac{4kt}{\pi}}\, e^{-\frac{x^2}{4kt}} - x\left(1 - \Phi\left(\dfrac{x}{\sqrt{4kt}}\right)\right)\right]$	$j(x,t) = q\left[1 - \Phi\left(\dfrac{x}{\sqrt{4kt}}\right)\right]$

Title	Heat flow through the surface $j(0, t)$	Total heat loss per m^2: $Q = \int_0^t j(0, t)\, dt$
1. Initial temperature u_0 Surface brought into contact with a body of temperature u_1	$j(0, t) = \lambda \dfrac{(u_1 - u_0)}{\sqrt{\pi k t}}$	$Q(t) = \dfrac{2\lambda(u_1 - u_0)}{\sqrt{\pi k}} \sqrt{t}$
2. Initial temperature $= u_0 - \alpha x$ otherwise as (1)	$j(0, t) = \lambda \left[\dfrac{u_1 - u_0}{\sqrt{\pi k t}} + \alpha \right]$	$Q(t) = \dfrac{2\lambda}{\sqrt{\pi k}} (u_1 - u_0) \sqrt{t} + \alpha \lambda t$
3. Surface preheated to temperature u_0 at the starting time t_0, then heated to temperature u_1 during time t	$j(0, t) = \lambda \left[\dfrac{u_1 - u_0}{\sqrt{\pi k t}} + \dfrac{u_0}{\sqrt{\pi k (t + t_0)}} \right]$	$Q(t) = \dfrac{2\lambda}{\sqrt{\pi k}} \left[(u_1 - u_0) \sqrt{t} + u_0(\sqrt{t + t_0} - \sqrt{t_0}) \right]$
4. Surface heated with constant power input/m^2 = q[kcal/h m^2]	$j(0, t) = q$	$Q = q \cdot t$

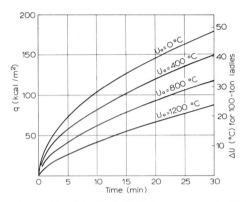

Fig. 19. Quantity of heat transmitted into the ladle lining on contact with liquid steel in relation to the time for various preheating temperatures of the lining. The left-hand ordinate shows the quantity of heat transmitted per square metre of contact area. On the right-hand ordinate the fall in temperature for 100-ton melts based on these figures is plotted. See also ref. 72.

tion of formulae for several special conditions is presented in Table 4. In the formulae the temperature of a thermal conducting material is designated u, the initial temperature u_0; x is used for the distance from the surface, t is the time, j the heat flow per hour and per unit of area, j_0 the heat flux through the surface into the adjacent refractory material, and q the quantity of heat transferred from the melt through unit area. As an example, the heat flux which occurs for one m² of conventional ladle lining is plotted in Fig. 19, if the ladle was not preheated or preheated for 1 hour to 400°, 800° and 1200°C. The scale shown on the right-hand side indicates the fall in temperature which occurs in a 100-ton ladle through heat transfer into the ladle lining.

In addition to these losses, which in practice occur in all degassing processes, temperature losses by radiation through the surface or by heat transfer to vessels to which the melt will be transferred or into which it will be sucked, must be taken into account. As these losses depend to a large extent on the operation of the individual vacuum processes, they will be discussed when these processes are described in detail. These temperature losses, besides the metallurgical efficiency and the various problems connected with the refractory linings, are an important characteristic in the appraisal of the individual processes.

1.2.4 Refractory lining of the vessels for the vacuum treatment

Vacuum treatment processes placed requirements on the refractory linings of the ladles and vessels which were until then unknown in steelmaking[73–78]. Side by side with the development of the individual vacuum treatment processes to industrial application, a development of new refractory materials took place[79]. The requirements placed on refractory materials for this purpose may be summed up as[80]:

(a) Resistance to liquid steel under the strongly reducing conditions of a vacuum.

(b) Chemical resistance to slags under both reducing and oxidizing conditions.
(c) Low vapour pressure and chemical stability under reduced pressure conditions.
(d) Favourable behaviour against tensile and compressive stresses at high temperatures.
(e) Good thermal shock stability.
(f) Resistance to mechanical wear (abrasion).
(g) Ease of handling during ramming, transport and storage.
(h) Favourable relation between price and service life.

So far no material is on the market which fulfils equally well all these conditions[81]. According to the vacuum process employed and to the working conditions, some of these requirements are more important than others so that one material may be more suitable in one process than in another. The specific-lining problems arising in the various vacuum processes will therefore be discussed during the description of these processes, but some general aspects for the selection of the refractory materials will now be considered.

(a) Resistance of the refractories to liquid steel

As mentioned earlier, metal oxides may be reduced in contact with liquid steel in a vacuum if the oxygen solution pressure of the oxide is higher than the solution pressure of the oxygen dissolved in the steel[63,69,81]. The thermodynamic data for the calculation of the oxygen pressure of the more important metal oxides and metal compounds of refractory materials are shown in Table 5, and are plotted in Fig. 20, in relation to the temperature. The quantities $RT \ln p_{O_2}$ are calculated from these thermodynamic data. The broken line for an oxygen partial pressure of 10^{-16} atm corresponds approximately to the conditions existing in a 1% carbon steel at 10 Torr. If the curves traverse the broken line the oxides will

TABLE 5

THERMODYNAMIC DATA FOR THE CALCULATION OF THE OXYGEN PRESSURE OF CERTAIN METAL OXIDES

$$\ln p_{O_2} = \frac{1}{RT}(H-TS)$$

Reaction equation	H (cal/mol)	S (cal/mol K)	Temperature range (K)
$4/3\ Al+O_2 = 2/3\ Al_2O_3$	−257,500	−44.30	930–2318
$4/3\ B+O_2 = 2/3\ B_2O_3$	−223,000	−35.50	298–1800
$2\ Ca+O_2 = 2\ CaO$	−307,100	−51.30	1124–1760
$4/3\ Cr+O_2 = 2/3\ Cr_2O_3$	−178,500	−41.40	298–1868
	−124,100	−29.90	843–1642
$2\ Fe+O_2 = 2\ FeO$	−103,950	−17.71	1642–1808
	−111,250	−21.67	1808–2000
$2\ Mg+O_2 = 2\ MgO$	−363,400	−102.60	1380–2000
$2\ Mn+O_2 = 2\ MnO$	−190,800	−39.25	1500–2051
$Si+O_2 = SiO_2$	−217,600	−48.79	1700–1986
$Ti+O_2 = TiO_2$	−217,500	−41.40	298–2080

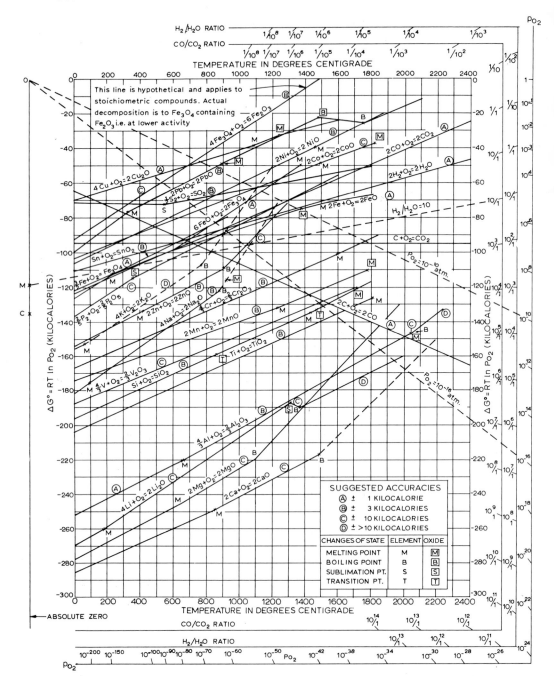

Fig. 20. Temperature dependence of the thermodynamic oxygen potential $RT \ln P_{O_2}$ of some metal oxides (according to Richardson and Jeffes).

be dissolved by the steel during vacuum treatment. It can also be seen that a dissolution of silica, chromium oxide and iron oxide may take place very quickly, whereas alumina, magnesite and lime are very stable indeed. It is therefore comprehensible that materials which belong to the latter group are used almost exclusively for the lining of vacuum vessels. Materials containing higher proportions of silica, chromium oxide, iron oxide or manganese oxide show, under most conditions, considerable wear.

(b) Resistance to slag reactions

As steel melts undergo an agitation during the vacuum treatment by the degassing action which is similar to the boiling action in converter refining, even small residual quantities of slag, which float on top of the melt, are brought into close contact with the refractory lining. As a result of the low interfacial tension between slag and refractory material these residual quantities of slag are easily absorbed and drawn into the pores of the lining. The slag reacts in many cases with the lining and changes the chemical composition of the refractory material to such an extent that low-melting-point mixtures are formed. As an example, the constitutional diagram of lime and silica is shown in Fig. 21 and that of lime and alumina in Fig. 22. In some cases the reacting slags form chemical compounds with the refractory material, which affect the mechanical properties of the lining. For instance, the thermal shock resistance may be reduced to such an extent that it becomes impossible to cool the lining without damaging it and so to use it for a second vacuum treatment.

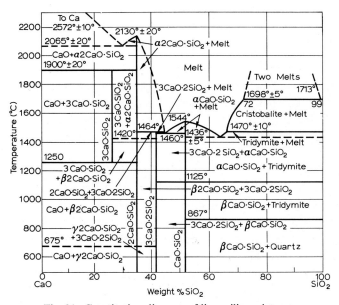

Fig. 21. Constitution diagram of lime–silica mixtures.

References pp. 507–515

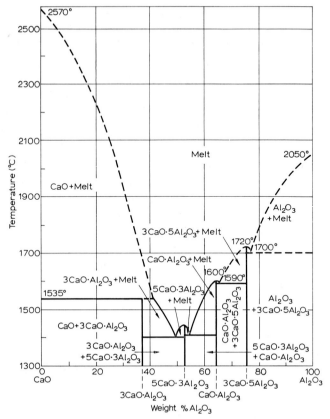

Fig. 22. Constitution diagram of lime–alumina mixtures.

(c) Low vapour pressure

Refractory materials may suffer marked losses by evaporation at low pressures. This loss may be attributed to three principal reactions. The first and most important is the decomposition of the oxide into its gaseous components. This phenomenon was the object of a study by Brewer[82], and the volatilization of lime according to the following reaction may be used to illustrate it:

$$CaO_{solid} = Ca_{gaseous} + \tfrac{1}{2}O_{2\,gaseous}. \qquad (7)$$

A similar mechanism has been observed for the evaporation of iron oxide, chromium oxide, magnesium oxide, aluminium oxide and silicon oxide as well as of manganese oxide. In other cases the decomposition of the oxides does not proceed to the metallic state, but goes only as far as the suboxide which may be easily volatile, for instance:

$$SiO_2 = SiO_{gaseous} + \tfrac{1}{2}O_{2\,gaseous}. \qquad (8)$$

The second mechanism involved is the decomposition of the oxide into

TABLE 6
WEIGHT LOSS OF REFRACTORY BRICKS IN A VACUUM AT 1650° C ACCORDING TO K. M. BONAR

	Weight loss (%)	Rate of weight loss $g/cm^2\, min \times 10^{-4}$ Measured	Chemical analysis (%)				
			MgO	Al_2O_3	Cr_2O_3	Fe_2O_3	CaO
Power pressed basic brick							
A. High purity magnesite (average of 4)	6.2	5.4	97	—	—	—	1.4
B. Direct bonded magnesite-chrome	6.6	5.2	73	10.0	9.3	5.1	1.1
C. Rebonded fused grain magnesite-chrome (A)	5.0	4.2	62	8.1	17.6	10.5	0.6
D. Rebonded fused grain magnesite-chrome (B)	6.9	5.9	52	10.3	22.7	11.8	0.9
E. Chrome brick	6.5	7.5	19.5	34.4	26.0	11.7	1.1
F. Spinel bonded magnesite	4.1	3.6	89	9.8	—	0.4	1.1
G. Lime brick (96% CaO)	1.0	0.6	2.7	0.3	—	0.3	96
H. High purity dolomite (99% MgO+CaO)	0.6	0.4					
Fusion cast brick							
I. Magnesite-chrome (average of 3)	14.0	12.0	57	9.5	19.5	10.3	1.5
J. Magnesite-spinel	4.8	3.2	80	16.0	—	1.0	0.8
High alumina brick			Al_2O_3	SiO_2	Fe_2O_3	TiO_2	Na_2O
K. High purity alumina	0.2	0.2	99.5	0.2	—	—	—
L. 90% alumina	0.8	0.6	89.3	10.2	0.2	—	—
M. Mullite	2.1	1.5	70.6	25.2	1.0	2.9	—
N. 70% alumina	5.2	3.5	71.3	24	1.3	2.9	—
O. 60% alumina	4.4	3.0	58.5	37.3	1.4	2.4	—
P. Fusion cast alumina	1.2	1.1	96	0.5	—	—	3.4
Zircon and zirconia brick			ZrO_2	SiO_2	CaO		
R. High purity lime-stabilized zirconia	0.15	0.17	96	—	4.0		
S. Self-bonded zircon	3.8	3.9	66	32.3			

oxygen and a solid suboxide, for instance:

$$Fe_2O_{3\,solid} = 2\,FeO_{solid} + \tfrac{1}{2}O_{2\,gaseous}. \qquad (9)$$

The third mechanism consists of the evaporation of a substance as a stable gas of the same composition as that of the solid, for instance:

$$P_2O_{5\,solid} = P_2O_{5\,gaseous}. \qquad (10)$$

Experimental results for the evaporation of refractory materials in a vacuum are reported by Bonar et al.[83] (cf. Table 6). Of the materials studied, zirconia and alumina showed lower weight losses by evaporation than did magnesite and lime, but the evaporation losses increased when the materials were combined with traces of iron oxide, silica and chromium oxide. This higher evaporation not only showed up in the weight losses, but also in an increase in pore volume. The weight loss per min of high-alumina bricks in relation to the silica content is demonstrated

References pp. 507–515

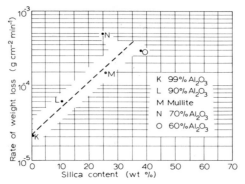

Fig. 23. Effect of alumina content on the evaporation rate of silica-containing alumina bricks in a vacuum at 1650°C[83].

as an example of this phenomenon in Fig. 23.

(d) Behaviour towards tensile and compressive stresses at high temperatures

As in most cases no allowance is made for a thermal expansion of the refractory lining in vacuum vessels, very much higher stresses are often set up in the refractory material of vacuum vessels than in conventional aggregates for the melting of steel. Moreover, the load-bearing capacity of refractory bricks sharply decreases with increasing temperature, as may be seen from Fig. 24. The suitability of a refractory material should therefore be tested not only at ambient temperature but also at 1600°C. The mechanical abrasion resistance behaves similarly. Wear occurs by both the vigorous turbulence of the steel during the vacuum treatment and by the addition of solid alloying elements.

Since longer downtimes between the individual vacuum treatments are unavoidable in steelmaking practice, the lining of the vessels for this purpose must be able to withstand frequent heating to the temperature of liquid steel and cooling

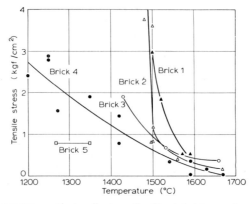

Fig. 24. Effect of temperature on the tensile strength of basic bricks produced by various methods.

to the ambient temperature without lowering of any properties. In many cases good thermal shock resistance can be obtained by changing to a coarse grain size of the refractory materials.

The decision on the applicability of a refractory material to vacuum processes cannot be arrived at in laboratory experiments, but can only be made after long-term experiments under industrial working conditions. It is customary practice in these studies to take various refractory materials and to use them at various positions on the same vessel and to observe their behaviour during several heats. In these pilot plant experiments the cost for the refractory material, for the transport and for the ramming or bricking up, must be determined with great accuracy. Only a comparison of the costs involved in several such operations can produce a true picture as to which refractory material is best suited for the particular vacuum process[63,84-86].

1.2.5 Insertion into the production schedule of steel works

It has already been pointed out that besides the metallurgical advantages, the costs for the vacuum treatment play a decisive part in the appraisal of a vacuum process. Besides the net costs for the operation of the plant, those expenses must be taken into account which occur as a result of additional work in the steelmaking process. The additional costs for the superheating of the melts in the steelmaking units have already been mentioned. The work necessary for the preparation of the degassing plant for the treatment of a heat represents another cost factor, and to this should be added any additional transport costs from the steelmaking unit to the vacuum plant and back to the casting pit. Moreover, a loss in time is incurred by the vacuum treatment which may reduce the production rate of a steel works. It is therefore important to take account of these additional costs in deciding which vacuum treatment process is best suited under the prevailing conditions. For this decision the number of the daily vacuum charges and the quality of the steel produced are of great importance. These factors must be thoroughly investigated in relation to the production programme of the individual steel works. The great variety of the degassing processes used in practice and the many modifications of the various processes must be attributed to these causes.

1.3 Ladle degassing processes

1.3.1 Ordinary ladle degassing

The simplest way to submit a heat of molten steel to a vacuum treatment is to place a vacuum-tight lid on the filled casting ladle or to insert the whole ladle into a vacuum vessel[87-97]. The first experimental arrangement used by Korber at the Škoda works during the years 1932–1935 is shown in Fig. 25[98]. The experiments were, however, soon discontinued, as no sufficiently powerful vacuum pumps were available to Korber. As pressures of between 30 and 60 Torr only

References pp. 507–515

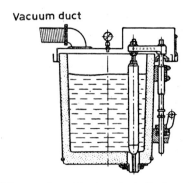

Fig. 25. Diagram of the experimental set-up for ladle degassing by Korber[98,99].

could be obtained, the degassing effect was small.

It was only after the war that Samarin[89,90] as well as Speransky[87,100] carried out similar experiments, but even then a vacuum no better than 30 and 50 Torr was attained. The treatment lasted between 10 and 20 min, and the experiments were concerned mainly with the treatment of transformer steel and structural alloy steels. The plant used is depicted in Fig. 26. After the ladle had been placed in the vacuum vessel, the lid, which was provided with a rubber seal, was placed upon the vacuum vessel and the vessel gradually evacuated. The behaviour of the melt during evacuation could be followed by looking through an observa-

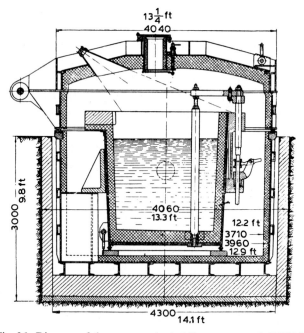

Fig. 26. Diagram of the vacuum plant at Dneprospetsstal, USSR[89].

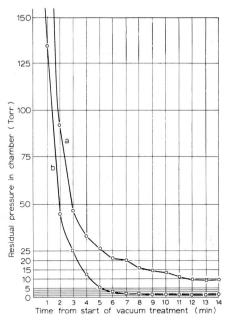

Fig. 27. Residual pressure during ladle degassing of 22-ton ladles. Experimental results obtained by Samarin et al.[102].

tion window provided in the lid, and the pressure was adjusted with the aid of a vacuum valve. As soon as the pressure in the vacuum vessel was lowered, the gases dissolved in the melt were liberated and brought about an intensive boiling action causing the level of the bath to rise about 30–40 cm in the ladle. The intensity of the boiling action varied with the temperature of the melt, the viscosity of the slag and type of steel treated. The most vigorous evolution of gas occurred with transformer steels, whereas it was lowest with chromium–nickel and ball-bearing steels. The wear of the ladle lining and of the stopper rod increased with increasing gas evolution and boiling action. The main risk in these experiments was that such a vigorous attack might occur on the stopper rod that it burned away. It was found necessary in these experiments to superheat the steel in the furnace before the vacuum treatment by 30–40°C. After completion of the vacuum treatment the vacuum vessel was opened to the atmosphere, the ladle removed and the steel teemed in the usual way.

This method, known as ordinary ladle degassing, was developed especially in Russian steel works[88–90,92,101–104]. In these experiments ladle capacities of 30 tons were rarely exceeded, and the initial results cannot therefore be applied to large melting units without caution, as will be shown later on. Samarin and his co-workers were the first to obtain pressures as low as 1 to 5 Torr for a melt of 22 tons[102]. Typical curves for the pressure during the vacuum treatment are shown in Fig. 27. Samarin also submitted rail steel to a vacuum treatment. The steel was

References pp. 507–515

TABLE 7

RESULTS OF LADLE DEGASSING PROCESS UPON
GAS CONTENTS OF RAIL STEEL (SAMARIN et al.[102])

Oxygen content in the molten metal		
Before treatment	57.7	ppm
After treatment	13.0	ppm
Relative reduction	77.5	%
Oxygen content of rolled stock	18.4	ppm
Relative reduction	68.1	%
Mean figures for	31	melts
Hydrogen content of the molten metal		
Before treatment	4.07	cm³ (S.T.P.)/100 g
After treatment	1.94	cm³ (S.T.P.)/100 g
Relative reduction	52.3	%
Mean figures for	38	melts

deoxidized during tapping with ferro-manganese and ferro-silicon, and during the vacuum treatment 50–100 g aluminium was added per ton of steel, but in some cases additions of vanadium and boron were also made. The oxygen and hydrogen levels obtained in these experiments are listed in Table 7. The elimination of the hydrogen during the vacuum treatment for various melts and its distribution in relation to the depth of the bath are plotted in Fig. 28[103]. The reduction of the hydrogen content in rail steel to final levels of between 2 and 2.4 cm³/100 g steel made it possible to produce rails free from flakes without a special heat-treatment. Samarin established that, in general terms, the quality of a converter steel could be raised by vacuum treatment to that of an open-hearth steel. For ladle degassing a treatment time of 14–15 min at a pressure of 5–10 Torr is required. It is even

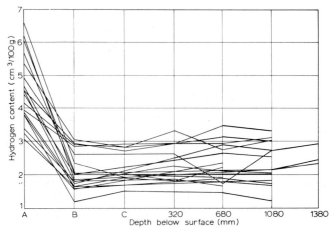

Fig. 28. Hydrogen content of ladle-degassed rail steel, also effect of depth from the surface on the hydrogen level. Results obtained by P. Ya. Sorkin[103]. Each curve represents one heat.
A = Hydrogen content at the surface before degassing.
B = Hydrogen content at the surface after degassing.
C = As B, second determination.

TABLE 8

OXYGEN AND NITROGEN CONTENT OF VARIOUS STEELS BEFORE AND AFTER DEGASSING[107]

Steel grade	Oxygen content		Nitrogen content	
	Before degassing ($10^{-3}\%$)	After degassing ($10^{-3}\%$)	Before degassing ($10^{-3}\%$)	After degassing ($10^{-3}\%$)
XC 35	10.2	6.5	9.2	7.1
25 CDV 4	9.5	4.6	8.1	7.4
XC 25	7.3	2.8	6.8	6.2
30 NCD 11	9.4	4.7	6.4	6.3
10 CD 9	10.0	5.3	7.6	6.5
32 NCD 8	8.2	4.9	7.5	6.8
XC 42	9.4	4.6	6.9	6.2
60 CD 4	9.9	2.8	8.1	6.5
XC 35	11.2	5.0	9.4	7.3
30 NCD 11	7.6	3.6	7.2	6.8

claimed that by special means the quality of converter steel could be raised by vacuum treatment to such a degree that it was superior to open-hearth steel with respect to weldability and technological properties.

Samarin also subjected melts which had not been previously deoxidized to a vacuum treatment. In this case the boiling action was so violent that the level of the bath was raised by 60–70 cm. During treatment of these melts the carbon content was lowered by 0.03–0.06%, but the content of the accompanying elements hardly changed.

The employment of pumps of higher pumping speeds brought about an improvement in the results of ladle degassing[94-96,105,106]. According to a French publication[107], hydrogen contents initially 5 cm³/100 g were reduced to 1.9 cm³/100 g during the vacuum degassing of 55-ton ladles. Oxygen and nitrogen levels obtained for several steel grades are shown in Table 8, and similar results were reported by Speith et al.[96,105]. These authors also established a relationship between reduction in hydrogen content and the final pressure during the vacuum

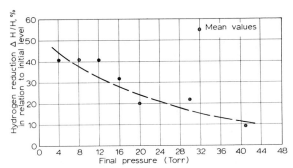

Fig. 29. Effect of pressure on the relative reduction of the hydrogen level during ladle degassing. According to Speith, vom Ende and Specht[105].

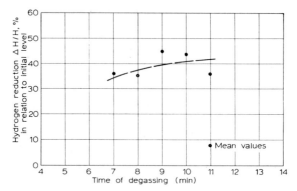

Fig. 30. Effect of degassing time on the relative reduction of the hydrogen level in ladle degassing at final pressures of 2–8 Torr[105].

treatment (Fig. 29). The duration of the degassing treatment seems, according to their results, to have no effect on the lowering of the hydrogen content, but to have a marked effect on the reduction of the oxygen level (Figs. 30 and 31). It should, however, be pointed out that the reduction of the oxygen content depends to a large extent on the initial oxygen level of the steel (Fig. 32).

The improvement in the number of oxide inclusions after ladle degassing is shown in Fig. 33. If, however, the vacuum applied was not adequate, a better cleanness was not attained. Usually the ladle degassing of semi-killed steels, which were produced under a black slag, showed a degree of purity similar to that of killed steels.

Levin and Umrichin[108] report on the heat losses during ladle degassing. Their results are listed in Table 9. They found that the main quantity of the hydrogen (to a total of about 90%) was evolved during degassing in the ladle in the first 5–8 min. It is therefore concluded that the hydrogen is removed principally by kinetic processes taking place in the melt. These authors express the opinion that

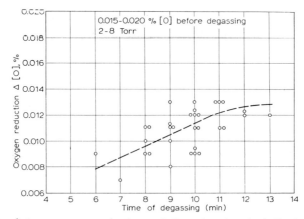

Fig. 31. Relation between oxygen reduction and degassing time in ladle degassing at final pressures of 2–8 Torr[105].

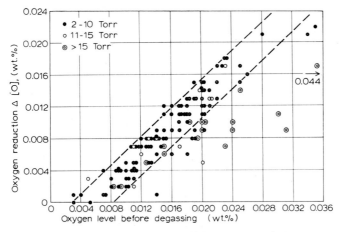

Fig. 32. Relation between oxygen reduction and initial oxygen level in ladle degassing of semi-killed steel melts[105].

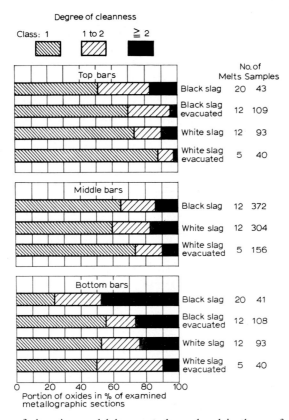

Fig. 33. Cleanness of chromium–molybdenum steels produced in the arc furnace by various methods, with and without application of vacuum degassing. Samples taken from the top, the middle and the bottom of the ingots[105].

References pp. 507–515

TABLE 9

REDUCTION Δt IN THE TEMPERATURE OF THE METAL (°C) DURING VACUUM TREATMENT IN THE LADLE (THE FIGURES IN PARENTHESES ARE LOSSES IN TEMPERATURE PER MIN)*

Works (or combine)	Steel** type	Q	P_{res}	τ (min)	Δt
Enakievo	B	15	70–120	12–14	30 (2.5–2.1)
	B	25	2***	12–15	30–40 (2.6)
Dneprospetsstal	Sh, LK	25	60–70⁺	10–12	35–45 (3.6)
	18KhNVA	25	13	12	30–35 (2.7)
	ShKh15	25	30	6–10	30 (3.7)
Red October	LK‡	20	10–15	7–10	30–50 (4.7)
	ShKh15	20	15–20	7–10	35–70 (2.3)
Kuznetsk	Sh, TR	—	15–20	8–12	20–30 (2.5)
V. I. Lenin (Permas)	K, Sh	60	10–15	12–13	40–50 (3.6)
A. K. Serov	Sh	90	4–8	15–20	27–30 (1.8)

* Q ladle capacity (tons); P_{res} residual pressure (Torr); τ vacuum treatment time.
** B converter, Sh ball bearing, LK alloy constructional, TR transformer, K constructional.
*** Reached in the 8th minute.
⁺ Reached in the 5th–6th minute.
‡ During the vacuum treatment the ladle was not full (12–16 tons, for LK steel, and 14 tons, for ShKh15 steel).

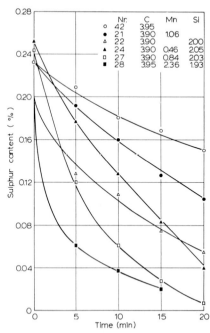

Fig. 34. Effect of carbon, manganese and silicon contents on the desulphurization in a vacuum[109]. Laboratory test results at 10^{-2}–5×10^{-2} Torr and 1300°C.

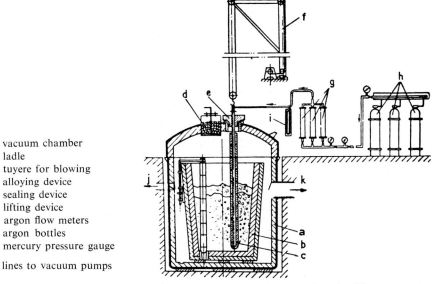

a vacuum chamber
b ladle
c tuyere for blowing
d alloying device
e sealing device
f lifting device
g argon flow meters
h argon bottles
i mercury pressure gauge
j ⎫
k ⎭ lines to vacuum pumps

Fig. 35. Diagram of a ladle degassing plant with argon injection[118].

the vacuum treatment should be discontinued as soon as the bath agitation ceases and therefore that the removal of the hydrogen takes place only by diffusion at the surface of the bath. If a further degassing effect is to be obtained additional devices should be applied, for instance, by injection of an inert gas into the melt, in order to improve the kinetic conditions of the melt.

Sorkin[103] arrives at a similar conclusion, which he derives from the data supplied by Samarin, and which are shown in Fig. 28. These data indicate the increase in hydrogen content with increasing distance from the surface of the melt after completion of the vacuum treatment. On the basis of this analysis an additional device for the agitation of the bath is regarded as being necessary. These requirements led to the development of equipment for degassing in the ladle with a scavenging gas or by use of a stirring coil which will be described in the next subsections.

Samarin[109] mentions that it should be possible to use the degassing in the ladle for the desulphurization of pig iron also. Samarin's experiments were, however, carried out only in a vacuum induction furnace[110]. His results are plotted in Fig. 34, and it may be seen that a favourable desulphurization can be obtained provided that both the silicon and manganese contents of the steel are high. Laboratory experiments by other workers carried out on cast iron have produced similar results[111-114]. It seems, however, that all attempts at desulphurization of pig iron or steel by vacuum treatment on a large industrial scale have failed.

1.3.2 Ladle degassing with injection of a purging gas

The often unsatisfactory results which were obtained by standard ladle de-

References pp. 507–515

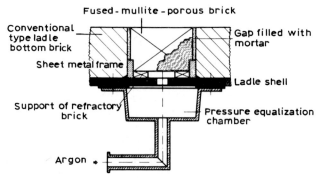

Fig. 36. Arrangement for the incorporation of a porous brick into the bottom lining of a ladle.

gassing in large ladles above 50 tons capacity induced the steel producers in the mid-fifties to change to other vacuum processes or to try to overcome these difficulties by the employment of additional devices. The method first used was to inject a gas into the melt in order to overcome the threshold of bubble formation, which is given by the ferrostatic pressure in the ladle, as well as to provide a sufficient degree of agitation of the melt[115-117]. In Fig. 35[118] an arrangement is de-

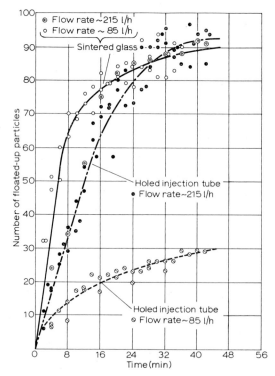

Fig. 37. Effect of blowing time on the number of floated-up paper particles for various flow rates and bubble sizes[122].

picted in which a stopper rod carrying a porous stopper at its lower end is immersed in the ladle. Another variant is shown in Fig. 36, in which a porous brick was incorporated in the bottom of the ladle lining[118,119]. The gas which is injected eccentrically into the ladle causes continuous circulation of the steel melt in the ladle and moves poorly degassed material from the lower part of the ladle to the surface[115,118,120,121].

This bath movement or circulation was studied by Speith and Steinhauer[122] on models in which the movement of particles suspended in water was registered. The number of particles which are transported to the surface on injection of air per unit of time was used as a measure of the movement of the bath. The results obtained are depicted in Fig. 37. It may be seen that bath agitation increases with increasing quantities of the scavenging gas. Bath agitation is greatest for the same gas flow if the injection nozzle is arranged laterally at the bottom of the ladle and it is lowest if the gas is injected at the centre of the bottom.

The gas most employed as purging gas is argon, but some American works have also used helium on an experimental basis[123]. A number of papers report on the hydrogen and oxygen levels obtained[95,123-125]. In an American works the average reduction of the hydrogen level from 4.3 to 1.75 ppm and of the oxygen level from 86 to 44.6 ppm were obtained on more than 100 melts[95], whereas a French report mentions hydrogen contents of between 2 and 3.2 cm^3/100 g steel for a similar number of heats[125]. The effect of the final pressure on these reductions in the hydrogen level is shown in Fig. 38.

Bath agitation causes an increased fall in temperature of the melt, and this heat loss is shown in Fig. 39 in relation to the time of purging[126] for 40- and 80-ton ladles.

On the other hand, bath agitation makes possible the addition of deoxidation agents or alloying elements during the degassing treatment. Ladle degassing plants with gas injection are therefore usually equipped with hoppers for alloying additions. The additions are, however, restricted to small quantities of silicon or aluminium, and for the correction of the manganese level.

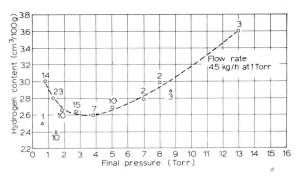

Fig. 38. Effect of the final pressure on the hydrogen content. The figures in the diagram indicate the treatment time in min[125].

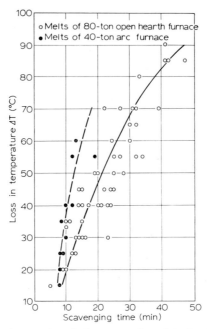

Fig. 39. Fall in temperature during scavenging of steel melts with argon.

1.3.3 Ladle degassing with stirring coil

On the basis of experience gained with induction furnaces, several ladle degassing plants were equipped with an induction coil for the agitation of the bath. The coil is arranged in the vacuum vessel in such a way that the ladle can be placed within it[127-132]. The use of a stirring coil results in a more vigorous bath agitation

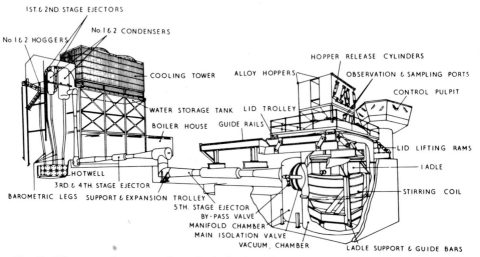

Fig. 40. Diagrammatic presentation of a ladle degassing plant with stirring coil for 110-ton melts[130].

compared with the injection of a scavenging gas. Stirring also results in lower falls of temperature than that produced with gas scavenging, as the boiling action of the steel, *i.e.* the disruption of the bath surface, is less and because the induced electrical current introduces additional heat to the melt. The increased agitation also makes it possible to add larger quantities of alloying elements even after completion of the vacuum treatment and to distribute them evenly throughout the melt. Figure 40 depicts the principle of such a plant. Vacuum pumps of adequate speed provide a good vacuum during the whole treatment; a final pressure of less than 0.1 Torr is maintained in modern plants. A typical pressure curve recorded during the degassing of a 110-ton ladle is shown in Fig. 41[130].

The employment of a stirring coil excludes the use of conventional steelworks ladles for the degassing treatment, as the iron shell shields the melt from the electromagnetic field. The iron shell of the ladle would be heated and the melt would be neither agitated nor heated[128,133]. Specially designed ladles constructed of non-magnetic steel must be used and suspension lugs for the transport of the ladle must be provided at a higher point than in conventional ladles. Furthermore, the refractory lining should be kept relatively thin in order to attain a favourable electrical efficiency in the coil. The coil is usually excited by a low-frequency current of $\frac{1}{2}$–1 Hz. In most cases the coil consists of three or four individual coils each with a phase shift. Each coil is constructed of a non-magnetic coil body which is insulated with such materials as asbestos or porcelain cement. It is also usual practice to incorporate thermocouples into the coil body in order to be able

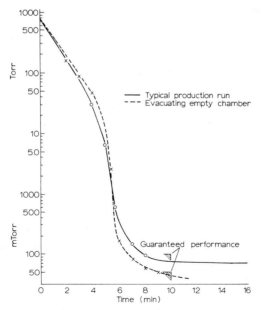

Fig. 41. Pressure/time diagram of a 110-ton melt during ladle degassing using a stirring coil[130].

References pp. 507–515

TABLE 10

COMPARISON OF AIR-MELTED AND LADLE-DEGASSED HEATS WITH INDUCTION STIRRING. DATA PUBLISHED BY PERRY[132]

	Grade 8620		Grade 52100	
Typical analysis (%)				
C	0.20		1.03	
Mn	0.80		0.40	
P	0.015		0.015	
S	0.015		0.010	
Si	0.30		0.30	
Cr	0.50		1.05	
Ni	0.55		—	
Mo	0.20		—	
Gas levels* (ppm)	Degassed	Undegassed	Degassed	Undegassed
Nitrogen	30–60	60–110	30–70	60–110
Oxygen	15–30	25–40	8–16	18–28
Cleanness rating**	Degassed	Undegassed	Degassed	Undegassed
Series A thin	2.0	1.0	1.5	1.0
heavy	0	0	0.5	0.5
Series B thin	1.5	2.0	1.0	2.0
heavy	0.5	1.0	0	0
Series C thin	0	0	0	1.0
heavy	0	0	0	0
Series D thin	1.5	1.5	1.5	1.5
heavy	0.5	1.5	0.5	0.5

* Vacuum fusion results on cold steel samples.
** Jernkontoret ratings.

to observe the temperature of the coil during operation and so to prevent overheating and destruction of the coil. According to the position of the individual coils and their electrical switching an agitation of the melt may be obtained in such a way that the melt at the surface moves either towards the centre or outwards from the centre. The power take-up of the coils may be varied between 50 and 600 kW. Plants of higher power input than this are so far not known.

Despite the power input into the melt by the induction coil it is not possible even for large ladles to manage the degassing without superheating the steel in the steelmaking furnace. Although exact data on the heat losses are not reported in the literature, a superheating of 40–60°C is regarded as necessary for the treatment of 110-ton melts[130, 133, 134].

As only a few plants of this type are in existence, the metallurgical data available are very scarce indeed. According to reports from an American steel producer[68] hydrogen levels of about 1 ppm have been obtained if the initial hydrogen content varied around 5 ppm, while oxygen and nitrogen levels were lowered by about 50%. The oxygen and nitrogen contents obtained on a low-alloy case-hardening steel (AISI 8620) and a ball-bearing steel (AISI 52 100) are shown in Table 10.

TABLE 11

EFFECT OF LOW PRESSURE ON OXIDE INCLUSION RATINGS OF
VACUUM-DEGASSED TYPE 52100 BALL-BEARING STEEL[131]

(a) Terminal pressure		Below 0.1 Torr	Between 0.10 and 0.16 Torr
Number of heats		263	110
Inclusion rating	Above avg. B	32	20 percent of heats
	Above avg. C	33	24
	Avg. C	28	45
	Below avg. C	3	6
	Below avg. D	4	5
(b) Heats degassed below 0.1 Torr			
Time below 0.1 Torr		Less than 5 minutes	5–10 minutes
Inclusion rating	Above avg. B	22	22 percent of heats
	Above avg. C	22	42
	Avg. C	49	29
	Below C	1	2.5
	Below D	6	4.5

The effect of degassing on the mechanical properties varies from grade to grade. Tensile strength and yield strength are usually slightly lowered, but the fatigue strength of ball-bearing steels is increased by about 20%. An improvement is usually also observed for elongation at fracture and in notch toughness.

Ladle degassing plants equipped with induction stirring are always provided with pumping systems of extremely high speeds, which make it possible to obtain pressures below 0.1 Torr during the degassing treatment[127,131,132]. As may be seen from Table 11, these low pressures bring about only a slight improvement in cleanness of ball-bearing steels compared with the treatment at higher pressures.

1.3.4 Ladle degassing with additional heating

During the last few years the various modifications of ladle degassing installations have been improved by equipping them, in addition to the stirring device, with a heating device. By this refinement ladle degassing was brought near to the modern vacuum melting processes with regard to its metallurgical potentialities, especially as regards the possibility of adding alloying elements after completion of the degassing process. Two principal variants have been tried in the steel works. In both, three-phase arc-heating is installed in the dome of the vacuum vessel; they are the Finkl–Mohr and the ASEA–SKF processes.

The principle of the Finkl–Mohr process[135–137] is shown in Fig. 42[136]. According to the claims of the inventors this process can be applied to ladles of 5–300 ton capacity. After the ladle has been placed in the vacuum vessel, the dome, which is operated by a hydraulic drive, is moved into the closing position and the vessel is then evacuated. At the same time purging gas is introduced through a porous stopper which is situated in the lower part of the ladle wall. About four

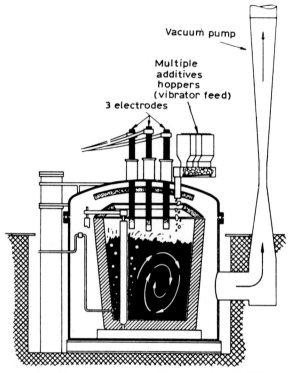

Fig. 42. Diagram of a ladle degassing plant with arc heating, Finkl–Mohr principle[138].

to seven minutes after the starting of the vacuum degassing the electrodes are brought into the working position above the melt and the arc is ignited. A power input of 5000 kW is installed in a plant for steel melts of 35–65 tons. It is reported that a power input of 3500 kW will keep the melt at a constant temperature during the degassing process, and that 5000 kW would be able to heat up the melt at a rate of 1–1.5°C/min. The expected temperature curve during the degassing treatment is shown in Fig. 43[138]. Data on the metallurgical results and on the life of the refractory lining have not been published up to now.

The other process, known as the ASEA–SKF process, has been developed in Swedish steel works[139,140]. It is depicted in Fig. 44[141]. In contrast with the Finkl–Mohr process the arc heating is not carried out under vacuum, but the installation is provided with two domes, one containing the vacuum line and the other the arc heater. Another difference is that the agitation of the bath is not effected by a scavenging gas but by induction. This process is in actual use for melts of 30 tons, but could also be applied to 60-ton melts. A peculiarity of this process is that the ladle does not carry a stopper rod but is provided with a special closure of the teeming nozzle. The stopper rod is inserted into the filled ladle only after completion of the degassing process and the heating up to the casting temperature.

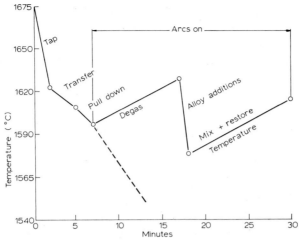

Fig. 43. Temperature changes in relation to time expected for the Finkl–Mohr process[138].

With this process it is possible to add alloying elements in any desired sequence and in any quantity after completion of the degassing process. It is, however, not known whether this process has been further developed so that the alloying additions may also be made under a vacuum. The ASEA–SKF process is similar to ladle degassing with induction stirring in that it too requires the employment of ladles fabricated from non-magnetic steel.

The usual procedure for the application of this process is as follows[139]: The melt is adjusted in the furnace to the required carbon level and temperature. The

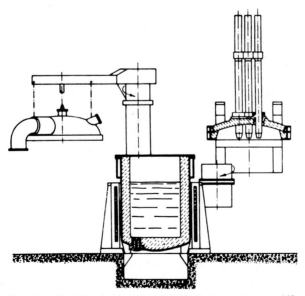

Fig. 44. Principle of the ASEA-SKF ladle degassing plant[140].

slag is then removed and the steel tapped into the special ladle mentioned above. The ladle is inserted into the induction coil, the lid of the vacuum vessel brought into position and lowered on to the ladle. The fall in temperature from the time slagging in the furnace was carried out to the beginning of the degassing process amounts to about 40°C. Bath agitation is observed by a closed circuit television camera via an observation window, and the pressure is lowered at such a rate that boiling-over of the steel does not occur. Degassing is usually continued for 10–15 min during which time the pressure is lowered to 0.2 Torr. During the degassing period the fall in temperature amounts to another 20–40°C. After completion of the degassing process the lid is removed from the vacuum vessel and the alloying elements or a reactive slag are introduced on to the bath of steel. Now the lid carrying the heating electrodes is brought into position and the steel is heated up again. This heating period usually lasts one hour, while the stirring coil is also in operation. Using stirring during heating is helpful to the heat transfer to the steel as the heated-up steel below the electrodes is drawn downwards, cooling at the walls of the ladle, and then rising. This bath agitation is said to be advantageous to the life of the ladle lining as the slags introduced are kept away from the lining. During the heating-up period the steel composition is adjusted. When the steel has attained the required temperature, the heating is discontinued, the stopper rod inserted, the ladle removed from the coil and the tapping hole opened from the outside. Teeming usually takes place under normal atmospheric conditions but it may also be carried out in a protective atmosphere.

A nominal rating of the stirring coil of 515 kVA at 1.75 Hz is mentioned for a 30-ton plant. The maximum voltage is 300 V and the maximum amperage 860 A[139]. The nominal transformer rating for the arc-heating is 3000 kVA, giving a power input to the steel melt of about 1400 kVA[139]. The steel degassed by this process is cleaner than that produced by conventional methods. The oxygen levels are lowered by up to 50% of that of the starting material and the hydrogen contents are reduced from about 5 to 2.5 cm^3 (s.t.p.)/100 g of steel.

Methods of heating other than arc-heating have been suggested for the ladles used in the degassing process[129]. For instance, it was proposed to attach a channel-type inductor to the ladle or to incorporate a resistance heater in the lid of the vacuum vessel or in a special lid which could be placed on top of the ladle in a vacuum-tight manner. Even plasma torches installed in the lid were thought of. But owing to too low a heat input or to other technological difficulties probably none of these methods will find its way into industrial application in the near future.

1.3.5 Refractory lining of the vacuum ladles

The refractory lining of the ladle plays a vital role in all ladle-degassing processes. The ladle lining conventionally used consisting of silica-base refractory bricks or silica-rich ramming mixes is only rarely employed, as in a vacuum violent reaction of the silica occurs with the carbon dissolved in the steel[142]. This reaction leads to heavy wear of the lining, to silicon pick-up by the melt, and also to an additional formation of carbon monoxide. Basic materials, on the other hand, cannot be used as they are too sensitive to thermal shock. Almost all degassing ladles are therefore lined with alumina-enriched fireclay bricks or with alumina bricks. The lower the silicon content of these bricks the less is the silica reaction, and the better is the service life of the linings. Unfortunately, the price of the bricks increases with increasing alumina contents. Results obtained by one steel works[143] with three materials are shown in Table 12. Although the alumina-rich lining could be used for more than twice as long as that consisting of normal bricks (up to 46 heats), the costs for the alumina-rich lining are substantially higher per ton of steel produced. Furthermore, alumina-rich bricks give a higher fall in temperature than do conventional fireclay bricks, and therefore higher melting costs, as may be seen from the bottom line of the table.

TABLE 12

COMPARISON OF 25%, 70%, AND 85% ALUMINA LADLE LINING FOR DEGASSING, AS EXPERIENCED AT A. FINKLE AND SONS CO.

Type	Bloating	70% Alumina	85% Alumina
Alumina content	25–30%	70%	83–85%
Ladle wall	8 in. thick	2½–4½ in. 70% Alumina 2½ in. – 0 IFB* 2½ in. Standard ladle brick	2½–4½ in. 85% Alumina 2½ in. – 0 IFB 2½ in. Standard ladle brick
Ladle bottom	11 in. thick	4½ in. 70% Alumina 2½ in. IFB 4 in. Standard ladle brick	4½ in. 85% Alumina 2½ in. IFB 4 in. Standard ladle brick
Cost 9 × 4½ × 2½	12 ¢	35 ¢	72 ¢
Cost including labour – 1963 35 ton ladle lining	$379.00	$837.00	$1,475.00
Number of heats per lining	18 average with 2 bottoms and 2 slag line patches	28 with 3 bottoms and 2 slag line patches	46 maximum to date with 4 bottoms and no slag line patch
Cost per 35 ton heat without slag line and bottom patches	60 ¢	85 ¢	92 ¢
Cost per 35 ton heat with slag line and bottom patches	86 ¢/ton	$1.21/ton	$1.37/ton
Temperature drop of steel before and after degassing without ladle pre-heating	x	$x+20$	$x+40$

* IFB = insulating fire brick.

References pp. 507–515

Attempts were also made to use chrome–magnesite bricks[93] for the lining of the ladles. Densely fired chrome–magnesite bricks have an excellent thermal-shock resistance and therefore allow the treatment of 18–30 melts with the same lining. However, owing to their good thermal conductivity even higher heat losses occur than by the employment of pure alumina bricks.

The ladle lining must be kept thin in ladle degassing processes which work with an induction coil for stirring. In order to prevent an excessive wear of the ladles, only bricks of a very high alumina content should be employed in these processes. In some instances even pure corundum bricks were used. Service lives of up to 100 heats are said to have been obtained if the melts are slagged with great care and if the upper part of the ladle lining, which comes into contact with the slag, is frequently repaired.

The stopper rod too is subjected to an increased wear in the ladle degassing process. Both the silica reaction mentioned above and the small quantities of slag remaining in the ladle attack the upper stopper-rod sleeves to a high degree. Although alumina-rich material withstands an attack by the molten steel quite well, it is more vulnerable to an attack by the slag, because the residues of the slag, which are in most cases rich in lime, form compounds of low melting-point with alumina. Stoppers made of chrome–magnesite are claimed to have proved their good resistance[93,144]. The use of stopper-rod sleeves of alumina which are coated with magnesite were also suggested[145]. Constant observation of the stopper-rod sleeves during the degassing process is regarded as necessary in any case.

1.4 Stream degassing processes

In the stream degassing processes, which are carried out during the teeming operation, the whole melt is not subjected to the vacuum treatment at one and the same time in contrast with the ladle degassing process. In stream degassing the stream enters an evacuated vessel and by the action of the vacuum the stream of steel is broken up into many small drops so that a large surface area is created[146–149]. The degassing rate is increased by this dispersion of the stream of steel to such an extent that the liberation of almost the whole quantity of gases takes place in the very short time in which the steel falls through the vacuum vessel into the collecting ladle. In the most simple variant—stream droplet degassing or ladle to ladle degassing—the steel is poured from one ladle into another which is contained in the vacuum vessel. In tap degassing the steel flows from the furnace into a ladle which carries a lid or dome so that it can be evacuated. In another variant the steel is teemed from a conventional ladle into an ingot mould which is placed in a vacuum vessel. This modification is known as vacuum ingot casting. All three variants were tried out by Gussstahlwerke Bochumer Verein A.G., Bochum, Germany, and were there developed to operational perfection on an industrial scale[150,151].

As vacuum ingot casting is rather similar to the ingot degassing process, it will be dealt with in Subsection 1.6. The so-called Therm-I-Vac process may also be regarded as a stream degassing process although it may be considered as well as a modification of the vacuum induction melting process for larger melts.

1.4.1 Stream droplet degassing

Stream degassing was first mentioned by Tix and his co-workers in 1955[50–55,152,153]. An installation for industrial application is depicted in Fig. 45[152,154–158]. A vacuum tank carrying a removable lid is placed in the casting pit and the casting ladle is inserted in the vacuum vessel. The lid is provided with observation windows and with an opening through which the steel enters the vacuum tank. This opening will be closed with an aluminium foil after the casting ladle has been inserted into the vacuum chamber and the lid placed into position (Fig. 46[159]). A tundish is then placed on top of the lid in a vacuum-tight manner, so that the tapping hole faces the opening in the lid which is closed by the aluminium foil. The vacuum tank is then evacuated and tested for leaks. For the degassing treatment the steel is tapped in the conventional way into a common ladle and the usual ladle additions introduced. This ladle is then positioned above the tundish and the steel teemed into it. When the tundish is half-filled with steel the stopper rod is lifted. The steel flowing out of it meets the aluminium foil, melts it and enters the vacuum tank, where the stream is dispersed into small droplets. Degassing takes place during the free fall of the droplets in the vacuum tank. The teeming rate from the tundish can be varied between 5 and 10 ton/min by choice of the run-out nozzle.

In many plants the tundish is omitted[160–163]. In these cases the transport ladle is provided with a flange at the bottom so that the ladle may be placed directly on the lid of the vacuum tank in a vacuum-tight fashion. In several in-

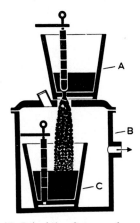

Fig. 45. Principle of stream degassing.
A = Intermediate ladle or holding ladle (pony ladle)
B = Vacuum tank; C = Casting ladle.

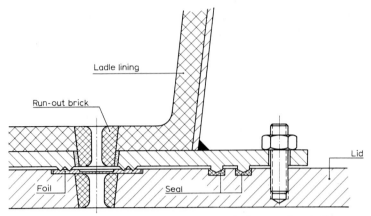

Fig. 46. Example of the sealing of the vacuum tank by the intermediate ladle or tundish and metal foil[159].

stallations another vacuum container is attached to the main vacuum tank for the addition of alloying elements.

In order to determine the effect of the vacuum treatment, the quantity and composition of the gas pumped off can be recorded. The quantity of gas liberated per unit time may be varied by controlling the stream of steel into the vacuum vessel and consequently the pressure during the degassing treatment. The lower the pressure the more intensive is the dispersion of the stream of steel, and therefore better degassing conditions are created since the surface area of the steel is increased. However, the teeming rate is somewhat limited, as too low a casting rate lowers the temperature of the steel excessively. Owing to the favourable kinetic conditions, an observation of the movement of the steel bath in contrast with ladle degassing is not necessary during stream degassing.

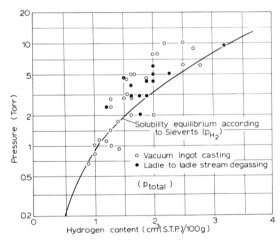

Fig. 47. Pressure dependence of the hydrogen content after stream degassing (samples taken from liquid steel) according to Tix[153].

Despite the shortness of the time during which the drops fall through the vacuum vessel, a high degree of degassing is obtained, as is shown by the results of Tix, Bandel, Coupette and Sickbert[150,152,164] as well as by the experimental studies of Tiberg[165]. Hydrogen levels after stream degassing in relation to the applied pressure are plotted in Fig. 47. As may be seen from these curves, the hydrogen levels obtained in both stream droplet degassing and vacuum ingot casting in the pressure range between 0.5 and 10 Torr are very close to the equilibrium curve derived from Sieverts' law. This highly effective removal of the hydrogen made it possible to omit the diffusion annealing of even large castings of flaw-sensitive steels for forging.

The deoxidizing action of a vacuum in stream degassing is shown in Fig. 48, where the carbon–oxygen product is plotted in relation to the pressure prevailing during treatment. According to these curves the oxygen contents are lowered to such an extent that at the end of the degassing treatment CO solubility products of about $2 \times 10^{-4} [\%]^2$ are obtained for mild steels. This value is about one order of magnitude lower than the value of $0.025 [\%]^2$ which applies to a carbon monoxide pressure of 1 atm. On the other hand, Fig. 48 also shows that the theoretical equilibria are not attained. This is in agreement with the findings obtained also on other vacuum degassing processes.

On the nitrogen transfer during stream degassing it is reported that the evolution of nitrogen may amount to up to 40% of the initial content for plain-carbon and medium-alloy steels, whereas for high-alloy steels and those killed with aluminium or titanium no noticeable change in the nitrogen level is observed.

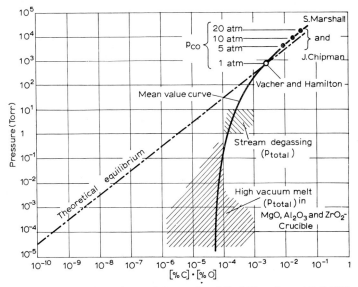

Fig. 48. Pressure dependence of the product [C]·[O] at about 1600°C[150].

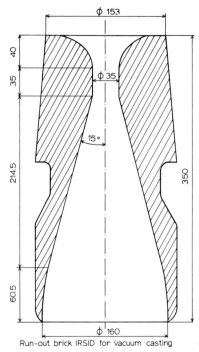

Fig. 49. Ladle outlet nozzle for stream degassing plants after IRSID[166].

The same considerations as those employed in ladle degassing apply to the refractory lining of the ladles used in stream degassing. In stream degassing, too, the carbon of the steel reacts with the silica of the lining and causes, besides the wear of the lining, a pick-up of silicon by the steel and an additional production of carbon monoxide. On the other hand, reactions of the refractory lining with slag

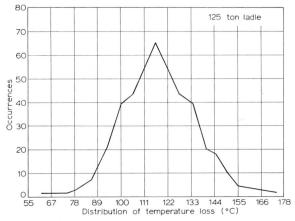

Fig. 50. Frequency distribution of the temperature loss during stream degassing determined on 1246 heats of 125 ton each[157].

do not occur in stream degassing, as the degassing process may be stopped as soon as the last portion of the steel leaves the intermediate ladle and is followed by slag. Usually conventional fireclay bricks or mullite bricks are employed for the ladle lining. Alumina bricks are not used since, owing to their good thermal conductivity, they would only increase the already high fall of temperature inherent in this process. In all stream degassing processes the ladle nozzle is subjected to extremely high wear. Instead of using the conventional fireclay nozzles, outlet bricks made of magnesite or zirconia are employed, and it is the usual practice to give these run-out bricks a special shape in order to impose a better control of the teeming conditions and of the dispersion of the stream. A run-out brick developed by IRSID, as an example of this, is shown in Fig. 49.

Installations for stream degassing exist in many steel works. The equipment is relatively simple in construction and can be used for vacuum casting, ladle to ladle degassing as well as for the ladle degassing process[101,104,164,167-169]. It is not known how large are the quantities which are degassed by each individual process in these installations, but from reports published in the literature it may be inferred that stream droplet degassing is nowadays not applied to the same extent as it was about a decade ago, owing to the high fall in temperature. One of the more recent evaluations of the loss of heat occurring during stream drople degassing is shown in Fig. 50. This high fall in temperature was probably also the reason for its inventors developing it to tap degassing.

1.4.2 Tap degassing

In tap degassing the steel running out of the melting furnace is immediately treated by the stream degassing process[144,151,170-176]. A diagram depicting this type of degassing plant is shown in Fig. 51. The tapping ladle forms the treatment

Fig. 51. Principle of tap degassing[144].

vessel and is therefore closed in a vacuum-tight manner by a lid. In order to provide a large enough reaction space in the ladle above the surface of the molten steel, this type of ladle has an increased capacity compared with that of conventional ladles. On top of the lid a rather smaller ladle is placed, which is called the entry funnel. Its capacity amounts to about 7 tons of steel for a tapping quantity of about 50 tons, and it is closed by a stopper rod which is operated by remote control. The entry hole is closed in a similar manner as in stream degassing by aluminium foil, so that the degassing ladle may be evacuated before tapping begins. The lid of the degassing ladle is provided with a port to which a flexible vacuum duct is connected. This duct leads to the permanent vacuum line system, the terminal of which is very near the melting furnace. In preparation for the degassing process the stopper rod is inserted into the ladle, the lid which is also fitted with a refractory lining is placed upon the degassing ladle, and finally the entry funnel is attached to the lid by a vacuum-tight joint. The ladle crane then picks up the whole assembly and sets it down in front of the furnace, where the ladle is connected to the flexible vacuum duct and so evacuated. During tapping the steel runs from the furnace into the entry funnel and from there directly into the evacuated degassing ladle. The flow rate of the steel is controlled by the electrically operated stopper rod. Since the capacity of the entry funnel is limited, and as different flow rates must be applied according to the quality of the steel to be degassed, a control of the tapping flow is essential. Hence tapping degassing can only be used in connection with tiltable melting units, in particular arc furnaces. The tapping and degassing time amounts, for instance, for a 30-ton furnace, to 3–5 min or a flow rate of 8–9 ton/min. The fall in temperature of the steel caused by the degassing process amounts to about 50°C[144, 151, 170, 176].

After completion of the treatment the ladle is opened to the atmosphere and the vacuum duct is disconnected from the ladle. The whole assembly consisting of ladle, lid and intermediate ladle is then transferred to the casting pit where the steel is teemed in the conventional manner.

As the steel flows directly from the melting furnace into the vacuum ladle, the fall in temperature will be less than in stream degassing. Moreover, additional transport within the steelworks is avoided, as the ladle has not to be transported to a special place for the vacuum treatment. One shortcoming is the relatively great height of the ladle assembly, which requires deep tapping pits and adequate heights of the crane gantry.

It is also possible in tap degassing to add deoxidation agents, such as silicon or aluminium, during the degassing process from a vacuum container attached to the lid of the degassing ladle. It is advisable to introduce these additions continuously during the degassing process, as a proper mixing of the steel in the ladle does not take place owing to the low directional energy of the stream of steel. Aluminium is usually immersed into the melt after completion of the degassing process.

In tap degassing, rimming, semi-killed and killed steels may be subjected to the degassing process. Studies were made of the metallurgical aspects of these

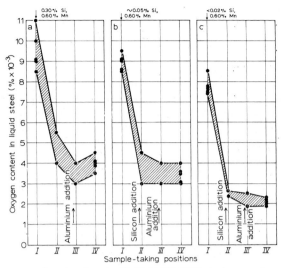

Fig. 52. Oxygen content of steels of grade Ck 35 (AISI 1035) during tap degassing after various degrees of killing[151].
(a) Killed in the furnace. (b) Partly killed in the furnace. (c) Rimming state in the furnace. Sample-taking positions: I Furnace before tapping; II Vacuum ladle after treatment; III Stream to mould; IV Ingot (liquid).

variants[170] and the effect of degassing was determined from immersion samples which were analysed for their oxygen and nitrogen contents. Simultaneously with immersion sample taking, samples were taken with a quartz pipette for the determination of the hydrogen content. The decrease in the oxygen level of steels of grade Ck 35 is plotted in Fig. 52. Part (a) shows the oxygen exchange of killed,

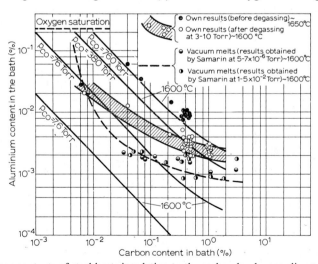

Fig. 53. Oxygen contents of steel heats in relation to the carbon level, according to Tix et al.[151], determined before and after tap degassing and compared with data obtained by Samarin on melts from a vacuum induction furnace.

References pp. 507–515

Part (b) that of semi-killed, and Part (c) that of rimming steel tapping. The mean values of the oxygen levels in the molten ingot were 0.0045% for killed, 0.0032% for semi-killed, and 0.0020% for rimming melts. Although the pressure of about 9 Torr during the degassing process of rimming melts was on the average 3 Torr higher than that of killed melts, an increased deoxidation effect was observed on the rimming steels. The oxygen of the rimming steels can therefore be separated more easily and faster than that combined with the silicon or other deoxidation agents of killed melts.

A comparison of the oxygen and carbon levels determined in various grades of steel with the theoretical carbon–oxygen equilibrium is shown in Fig. 53. The solid black dots represent the oxygen contents of the melt in the entry funnel before the degassing operation. They agree fairly well with the equilibrium line for 760 Torr. Despite the very short time of 0.5 sec during which the steel is falling through the degassing vessel, oxygen levels are attained which correspond to equilibrium pressures between 76 and 500 Torr (small circles). Also plotted in this diagram are data obtained by Samarin[167] in a vacuum induction furnace at pressures of 10^{-2} and 10^{-6} Torr (half-shaded dots and broken line). Although the pressure in Samarin's work is considerably lower than in tap degassing, his values are only slightly lower. Both results, however, still deviate greatly from the carbon–oxygen equilibria. It can be assumed that in technological processes the poor approach to the theoretical equilibrium conditions is attributable to the diffusion conditions for the carbon and oxygen in the liquid steel and to the so-called crucible reaction, *i.e.* pick-up of oxygen from the refractory lining. It appears therefore that in most cases in stream degassing it will not be necessary to operate at pres-

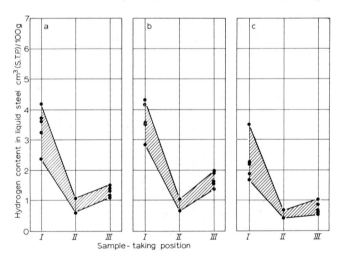

Fig. 54. Lowering of the hydrogen level during tap degassing for various degrees of pre-deoxidation of steel Ck 35[176].
(a) Killed in the furnace. (b) Partly killed in the furnace. (c) Rimming state in the furnace (see Fig. 52). Sampletaking position: I. Furnace before tapping; II Stream to mould; III Ingot (liquid).

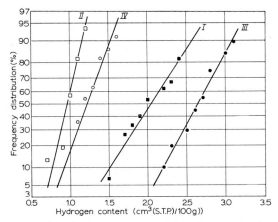

Fig. 55. Frequency curve of the hydrogen contents in the furnace and in the ingot while still liquid after tapping degassing of steels 41 Cr 4 and 100 Cr 6[151].
16 Heats of steel 41 Cr4. Sample-taking position: I Furnace before tapping; II Ingot (liquid).
22 Heats of steel 100 Cr6. Sample-taking position: III Furnace before tapping; IV Ingot (liquid).

sures markedly lower than 1–10 Torr, and that the capital investments for obtaining a better vacuum are not justified.

The hydrogen elimination during the tap degassing of steels of grade Ck 35 is plotted in Fig. 54 for killed, semi-killed and rimming melts. Lower hydrogen levels are obtained in the degassing of rimming steel than with killed steel owing to a lower hydrogen partial pressure. The hydrogen levels, which are in all cases

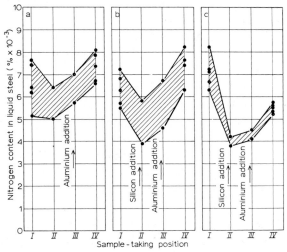

Fig. 56. Lowering of the nitrogen content during and after tap degassing for various types of pre-deoxidation. Steel Ck 35[151].
(a) Killed in the furnace. (b) Partly killed in the furnace. (c) Rimming state in the furnace (see Fig. 52). Sample-taking position: I Furnace before tapping; II Vacuum ladle after treatment; III Stream to mould; IV Ingot (liquid).

References pp. 507–515

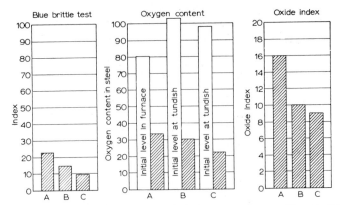

Fig. 57. Comparison of the blue-brittleness appraisal, the oxygen content (% × 10⁻⁴) in the solidified ingot and the oxide inclusion index of steel of type 41 Cr4 produced by various steelmaking processes (A, B, C)[151].

A: Oxidized, refined, killed in the furnace, without vacuum deoxidation.
B: Partly oxidized, not refined, vacuum deoxidized in the rimming state.
C: Oxidized, not refined, vacuum deoxidized in the rimming state.

around or below 2 cm³ (s.t.p.)/100 g steel, are sufficiently low to guarantee freedom from flaws during the cooling of the cast ingots from the forging or rolling temperature in air[177]. The frequency distribution of the hydrogen contents of steel grades 41 Cr 4 and 100 Cr 6 before degassing and after teeming is plotted in Fig. 55. The melts of grade 41 Cr 4 were not killed before vacuum-degassing, those of grade 100 Cr 6 were killed. All ingots were bottom-cast in air. It is quite possible that a certain amount of hydrogen was picked up from the moisture of the atmosphere and from the runner bricks after the degassing treatment[151].

The lowering of the nitrogen level for steel grade Ck 35 is plotted in Fig. 56. The difference between melts which were degassed in the killed and in the untreated states can easily be seen.

As an appraisal of cleanness and oxide inclusions, blue brittleness coefficients, oxygen content and oxide inclusion coefficients were determined on grade 41 Cr 4 for three types of melting practice, and the results obtained are shown in Fig. 57. Group A represents steels oxidized and refined in a conventional way, killed in the furnace and tapped without subsequent vacuum treatment. For Group B the oxidizing period was shortened, the melts were not refined but vacuum-treated, while steels of Group C were oxidized in the conventional way, not refined but vacuum-treated. The steels of this group show undoubtedly a higher degree of cleanness than do those produced in the usual way. Oxygen levels and blue brittleness coefficients show a similar trend. The size distribution of the inclusions is plotted in Fig. 58. The oxide inclusions are not only reduced with regard to their number, but they are also dispersed in a finer form and are therefore less detrimental.

In tap degassing, too, the reaction of the carbon in the steel with the lining

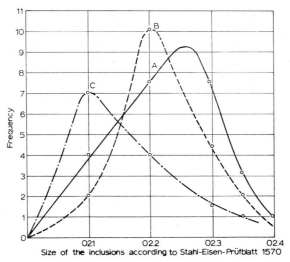

Fig. 58. Frequency distribution of the size of the oxide inclusions for steel 41 Cr 4 according to Tix et al.[151].
A: Oxidized, refined, killed in the furnace, without vacuum deoxidation.
B: Partly oxidized, not refined, vacuum deoxidized in the rimming state.
C: Oxidized, not refined, vacuum deoxidized in the rimming state.

of the degassing ladle plays an important role with regard to the ultimate oxygen contents obtainable. As the ladle is allowed to cool down after every vacuum treatment and is equipped with a fresh stopper rod, only shock-resistant lining materials may be used. These refractory materials usually contain a higher proportion of silica which is easily reduced in a vacuum. In many cases, therefore, no loss in silicon was found in the steel but a greater increase in the silicon content than would be expected from the quantity of ferro-silicon added. If the degassing ladle was lined with the usual refractory bricks containing about 45% silica and 48% alumina, boiling in the ladle was observed especially with rimming steel. This boiling must be attributed to the reaction mentioned. An observation carried out on six heats showed an average gain in silicon of 0.04%[170]; this means a silicon pick-up from the ladle lining of about 20 kg for a 50-ton melt. From purely stoichiometric considerations the quantity of oxygen evolved should cause a lowering of the carbon level of 0.034%. However, since the loss in carbon cannot be determined experimentally with sufficient accuracy, it is not possible to follow the reaction via the carbon reduction, but attempts were made to evaluate this reaction from the gas phase, *i.e.* from the quantity of the carbon monoxide pumped off. The findings confirmed the great effect of the crucible reaction, which was about four times as much as that of the degassing reaction itself. The pumping speeds of the vacuum pumps must therefore be very much higher in tap degassing than in mould degassing.

1.4.3 *Therm-I-Vac process*

The Latrobe process, also called the Therm-I-Vac process, is a stream de-

References pp. 507–515

gassing process which uses a vacuum-induction furnace instead of the receiving ladle[126,131,178]. The principle of this process is depicted in Fig. 59. The plant is laid out for the vacuum treatment of 30 tons of molten steel. Degassing practice in this plant is as follows: The steel is melted in a conventional arc furnace and refined. The steel is tapped into a casting ladle and transported to the vacuum plant, where the ladle is placed on top of the vacuum vessel in a vacuum-tight manner. The vacuum vessel is evacuated and the melt is poured into the induction furnace via a gate valve. During the teeming operation the gases liberated are removed by steam ejector pumps. As soon as the casting ladle has been emptied, the gate is closed and the vacuum vessel evacuated to lower pressures. Oil booster pumps which take over the evacuation maintain a pressure of between 1 and 5×10^{-3} Torr during the succeeding operation. As soon as the melt is in the induction furnace, the induction current is switched on, which heats and stirs the melt. The plant now operates as a vacuum induction furnace. Alloy additions may be made or reactive substances introduced. Samples of the steel may be taken via a lock system. When the steel has attained the desired composition or the required casting temperature, it is vacuum-poured into moulds which are also contained in the vacuum vessel.

For the pouring operation the crucible is tilted and the steel runs into a tundish which is arranged for the simultaneous filling of two moulds. (One mould may also be filled on its own.) The moulds are placed on a carriage which can be moved hydraulically. A lock system makes possible the changing of mould carriages while the vessel is evacuated.

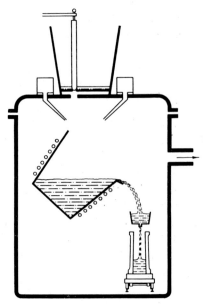

Fig. 59. Diagram showing the principle of the Therm-I-Vac process.

TABLE 13
CHARACTERISTICS OF THE THERM-I-VAC MELTING CHAMBER AND MOULD TUNNEL

Melting chamber		
Design		Vertical
Diameter × height		22 ft. o.d. × 24 ft.
Top opening, diameter		16 ft.
Cover operation		Lift and roll away
Top valve, diameter		36 in.
Roughing manifold, diameter		30 in.
Vacuum manifold, diameter		72 in.
Power port, diameter		22 in.
Manhole, diameter		24 in.
Number of sight ports		6
Cooling		Water jacket, upper part
Mould tunnel		
Width × height × length		8 × 18 × 88 ft.
Doors		One, on front
Mould carts		3
Cart length		14 ft.
Mould height, maximum		15 ft.
Weight capacity per cart		85 tons
Cooling		Water jackets in critical areas
Volume of chamber + tunnel		Approx. 20 500 cu. ft.
Auxiliary equipment		
Bulk charger:	Design	Rectangular, over melt, on rails
	Operation	Electric winch
	Chamber size	36 sq.in. × 96 in.
	Valve diameter	36 in.
	Bucket vol. capacity	Approx. 30 cu. ft.
Tundishes		2
Spray sleeve		3 sections
	Operation	Pneumatically, swing aside

The plant may also, however, be operated as a vacuum induction furnace using solid charges. Owing to the restricted heating input the melting capacity is limited to 15 tons. The plant which was installed by Latrobe Steel Co. in 1964 is laid out generously. The vacuum vessel has a capacity of 650 m³ (22,500 ft³) and the mould tunnel is 27 m (90 ft) long. Further technical details of this plant are compiled in Table 13, while Table 14 shows details of the pumping system installed.

The Therm-I-Vac plant at Latrobe is intended mainly for the production of high-alloy special steel for aircraft and space technology. Metallurgical results have so far not been revealed. It cannot, therefore, be estimated whether the high expenditure for the installation and the high running costs, as compared with those of other steel degassing plants, are still economically feasible for the production of conventional high-alloy steels. For the highest requirements with respect to quality without any consideration of production costs this process is certainly promising.

References pp. 507–515

TABLE 14
CHARACTERISTICS OF THE THERM-I-VAC VACUUM-PUMPING SYSTEM

	App. steam consumption (kg/h)	Cut-in pressure (Torr)
Steam ejector pumping system		
Number of stages		4
Steam pressure, dry and saturated		10–13 atm
Steam temperature		180° C
First stage	240	<5
Second stage	1460	<35
Third stage	580	<100
Fourth stage	440 ⎫ Parallel	760
Hogging jet	5350 ⎭	
Water requirements: First intercondenser		2.3 m³/min
Second intercondenser		0.7 m³/min
Pumping capacity, dry air at 5×10^{-1} Torr		82 kg/h
Pumping-down time to 5×10^{-1} Torr		20 min
Blank-off pressure, approx.		1×10^{-1} Torr
Ring-jet oil-booster pumping system		
Number of booster pumps		20
Size of pumps, diameter		16 in.
Cut-in pressure		$<25 \times 10^{-2}$ Torr
Cold tank pressure		1×10^{-3} Torr
Furnace operating pressure (30 ton heat)		1×10^{-2} Torr
Pump-down time to 5×10^{-2} Torr		45–50 min
Combined pumping speed at 10^{-2} Torr, approx.		5×10^4 l/s
Power requirement		850 kW
Mechanical holding pumps		
Type		Rotary, piston, oil sealed
Number of pumps		2
Blank-off pressure		5×10^{-1} Torr
Combined pumping speed at 5×10^{-1} Torr, approx.		250 l/s

1.5 Cycling and circulation degassing processes

In the cycling and circulation degassing processes the whole melt is not vacuum-treated at the same time, but portions of the melt are taken and are introduced into a special vacuum vessel where each portion remains for a certain time. After the degassing, the portion is discharged either into the original ladle or into another. The individual processes differ in the type and manner of transport of the steel through the vacuum vessel as well as in their metallurgical and technological potentialities.

1.5.1 The DH process

The DH process, also called the vacuum lifter process, was mentioned for the first time in 1956[61]. It was developed to full operational reliability by Dortmund–Hörder Hüttenunion of Germany in a relatively short period of time[179-184] and

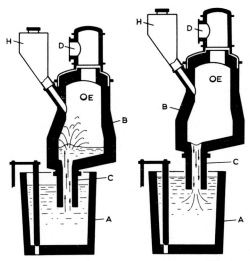

Fig. 60. Principle of the DH process[190].

is being applied in more than 60 degassing plants for melts of 10–400 tons[185–194].

The principle of the DH process is shown in Fig. 60. A portion of steel in a conventional casting ladle is sucked into a vacuum vessel and is returned to this ladle after a short degassing period. The steel is sucked through a refractory-lined pipe, which dips into the molten steel, into the vacuum vessel, which is also refractory-lined. Immediately after the suction pipe is dipped into the molten steel the vacuum pumps are switched on so that the vacuum vessel is evacuated. Owing to the difference in pressure, the steel rises in the suction pipe until the difference in height between the steel in the ladle and that in the vacuum vessel is about 1.4 m. On further lowering the vessel it will be filled with steel, the difference in height of 1.4 m being still maintained. The atmospheric pressure therefore forces a certain quantity of steel into the vacuum vessel, this quantity depending upon its width. As soon as the steel enters the vacuum vessel, the degassing reaction sets in, causing vigorous bubble formation and spattering of the steel[182,195–197]. When the reaction quietens down and the degassing process is almost completed, the vacuum vessel is lifted so that the treated steel is forced to flow back into the ladle. Lowering and lifting are repeated as often as required until a large enough quantity of steel has passed through the vacuum vessel and the desired degree of degassing has been obtained.

The vigorous degassing reaction which sets in as soon as the steel enters the vacuum vessel spontaneously liberates large quantities of gas[72,198], this gas evolution resulting in a characteristic change in the pressure existing in the vacuum vessel during the vacuum treatment. Such a pressure curve is shown in Fig. 61 and the periodical pressure fluctuations which correspond to the lowering and lifting cycle are clearly evident. Owing to the reduction of the gas content of the

References pp. 507–515

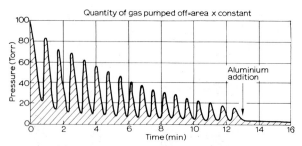

Fig. 61. Relation between pressure and time during the DH vacuum treatment[71].

melt with progressing time of the vacuum treatment, the pressure, as well as the rise in pressure on the entry of a fresh portion into the vacuum vessel, will become lower and lower, until the whole melt is almost completely degassed so that practically no pressure variations will be observed when a fresh portion enters the vacuum vessel.

On completion of the vacuum treatment the composition of the steel is corrected and adjusted, and the vacuum vessel pressurized to atmospheric pressure (refs. 199–201). After the suction pipe has been removed from the steel, the ladle is transported to the casting pit and teemed in the conventional way. According to the quantity and the quality of the steel, the time for degassing varies between 5 and 15 min[202–204]. It is of no significance whether the ladle or the vacuum vessel is lowered and lifted. In most plants this operation is carried out by hydraulic means, but electro-mechanical operation is also applied, though to a lesser extent. The lowering and lifting operation is usually automatic after the lower and upper reversing points at the beginning of the vacuum treatment have been adjusted according to the size of the ladle and the level of the steel in it. Most of the degassing plants are stationary, *i.e.* the ladle is placed on a carriage and moved under the degassing vessel, but there are also mobile plants in existence. In these the ladle is placed at some suitable location, usually in a pit, and the degassing plant is moved on to the top of the ladle.

All DH vacuum-degassing plants are equipped with means of heating the vacuum vessel, so that the vessel may be preheated to the temperature of the liquid steel. During a campaign the vessel is maintained at this temperature[71, 205–210]. According to local conditions, electric heating, using a graphite rod as the resistor, or fuel-firing may be employed. In the latter, recuperative systems with air preheating and oil–oxygen or gas–oxygen heating systems are both used. The costs for heating are usually lower for fuel-fired systems than for electric systems, but fuel-fired systems place far higher requirements on the refractory material which is exposed during the vacuum treatment to the reducing action of the vacuum and during the firing cycle to the oxidizing atmosphere of the waste gases. Electric heating therefore results in a longer life of the lining and has also the additional advantage that heating can be applied during the degassing treatment.

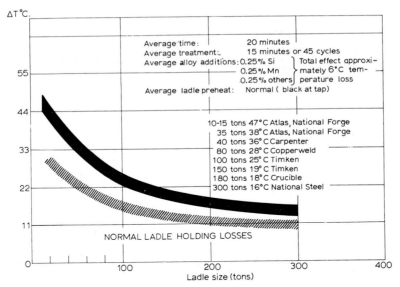

Fig. 62. Fall in temperature with and without DH vacuum treatment in relation to the capacity of the ladle for a holding time of 20 min between tapping and pouring[211].

Heating of the vacuum vessels ensures that the steel does not suffer any fall in temperature inside the vacuum vessel either by radiation or by direct contact with the refractory materials. The fall in temperature during a DH vacuum treatment cycle is therefore practically that which the steel would experience during the holding time between tapping and teeming, *i.e.* a period of time made up of the duration of the vacuum treatment and the additional time for transport to and from the degassing plant. The fall in temperature in relation to the size of the melts observed in 15 plants is plotted in Fig. 62. For the sake of comparison, the fall in temperature occurring through the lining of the ladle at equally long holding times is also shown.

The inventors of the vacuum lifter process had already pointed out in their first publication that the DH process was developed not only for vacuum degassing but also for the application of vacuum-metallurgical reactions[61, 212–214]. One of the main objects of this development was to degas the steel in the unkilled condition, *i.e.* before the oxygen-fixing elements were added, and to add the alloying elements and to adjust the composition after the removal of the oxygen and the hydrogen[12, 215, 216]. The reason for this procedure is the avoidance of the formation of most of the primary deoxidation products which occurs if the oxygen-fixing elements, such as aluminium, silicon and manganese, are added after the steel has been degassed[211, 217]. The DH plants were therefore equipped right from the beginning with a generous alloy addition system, which has been further and further perfected over the years[218].

Whereas the first plants carried several hoppers which were charged with the

References pp. 507–515

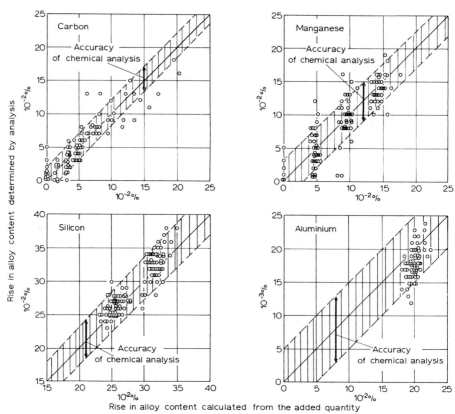

Fig. 63. Yield of alloying elements in carbon steels containing carbon 0.2–0.5% and manganese 0.6–1.5% after composition adjustment by the DH process.

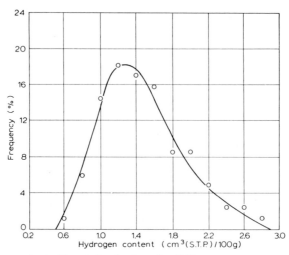

Fig. 64. Hydrogen content of ingots for forging after DH vacuum treatment. Frequency distribution[215].

alloying elements before the vacuum treatment was started, recent installations have preferred a system in which the alloying elements are automatically weighed while operating the plant and introduced into the vacuum vessel via a lock system. The addition of alloying elements to the batch of steel in the vacuum vessel produces a type of liquid master alloy under vacuum. During lifting of the vacuum vessel this master alloy is introduced into the melt with exclusion of air and as a result of the high speed of the flowing steel is well mixed with the steel in the ladle. The loss in alloying elements by burning in air or by reaction with the slag is therefore avoided and the yield in the application of alloying elements greatly increased[217,219,220]. At the same time the accuracy of alloy additions and therefore of the composition of the steel is assured. Figure 63 shows some results of the precision with which a required analysis can be obtained as well as the yield in alloying elements. The theoretical increase in alloy content, *i.e.* if the yield of the added element should be 100%, is plotted on the abscissa, whilst the levels obtained by actual analyses are plotted on the ordinate. This type of alloying element addition and composition adjustment under a vacuum after vacuum degassing originally developed for the DH process has proved so successful that the attempt has been made to make use of it for other types of steel degassing processes.

DH plants are employed for a wide range of steel grades[187,221-225], for instance, for large plain-carbon and alloy steel forgings, low- and medium-alloy heat-treatable and case-hardening steels, stainless steels, ball-bearing steels, die-forging steels, tube steels, and steels of extremely low carbon levels.

As a result of the wide range of application the number of publications on this process is extremely large, therefore only a few examples can be quoted here. Figure 64[221] shows the hydrogen levels obtained in ingots for forgings of vacuum treated steel, while the oxide residue levels of this steel before and after vacuum-treatment are plotted in Fig. 65.

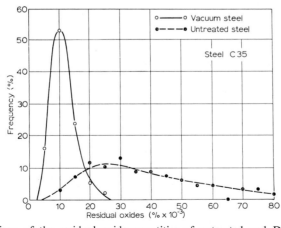

Fig. 65. Comparison of the residual oxide quantities of untreated and DH vacuum-treated steels[222].

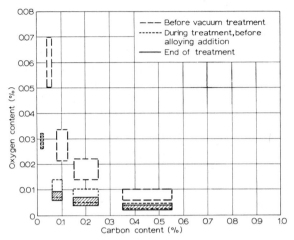

Fig. 66. Oxygen contents during and after a DH vacuum treatment[217].

Figure 66 shows a survey of the oxygen contents before vacuum treatment, after the degassing period, *i.e.* before addition of the alloying elements, and also at the end of the vacuum treatment in relation to the carbon level. Figure 67 shows the effect of a DH treatment on silicon-killed open-hearth steels. In this diagram the oxygen levels of the same steel before and after vacuum treatment are plotted. In order to create a correct basis for comparison, the steel was tapped into two ladles. The first ladle was vacuum-treated without any addition of oxygen-fixing elements. The silicon was added only after completion of the degassing treatment. The steel in the second ladle received ferro-silicon and calcium–silicon during tapping and was teemed without any other treatment.

The quantity and distribution of oxide inclusions in blooms of silicon-killed steel are plotted in Fig. 68. In order to determine these data over the whole length

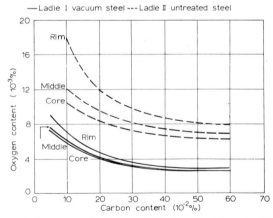

Fig. 67. Comparison of the oxygen levels in solidified steel with and without a DH vacuum treatment in relation to the carbon content[119]. Parameter: Location of the samples in the ingot.

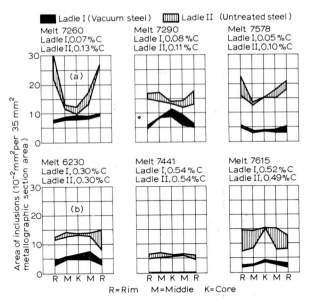

Fig. 68. Comparison of the cleanness of untreated and vacuum-treated steels of various carbon contents. Parameter: Location of the samples in the ingot.

of the bloom, samples were always taken from sections 2 m apart and averaged over the total length of the bloom. Heavy precipitation of the inclusions was found in the rim of non-vacuum treated blooms of mild steel, but this accumulation of inclusions was not observed in steels of higher carbon contents as the proportion of primary deoxidation products is lower in these steels. The vacuum-treated mild steels showed no increase in the number of oxide inclusions towards the rim, although the carbon levels were always lower than those of corresponding non-treated steels. In all cases there was a distinct difference in the quantity of inclusions, which was always lower with the vacuum-treated blooms. The cleanness appraisal of these melts, which is shown in Fig. 69, confirms this result. The size and number of the inclusions were measured and counted on a 30-mm-square sample under the microscope. As a result of the vacuum treatment large-size inclusions are almost completely absent and the number of small ones is greatly reduced.

Similar improvements are also reported after a vacuum-treatment of alloy steels. In Table 15 the hydrogen, oxygen and nitrogen levels of open-hearth steels which have been subjected to a DH vacuum treatment are listed.

In many cases it is possible to replace the conventional two-slag practice in the arc furnace by a refining of the steels under one slag and a subsequent DH vacuum treatment. A comparison of several grades of steels treated in both ways is shown in Fig. 70. Besides the obvious improvement in the cleanness of the steel obtained by the vacuum treatment, the one-slag practice followed by DH vacuum treatment results in a shortening of the furnace time and therefore in an increase in produc-

References pp. 507–515

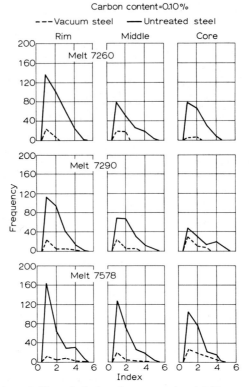

Fig. 69. Size distribution of silicate inclusions in untreated and DH vacuum-treated steels containing 0.10% carbon. Index after Diergarten (1 = very fine; 2 = medium; 4 = coarse; 6 = very coarse inclusions)[119].

tivity. Moreover, DH vacuum treatment brings about a marked improvement in the surface quality of medium-alloy steels as may be seen from Table 16[227]. Similar good results are reported on alloy steels for heavy and medium plates[224]. Figure 71 shows the results of macro-etchings of heavy plate of steel AISI 4140 of conventional and of vacuum-treated melts obtained during routine quality

TABLE 15

GAS CONTENTS OF ALLOY STEELS MELTED IN THE OPEN-HEARTH FURNACE AND DEGASSED BY THE DH PROCESS

Grade	Hydrogen cm^3 (S.T.P.)/100 g	Oxygen 10^{-4}%	Nitrogen 10^{-4}%
100 Cr 6	0.50–1.00	10–27	25–32
105 Cr 4	0.50–1.15	18–21	26–28
55 NiCrMoV 6	1.40–2.10	15–26	55–67
42 CrMo 4	1.15–1.90	16–25	25–30
17 CrNiMo 6	1.10–1.24	20–34	36–41
16 MnCr 5	0.80–2.20	7–35	25–40
8 · Si 7	1.60–2.20	48–92	28–36

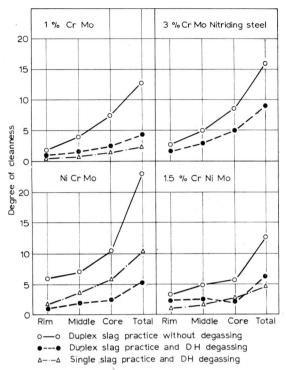

Fig. 70. Cleanness of arc-furnace steels of various types of steelmaking practice[217, 226].

controls. The improvement of the cleanness of ball-bearing steels after a vacuum-treatment by the DH process may be seen from Table 17[227]. Besides a more uniform quality of the product, the improvement of the cleanness also results in an increase in the service life of the ball bearings as will be demonstrated later.

The DH vacuum treatment may also be applied to the production of steels of less than 0.01% carbon as it is possible to obtain an effective degassing of rimming grades of high oxygen content[228]. The typical carbon exchange during the decarburization treatment of a 170-ton melt is shown in Fig. 72. The effect of the DH treatment on the decarburization depends mainly on the carbon content

TABLE 16

BLOOM SURFACE SCRAP DURING PERIODS OF FULL CAPACITY OPERATIONS

Period	May to October 1963	October 1964 to March 1965	May to October 1965
Remarks	Last peak period before degassing	Last period of one D-H unit operation	Since operation of two D-H units
Number of months	6	6	6
Average tons bloomed per month	43 287	44 217	45 173
Per cent of tonnage degassed	0	47.9	86.1
Bloom scrap index	Normal	54 per cent of normal	27 per cent of normal

References pp. 507–515

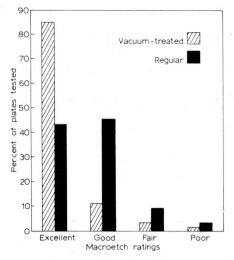

Fig. 71. Comparison of macro-etch quality of heavy gauge AISI 4140 plates.

TABLE 17

NON-METALLIC OXIDE INCLUSION RATING BY JERNKONTORET CHART OF AIR MELTED vs. DH VACUUM DEGASSED 52100 STEEL[227]

Type	E52100 (air melted)	DE52100 (DH vacuum degassed)
Heats studied	11	11
Samples examined	66	66
Series B (alumina type)	Per cent of 66 samples	Per cent of 66 samples
1.0 thin	36.5	63.6
1.5 thin	43.6	39.4
2.0 thin	9.1	18.2
2.5 thin	10.6	3.0
3.0 thin	0	0
1.0 thick	15.2	3.0
1.5 thick	9.1	1.5
2.0 thick	3.0	0
2.5 thick	0	0
Series C (silicate type)	No rateable fields	No rateable fields
Series D (globular-type oxides)		
1.0 thin	100.0	100.0
1.5 thin	82.0	9.1
2.0 thin	0	0
2.5 thin	0	0
1.0 thick	42.5	3.0
1.5 thick	3.0	0
2.0 thick	0	0

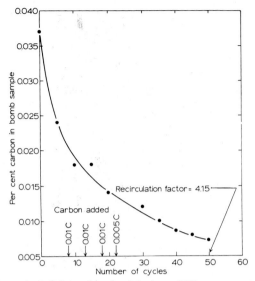

Fig. 72. Lowering of the carbon level during a DH vacuum treatment[202].

of the melt and on the vacuum conditions. Figure 73 shows the theoretically attainable final carbon contents in relation to the starting contents. The lower the initial carbon contents are and the higher the oxygen content is, the lower are the final carbon levels after the vacuum treatment. As may be seen from the diagram, the production melts closely follow the broken line which represents the calculation assuming stoichiometric carbon and oxygen removal. By addition of iron ore during the vacuum treatment with the aid of the above-mentioned alloying element charging system even greater reduction in the carbon level may be obtained, *i.e.* oxidizing of the melts under a vacuum is therefore possible. Vacuum-decarburized melts are characterized by the fact that they solidify in a quiescent fashion without

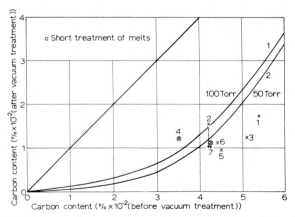

Fig. 73. Theoretical carbon contents and carbon contents after a decarburization treatment by the DH process in relation to the carbon content before the treatment[229].

References pp. 507–515

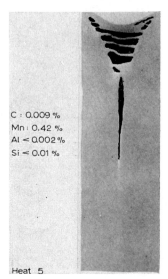

Fig. 74. Section through a 5-ton ingot decarburized by the DH process. No deoxidation agents were added to the steel at any stage.

any addition of deoxidation agents, either before or after the vacuum treatment. The section of a vacuum-killed 5-ton ingot is shown in Fig. 74. The analysis of this steel is carbon 0.009 % and manganese 0.42 % with traces of aluminium and silicon. This vacuum-decarburized steel shows a high cleanness, a good surface quality, and a structure which is almost free from segregation.

The DH process offers a number of additional possibilities in steelmaking. In order to utilize them, melting plant and vacuum installation should be regarded as one processing line. It is in many cases possible and even advantageous to introduce simplifications in furnace practice and to adjust the required quality grade in the vacuum installation. The addition of alloying elements towards the end of the vacuum degassing process and the excellent distribution of the alloying elements in the melt make it possible to produce steels within very narrow analytical limits. Besides silicon and aluminium, carbon, chromium, manganese, nickel and other elements may be added to the exact levels required. Also trace elements, such as zirconium, boron, niobium and rare-earth metals may be introduced with a precision that is within the accuracy of the analytical methods.

The short time required for the treatment, the heating of the vacuum vessel and the protection of the steel surface in the ladle against radiation losses by a cover of slag, ensure that the fall in temperature of the melt is small. In preparation for vacuum treatment by the DH process only the addition hoppers require checking or topping up. It is also advisable to attach a slag separator to the mouth of the suction pipe, so that slag is prevented from entering the vacuum vessel on dipping the pipe into the steel. As a result of the very short time required for the preparation of the vacuum treatment plant it is possible to treat several ladles,

TABLE 18
PERFORMANCE OF REFRACTORY LININGS USED IN DH VESSELS AND SNORKELS

Lining section	Refractory material	Number of linings	Heats per lining (range)	Heats per lining (average)
Vessel bottom and lower wall	Corhart 104 RFC	18	220–710	382
Vessel upper wall	Harklase (97 per cent MgO)	4	542–1373	919
Vessel roof	Harklase (97 per cent MgO)	8	351–1018	807
Snorkel	90 per cent aluminium castable with 6-in. top ring of Corhart 104 RFC	Last 80	52–128	88

which are tapped from one or several melting units, immediately one after the other by the DH process, without the slightest risk of a disruption of the usual routine of a steel works[190]. On the other hand, it is also quite possible to produce several grades of steel from one melting charge by tapping it into several ladles which are then treated differently.

Constant heating of the refractory lining of the DH vacuum vessel to the temperature of the liquid steel provides favourable conditions for the durability of the lining. Basic bricks which show a low tendency to react with the liquid steel may be used for the DH process, and in most cases magnesite bricks or spinel bricks on a magnesite–chrome basis are employed. During the 14 years in which the process has been applied on an industrial basis the durability of the lining of the vacuum vessel showed steady improvement[82, 227], and service lives of up to 700 heats have frequently been reported. A survey of the service life obtained in a steel works over a period of 18 months is shown in Table 18[227].

1.5.2 The RH process

The RH process, or circulation degassing process, was developed for application on an industrial scale by Ruhrstahl A.G. Hattingen of Germany. The principle of this process is depicted in Fig. 75. In contrast with the DH process, the refractory-lined vacuum chamber of the RH process has two tubes attached to the bottom; one of these is used for the inlet and the other for the outlet of the steel[230–233]. After both tubes have been dipped into the molten steel the vacuum vessel is evacuated, so that the steel rises to the barometric height of 1.40 m above the level of the steel surface. Now, at the lower third of the inlet tube, close to the surface of the steel in the ladle, a conveyor gas is injected. Argon is usually employed for this purpose. The gas rises with great increase in volume thus accelerating the speed with which the steel flows upwards. This upward movement is further helped by the gases liberated from the steel in the upper part of the entry tube as a result of the lowering in the pressure. In agreement with results on model tests it was found that a mixture consisting of one part of steel and ten parts of gas by volume flows at such a rate through the entry tube that the liquid steel reaches

References pp. 507–515

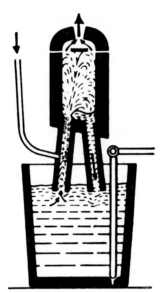

Fig. 75. Principle of the RH degassing process.

a height of about 1 m above the barometric height, and in some instances splashes of steel were thrown on to the refractory baffle provided at the upper part of the vacuum vessel[231]. The degassed steel collects at the bottom of the vacuum vessel and returns through the outlet tube into the ladle. The circulation speed of the steel is governed by the elevating capacity of the inlet tube, which acts as a pump, and by the level of the bath in the degassing vessel.

If the distance between the upper end of the inlet tube at the bottom of the

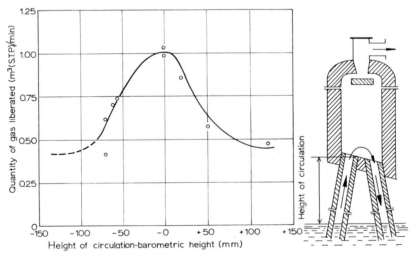

Fig. 76. Effect of height of circulation on the degassing rate in the RH process, after Thielmann et al.[231].

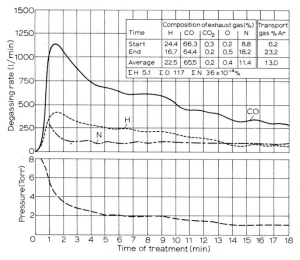

Fig. 77. Effect of time of treatment on the degassing rate in the RH degassing process[231].

vacuum vessel and the level of the bath in the ladle is considered as the height of circulation, the circulation velocity and therefore the degassing speed is a maximum, if this distance equals the barometric height (Fig. 76)[231].

The pressure and composition of the pumped-off gases are constantly re-

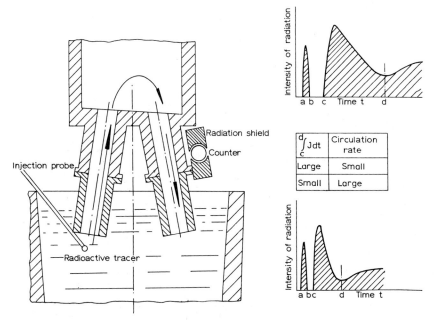

Fig. 78. Determination of the circulation rate by injection of radioactive gold into the melt, after Maas[234].

TABLE 19

RESULTS OF CIRCULATING SPEED MEASUREMENTS[235]
up-leg diam. 200 mm; down-leg diam. 150 mm

Test No.	Steel in ladle (ton)	Argon (l/ton)	Circulating speed (ton/min)	Tracer injection 198 Au, mc	Position
1	101.5	5.6	17.8	82	Up-leg
		6.7	17.9	82	
2	98.3	3.5	14.1	90	Up-leg
		6.2	20.8	90	
3	100.8	4.8	20.7	82	Up-leg
4	106.5	4.5	20.2	85	Alloy hopper

corded during the treatment, and a typical curve obtained for AISI 1022 steel is shown in Fig. 77. The quantity of steel passing through the vacuum vessel during the actual treatment cannot be measured directly[234], but measurements were carried out on batches to which radioactive gold was added to the steel as a tracer, as indicated in Fig. 78[234]. For a diameter of 20 cm of the inlet tube circulation velocities of 10–20 ton/min were measured according to the quantity of conveyor gas used. More recent results obtained by Hirata[235] are shown in Table 19.

It is the customary practice in the RH process to subject the steels to the degassing treatment in the semi-killed condition and to adjust the final silicon or

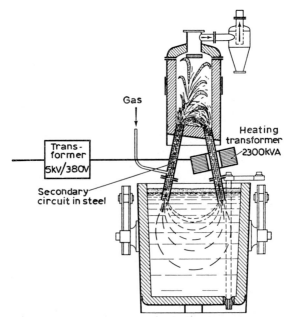

Fig. 79. Suggestion for heating during circulation degassing[231].

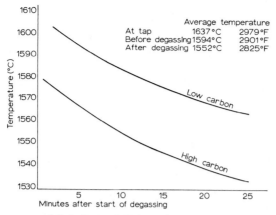

Fig. 80. Temperature fall during an RH degassing cycle. Ladle size = 120 tons[233].

aluminium levels in the degassing plant[236-238]. However, this practice may be changed and the steel treated in the fully killed condition, in which case only a correction of the carbon and aluminium levels is required at the end of the degassing treatment[233]. More recently some plants have been equipped with hoppers for the addition of other elements; these hoppers are evacuated together with the vacuum vessel. Vibration channels below the hoppers provide the transport of these elements into the vacuum vessel. However, no publication exists to date on the actual employment and the usefulness of these addition devices.

The vacuum vessel used for the RH process is customarily preheated by gas

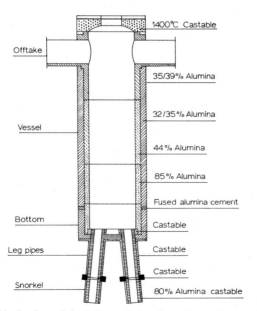

Fig. 81. Section of the RH vessel showing refractories used[233].

References pp. 507–515

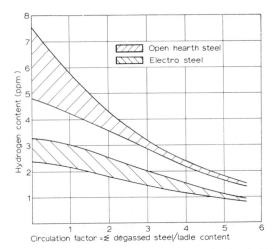

Fig. 82. Lowering of the hydrogen content during degassing by the RH process[240].

or oil burners and temperatures of 800–1200°C are attained in the interior of the vacuum vessel. These low preheating temperatures often lead to break-downs in the operation as the metal deposits at the walls of the vacuum vessel in the form of skulls. The vessel is therefore cooled down after about 25 heats and the skulls removed[233]. In order to prevent the formation of these skulls, attempts were made quite early in the development of this process to install higher heat inputs. A proposed induction heating using a transformer around the return tube is depicted in Fig. 79[231]. The closed secondary winding in this arrangement consists of the steel in the vacuum vessel, in the ladle and in the two tubes. It has also been suggested that plasma torches be employed for the preheating of the vacuum vessel. It is not known, however, whether these types of heating have actually been used.

As a result of the low degree of preheating of the vacuum vessel the fall in temperature observed during the degassing operation by the RH process is higher

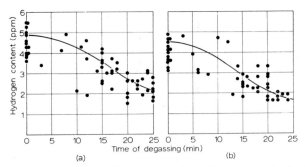

Fig. 83. Effect of degassing time on the hydrogen contents during degassing by the RH process. a = Open-hearth steel (black slag); b = arc-furnace steel (black slag), after Wahlster et al.[240].

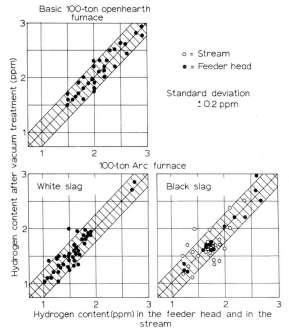

Fig. 84. Comparison of the hydrogen contents at the end of the vacuum treatment with those in the casting stream and the feeder head of the mould for various types of steelmaking practice[240].

than that occurring in the DH process. The temperature changes during an RH degassing operation are shown in Fig. 80[233]. Falls in temperature of about 1.7°C/min for 150-ton melts and of about 2.8°C/min for 70-ton melts have been reported elsewhere[239].

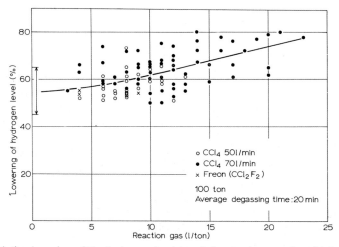

Fig. 85. Relative lowering of the hydrogen level in relation to the quantity of injected reaction gases[240].

References pp. 507–515

Owing to the required high thermal-shock resistance of the refractory material, the vacuum vessels for this purpose are usually lined with alumina-rich fireclay. The types of refractory used at the various parts of the vacuum vessel are indicated in Fig. 81. A service life of the lining of several hundred melts in the upper part of the vessel was attained, whereas that in the bottom part varies between 40 and 80 heats. The durability of the entry tubes is about 25 melts[233].

The degree of degassing in the RH process depends to a large extent on the degassing time and therefore on the quantity of steel passing through the vacuum vessel in unit time. The relation of hydrogen removal to the circulation factor, *i.e.* the ratio of quantity treated to ladle content, is depicted in Fig. 82[240]. The hydrogen contents are, for instance, reduced to one-half the initial level if a circulation factor of at least 3 is attained, *i.e.* if the whole ladle content has passed three times through the vacuum vessel. If the circulation factor is increased, the degassing

TABLE 20

OXYGEN CONTENTS (MEAN) BEFORE AND AFTER RH-DEGASSING USING ALUMINIUM ADDITION BEFORE OR DURING THE DEGASSING PROCESS[233,240]

	Wahlster et al.[240]*		Forster[233]	
	Before	After	Before	After
Killed during degassing	137	70	192	90 ppm
Killed after degassing	70	32	104	47 ppm

* Steel composition: 0.16–0.20% C; 0.25–0.40% Si; 1.30–1.50% Mn; 0.020–0.040% Al.

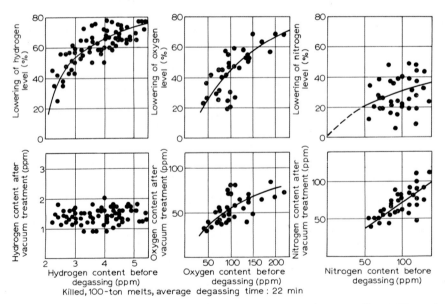

Killed, 100-ton melts, average degassing time: 22 min

Fig. 86. Gas content of killed melts before and after an RH degassing, according to Watanabe *et al.*[241].

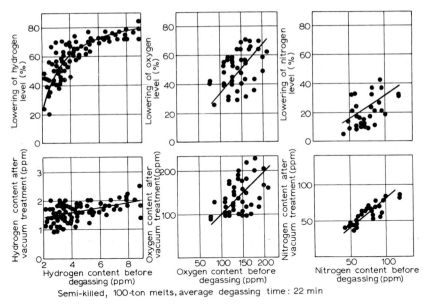

Fig. 87. Gas content of semi-killed melts before and after an RH degassing[241].

effect is also increased. The hydrogen exchange during a degassing treatment of steel containing initial hydrogen levels of 4–5 ppm is shown in Fig. 83, and as may be seen, levels between 1.8 and 3 ppm are obtained for a treatment time of 20 min. As is plotted in Fig. 84, no hydrogen pick-up occurs by the stream of molten steel

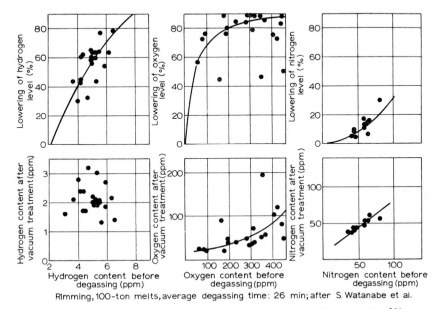

Fig. 88. Gas content of rimming melts before and after an RH degassing[241].

References pp. 507–515

during teeming or by the feeder head. The hydrogen removal may be slightly improved by additives to the conveyor gas which enter into a chemical reaction with the hydrogen. Such agents are carbon tetrachloride, Frigen and Freon. The increased hydrogen removal in relation to the type and quantity of the reaction gas is shown in Fig. 85. The removal of the oxygen during degassing by the RH process is shown in Table 20. Watanabe et al.[241] compared the degassing effect in the treatment of killed, semi-killed or rimming steels. The results are shown in Figs. 86, 87 and 88 for the hydrogen, oxygen and nitrogen levels before and after RH treatment and the reduction of the gas contents of 100-ton melts attained after a degassing time of 22–26 min. From these results it appears that in order to obtain low final gas contents in the RH process, killing of the melt with aluminium is required before the vacuum treatment.

A decarburization of rimming steel may also be carried out by the RH process. Studies are reported on the change of carbon, oxygen, manganese and silicon contents with time[242]. In these studies samples were taken from the ladle directly below the nozzle of the return tube. The results obtained are shown in Fig. 89. While a loss in manganese occurred, a considerable increase in the silicon content took place during the treatment. Vacuum-decarburized melts similar to those treated by the DH process showed solidification in a quiescent fashion and were free from rim holes as well as from noticeable sulphur, carbon or phosphorus segregation. It is assumed that the iron reduced the silica from the refractory lining or from the slag and that this reaction might even lead to still higher silicon

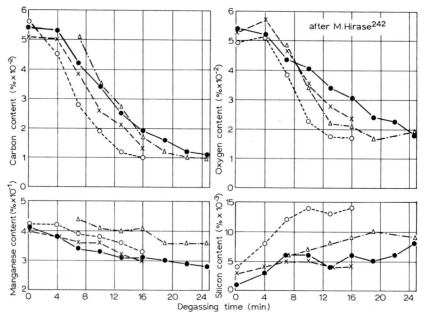

Fig. 89. Effect of an RH degassing on the composition of rimming mild steel, after Hirase[242].

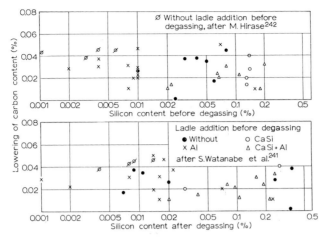

Fig. 90. Lowering of the carbon level during an RH degassing of mild steel in relation to the silicon content.

contents of the steel[240]. That in fact these melts solidify without any boiling action is ascribed by Wahlster and Reichel[240] to the deoxidation effect of the silicon. This is in agreement with the type of non-metallic inclusions shown by Hirase[242]. It is noteworthy that a decarburization also occurred in melts which had received considerable quantities of silicon in the form of CaSi before the vacuum treatment. Even additions of aluminium of up to 0.5 kg/ton steel have little effect on the lowering of the carbon levels. The results obtained by Hirase[242] and by Watanabe et al.[241] mainly on mild steel are shown in Fig. 90. It is evident that there is no clear relationship between the reduction of the carbon level and the silicon content[240].

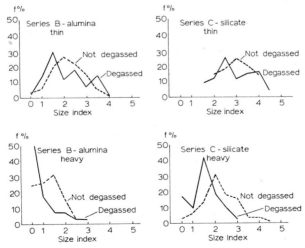

Fig. 91. Effect of an RH degassing on cleanness of the steel, evaluated by the ASTM method[233].

References pp. 507–515

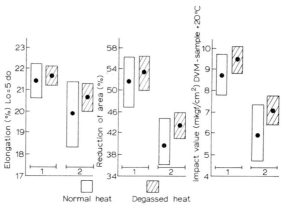

Fig. 92. Comparison of the technological properties of low-alloy steel before and after degassing[236].

This metallurgically inexplicable fact might be caused by the special conditions prevailing in the RH process and renders it difficult to maintain the margin in carbon content of $\pm 0.02\%$ usually required if rimming or semi-killed steels are degassed. For the RH process it is therefore recommended to add: (1) all the silicon in the form of FeSi or CaSi to the ladle during the tapping operation, and aluminium alone during the vacuum treatment, or (2) all deoxidation agents to the ladle during the tapping operation and nothing during the vacuum treatment.

Besides a reduction of the gas content in the RH process, a reduction in the non-metallic inclusions has also been reported. A comparison of the cleanness of degassed and non-degassed steel is shown in Fig. 91[235]. Moreover, a slight improvement of the mechanical properties after degassing of certain grades of steel has been observed as is evident from Fig. 92[236]. The RH treatment also brought about a reduction in the rate of rejects and an improvement of the service properties of the steels.

Attempts were also made to carry out RH degassing not in the ladle but in the arc furnace, as Fig. 93 illustrates. But this modification never got beyond the experimental stage.

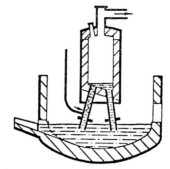

Fig. 93. Principle of degassing in the furnace.

1.5.3 Thermo-Flow process

The so-called Thermo-Flow process[243-245] is a modification of the circulation degassing process, which consists of a replacement of the proposed heating device shown in Fig. 79 by a large induction coil and therefore it secures not only a heating effect but also the transport of the metal into, or out of, the vacuum vessel. The principle of this process is depicted in Fig. 94. The vacuum vessel is also lined with a refractory material and the two tubes are each surrounded by an induction coil, and act as electromagnetic pumps. Reaction gases may also be injected into the inlet tube. As a result of this special arrangement of the two induction coils a considerably more intensive circulation of the molten steel than in the RH process is achieved so that much larger quantities of alloying elements may be introduced into the vacuum vessel than in the RH process. The Thermo-Flow process is said to be applicable to melt capacities of between 25 and 400 tons. Similarly to the DH process, it is intended to treat the melts in the unkilled condition and to carry out the whole alloying addition procedure in the vacuum plant. For this purpose a sample of the steel is taken from the ladle after tapping and this is analysed while the ladle is being transported to the degassing plant. Another aim of this type of degassing plant is to obviate superheating of the melt in the furnace in order to save production costs and to increase productivity. Furthermore, it is said that the yield in alloying elements rises to 98% and that and accurate analysis adjustment of the final product is attainable.

The process was installed in an American steel works for the degassing of 100-ton melts some years ago, but no further information has been made available

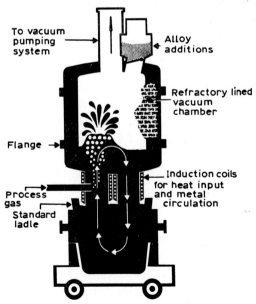

Fig. 94. Principle of the Thermo-Flow process[245].

References pp. 507–515

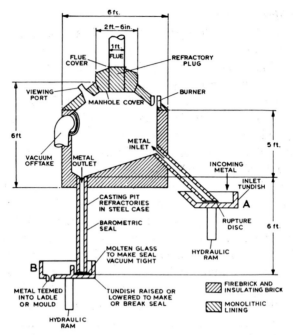

Fig. 95. Principle of transfer degassing.

since its installation so that it is not possible to describe its mode of operation and other technical details. It can be assumed that this process still requires further modification and extensive technological development before it can be regarded as suitable for routine industrial application. It cannot therefore be expected that this process will be used on an industrial scale in the near future.

1.5.4 Transfer degassing

This degassing method which was developed by the British Iron and Steel Research Association, London[246-249] is depicted in Fig. 95. It resembles the cycling and circulation degassing processes so far discussed only externally. Technologically it is a combination of these processes with the stream degassing process. In contrast with the cycling and circulation processes the steel sucked into the vacuum chamber is not returned to the original ladle but is discharged into another vessel. The advantage of this process lies in the fact that it may be inserted between the casting ladle and the tundish of continuous casting plants. The steel is poured from the casting ladle into the inlet tundish A and flows into the vacuum chamber through the entry tube which is arranged with an upward slope of 45°. After degassing, the metal is discharged through the outlet tube into a second tundish B from which the steel may flow directly into the mould of a continuous casting machine. In order to allow the plant to be started up, the vacuum vessel is evacuated. A metal foil at the mouth of the inlet tube and a vacuum seal of

TABLE 21

TEMPERATURE LOSSES IN CONTINUOUS VACUUM DEGASSING[250]

Source of loss	Temperature drop (°C)
Furnace to furnace ladle	45
Furnace ladle to inlet tundish	37
Inlet tundish to outlet tundish	30
Outlet tundish to casting ladle	61
Total, furnace to casting ladle	173

molten glass at the end of the outlet tube in tundish B provide the sealing of the vacuum vessel. When the steel is teemed into tundish A the metal foil melts and the steel is sucked into the vacuum vessel. As soon as the vacuum vessel is filled sufficiently with steel and the outlet tube is full, tundish B is lowered so that the molten glass seal is broken. The steel now flows into this tundish.

A pilot plant, which has been developed in the laboratories of BISRA, was, after several modifications, installed in a steelworks for further tests. According to the furnace sizes in this steelworks melts of 2.5–5 tons were treated. The degassing vessel is preheated with a burner to 1300–1350°C. A pressure of 1.5–8 Torr was usually attained in the vacuum vessel with the aid of mechanical pumps during the degassing process. A fall of temperature of about 100°C between furnace ladle and casting ladle was observed over a number of melts. The fall in temperature measured at the individual stages is shown in Table 21. It should be expected that this fall in temperature will be lower for larger melts. As the melts were alloyed and deoxidized before vacuum treatment, no noticeable reduction in oxygen and nitrogen contents was observed. The loss in alloying elements was relatively high for manganese, which could amount to up to 17.8% for steels containing 2% manganese. The manganese losses in relation to the initial manganese content are plotted in Fig. 96. An average reduction in the hydrogen level

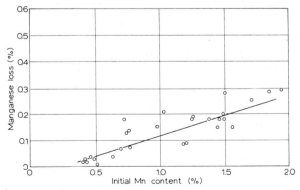

Fig. 96. Manganese loss during transfer degassing in relation to the manganese content of the melt before degassing[250].

References pp. 507–515

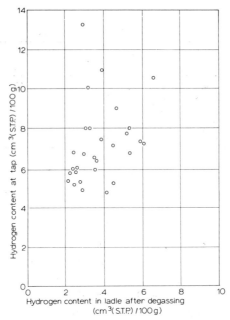

Fig. 97. Hydrogen levels in the steel before and after transfer degassing[250].

from an initial 7.3 to 3.8 cm³ (s.t.p.)/100 g, *i.e.* a reduction of 48%, could be obtained. The highest reduction in the hydrogen content observed was 78% (from 13.2 to 2.9 cm³/100 g). The hydrogen levels before and after degassing of 30 heats are shown in Fig. 97. In the tests it was possible to manage the melting process with one slag only instead of the usual two. Advantages of this practice are lower furnace operating costs and increased furnace production. Furthermore, high losses of deoxidation agents are prevented and savings in deoxidation agents are obtained.

Improvements in quality were particularly observed on stainless steels of grade EN52 containing carbon 0.4–0.5%, manganese 0.3–0.6%, silicon 3.0–3.75% and chromium 7.5–9.5% with maximum levels of 0.5% nickel and 0.04% each of sulphur and phosphorus. It was usual with this type of steel before vacuum treatment for rejects to occur after an incubation period of several days owing to poor cleanness, rim holes, slag inclusions and non-metallic inclusions from the refractory material. The number of these defects greatly depended on the hydrogen content of the steel, and as the hydrogen level could be lowered, these defects were prevented from occurring[248].

Despite these reported improvements no further plant according to this principle has been installed in a steel works so far. It must therefore be assumed that the preparatory work for the degassing of larger melt quantities is so elaborate that at present this process cannot compete with other processes with regard to operating costs.

1.6 Mould degassing processes

Mould degassing processes are here held to comprise all those processes in which the vacuum treatment is applied immediately before or during the solidification process of the steel.

1.6.1 Vacuum casting inside a vacuum tank

Vacuum ingot casting which was developed at the beginning of the fifties to industrial application and reliability[50-55,57,58,146,147,148,156,251] is a stream degassing process, the principle of which is shown in Fig. 98. The ingot mould in which the steel is to solidify is placed in a vacuum tank. On top of the vacuum tank an intermediate ladle, or in more recent installations the casting ladle itself, is placed in a vacuum-tight fashion. The nozzle through which the steel will enter the vacuum chamber is closed by metal foil and the vacuum tank is then evacuated. After the casting ladle (or the intermediate ladle) has been placed on top of the vacuum chamber, the stopper rod is lifted and the stream of steel melts away the sealing foil. The stream of molten metal then enters the vacuum chamber and is split up into very fine droplets by the liberation of the gases. This dispersion of the stream contributes to the acceleration and effectiveness of the degassing process; furthermore, any oxidation of the steel by air during filling of the mould is excluded with certainty. Another advantage is that the mould does not need to be lacquered without impairing its service life so that there is no possibility of any hydrogen pick-up from the lacquer. Splashing on to the mould walls by the dispersed stream of molten metal is, in vacuum casting as opposed to casting in air, of no great importance provided that the melt is hot enough, as no oxidation of steel can occur. By special means, such as the insertion of a refractory tube (Fig. 99), the angle of scattering of the steel droplets may be reduced. Small quantities of deoxidation elements may still be added to the ingot mould during

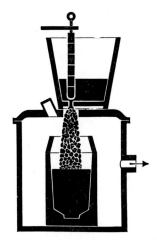

Fig. 98. Principle of vacuum ingot casting[153].

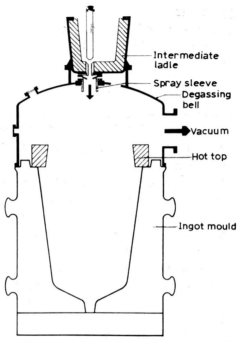

Fig. 99. Vacuum ingot casting with spray sleeve[154].

the filling process. After completion of the casting process, the casting ladle and the lid of the vacuum vessel are removed and the steel in the ingot mould solidifies under atmospheric pressure.

This process found wide application, particularly for heavy castings for forging, all the more so as it is relatively simple[92,101,149,154,161,164,169,252-265]. Ingots of up to 250 tons in weight were produced by this method; of course such sizes require rather large vacuum tanks. Besides the removal of gases from the steel, the vacuum casting process offers the additional advantage that residual moisture is removed to a large extent from the refractory-lined feeder heads of the ingot mould itself during the evacuation of the vacuum vessel. Furthermore, the casting operation may be carried out without exposure of the casting personnel to fumes or heat, although the casting process may be observed through an observation window provided in the lid of the vacuum vessel so that it may be controlled.

As the steel begins to solidify in the mould immediately after casting, no metallurgical process, such as decarburization or adjustment of the composition, can be carried out in vacuum ingot casting. In almost all cases, therefore, the steel is cast and degassed in the fully killed condition. A typical treatment cycle is plotted in Figs. 100 and 101[255,258] on a Cr–Ni–Mo steel as an example. After the oxidizing period, which takes about 1½ hours, ferro-molybdenum is added to the charge in the furnace, which is slagged about one hour later. After addition of

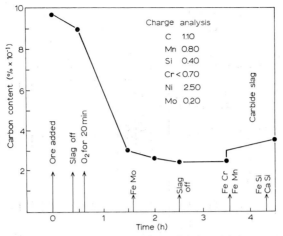

Fig. 100. Typical melting process of a chromium–nickel–molybdenum steel, according to Scalise[258]. (Carbon content in arbitrary units.)

ferro-chromium and ferro-manganese to the desired level, the carbide-containing second slag is charged and ferro-silicon and calcium–silicon are added. It is the usual practice to tap the melt at a temperature which is 20–30°C above that for casting in air. The ladle is then transported to the vacuum unit. The pressure in the vacuum tank, which has been maintained below 0.1 Torr for about 2 hours, is raised to about 15 Torr. The reason for increasing the pressure is that the stream of molten metal assumes too wide a dispersal cone if the pressure is too low. Particularly at the beginning of the casting process the stream should be slightly kept together, as splashing to the mould wall at this time may lead to scab formation and therefore to poor surface quality. Observation of the casting stream later

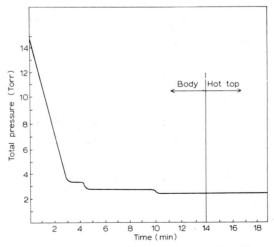

Fig. 101. Pressure in relation to time during the casting of a 100-ton ingot of a steel according to Fig. 100[258].

References pp. 507–515

TABLE 22

COMPOSITION OF GASES COLLECTED DURING VACUUM CASTING[146]

Ingot no.	Grade	Casting time (min)		Sample no.	Time sampled after start of pour (min)	Absolute pressure (10^{-3} Torr)	Gas composition (mol%)						
		Ingot body	Total				CO_2	A	O_2	N_2	CO	H_2O	H_2
11	Ni-Mo-V	12.5	21.0	1	0	80	1.6	0.5	12.9	50.1	—	34.9	—
				2	5	320	1.9	0.2	1.5	30.9	21.7	3.5	40.3
				3	10	360	0.6	0.1	0.5	11.5	26.0	2.3	59.0
				4	15	340	2.5	0.3	0.8	25.0	38.4	1.7	31.4
12	Ni-Cr-Mo-V	16.0	25.0	1	0	175	3.2	0.6	9.0	68.2	18.5	6.7	1.6
				2	5	400	0.7	0.1	0.08	10.8	26.1	0.08	62.1
				3	10	410	0.4	ND	0.2	11.9	19.6	0.6	67.2
				4	16	300	2.1	ND	4.4	14.7	27.0	1.8	50.0
				5	24	—	4.3	0.4	0.06	32.6	38.2	0.3	24.0

HYDROGEN, OXYGEN AND NITROGEN CONTENTS OF HEATS BEFORE AND AFTER VACUUM CASTING

Ingot no.	Grade	Ingot diam (in.)	Avg. gas content (ppm)					
			Hydrogen		Oxygen		Nitrogen	
			Before	After	Before	After	Before	After
1	Ni-Mo-V	72	3.3	1.0	40	25	30	30
2	Ni-Mo-V	72	5.1	0.8	20	15	60	55
3	Ni-Mo-V	72	4.7	0.8	25	15	70	65
4	Ni-Cr-Mo-V	72	4.9	0.8	20	10	80	80
5	AISI-4340	28*	3.7	0.8	50	40	95	95
6	AISI-52100	28*	4.0	0.6	30	20	80	30
7	Mn-Ni-Mo-V	72	4.5	1.7	—	—	—	—
8	Mn-Ni-Mo-V	72	5.0	2.0	20	—	70	—
10	Type 422	39	5.8	1.4	70	60	200	190
11	Ni-Mo-V	72	4.1	—	10	—	60	—
12	Ni-Cr-Mo-V	77	3.3	0.42	10	—	70	—
14	Ni-Mo-V	95	3.8	0.27	10	—	65	—
15	Cr-Mo-V	95	3.5	0.29	10	—	90	—
16	Ni-Mo-V	95	4.2	0.43	<10	—	55	—
17	Cr-Mo-V	95	4.6	0.81	30	—	75	—
18	Cr-Mo-V	95	4.7	0.38	<10	—	70	—
19	Ni-Mo-V	95	4.7	0.57	20	—	75	—
20	Ni-Mo-V	72	4.0	0.61	—	—	60	—

* Square ingot.

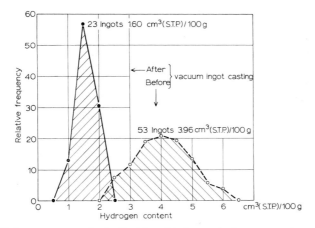

Fig. 102. Hydrogen contents in the liquid steel before and after vacuum ingot casting[153].

makes possible the gradual reduction of the pressure until an optimum dispersion cone or the lowest attainable pressure is reached. The pressure will then be kept constant either by the use of a butterfly valve or by the pumping speed itself.

The behaviour of the vacuum-cast steel in the mould is similar to that of rimming steel cast in air, as the vacuum-cast steel still gives off gases and therefore boils. Analyses of the gases liberated are shown in Table 22[146]. On the whole the pumped-off gas consists half of hydrogen and half of equal parts of nitrogen and carbon monoxide. The hydrogen level of the ingots is therefore reduced to a greater extent than that of the other gases. The hydrogen contents attainable are mainly governed by the pressure existing during vacuum casting. Comprehensive studies by Tix et al.[102,152,153] showed that the hydrogen contents obtained obey quite closely Sieverts' law on the hydrogen solubility (see also Fig. 47). The statistical

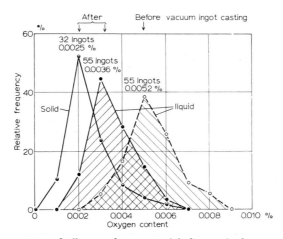

Fig. 103. Oxygen contents of alloy arc-furnace steel before and after vacuum ingot casting. Also oxygen content of samples taken after solidification. Ingot sizes 60–150 tons[153].

References pp. 507–515

distributions of the hydrogen levels before and after vacuum casting are shown in Fig. 102, and the statistical distributions of the oxygen contents in the casting ladle and in the solidified ingot after vacuum treatment are plotted in Fig. 103.

As a result of the effective removal of hydrogen during vacuum casting the danger of flaking is almost eliminated for all steel grades which are used for forging. Heat-treatment of the ingots can therefore be simplified and the annealing times can be greatly reduced. A comparison of the annealing times and annealing temperatures of forging ingots cast in air and cast in a vacuum is shown in Table

TABLE 23

HEAT TREATMENT OF AIR-CAST AND VACUUM-CAST INGOTS FOR FORGING, ACCORDING TO SCALISE[257]

INGOTS CAST IN AIR

Ingot size (cm)	Forging scheme	Efficient annealing time (h)	Billet size (mm)	Intermediate annealing time (efficient), h		Final diam. (mm)	Final annealing time, h		Overall annealing time, h	
				650° C	Total		650° C	Total	650 °C	Total
120	A 2	750	—	—	—	1,000	750	990	750	990
136	A 2	750	—	—	—	1,000	750	990	750	990
153	A 2	900	—	—	—	1,000	900	1,140	900	1,140
	A 2	900	—	—	—	1,200	1,300	1,588	1,300	1,588
	A 3	900	1,400	885	1,221	1,200	650	938	1,535	2,159
173	A 3	1,000	1,600	1,280	1,616	1,000	500	740	1,780	2,356
	A 3	1,000	1,600	1,280	1,616	1,200	720	1,008	2,000	2,624
	A 4	1,000	1,600	855×2	1,291×2	1,400	655	991	2,365	3,573
204	A 3	1,000	1,800	1,620	1,956	1,000	500	740	2,120	2,696
	A 4	1,000	1,800	1,080×2	1,416×2	1,400	655	991	2,815	3,823
>204	A 4	1,100	2,000	1,465×2	1,501×2	1,000	367	607	3,297	3,609
	A 4	1,100	2,000	1,465×2	1,501×2	1,500	825	1,161	3,755	4,163

INGOTS CAST IN VACUUM

Ingot size (cm)	Forging scheme	Efficient annealing time (h)	Billet size (mm)	Intermediate annealing time (efficient), h		Final diam. (mm)	Final annealing time, h		Overall annealing time, h	
				650° C	Total		650° C	Total	650° C	Total
120	V 1	400	—	—	—	1,000	400	592	400	592
136	V 1	400	—	—	—	1,000	400	592	400	592
153	V 1	400	—	—	—	1,000	400	592	400	592
	V 1	400	—	—	—	1,200	576	792	400	792
	V 2	400	1,300	720	936	1,200	240	456	960	1,392
173	V 2	400	1,100	486	702	1,000	240	432	726	1,134
	V 2	400	1,300	680	896	1,200	240	456	920	1,352
	V 2	400	1,500	900	1,140	1,400	240	480	1,140	1,620
204	V 2	400	1,100	486	702	1,000	240	432	726	1,134
	V 2	400	1,500	900	1,140	1,400	240	480	1,140	1,620
>204	V 2	400	1,100	486	702	1,000	240	432	726	1,134
	V 2	400	1,600	1,028	1,268	1,500	240	480	1,238	1,748

23[257]. According to ingot size the saving in annealing time may amount to up to 70%. Besides the shortening of the annealing times a beneficial effect of vacuum casting on the technological properties of the ingots for forging is mentioned by a number of authors[146,149,257]. This effect is particularly pronounced in respect of elongation and necking.

In contrast with the vacuum treatment of forging steel, the requirement for the vacuum treatment of mild (*i.e.* low-carbon) steel is that it should be cast in ingot mould of 5–10 ton capacity from large melting units. Owing to the low carbon contents of these mild steels and the resulting oxygen contents considerably more carbon monoxide is produced during the degassing of these steels than during the degassing of steel for forgings. In order to provide a vacuum treatment of the casting stream of these steels the method of vacuum ingot casting described in the preceding paragraphs must be modified in several important particulars.

The principle of the vacuum stream degassing process used for plain-carbon steel of low carbon content is shown in Fig. 104[266,267]. Owing to the large quantities of carbon monoxide liberated, the casting stream disperses immediately below the ladle nozzle on entering the vacuum chamber to such an extent that a dispersion paraboloid of small droplets is created which assumes a diameter of several metres a very short distance below its point of entry. The process therefore requires a special tundish which is interposed between casting ladle and ingot mould. This tundish makes possible effective degassing during dispersion of the casting stream and it also collects the steel again so that it is run as a coherent stream into the mould. After completion of the pouring operation, the vacuum vessel is usually opened to the air at once. The cast slabs solidify quietly, but if the vacuum is maintained the steel boils for a short time.

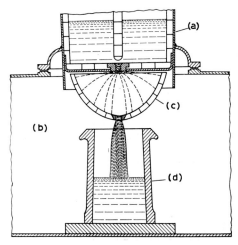

Fig. 104. Schematic representation of pouring jet vacuum degassing equipment for rimming low-carbon steel. a, pouring ladle; b, vacuum chamber; c, dispersion tank; d, ingot mould.

References pp. 507–515

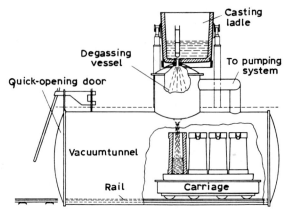

Fig. 105. Stream degassing of mild steel. Principle of a plant for top or bottom pouring and rapid mould change[266].

As mild steel of deep-drawing quality is produced in huge quantities, and as the vacuum treatment plant is therefore laid out for large throughputs, it is planned to develop the vacuum vessel, which contains the ingot mould, as a vacuum lock, so that the ingot moulds may be moved into the vacuum vessel on carriages. Such a plant is depicted in Fig. 105, but it is not yet in operation.

Provided that the vacuum practice is carried out correctly, the ingots are entirely free from pipes and rim holes. Ultrasonic tests of the cast slabs gave no indications which could be interpreted as non-metallic inclusions of more than marginal size[266]. Moreover, the cast ingots do not show any sulphur segregations, which are always present in rimming deep-drawing steel if cast by conventional

TABLE 24

CHEMICAL COMPOSITION AND MECHANICAL PROPERTIES OF A COLD-FORMED STRIP OF POURING JET DEGASSED LOW-CARBON STEEL (SAMPLES FROM THE BEGINNING, FROM THE MIDDLE AND FROM THE END OF THE STRIP)[267]

Distance of sample from the beginning of the strip in % strip length	Chemical composition				Mechanical properties					
	% C	% Mn	% S	% O[a]	Tensile strength (kgf/mm^2)	Yield strength (kgf/mm^2)	Elongation $L_0 = 800$ mm (%)	Hardness HRB	Cupping number[b] (%)	Limiting blank diameter[c] (mm)
3 (Beginning)	0.06	0.31	0.033	0.012	32.6	19.9	41	46	115	70
6 (Beginning)	0.06	0.32	0.032	0.011	32.4	19.5	41	46	113	71
9 (Beginning)	0.06	0.32	0.032	0.011	32.2	18.9	43	45	113	70
50 (Centre)	0.06	0.32	0.030	0.011	32.4	19.4	39	48	112	70
97 (End)	0.05	0.31	0.028	0.010	31.9	18.6	42	45	113	71

[a] Hot extraction method at 1800° C. [b] Related to minimum values for UST 13 acc. DIN 1623. [c] Cup forming test with strip thickness of 1.25 mm.

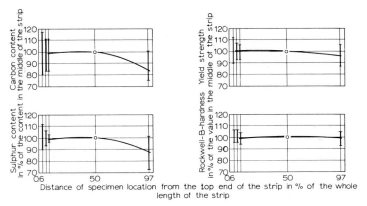

Fig. 106. Carbon and sulphur contents, hardness and yield stress of cold-rolled strip of vacuum-cast rimming steel of low carbon content measured along the strip. The results always refer to the middle of the coil. The "I" in the diagrams indicates the scatter[267].

methods and which often give rise to trouble. The excellent consistency of the composition and of the mechanical properties over the whole length of the strip produced from these ingots is evident from Table 24. The high uniformity of the steel degassed by this process may also be seen from Fig. 106, in which the mean values of the carbon and sulphur contents as well as yield strength and hardness are plotted in relation to the position in the strip. As a reference the value obtained in the centre of the strip is used and the small vertical lines in the form of an "I"

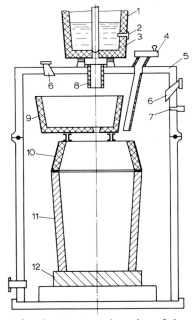

Fig. 107. Principle of a plant for the two-stage degassing of the casting stream, after Zverev et al.[158].

References pp. 507–515

TABLE 25

COMPARISON OF VACUUM CASTING USING AN INTERMEDIATE LADLE WITH CONVENTIONAL VACUUM CASTING[158]

	Duplex vacuum treatment		Conventional vacuum casting	
Melt No.	7	8	9	10
Casting rate in tons per minute	2.37	1.96	2.40	3.56
Pressure in the chamber on entry of the metal into the intermediate ladle, Torr	6	5	6	4
Quantity of gases liberated, l/ton	170	202	144	156
Quantity of hydrogen liberated from the metal, $cm^3/100$ g	2.26	3.05	1.66	1.22
Quantity of carbon monoxide liberated, l/ton	104	100	70.6	56.8

indicate the scatter of the values.

The modification of inserting an intermediate tundish between casting ladle and mould was also applied to the degassing of large forging ingots[158], and such an arrangement on a pilot-plant scale is shown in Fig. 107. In the experiments the depth of the pool of molten steel in the tundish was about 30–40 mm. Intensive degassing took place in this tundish so that the stream of steel leaving the tundish remained coherent. The arrangement prevented the formation of scabs which are common in direct vacuum casting. Boiling of the metal in the mould, too, was no longer observed. This method also makes possible the addition of small amounts of deoxidation agents after degassing. As shown by the results of measurements the quantity of gas liberated in this process is greater than in the straight vacuum ingot casting process (see Table 25). If, however, a powerful deoxidation agent—in the tests lanthanides were used—is added to the steel in the tundish or in the mould, development of gas is considerably lower than in direct vacuum casting.

1.6.2 Vacuum casting without an outer vacuum chamber

In order to apply ingot degassing to small ingot sizes as well, a process was developed in which the ingot mould itself was sealed and connected to the casting ladle in a vacuum-tight manner[48,99,159,268–272]. An interesting method was used in 1941 on an experimental scale, the principle of which is depicted in Fig. 108[48]. An intermediate chamber is connected to the casting ladle in a vacuum-tight manner and this assembly is connected, also vacuum-tight, to the ingot mould. The vacuum in the mould is created by the suction action of a Venturi tube. The suction ejector is operated by a small air compressor. In later modifications of this process the suction ejector was replaced by a proper vacuum pumping system. The first arrangement of this type was probably operated on an experimental scale by Eminger et al.[99,268,273]. This arrangement is shown in Fig. 109[273]. The individual parts of the mould were joined by special sealing materials so that they were vacuum-tight. The upper part of the ingot mould, which contains

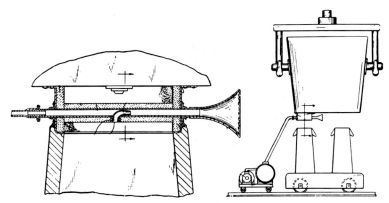

Fig. 108. Proposal for the exhaustion of ingot moulds by an ejector operated by compressed air[48].

the refractory-lined head, was extended and provided with a suction port and an observation window. An intermediate ladle was then attached to the head. In more recent units this arrangement has been modified by an enlargement of the chamber which rests on the ingot mould. A protection against splashing was also provided. This modification which is currently used in a number of steelworks is depicted in Fig. 110[271, 272]. The pumping system is movable along the casting pit and may be attached to the moulds by a quick-closing lock. It is the usual practice in this process to treat only fully killed steel, which is tapped from the furnace at

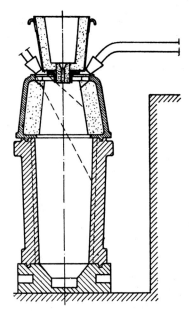

Fig. 109. Experimental set-up of Eminger et al. for the degassing of steel during pouring and solidification[99].

References pp. 507–515

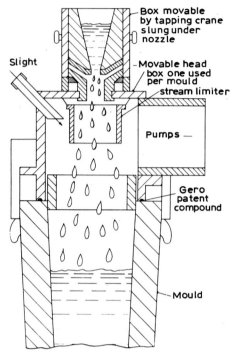

Fig. 110. Principle of the Gero process[272].

the conventional temperature without superheating. After the filling of the mould the vacuum system is detached and connected to the next mould. The ingot solidifies under atmospheric pressure.

Using this method, the hydrogen levels are said to be lowered by 50–75% of the initial level and the quantity of oxide inclusions by 25–35%[272]. Otherwise, the results obtained with this method differ very little from those obtained by vacuum casting in a vacuum tank.

1.6.3 Ingot degassing

Under the term "ingot degassing" are grouped all methods and modifications that involve a vacuum treatment of steel during the solidification process. As far back as 1957, Eminger et al.[99] described experiments in which a vacuum was maintained for 20 and 60 min after completion of the mould-filling operation. The arrangement used for these experiments was that depicted in Fig. 109. As may be seen from Table 26, the hydrogen level of the steel could be further lowered by this arrangement. Solidification under vacuum has almost no influence on the distribution of the alloying elements, as is illustrated in Fig. 111 for two 4.7-ton ingots of chromium–nickel–molybdenum steel. Ingots solidified under a vacuum are usually more fine-grained. The granular rim zone grows wider the longer the block is exposed to the vacuum. The centre of the ingot is also markedly more

TABLE 26

EFFECT OF TIME UNDER VACUUM ON THE HYDROGEN CONTENT[99]
(vacuum = constant)

Ingot no. (weight 4700 kg)	Type of casting	Time under vacuum (s)	Hydrogen content (%)
1	Conventional	0	100
2	Under vacuum	200	77.2
3	Under vacuum plus 20 min. during solidification	1200	53.7

Remarks: Hydrogen content of air-cast ingots = 100%.

fine-grained and is predominantly granular. A comparison of the segregation found in air-cast, vacuum-cast, and vacuum-solidified ingots is shown in Fig. 112. The depth of the vacuum pipe decreases in ingot degassing as well as the segregations. These effects are the more pronounced the longer the steel is exposed to the vacuum. The ghost lines are shifted from the centre of the ingot towards the rim by the action of the vacuum. The application of a vacuum also causes a considerable enlargement of the ghost lines and a change in their slope in the ingot. They are not so strong, however, as those of air-cast ingots.

Besides these ghost lines, numerous blow holes, both in the rim zone and in the centre, are observed, particularly in the upper half of the ingot, which indicates

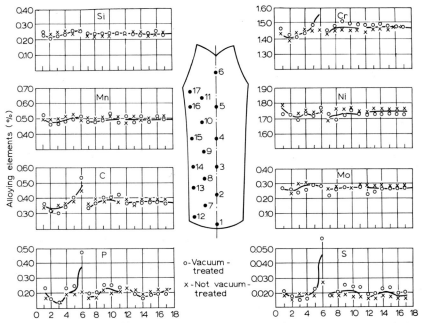

Fig. 111. Distribution of the elements over the cross-section of 4.7-ton ingots of a chromium–nickel–molybdenum steel cast in air and vacuum-cast[99].

References pp. 507–515

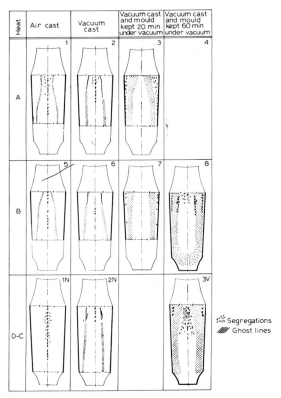

Fig. 112. Position of segregations and ghost lines in 4.7-ton ingots of chromium–nickel–molybdenum steel cast under various conditions[99].

that the steel, even with the use of deoxidation agents, still shows some rimming behaviour if solidified in a vacuum.

The occurrence of blow holes and shrinkage cavities leads to the conclusion that, despite an improvement in some specific properties, a vacuum treatment of killed steels after casting is not a practical proposition. However, Henke et al.[266], as well as Schulte[274], showed that advantages are obtained by vacuum treatment during the solidification of mild rimming steel. These authors carried out experiments of a type depicted in Fig. 113. The idea of these trials was based on the observation that the ferrostatic pressure greatly impairs the scavenging and purifying boiling process in the ingot bottom. A lowering of the ambient pressure of 760 to less than 100 Tor should be sufficient to create a boiling action and therefore a scavenging effect even in the lowest layers of the ingot. By controlling the boiling action with regard to intensity and time, pronounced segregation should be avoided and good uniformity of the mechanical properties in the final product obtained.

During the tests it was found that a careful time-dependent lowering of the pressure, which varies with both ingot size and composition of the steel, is an essential requirement for an adequate separation of the slag, a favourable segre-

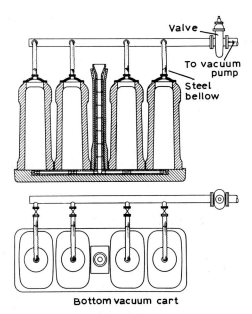

Fig. 113. Arrangement for the bottom casting of low-carbon, rimming steel according to Henke and Hess[266].

gation picture and correct formation of the ingot head, which is very important for the subsequent rolling operation. The lowering of the pressure in relation to time during the simultaneous bottom pouring of four 7.5-ton slabs is plotted in Fig. 114. The steel used contained about 0.045% carbon and 0.030% manganese. The vacuum was maintained for about 15–20 minutes, when the pressure was increased to ambient. When the pressure is being increased the boiling action of the steel in the mould ceases instantly, the surface level of the still liquid steel in the centre falls a little and the ingot gives the impression of a killed steel. But even in this condition the steel still contains dissolved oxygen which may lead to blow hole formation at the solidification front. If this residual oxygen is combined with aluminium the ingot head shows a well-formed rim and inside this rim a uniform

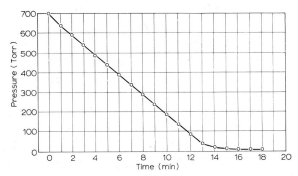

Fig. 114. Adjustment of the vacuum during vacuum bottom casting according to Fig. 113.

References pp. 507–515

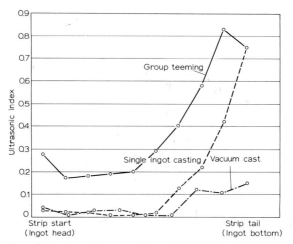

Fig. 115. Distribution of the impurities over the whole length of the finished cold strip for various types of casting of the starting ingots[266].

microstructure and only little segregation so that the head parts lead to far fewer rejects during the subsequent rolling process and sheet manufacture than those of conventionally cast steels. The favourable effect of a vacuum treatment on the quantity and distribution of the slag inclusions in the finished cold-rolled strip is shown in Fig. 115. In a similar way the beneficial effect of a vacuum treatment on the yield point of cold-rolled strip may be seen from Fig. 116. From these data it can be concluded that the solidification picture of the rimming ingots which have been subjected to a vacuum treatment resembles that of killed ingots whereby, however, the higher yield of the rimming ingots is obtained.

1.7 Discussion of the vacuum processes

The description of the individual vacuum processes in the preceding subsections will have led the reader to appraise and compare the individual processes and to consider their comparative values. Some of these processes are closely related, while others are entirely different. It would appear simple to make an immediate appraisal of the advantages and shortcomings of the various vacuum processes, but it is in fact extremely difficult to make such a comparison objectively, as their most advantageous application to industrial practice is governed by many factors, which depend to a large extent on local conditions and may therefore change from one works to another. For instance, excessive heat losses during the vacuum treatment and consequently the use of higher melting temperatures result in considerably less financial and technological disadvantages in steel works equipped with arc furnaces than in those employing large open-hearth furnaces. Moreover, the individual processes are at present still subject to constant technical modifications, so that the relation between the various processes must change in time.

It is therefore not surprising that papers are published at periodic intervals

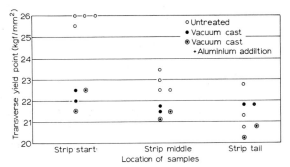

Fig. 116. Change of the yield point of cold strip over the whole length of the strip determined on strip rolled from ingots cast by various methods.

which report on the state of development of the individual processes and point out the differences existing between them*.

Most of these papers are restricted to a description of the state of the development of the installations and to a report on some important results which have been obtained in these plants. Other papers place the emphasis on the degassing effects or on the degree of purity or cleanness of the treated steels. Some studies are devoted more to the usefulness in, and suitability for, the particular steelworks, or to the heat losses during treatment, or to the costs of the vacuum treatment. As various as the individual studies may be, they all show that, besides the kinetics of degassing, *i.e.* the elimination of hydrogen, oxygen and nitrogen, those factors are also of great importance for a comparison of the various processes, which are governed by the production programme of the works, their type of melting units and their size and output. A comparison may therefore well arrive at different results, according to the manner in which one and the same process is used in the various steelworks. It is for these reasons that it is not possible to make generally applicable statements in the subsequent subsections. The following considerations will therefore stress only those aspects which are of importance in the appraisal of the individual degassing process from the practical, technological point of view, so that making a decision in favour of one or other process will be simplified. It is, therefore, to be expected that the present author cannot be wholly objective, and that his own opinions must play some part in appraising the importance of the individual factors, even though he will endeavour as far as possible to remain unbiased.

1.7.1 Gas content and the kinetics of degassing

An appraisal of the various degassing processes with respect to the minimum gas content of the steel obtainable is immediately involved in great difficulties. As

---

* Comprehensive reports on the state of development of vacuum-treatment plants and appraisals of the various types of plant are contained in the papers in references 98, 180, 259, 263 and 275–303.

References pp. 507–515

the processes operate in the various steelworks under different conditions, the conditions for the determination of the gas contents before, during or after the vacuum treatment are not uniform. Furthermore, the practice of sample taking from the steel heat as well as the method of analysis employed may be different. For this reason, attempts have been made for many years to lay down rules for a standardized sample-taking process by the iron and steel institutes of the various steel-producing countries. However, even in those countries in which standardized sample-taking methods have been introduced, most steelworks do not use them. Finally, the results obtained from the same samples show a scatter to a large degree if different analytical methods and pieces of analytical apparatus are employed, particularly if the gas contents are low, as is the case after a vacuum treatment.

Different sample-taking techniques and different analytical methods must be employed for different gases present. Sample taking for an evaluation of the hydrogen content of the steel is carried out mainly by one of three techniques. In the first, the sample is taken with a spoon from the steel heat, the sample is killed with aluminium and cast in a copper mould. The solidified sample is removed from the mould and quenched in water. The sample is then stored at a low temperature—usually in a mixture of dry ice and acetone—before analysis in the laboratory. In the second technique, the steel sample is sucked into a quartz tube and cooled as quickly as possible. In the third variation, small evacuated quartz tubes are immersed in the molten steel. The steel enters the quartz tube at one end and solidifies in the tube. This method has the advantage that not only the hydrogen contained in the solidified sample is measured, but also that portion of the hydrogen that is liberated during solidification.

The analysis for the hydrogen content of the samples is also carried out mainly by one of three methods. In the first method, the sample is heated to 600–800°C in a vacuum and the hydrogen diffused from the sample at this temperature is determined. In the second method, the tin fusion method, the sample is melted in a tin bath and the quantity of hydrogen liberated is measured. In the third variation, the sample is melted in a vacuum at 1600°C and the quantity of hydrogen liberated is determined. How much the results obtained by the individual methods differ from each other depends on the details of the analytical apparatus used and on the size or weight of the sample. Comprehensive studies have been carried out under the auspices of the appropriate research institutes of the iron and steel industry of the various countries. The results of these studies will not be discussed here, but variations of the order of ± 0.5 ppm hydrogen are quite possible depending on the type of analytical method employed.

An even greater variety in sample-taking methods is used in the determination of the oxygen and nitrogen contents of steel. In some cases samples are of the order of 50–200 g (killed with aluminium); sometimes samples of 1–2 kg are taken and test-pieces, the weight of which may be only 2–10 g according to the type of analytical apparatus used, are then prepared from these samples. These pieces are melted in graphite crucibles in a vacuum and are then heated to 1700–1900°C and the quantities of gases liberated are measured. If the oxygen levels are high, above 400 ppm, errors in the results are introduced by the sample-taking method; on the other hand, considerable scatter of the results is often caused by the method of analysis at low oxygen levels of the steel. However, the determination of the nitrogen content of steel is in general independent of both the method of sample taking and that of analysis (see also Chapter VI).

The scatter of the results in the determination of the oxygen and hydrogen contents, which was carried out in different types of apparatus at the different steelworks, does not, as a rule, allow direct comparison of the results[281, 286]. In general, an exact statement on the lowering of the gas content as a result of a vacuum treatment of the steel can only be made if the samples before and after treatment are taken by strictly the same procedure and analysed by the same method.

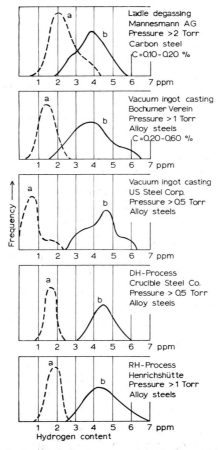

Fig. 117. Frequency distribution of the hydrogen contents before and after a vacuum degassing. Comparison of various processes according to Edström[280]. (a) After degassing; (b) Before degassing.

References pp. 507–515

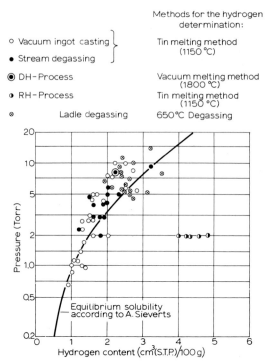

Fig. 118. Comparison of the final hydrogen contents for various degassing processes[96, 303].

Unfortunately, descriptions of the methods used for these determinations are often omitted from the results reported in the literature, so that these values cannot be used for a comparison of the various methods. Taking these uncertainties into consideration a comparison of the various processes for the vacuum treatment of steel, as well as a summing-up of the many publications available on this subject, makes it possible to draw the following conclusions.

The hydrogen content can be lowered to levels of 1–2.5 ppm by nearly every method, independently of time of treatment and quality of the steel. The reduction in the hydrogen level in these processes amounts to between 60 and 80% of that initially present. This is demonstrated in a study by Edström[280] and is shown in Fig. 117. It has also been found that the results obtained on killed steels agree well with the theoretical equilibria derived from Sieverts' law for the hydrogen content of steels, if the results are related to the total pressures employed in the vacuum-treatment process (Fig. 118). In the degassing of semi-killed or rimming steels lower final hydrogen levels may usually be obtained by most processes than in the degassing of killed steels. The reason for this phenomenon is that the hydrogen partial pressure of semi-killed or rimming steels is lower at the same total pressure and that the carbon monoxide given off simultaneously by these steels acts as a scavenging gas.

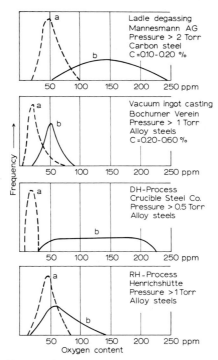

Fig. 119. Frequency distribution of the oxygen contents before and after a vacuum degassing. Comparison of various processes according to Edström[280]. (a) After degassing; (b) Before degassing.

As the comparison depicted in Fig. 119 demonstrates, the oxygen content of the steels will also be lowered to almost the same level by any of the vacuum processes in current use. According to the steel grade and the type of degassing used, reductions in the oxygen level of between 50 and 85% are obtained. The highest degrees of lowering of the oxygen content are observed on steels to which no deoxidation agents, such as aluminium or silicon, have been added before the vacuum treatment. In all vacuum-treatment processes, however, the lowering of the oxygen content depends greatly on the carbon level of the steels. Figure 120 shows the oxygen content of rimming steel melts before and after a vacuum treatment by various degassing processes in relation to the carbon content of the steel. As may be seen, the results after vacuum degassing—but before the addition of deoxidation agents—are grouped along the equilibrium curve, which corresponds to a carbon monoxide pressure of about 0.1 atm. They are therefore considerably higher than the equilibrium values corresponding to the pressure of 0.5–10 Torr applied during the vacuum treatment.

This phenomenon has frequently been discussed in the literature[62,304–313], and the deviation of the carbon monoxide reaction from the equilibrium line is

References pp. 507–515

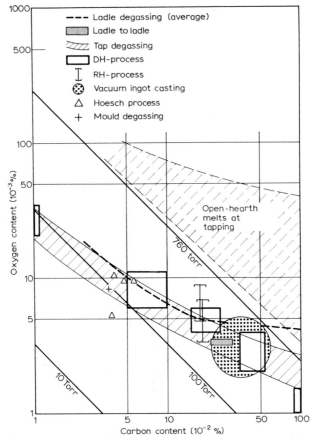

Fig. 120. Effect of the carbon content of steels on the ultimate oxygen contents obtainable by the various degassing processes.

attributed mainly to the following four causes:

(i) The carbon monoxide reaction is markedly slower than the elimination of the hydrogen from the molten steel, as the diffusion rate of the dissolved carbon and oxygen towards the surface of the steel melt or towards the surface of a gas bubble inside the melt is slower than that of the dissolved hydrogen. This explains why the elimination of the hydrogen proceeds almost to completion in the time available for the vacuum treatment, whereas the carbon monoxide is removed only partially.

(ii) The oxygen present in the steel is to a large extent combined with strong deoxidation agents and is therefore, in fact, present in the form of oxide inclusions. These oxide inclusions are only partly dissociated during the vacuum treatment.

(iii) The gas bubbles formed in the melt, and into which most of the excess in oxygen and carbon diffuses, are under a pressure which is higher than the pressure above the melt, as both surface tension and the hydrostatic pressure of the molten steel above the bubble create an additional pressure.

(iv) A constant supply in oxygen may be available during the vacuum treatment, for instance from the refractory lining of the ladle or the vacuum vessel or even from the slag in the ladle, which prevents the attainment of equilibrium.

These four factors assume different importance in the various degassing processes and it is therefore to be expected that the degree of deviation from equilibrium of the oxygen levels observed will be different for the different degassing processes. There are a number of papers in which attempts have been made to explain this discrepancy by theoretical considerations. For instance, the kinetics of degassing have been dealt with by Kraus[244], Levin[123,124] and Knüppel et al.[270]. Their studies show that for a rapid degassing reaction a high ratio of the free surface area to the volume of the steel is required. According to the particular degassing process employed this large specific surface area is produced either by a dispersion of the stream of steel into discrete droplets or by formation of bubbles in the melt. As the effect is, in principle, the same, almost identical results are obtained in these types of degassing process.

The lowering of the nitrogen level obtained is also almost identical in the various degassing processes for the same grade of steel. For plain-carbon steels, final contents of 25–30 ppm nitrogen are obtained from initial values of 30–50 ppm. The degree of reduction of the nitrogen level is higher for alloy steels because of the higher initial nitrogen contents.

1.7.2 Cleanness

The appraisal of the cleanness of steels is also carried out by a variety of test methods, but in general, one of two variants is usually employed. In the first method, a solid steel sample is dissolved by electrolysis and the insoluble impurities are filtered off from the electrolyte and their composition is determined by chemical and microscopical methods. In the other method a metallographic section is prepared from the steel sample, and the size and shape of the inclusions are determined under the microscope at magnifications of × 100 or × 200 and the inclusions are counted according to an established process. The best known counting methods are those of the ASTM or Diergarten. More recently, automatic counting methods have also been employed, for instance, the QTM method.

An improvement in the cleanness of the steels is not obtained by many vacuum-degassing processes to the extent corresponding to the degree of lowering of the oxygen content. Large differences in the improvement in cleanness are found, as is demonstrated by Figs. 69 and 91, even where similar degassing processes are employed. From published results it may be concluded that the improvement in cleanness is the greater the less the steel has been deoxidized by deoxidation agents before the vacuum-degassing process[119,190,202,203,314–316]. The most likely reason for this behaviour is the fact that the deoxidation products are not always quantitatively dissolved during the vacuum treatment and that therefore oxide inclusions, particularly large-size inclusions, may remain in the steel after the vacuum treatment. If, on the other hand, the deoxidation treatment is carried out after the vacuum treatment, the formation of large-size oxide inclusions is suppressed, as the inclusions are usually formed only during the solidification stage, so that their size does not exceed 5 μm. From these facts the conclusions may be drawn that from the point of view of a high cleanness the vacuum-degassing processes that are particularly well suited are those that permit the steel to be treated in the rimming condition and the deoxidation agents, in particular aluminium and silicon, to be added only after completion of the degassing process.

1.7.3 Vacuum-metallurgical potentialities

The metallurgical potentialities of the individual vacuum-degassing processes should also be borne in mind when comparing the various degassing processes. Such potentialities are particularly realized when the degassing plant is provided with a device for the addition of alloying elements, which makes it possible for the alloying procedure to be carried out under a vacuum after completion of degassing. This practice has the principal advantage over the conventional method of alloy addition in the furnace or in the ladle that practically no burn-off of the alloying material occurs[190, 229, 317]. Not only is yield from alloying additions high, but at the same time a greater accuracy in the steel composition is achieved, superior to that after alloying addition in air (see also Fig. 63). By obtaining this high accuracy in the predetermined composition not only can rejects of whole charges be avoided in the steel works, but the extremely high requirements for correct composition in high-grade steels can consistently be met. This alloying technique also makes it possible to add the various alloying elements separately in a metallurgically logical sequence, so that the few remaining oxide-type impurities will be present in a form in which they will not impair the fabrication of the steels[217]. In order to be able to add rather large quantities of alloying elements in a vacuum, a homogeneous and thorough mixing of the melt must be guaranteed. The best results achieved so far have been obtained with the DH process in which the degassed or alloyed portion of the melt is returned batchwise to the bath. A high mixing impulse and rapid thorough mixing are therefore obtained, as may be seen from Fig. 121.

Another important metallurgical application of vacuum degassing is the possibility of lowering the carbon level in mild steels, if the melts can be treated in the rimming condition before deoxidation. By this practice carbon contents of less than 0.008% may be obtained even in large units. This lowering of the carbon contents is even possible for stainless steels, if the vacuum treatment is carried out at steel temperatures of between 1650° and 1800°C according to the required ultimate carbon level. Lowering of the carbon content from 0.15% to 0.01% is possible in many plants.

In summing up it may be said that a vacuum process that is capable of making

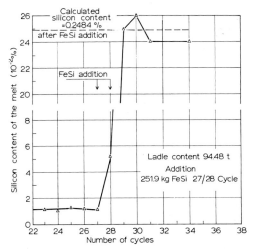

Fig. 121. Efficiency of the mixing of ferro-silicon to the ladle during operation of the degassing process. Duration of a lifting cycle, 20 sec.[190].

use of vacuum metallurgical possibilities to the fullest extent must provide the means for degassing steel in the rimming condition, must make possible the quantitative addition of the alloying agents under a vacuum and finally must guarantee a rapid and thorough mixing of the alloying elements with the melt.

1.7.4 Incorporation in the production process

It is imperative in the course of the rationalization of the steel works to choose steel degassing processes which fit in well with the whole steelmaking procedure[190,193,217,227]. The treatment times for vacuum degassing should be short, no material heat loss should occur during the vacuum treatment, the transport of the steel to and from the degassing plant should take up little time and must not interfere with the conventional materials flow of the steel works, the running costs should be low, the degassing plant must be available for use at any time and, finally, it should permit a high throughput. These stringent requirement have led to the consequence that only very few degassing processes have succeeded in large-scale steel production. On the other hand, these requirements have also caused the introduction of the various modifications of the individual processes at the various steelworks and the equipment with different additional auxiliary devices of these plants.

The incorporation of a degassing plant into an existing steel works requires an intimate knowledge of the production programme of the works, the existing space availability and the transport possibilities, the rating of the various steelmaking units and the established metallurgical practice. The incorporation and installation of a vacuum degassing plant is therefore always a case for close collaboration between manufacturer and buyer of the plant. The manufacturers of vacuum degassing plants can also always provide detailed information on the heat losses during treatment, on the expected increase or decrease in output, and on the operating costs for the vacuum treatment under the existing individual conditions.

These operational requirements restrict the choice of vacuum degassing plant and they are indeed mainly responsible for the fact that only about one-third of the 15 processes described in Subsections 1.3 to 1.6 have succeeded in industrial application on a broad basis.

The straight type of ladle degassing—widely used in the early stages of the introduction of the degassing of steel—is no longer in use for large units because of the high heat losses incurred and the poor degassing effect. The existing plants have all been subsequently equipped with a gas scavenging device or they have been so modified that the ladle-to-ladle degassing process or the vacuum ingot-casting process may be applied. Vacuum casting is still used to a large extent for the production of ingots for forging, whereas stream degassing from ladle-to-ladle is employed only in exceptional cases because of the high heat losses and the large requirement in man-power. Ladle degassing incorporating a stirring coil has not succeeded, compared with ladle degassing with a scavenging gas, although the two modifications of the ladle degassing process give almost identical results with respect to the removal of gases, the metallurgical possibilities and the heat losses, but the expenditure in installation and the operating costs of the former are much higher as a result of the installation of the stirring coil and the necessity for employing ladles fabricated from costly non-magnetic steel and lined with high-quality, expensive refractories. Ladle degassing with additional arc-heating is in operation only in a few high-grade steel works owing to the high capital expenditure involved.

Of the stream degassing processes, tap degassing has replaced the ladle-to-ladle degassing process, as tap degassing causes markedly lower heat losses and because it is also more flexible in application. The Therm-I-Vac process requires high capital expenditure and is costly to operate and is therefore, so far as is known, in operation in only one high-grade steel works for the production of special steels of extremely high purity.

Of the cycling and circulation degassing processes, the DH and the RH processes have succeeded. The Thermo-Flow process and the transfer degassing process are each employed in only one steel works. As far as is known neither process has yet been developed to full operational reliability.

The mould degassing processes, with the sole exception of the vacuum ingot casting process, have not succeeded in industrial application. The reason for this failure has been the fact that they put so great a load on the casting pits, which in many cases even under normal condition represent a bottleneck in the steel works, so that an installation of any of the mould degassing processes requires enlargement of the casting bay, the mould store and the labour force. As these additional capital-investments and the internal operating costs must be added to the cost of the mould degassing processes, they simply cannot compete with the other methods of degassing.

If we now consider those degassing processes which are in actual industrial operation at a large number of plants, say in more than ten or so, we find that only the group of cycling and circulation degassing processes exists in two modifications, *i.e.* as the DH process and as the RH process. All other degassing processes are represented by only one type, *viz.*: ladle degassing with a scavenging gas represents the ladle-degassing process, tap degassing the stream-degassing process, and vacuum ingot-casting the mould-degassing process. Vacuum ingot-casting is used in almost all steel works which produce large ingots for forging; on the other hand this process is restricted to this particular application. These plants are therefore also used for the degassing of other grades of steel by the ladle-degassing process with a scavenging gas. Ladle degassing with a scavenging gas is also used by those steel works which subject only part of their production to a vacuum treatment. The shortcomings of the ladle-degassing process—slagging of the melts, increased tapping temperatures, etc.—can be tolerated as the installation costs of the plant are low. Nevertheless, more recently a number of steel works have turned to tap degassing, provided that the furnaces are such as to make possible control of the stream of the tapped steel, as slagging of the melt is not necessary and as the heat losses are lower than in ladle degassing. The cycling and circulation processes are given preference in those steel works which subject a large proportion of their steel production to a vacuum treatment since these processes fit in well with the general flow scheme of the materials. About twice as many plants are in operation according to the DH process than according to the RH process. Mainly the large steel works and those which treat almost all their production by a vacuum process, prefer the DH process. Besides the longer life of the refractory lining, the superior alloying technique with the resulting improvements in steel quality is most probably responsible for the lead of the DH process over the other.

1.7.5 Effect of degassing processes on the properties of steels

The following considerations cannot claim to be exhaustive, as they are only mentioned to demonstrate the effect of vacuum treatment of the molten steel on its service properties, separately for the various types of steel but independent from the degassing process employed. A warning should be given first of all about the belief that simple application of a vacuum-metallurgical process subsequent to the steelmaking procedure will in all cases result in an improvement of the properties of the steel. It is in fact the case that faulty application of vacuum treatment may even have a detrimental effect[318]. Finally, there are also cases of a deceptive improvement of the properties as a result of a vacuum treatment. In such instances, it usually happens that a steel has been used as starting material

References pp. 507–515

for the vacuum treatment that shows poor properties as a result of faulty steelmaking technique compared with a correctly melted steel. In other words, improvements appear to have been obtained by the vacuum treatment that could in fact easily have been attained in the first place by careful steelmaking practice in air. Indeed, the usual great care that is commonly taken in the practice of making steel later to be subjected to a vacuum treatment may by itself lead to an improvement in quality.

In contrast with the vacuum melting of steel, a vacuum degassing of finished liquid steel does not in general have a noticeable effect on the technological properties, such as ultimate tensile strength, elongation and notch-impact toughness, as these are usually governed by chemical composition and structural characteristics. The latter depend—apart from the nature of the steel—mainly on solidification conditions and the subsequent working and heat-treatment practice. These are either not at all or only slightly changed in vacuum-treated steels as compared with steels of the same composition not subjected to a vacuum treatment. Despite this, a number of improvements are obtained by the vacuum treatment, which cannot be indicated by technological data alone. These improvements show up during the working and fabrication of the steels and in their behaviour during applications and use and may be summarized as follows:

Low oxygen and hydrogen contents.
Excellent and consistent cleanness from one heat to another with respect to oxide inclusions.
Consistency in composition from one heat to another.
High accuracy in the melting of steels of predetermined grain size.
Great uniformity in grain size within one heat.
Reduced fluctuations in transformation behaviour from one heat to another, in particular in plain-carbon steels.
No susceptibility to flake formation.
Uniform behaviour during working or shaping from one heat to another.
Reliable introduction of trace elements with a high affinity for oxygen into the melt.
Reliable adjustment of the casting temperature and consequently high surface quality of the ingots.

From this it follows that vacuum-treated steels are applied with advantage in cases where fabrication conditions require a high degree of uniformity and consistency within one heat as well as from one heat to another.

A vacuum treatment should therefore be applied particularly to large ingots for forging, to ball-bearing steels subjected to high loads, to low- and high-alloy high-grade steels as well as to high-quality steels with special properties.

(a) Steels for forging
The production of large forgings of alloy steels without the application of a

TABLE 27

DATA COMPARING THE NUMBER OF REJECTED FORGINGS THAT WERE PRODUCED FROM ELECTRIC-FURNACE OR ELECTRIC-FURNACE AND VACUUM-DEGASSED HEATS[322]

Valve block cross-section size (in.)	Air-melted material		Vacuum-degassed material	
	No. of valve blocks fabricated	No. of valve blocks rejected	No. of valve blocks fabricated	No. of valve blocks rejected
$17\frac{3}{4} \times 25$	24	6	20	0
$21 \times 22\frac{3}{4}$	25	4	20	0
$21 \times 23\frac{3}{4}$	17	2	17	0
$25 \times 26\frac{1}{2}$	17	6	18	0
Totals	83	18	75	0

vacuum treatment is extremely difficult and uneconomical particularly if sophisticated testing methods and high quality requirements are to be met. A vacuum treatment makes the production of the most exacting forging of high operational reliability possible[156, 221, 254, 269, 301, 319–321].

A typical example is shown in Table 27, in which the results of an evaluation of the rejects of 158 ingots for forging are listed. The composition of the ingots

TABLE 28

COMPARISON OF PRODUCTION COSTS OF A TURBINE SHAFT FROM AIR- AND VACUUM-CAST INGOTS

Cr–Mo–V–steel		Air-cast ingot with casting pit costs	Vacuum-cast ingot without casting pit costs
Production costs	S*/t	4000.50	4000.50
Casting pit costs	S/t	330.00	—
Costs for vacuum casting	S/t	—	300.00
Total	S/t	4330.50	4300.50
Cost of storage and sundries	S/t	202.00	199.00
Ingot production costs			
(As cast)	S/t	4532.50	4499.50
(Finished)	S/t	4847.90	4764.50
Forging and heat treatment costs of a turbine shaft of 18 500 kg service weight (starting with an ingot weight of 33 800 kg)			
Ingot as-cast	S	125 520.00	122 470.00
Fabrication costs in the forge	S	18 510.00	18 510.00
Isotherm annealing (S/t 450.00)[1]	S	9 090.00	—
Soft annealing (S/t 330.00)[2]	S	—	6 105.00
Total production costs	S	153 120.00	147 085.00
Saving obtained by use of a vacuum-cast ingot	S		6 035.00

[1] Costs for furnace plant for a production of 500 ton/month.
[2] At a furnace throughput of 280–300 kg/h.
* S = Austrian schillings

was: C 0.40, Mn 0.60, Cr 1, Si 0.2 and V 0.10% (AISI 4110). From a total of 83 ingots produced by conventional techniques 18 were rejected, whereas there was not one reject from the 75 ingots which were vacuum-treated by the DH process[322].

Equally good results have been reported from vacuum ingot-casting[151,153,182,199,etc.]. The technique of vacuum casting has been perfected to such an extent that the vacuum-cast ingots now show the same high surface quality as do the air-cast ingots. The highest quality of forging is obtained with arc-furnace steel as with this type of steel the lowest levels in phosphorus and sulphur are attainable. The mechanical properties of forgings made of vacuum-treated steel are relatively superior and more consistent than those of forgings cast from conventional steels. The quality of inductor shafts made of vacuum-treated steel was so excellent in ultrasonic testing, that axial trepanning could be omitted[323]. Furthermore, the long-term annealing treatment of the ingots for forging for the removal of the hydrogen can be omitted and the final annealing treatment of the finished forgings may substantially be shortened.

Better internal homogeneity of the ingots also in many cases makes it possible to simplify the forging procedure. The savings in production costs obtained as a result of the use of vacuum-treated steel are listed in Table 28. In this table the production costs are compared for turbine shafts of a chromium–molybdenum–vanadium steel, 18.5 tons in weight forged from air-cast and from vacuum-cast ingots[264].

(b) Low-alloy steels

Vacuum treatment has usually been particularly successful for low-alloy steels in production of ball-bearing steels[131,227,317,324–326]. The reduced quantity of inclusions and the consistency and uniformity of the technological properties have a beneficial effect on the service life of the ball bearings[132,187,199,225,327–330]. A frequency curve of the fatigue life of ball bearings manufactured from conventional and from vacuum-treated steel is shown in Fig. 122[225]. For the tests a steel containing C 1.0, Mn 0.4, Cr 1.05 and Si 0.3% was used and the finished ball bearings were loaded at a pressure of 100 kgf/mm^2. As may be seen from the diagram the service life of the ball bearings was increased three to five times as the result of the vacuum treatment of the steel.

These curves also quite clearly demonstrate that the manner of applying the vacuum treatment of the steel has an effect on the service life of the ball bearings. Curve b was obtained with steels to which an addition of the silicon was made before the vacuum treatment, whereas curve c depicts the fatigue life of ball bearings to which the addition of silicon and aluminium to the steel was made after the vacuum treatment.

Similar improvements in service life are reported for case-hardened ball-bearing steels of grade 8620 (composition: C 0.20, Mn 0.80, Cr 0.50, Si 0.30, Ni 0.55 and Mo 0.20%)[131,227,328]. A comparison of the service life between con-

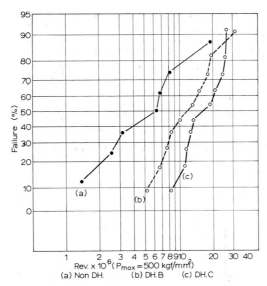

Fig. 122. Comparison of the fatigue life of vacuum-treated and non-treated ball-bearing steel[225]. (a) Not treated; (b) Semi-killed, vacuum-treated by the DH process; (c) Rimming, vacuum-treated by the DH process.

ventional and vacuum-treated steels is plotted in Fig. 123[131,328].

A marked improvement in the micro-cleanness of low- and medium-alloy case-hardening and heat-treatable steels may be obtained by vacuum treatment as is evident from Figs. 124 and 125[321,322,331,332]. Even steels produced by the duplex-slag process in the arc furnace share this improvement in cleanness as may be seen from Fig. 70. The cleanness of the steels may even be increased if they are melted

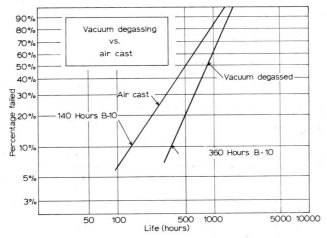

Fig. 123. Comparison of bearing fatigue life for carburized type 8620 steel produced by electric furnace practice and by the ladle degassing process with induction stirring[328].

References pr 507–515

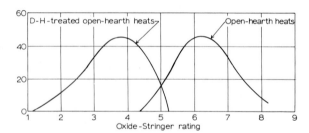

Fig. 124. Oxide-stringer ratings of carbon and alloy steels[322].

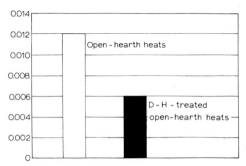

Fig. 125. Total weight of oxide-type inclusions in steels open-hearth melted with and without vacuum degassing[322].

TABLE 29

EFFECT OF VACUUM DEGASSING ON CLEANNESS RATING

AISI – 3312	Number of heats	Average rating*
Not degassed	21	25.6
Degassed – Practice No. 1	19	22.1
Degassed – Practice No. 2	22	9.3

* Atlas rating method – magnafluxed stepdown tests.

AISI – 8620	Number of heats	Average frequency	Average severity
AMS 2301B Limit	–	0.40	0.35
Not degassed	5	0.38	0.35
Degassed – Practice No. 1	8	0.36	0.35
Degassed – Practice No. 2	4	0.06	0.04

AISI – 4340	Number of heats	Average frequency	Average severity
AMS 2301B Limit	—	0.37	0.32
Not degassed	32	0.38	0.30
Degassed – Practice No. 1	21	0.20	0.14
Degassed – Practice No. 2	6	0.07	0.05

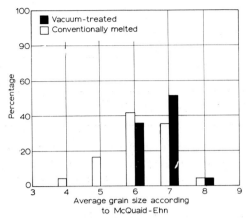

Fig. 126. Average grain-size distribution of steels with and without vacuum degassing. Steel grade 16 Mn Cr5—fine grain[282].

in the arc furnace with the one-slag practice and if the addition of the deoxidation agents—in this case of the highly active type—is made towards the end of the vacuum treatment. Further examples of the improvement in cleanness are compiled in Table 29[333].

Improved cleanness of a steel lowers the rate of rejects during fabrication and improves the service behaviour. Furthermore, the hydrogen content of the steel is reduced by the vacuum treatment to such levels that there is no longer any danger of flake formation and the ingots may safely be cooled in air[67, 224, 334]. Slow cooling rates are used for technological reasons for air-hardening steels only.

As the addition of the aluminium to the steel can be made in the theoretically correct quantity in the vacuum process an exact adjustment of the grain size and consequently a definite control of the hardenability of these steels is possible. Even with fine-grain steels the grain-size may be controlled within narrower limits in vacuum-treated heats than in conventional heats, as may be seen from Fig. 126. Moreover, as a result of the accurate adjustment of the boron level in boron-alloyed case-hardening steels in vacuum-treated heats an improvement in the case-hardenability is obtained. In many instances an improvement in tensile strength and notch-impact toughness transverse to the rolling direction has been reported[199, 224, 227, 236, 322].

Vacuum treatment also has a beneficial effect on the properties of high-silicon or vanadium-bearing spring steels, of alloy steels for tube production and of steels for plate and heavy sheet rolling. In these grades of steel this effect is particularly pronounced with respect to a reduction in rejects during working and fabrication.

(c) High-alloy steels

For high-alloy steels improvements were observed mainly in austenitic and ferritic stainless steels, as well as in high-nickel alloy steels.

References pp. 507–515

TABLE 30

EFFECT OF DH TREATMENT IN REDUCING EDGE CRACKING IN PLANETARY HOT ROLLED COILS[333]

Grade	Treatment	% Coils with "severe" cracks
Stainless 301	Not degassed	11.1
	Degassed	1.6

The improvements in ferritic stainless steels are mainly the results of the reduction of the hydrogen and the oxygen contents. The prevention of subsurface blow holes as well as the suppression of the tendency to "wildness" lead to a better surface quality[78,335].

The carbon level of austenitic stainless steels may further be lowered by a vacuum treatment, provided that this is carried out at higher steel temperatures. Carbon contents below 0.02% have been attained by this process. As a result of the improved cleanness of the vacuum-treated steels the hot workability of both strip and sheet products is raised, for instance a marked reduction of edge cracks in hot strip rolling has been reported[78]. As may be seen from Table 30, edge cracking could be reduced to one-seventh of the former level.

Titanium-stabilized stainless steels, to which the titanium addition was made under a vacuum, show a marked reduction of the so-called "titanium strakes". Also in titanium-free austenitic stainless steels the surface quality of cold-rolled strip was improved as a result of the improvement of the cleanness, for instance the reject rate could be reduced at one steel works from 9.5% to zero[78].

In some works high-nickel alloy steels, e.g. 18 Ni maraging steel, containing C 0.017, Si 0.15, Ni 18, Co 8, Mo 5 and Ti 0.5%, have been vacuum-treated, with excellent results. Tool steels of grade Cromo-V (H-11), containing C 0.41, Si 1, Cr 5, V 0.5 and Mo 1.26%, also showed an improvement after a vacuum treatment. Uniform, consistent and improved technological properties are mentioned as particular advantages[322].

(d) High-grade steels

In the group of high-quality steels a vacuum treatment is usually applied to plain-carbon heat-treatable and case-hardening steels, rail steels and transformer steels, as well as to high-strength structural steels with and without the addition of trace elements.

Consistency in composition and cleanness and, in particular, uniformity of the grain size are obtained from one heat to another by the vacuum treatment of plain-carbon heat-treatable and case-hardening steels. These improvements have a beneficial effect on working and fabrication. As a result of the lower oxygen contents of the vacuum-treated steel an increase in austenite grain size is obtained, as compared with conventional steel, as may be seen from Fig. 127. Increased

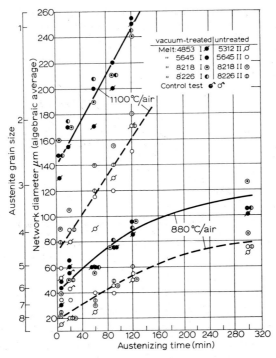

Fig. 127. Effect of a vacuum treatment on the grain size of the austenite of steel Ck 35 for various austenizing temperatures and holding times[336].

austenitic grain size enhances the hardenability of these steels, and by addition of aluminium, which acts as a grain-refining agent, the grain-size distribution may be closely controlled provided that a high enough level of nitrogen is present in the steel.

The degree of lowering of the number of oxide inclusions as a result of the vacuum treatment is depicted in Fig. 128. An extension of the times between regrinding of tool tips in machining operations is also found for vacuum-treated steels, as may be seen from Fig. 129.

Rail steels and wheel steels are both sensitive to flaking at elevated manganese and carbon contents, which are both necessary to meet the requirements for strength. By employing a vacuum treatment, *i.e.* by removing the hydrogen from the steel, it has been possible to suppress the sensitivity to flaking even at high carbon and manganese contents and to omit the annealing treatment or the slow cooling in cooling beds. Steels for high-strength railway rails are therefore almost all vacuum-treated[150,337]. Side by side with the lowering of the hydrogen content the cleanness of the steel is improved by the vacuum treatment. It has been reported that rail and wheel steels have been produced of a cleanness that compares with that of high-quality alloy steels[316].

References pp. 507–515

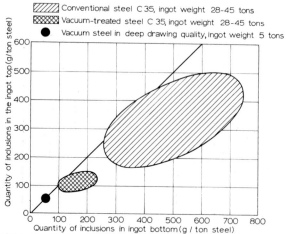

Fig. 128. Total weight of oxide-type inclusions in steel C 35 with and without vacuum degassing and a vacuum steel of deep-drawing quality according to ref. 222. (Scatter range of 68% of the test results.)

The users of high-strength structural steels often demand the adherence to the so-called "carbon equivalent" because of the required weldability of the steels. In the most simple case the carbon equivalent means that the sum of the carbon content and of one-sixth of the manganese content must not exceed a certain quantity, for instance 0.42. In the other hand, high carbon and manganese contents are necessary because of the required high strength of the steel. This means for the steelmaker the restriction to a very narrow tolerance in the composition of the steels, which is for a number of grades limited to 0.02% for the carbon and 0.10% for the manganese content. This limit in analysis is so close that it cannot even be met in open-hearth practice with reliability so that a large portion of the heats had to be relegated to lower grades. A vacuum process which makes possible an accurate adjustment of the steel analysis may in these cases eliminate all difficulties.

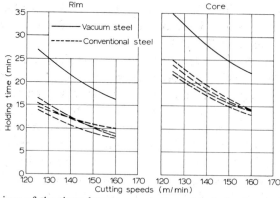

Fig. 129. Comparison of the times between regrinding of tool tips in machining of vacuum-treated and non-treated steel Ck 45[203].

There is also a growing tendency to obtain an increase in the strength of high-strength structural steels not by raising the carbon and manganese levels, but by addition of strength-raising elements, such as niobium, vanadium and titanium[338]. Additions of titanium are particularly important for heavy-plate rolling, when high technological values are required transverse to the rolling direction. Figure 130 demonstrates that a titanium recovery of 80–90% at a very low scatter of the analytical results may be obtained, if the titanium is added to the steel in a vacuum. This higher yield in the titanium addition in itself may save so much of the cost of the added element that the additional costs for the vacuum treatment are met. A similar saving in costs may also be secured for alloying with vanadium in a vacuum. Apart from these improvements, vacuum treatment of the structural steels brings about an improved cleanness[339], which shows up in a reduction of the anisotropy of the technological properties and in a sharpening of the ultrasound pattern.

The high degree of cleanness of vacuum-treated steels also offers advantages in the production and fabrication of seamless tubes and welded pipes as well in the production of high-strength steel wire[219,337]. An improved cleanness and a better surface quality are even found in semi-killed steels, because the degree of partial killing can be met more accurately and consistently[340]. Aluminium-killed deep-drawing steel, as used for the manufacture of non-ageing thin sheet, is also vacuum-treated with advantage[185].

Vacuum-decarburized steel offers interesting possibilities from both the metallurgical and the technological aspects as it solidifies in the killed condition without addition of deoxidation agents. Its technological behaviour resembles that of a rimming steel and its excellent surface quality and cleanness make possible its use for thin sheet with the highest requirements for surface properties, *i.e.* for use in one-coat white enamelling. The possibility of enhancing the technological properties of a vacuum-decarburized steel by addition of trace elements opens

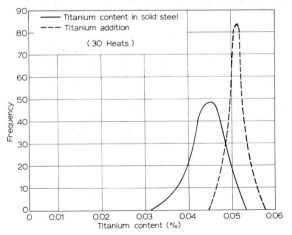

Fig. 130. Titanium recovery in titanium-stabilized, DH vacuum-treated structural steels.

References pp. 507–515

up new ways for the production of special-purpose steel. In this connection the strength-enhancing effects of niobium and vanadium may be mentioned, which allow the production of high-strength steels having particularly good welding and ductility behaviour. By addition of small amounts of boron and titanium non-ageing steels are obtained; at the same time the lower yield point and the deep-drawing properties are improved. A lowering of the carbon content of aluminium-killed steels gives a particularly soft and ductile sheet of low proof stress without the formation of "Lüders" lines. Vacuum-decarburized steel may therefore be used for purposes which up to now have been the domain of the substantially more expensive non-ferrous metal alloys.

Low carbon levels coupled with high cleanness are most important for obtaining low power losses in transformer sheet. Samarin as well as Oiks and his co-workers[91, 341-343] have reported on the beneficial effect of a vacuum treatment on transformer steel. In their experiments the yield in top-class sheets, that previously never exceeded 55%, could be raised to more than 90% on the annual average[91]. Analyses and magnetic data for transformer steel, which had been decarburized during the vacuum treatment and alloyed with silicon, also under a vacuum, are listed in Table 31[223]. This practice is therefore employed by many manufacturers of transformer sheet for the production of high-grade transformer steels.

Vacuum-treated steel may also be used with success in the production of castings[86, 334, 344, 345]. Of the defects occurring in steel castings, gas bubbles deserve special attention as they are very common. This type of bubble is caused by the increased gas content of the steel as a result of the usually very high casting temperatures. So far, however, the vacuum treatment of cast steel has been applied to only a small extent.

After discussing these examples from practice, this brief survey of the properties of vacuum-treated steels may be concluded. It must, of course, be expected

TABLE 31

ANALYSIS AND PROPERTIES OF AN ELECTRIC STEEL SHEET MADE FROM VACUUM TREATED STEEL[223]

Analysis (%)	C	O	Si	Al
Before vacuum treatment	0.04	0.11	—	—
Vacuum treatment				
Before Al addition	0.011	0.07	—	—
After Al and Si additions	0.011	0.005	1.5	0.11
Hot strip	0.011	0.004	1.5	0.11
Cold strip (annealed)	0.004	0.004	1.5	0.11
Wattage loss/kg at 10 000 G (Eppstein test) 2.06–2.16				
at 15 000 G 4.60–4.75				
Magnetic reduction, G				
B5 B10	B25	B50		B100
13 800 15 200	16 300	17 300		18 400

that other important fields of application of the vacuum treatment of steel will emerge in the future. The crucial point of all cases, in which the application of vacuum-treated steels is under consideration, is, of course, that of economics. The evidence for the attainment of improved properties without taking into account the cost of the process is usually not sufficient to justify an application of the vacuum-degassing process, disregarding of course such special fields as aircraft construction and the space industry in which increased safety considerations are overriding and in which the materials are stressed to the extreme. The point in all other practical cases is whether the additional expenditure caused by the application of the vacuum treatment is recompensed by a higher output, a lowering of the reject rate or by other savings in the fabrication of the steels and by an increase in the service life of the finished parts. In so far as the vacuum-treated steels show these properties, they have succeeded and have gained a firm footing in many branches of engineering and in new spheres of application which are continuously opened up.

2. Vacuum Degassing in the Solid State

2.1 Introduction

Gases may be present in metals in various forms. The originally gaseous molecules can be dissociated and dissolved in the metal lattice as atoms or ions. Owing to their small size they usually act as interstitials and it can be assumed that accumulation of the dissolved gases will occur at dislocations and small subgrain boundaries.

In contrast with this behaviour, compounds of metals and gases form a separate phase at the surface of the solid body or in its interior (inclusions). These compounds are in many cases insulators or semi-conductors, but some suboxides or subnitrides having a metallic character are known. Metal–gas compounds are formed in the interior of the solid metallic body, for example, from supersaturated solutions during cooling.

Where no stable or metastable metal–gas compounds exist, gas in the molecular form separates from the supersaturated solution during cooling. This separation takes place on external and internal surfaces, such as shrinkage cavities, pores, microcracks, etc., sometimes building up an internal pressure of many atmospheres.

Gas is also adsorbed by the external and internal faces of the metallic body— physically, under the application of Van der Waals' forces, or chemically[*]. This surface adsorption layer even continues in an ultra-high vacuum. It is absent only

---

[*] Rough models are used here to differentiate between physical adsorption and chemisorption. In reality, however, a whole range of different binding states occur at the surface of the solid metallic body.

References pp. 507–515

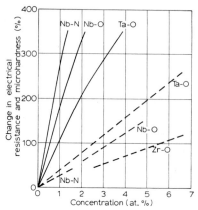

Fig. 131. Effect of dissolved gases on the mechanical and electrical properties of recrystallized, high-melting-point metals[346–349].
———— Microhardness.   - - - - - Specific electrical resistance.

at the surface of fresh fractures which occur in the interior of the metal.

The properties of a metal are influenced by the gases in it, to a greater or lesser extent, depending on the content and state of these gases. Gases dissolved in solid metal change the electrical resistance, thermal conductivity, e.m.f., magnetic susceptibility, hardness, yield point, ultimate tensile strength, elongation and reduction in area, the internal friction (damping), and many other properties. To emphasize the importance of this point, it may be mentioned that as little as 1 at.% nitrogen, *i.e.* only 0.15 wt.% or 1500 ppm, dissolved in niobium, changes its hardness by about 300% and its electrical resistance by about 30% (see Fig. 131). Furthermore, the notch impact strength of titanium is reduced to one-tenth of that of the pure metal by as little as 2 at.% hydrogen, *i.e.* only 0.085 wt.% or

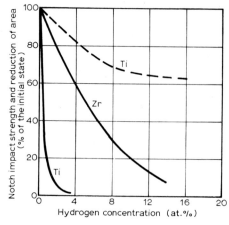

Fig. 132. Effect of dissolved gases on the mechanical properties of quenched metals[350, 351].
———— Notch impact strength.   - - - - - Reduction in area.

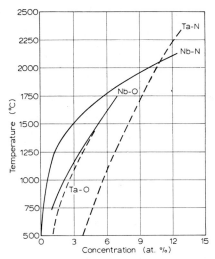

Fig. 133. Saturation concentration of gases in high-melting-point metals[346, 347, 352, 353].

850 ppm (Fig. 132). In fact, the solubility of gases in metals is, to some extent, quite considerable (Fig. 133), and in some instances even very high, as in α-titanium (7–8 at.% hydrogen), β-titanium (about 50 at.% hydrogen), α-zirconium (6 at.% hydrogen, 33 at.% oxygen), β-zirconium (50 at.% hydrogen) to name just a few. From these examples both the technological potentialities as well as the damage which could result from insufficient consideration of these effects may well be imagined.

Metal–gas compounds also have a marked effect on the properties of metallic bodies. Stable compounds on the surface can form a highly corrosion-resistant layer or a protective layer against scaling (chemical properties). On the other hand, as is evident from Fig. 134, they are able to change the optical properties of the metal entirely. Stringer-like inclusions (gas–metal compounds) within the metal adversely affect the fatigue strength and its machining properties, etc.

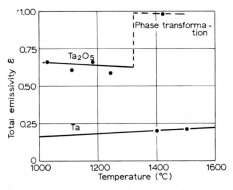

Fig. 134. Relationship between the total emissivity of tantalum and tantalum oxide and the temperature[346, 354].

References pp. 507–515

Gases in molecular form trapped in cavities, such as shrink holes, cracks, etc., may give rise to crack formation during cooling (flakes in steel containing hydrogen) or to long-term failure. Finally, a catalytic acceleration of chemical reaction can be set up by adsorbed gases, and they are even able to cause a change in the electrical resistance of thin metal films[355].

These examples of the diverse effects of gases on the properties of solid metals emphasize the significance of the type of gases and their state in metals and the importance of controlling the gas content. Besides the degassing reactions of liquid metals which are frequently employed, degassing reactions in the solid state are also important for attaining favourable metal–gas equilibrium.

These degassing processes will now be considered, but the question of selecting the temperatures and pressures necessary for obtaining the required final state must first be studied (Subsections 2.2 and 2.3). Kinetics are discussed in Subsection 2.4. In accordance with the aims of this book, only those degassing processes are dealt with which are carried out in a vacuum, or in which the application of a vacuum offers considerable advantages compared with a protective atmosphere.

For reasons of economy, degassing in the solid state is nearly always carried out in conjunction with other production or manufacturing processes. This is particularly true in such processes as sintering, annealing and in chemical reactions. There are, however, special cases in which degassing in the solid state is carried out as a separate process. For instance hydrogen, picked up when pickling titanium, can be extracted under high vacuum at 550–800°C. Another example is the annealing of metal powder—tantalum, niobium, etc.—in a high vacuum before sintering, in order to remove at least part of the adsorbed gases. After the vacuum-annealing treatment the powders are more ductile and easier to compact. The degassing of structural parts for vacuum plants, in particular those for ultra-high vacuum plants, of radio valves, etc., is a process that is often employed and which has been comprehensively reviewed elsewhere[356].

Since, however, degassing in the solid state is usually conducted during other processes, and because, except in the case of titanium, no special plants are needed for the degassing process, only the controlling equilibria and degassing processes which take place during these combined treatments are discussed. The commercial plants which are used for these processes are described elsewhere in this book.

2.2 Thermodynamic equilibria

2.2.1 Introductory remarks

The pressure, temperature and duration which have to be chosen for carrying out a degassing reaction must be known. Pressure and temperature should be based on the final state of the reaction being aimed for, taking into account the controlling thermodynamic equilibria and stationary states. On the other hand, the necessary process duration, which depends on the temperature, is a question of kinetics.

To illustrate the thermodynamic equilibria of metal–gas systems the conventional phase diagrams used in metallurgy are well suited. If Gibbs' phase rule is applied to metal–gas systems it should be borne in mind that, in this case, in contrast with the pure solid-state and fusion reactions, the gas phase and consequently the pressure play a decisive role. Gibbs' phase rule contains therefore an additional degree of freedom, f, and may be written as

$$P + f = c + 2 \qquad (1)$$

where P is the number of phases and c the number of components.

In consequence of this extension a binary system consisting of a pure metal and a gas ($c = 2$) has to be represented three-dimensionally or, if two-dimensional illustration is employed, the use of isothermal, isobaric and isosteric sections is mandatory.

2.2.2 Binary systems

The equilibrium conditions important for solid-state degassing processes are depicted in Figs. 135–139. They refer to the frequent case in which the gaseous component vaporizes in the molecular form and not as a compound with the metallic component. Simultaneously, metal atoms enter the gas phase; in most cases, however, in far smaller quantities. This frequently occurring type of degassing, in which metal–gas compounds are insignificant with regard to the degassing process, has been designated "type A". It is first dealt with generally, then illustrated by some examples.

According to the isothermal section in Fig. 135, a reduction in pressure on reaching the ternary-phase equilibrium E′–G′–H′ leads to decomposition of com-

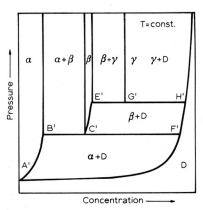

Fig. 135. Isothermal section through the three-dimensional equilibrium diagram of a metal–gas system of degassing type A with intermediate compounds β and γ.

References pp. 507–515

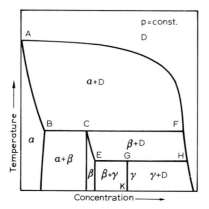

Fig. 136. Isobaric section for a metal–gas system as shown in Fig. 135.

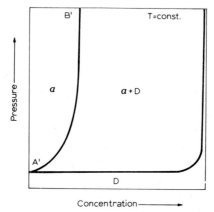

Fig. 137. Isothermal section through a three-dimensional equilibrium diagram of a system of degassing type A not forming intermediate compounds.

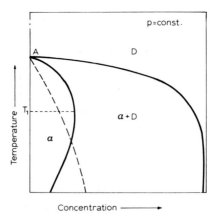

Fig. 138. Isobaric section for a system as shown in Fig. 137.

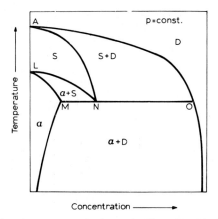

Fig. 139. Isobaric section for a system as shown in Fig. 137 with the presence of melt S.

pound γ, while at B′–C′–F′ to a decomposition of compound β. Afterwards the solid metallic body consists only of the α-solid solution phase together with the saturation concentration $C_{\alpha(max.)}$ (Point B′) characteristic of the momentary temperature T.

If the pressure* is lowered any further, the amount of gas in solution diminishes along the line B′–A′. In systems where no compound is formed (Figs. 137–139) there are, of course, no decomposition reactions, and the degassing of the homogeneous metal takes place in exactly the same way, along the curve B′–A′ (Fig. 137).

While a reduction in pressure always leads to some degassing, and an increase in pressure always to gas adsorption, the effect of a change in temperature at constant pressure is not obvious. Depending on the system, temperature range, etc., the α-solid solution may pick up gas when the temperature is raised (Fig. 136 below B–C–F, Fig. 138 below T_1, Fig. 139 below M–N–O) or it may release gas (Fig. 136 along B–A, Fig. 138 broken line and above T_1, Fig. 139 line M–L). In systems according to Fig. 136, it should be noted that below B–C–F the α-solid solution will pick up gas in the manner described above if the temperature is raised. The whole sample, however, in equilibrium with gas phase D, continually releases gas along the line K–G–E–C–B. Phase transformations of the basic metal in the solid or liquid phase (Fig. 139) make the picture appear even more complex.

The systems Ti–H (Fig. 140), Zr–H (Fig. 141), Fe–H (Figs. 142 and 143), Nb–N (Figs. 144 and 145) and Ta–N (Fig. 146) have been selected to demonstrate this complexity. Other systems are described in the relevant literature (refs. 359–361). During degassing in the solid state conditions simplify very much since the

---

* The partial pressures of the gaseous and metallic components are of primary importance, while the partial pressures of inert substances, *i.e.* inert gases, not involved in the equilibrium, as well as the total system pressure, are of secondary importance.

References pp. 507–515

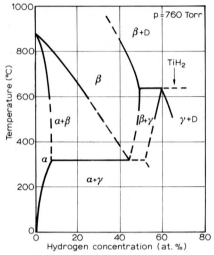

Fig. 140. Isobaric section through the constitution diagram for titanium–hydrogen[357].

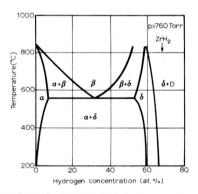

Fig. 141. Isobaric section through the constitution diagram for zirconium–hydrogen[357].

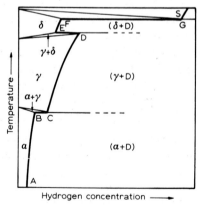

Fig. 142. The iron–hydrogen system, isobaric section through the constitution diagram (schematic).

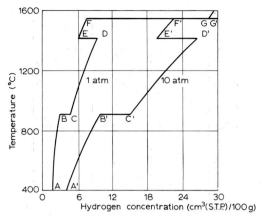

Fig. 143. Experimentally determined solubility (curve A–B–C–D–E–F–G– of Fig. 142) for isobaric sections at 1 and 10 atm[358].

relationship between the metal's gas content and pressure is, for all binary systems of degassing type A, as simple as that shown in Figs. 135 and 137. In the following these relationships will be discussed for the Nb–N system as this is a simple and particularly well studied example[352] (Figs. 144 and 145). Similar results were also obtained for Ta–N (Fig. 146).

The lines A′–B′ are temperature-dependent straight lines on log–log co-ordinates (Fig. 145). The intersection points of the horizontal lines with the two inclined straight lines represent, on the left-hand side, corresponding to point B′, the saturation concentration of the gas in the metal at the momentary temperature

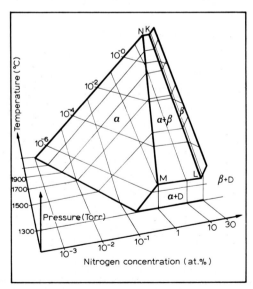

Fig. 144. Three-dimensional illustration of the niobium–nitrogen[352] system.

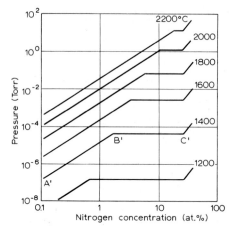

Fig. 145. Isothermal section through the niobium–nitrogen system in the range of the solid solution and subnitrides[352] (log–log presentation).

and on the right-hand side, corresponding to point C′, the composition of the subnitride β at the decomposition point. Figure 144 shows these relationships for the Nb–N system in the form of a three-dimensional illustration. On increasing the temperature and/or reducing the pressure the subnitride Nb–N decomposes along the line K–L (univariant equilibrium) into the saturated solid solution α and the molecule N_2. As soon as the solid body consists only of the α-phase, its gas content (maximum solubility) is completely governed by the temperature and pressure (bivariant equilibrium), corresponding to the area designated α in Fig. 144.

This area, which is important for the degassing of the α-solid solution, may

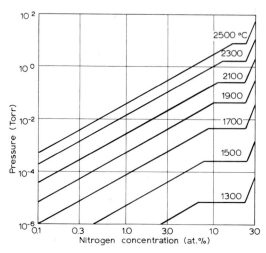

Fig. 146. Isothermal section through the tantalum–nitrogen system in the range of the solid solution and subnitrides[353] (log–log presentation).

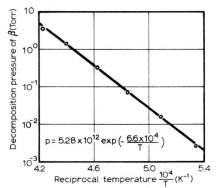

Fig. 147. Effect of temperature on the decomposition pressure of Nb–N[352]. (Projection of the straight lines B–C–F and B'–C'–F' from Figs. 135 and 136 on the p–T plane.)

be expressed for type A as:

$$C_\alpha = p^x a \exp\left[-\frac{\Delta H_1^0}{RT}\right]. \tag{2}$$

If the coefficient x assumes the value $\frac{1}{2}$, which, as an approximation, it often does for di-atomic gases, an ideal solution of the gas in the metal lattice exists and at constant temperature, the well-known $\sqrt{p}$ law evolves. In the p–T plane the α-field merges into the vapour pressure curve of the pure metal (points A and A' in Figs. 135–139). On the other side this field is bounded by the line M–N (Fig. 144) which indicates the temperature- and pressure-dependent saturation concentration (univariant equilibria). It is expressed approximately by the following equation:

$$C_{\alpha\,(\text{max.})} = b \exp\left[-\frac{\Delta H_2^0}{RT}\right]. \tag{3}$$

Finally, the $\alpha+\beta$ area which is defined by the lines M–N and K–L in Fig. 144 always runs parallel to the concentration co-ordinate. When projected to the T–p plane, a curve results which represents the decomposition pressure of the nitride β as a function of the temperature (Fig. 147). This relationship may be expressed by:

$$p_\beta = d \exp\left[-\frac{\Delta H_3^0}{RT}\right], \tag{4}$$

where ΔH_3^0 is the molar enthalpy for the formation of the compound, consisting of saturated α-solid solution and the molecule from the gas phase—thus in the Nb–N case, for the reaction:

$$4\,[\text{Nb}_{\text{sat.}}] + (\text{N}_2) \rightarrow 2 <\text{Nb}_2\text{N}> . \tag{5}$$

Data on the decomposition pressure and the free enthalpy H_3^0 may be found

References pp. 507–515

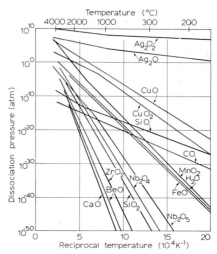

Fig. 148. Decomposition pressure of oxides calculated according to values from ref. 363.

in various publications (see, for instance, refs. 354, 360, 362–364). Figs. 148 and 149 show a number of such values for the decomposition pressure of oxides and nitrides taken from data by Elliott and Gleiser[363].

On the whole it may be said that many hydrogen- and nitrogen-containing systems, and a few oxygen-containing systems, are susceptible to degassing in a vacuum. The decomposition pressures of most oxides are so low that degassing by a change in pressure and temperature alone is no longer possible.

In many metal-gas systems of type A, compounds occur of the kind indicated in Figs. 135 and 136, for instance Nb–N, Ta–N, Fe–N, Fe–O, Nb–H etc. In these cases, the decomposition pressure and the saturation concentration being known, the α-field—the equilibrium field between saturated α-solid solutions and the gas phase—can be constructed as a first approximation. For this construction, coefficient x is assumed to be $\frac{1}{2}$, and the governing relationships between eqns. (2), (3) and (4) taken into account:

$$a = \frac{b}{d^x}; \quad \Delta H_1^0 = \Delta H_2^0 - x\Delta H_3^0 \tag{6}$$

Isothermal sections can also be constructed from various paired values p_β, $C_{\alpha(max)}$.

From these α-fields—whether they be determined by experiment or approximate calculation—it is possible to derive the paired values of pressure and temperature, which should be applied in the degassing process to obtain the desired final state. On the other hand, when this is known, the equilibrium gas content can be derived which will probably exist after sufficient time under the given pressure and temperature conditions prevailing during sintering, annealing or other manufacturing processes.

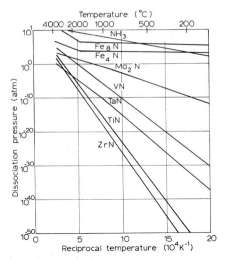

Fig. 149. Decomposition pressure of nitrides calculated according to values from ref. 363.

Such considerations and deductions, however, only lead to reasonable approximations if the system being considered really is of the simple degassing type A. This method is impracticable for some oxygen-containing systems, as, for instance, the systems Ta–O, Nb–O, V–O, Mo–O and W–O, of which Ta–O and Nb–O have been studied in great detail. From Fig. 148 at 1700°C a decomposition pressure of less than 10^{-8} Torr for the oxide Nb_2O_5 can be taken. Using a saturation concentration value $C\ \alpha(max.)$ of 7 at.% extrapolated from Fig. 133, the maximum solubility in equilibrium is obtained, being 0.7 at.% at 10^{-10} Torr, and 0.07 at.% at 10^{-12} Torr. These equilibrium pressures are still lower if, as shown in Figs. 135 and 136, a metal-rich suboxide takes part in the equilibrium instead of the oxygen-rich oxide Nb_2O_5. These considerations show that removal of oxygen from niobium below a residual gas level of 1 at.% is not possible at 1700°C and pressures above 10^{-9} Torr.

Nevertheless, Gebhardt, Fromm and Jakob[365] attained an effective reduction in the oxygen content to levels below 1 at.% after a degassing time of 200 min at a pressure of 10^{-5} Torr. Strangely enough, even in a closed vessel, no noticeable increase in pressure was observed. The explanation for this behaviour is that Nb–O does not represent a simple type A system, but is a complex one having a phase diagram as depicted in Fig. 150*. Such behaviour, which is also shown by Ta–O (Fig. 151), will be referred to as degassing type B in the following paragraphs.

The vital difference between types A (Fig. 135) and B (Fig. 150) is revealed

---

* To date it is not known with certainty whether Ta or Nb is in equilibrium with TaO or NbO (Figs. 150a, 151a) or whether Ta or Nb is in equilibrium with Ta_2O_5 or Nb_2O_5 (Figs. 150b, 151b).

References pp. 507–515

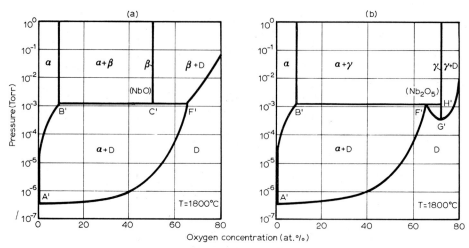

Fig. 150. Isothermal section through the Nb–O system at 1800°C[366].
(a) NbO in equilibrium with the solid solution.
(b) Nb_2O_5 in equilibrium with the solid solution.

by the shape of curve A'-F', and by point F' itself, which indicates the composition of the gas phase in equilibrium with the saturated solid solution.

For type A (Fig. 135) this curve originates at A' and runs to a point F', which lies at almost 100% gas content.

For type B, on the other hand, curve A'–F' in Figs. 150 and 151 clearly indicates that the gas phase consists of both the metallic element and the gaseous component.

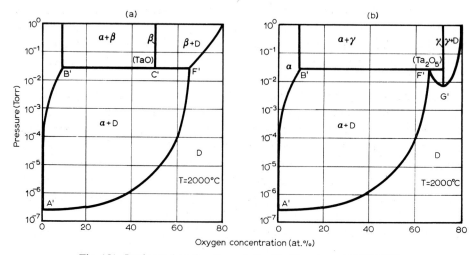

Fig. 151. Isothermal section through the Ta–O system at 2000°C[366].
(a) TaO in equilibrium with the solid solution.
(b) Ta_2O_5 in equilibrium with the solid solution.

During degassing, therefore, it is not the gaseous component which will escape from the metal, but a compound or a mixture of compounds whose composition or ratio of components not only depends on the temperature but also on the momentary gas content of the metal. Hence a mixture of NbO and NbO_2 evaporates from niobium and TaO and TaO_2[366] from tantalum.

These temperature- and pressure-dependent component ratios and their experimental determination have been carefully studied[366,367]. The α-equilibrium field for maximum solubility no longer obeys eqn. (2), but must be expressed by the much more complex equation[366]:

$$p = c \cdot T^{\frac{1}{2}} \cdot g \exp\left[-\frac{\Delta H_4^0}{RT}\right] \cdot \left\{1 + c \cdot h \cdot \exp\left[-\frac{\Delta H_5^0 - \Delta H_4^0}{RT}\right]\right\} + (1-c)i \exp\left[-\frac{\Delta H_6^0}{RT}\right] \quad (7)$$

where g, h and i are constants, ΔH_4^0 is the molar enthalpy for the formation of the gaseous suboxides TaO or NbO, ΔH_5^0 is the enthalpy for the oxides TaO_2 and NbO_2 and ΔH_6^0 the atomic enthalpy for the evaporation of the metals tantalum or niobium. Equation (7) determined for Nb–O and Ta–O applies generally for the case in which the gas phase consists of a mixture of two compounds in an atomic ratio of 1:1 and 2:1 and the metal.

Type A systems can contain intermediate compounds (Figs. 135, 136), but on a lowering of the pressure these will always decompose before degassing of the homogeneous solid solution begins (Fig. 135)*. The same may also occur with type B systems (Figs. 150a and 151a). It is, however, quite possible for highly stable compounds to exist down to very low pressures and then to evaporate without

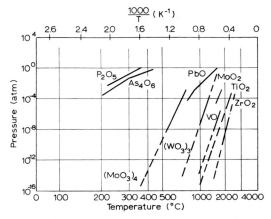

Fig. 152. Vapour pressures of various oxides[363].

---

* This only applies if the decomposition is not inhibited.

decomposition, *i.e.* congruent evaporation (Figs. 150b and 151b). In such cases, therefore, the vapour pressure of the compound (Fig. 152) should be taken into account during the degassing process.

When the vacuum plant is being laid out and the degassing reaction conducted, it is important to remember that in type B systems the gas phase consists substantially of compounds or elements which normally being solid at room temperature will therefore condense on the cold surfaces of vacuum vessels or in cooling traps. As mentioned previously when discussing the degassing of oxygen-containing tantalum and niobium, no noticeable pressure increase is observed during degassing. In these cases no true equilibrium conditions are attained during the degassing process; instead, only the stationary states described in Subsection 2.3 set in.

Finally, it should be pointed out that homogeneous alloys consisting of several metallic components behave identically with or similarly to pure metals during the degassing process. These may therefore be regarded as a quasi-binary system if in contact with a single gaseous component.

2.2.3 Multi-component systems having at least two non-metallic (gaseous) components

The binary systems described here are of great value for the understanding and theoretical treatment of this subject. They rarely occur in practice, however. On the contrary systems consisting of two or more non-metallic components, for instance, those present in the residual-gas atmosphere of an evacuated vessel or in conventional technological protective gases are most often met. It is not the total number of components present that is decisive, however, but only the number of components actually taking part in the degassing reaction.

These conditions are least complex in multi-component systems in which the non-metallic components do not form compounds among themselves. The degassing reactions therefore follow a pattern similar to that with corresponding

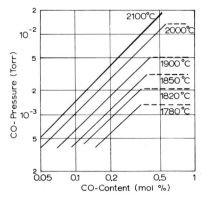

Fig. 153. Isothermal section in the system Ta–C–O for a C:O ratio 1:1[368] (log–log presentation).

binary systems. This is the case, for instance, when degassing tantalum and niobium which contain both nitrogen and oxygen. As is known from binary systems, nitrogen also escapes here in molecular form during the degassing process, whereas oxygen forms the suboxides TaO and TaO_2, or NbO and NbO_2, which evaporate from the surface of the solid body. The interaction of dissolved nitrogen and dissolved oxygen results only in a displacement of the values of maximum solubility and saturation concentration at given partial pressures.

An entirely different situation is found, however, if chemical reactions take place between the non-metallic components. This will be demonstrated by the well studied systems Ta–C–O[368] and Nb–C–O[369].

The relationship between the sample's CO content and the CO partial pressure at various temperatures is shown for tantalum in Fig. 153. Similar to Figs. 145 and 146, the steep straight lines on the left represent the homogeneous solid solution in which carbon and oxygen are atomically dissolved in equal proportions. The flat end of the curve represents the three-phase region (solid solution–oxycarbide–gas phase). While these parts of the curves in the binary system (Figs. 145 and 146) are strictly parallel to the abscissa for the heterogeneous range, *i.e.* the mixture ratio of solid solution and compounds is independent of the pressure, this does not apply to the ternary system. The gradient may vary between zero and the slope of the first part*, which represents the solid solution. Here, too, the kinks indicate the saturation concentration of the solid solution.

The relationship between pressure and temperature at constant gas content of the solid metallic body is shown in Fig. 154 for the same system. The less-steep lines correspond to homogeneous solid solutions of various C+O contents,

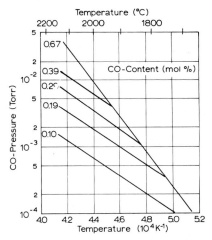

Fig. 154. Isosterics in the system Ta–C–O for a C:O ratio of 1:1[368].

---

* In the figure a slope of zero was plotted for the system Ta–C–O in accordance with experimental results.

References pp. 507–515

whereas the steeper straight line represents the ternary phase range (solid solution + oxycarbide + gas phase).

The two-phase equilibrium between solid solution and gas phase according to the reaction:

$$[O]_\alpha + [C]_\alpha \rightleftharpoons CO \tag{8}$$

may be expressed by the equation:

$$P_{CO} = a_C \cdot a_O \cdot b \cdot \exp\left[\frac{-\Delta H^0}{RT}\right], \tag{9}$$

where a_C is the activity of the dissolved carbon, a_O that of the dissolved oxygen, b is a constant and ΔH^0 is the molar enthalpy for CO.

By comparing Figs. 151 and 153 the degassing influence of carbon becomes apparent. If, at the same time, a large amount of dissolved carbon is present, the CO partial pressure above the solid solution is several powers of ten greater than the equilibrium pressure of the suboxide in the binary system Ta–O.

The reducing and degassing effect of carbon and hydrogen or their compounds is of extreme importance for oxygen systems whose equilibrium pressures are frequently far below the pressure range attainable under technological conditions.

From the kinks in the curves in Figs. 153 and 154, the saturation concentration for a carbon : oxygen ratio of 1 : 1 is found to be:

$$C_{max} = d \exp\left[-\frac{\Delta H^0}{RT}\right]. \tag{10}$$

This relationship is plotted in Fig. 155 for the system Ta–C–O. The experimental results show heavy scattering and data obtained from two different test methods

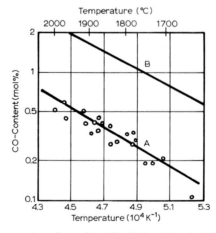

Fig. 155. Saturation concentration C_{max} of the Ta–C–O solid solution in equilibrium with CO. Curve A according to ref. 368. Curve B according to ref. 370.

deviate in the ratio of 1:3. This high inaccuracy is caused by the great experimental difficulties attached to such studies, and comprehensive, exact investigation of these multi-component systems is still only beginning.

For these reasons a rough approximation had to suffice in most cases. In such estimates mutual solubility was neglected, and it was mainly the equilibrium pressure of oxides, hydrides, etc., that was considered. Thus for the reduction of the iron oxide, $Fe_{0.95}O$, which has the lowest oxygen content according to the reaction:

$$Fe_{0.95}O + CO \rightleftharpoons 0.95\ Fe + CO_2 \tag{11}$$

the following approximate relation is obtained:

$$\frac{P_{CO_2}}{P_{CO}} = K_p. \tag{12}$$

From tabulated thermodynamic data[360] for $Fe_{0.95}O$, CO and CO_2 and using the equation:

$$\Delta G° = -RT \ln K_p, \tag{13}$$

the following result is obtained for the case in question:

$$\log \frac{P_{CO_2}}{P_{CO}} \approx \frac{800}{T} - 0.66, \tag{14}$$

where P is in atm and T in °K.

Since such calculations are described in great detail in relevant literature, e.g. ref. 360, and since they are equally applicable to degassing of metals in the liquid state, they need not be discussed any further here.

2.3 Stationary states

The attainable ultimate state of a degassing process can only be derived from thermodynamic considerations if equilibrium conditions have, in fact, been approached during the process. There are, however, cases where no equilibria are attained, even after any length of time, but where only so-called stationary states are achieved. These occur preferably in systems of type B if the walls of the reaction vessel are cooled. As examples of these phenomena the systems Ta–O, Nb–O[371] and Ta–H–O, Nb–H–O[372] will be discussed.

As mentioned in Subsection 2.2, the degassing of oxygen-containing niobium and tantalum is effected by evaporation of the oxides NbO and NbO_2[366] and TaO and TaO_2[366]. The bulk composition of the oxide mixture is therefore NbO_x and TaO_x, where, according to Subsection 2.2, the factor x depends on both the temperature and pressure. In general, the degassing reaction may be written as:

$$\alpha\text{-solid solution} \rightarrow \text{metal} + \text{volatile oxide} \qquad (15)$$

If the volatile oxide is deposited on the cold parts of the plant, the reverse reaction is almost inhibited and degassing leads to very low oxygen values, *e.g.* below 0.01 at. % for tantalum and niobium[346, 365]. Total removal of oxygen does not result because gas is simultaneously picked up from the environment according to the reaction:

$$\tfrac{1}{2}O_2 \rightarrow (O)_{\text{dissolved}} \qquad (16)$$

The constant residual oxygen level is reached when the temperature-dependent rates of the reactions, according to eqns. (15) and (16), are equal. Since, in addition, reaction (15) is predominantly dependent on the concentration and (16) on the pressure, the residual oxygen level in niobium and tantalum is governed by the pressure and temperature as depicted in Figs. 156 and 157. These diagrams enable the expected ultimate state of a degassing reaction to be determined and, in the case of stationary states, replace the phase diagrams shown in Subsection 2.2.

Stationary states may also occur in systems having several gaseous components

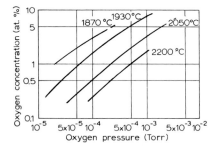

Fig. 156. Pressure/concentration isotherms of the stationary state during annealing of niobium in oxygen[371].

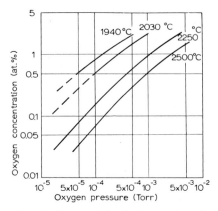

Fig. 157. Pressure/concentration isotherms of the stationary state during annealing of tantalum in oxygen[371].

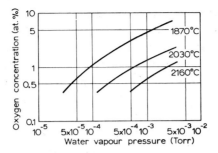

Fig. 158. Pressure/concentration isotherms of the stationary state during annealing of niobium in water vapour[372].

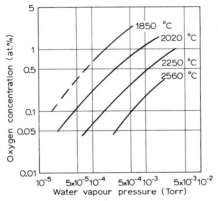

Fig. 159. Pressure/concentration isotherms of the stationary state during annealing of tantalum in water vapour[372].

if volatile compounds evaporate and effect degassing. Reaction (16) for an oxygen pick-up is then replaced, for example, by:

$$\text{water vapour} \rightarrow \text{dissolved oxygen} + \text{gaseous hydrogen} \tag{17}$$

On annealing tantalum and niobium in water vapour a stationary state according to reactions (15) and (17) is attained. The oxygen concentration in the solution depends on the water-vapour pressure and the temperature[372], according to Figs. 158 and 159.

2.4 Kinetics of degassing

2.4.1 Theory

The macroscopically observable entire process of degassing in the solid state is always made up of individual atomic processes, the theoretical elucidation of which is essential for understanding the degassing process. Such unit processes, also called partial or elementary steps, are, for instance, diffusion, transition from the metal lattice to the surface, recombination or transfer into the gas phase.

References pp. 507–515

In degassing, just as with other kinetic processes, certain elementary steps follow each other in strict sequence, while others may take place simultaneously. For instance, a dissolved atom must first diffuse to the surface from the interior of a solid metal body before it can pass into the state of chemisorption. In this case, diffusion and transition, therefore, follow each other. On the other hand, dissolved atoms may diffuse to the surface either via grain boundaries or via lattice interstitials, as both processes are independent of each other. Grain boundary diffusion and diffusion through the metal lattice may take place together.

It is obvious that in equivalent, simultaneous elementary steps the fastest step is rate-determining, while in sequential partial steps the slowest one is the governing factor of the whole degassing process. This fact is very important, as it greatly simplifies the quantitative treatment of the overall process, it being sufficient to find the slowest elementary step and to calculate its magnitude.

This simplified method fails only if several individual steps proceed at a similarly slow rate, but such a coincidence will seldom occur. In such a case the kinetics equations of the elementary steps must be coupled and the system of coupled differential equations then solved[373].

(a) Elementary steps of the degassing process

The most important partial steps in degassing processes are briefly enumerated here, but will be discussed in detail afterwards.

I. Gases dissolved in solid metal bodies are positioned as interstitials in the metal lattice and during degassing have to diffuse to the surface. Diffusion takes place via interstitials, along grain boundaries, or along linear or facet-shaped lattice defects, such as dislocations, low-angle grain boundaries, etc. In most cases diffusion via interstitials progresses the fastest and is therefore rate-determining for the complete process.

II. Atoms or ions diffused from the interior of the metal to its surface leave the metal lattice and pass into the state of adsorption at the surface, preferentially at the so-called active sites.

III. The adsorbed atoms recombine and leave the surface. The simplest conditions exist during the recombination of homogeneous atoms, for instance:

$$[H]_{ads} + [H]_{ads} \rightarrow (H)_2 \qquad (18)$$

This process is similar to the Langmuir and Hinshelwood mechanism[374].

The recombination of heterogeneous atoms should also be considered, for instance:

$$[C]_{ads} + [O]_{ads} \rightarrow (CO) \qquad (19)$$

or

$$[C]_{ads} + 2[O]_{ads} \rightarrow (CO_2) \qquad (20)$$

and the reaction of an adsorbed atom with one from the host lattice, as

$$\langle Ta \rangle + [O]_{ads} \rightarrow (TaO) \tag{21}$$

IV. Molecules in the state of physical adsorption leave the solid metal surface without changing their chemical composition.

This elementary process plays a role only in pure surface degassing, as for example in pumping off moisture from the walls of a vacuum chamber. It may be disregarded when gases are being eliminated from the interior of the metal.

V. Finally, the molecule evaporated from the surface is pumped out of the vacuum chamber, a process which is termed "transport in the gas space". The type of gas flow during this transporting process is governed by the mean free path in relation to the geometrical dimensions of the vacuum tank.

VI. If the component which is to be removed is not dissolved in the metal, but is present as a metal compound, dissociation of the compound can occur during the degassing process.

VII. As described in Subsection 2.2.2, a compound can evaporate directly from the surface of the metal without undergoing decomposition or any other chemical reaction or conversion.

VIII. If a gas is trapped in shrinkage cavities, cracks, etc., in the interior of the metal, it must first be dissolved in the metal before it can be transported to the surface by a diffusion process. In this process the following additional elementary steps occur:
(a) Physical adsorption at the surface of the shrinkage cavities, etc.
(b) Chemisorption and dissociation
(c) Transfer from the chemisorbed state into the metal lattice.

IX. Finally, a chemical reaction with a component present in the gas phase can lead to degassing of the metal, for instance, the reduction of a compound formed at the surface of the metal by hydrogen, or the reaction of a dissolved component with a gaseous reduction agent. In these cases the gaseous component must carry out the following elementary steps:
(a) Transport in the gas space
(b) Physical adsorption at the surface of the metal or the compound
(c) Chemical adsorption and reaction.

For the simplest but at the same time the most important case of removing dissolved gas from the metal lattice, the energy relationships are shown in Fig. 160 and Table 32, using the systems Ta–N[375] and Nb–N[376] as examples. From the different values of activation energy Q_D (elementary step I), Q_{LA} (II) and Q_{AG} (III), there may be a different temperature dependence of each of these three elementary steps. According to Fig. 161, elementary step B may be the slowest in Range a at low temperatures and could well become rate-determining for the total degassing reaction A. On the other hand in Range c at higher temperatures the elementary step C is rate-determining. In the intermediate temperature Range b both elemen-

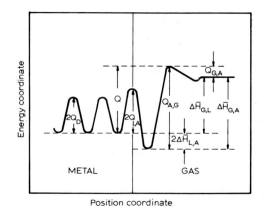

Fig. 160. Energy diagram of the degassing of metal–gas solid solutions of degassing Type A[375–377]. For keys see Table 32.

tary steps, which follow one another, proceed at about the same rate so that the whole degassing process becomes blurred, rendering a mathematical treatment therefore much more difficult.

Besides the temperature, numerous other parameters play a role, such as the dimensions of the sample, the pressure, pumping rate of the vacuum pumps,

TABLE 32

ENERGY QUANTITIES IMPORTANT FOR THE DEGASSING REACTION $2N_{diss} \rightarrow N_{2\,gas}$ FOR NIOBIUM AND TANTALUM[375,377]

Quantity	Meaning	Reaction	Niobium		Tantalum	
			Energy	Temperature range (°C)	Energy	Temperature range (°C)
Q_D	Activation energy for diffusion of N in the metal lattice	—	38.8 kcal/g-atom	800–1600	37.8 kcal/g-atom	100–300
Q_{LA}	Activation energy for transfer	$N_{diss} \rightarrow N_{ad}$	—	—	—	—
Q_{AG}	Activation energy for recombination and desorption	$2N_{ad} \rightarrow N_{2\,gas}$	—	—	187 kcal/mol*	—
Q	Activation energy for degassing	$2N_{diss} \rightarrow N_{2\,gas}$	124 kcal/mol	1600–2000	134 kcal/mol	1700–2200
Q_{GA}	Activation energy for adsorption	$N_{2\,gas} \rightarrow 2N_{ad}$	35 kcal/mol*	1600–2000	47 kcal/mol*	1700–2200
$\overline{\Delta H}_{GL}$	Solution enthalpy for the reaction	$N_{2\,gas} \rightarrow 2N_{diss}$	−85 to −92 kcal/mol	1600–2100	−87 kcal/mol	1600–2400
$\overline{\Delta H}_{GA}$	Adsorption enthalpy for the reaction	$N_{2\,gas} \rightarrow 2N_{ad}$	—	—	−140 kcal/mol	Room temperature
$\overline{\Delta H}_{LA}$	Adsorption enthalpy for the reaction	$N_{diss} \rightarrow N_{ad}$	—	—	−26.5 kcal/g-atom*	1700–2200

* Calculated values.

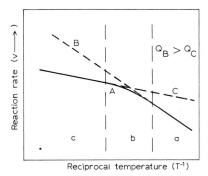

Fig. 161. Rate of a reaction (A) consisting of two partial reactions (B and C) which are both rate-determining.

size of the vacuum tank, etc. For a theoretical treatment, the rate-determining elementary step prevailing under the momentary existing conditions should always be found, as a number of factors can then be immediately eliminated from the analysis, thereby significantly simplifying the calculation.

The most important elementary steps stated from I to IX will be discussed in detail in the following Subsections[373,376].

(b) Diffusion (Step I)

With regard to diffusion via interstitials—by far the most important mechanism with respect to the diffusion of gases in solid metals—the well known Fick's laws[378] apply, in particular Fick's second law:

$$\frac{\partial C}{\partial t} = \frac{\partial}{\partial x}\left(D\frac{\partial C}{\partial x}\right) \qquad (22)$$

In the range of low gas contents up to about 5 at. %, the diffusion coefficient may be regarded as independent of concentration so that eqn. (22) may be written in a simplified form as

$$\frac{\partial C}{\partial t} = D\frac{\partial^2 C}{\partial x^2}. \qquad (23)$$

Taking into account the boundary and initial conditions, the change in concentration with degassing times may be calculated with the aid of the diffusion coefficient (Figs. 162 and 163). The various quantities for the diffusion coefficient D are listed in Table 33.

If the diffusion process progresses substantially faster than all other partial steps, an almost homogeneous distribution of the dissolved gas in the metal prevails throughout the degassing process (Fig. 162). If, on the other hand, diffusion is rate-determining for the whole process, the gas concentration at the surface zone of the metallic body decreases very rapidly towards zero (Fig. 163). If this is true

References pp. 507–515

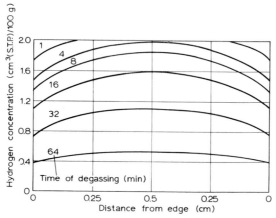

Fig. 162. Hydrogen concentration during the degassing of a sheet iron sample of 1 cm thickness. The degassing rate is determined by the boundary interface reaction (400°C, $D = 1.3 \times 10^{-4}$ cm²/s).

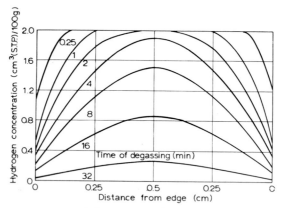

Fig. 163. Hydrogen concentration during the degassing of a sheet iron sample of 1 cm thickness. The degassing rate is determined by the diffusion in the solid solution. (The transfer rate at the surface is higher by a factor of 12.5 than that of Fig. 162 under otherwise equal conditions.)

then the degassing equation

$$\frac{\bar{C}}{C_0} = k' \exp(-kt) \tag{24}$$

is valid for sufficiently long test durations t.
$\bar{C}$ is the mean gas concentration across the section of the sample,
C_0 the initial gas concentration and
k, k' are constants.

The rate constant k obeys the Arrhenius Law:

$$k = \frac{D_0}{F_p} \cdot \exp[-Q_D/RT], \tag{25}$$

TABLE 33

DIFFUSION COEFFICIENT OF GASES IN METALS, $D = D_0 \exp[-Q/RT]$[379]

Gas	Metal	D_0 (cm²/s)	Q (kcal/mol)	Temperature range (°C)
H	Al	0.21	10.9	470–590
	Au	0.56×10^{-3}	24.4	500–940
	Cu	0.011	9.2	430–650
	α-Fe	1.4×10^{-3}	3.2	200–780
	Nb	0.0215	9.4	795–870
	Ni	4.5×10^{-3}	8.1	380–990
	Pd	4.3×10^{-3}	5.6	200–700
	Si	9.4×10^{-3}	11.0	1090–1200
	α-Ti	1.8×10^{-3}	12.4	500–825
	β-Ti	1.95×10^{-3}	6.6	600–1000
	α-Zr	4.15×10^{-3}	9.5	450–700
	β-Zr	5.32×10^{-3}	8.3	760–1010
N	α-Fe	7.8×10^{-3}	18.9	500–850
	γ-Fe	0.91	40.3	950–1350
	Nb	0.061	38.8	800–1600
	α-Ti	0.012	45.2	500–825
	β-Ti	0.035	33.8	900–1570
	β-Zr	0.015	30.7	920–1640
O	Ag	3.66×10^{-3}	11.0	412–862
	Cu	0.075	46.0	600–950
	Ta	0.015	26.7	700–1400
	α-Ti	5.08×10^{-3}	33.5	700–850
	β-Ti	1.6	48.2	950–1414
	α-Zr	9.4	51.8	400–600

where F_p is the surface area of the sample and D_0 is the temperature-independent part of the diffusion coefficient (see also Table 33). After differentiation of eqn. (24) the formula

$$\frac{d\bar{C}}{dt} = -kC_0 k' \exp(-kt) = -k\bar{C} \tag{26}$$

is obtained. Equation (26) shows that, in this case, the degassing process is similar to a first-order reaction.

(c) Transition from the dissolved to the chemisorbed state (Step II)

The transition of a gas atom G_{diss} from an interstitial site Z just being released to a free adsorption site A into the absorbed state G_{ads} takes place according to the reaction equation:

$$G_{diss} + A \underset{v_-}{\overset{v_+}{\rightleftharpoons}} G_{ads} + Z \tag{27}$$

References pp. 507–515

In order to be able to express the reaction rates v_+ and v_-, not only is the concentration of dissolved gas c required but also measures of the concentration for the remaining states.

This knowledge may be obtained from the degree of coverage, θ, which relates the number of the occupied surface sites (G_{ads}) to the total number of surface sites available ($G_{ads} + A$). Since the number of interstitial sites is very high, only a few can be occupied. Therefore, the concentration of the free interstitial sites Z is always approximately one.

From the expressions

$$v_+ = k_+ \cdot C(1-\theta) \tag{28}$$

and

$$v_- = k_- \cdot \theta \tag{29}$$

it follows that:

$$\frac{dC}{dt} = -k'' [k_+ C(1-\theta) - k_- \cdot \theta] \tag{30}$$

Assuming that the transition is by far the slowest partial step and therefore rate-determining for the total reaction, the degree of coverage θ becomes very small, so that a first approximation is:

$$\frac{dC}{dt} = -kC \tag{31}$$

This is a reaction of the first order, which leads to the degassing law as:

$$\frac{C}{C_0} = \exp[-kt] \tag{32}$$

where

$$k = k_0 \cdot \exp[-Q_{LA}/RT] = k_0' \frac{F_p}{V_p} \cdot \exp[-Q_{LA}/RT], \tag{33}$$

where F_p is the surface area of the sample and V_p is the volume of the sample. (See also Table 32.)

(d) Recombination and evaporation from the surface (Step III)

In the simplest case, a reaction takes place between two adsorbed gas atoms, G_{ads}, forming as a result the gaseous molecule, G_2, and liberating two surface sites, A, according to:

$$2\,G_{ads} \underset{v_-}{\overset{v_+}{\rightleftharpoons}} 2\,A + (G_2), \tag{34}$$

at a reaction rate which may be expressed as:

$$v = k_+ \theta^2 - k_- \cdot C_{(gs)} \cdot (1-\theta)^2, \tag{35}$$

where $C_{(gs)}$ is the concentration of the gas molecules in the gas space.

If this recombination is rate-determining for the whole degassing reaction and if, at the same time, the pumping rate is sufficiently large, $C_{(gs)}$ will be very much smaller than unity and θ in equilibrium with the solid solution. Therefore, an equilibrium isotherm may be introduced, for instance that derived by Langmuir:

$$\theta = \frac{K \cdot C}{1 + K \cdot C}. \tag{36}$$

It follows that

$$\frac{dC}{dt} = -k \left(\frac{K \cdot C}{1+K \cdot C}\right)^2 = -k \left(\frac{1}{1+\frac{1}{K \cdot C}}\right)^2. \tag{37}$$

The two limiting cases, $\theta \approx 1$, i.e. $K \cdot C \gg 1$, and $\theta \ll 1$, i.e. $K \cdot C \ll 1$, give then the following results:

$$\text{Case (i): } \theta \approx 1 \qquad \frac{dC}{dt} = -k \tag{38}$$

$$\text{Degassing law } \frac{C}{C_0} = 1 - \frac{k}{C_0} \cdot t, \tag{39}$$

where C_0 is the initial concentration of the solution and

$$k = k_0 \cdot \exp\left[-Q_{AG}/RT\right] \tag{40}$$

$$\text{Case (ii): } \theta \ll 1 \qquad \frac{dc}{dt} = -k \cdot K^2 \cdot C^2 = k'C^2 \tag{41}$$

$$\text{Degassing law } \frac{C_0}{C} = k'C_0 \cdot t + 1, \tag{42}$$

where $k' = k'_0 \cdot \exp\left[-(\Delta 2\bar{H}_{LA} + Q_{AG})/RT\right]$. \hfill (43)

It should be noted that in case (ii) the temperature dependency of the reaction is governed jointly by the true activation energy Q_{AG} and the heat of solution $\Delta \bar{H}_{LA}$.

Instead of eqn. (37) an approximation is often used between $1 \gg \theta \approx 1$, this being:

$$\frac{dC}{dt} = -a \cdot C^n, \tag{44}$$

where $0 < n < 2$.

The application of the Freundlich adsorption isotherm

$$\theta = b \cdot C^m, \tag{45}$$

where $0 < m < 1$, gives immediately eqn. (44) instead of eqn. (37). This means that eqn. (44) certainly covers a wide range of practical cases.

The simplest case has been considered so far in which two homogeneous atoms recombine. The other cases can be calculated similarly by introducing different degrees of coverage $\theta_1, \theta_2 \ldots$ for the various types of atoms.

(e) Transport in the gas space

The pumping rate of vacuum plants is usually adequate, so that one of the elementary stages already dealt with will become rate-determining for the degassing process. However, during sudden gas eruptions, when pumping is throttled in narrow heating pipes incorporating radiation shields, or during very rapid heating, it may occur that the removal of the gases liberated from the sample could become rate-determining for the whole degassing process. In this case

$$v = k \cdot C_{(\text{gs})}, \tag{46}$$

where among other things the constant k incorporates the pumping rate of the pumping system and where $C_{(\text{gs})}$ expresses the concentration of the liberated gas in the vicinity of the sample.

Under the above-mentioned assumption that all other elementary steps during degassing proceed markedly faster than the pumping-off rate, a state of equilibrium is almost reached between the solution/chemisorption layer and the recombined gas in the gas space. Using the relevant equilibrium constants an expression for the relationship between the gas content in the gas space, $C_{(\text{gs})}$, and the concentration of the dissolved gas, C, may be derived. For the gas concentration of the sample, a formula is obtained which resembles a reaction law of the second order

$$\frac{dC}{dt} = -k \cdot C^2. \tag{47}$$

This leads to a degassing law of the following form:

$$\frac{C_0}{C} = k \cdot C_0 \cdot t + 1, \tag{48}$$

where the rate constant k incorporates both the pumping rate, which is independent of the temperature of the sample, and the two equilibrium constants. The temperature dependency may be expressed as:

$$k = k_0 \cdot \exp\left[-(2\Delta \bar{H}_{\text{LA}} - \Delta \bar{H}_{\text{GA}})/RT\right]. \tag{49}$$

It should be noted that—according to Fig. 160—the two adsorption enthalpies $2\Delta \bar{H}_{LA}$ and $\Delta \bar{H}_{GA}$ represent the solution enthalpy $\Delta \bar{H}_{GL}$.

2.4.2 Results of experiments

The degassing of solid metallic materials in a vacuum at high temperatures plays an important role in technology, for instance in sintering and annealing processes. The degassing process has therefore been studied on various occasions in a series of carefully planned and very skilfully conducted experiments. The evaluation of these experiments has been greatly assisted by the theoretical considerations dealt with in Subsection 2.4.1.

None of the degassing laws could be exclusively co-ordinated with any one metal–gas system, as was expected. On the contrary, the type of degassing law for any binary or multi-component system is dependent on the temperature, the size of the sample, the pumping rate in the vacuum chamber and also, possibly, on the gas concentration inside the sample, as is shown in Table 34. For instance, the

TABLE 34

FACTORS WHICH MAY CAUSE A CHANGE IN THE RATE-DETERMINING DEGASSING LAW

Factor	Indication of the type of effect
Temperature	See Fig. 161
Concentration	Effect not readily determined
Section size of sample	Transport rate of the gas atoms decreases with increasing section size
Surface area: volume	Surface reactions (see Subsections 2.4.1 b and c) become more important for high surface area: volume ratios
Pumping rate	See example shown in Figs. 171 and 172

transport of gas from the interior to the surface of the sample is made more difficult with increasing thickness of the sample. It is therefore quite possible that, for a thin sample, a surface reaction could be the slowest step, becoming therefore rate-determining for the kinetics of the whole degassing process, while under otherwise identical conditions the overall kinetics of a thick sample are determined by the diffusion of the gases in its interior.

From practical experiments, some characteristic examples of which will subsequently be described, important factors may be derived such as the diffusion coefficient, specific rate constants, etc. Knowing these and using the equations quoted in Subsection 2.4.1, the individual reactions involved in the degassing process can be calculated.

(a) Single-phase binary systems

The degassing of homogeneous, metallic materials, having a bright surface and containing only one type of gas, takes place by a process either of Type A or

Type B, which have been considered extensively with regard to their thermodynamic properties in Subsection 2.2. It is immaterial whether only the two components of the binary system or a number of other elements are present, so long as they do not take part in the degassing reaction to any noticeable extent. It is for this reason that there is no qualitative difference between a pure metal and an alloy in the degassing mechanism, provided that the alloying elements do not cause any major change in the chemical behaviour, in which case they would, of course, significantly affect the degassing reaction.

At high temperatures and low pressures, *i.e.* under conditions preferable for a degassing reaction, a marked degree of evaporation of the metal usually occurs in addition. As this process is entirely independent of the degassing process and as it takes place side by side with the latter, it will not be considered in the following (see Chapter I).

Evaporation of the gas molecules—Degassing Type A. Particularly careful studies were carried out with thin wire samples of nitrogen-bearing niobium and tantalum[375, 380]. In these experiments the gas-free samples were annealed under the subsequent degassing conditions in a high vacuum and the kinetics of metal evaporation were determined from the weight loss. The vacuum chamber was then closed and the sample charged with nitrogen by introducing a predetermined quantity into the closed vacuum chamber (time t_1 of Fig. 164).

Immediately afterwards the vacuum chamber was reconnected to the pumping system and pumping began. The degassing that took place led to a reduction in the weight of the sample. The loss of weight, m, as shown in Fig. 164, was caused by the quantity of nitrogen liberated, m_{N_2}, and by loss of the evaporated metal, m_{Me}. After completion of the degassing process (time t_2) the rate of evaporation of the metal was again checked. The electrical resistance of the sample, which is greatly affected by the content of dissolved nitrogen, was also measured. Other quantities too, such as the hardness, the magnetic susceptibility and the internal friction,

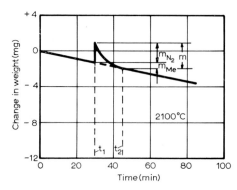

Fig. 164. Change in weight of niobium during the pick-up (t_1) and liberation (t_1 to t_2) of nitrogen[381].

were determined, but in these experiments the use of those quantities is recommended which may be determined continuously without removal of the sample from the vacuum chamber.

The course of the degassing process of nitrogen-bearing niobium and tantalum is depicted in Figs. 165 and 166. In these experiments the pressure in the reaction chamber was below 10^{-4} Torr a few minutes after the start of the reaction. It reached values below 1×10^{-6} Torr towards the end of the reaction.

Fig. 167 and 168, in which the quantities measured have been plotted differently, show that the degassing process conforms to an equation ((42) or (48)) derived in Subsection 2.4.1:

$$C_0/C = k \cdot C_0 \cdot t + 1$$

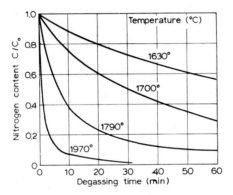

Fig. 165. Lowering of the nitrogen content, C, during the isothermal degassing of niobium wire[380]. $C_0 \approx 3$ at.%. Diameter 0.5 mm.

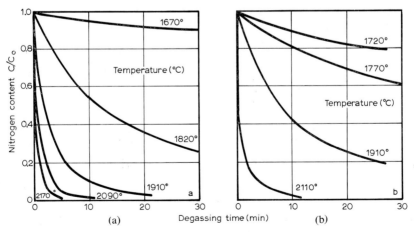

Fig. 166. Lowering of the nitrogen content, C, during the isothermal degassing of tantalum wire[375]. (a) Initial concentration $C_0 \approx 5.7$ at.%; (b) Initial concentration $C_0 \approx 4.3$ at.%.

References pp. 507–515

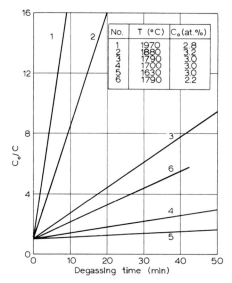

Fig. 167. Plot of C_0/C of the results shown in Fig. 165.

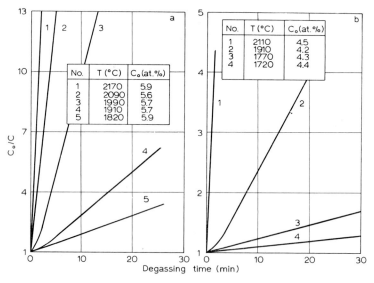

Fig. 168. Plot of C_0/C of the results shown in Fig. 166. (a) Initial concentration $C_0 \approx 5.7$ at.%; (b) Initial concentration $C_0 \approx 4.3$ at.%.

This is therefore a reaction of the second order (eqn. (41) or (47)):

$$\frac{dC}{dt} = -k \cdot C^2$$

In these experiments the pumping rate was high (about 60 l/s) and the pressure in the vacuum chamber kept low. Transport in the gas space may therefore be excluded as the rate-determining step (eqns. (47) and (48)). From the fact that eqn. (42) is obeyed, it may be concluded that the rate of nitrogen liberation from tantalum and niobium is, under the prevailing conditions, determined by the recombination of nitrogen atoms and the subsequent transfer of the freshly formed molecules into the gas space. That this is in fact so was confirmed by other experiments and theoretical considerations reported in the original papers.

The constant k in eqn. (42) may be expressed as

$$k = \frac{F}{M} k^*, \qquad (50)$$

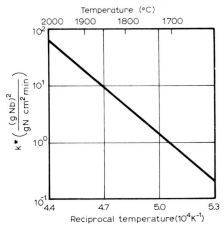

Fig. 169. Temperature dependence of the specific rate constant, k^*, for the liberation of nitrogen from niobium[380].

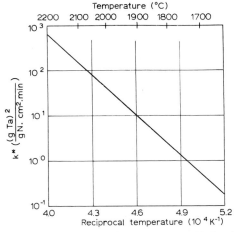

Fig. 170. Temperature dependence of the specific rate constant, k^*, for the liberation of nitrogen from tantalum[375]. Initial concentration C_0 2.7–6.5 at.%.

where F is the surface area of the sample in cm^2, M is the mass of the sample in g, and k^* is the specific-rate constant in

$$\frac{\text{g sample} \cdot \text{g sample}}{\text{cm}^2 \cdot \text{g N} \cdot \text{min}}.$$

The specific rate constant k^*, as may be seen from Figs. 169 and 170, obeys an Arrhenius Law of the form:

$$k^* = k_0^* \cdot \exp[-Q/RT], \tag{51}$$

i.e. for niobium

$$k^* = 5.5 \times 10^{13} \cdot \exp[-(6.25 \times 10^4)/T] \tag{52}$$

and for tantalum

$$k^* = 3.58 \times 10^{14} \cdot \exp[-(6.76 \times 10^4)/T]. \tag{53}$$

The values used for the activation energy Q were taken from the diagrams Figs. 169 and 170, these being:

for Nb–N $Q = 124$ kcal/mol N$_2$ and
for Ta–N $Q = 134$ kcal/mol N$_2$.

The energy relationships determined so far are shown in Table 32.

In summing up it may be said that the degassing reaction for the binary systems of degassing Type A can be calculated quantitatively with the aid of eqns.

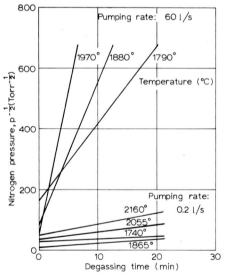

Fig. 171. Effect of pumping rate on the nitrogen pressure in the vacuum chamber during the degassing of niobium.

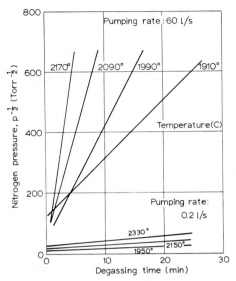

Fig. 172. Effect of pumping rate on the nitrogen pressure in the vacuum chamber during the degassing of tantalum.

(42), (50) and (51), provided that the specific-rate constant k^* is known and provided—this is most important—that the recombination of the gas atoms is rate-determining for the reaction. This assumption is, as mentioned earlier, not always true for any given binary system, as it is affected by numerous other factors (Table 34).

The degassing reaction of the systems Nb–N and Ta–N described earlier was also studied under throttled pumping rate conditions, *i.e.* 1/300 of the original value[375,380]. The relationship between pressure and pumping time is compared with that obtained at a high pumping rate in Figs. 171 and 172. In both cases straight lines are obtained for $p^{-\frac{1}{2}}$ versus t. While at a high pumping rate recombination, and therefore eqns. (41) and (43), were decisive, transport in the gas space is rate-determining at a low pumping rate and therefore eqns. (47)–(49) apply. This difference is clearly shown in the values obtained for the activation energy (Table 35).

TABLE 35

ACTIVATION ENERGY OF THE LIBERATION OF NITROGEN FROM NIOBIUM AND TANTALUM

Metal	Pumping rate (l/s)	Rate-determining step	Activation energy (kcal/mol N_2)
Nb	60	Recombination	124
	0.2	Transport in the gas space	78
Ta	60	Recombination	134
	0.2	Transport in the gas space	82

References pp. 507–515

Evaporation of metal-gas compounds—Degassing Type B. Degassing according to Type B progresses quite differently with the formation of volatile compounds between metal atoms and gas atoms at the surface of the metal. The difference from that of Type A discussed earlier is obvious from a comparison of the weight loss during a gassing–degassing experiment (Figs. 164 and 173). If the experiment on which the diagram Fig. 164 is based is conducted in a similar way, with the exception that instead of using nitrogen (m_{N_2}) an equally large quantity of oxygen, m_{O_2}, is introduced, the sample will lose considerably more weight (m) during the subsequent degassing period $t_2 - t_1$. The major part comes from the quantity of metal m_{Me}, which in Type B degassing combines with the gas, m_{O_2}. The compound, or the mixture of compounds subsequently formed, evaporates.

Such cases have so far been studied in detail only in the systems Nb–O[381], Ta–O[381] and V–O[382]. In these experiments the existence of the compounds NbO, NbO_2, TaO, TaO_2 and VO was established by gravimetry and mass spectrometry.

As a result of their low vapour pressure these suboxides condense on the cooler parts of the vacuum chamber and are therefore not pumped off. The Type B degassing process is therefore not characterized by a noticeable rise in pressure. Owing to the freezing-out of the suboxides equilibrium is not reached at the end of the degassing process, but a stationary state is set up, as described in Subsection 2.3, in which the oxygen pick-up from the residual atmosphere is equally as large as the loss in oxygen through the formation of suboxides.

At very high temperatures in the vicinity of the melting point of the metal, and even more pronounced in the liquid state, evaporation of the metal may be

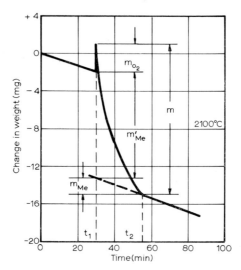

Fig. 173. Change in weight of niobium during the pick-up (t_1) and liberation (t_1 to t_2) of oxygen[381].

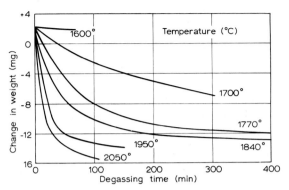

Fig. 174. Change in weight of a niobium sample during the liberation of oxygen[381].

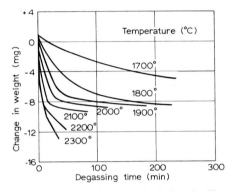

Fig. 175. Change in weight of a tantalum sample during the liberation of oxygen[381].

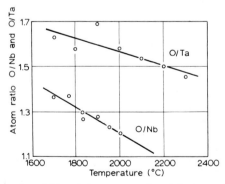

Fig. 176. Concentration ratio of liberated oxygen and metal atoms during the degassing of niobium and tantalum[381].

very considerable. The manner in which the gas content of the sample changes—in other words whether an increase or a decrease in the gas concentration occurs—depends on the ratio of the rates of oxide evaporation, metal evaporation and oxygen pick-up from the residual gas. Under adverse conditions, an increase in

References pp. 507–515

gas content may even occur despite the very low pressure and high temperature. Concentration changes of this type were studied in detail by Hörz[383,384] and extended to Type A degassing.

Figures 174 to 181 show experimental results for the systems Nb–O, Ta–O and V–O carried out on thin wire and hollow cylindrical samples in which the diffusion of the oxygen to the surface proceeded rapidly. Similarly to Fig. 173 the atom ratio of oxygen to metal in the volatile compound may be calculated from the ratio $m_{O_2}:m_{Me}$ of Figs. 174 and 175. The result of this calculation (Fig. 176) shows that a mixture of compounds evaporates during the liberation of oxygen from niobium and tantalum. These were identified by mass spectrometry as NbO and NbO_2 and TaO and TaO_2, respectively. As can be seen from Fig. 176, the ratio in the mixture of the two compounds is temperature-dependent.

Fromm[367] worked out the theoretical fundamentals regarding the evaporation of two compounds from a mixed phase and so provided the basis for evaluating experimental results. He derived the following expressions for the evaporation rate during oxygen removal from niobium and tantalum:

$$v_{NbO_x} = C \cdot 9.55 \times 10^{11} \cdot \exp[-6.53 \times 10^4/T]$$
$$\cdot \{1 + C \times 10^{-2} \cdot \exp[9.26 \times 10^3/T]\} \tag{54}$$

and

$$v_{TaO_x} = C \cdot 7.80 \times 10^{11} \cdot \exp[-6.65 \times 10^4/T]$$
$$\cdot \{1 + C \cdot 2.80 \times 10^{-2} \cdot \exp[1 \times 10^4/T]\} \tag{55}$$

in which v is in mg/min cm^2, C is in at.–% and T is the absolute temperature.

The total oxide evaporation is therefore made up of the sum of two expressions, which characterize the evaporation of MeO and MeO_2. The first one is directly proportional to the oxygen concentration, C, in the solid solution, the other to the square of this concentration. The mixture ratio of these two oxides is therefore dependent on the momentary solid solution concentration. During removal of oxygen from tantalum and niobium the mixture ratio changes in such a manner that the monoxide portion increases steadily as degassing proceeds. At very low oxygen concentrations, as is present at the end of the degassing process, the monoxide alone, or almost alone, is evaporated*.

The concentration dependence of the mixture ratio affects the overall ratio X—derived from the total quantity of oxygen liberated and from the quantity of metal combined with it—as X depends also on the initial concentration (Figs. 177 and 178).

Considering eqns. (54) and (55) no simple degassing law can be expected. This is confirmed by Figs. 179 and 180 which were derived from Figs. 174 and 175,

---

* The pure metal always evaporates as well, of course.

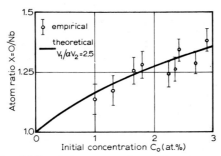

Fig. 177. Atom ratio $X = O/Nb$ of the evaporating oxide mixture during the degassing of niobium in relation to the initial concentration C_0 of the solid solution[366].

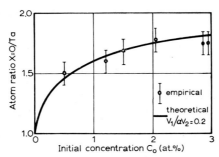

Fig. 178. Atom ratio $X = O/Ta$ of the evaporating oxide mixture during the degassing of tantalum in relation to the initial concentration C_0 of the solid solution[366].

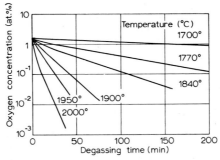

Fig. 179. Change of the oxygen concentration in a niobium sample during degassing[381].

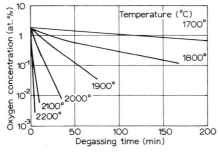

Fig. 180. Change of the oxygen concentration in a tantalum sample during degassing[381].

References pp. 507–515

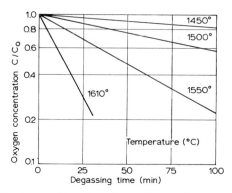

Fig. 181. Oxygen liberation of vanadium during degassing[382].

since by plotting the experimental results on logarithmic co-ordinates, curved lines are obtained, particularly for short degassing times, high gas concentrations and low temperatures.

At low concentrations and high temperatures, on the other hand, where according to Figs. 176 to 178 mainly monoxide evaporates, the term representing the quadratic concentration effect may be disregarded in eqns. (54) and (55). This results in a reaction of the first order for the reduction of the oxygen concentration C in a sample of surface area F and mass M, which may be written as:

$$\frac{dC}{dt} = -\frac{Fk^x}{M} \cdot C. \tag{56}$$

TABLE 36

DEGASSING OF THIN SAMPLES OF THE SYSTEMS Nb–O, Ta–O, AND V–O[381,382,384]

System	Applicable law of degassing	Activation energy (kcal/mol) MeO)	Range of validity
Nb–O	$\ln \dfrac{C}{C_0} = -\dfrac{F}{M} \cdot t \cdot 1.35 \times 10^{11} \exp\left[-\dfrac{6.53 \times 10^4}{T}\right]$	130	$T = 1700\text{--}2050°C$ $C < 0.8$ at.% O
Ta–O	$\ln \dfrac{C}{C_0} = -\dfrac{F}{M} \cdot t \cdot 2.34 \times 10^{11} \exp\left[-\dfrac{6.65 \times 10^4}{T}\right]$	132	$T = 1700\text{--}2200°C$ $C < 0.3$ at.% O
V–O	$\ln \dfrac{C}{C_0} = -\dfrac{F}{M} \cdot t \cdot 2.55 \times 10^{13} \exp\left[-\dfrac{6.88 \times 10^4}{T}\right]$	136	$T = 1450\text{--}1610°C$ $C < 0.7$ at.% O

C: Gas concentration.
C_0: Initial gas concentration.
F: Sample surface in cm².
M: Sample mass in g.
t: Degassing time in min.
T: Absolute temperature.

The degassing law may therefore be expressed as:

$$\ln \frac{C}{C_0} = -\frac{F}{M} k^x \cdot t \qquad (57)$$

Graphical presentation of this equation leads to straight lines.

A similar degassing law was found to be obeyed by the V–O[382] system for various initial concentrations between 0.15 and 0.70 at. % O in thin wires (Fig. 181), where only vanadium monoxide is evaporated.

The degassing laws for the simple case that only monoxide evaporates are compiled in Table 36. The formation and evaporation of the monoxide is the rate-determining step in these reactions. Outside the quoted boundary conditions, evaporation of a mixture of compounds, involving complex degassing laws, should be expected[367]. With increasing section size of the samples diffusion may become rate-determining. In this case the laws derived in Subsection 2.4.1a apply, in fact, for all degassing types.

(b) Multi-phase binary systems

The considerations discussed so far have concerned single-phase samples with one gas component in solution. The problem now arises as to what changes take place during a degassing process, if the metal contains compounds, either in the interior or at the surface, before the degassing process starts. Inclusions in the interior of the sample are dissolved when the concentration of dissolved gas falls below the saturation concentration. Compounds on the surface are decomposed by entirely different mechanisms: by congruent evaporation; by dissociation into gas and sub-compounds; by decomposition into metal and gas.

TABLE 37

BEHAVIOUR OF OXIDES AND NITRIDES DURING EVAPORATION IN A HIGH VACUUM[385]

Reaction	Compound
Evaporation without change in composition	BaO, VO, NbO
$[Me_xO_y] \rightarrow (Me_xO_y)_{gas}$	B_2O_3
$[Me_xO_y] \rightarrow (Me_xO_y)_{n\ gas}$	$MoO_3, WO_3, UO_3\ (?)$
Evaporation of a mixture	BeO
$[Me_xO_y] \nearrow (Me_xO_y)_{gas}$	TiO_2, ZrO_2, HfO_2
$\searrow (Me_xO_{y-a})_{gas} + \frac{a}{2}O_2$	UO_2
Evaporation of a suboxide	$Sc_2O_3, Y_2O_3, La_2O_3,$
$[Me_xO_y] \rightarrow (Me_xO_{y-a})_{gas} + \frac{a}{2}O_2$	$SiO_2, GeO_2, SnO_2,$ Ta_2O_5, Nb_2O_5
Decomposition into metal and gas	MgO, CaO, SrO, Al_2O_3,
$[Me_xO_y] \rightarrow xMe + \frac{y}{2}O_2$	BN, AlN, GaN, Nb_2N

References pp. 507–515

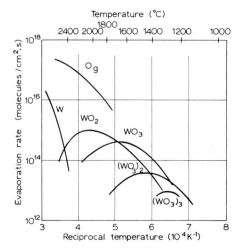

Fig. 182. Evaporation from a tungsten surface at an oxygen pressure of 2.1×10^{-4} Torr[387].

Investigations into the actual behaviour of compounds may be very conveniently carried out by evaporation in Knudsen cells with subsequent analysis in a mass-spectrometer[385-387]. Results obtained in this way are shown in Table 37 for various oxides. In these investigations the vapour pressure of the compound and the decomposition energy may be determined at the same time.

As may be seen from Table 37, WO_3, for instance, evaporates without decomposition, but in the mass-spectrometer the associated complexes $(WO_3)_2$, $(WO_3)_3$ and $(WO_3)_4$ are also found.

For tungsten, the evaporation of the oxide proceeds to completion leaving behind a bright metal surface. Additional oxide may be formed at the tungsten surface from the oxygen in the residual atmosphere, which also evaporates immediately. This behaviour is depicted in Fig. 182 for an oxygen partial pressure of 2.1×10^{-4} Torr. The evaporation rate increases with increasing temperature, but the size of the molecule decreases[387].

The particularly stable oxides BeO, TiO_2, ZrO_2, etc., also evaporate without decomposition; here, too, suboxides were detected[385].

If a piece of tantalum, which is covered with an oxide film, is degassed in a high vacuum, Ta_2O_5 is decomposed and the suboxides TaO and TaO_2 are formed, and at the same time the tantalum solid solution is degassed.

During the degassing of magnesium, aluminium and similar metals a decomposition of the oxide film into metal and oxygen takes place in the Knudsen cell at high temperature and low pressures.

(c) Multi-component systems

Numerous materials consisting of several metallic components behave during degassing qualitatively like the corresponding pure metals, but the quantitative values of the degassing rate, the activating energy, etc., are usually slightly changed by the alloy constituents. This behaviour changes, of course, if the chemical state has been fundamentally altered by alloying, as for instance in stainless steels.

If metallic materials contain several gaseous components, which mutually interfere or even react with one another, the behaviour during degassing is entirely different from that of the corresponding binary systems. The various gaseous components in this case diffuse separately to the surface where they may form volatile compounds. For instance, in the degassing of solid iron containing oxygen, hydrogen and carbon the formation of carbon monoxide, carbon dioxide and methane is observed. These additional possibilities not only affect the corresponding equilibria, as described in Subsection 2.2.3, but also the kinetic process to a marked degree.

Similar major changes may be brought about by the addition to the residual gas atmosphere of a reactive component which was not originally present in the solid metallic sample. For instance, by introducing a low oxygen partial pressure, decarburization of bright steel may be obtained as a result of the formation of carbon monoxide and carbon dioxide. This type of process has only recently been studied theoretically and the results applied to the purification of niobium by controlled admission of oxygen and checking of the carbon monoxide pressure[388]. Similar purification effects are obtainable by the use of hydrogen–water vapour mixtures. As this effect is not degassing in the strict sense, but a purification by chemical reaction, the reader is referred to the appropriate sections in Chapter III.

REFERENCES

Section 1
1. H. THOLANDER, *Ger. Pat. 164,479* (30.7.1881).
2. R. AITKEN, *Ger. Pat. 22,170* (12.11.1882).
3. Compressed Steel Comp., *Ger. Pat. 33,710* (9.6.1885).
4. E. TAUSSIG, *Ger. Pat. 52,650* (3.12.1889).
5. E. TAUSSIG, *Ger. Pat. 58,908* (26.2.1891).
6. E. MAY, *Ger. Pat. 64,831* (5.5.1891).
7. E. TAUSSIG, *Ger. Pat. 65,592* (10.12.1891).
8. E. TAUSSIG, *Ger. Pat. 65,592* (10.12.1891).
9. W. S. SIMPSON, *Ger. Pat. 68,483* (27.7.1892).
10. E. MAY, *Ger. Pat. 96,836* (31.1.1897).
11. DELLWIK (Fleischer Wassergas-Gesellschaft), *Ger. Pat. 250,999* (1.11.1911) and *Ger. Pat. 258,981* (12.10.1912).
12. AEG, *Ger. Pat. 316,047* (17.3.1918).
13. D. LARNER, *Ger. Pat. 479,627* (18.12.1925).
14. W. BÖHM, *Ger. Pat. 569,699* (24.2.1927).
15. T. STROHMEIER, *Ger. Pat. 638,931* (24.6.1932).
16. A. WACKER, *Ger. Pat. 687,720* (2.12.1934).
17. DHHU, *Ger. Pat. 660,094* (18.8.1935).
18. E. BREUER, *Ger. Pat. 736,603* (6.10.1940).
19. KNAPPSACK, A.G., *Ger. Pat. 862,220* (4.7.1943, ÜLG).
20. H. WINTER, *Ger. Pat. 859,801* (18.2.1945, ÜLG).

21 F. Jansen, *Ger. Pat. 837,707* (23.12.1950, ÜLG).
22 K. Daeves, *Ger. Pat. 832,903* (25.6.1950, ÜLG).
23 W. Coupette, *Ger. Pat. 866,231* (29.4.1951, ÜLG).
24 H. Bessemer, *U.S. Pat. 49,051* (25.7.1865).
25 R. H. Gordon, *U.S. Pat. 287,273* (23.10.1883).
26 E. S. Hitt, *U.S. Pat. 916,314* (16.9.1907).
27 J. O. Betterton, *U.S. Pat. 2,054,923* (12.11.1923).
28 C. E. Williams, *U.S. Pat. 2,587,793* (23.3.1931).
29 V. C. Doerschuk, *U.S. Pat. 1,944,733* (22.11.1932).
30 W. K. Thomson, *U.S. Pat. 2,140,607* (19.10.1935).
31 L. Maré, *U.S. Pat. 2,253,421* (Sweden 20.9.1938).
32 C. E. Logan, *U.S. Pat. 2,372,628* (16.2.1944).
33 F. B. Waldron, *U.S. Pat. 2,587,793*, corresponds to *Brit. Pats. 653,965 and 702,225* (19.7.1949).
34 C. F. Bennet, *U.S. Pat. 2,568,578* (23.12.1949).
35 W. Coupette, *U.S. Pat. 2,784,961* (Germany 5.12.1953).
36 J. W. Rozian, *U.S. Pat. 2,837,790* (28.12.1953).
37 L. Southern, *U.S. Pat. 2,734,240* (22.11.1954).
38 T. Wainright, *Brit. Pat. 2,293* (28.1.1898).
39 R. Krulla, *Austrian Pat. 138,020* (19.7.1933).
40 H. M. Ridge, *Brit. Pat. 425,369* (19.11.1933).
41 F. Thomas, *French Pat. 992,654* (25.7.1944).
42 *U.S. Pat. 1,068,643* (2.3.1906).
43 E. F. Lake, *Stahl Eisen*, 32 (1912) 1062.
44 E. Améen and H. Willners, *Jernkontorets Ann.*, 112 (1928) 195.
45 W. Rohn, *Z. Metallk.*, 21 (1920) 12.
46 H. T. Reeve, *Metal Ind. (London)*, 38 (1931) 261.
47 C. G. Carlsson, *Jernkontorets Ann.*, 124 (1940) 686.
48 Anon., *Steel (USSR)*, 108 (1941) 82.
49 B. Lunn, *Giesserei, Tech. Wiss. Beih. Giessereiw. Metallk.*, (1953) 485.
50 A. Tix, *Hütten-Ztg. (Bochumer-Verein)*, (1954) 2.
51 Anon., *Iron Age*, 175 (1955) 91.
52 Anon., *Iron & Coal Trades Rev.*, 170 (1955) 246.
53 Anon., *Steel (USSR)*, 138 (1956) 112.
54 A. Tix, *Stahl Eisen*, 76 (1956) 61.
55 P. M. Unterweiser, *Iron Age*, 177 (1956) 67.
56 A. Tix, *J. Metals*, 8 (1956) 420.
57 A. Tix, *Metal Progr.*, 69 (1956) 81.
58 Anon., *Iron & Coal Trades Rev.*, 174 (1957) 274.
59 Anon., *Steel (USSR)*, 141 (1957) 70.
60 W. A. Fischer, *Arch. Eisenhüttenw.*, 31 (1960) 1.
61 F. Harders, H. Knüppel and K. Brotzmann, *Stahl Eisen*, 76 (1956) 1721.
62 Hsu Chia-Lung, A. J. Polyakov et al., *Izv. Akad. Nauk SSR, Met. i Gornoe Delo*, (1964) 17.
63 L. Tiberg, *Jernkontorets Ann.*, 144 (1960) 771.
64 A. S. Darling, *Metallurgia*, 64 (1961) 7, 71.
65 J. I. T. Green and D. M. Hayes, *Iron & Coal Trades Rev.*, 181 (1960) 827.
66 W. A. Fischer and A. Hoffmann, *Arch. Eisenhüttenw.*, 31 (1960) 411.
67 H. Schenk and H. Domalski, *Arch. Eisenhüttenw.*, 32 (1961) 753.
68 R. G. Ward, *J. Iron Steel Inst. (London)*, 201 (1963) 11.
69 T. R. Meadowcroft, *Ind. Heating*, 33 (1966) 1285.
70 W. Coupette, *Die Vakuumbehandlung des flüssigen Stahles*, R. A. Lang Verlag, Wiesbaden-Esch/Taunus, 1967, p. 72.
71 H. Knüppel and K. Brotzmann, *Stahl Eisen*, 79 (1959) 272.
72 K. Brotzmann, K. Rüttiger et al., *Stahl Eisen*, 79 (1959) 410.
73 J. R. Lakin and G. Payne, *Steel Times*, 192 (1966) 83.
74 W. H. Schaefer, Jr., *J. Metals*, 18 (1966) 69.

REFERENCES

75 R. H. Venable, *J. Metals*, *18* (1966) 495.
76 K. Granitzki, *Rheinstahl-Tech.*, *3* (1966) 91.
77 A. M. Sage, *Iron & Steel (London)*, *39* (1966) 124.
78 S. Eketorp, *Jernkontorets Ann.*, *150* (1966) 585.
79 K. Brotzmann, *Arch. Eisenhüttenw.*, *31* (1960) 67.
80 K. Rüttiger and P. G. Mantey, *Tonind. Ztg. Keram. Rundschau*, *92* (1968) 251.
81 T. H. Turner, *Engineer*, *221* (1966) 211.
82 L. Brewer, *Chem. Rev.*, *52* (1953) 1.
83 K. M. Bonar et al., *Trans. Vacuum Met. Conf.*, Boston, Mass., 1966, p. 437.
84 H. O. McIntire, *Foundry Trade J.*, *103* (1957) 543.
85 A. Eichner, *Hutnik Praha*, *12* (1962) 586.
86 J. A. Rassenfoss and R. K. Matuschkowitz, *J. Metals*, *17* (1965) 1362.
87 W. G. Speransky, *Metallurg*, (1956) 12.
88 G. A. Garnik and A. M. Samarin, *Stal'*, *16* (1956) 514.
89 A. M. Samarin, L. M. Novik et al., *Stal'*, *16* (1956) 700.
90 A. M. Samarin and L. M. Novik, *Tr. Inst. Met. im. A. A. Baikova, Akad. Nauk SSSR*, (1957) 39.
91 A. M. Samarin et al., *Neue Hütte*, (1957) 582; see also *Stal'*, *16* (1956) 700.
92 G. A. Sokolov, G. N. Oiks et al., *Metallurg*, (1958) 10.
93 P. N. D'yachkov and S. S. D'yachkova, *Ogneupory*, *23* (1958) 446.
94 C. W. Finkl, *Elec. Furnace Conf. Proc.*, *16* (1958) 93.
95 C. W. Finkl, in R. F. Bunshah (Ed.), *Trans. Vacuum Met. Conf. 1959*, New York University Press, New York, 1960, p. 93.
96 K. G. Speith and H. vom Ende, *Stahl Eisen*, *80* (1960) 737.
97 B. A. Pogulev, *Metallurg*, (1966) 23.
98 W. Küntscher, *Neue Hütte*, *5* (1960) 383.
99 Z. Eminger, F. Kinsky and Z. Kletecka, *Hutnicke Listy*, *12* (1957) 755.
100 W. Speransky and G. Pletschun, *Vakuumnaja Obrabotka Staly*, Moscow, 1958, p. 1.
101 G. A. Sokolov and G. N. Oiks, *Metallurg*, (1958) 16.
102 A. M. Samarin, L. M. Novik et al., *Stal'*, *19* (1959) 231.
103 P. Ya. Sorkin, *Stal'*, *20* (1960) 988.
104 K. M. Petrov, V. I. Djakonov et al., *Steel (USSR)*, (1962) 98, 113.
105 K. G. Speith, H. vom Ende et al., *Stahl Eisen*, *81* (1961) 1449.
106 W. Wilson, *J. Metals*, *13* (1961) 350.
107 J. C. Paurise, *Circ. Inform. Techn.*, *22* (1965) 2655.
108 E. S. Levin and P. V. Umrichin, *Stal'*, *25* (1965) 125.
109 A. M. Samarin, *Rev. Universelle Mines*, *101* (1958) 511.
110 J. M. Afanasev and B. V. Lincevsky, *Izv. Akad. Nauk SSSR, Otd. Tekhn. Nauk*, (1963) 76.
111 V. P. Chernobrovkin, V. S. Beljaev et al., *Fiz. Metal. i Metalloved.*, *8* (1959) 747.
112 V. T. Burtsev and R. A. Karasev, *Izv. Vysshikh Uchebn. Zavedenii Chernaya Met.*, *5* (1962) 86.
113 V. D. Sehgal and A. Mitchell, *J. Iron Steel Inst. (London)*, *202* (1964) 216.
114 B. Lux and W. Kurz, *Giessereiforschung*, *19* (1966) 43.
115 Anon., *Steel (USSR)*, *145* (1959) 106, 110.
116 A. Robens, *Iron & Coal Trades Rev.*, *182* (1960) 717.
117 J. Duflot, J. Verge et al., *J. Metals*, *13* (1961) 417.
118 N. M. Chuiko, V. B. Rutkovski et al., *Stal'*, *22* (1962) 809.
119 H. Knüppel, A. Wicher et al., *Stahl Eisen*, *82* (1962) 604.
120 Anon., *Metal Progr.*, *76* (1959) 111.
121 Anon., *Foundry*, *87* (1959) 130.
122 K. G. Speith and O. Steinhauer, *Stahl Eisen*, *83* (1963) 75.
123 E. S. Levin, *Izv. Akad. Nauk SSSR, Otd. Tekhn. Nauk*, (1963) 68.
124 E. S. Levin, K. T. Kurockin et al., *Izv. Vysshikh Uchebn. Zavedenii Chernaya Met.*, *6* (1963) 43.
125 A. le Moyne, J. Houis et al., *Circ. Inform. Tech.*, *22* (1965) 2473.
126 Anon., *Iron Steel Eng.*, (1963) 167.

127 C. L. KOBRIN, *Iron Age*, *192* (1963) 90.
128 T. E. PERRY, *Metal Progr.*, *84* (1963) 88.
129 E. HAMMARLUND AND P. SAMUELSSON, *ASEA J.*, *9* (1964) 48.
130 A. THORPE AND G. F. W. JACKSON, *Steel Times*, *189* (1964) 851.
131 T. E. PERRY, *Iron Steel Eng.*, *42* (1965) 89.
132 T. E. PERRY, *J. Iron Steel Inst. (London)*, *203* (1965) 977.
133 Anon., *Engineer*, *218* (1964) 1034.
134 G. W. S. JACKSON, A. THORPE et al., *Iron Steel Inst. (London) Spec. Rept. No. 92*, 1965, p. 111.
135 Anon., *Steel (USSR)*, April (1966) 72.
136 A. M. SAGE, *Tidsskr. Kjemi, Bergvesen Met.*, (1966) 85.
137 Anon., *Klepzig Fachber.*, *74* (1966) 476.
138 Anon., *J. Metals*, *18* (1966) 549.
139 H. STICKLER, *Elektrowärme*, *24* (1966) 376.
140 M. TIBERG, T. BUHRE et al., *ASEA J.*, *11* (1966) 47.
141 M. TIBERG, T. BUHRE et al., *Iron Steel Eng.*, (1966) 153.
142 C. B. GRIFFITH AND M. P. FEDOCK, *Trans. Vacuum Met. Conf.*, *1968*, Am. Vac. Soc., New York, 1968, p. 463.
143 Anon., *Ind. Heating*, *33* (1966) 920.
144 A. SICKBERT AND P. SCHWARZFISCHER, *Tech. Mitt. Krupp*, (1965) 117.
145 A. G. ZUBAREV, *Ogneupory*, 29 (1964) 442.
146 J. N. HORNAK AND M. A. OREHOSKI, *J. Metals*, *10* (1958) 471.
147 J. H. STOLL, *Blast Furnace Steel Plant*, *46* (1958) 595, 605.
148 Anon., *Iron Age*, *182* (1958) 81.
149 H. C. BIGGE, *Iron Steel Eng.*, *36* (1959) 128.
150 A. TIX, G. BANDEL et al., *Stahl Eisen*, *79* (1959) 472.
151 A. TIX, G. BANDEL et al., *Stahl Eisen*, *85* (1965) 1033.
152 A. TIX AND W. COUPETTE, in R. F. BUNSHAH, *Trans. Vacuum Met. Conf.*, *1959*, New York University Press, New York, 1960, p. 98.
153 A. TIX, *Tech. Mitt. Krupp*, *19* (1961) 122.
154 K. C. TAYLOR, *Iron Steel Eng.*, *40* (1963) 111.
155 J. D. SHARP, *Iron Steel Inst. (London) Spec. Rept. No. 92*, 1965, p. 50.
156 K. KNAGIGS AND P. H. BROXHAM, *Iron Steel Inst. (London) Spec. Rept. No. 92*, 1965, p. 170.
157 L. H. WLSON AND T. F. UNICK, *Blast Furnace Steel Plant*, *53* (1965) 823.
158 B. F. ZVEREV, I. I. ANSHELES et al., *Izv. Vysshikh Uchebn. Zavedenii, Chernaya Met.*, (1966) 64.
159 Z. MOTLOCH, *Hutnik Praha*, *8* (1958) 396.
160 J. D. HOBSON, *Iron Steel Inst. (London) Spec. Rept. No. 73*, 1962, p. 30.
161 A. G. MCMILLAN AND R. L. P. SALMOND, *Iron Steel Inst. (London) Spec. Rept. No. 73*, 1962, 41.
162 J. HEWITT, *Iron Steel Inst. (London) Spec. Rept. No. 77*, 1963, p. 70.
163 J. D. SHARP, *Iron and Steel (London)*, *36* (1963) 554.
164 G. BANDEL, *Berg-Hüttenmänn. Monatsh. Montan. Hochschule Leoben*, *105* (1960) 142.
165 L. TIBERG, *Jernkontorets Ann.*, *144* (1960) 757.
166 J. CHEDAILLE, M. SIVET et al., *Rev. Met. (Paris)*, *60* (1963) 663.
167 R. F. BUNSHAH (Ed.), *Trans. Vacuum Met. Conf., New York, 1957*, Reinhold, New York, 1958.
168 G. E. DANNER AND E. DYBLE, *Metal. Progr.*, *79* (1961) 74.
169 L. SZÖKE, *Kohasz. Lapok*, *94* (1961) 135, 151.
170 Anon., *Steel (USSR)*, *154* (1964) 32.
171 A. SICKBERT, *Iron Steel Inst. (London) Spec. Rept. No. 92*, 1965, p. 95.
172 H. MORIKAWA, T. IKEMI et al., *The Japan Steel Works Co. J.*, 1966.
173 A. SICKBERT AND P. SCHWARZFISCHER, *Rev. (ILAFA) Latinoam. Siderurgia*, 1966, p. 37.
174 H. MORIKAWA et al., *Tetsu To-Hagane*, *52* (1966) 443.
175 P. SCHWARZFISCHER, *Hutnicke Aktuality*, (1966) 11.
176 A. SICKBERT AND P. SCHWARZFISCHER, *V. Congreso Latinoam. Siderurgia, Santiago, 1966*, p. 71.

REFERENCES

177 P. Dewsnap and G. Hoyle, *J. Iron Steel Inst. (London)*, 203 (1965) 988.
178 Anon., *Iron Steel Eng.*, 43 (1966) 223.
179 F. Harders, *J. Iron Steel Inst. (London)*, 190 (1958) 306.
180 F. Harders, *Klepzig Fachber.*, 66 (1958) 171.
181 H. J. Därmann, *Berg-Hüttenmänn. Monatsh. Montan. Hochschule, Leoben*, 105 (1960) 144.
182 Anon., *Iron Age*, 187 (1961) 104.
183 Anon., *Steel (USSR)*, 149 (1961) 76, 78.
184 Anon., *Iron Steel Eng.*, 38 (1961) 183.
185 A. J. Park and R. F. Kowal et al., *J. Metals*, 17 (1965) 897.
186 F. Cousin, *Met. Constr. Mecan.*, 98 (1966) 79.
187 K. Matsuda, K. Oba et al., *Yawata Iron & Steel Co. J.*, 1966.
188 H. Matsuoka and R. Kato, *Sumitomo Metal Industries Co. J.*, 1966.
189 K. Ushijima and Ta. Ikeda, *Sumitomo Metal Industries Co. J.*, 1966.
190 F. Harders, F. Oeters et al., *Hoesch Ber.*, 4 (1966) 7.
191 K. Kawamoto, H. Matsuoka et al., *Sumitomo Metals*, 1966, p. 148.
192 E. G. Schempp, *AISE Pittsburgh Sect. Meeting*, 1966.
193 W. Sieckmann, *V. Congr. Latinoam. Siderurgia*, Santiago, 1966, p. 150.
194 H. Matsuoka and R. Kato, *Tetsu To Hagane*, 52 (1966) 446.
195 P. J. Wooding and W. Sieckmann, *Metal Progr.*, 77 (1960) 116.
196 F. Oeters, K. Rüttiger et al., *Hutnicke Aktuality*, No. 11, 1966.
197 U. Kiyote and T. Ikeda, *Tetsu To Hagane*, 52 (1966) 448.
198 W. L. Finlay and C. D. Preusch, *Yearbook Am. Iron Steel Inst.*, 1961, p. 133.
199 W. L. Finlay and C. D. Preusch, *Blast Furnace Steel Plant*, 49 (1961) 867.
200 P. J. Wooding and W. Sieckmann, *Trans. Vacuum Met. Conf., New York, 1960*, Interscience, New York, 1961, p. 243.
201 Anon., *Ind., Heating*, 28 (1961) 2201.
202 P. L. Jackson and W. M. Hyams, *AISI-Reprint*, (1963) 31.
203 H. Kutscher, *Tech. Rundschau Bern.*, 54 (1962) 2.
204 R. Hentrich, *Stahl Eisen*, 82 (1962) 899.
205 E. J. Fitzgerald, *J. Metals*, 13 (1961) 431.
206 C. D. Preusch, *Reg. Tech. Meetings Am. Iron Steel Inst.*, 1961, p. 63.
207 W. C. Kollmann and C. D. Preusch, *Elec. Furnace Conf. Proc.*, 19 (1961) 23.
208 W. L. Finlay and C. D. Preusch, *Blast Furnace Steel Plant*, 49 (1961) 867.
209 C. D. Preusch, *Metal Progr.*, 81 (1962) 80, 116.
210 K. Brotzmann, *Klepzig Fachber.*, 69 (1961) 168.
211 E. G. Schempp, *Iron Steel Eng.*, (1967) 89.
212 F. Harders, H. Knüppel et al., *Stahl Eisen*, 79 (1959) 267.
213 G. J. McManus, *Iron Age*, 184 (1959) 65.
214 R. Sevin, *J. Four Elec.*, 71 (1966) 203.
215 H. Knüppel, A. Drevermann et al., *Stahl Eisen*, 79 (1959) 414.
216 W. Sieckmann and E. G. Schempp, *Iron Steel Eng.*, 40 (1963) 117.
217 F. Oeters, K. Rüttiger and H. Kutscher, *Rev. Met. (Paris)*, 64 (1967) 711.
218 H. Kutscher, F. Oeters et al., *Tech. Forsch.*, 41 (1966) No. 170.
219 I. K. Maclean and J. M. Lupton, *Iron Steel Inst. (London) Spec. Rept.* No. 92, 1965, p. 176.
220 H. Mori, K. Tanizawa et al., *Yawata Iron & Steel Co. J.*, 1966.
221 C. H. Pottgiesser and H. J. Därmann, *Stahl Eisen*, 79 (1959) 463.
222 C. H. Pottgiesser, H. A. Wicher et al., *Stahl Eisen*, 79 (1959) 468.
223 K. Brotzmann, *Iron Steel Inst. (London) Spec. Rept.* No. 92, 1965, p. 148.
224 E. L. Fogleman and H. W. Wiltna, *J. Metals*, 18 (1966) 623.
225 K. Matsuda, K. Oba et al., *Tetsu To Hogane*, 52 (1966) 451.
226 J. C. C. Leach and W. Sorby, *Iron Steel Inst. (London) Spec. Rept.* No. 87, 1965, p. 52.
227 A. Capuano, J. K. Preston et al., *Iron Steel Eng.*, 43 (1966) 85.
228 K. Rüttiger and A. Diener, *Trans. Vacuum Met. Conf., New York, 1966*, Am. Vac. Soc., Boston, Mass., 1967, p. 384.
229 I. K. Maclean and H. G. Trotter, *Steel Times*, 192 (1966) 658.

230 F. W. STARRATT, *J. Metals*, *10* (1958) 465.
231 H. THIELMANN AND H. MAAS, *Stahl Eisen*, *79* (1959) 276.
232 Anon., *Steel Times*, *191* (1965) 328.
233 G. B. FORSTER, *J. Metals*, *18* (1966) 628.
234 H. MAAS, *Berg-Hüttenmänn. Monatsh. Montan. Hochschule, Leoben*, *165* (1960) 151.
235 R. HIRATA, *J. Metals*, *18* (1966) 617.
236 M. WAHLSTER, A. DAMMER et al., *Iron Steel Inst. (London) Spec. Rept.*, No. 92, 1965, p. 129.
237 A. DAMMER, *Intern. Conf. Quality Improvement of Steel, Prague, 1965*, Ref. No. 11.
238 M. WAHLSTER AND H. H. REICHEL, *Trans. Vacuum Met. Conf., New York, 1966*, Am. Vac. Soc., Boston, Mass., 1967, p. 415.
239 E. D. SCHERRER, *J. Metals*, *18* (1966) 977.
240 M. WAHLSTER AND H. REICHEL, *Rheinstahl-Tech.*, *3* (1966) 80.
241 S. WATANABE, H. WATANABE, K. ASANO, K. MATSUI AND N. NAKAYAMA, *Tetsu To Hagane*, *50* (1964) 456.
242 M. HIRASE, *Iron Steel Inst. (London) Spec. Rept.* No. 92, 1965, p. 154.
243 Anon., *Iron Age*, *196* (1965) 90.
244 T. KRAUS AND O. WINKLER, *Iron Steel Inst. (London) Spec. Rept.* No. 92, 1965, p. 45.
245 Anon., *Iron Age Metalworking Int.*, *5* (1966) 27.
246 G. HOYLE, *Iron Steel Inst. (London) Spec. Rept.* No. 73, 1962, p. 24.
247 G. HOYLE, *J. Iron Steel Inst. (London)*, *200* (1962) 605.
248 Anon., *Brit. Steelmaker*, *31* (1965) 188.
249 Anon., *Engineer*, *219* (1965) 873.
250 G. HOYLE AND M. A. RAMSBOTTOM, *Iron Steel Inst. (London) Spec. Rept.* No. 92, 1965, p. 135.
251 Anon., *Metal Progr.*, *74* (1958) 142, 144, 145, 146.
252 J. LEVAUX AND M. NEPPER, *J. Iron Steel Inst. (London)*, *192* (1959) 77.
253 P. BASTIEN, *Rev. Met. (Paris)*, *57* (1960) 815.
254 A. A. MARKARYANTS, J. D. SMIRNOV et al., *Stal'*, *20* (1960) 148.
255 V. SCALISE, *Iron & Coal Trades Rev.*, *181* (1960) 717.
256 V. SCALISE, A. DE NEGRI AND P. PICARDI, *Metallurgia Ital.*, *52* (1960) 237; see also *J. Iron Steel Inst. (London)* *195* (1960) 260.
257 V. SCALISE, *J. West Scot. Iron Steel Inst.*, *68* (1960/61) 127.
258 V. SCALISE et al., *J. Iron Steel Inst. (London)*, *195* (1960) 260.
259 I. MYDLARZ, *Neue Hütte*, *6* (1961) 259.
260 M. STANKIEWICZ AND J. OBREBSKI, *Hutnik*, *28* (1961) 37.
261 G. E. DANNER AND G. TAYLOR, *Trans. Vacuum Met. Conf., New York, 1960*, Interscience, New York, 1961, p. 225.
262 A. WOJCIK, *Hutnik*, *30* (1963) 407.
263 G. J. MCMANUS, *Iron Age*, *194* (1964) 127.
264 A. SCHÖBERL AND H. RAISKY, *Radex Rundschau*, *(1964)* 341.
265 A. FIALA, *Sb. Praci Vyzkum. Zkusebniho Ustavu Leninovych Zadovu v Plzni*, (1965) 250.
266 G. HENKE AND W. HESS, *Stahl Eisen*, *79* (1959) 405.
267 H. J. KOPINECK, E. SCHULTE et al., *Stahl Eisen*, *82* (1962) 846.
268 Z. EMINGER AND F. KINSKY, *Hutnicke Listy*, *11* (1956) 345.
269 Z. MOTLOCH, *Hutnicke Listy*, *13* (1958) 1037.
270 H. KNÜPPEL AND F. OETERS, *Arch. Eisenhüttenw.*, *33* (1962) 729.
271 A. E. NEHRENBERG, *Iron Age Metalworking Int.*, *2* (1963) 28.
272 J. D. SHARP, *Iron & Steel (London)*, *39* (1966) 466.
273 H. ZAKOWA, *Hutnik*, *24* (1957) 234.
274 E. SCHULTE, *Klepzig Fachber.*, *69* (1961) 179.
275 Anon., *Iron & Coal Trades Rev.*, *179* (1959) 1183.
276 A. M. AKSOY, *J. Metals*, *11* (1959) 468.
277 J. VERÖ, *Kohasz. Lapok*, *92* (1959) 569.
278 K. G. LEWIS, *Iron & Steel (London)*, *33* (1960) 3.
279 Anon., *Klepzig Fachber.*, *69* (1961) 107.
280 J. O. EDSTRÖM, *Jernkontorets Ann.*, *146* (1962) 549.

281 A. Mund, *Stahl Eisen*, 82 (1962) 1485.
282 A. Randak and R. Hentrich, *Tech. Ber. Stahlwerk Südwestfalen*, July 1963.
283 C. Belvedere and A. Concina, *Met. Ital.*, 54 (1962) 131.
284 L. Colombier, *Rev. Met. (Paris)*, 60 (1963) 115.
285 W. Coupette, *Vakuum-Tech.*, 13 (1964) 80, 185.
286 H. Hellbrügge and A. Tix, *Stahl Eisen*, 84 (1964) 1723.
287 J. Verge, *Rev. Met. (Paris)*, 61 (1964) 755.
288 J. H. Flux, *Iron Steel Inst. (London) Spec. Rept. No. 92*, 1965, p. 1.
289 T. E. Perry, *Blast Furnace Steel Plant*, 53 (1965) 1017.
290 T. W. Perry, *Rev. Latinoam. Siderurgia*, 6 (1965) 29.
291 K. C. Barraclough, *Steel Times*, 191 (1965) 96.
292 C. J. Hunter and J. H. Hornak, *Blast Furnace Steel Plant*, 53 (1965) 699.
293 H. Schenck, *Rev. Latinoam. Siderurgia*, 6 (1966) 28.
294 F. Sperner, *Chemiker Ztg.*, 96 (1966) 43.
295 S. Eketorp, *Vakuumbehandlingens Möjligheter*, Almqvist & Wiksell, Uppsala, 1966.
296 F. Sperner, *Vakuum-Tech.*, 15 (1966) 29.
297 K. Narita, *Kobe Steel Eng. Rept.*, 15 (1965) 200.
298 K. Narita, *Kobe Steel Eng. Rept.*, 15 (1965) 261.
299 Anon., *Kobe Plant Steel Co. J.*, 1966.
300 J. C. C. Leach, *Steel Times*, Oct. (1966) 197.
301 H. Stickler, *ASEA J.*, 39 (1966) 90.
302 M. A. Orehoski, *Trans. Conf. Vacuum Metallurgy, New York, 1966*, Am. Vac. Soc., Boston, Mass., 1967, p. 321.
303 N. Miholich, *Vakuumbehandling av flytande Stal*, Stallaval, Linköping, Sweden, 1961.
304 S. Maekawa, Y. Nakagawa et al., *Tetsu To Hogane*, 46 (1960) 1233.
305 J. Berve and H. Gravenhorst, *Stahl Eisen*, 82 (1962) 1036.
306 A. M. Samarin, in J. F. Elliot (Ed.), *Steelmaking: The Chipman Conference*, M.I.T. Press, 1965, p. 269.
307 A. V. Bradshaw and F. D. Richardson, *Iron Steel Inst. (London) Spec. Rept. No. 92*, 1965, p. 24.
308 T. F. Meadowcroft, *Steel Plant Furnaces*, 1966, p. 1285.
309 I. Dragomir, *Metallurgia*, (1966) 157.
310 L. Pöcze and O. Fölkl, *Kohasz. Lapok*, 97 (1964) 153.
311 A. G. Shalimov, G. N. Okorokov et al., *Izv. Akad. Nauk SSR*, (1964) 35.
312 Z. Klisiewicz, *Hutnik*, 33 (1966) 357.
313 Z. Klisiewicz, *Prace Inst. Hutniczych*, 18 (1966) 321.
314 K. Brotzmann, *Schweiz. Arch. Angew. Wiss. Tech.*, 28 (1962) 473.
315 G. H. Ockenhouse and J. E. Werner, *J. Metals*, 18 (1966) 52.
316 T. Mazanek and K. Mamro, *Hutnik*, 32 (1965) 309.
317 G. N. Oiks, P. P. Matewossjan et al., *Stal'*, 20 (1960) 308.
318 H. Knüppel, K. Brotzmann et al., *Stahl Eisen*, 85 (1965) 675.
319 H. C. Myers, *Iron Steel Inst. (London) Spec. Rept. No. 92*, 1965, p. 73.
320 G. B. Boccone and R. De Martini, *Met. Ital.*, 59 (1967) 64.
321 J. C. C. Leach, *E.S.C. Rev.*, (1965) 15.
322 R. F. O'Railly, *D. M. I. C. Battelle Inst.*, DMIC-Rept. 200, 1964.
323 J. Zlatnikova and I. Zlatnik, *Neue Hütte*, 10 (1965) 649.
324 I. A. Baranov, G. N. Oiks et al., *Izv. Vysshikh Uchebn. Zavedenii Chernaya Met.*, 5 (1962) 78.
325 O. C. Fatkullin, *Metallurg*, 10 (1965) 20.
326 C. D. Church, *J. Metals*, 18 (1966) 62.
327 G. Riedel, *VDI Z.*, 107 (1965) 1407.
328 R. D. Smith, *J. Metals*, 18 (1966) 59.
329 S. Yuki, K. Kajikawa et al., *Sanyo Special Steel Co. J.*, 1966.
330 A. A. Conrad, C. W. Darby et al., *S.A.E. J.*, 74 (1966) 53.
331 J. C. C. Leach and W. Sorby, *Iron Steel Inst. (London) Spec. Rept. No. 92*, 1965, p. 119.
332 J. Skala, *Hutnicke Listy*, 21 (1966) 311.

333 J. R. BARRICK, *Trans. Vacuum Met. Conf. New York, 1966*, Am. Vac. Soc., Boston, Mass., 1967, p. 370.
334 J. B. DABNEY, M. C. FLEMINGS et al., *Trans. Am. Foundrymens' Soc.*, 69 (1961) 778.
335 W. M. P. HERBERT, *Iron Steel Eng.*, (1966) 131.
336 A. ROSE AND A. WICHER, *Arch. Eisenhüttenw.*, 34 (1963) 617.
337 J. BECVAR, *Hutnik Praha*, 14 (1964) 271.
338 J. KERMES AND J. ELLINGER, *Sb. Praci Vyzkum. Zkusebniho Ustavu Leninovych Zavodu v Plzni*, (1965) 315.
339 YU. A. STAROVOITOV AND A. B. KUSLITSKY, *Fiz.-Khim. Mekhan. Mat.*, 2 (1966) 637.
340 T. KATO, K. MATSUDA et al., *Iron Steel Inst. (London) Spec. Rept. No. 92*, 1965, p. 164.
341 G. N. OIKS et al., *Stal'*, 25 (1965) 711.
342 I. M. ZUEV, G. A. SOKOLOV et al., *Izv. Vysshikh Uchebn. Zavedenii Chernaya Met.*, (1966) 39.
343 I. M. ZUEV, *Izv. Vysshikh Uchebn. Zavedenii Chernaya Met.*, 9 (1966) 39.
344 W. J. JACKSON, *B.S.C.R.A. J.*, (1966) 15.
345 W. COUPETTE, *Giesserei*, 53 (1966) 224.

Section 2
346 E. GEBHARDT AND H. D. SEGHEZZI, *Z. Metallk.*, 50 (1959) 521.
347 E. GEBHARDT AND R. ROTHENBACHER, *Z. Metallk.*, 54 (1963) 623.
348 E. GEBHARDT, W. DÜRRSCHNABEL AND G. HÖRZ, *J. Nucl. Mater.*, 18 (1966) 119.
349 E. GEBHARDT, H. D. SEGHEZZI AND W. DÜRRSCHNABEL, *J. Nucl. Mater.*, 4 (1961) 241.
350 P. COTTERILL, Hydrogen embrittlement of metals. In B. CHALMERS (Ed.) *Progress in Materials Science*, Vol. 9, Pergamon, Oxford, 1962.
351 L. S. MOROZ AND B. B. ČEČULIN, *Vodorodnaja chrupkost' metallov*, Izd. Metallurgija, Moscow, 1967.
352 E. GEBHARDT, E. FROMM AND D. JAKOB, *Z. Metallk.*, 55 (1964) 423.
353 E. GEBHARDT, H. D. SEGHEZZI AND E. FROMM, *Z. Metallk.*, 52 (1961) 464.
354 C. J. SMITHELLS, *Metals Reference Book*, Vol. II, Butterworth, London, 1962.
355 E. GEBHARDT, H. D. SEGHEZZI AND H. KEIL, *Z. Metallk.*, 53 (1962) 524.
356 W. ESPE, *Werkstoffkunde der Hochvakuumtechnik*, Vol. 1, VEB Deutscher Verl. d. Wissensch., Berlin, 1959.
357 N. A. GALAKTIONOVA, *Vodorod v metallach*, Izd. Metallurgija, Moscow, 1967.
358 G. V. KARPENKO AND R. I. KRIPYAKEVICH, *Vljanie Vodoroda na svojstra stali*, Izd. Metallurgizdad, Moscow, 1962.
359 M. HANSEN, *Constitution of Binary Alloys*, McGraw-Hill, New York, 1958.
360 J. D. FAST, *Interaction of Metals and Gases*, Vol. 1, Philips Technical Library, Eindhoven, 1965.
361 E. GEBHARDT AND E. FROMM, *Metall-Gas-Systeme*, Springer, Berlin, in preparation.
362 O. KUBASCHEWSKI AND E. U. EVANS, *Thermochemical Data of Alloys*, Pergamon, London, 1967.
363 J. F. ELLIOTT AND M. GLEISER, *Thermochemistry for Steelmaking*, Vol. 1, Addison-Wesley, Reading, Mass., 1960.
364 W. C. HERAEUS, *Vakuumtechn. Tabellen und Diagramme*, p. 17 (company publication).
365 E. GEBHARDT, E. FROMM AND D. JAKOB, *Z. Metallk.*, 55 (1964) 432.
366 E. FROMM, *Z. Metallk.*, 57 (1966) 540.
367 E. FROMM, *Z. Metallk.*, 57 (1966) 477.
368 E. FROMM AND O. HEINKEL, *Z. Metallk.*, 58 (1967) 805.
369 E. FROMM AND G. SPÄTH, *Z. Metallk.*, 59 (1968) 65.
370 E. GEBHARDT, E. FROMM AND U. ROY, *Z. Metallk.*, 57 (1966) 682.
371 E. FROMM AND H. JEHN, *Z. Metallk.*, 58 (1967) 61.
372 E. FROMM AND H. JEHN, *Z. Metallk.*, 58 (1967) 120.
373 G. HÖRZ, *Z. Metallk.*, 57 (1966) 703.
374 K. J. LAIDLER, *Chemical Kinetics*, McGraw-Hill, New York, 1950.
375 G. HÖRZ AND E. GEBHARDT, *Z. Metallk.*, 57 (1966) 812.
376 E. GEBHARDT, W. DÜRRSCHNABEL AND G. HÖRZ, *J. Nucl. Mater.*, 18 (1966) 149.

377 G. Hörz, *Metall*, 22 (1968) 1201.
378 W. Jost, *Diffusion in Solids, Liquids, Gases*, Academic Press, New York, 1952.
379 C. J. Smithells, *Metals Reference Book*, Vol. II, 4th edn., Butterworth, London, 1967.
380 G. Hörz and E. Gebhardt, *Z. Metallk.*, 57 (1966) 737.
381 E. Gebhardt, E. Fromm and D. Jakob, in F. Benesovsky (Ed.), *Metalle für die Raumfahrt*, (5. Plansee-Seminar), Springer Verlag, Vienna, 1965.
382 G. Hörz, *Z. Metallk.*, 60 (1969) 50.
383 G. Hörz, *Z. Metallk.*, 60 (1969) 115.
384 G. Hörz, *Z. Metallk.*, 60 (1969) 121.
385 J. Drowart, A. Pattoret and S. Smols, *Proc. Brit. Ceram. Soc.*, 8 (1967) 67.
386 J. Drowart and P. Goldfinger, *Angew. Chem.*, 79 (1967) 589.
387 P. O. Schissel and O. C. Trulson, *J. Chem. Phys.*, 43 (1965) 737.
388 G. Melchior, *Vakuum-Tech.*, 18 (1969) 1.

Chapter V

Vacuum Melting

Sections 1–3

H. C. CHILD AND G. E. OLDFIELD

Section 4

R. BAKISH

Section 5

A. LAWLEY

Introduction to Sections 1–3

Vacuum melting started the transition from the laboratory to the shop floor in 1917, when Rohn[1] melted nickel-base alloys by resistance heating. By 1923, Heraeus Vacuumschmelze in Germany had been founded, operating vacuum induction furnaces. By 1926, two 4-ton units with a 350-kW mains frequency power supply were producing mainly thermocouple materials and resistance heating alloys.

The economic factors were against the widespread growth of the process. The premium of 20 cents per kg at the ingot stage was stated by Rohn[2] to be justified only for applications involving thin wire. The high cost was due to the limitations of the vacuum pumps causing prolonged heat times of 14–15 hours, and to short hearth lives.

Consequently developments languished, although Telegraph Construction and Maintenance Co. in the U.K. started production of electric and soft magnetic alloys in the late 1930's.

The next stimulus to development, aided by the availability of much improved vacuum pumps originating from the atomic energy industry, was the advent of titanium in the U.S.A. in the early 1950's. "Home-made" consumable electrode furnaces of several tons capacity were soon in production and were quickly followed by more sophisticated units designed and manufactured by established vacuum engineering and electric furnace concerns.

A further stimulus came rapidly from the need for vacuum induction melting of wrought nickel-base alloys for turbine blades. This led to the installation of many medium-sized vacuum induction installations of up to 1 ton in size.

By about 1956 the potential of vacuum arc remelting for steels and nickel-base alloys had been realised, the extremely sound homogeneous ingot produced by this process having resulted in forgings for gas turbine discs, shafts and casings of much improved quality and properties compared with their air-melted counterparts. These early developments have been described by, among others, Dyrkacz, De Vries and Pitler[3] and Child and Harris[4].

The growing demand for large forgings in nickel-base alloys led to the realisation about 1960 that better control of analysis and quality could be achieved if the electrodes for the vacuum arc process were cast in a vacuum induction melting furnace. This so-called "double vacuum melting" technique is widely used and has stimulated the development of very large vacuum induction furnaces to produce the electrodes most economically.

At the present time, vacuum induction melting furnaces of up to 30-tons capacity are operational, and a 100-ton unit is already under construction. Details of the design and operation of the 30-ton unit at Latrobe have been described by Schlatter and Simkovich[5].

The largest vacuum arc furnace installation is that at Midvale–Heppenstall described by Busby[6]. It can melt ingots of up to 50 tons in weight and 1.5 m diameter. As far as is known to the authors, no larger units are at present operational though furnaces of up to 100–200 tons are under consideration.

The production capacity per annum of the two processes in the Western World in 1967 has been reviewed by Nisbet[7] and Child[8].

	W. Europe[8]	U.S.A.[7]
Vacuum arc remelting	90,000 tons	250,000 tons
Vacuum induction melting	15,000 tons	250,000 tons

The equivalent capacity of the two processes in the U.S.A. reflects the preponderance of double vacuum melted nickel-base alloys. By contrast, in Europe much of the production is of single vacuum arc melted steels using the basic electric arc or air induction process for electrode casting.

1. Vacuum Melting in Resistance Furnaces

Although resistance-heated melting furnaces were used initially by Rohn[1], the technique has never found favour for production melting. This is due to several factors.

Supplying the heat from an external source is thermally inefficient, but a more important aspect is that the crucible material cannot be insulated and must withstand temperatures in excess of that of the bath throughout its wall thickness. Indeed, outer wall temperatures several hundred degrees in excess of the bath temperature are incurred when effective heating rates are used with refractory oxide crucibles.

Refractory limitations have always been a major problem in vacuum melting and the selection of induction heating instead of resistance heating enables temperatures to be minimised and effective insulation to be used at the outer wall.

Resistance heating by means of a graphite spiral or other suitable configuration is used in conjunction with a graphite crucible for melting small samples for vacuum fusion analysis of gas content. For other laboratory work, indirect heating of ceramic crucibles has been used, with resistors of the refractory metals molybdenum, tungsten and tantalum, and of graphite.

2. Vacuum Induction Melting

In the vacuum induction melting process virgin alloys or scrap are melted down under vacuum or inert atmosphere, to avoid oxidation and contamination from the atmosphere. The molten charge is then refined by exposure to the vacuum and subsequently cast, mostly under vacuum. The product may take the form of ingots for subsequent forging, rolling or extrusion, cast electrodes for vacuum arc remelting, castings, or cast stock bars. The last named are themselves used as the charge for remelting in vacuum precision casting furnaces, generally induction heated.

2.1 Design features of modern plant

2.1.1 General layout of vacuum chambers (considered with respect to pouring technique)

A noteworthy feature of vacuum induction melting furnaces has been the lack of standardisation of design, particularly in the case of the larger units. This is due to the widely varying requirements of the individual producers and to the limited number of furnaces built per year, the latter factor making obsolete a design before an appreciable number of units can be produced.

It is appropriate, however, to consider the basic layouts of vacuum chambers and how the choice is influenced by the producer's requirements, particularly the pouring technique to be used.

The simplest design of melting chamber, as shown in Fig. 1(a), is a small vertical tank with a fixed, bottom-poured crucible situated directly above the ingot mould. The use of such units has been severely restricted on account of the difficulty of designing and operating bottom-poured tapping devices. The technique has, however, found favour for those metals which permit the use of graphite crucibles, notably uranium and its alloys. In the case of bottom pouring it is practical to have the induction coil outside of the vacuum chamber. Figure 2 shows such a unit for uranium melting based on a silica sleeve with a susceptor, crucible and mould situated inside.

Where a lip-tilted crucible is necessary, Fig. 1(b) shows the simplest type of layout. It has the advantage of compactness and has been widely used for small units of up to a few hundred pounds capacity. Access to the crucible and chamber

References pp. 642–647

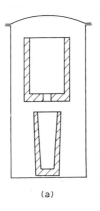
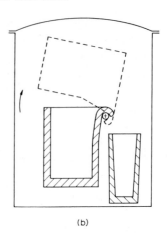
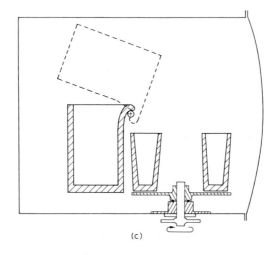
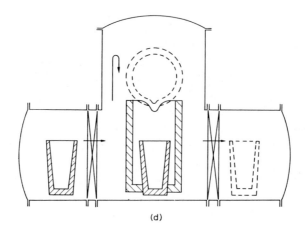

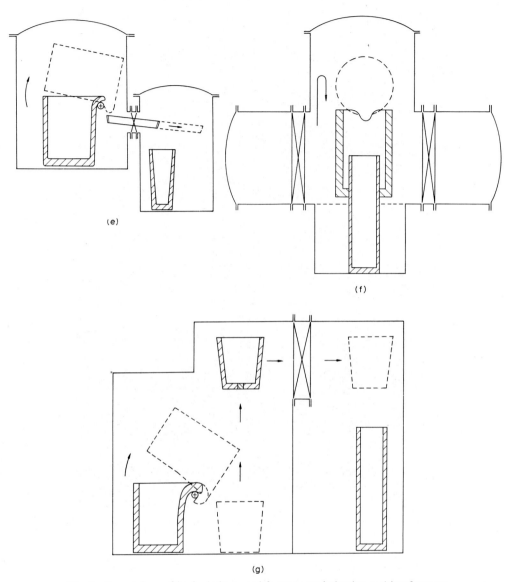

Fig. 1. Comparison of basic designs used for vacuum induction melting furnaces.
(a) Fixed crucible, bottom-poured; (b) Vertical tank with lip-poured crucible; (c) Horizontal tank with rotating mould table; (d) Use of twin mould locks; (e) Unit with separate mould chamber, launder and launder lock; (f) Modification to type (d) for long electrode mould; (g) Use of ladle with hoist, transfer and appropriate vacuum lock.

is restricted, but this presents no problem in small units. The various services must be introduced through the lid, a feature which is acceptable for such small units but is undesirable on large plant.

The horizontal design of tank shown in Fig. 1(c) is the most widely used. The

References pp. 642–647

Fig. 2. Vacuum induction quartz tube furnace used for melting nuclear fuels and subsequent casting in moulds by bottom pouring. The moulds are pre-heated by the lower induction coil.

axis of the tank may be parallel to or at right angles to the tilting axis of the crucible, depending on the preferred layout of the ancillaries. The various service devices can be fixed to the main tank and the "door" used only for access. It is now possible to elongate the tank to accommodate such devices as circular mould tables, and a typical 300 kg unit of this basic design is illustrated in Fig. 3. Where provision for more than one mould is required in the basic design shown in Fig. 1(b), the moulds must be arranged round the periphery of the tank or a tank of unnecessarily large volume is required.

The use of a horizontal tank enables an important design feature, common in modern European furnaces, to be incorporated. Either the furnace shell, or that

Fig. 3. 600-lb Wild Barfield vacuum induction melting furnace at Jessop–Saville Ltd.

part of the structure carrying the coil, is movable, exposing the crucible to facilitate exchange or repairs.

These relatively simple designs are possible where a mould lock is not used. In practice the provision of a mould lock is normally considered desirable for units of 1-ton capacity and above.

In its simplest form the mould lock is an extension to the tank of the necessary dimensions, as shown in Fig. 1(d), and is isolated from the tank by a large gate valve. A typical unit is shown in Fig. 4. Where necessary two mould locks can be provided to permit the continuous movement of moulds from one side of the furnace to the other. This facility expedites the teeming process where more moulds are involved than can reasonably be accommodated on one mould trolley. The furnace can be operated continuously without breaking the vacuum in the melting chamber.

On account of the high cost of gate valves long enough to pass mould trolleys, several designs have been evolved whereby the hot metal may be teemed through a tundish to a separate casting chamber. The melting chamber and the mould chamber are separated by a small lock, which is sometimes part of the tundish arrangement. In one form of the arrangement, the launder connecting the two chambers is coaxial with the tilting axis of the crucible. Casting is carried out by pivoting the complete melting tank and crucible about this axis. This permits the

References pp. 642–647

Fig. 4. 2-ton G.E.C.–V.I.A. vacuum induction furnace at Jessop–Saville Ltd.

launder to remain in the casting position. A disadvantage of this design is the difficulty of providing services to the main tank.

A more versatile alternative, as shown in Fig. 1(e), is to cast through a small horizontal launder which can be moved in and out of a vacuum lock. A further alternative is to cast vertically through a lined vacuum lock into a mould chamber situated below. In both cases considerable care in design and operation is required to avoid heat, and hot metal from damaging the lock. A schematic illustration of such a design is shown in Fig. 5.

As will be discussed later, a major use of vacuum induction furnaces is the production of very long electrodes (*e.g.* 4.6 m) for vacuum arc remelting, and it is

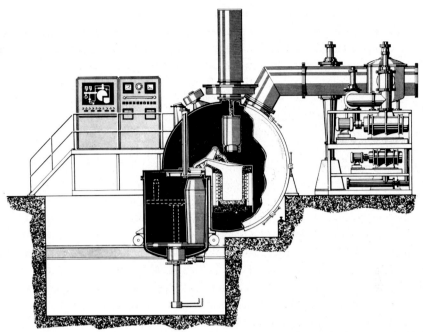

Fig. 5. Schematic view of vacuum induction furnace with pouring lock. (Courtesy Leybold–Heraeus)

the problem of designing a unit for this purpose most economically in terms of cost and space that has influenced thinking on the most recent vacuum induction plant. The provision of a mould chamber, as shown in Fig. 1(d), of adequate height over its whole length to cast such long electrodes is expensive in both space and capital cost. Several variants have accordingly been used or proposed. As normally only one or two such large electrodes are cast per heat, an obvious approach is to provide a deep well locally in front of the melting unit, as shown in Fig. 1(f), to enable the rest of the mould chamber to be smaller and to be used only in the case of the production of a heat split into many small ingots. In this case the moulds are normally charged through a lid in the mould chamber. It will be appreciated that designs such as that shown in Fig. 1(e) and Fig. 5 lend themselves to this approach, as the mould chamber is completely independent of the furnace chamber and the furnace chamber can be isolated during the charging of the long moulds. Such a procedure is not possible in the case of designs such as Fig. 1(f).

An alternative approach to minimising the height of unit required for these large moulds is the provision of a ladle within the vacuum system capable of taking the full crucible weight of liquid metal[9]. If this is provided with a hoist arrangement the overall height of units of the basic design shown in Fig. 1(d) can be minimised. Alternatively, both lift and transport of the ladle can be used as shown in the basic design, Fig. 1(g).

References pp. 642–647

Another feature which has a major influence on the basic design of the vacuum chamber is whether or not hot metal charging is to be used. Most of the basic designs described above can be fitted with a bulk charging device operating through a vacuum lock of comparable dimensions with the crucible diameter. In the case of hot metal charging, however, there are a number of other features to be considered.

If the metal is to be introduced at atmospheric pressure it is important that an access lid be provided allowing for the minimum free drop of hot metal into the crucible. This is most easily provided by the use of a vertical melting chamber with a separation line near the top of the crucible. Designs have also been considered where the entire crucible assembly can be moved forward on rails and through a vertical access door to a charging position outside the vacuum chamber. The design incorporating both lift and travel of a ladle (Fig. 1(g)) can be used equally well for charging the crucible with hot metal as for teeming.

If it is considered desirable to stream degas the hot metal charge there must be provision for either using a pony ladle situated directly above the melting crucible, or alternatively accommodating the charging ladle on a vacuum seal in a similar position. Here again, it is important that the free fall should not be too high owing to the difficulty of constraining the "exploding" stream of degassed metal.

In the early days of construction of vacuum induction furnaces, cylindrical melting tanks were favoured using initially double wall construction and later coils of copper tubing brazed in position for cooling. Lids and doors were circular or rectangular with rounded ends in order to facilitate the use of conventional grooved "O" ring seals.

Stainless steel is to be preferred in view of its corrosion resistance and reduced tendency to heat in the presence of stray magnetic fields, the latter feature enabling tank volumes to be minimised. On the other hand, cost, higher thermal conductivity and melting point favour the use of mild steel. The last-named attributes reduce the risk of furnace puncture according to Taylor[10].

In recent years novel seals such as the "Delta-shaped seal"[11] have been developed for awkward sealing faces, *e.g.* square and rectangular flanges. The main advantage of the seal is that it can be applied to any sealing surface and eliminates expensive machining of the conventional dovetail groove, into which a trapped seal is inserted. These seals were made from round section neoprene material and only a very small percentage of the rubber protruded above the flange. In the Delta-shaped seal, two round section steel bars are bent to the configuration of the flange and tack welded to the flange with a constant gap. A neoprene seal of ∩ section is trapped in the space between the bars, with approximately 30% of the rubber section protruding. This form of sealing has facilitated the manufacture of very large chambers, as the sealing flange may be prefabricated by machining in sections and then welded to the vessel on site. The feature also allows for considerably more flange distortion than conventional grooved seals.

With such a seal, cheaper forms of construction involving rectangular tanks and doors strengthened by ribbing may be used. This has given added flexibility to furnace design and these factors have shifted some of the emphasis in selection of layout, particularly in that they enable better use to be made of the volume of the melting and casting chambers.

2.1.2 *Charging and ancillary devices*

The various services which may be required in a vacuum induction furnace are as follows:
(a) Bulk charger
(b) Hot metal charging
(c) Alloy charging
(d) Liquid metal sampling
(e) Bath temperature measuring
(f) Bridge breaker
(g) Crucible lid
(h) Mould charging
(i) Tundish charging
(j) Casting ladle facility
(k) Hot topping facility.

(a) Bulk charger

Although on small units it is not uncommon to hand-charge the crucible under atmospheric conditions, on larger units it is the normal procedure to operate with the melting chamber under continuous vacuum and to bulk charge through a vacuum lock. This lock is normally situated directly above the crucible and has a diameter at least as great as that of the crucible. The material to be charged, be it virgin alloy or scrap, is contained in a drop-bottom basket or a consumable can fabricated from the base alloy being melted, as illustrated in Fig. 6.

On many units the bulk charging lock is also used for some of the other services described below (see Fig. 7).

(b) Hot metal charging

Where hot metal is to be charged as an alternative to cold metal, this may be conveniently carried out by removing the bulk charging vessel and substituting a small pony ladle over the charging lock. Alternatively, the transfer ladle itself may be located on a suitable seal on the furnace chamber (Fig. 8). It is not usual for equipment to be built solely for taking hot metal charges, although many units have this additional facility. The metallurgical and operational considerations of hot metal charging have been described by Schlatter and Simkovich[5,9,12].

An important feature with hot metal charging is suitable thermal shielding of the seal of the vacuum lock and the use of a constrictor (see Fig. 8), normally refractory-lined, to contain the spray of metal which occurs when stream degassing

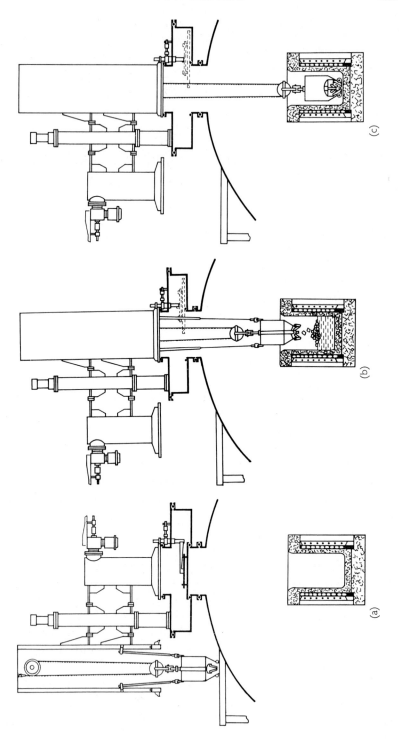

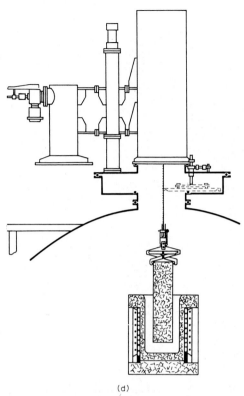

(d)

Fig. 6. Bulk charging facility for continuous operation of large vacuum induction melting furnace.
(a) Charging basket on the charging platform prior to introduction into a rotary container; (b) Additional charging during melting from a charging basket opening at any desired height; (c) Introduction of melting material from a charging basket which opens spontaneously on striking the crucible floor; (d) Charging of ingots by means of gripping tongs. (Courtesy Leybold–Heraeus)

alloys (particularly iron-base) into a vacuum.

The spray constrictor must be able to be mechanically removed from the top of the crucible to facilitate alloy charging and other services at a later stage in the process. Because of the complexity and cost that this represents, the alternative approach of hot metal charging under atmospheric conditions is sometimes preferred, as discussed in the previous section.

(c) Alloy charging

It is not normally considered desirable to charge small quantities of alloy additions through the bulk charger on account of lack of control of the feed rate and of the unnecessary complexity of the operation. An alloy charger is, therefore, normally provided capable of holding from 2 to 10% of the bath weight in the form of graded lumps not normally exceeding about 50 mm. These must be contained in either segmented compartments above the melt chamber, or alternatively

References pp. 642–647

Fig. 7. Large vacuum induction furnace with bulk charger in position, casting electrode moulds at Latrobe Steel Co. (Schematic)

introduced in charging pans or skips which may be mechanically discharged—after entering the vacuum chamber through a vacuum lock—on to a vibratory hopper discharging into the crucible. A typical design is illustrated in Fig. 3.

(d) *Liquid melting sampling*

It is necessary to be able to sample the liquid bath at appropriate stages of the melt to determine the correct analysis. While yields of metallic elements are very consistent, the use of carbon deoxidation to purify the charge invariably necessitates analysis for this element, if for no other. Small metallic or refractory moulds are introduced on the end of a rod through a small vacuum lock, or alternatively the bulk charge lock is used (Fig. 9(a)). If a small sampling lock is

Fig. 8. Hot metal charging of 30-ton vacuum induction furnace at Latrobe Steel Co. (Schematic)

used, care must be taken to avoid overheating of the seals, as the sample must be withdrawn while still very hot in the interests of speed.

(e) *Bath temperature measurement*

Radiation pyrometers must be used with discretion because of the varying emissivity of the bath and the problem of maintaining film-free windows. Careful location of windows in positions of minimum deposition of metal vapour, the use of suitably placed condensing areas, and the ability to change windows during the melt, overcome the latter problem.

References pp. 642–647

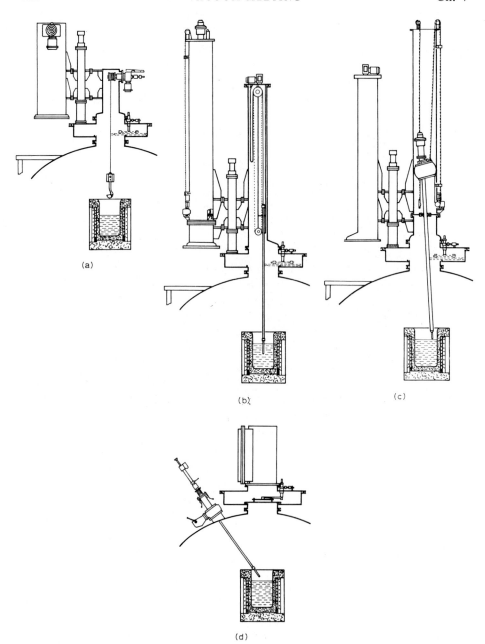

Fig. 9. Sampling, temperature measurement and bridge breaking on vacuum induction furnace. (Courtesy Leybold–Heraeus)
(a) Device for taking melt samples before pouring; (b) Immersion thermocouple for measuring pouring temperature. During the melting process, temperature is monitored with a pyrometer; (c) Poker for breaking up metal bridges on melting bath surface; (d) The instrument required is applied to a connection which can be separated from the tank by a slide valve. Instruments not required are laid aside on the furnace platform.

It remains standard practice, however, to determine the bath temperature prior to casting by introducing a dip thermocouple through a sliding vacuum seal. The dip thermocouple can be removed for repairing, etc., by means of a small vacuum lock (Fig. 9(d)). As it permits such maintenance, this arrangement is widely preferred in comparison with having a permanent couple inside the melting chamber. Then again use may be made of the vacuum lock used for bulk charging (Fig. 9(b)).

(f) *Bridge breaker*

While charge selection and care in melt-down can minimise "bridge" formation, it is a recurring problem. A heavy bar is often introduced through the thermocouple lock to act as a bridge breaker or a device may be permanently fitted. Counterbalancing is necessary on larger units. A heavy "piercing cone" can be operated through the bulk charger (Fig. 9(c)).

(g) *Crucible lid*

A removable lid, generally refractory-lined, is frequently used to protect the top of the furnace casing from undue radiation during any lengthy refining periods.

(h) *Mould charging*

The provision of a mould lock, as mentioned earlier, is normally confined to the larger units of 1-ton capacity and above. In the case of smaller units, the added cost and complexity of this facility is not always justified. In practice it is not considered desirable to open the melting chamber to atmosphere until the crucible has cooled to a temperature at which any residual metal will not severely oxidise, and this time limitation is only significant on large units. Similarly, small moulds may be removed to the atmosphere after a minimum time period, while it may be considered desirable to cool larger moulds in the mould lock for a more extended period and the provision of the mould lock enables the next melt to be started simultaneously. The time taken in handling the larger assemblies of moulds used with larger units is inevitably extended, and again justifies the provision of a mould lock.

The types of lock used have been considered in the previous subsection (2.1.1)

(i) *Tundish charging*

Direct teeming into moulds is not satisfactory on account of excessive splashing of the casting stream which results in poor ingot skins.

While most furnaces have used a fixed tundish to give a controlled vertical stream into the mould, there is a growing tendency to provide a tundish lock to enable a pre-heated tundish to be introduced at the appropriate stage in the process.

(j) *Casting ladle facility*

Where a ladle is to be used as an intermediate stage between the crucible and the ingot moulds, it is essential that it should be capable of being pre-heated. It is the practice, therefore, to introduce the ladle into the furnace at the end of the refining cycle. This procedure necessitates the use of a very wide mould lock.

Design refinements are possible if the ladle is introduced to the mould chamber through a ladle lock, as in Fig. 1(g). In this case no mould lock is necessary as the melting and mould chambers are completely separate. A disadvantage of this design is the inability to cast directly from the crucible through a fixed tundish into the ingot moulds as an alternative to ladle casting.

(k) *Hot topping facility*

The metallurgical aspects of this problem are considered later. An electric arc topping facility can be provided in the mould chamber. This consists of tungsten tipped electrodes operating through sliding seals and using d.c. welding generators as the heat source.

2.1.3 Vacuum requirements, selection of pumping systems and vacuum measurement

The selection of vacuum pumping equipment is governed by a number of factors:

(a) *The initial evacuation of the melting chamber and of the various service locks*

This aspect is particularly important on large units if the evacuation time is not to become a limiting feature in determining the melting cycle of the furnace. It is normally accepted that a time period of the order of 15 minutes is acceptable for evacuating the main melting chamber to a pressure of less than 100 millitorr to enable melting down to commence, but a much shorter time period of the order of a minute or two is necessary for most of the service locks. In the case of a large mould lock, a time period of 10 minutes might be considered acceptable.

(b) *The total gas load occurring during the refining period*

The total gas load varies considerably with the selection of raw materials, but an average gas load of 10–100 Torr l/s ton is encountered when melting steel and nickel-base alloys. It should be noted that the maximum gas load occurs when the material is first melted and is significantly greater by a factor of 3 or 4 than at a later stage in the purification cycle. It is the smaller value which is of importance when considering the pumping characteristics required to obtain a specific ultimate pressure.

(c) *Pressure range necessary for refining*

For steels and nickel-base alloys most of the refining is by a carbon deoxidation process and is found to proceed rapidly at relatively high pressures of the order of 100 millitorr or more. It is not necessary, therefore, to attain very low pressures to achieve most of the refining[10]. It is found, however, that the ultimate purity of the charge with respect to volatile impurities such as lead is influenced considerably by the ultimate pressure. For example, in the case of lead it has been shown by Wood and Cook[13] that prolonged exposures to pressures of less than 10 millitorr are necessary to achieve maximum refining (see Fig. 10). Precise evidence such as this for the necessary refining conditions is limited but it is normal to specify equipment capable of an ultimate vacuum over hot metal in the pressure range 1–10 millitorr.

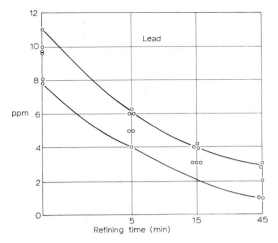

Fig. 10. Effect of vacuum treatment on the lead content of nickel base alloy[13] held at 0.5–1.0 millitorr at 1500°C.

(d) Extraneous gas loads

Particularly in large equipment it must be recognised that there will be significant gas load arising from leaks in the various vacuum seals and outgassing from the large quantity of refractories associated not only with the crucible and its insulation but with hot tundishes, launders or indeed hot ladles. It is also standard practice to use materials such as asbestos insulation for the furnace construction and refractory insulation for the crucible lid, which again add to the gas load.

While a very small leak and outgassing rate, corresponding to a pressure rise of 15 millitorr/h, is normally specified from the point of view of equipment acceptance, this applies only to a cold empty chamber. A much higher figure is generally considered acceptable as a specification imposed by the melter before commencing melting. This may be of the order of 5×10^{-4}–1×10^{-3} Torr l/s kg (kg being the kilogram capacity of the crucible).

It is important to bear in mind that even such high leak and outgassing rates as this may be difficult to achieve if the melting chamber has been exposed to atmospheric pressure for furnace fettling, etc., with significant quantities of volatilised metal deposited and condensate formed on the cooler surfaces of the melting chamber.

It is important that such leak checks are carried out over the whole vacuum system with the various service locks open, as otherwise harmful pressure rises can occur at critical stages of the melt.

The adverse effect of the deposition of metal vapours can be minimised by providing a water-cooled condensing plate near the exit manifold. This may be removed and cleaned more readily than dealing with the whole of the furnace shell. Condensation of water vapour during periods when the furnace is open can be avoided by using warm water for cooling purposes.

References pp. 642–647

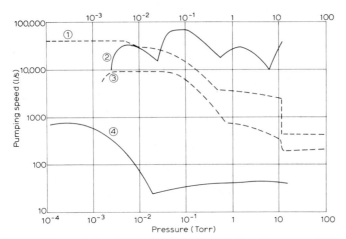

Fig. 11. Characteristics of typical pumping installations for vacuum induction melting furnaces. 1. Oil booster (4×30 B5.A)+Roots pumps+oil-sealed mechanical pumps for 5-ton unit; 2. 6-Stage steam ejectors with 5-millitorr design point for 5–10-ton unit; 3. Oil booster (30 B4)+ Roots pumps+oil-sealed mechanical pumps for 250–400 kg unit; 4. Oil diffusion pump+2-stage oil-sealed mechanical pump for 10-kg laboratory furnace.

It is not intended to deal in detail in this section with the characteristics of the available pumping systems as these are dealt with more fully in Chapter II. However, in Fig. 11, several typical pumping systems which have been chosen for vacuum induction melting furnaces are shown. It is interesting to note that steam ejectors are to be preferred for the largest units, as they minimise the initial pumpdown time of the melting chamber and mould lock, although they are restricted with respect to the ultimate vacuum which they can achieve.

The use of steam ejectors also enables filter systems to be dispensed with. These are considered essential for most metallurgical furnaces using any other pumping system.

Selection of vacuum gauges is dealt with in detail in Chapter II, but some general comments are appropriate. It is important that each individual vacuum chamber comprising the unit should have a suitable gauge. At the least, this should be of the Bourdon type to enable the operator to follow the pump-down of the auxiliary chamber.

If a more precise gauge is fitted, the leak rate of each individual chamber may be measured. This can be of considerable importance from an operational and leak-chasing aspect. The use of interlocks based on vacuum pressure are worthy of consideration to ensure that auxiliary chambers are not opened to the main chamber until a suitable pressure is reached.

2.1.4 Selection of power supply

As vacuum induction furnaces represent a high capital cost, it is important that factors restricting their output be avoided. It is the tendency, therefore, to ensure that they are adequately powered. As one limitation in the heat cycle is the refining time, governed solely by metallurgical considerations, and as this may take several hours, an overall heat time of 4–6 hours is normally considered acceptable. This means that the melt-down time must be limited to about two hours.

To achieve such a melt-down time one requires between 300 and 500 kW per ton, depending on the charging facilities and degree of coupling achieved. Some typical power supplies for various sized units are shown in Table 1. As important

TABLE 1

TYPICAL POWER SUPPLIES FOR VACUUM INDUCTION MELTING INSTALLATIONS

	Furnace size					
	80 kg	300 kg	1 ton	3.5 ton	5 ton	30 ton
Melting supply:						
Power	100 kW	300 kW	600 kW	1200 kW	1600 kW	2400 kW
Voltage	250 V	400–600 V	400–600 V	400–600 V	400–600 V	600 V
Frequency	2–4 kHz	1–2.5 kHz	0.5–1 kHz	0.5–1 kHz	0.15–0.5 kHz	60 Hz
Stirring supply:						
Power	—	—	700 kW	1800 kW	2500 kW	—
Voltage	—	—	20–65 V	20–65 V	20–65 V	—
Frequency	—	—	approx. 50 Hz	approx. 50 Hz	approx. 50 Hz	—

as the kW rating is the frequency chosen for the power supply, as this affects the stirring of the bath. Without effective stirring the melt times are prolonged on account of retardation of the refining reactions and of solution of alloy in the bath. The most effective frequency for a variety of furnace sizes is shown also in Table 1.

Because of insulation problems and resultant discharge phenomena many of the earlier furnaces restricted the supply voltage to a maximum of 250. The development of more effective coil insulation techniques has led to the introduction of supply voltages of up to 800.

As will be seen from Table 1, up to a furnace size of two or three tons the frequency has been 500 Hz or over, and motor generators have been used. With the advent of still larger units, triplers have been preferred because of initial cost and ease of maintenance. For small installations of up to about 100-kg capacity, valve sets of medium or variable frequency can be used. These may eventually be superseded by designs incorporating silicon-controlled rectifiers.

2.1.5 Power feed and insulation problems

The earliest furnaces were small in capacity and had a power supply in the 3–10 kHz range. To avoid unnecessary power losses coaxial power input designs were generally used. In some cases these gave rise to considerable trouble[14] as the furnace coil could only be removed for ramming new linings by removing the insulation at some point. Replacing this insulation led to practical difficulties in ensuring effectiveness. In other cases, by employing a very low coil voltage of

Fig. 12. Water-cooled flexible cables supplying melt power to crucible.

about 200–220 V, the coil and a coaxial input were used without external insulation and the problem was avoided.

As the size of units grew and the associated power supply was of lower frequency, and hence power losses less of a problem, there was a tendency to use rigid water-cooled connections inside the furnace. These proved cheaper and simpler to engineer at the higher amperages involved than the earlier coaxial designs. They suffered from the same problem of re-insulation when the coil had to be removed from the melting chamber.

More recently improved water-cooled flexible cables have been used inside the furnace. These enable the design to be such that the power supply to the coil can be disconnected outside the vacuum chamber, and the coil with its flexible cables removed, without interfering with any of the insulation exposed to vacuum. The use of such units has greatly simplified the maintenance of satisfactory insulation and improved the utilisation of the equipment.

A modern unit using water-cooled flexible cables is shown in Fig. 12.

2.1.6 Auxiliary stirring

As discussed under "Power supply" effective stirring can be achieved in the melt provided that the frequency of the generator is correctly chosen and the coil and crucible are of suitable design. In certain cases standard generators have been chosen for installations because of cost and availability, and these have not given adequate stirring.

Apart from this, there is a fundamental problem in that while a unit may stir effectively using the maximum available power, it may prove impossible to control the temperature of the melt under these conditions. With certain alloys it is important that the refining be carried out at a certain maximum temperature and with effective stirring, and this may mean that auxiliary stirring is required.

A number of production units have been fitted with independent 50-cycle single-phase stirring applied to the melting coil. Typical power ratings of such auxiliary stirring are shown in Table 1.

In other cases 3-phase separate stirring coils have been used. This system has the advantage of giving complete stirring of the bath as distinct from the quadrant stirring obtained with single-phase supply. In practice, the necessity for two independent coils and the danger of interference between the two have led to single-phase auxiliary stirring being preferred. Stirring is improved and overheating minimised if the crucible is tilted at this stage of the melt.

2.2 Process techniques

2.2.1 Selection and preparation of the crucible

With small furnaces of up to about 25-kg capacity it is standard practice to use pre-fired crucibles. Although it is possible to use rammed linings, the time and labour cost compared with pre-fired crucibles is unfavourable. The upper size limit of pre-fired crucibles is about 250 kg, but towards the upper end of the range the economics of rammed linings become more favourable.

Alumina and magnesia base crucibles of varying degrees of purity are most commonly chosen depending on the chemistry of the alloy to be melted. High residual oxygen contents will occur in the melt if crucibles such as commercial magnesite with high iron oxide and silica contents are used. In the case of the more reactive elements that are induction melted (*e.g.* chromium), lime- or yttria-stabilised zirconia is preferred. Lime-stabilised zirconia has also been favoured for the remelting of nickel base casting alloys because of its greater stability and non-wetting characteristics.

In the case of lower melting-point alloys, based for example on Cu and U, which are non-reactive to carbon, plumbago and graphite crucibles give excellent life with minimum contamination.

References pp. 642–647

The larger vacuum induction units which are used for ferrous and nickel-base alloy production generally use rammed linings of the 70% magnesia, 30% alumina type, or less commonly 90% magnesia, 10% alumina, or pure MgO. A small quantity (2–5%) of bond is common to promote fritting. This may be of a suitable clay such as bentonite or of the borax type.

Details of the composition of some typical refractories used for pre-fired crucibles and rammed linings are given in Table 2.

TABLE 2
TYPICAL REFRACTORY COMPONENTS USED IN VACUUM INDUCTION MELTING

Application	Material	Chemical analysis (%)							
		SiO_2	Al_2O_3	Fe_2O_3	CaO	MgO	ZrO_2	B	Na_2O
Rammed furnace linings	Magnesia (10 kg lab. size)	Tr.	Tr.	Tr.	Tr.	99	—	1	—
	Commercial Mix A (U.K.)	5	26	Tr.	Tr.	68	—	—	—
	Commercial Mix B (U.K.)	4	27	Tr.	Tr.	68	—	—	—
	Commercial Mix C (U.S.A.)	6	25	Tr.	Tr.	67	—	—	—
	Commercial Mix D (U.S.A.)	3.5	20	Tr.	Tr.	75	—	1	—
Prefired crucibles	Alumina (10 kg lab. size)	4	95	Tr.	Tr.	Tr.	—	—	—
	Alumina (large–commercial–U.K.)	2	96	Tr.	Tr.	Tr.	—	—	—
	Magnesia (large–commercial–U.K.)	2	Tr.	1	Tr.	96	—	—	—
Crucible capping ring	(Lining material + calcium aluminate cement)	15	41	Tr.	5	52	—	—	—
Sample moulds	Zircon	34	Tr.	Tr.	Tr.	Tr.	65	—	1
Pouring tundishes	(1) Firebrick	56	40	2	Tr.	Tr.	—	—	—
	(2) Sillimanite	41	57	1	Tr.	Tr.	—	—	—
	(3) Zircon	32	1	1	Tr.	Tr.	65	—	—

Tr = Trace.

	Sieve grading (B.S.S.) (%)											
	+8	−8 +10	−10 +16	−16 +22	−22 +30	−30 +44	−44 +60	−60 +72	−72 +85	−85 +100	−100 +150	−150 +200	−200
Magnesia (10 kg lab. size)		45				33				22			
Commercial Mix A	35	6	15	4	1	1	1	1	3	1	5	4	21
Commercial Mix B	34	6	12	4	2	1	1	2	3	2	8	4	21
Commercial Mix C	30	7	12	4	6	6	6	1	2	1	2	1	23
Commercial Mix D	37	7	6	5	4	3	3	2	2	1	3	11	15
Lining material for crucible capping ring	25	4	9	3	2	1	1	2	2	1	6	3	41

The grading of the material used for a rammed lining is important, and is dependent on the size of the crucible, although in the range $\frac{1}{2}$–5 ton melt size similar gradings are standard (see Table 2).

The crucible material is normally dry rammed against a solid graphite or hollow metal former using a pneumatic hammer to ensure uniform packing. The initial fritting is carried out in air to about 750°C to facilitate degassing and then fired at a moderate vacuum at about 1540°C. When using a graphite former, the furnace is allowed to cool to about 1100°C, opened to atmosphere, and the graphite plug stripped at about 1000°C.

Where a steel- or nickel-base former is used, the desired fritting temperature is achieved by a washout heat.

A finer grained mixture, generally bonded with sodium silicate, is used for the top ring of the furnace. The pouring spout may for convenience be pre-fired.

Behind the densely sintered crucible face there must be a loose unsintered zone to accommodate movement of the crucible face due to thermal or mechanical causes, to act as a thermal insulator, and to mitigate the effects of metal penetration from cracks in the sintered layer.

So far there has been little satisfactory experience using pre-fired bricks for the furnace lining, although it is important to note that one of the largest induction furnaces of 30-ton capacity at Latrobe Steel Co. uses this principle.

Schlatter and Simkovich[9,12] have described the development of a new mullite-bonded Korundal-XD type refractory which, because of high purity and high density, has excellent structural stability and resistance to thermal cycling. Close size tolerances, good bricklaying and improved mortars resulted in a very successful working lining. All crucibles are lined with double layers of firebrick. In all cases Korundal-XD is used as the back-up lining, but the working face has varied between this and high purity magnesia bonded with spinel, *i.e.* 70–80% magnesia, 20–30% alumina type.

The life of the brick linings has in this instance exceeded that of conventionally rammed units.

Where a metal former has been used it is invariably the practice to regard the fritting heat as a washout heat to condition suitably the furnace lining. Where graphite fritting is employed, a washout heat may or may not be deemed necessary dependent on the product.

After several heats it is generally necessary to remove spatter rings and other incrustations from the top of the crucible, patching where necessary with a cement of the same general nature as that of the furnace lining, but with a higher content of bonding material (frequently sodium silicate). Extreme care is necessary with patching if undue erosion and contamination of subsequent heats are to be avoided.

2.2.2 Charge material, including hot metal

The charge material may be considered in three categories:

(a) *Virgin alloy charge*

Where a raw material charge is to be used it is standard practice to apply a more demanding specification with respect to certain impurities than with air melting. This particularly relates to such detrimental impurities as phosphorus and sulphur, which cannot be removed by normal vacuum induction melting procedures. In the case of gaseous impurities such as oxygen and nitrogen, it is generally more economical to purchase raw materials of a low content rather than carry out unnecessary purification in the vacuum induction furnace. The same consideration should also be given to the selection of raw material containing any significant content of volatile impurity.

(b) *Scrap*

In the interest of economics it is desirable to re-cycle scrap. This must be clean and free from scale, oil, grease, or in the case of castings, from moulding materials.

(c) *Hot metal*

Where a hot metal charge is used it is desirable that this should be free from reactive elements such as titanium and aluminium. These should be added after initial refining under vacuum. In most cases it is desirable to allow carbon deoxidation to proceed under the vacuum conditions, so the hot metal charge should be un-deoxidised or in the semi-killed condition.

2.2.3 Melting and refining

The initial charge will normally contain all the non-reactive elements with sufficient carbon to enable carbon deoxidation to proceed during melt down. After the initial charge has been refined, the bath is generally checked for carbon. Provided that residual carbon is still present, a pressure rise check may then be carried out to assess whether deoxidation is complete. If this proves to be the case, reactive element additions are made. At this stage it is important that the bath temperature should not be allowed to rise above a predetermined maximum, as otherwise metal/crucible reactions are encouraged.

It is important at this stage too that effective stirring is applied, otherwise pronounced stratification can occur with elements such as aluminium, which are of low melting point and low density. After the reactive elements, volatile elements are added, generally under a reduced pressure of inert gas. The melt is then adjusted to the casting temperature.

Where scrap is being melted it is simpler if 100% scrap charges can be used. In this instance the melting procedure may be regarded merely as an avoidance of oxidation of reactive elements in the charge followed by the limited refining which can be expected in the presence of reactive elements. A certain amount of "topping-up" may be necessary to allow for losses in volatile elements such as Mn. Where mixed virgin alloy/scrap heats are to be made, the non-reactive virgin charge should be refined, the clean scrap added, followed finally by reactive elements.

As described above, hot metal is normally added as a melting base of non-reactive elements, and after carbon deoxidation, the heat is finished in a similar way to that described for virgin alloy charges. The use of hot metal has certain advantages:

(a) The heat time can be substantially reduced with economic advantages.

(b) Less pure raw materials can be purified in air using standard slag refining technology, again with economic advantages.

(c) Scrap which might be unsuitable for vacuum induction melting because of excessive contamination can be refined by standard air-melting practices. Oxygen blowing as well as slag reactions can be employed to give a suitable and cheap melting base.

2.2.4 Casting techniques

(a) Bottom pouring

As mentioned earlier, bottom pouring is an excellent technique to achieve complete separation of any surface dross which may be present. In practice its use has been limited because of the difficulty of maintaining a satisfactory pouring orifice at the base of a melting crucible. While satisfactory procedures exist for ladles, these necessitate the replacement of nozzles after every heat, a procedure generally deemed impractical with rammed melting crucibles. In the case of the graphite or plumbago crucibles used for the lower melting-point elements, copper and uranium, this technique is possible. A graphite stopper may be used, or alternatively a water-cooled copper cone which is applied below the tapping orifice,

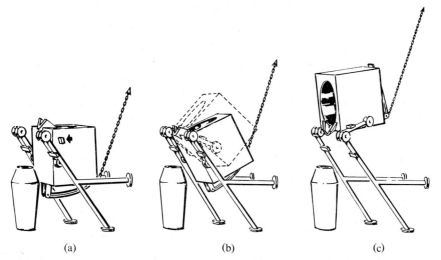

Fig. 13. Change of axis of rotation during casting to minimise throw of metal stream. (Courtesy Leybold–Heraeus)
(a) Melting coil and crucible in the melting position. (c) Melting coil and crucible in the pouring position.

References pp. 642–647

Fig. 14. Photograph of tundish showing use of slag traps.

as described by Winkler[15]. During melting a solid plug is formed in the nozzle but this melts within 30 seconds of removing the water-cooled cone.

The use of electromagnetic throttling has been suggested but practical difficulties have so far prevented its adaptation to production units.

(b) *Direct lip pouring*

This is rarely used, partly on account of the changing throw of the metal stream as the liquid metal level in the crucible changes. Where this is considered desirable, movement of the casting stream can be minimised by shifting the axis of rotation (see Fig. 13).

(c) *Fixed tundish*

A fixed tundish is normally used and overcomes the problem of metal throw. The design of the tundish is important in the case of melts having a residual dross cover. The provision of one or two bridges through which the metal must flow is common, and these effectively hold back dross (Fig. 14). For certain designs a launder may be used instead of a fixed tundish.

In the case of both the launder and the tundish, a secondary tundish may be used, particularly in the case of the filling of multiple small moulds. Where only one large ingot mould is to be filled a secondary tundish is not normally used.

(d) Ladles

With the demand for higher and higher standards of cleanness in vacuum induction melted metal there has been a growing consideration for the use of ladles. The use of ladles of the same capacity as that of the crucible is not common, but there are several instances of large furnaces using pony ladles. This technique, in effect, has all the advantages with none of the disadvantages of the bottom-poured crucible. Control of the stopper of the ladle in a sensitive manner under vacuum is tackled by using either hydraulic or pneumatic actuation.

(e) Hot topping

On account of the absence of atmospheric pressure, which leads to a lower solidification rate because of the reduced heat transfer to the mould, the "feeding" of ingots during cooling under vacuum is more of a problem than with conventional air casting. Added to this is the desirability of avoiding exothermic reactions so as to maintain the clean conditions necessary in a vacuum melting unit. Because of this, the use of insulated tiles for the hot-top has been preferred, and, although attempts have been made to use an electric arc for hot-topping, the success has been strictly limited.

This difficulty of obtaining effective hot-topping of vacuum induction cast ingots has not yet been satisfactorily solved and the yield of such ingots is generally low compared with the air-melted equivalent. It should be noted, however, that in practice the bulk of the output from vacuum induction furnaces is not in the form of forging ingots.

2.3 Metallurgical effects

2.3.1 Purification reactions

The main impurities of interest to the metallurgist are the gases (O_2, N_2, H_2), certain non-metallic impurities such as S and P, and metallic impurities such as Pb and Sn.

One of the main virtues of the vacuum induction melting process is that time, temperature and pressure can be varied independently and over wide ranges, thus allowing certain purification reactions to be expedited. This is in contradistinction to the vacuum arc melting process, where the dwell time of metal in the arc and in the molten pool is limited, and where the time and pressure cannot be controlled except to a limited extent through adjustment of the melting parameters.

It should be emphasised too that an essential feature of the vacuum induction melting process is the presence of a crucible which is never completely inert to the molten charge. The ensuing reactions are of the utmost importance and limit the potential of the process.

References pp. 642–647

The main purification reaction in most vacuum melting processes is the removal of oxygen by reaction with carbon, this being more effective than thermal dissocation.

$$C + MO \rightleftharpoons CO + M$$

The thermodynamics and kinetics of this reaction in oxide crucible have been studied by several investigators[16,17]*. Even in the case of pure iron the reaction never reaches the equilibrium expected under reduced pressures. The oxygen content of the melt is determined by a balance of the supply of oxygen, mainly from carbon reduction of the crucible, and the removal of oxygen as carbon monoxide at the metal surface.

It should be noted that the full advantage of reduced pressure is only operative at the surface of the melt as the ferro-static head inhibits carbon deoxidation in the body of the melt. This factor is of considerable importance in larger units. The kinetics of removal of oxygen at the melt surface are therefore important and affected by (a) crucible geometry and (b) bath stirring.

Apart from crucible reactions, oxygen may also be picked up by the melt from leakages in the plant, and out-gassing of the plant components, particularly the crucible lining and insulation.

Reaction with the lining is often of the nature of carbon reduction.

$$[C]_{Fe} + \langle MO \rangle \rightleftharpoons (CO)_g + [M]_{Fe}$$

It is observed that the loss in carbon during refining is much greater than can be accounted for by reaction with oxygen in the melt, as reported for example by Graham and Argent[18]. These authors and Bennett et al.[19] all report the lowest residual oxygen levels in vacuum induction melted 0.2% C steel to be as high as 10 ppm for aluminosilicate and magnesite type crucibles. The equilibrium oxygen content in an inert crucible would be less than 1 ppm for the pressures studied (approximately 2 millitorr).

Hydrogen can also be used to reduce oxides in the melt, and indeed in the case of very low carbon alloys this technique is essential. In this instance the reaction

$$MO + H_2 \rightleftharpoons H_2O + M$$

is not favoured by low pressures. For effective reaction hydrogen is sprayed on to the metal surface or injected into the bath. The hydrogen used may be readily removed from the melt by subsequent vacuum treatment.

Deoxidation can also be achieved by the addition of metallic deoxidants such as Ca and Mg but it must be noted that these result in solid reaction products which can be entrapped in the melt and form harmful non-metallic inclusions in

---

* See also Chapter I.

the product. Where this technique is used Mg is sometimes favoured, as excess may be added and distilled off because of the high vapour pressure.

In recent years the possible advantages of less common deoxidisers such as germanium[20] and yttrium[21] have been investigated. So far no production use is made of such additions, although yttrium in particular has proved most effective in reducing the oxygen and nitrogen levels of chromium-base alloys.

In some cases the purification reaction is dependent on the thermal dissociation of stable phases. This is particularly true in the case of nitrogen, as the conditions for the dissociation of the gas–metal compound are much more favourable than in the case of oxygen, the heats of formation of the nitrides being markedly lower than those of the corresponding oxides. Simkovich[22] has studied the effect of composition, chamber pressure and temperature on the removal of nitrogen from iron and nickel-base alloys, and compared the results obtained from laboratory size furnaces with those from production heats made in 15- and 30-ton crucibles. The results show that aluminium and titanium in an alloy are particularly effective in retaining nitrogen, but that after vacuum refinement alloys containing up to 15% chromium do not possess higher residual levels than do chromium-free alloys. Increased temperature improved the rate of removal of nitrogen owing to improved kinetics.

The non-metallic impurities S and P are not effectively removed by vacuum treatment. Sulphur can be removed to some extent by the addition of elements of high affinity such as Ce[23] and Ca. As these elements are also deoxidisers, effective deoxidation must precede desulphurisation.

Volkov et al.[24] and Ward and Hall[25] have described how desulphurisation may also be achieved by using slag reagents based on lime. According to Volkov, the degree of vacuum increases the desulphurising capacity of the slags particularly when melting low carbon steels and alloys. A 90%–10% mixture of finely divided freshly burned lime and fluorspar is placed in the crucible before charging. The desulphurisation occurs during melt-down and is most effective at pressures under one Torr and for carbon contents above 0.05%. The use of such a technique on a production scale would, however, necessitate the handling of slag volumes of 2–3% of the bath volume, and so far no production use is made of this technique to the authors' knowledge.

The use of reactive gases, *e.g.* oxygen, hydrogen, methane and carbon tetrachloride, for vacuum treatment has been described by Coupette[26]. Such a technique enables certain desirable reactions to be controlled and intensified without resorting to the use of slags.

The removal of volatile impurities such as lead by evaporation is a particularly important practical function of vacuum induction melting. The degree of purification achieved under different conditions has been studied by Wood and Cook[13] with the results shown in Fig. 10 for lead in nickel-base alloys. As a general guide to the ease of removal of volatile impurities, one should consider the vapour

References pp. 642–647

pressure of the impurity in question at the melting temperature as shown in Fig. 15. However, on account of the varying activity of solutes in dilute solutions, experience has shown that such information is only a guide. For example, manganese and aluminium which have very similar vapour pressures behave very differently in dilute solutions in iron and nickel, manganese being removed much more readily than aluminium.

Turillon[27], in a study of the evaporation during vacuum induction melting of a wide range of elements at the 0.1% level from an 80% Ni–20% Cr base at 1565°C, showed that Te, Pb, Bi, Se and Cu were volatilised, but Sb, As and Sn were not. The failure of Sb and As to volatilise was due to a large negative deviation from ideality (Sieverts' Law), whereas Sn showed a positive deviation.

The removal of more substantial quantities of impurities (*ca.* 1%) has been reported by Duckworth and Appleby[28]; Zn, Mg, Ca, Bi, Sb, Pb, Mn, Sn and Cu were readily removed, but the rate of loss decreased with time to a greater extent than predicted theoretically. Consequently, the results should not be applied to the production of alloys with impurity levels of 1–10 ppm.

It is important to note that there is a physical aspect to the purification process in vacuum induction melting. Slag or dross may form on the surface of

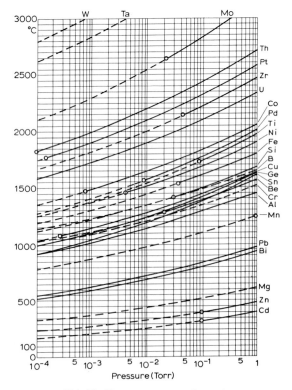

Fig. 15. Vapour pressure of metals.

the melt owing to non-metallic impurities present in the raw material or scrap, or owing to reactions with, or erosion of, the furnace lining. This oxide material can be reduced by carbon in the melt, but the kinetics are unfavourable. It is often experienced that this dross adheres to the furnace lining, particularly if superheat is used. If not, it may be separated from the purified metal by casting with the power on, under which conditions the scum tends to be held against the crucible wall. As discussed earlier, use of trapped launders or ladles results in further separation.

2.3.2 Compositional control

As will be appreciated from the discussion of purification reactions, carbon is lost in substantial quantities owing to reaction with oxide impurities in the melt and with the furnace lining. It is desirable, therefore, to sample for this element at several stages in the melt.

When reactive elements are added to a well-deoxidised bath, virtually quantitative yields are obtained. Notable exceptions are the very reactive elements such as yttrium, which can itself reduce the crucible materials used.

The yield of volatile materials is dependent on the time delay between the addition and the tapping of the melt. While in practice the more common volatile elements such as manganese and nitrogen present no problem, strictly controlled melting procedures are essential.

A primary function of vacuum induction melting is the precise control of chemistry. The degree to which this can be achieved has been illustrated by Boesch and Radavich[29], who surveyed 123 heats of Udimet 700 as shown in Table 3.

TABLE 3

ILLUSTRATION OF THE PRECISE DEGREE OF CHEMISTRY ACHIEVABLE WITH VACUUM INDUCTION MELTING[29] FOR 123 HEATS OF UDIMET 700

Element	Max. (%)	Min. (%)	Avg. (%)
Aluminium	4.45	4.29	4.36
Titanium	3.55	3.42	3.47
Boron	0.032	0.026	0.029
Molybdenum	5.32	5.00	5.14
Carbon	0.08	0.06	0.07
Chromium	15.85	14.83	15.32
Cobalt	19.05	18.23	18.70
Iron	0.33	0.11	0.15
Nickel			Remainder

The consistency, particularly in the case of titanium and aluminium which show notorious variations in air melting, is remarkable.

2.3.3 Cleanness

Cleanness, concomitant with precise control of chemistry, is the main aim of vacuum induction melting.

References pp. 642–647

The measurement of cleanness in the case of vacuum melted products is no easy task. The majority of the inclusions present are found to be less than 10^{-2} mm in size and any larger inclusions occur in a random manner and are not found in the few square inches of sample normally used for cleanness assessment. They are found, however, when the whole product of a cast is examined, say in the form of thin sheet, or tested comprehensively as on an actual product.

It is not surprising therefore that published information dealing comprehensively with the distribution of inclusions in vacuum induction melted alloys is scant. The occasional occurrence of large random inclusions is accepted as a limitation of the process. A typical report by Leinbach and Hamjian[30] on improved cleanness of 1% C, 1.3% Cr ball race steel (AISI 52,100) is summarised in Table 4.

TABLE 4

IMPROVEMENT IN CLEANNESS OBTAINED BY VACUUM INDUCTION MELTING 1% C, 1.3% Cr BALL BEARING STEEL[30] (AISI. 52100). TESTING METHOD ASTM. E.45.A – MODIFIED

Inclusion type	A	B	C	D
	Sulphides	Alumina	Silicates	Oxides
Maximum J–K rating of any one sample, thick and thin series combined	1	1	0	1
Avg. no. of max. J–K readings per standard specimen, vacuum induction melted	2.14	0.86	–	–
Avg. no. max. J–K readings per standard specimen, air melted	2.5	10.00	–	–

As carbon deoxidation is normally used to reduce the oxygen content because it is more reliable than deoxidation by metallic additions with solid reaction products from the aspect of cleanness, it is not surprising to note that considerable difficulty is encountered in producing very low carbon alloys to cleanness standards comparable with those for carbon-containing alloys.

2.3.4 Improvement in physical and mechanical properties

The improvement in physical and mechanical properties, it should be emphasised, arises from the factors already considered, *i.e.*:
(a) Control of chemistry
(b) Purity with respect to residual elements
(c) Freedom from non-metallic inclusions.

As an example of physical property improvement due to all three factors, Leinbach and Hamjian[30] have shown the dependence of magnetic permeability distribution upon melting technique (Fig. 16). The average initial permeability of Permalloy may be increased by about 50% by changing from the best air melting practice to vacuum induction melting.

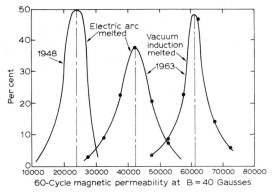

Fig. 16. Dependence of magnetic permeability distribution upon melting technique (Leinbach and Hamjian)[30].

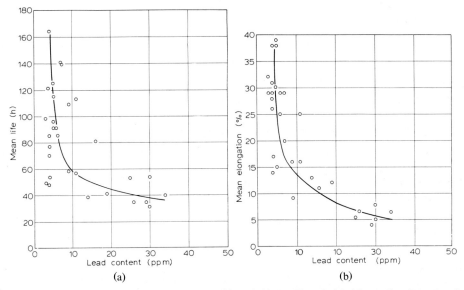

Fig. 17. Effect of lead content on (a) rupture life and (b) ductility of nickel-base alloy (Wood and Cook[13]).
(a) Rupture life *vs.* lead content for wrought alloy "A" tested at 11 kgf/mm² and 940°C; (b) Elongation at fracture *vs.* lead content for wrought alloy "A" tested at 11 kgf/mm² and 960°C.

In nickel-base superalloys, the creep strength and rupture ductility are improved. The main factor is believed to be higher purity with respect to residual elements as illustrated with respect to lead by Wood and Cook[13] (Fig. 17a and b).

The fatigue strength is significantly raised on account of the reductions in inclusions. This may be exemplified by the case of ball race steel as described by Child[31], who reported improvements of approximately 50% in the fatigue limit at 10^8 cycles for both 1% C and 1.3% Cr and 18% W ball race steels.

References pp. 642–647

2.3.5 Brief description of typical products

The earliest use of vacuum induction melting some 40 years ago was the production of wire products in alloys for resistance heating. The improved cleanness and purity facilitated the wire drawing in a material otherwise difficult to handle. Another product made at that period was the soft magnetic alloys.

Further development of applications came 20 years later owing to the stringent needs of the gas turbine. The nickel-base superalloys for rotor blades were produced to improved standards of analysis, and mechanical properties, by vacuum melting. Most important of all, the forging characteristics of the alloys were substantially improved on account of the absence of inclusion bands, the control of critical residuals such as boron, and the improved purity[32].

In later years the process has also been used to produce cast stock and precision casting for turbine rotor and stator blades. To a lesser extent it has been used for bearing steels, because here the other vacuum melting processes are generally preferred.

By far the most important use of vacuum induction melting is the production of cast electrodes for subsequent remelting by the vacuum arc or electroslag process. The ultimate end products are turbine and compressor discs and casings in nickel-base alloys, high strength forgings in Marage and other alloy steels, and stainless steels for fuel element cladding in atomic reactors.

2.4 Special requirements for precision casting furnaces

As mentioned earlier, considerable use is made of the vacuum induction melting process to produce precision castings[33]. The production is normally carried out in two stages. The first, the production of cast bar-stock 50–100 mm diam., is done in conventional vacuum melting units as described. The larger the scale of operation the cheaper the product, as the extensive cost of analysis and testing is spread.

The second stage is the remelting of this stock in small units to produce vacuum precision castings. The lost wax process is normally used to produce moulds, and at the current stage of development they rarely exceed 15–20 kg in weight. Most of the current production facilities are single-shot furnaces where one such mould is filled per melt.

The furnace must be capable of receiving through a vacuum lock the pre-weighed, pre-alloyed charge, melting this very rapidly to maximise output, and accepting a mould (pre-heated to about 1000°C) through a vacuum lock to receive the molten charge at the appropriate time. Rapid cycle times and freedom from down-time are essential, as the melting programme must dovetail into a mould production programme. Units are typically of 25 kg capacity and have a power rating of up to 100 kW. A typical installation is shown in Fig. 18. For certain castings, the provision of a mould spinning table is considered desirable.

It is also possible to fill moulds from a larger furnace of, say, 250-kg capacity.

Fig. 18. Typical vacuum precision casting furnace. (Courtesy of Rolls–Royce Ltd. and Balzers AG)

In this case, scrap or virgin alloys can be melted, *i.e.* the production of stock carried out in the same unit. The process is most suited where extensive mechanical testing is not considered necessary and the consequent risk of expensive scrap is at a minimum. This thinking applies to certain nozzle guide vanes, and also to gas turbine casings.

As it is necessary to fill each mould rapidly with a pre-determined volume of metal, some form of metering ladle is essential. The pre-heated moulds can be introduced through a mould lock or heated *in situ*. So far only one unit of this type is known to be operational and there are no published details.

3. Vacuum Arc Melting

In this process, heat is generated by the application of a low-voltage, high-current electric arc to the surface of the molten metal from an electrode which may be either consumable or non-consumable. In the latter case—also known as the "permanent electrode" method—the electrode is to all intents inert and plays no part in the melting other than the supply of heat for the fusion of the raw material which must be fed into the pool by suitable means. This electrode is usually of a highly refractory metal such as tungsten (thoriated) or of graphite or metal carbide,

References pp. 642–647

and in order to assist ionisation and also to minimise vaporisation of the electrode, melting is usually carried out under a low pressure of inert gas as distinct from high vacuum.

For this latter reason, and since uniform ingot structure is difficult to achieve when the raw material is fed into the pool by external means, the process has never been widely used for the production of large ingots on an industrial scale although still finding application in the laboratory for the preparation of small quantities of metal for experimental purposes. The consumable arc method of melting on the other hand now finds widespread use, not only for the production of the reactive and refractory metals for which it was originally developed, but also for steels and nickel-base superalloys in ever increasing quantities.

In this form of melting the material to be melted is preformed into a bar or electrode which is used as one pole of the arc and is itself consumed in the process. In the earlier designs of furnace, the high current melting power was fed to the furnace via sliding contact shoes which pressed on to the electrode surface as it was progressively fed downwards during the melting. The whole of the electrode, current feed and control mechanism were contained in a vacuum-tight enclosure. Difficulties with this method of current feed because of variations in contact resistance, however, eventually led to the development of the modern design of consumable arc furnace in which the power connection is outside the furnace.

3.1 Design features for modern plant

3.1.1 Basic designs

The principle of the modern consumable arc melting furnace is illustrated schematically in Fig. 19. The furnace consists of a copper crucible immersed in a water-cooling jacket and connected to the positive of the melting supply. The negative of the supply is connected to the ram or "stinger" rod which passes into the furnace body through a sliding vacuum seal. The material to be melted (the electrode) is clamped to the base of the control ram and after the evacuation of the chamber, is lowered to strike an arc with a metal pad of similar material on the base of the crucible. As the electrode melts owing to the arc power the ram is lowered by an electric or hydraulic control system so as to preserve a constant spacing between the electrode and the resulting molten pool. As melting proceeds, an ingot is progressively built up in the crucible.

This basic principle applies to most vacuum arc furnaces ranging in size from the small experimental units used to melt laboratory ingots up to the largest commercial furnaces at present available (50 tons), though naturally variations are introduced depending upon the particular melting application.

In order to introduce the electrode into the furnace and also to remove the completed ingot it is necessary to separate the crucible from the upper furnace body in one of two ways. Either the crucible may remain fixed and the upper

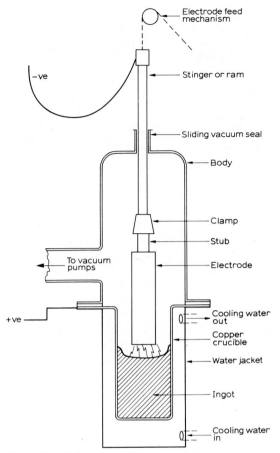

Fig. 19. Schematic illustration of the principal features of the vacuum consumable arc furnace.

furnace body removed from it or vice versa. Earlier designs utilising the former principle relied on the entire removal of the furnace body and superstructure by means of a crane, but modern designs based on a lifting and a pivoting type superstructure covering dual melting stations have been developed in order to minimise down-time between melts. A typical layout is shown in Fig. 20; unloading of the ingot and loading of the next electrode may proceed in one station while melting is in progress in the other.

In an alternative design the furnace body and superstructure remain fixed while movable crucibles may be alternately positioned under the furnace for melting. In one case the two crucibles are diametrically opposed and are rotated into position and offered up to the body by a hydraulic ram, and in another method a single crucible is moved into, or out of, position on a movable trolley. The fixed furnace top design is particularly favoured for the melting of reactive metals on account of the smaller area occupied, hence minimising the blast protection required.

References pp. 642–647

Fig. 20. Typical vacuum arc melting installation showing two melting furnaces of the dual station type. (Courtesy Lectromelt)

With the increase in numbers of vacuum arc furnaces now being manufactured some degree of standardisation and preassembly has been achieved so that expensive installation time on site is reduced. Considerable attention has also been paid to kinematic design so that the turn-around time (from end of one melt to commencement of next) is reduced to a minimum. Both vacuum and electrical connections are so terminated that time-consuming coupling and uncoupling operations are unnecessary between melts. Connection to the stinger rod and electrode from the negative bus bars is by flexible leads to allow the necessary traverse.

The consumable electrode is connected to the stinger rod by a stub (usually of similar material) and clamping device. The weld between stub and electrode may be formed inside the furnace using an arc under vacuum or inert gas conditions, or alternatively welding may take place externally so as not to occupy valuable furnace time. The weld must be capable not only of supporting the weight of the electrode but also of carrying the ultimate melting current.

Crucibles are constructed from copper tube welded to a flange which together with the water jacket is clamped to the furnace body to position the crucible and form the vacuum seal. Since this flange also completes the electrical connection and carries the full melting current it is frequently of bronze or copper but may be backed by a steel flange for added strength. Although square or octagonal sections have been used, crucibles are invariably circular and ideally are constructed from seamless high conductivity copper or are manufactured by electrodeposition. Earlier crucibles had a solid integral bottom necessitating inversion and sometimes considerable "bouncing" to remove the ingot but the modern tendency is toward removable bottoms. Ingot stripping is facilitated and inspection, cleaning and repairs are greatly simplified.

The cost of copper is high and since a wide range of crucible sizes may be utilised in conjunction with a given furnace this can form a considerable proportion of the total plant capital outlay. With care, however, a life of many hundred melts may be obtained.

3.1.2 Vacuum requirements

Consumable vacuum arc melting normally operates in the range from about 1 to 10 millitorr and the vacuum system utilised must fulfil two requirements. In the first instance it is required to evacuate the furnace volume to the appropriate operating range in a period of the order of 10–20 minutes and secondly it must be capable of handling the gas volume evolved during the melting of the electrode. The latter will be a function of the electrode material and size and will usually dictate the particular system chosen.

A widely used configuration for the high vacuum side consists of a mechanical booster (Roots blower) type pump in parallel with a vapour booster. The mechanical booster has a high pumping capacity in the 10^{-2}–10 Torr range and speeds evacuation of the furnace during initial pump-down. It also assists in minimising the effect of sudden gas evolutions during the melting of the electrode. The pressure is quickly restored to optimum before diffuse or misplaced arcs can affect the heat balance to the pool or cause damage to the crucible walls. The parallel vapour booster pump can handle large quantities of gas over the normal operating range of the furnace and offers the additional bonus that the pumping speed for hydrogen is about double that for air at the lower pressures. This is of particular value in the melting of titanium. Backing this high vacuum combination would be the usual mechanical forepumps. A typical configuration is shown in Fig. 21.

For large installations and where space permits, steam ejector pumps may be used and also water ring pumps.

In most modern installations all the vacuum pumps are remotely operated from the control desk by pneumatic or magnetically operated valves. Indeed the pumping programme may be rendered automatic by operating the appropriate valves and pump contactors by signals from the normal vacuum gauges and the state of the furnace evacuation programme may readily be seen on a "mimic" diagram at the control desk. A further essential feature is the inclusion of automatic interlocks to obviate misuse and to render the furnace safe (or relatively so)

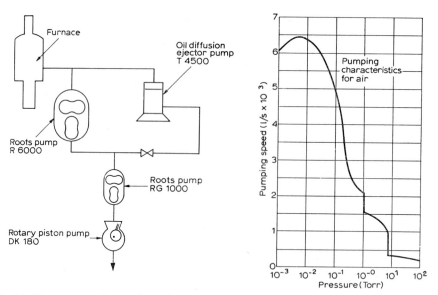

Fig. 21. Vacuum pumping configuration for production size consumable arc furnace. (Courtesy Leybold–Heraeus)

in the event of a dangerous situation developing. For instance the melting current may automatically be disconnected if the furnace pressure rises above a pre-set level or in the event of a water failure to the cooling system of the vacuum pumps.

As in all vacuum systems, conductance of the connecting pipework is maximised by mounting the pumps as close to the furnace as possible and utilising large diameter manifolds.

3.1.3 Power supplies

With the exception of a special design of furnace, which will be referred to later, d.c. power is required for the melting supply during consumable arc melting, thus necessitating rectification of the normal mains supply either by rotary or static devices.

Designers of the earliest consumable arc furnaces—designed largely for reactive metal melting—turned quite naturally to existing equipment in the form of motor-generator sets already widely used in an allied process, namely d.c. arc welding. These welding generators represented a readily available package offering best value for money in terms of amperes per unit of capital outlay and were assembled in large banks of up to a total power of many thousands of amps utilising common control units. Although later supplanted by the static type rectifier, they are still widely used for the smaller laboratory or development unit on account of their low initial cost and their portability. A further advantage of the rotary welding generator for use with small experimental furnaces is the absence of any mains harmonic frequency ripple on the supply. This is a very desirable feature when high-speed recordings are being made to study the effect of different melting techniques or electrode control systems on the arc phenomena.

Adjustment of the output of this type of generator is by two controls: one which increases or decreases the open circuit voltage, and one which adjusts the short circuit current while maintaining a constant voltage. (See Fig. 22(a) and (b).)

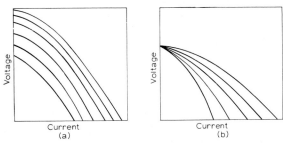

Fig. 22. Typical rotary generator characteristics showing range of variation obtainable with (a) voltage control, and (b) current control.

Since the arc established during melting is basically a constant voltage device, a factor further augmented by the electrode position control system, what is required of the power supply is a "steep" characteristic to ensure that the current and hence the power is also maintained as constant as possible. In practice, a setting which gives an open circuit voltage of about 70–90 V is chosen, and large changes in power are obtained by switching in or out more generators.

With the growth in the use of arc melting the static rectifier increased in favour because of its higher operating efficiency, better space utilisation and ease of control, and the motor generator was replaced by the selenium rectifier unit. This in turn was superseded by the silicon unit on account of its higher current rating for a given size. The germanium rectifier has the highest efficiency of all but is limited in temperature of operation to about 65°C and has therefore not been widely used. Control of the current to give the required steep characteristic is by reactors usually in the primary circuit of the mains transformer[34]. These are normally of the "saturable" or d.c. control type in which the reactance is varied by passing a small direct current through a control winding. In this way it is possible to adjust the rectifier smoothly from 100% down to about 10% of its full rating by means of a small control on the instrument panel. Figure 23 shows schematically a typical circuit and the resulting current voltage characteristic is shown in Fig. 24(a).

It can be shown that the maximum power output from a unit of this characteristic occurs at about 55–60 V, whereas arc operating characteristics of most steels and alloys are typically about 25–35 V. The need for constancy of current in this region coupled with high open circuit voltage in order to strike the arc is therefore obtained uneconomically since such a unit is only operating at about

References pp. 642–647

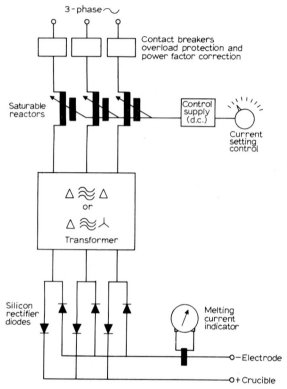

Fig. 23. Schematic diagram of d.c. melting power supply unit for consumable arc melting.

one-half its nominal rating. Since very few metals require an operating arc voltage much above about 40 V the type of characteristic shown in Fig. 24(b) is now usually demanded.

In this case, the main rectifier is designed with a lower open circuit voltage (say 60 V), thus effecting considerable saving in initial cost. A small low-current rectifier known as the "tickler" supply is incorporated with an open circuit voltage of about 80 V in order to facilitate striking the arc at commencement of the melt or to restrike the arc in the event of an open circuit during the melt. Constancy of current over the operating range is effected by the use of a closed loop control system in which a signal proportional to the rectified load current is compared with a stabilised pre-set reference voltage and any error amplified by a d.c. amplifier. This amplified error signal is then used to control the saturable reactors and adjust the current level in the appropriate direction so as to reduce the error to zero. The operating signal for such a system is usually derived from the transductor supplying the current indicating meter and an overall stability of $\pm 2\%$ for widespread changes in loading and mains fluctuation may be achieved by this type of system. Without closed loop control of current the stability from all causes (mains variation or short circuits during melting) may be of the order of -5% to $+10\%$.

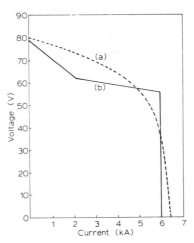

Fig. 24. Voltage–current characteristics for d.c. power supplies (a) saturable reactor controlled type, (b) later type with closed-loop feed-back control system.

The latest step in the development of power supplies came with the advent of the silicon-controlled rectifier or SCR, a development of the silicon diode but incorporating a third "electrode" used to control the particular instant at which the device commences to conduct during the appropriate half-cycle of the supply frequency. If "triggered" at the start of the half-cycle the device conducts over the whole of that half-cycle and the full load current flows. As the trigger point is delayed to progressively later stages of the half-cycle so the load current is progressively diminished. Such a device offers many advantages; control signal requirements are several orders of magnitude less than with the previously described systems, and since no saturable reactors are required some saving in space is effected. The SCR device is, however, very sensitive to transient overloads and adequate protection in the form of special fuses across each rectifier is necessary. The application of such a power supply to consumable arc melting was first described by Richardson in 1964[35]. A 2000 A supply was successfully developed for a small laboratory-size arc furnace and was later followed by larger units for production use.

The rectifier characteristics such as these typified in Figs. 24(a) and (b) indicate those appertaining under static conditions and it is important to bear in mind that the dynamic or transient characteristics of the unit are equally important. In the event of a short circuit occurring between electrode and pool, the current established—assuming the condition to be sustained for several seconds—is that indicated at zero volts on the static characteristic. Instantaneously, however, the current may rise to a considerably higher value than this owing to sluggishness on the part of the closed loop control system and also reactive components normally present in the circuitry. Examples of typical and extreme response time are shown in Fig. 25, and it will be seen that in the worst case the overshoot can in-

References pp. 642–647

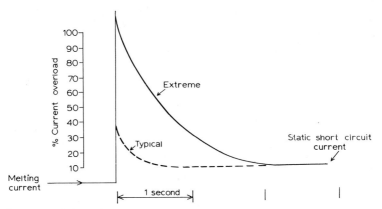

Fig. 25. Examples of dynamic short circuit characteristics of a melting supply showing typical and extreme cases.

stantaneously be over 100% and may last for several seconds. Apart from any adverse effects on the actual melt this can also cause severe strains to be imposed on the bus bars and stinger rod owing to the high electromechanical forces obtaining. Careful attention to circuit design, however, can reduce these transient conditions to acceptable proportions.

A further factor of importance which has received considerable attention in recent years is the mode of current supply to the crucible. The view is widely held that spot segregation, an undesirable feature of some vacuum arc melted steels and alloys, is aggravated by disturbances during ingot solidification.

Uncontrolled movement of the molten metal pool due to electromagnetic reactions between the field from the melting current flowing and any unbalanced magnetic fields in the neighbourhood of the crucible is thought to be a contributory factor. In this connection therefore attention is directed to methods of arranging the bus bar runs and connections to the crucible so as to keep the input as symmetrical as possible. Connections to the base of the melting crucible have been suggested instead of the top flange coupling, and modifications to the extent of constructing a "bifilar" bus bar supply and circular distribution around the crucible have been described[36].

The extent of the power supply requirements for a given furnace is dictated by the maximum ingot size considered, and since most furnaces have been designed to operate with a range of crucible sizes it was not unusual in the early days of development for organisations entering this particular field of manufacture to moderate their power requirements initially until experience had been gained in the melting of the smaller size ingots. As experience was accumulated and production requirements grew, extra "blocks" of rectifiers were added to increase the ingot capacity and it has been observed that "hybrid" combinations of the different types of rectifier can usually be operated together quite successfully.

Finally, reference must be made to the special design of furnace which utilises alternating current for the melting supply. The furnace and its mode of operation are described in subsection 3.2, but it will be apparent that the power and electrode control arrangements are somewhat similar to those of the normal electric arc furnace, the absence of rectifiers effecting some economy in capital outlay.

3.1.4 Electrode position control

Great stress has been laid in recent years on methods of arc control since homogeneity of the resulting ingot has been found to be profoundly dependent upon uniformity of arc conditions throughout the melt.

The commonest and most widely used principle is the voltage-dependent system illustrated schematically in Fig. 26. The voltage between electrode and

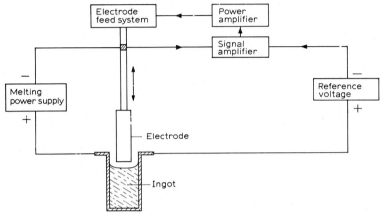

Fig. 26. Principle of the simple "voltage-dependent" electrode control system.

molten pool is compared with a pre-set stable reference potential and any error is amplified and used to feed the electrode either up or down so as to reduce the error to zero.

The earliest systems were based on similar methods to those already in existence on air melting electric arc furnaces and used the rotating amplifier or amplidyne to generate the required power to operate the feed motor in response to the error signal. These and later versions using the static magnetic amplifier have been described by Borrebach[37].

Developments on this basic principle have been mainly directed at improving sensitivity and speed of response of the system as a whole and have concentrated on the method of power amplification and the particular type of motive system employed to feed the electrode.

The rotating amplifier initially used has given way to the thyratron valve unit or magnetic amplifier or latterly the silicon-controlled rectifier. In the earlier furnaces the drive system was a simple reversible d.c. motor and suffered several limitations. The inertia of the armature was appreciable, limiting the speed of

References pp. 642–647

corrective action attainable and the range of effective speed control was limited to something of the order of 30 to 1. These disadvantages were obviated by the introduction of the two motor system which controlled the electrode through a differential gear box. This in its turn was superseded by the double differential system illustrated in Fig. 27[9]. The differential gear for the electrode feed is driven

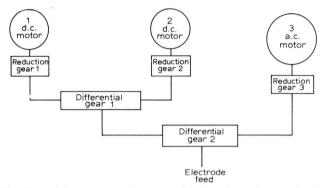

Fig. 27. Double-differential gear system for comprehensive electrode control (after Scheidig[38]).

by two d.c. motors and has a transmission ratio such that the speed differential of both motors is considerable and can be controlled or reversed easily. The second differential gear utilises an a.c. motor and brake and during melting comes into operation only in the event of an actual short circuit between electrode and pool by retracting the electrode as quickly as possible. The same motor is also used for the manual control of the stinger rod during the charging of the electrode into the furnace. This system has been described by Scheidig[38].

The systems so far considered are voltage-dependent and rely on the relationship between arc voltage and arc length. Typical arc gradient characteristics for titanium and steel in vacuum are shown in Fig. 28, and it will be apparent that in the case of steels particularly a large change in arc length can occur without a significant change in voltage except at near short-circuit conditions. The situation

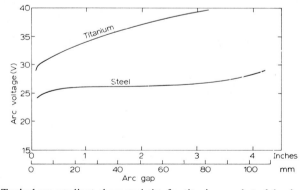

Fig. 28. Typical arc gradient characteristics for titanium and steel (under vacuum).

is further aggravated by the fact that changes in pressure too can affect the slope of the characteristic. Under certain conditions of glow discharge or diffused arc the relationship may change slope and an *increase* in arc gap be associated with a *fall* in voltage. As a result, the control system would respond to the reduction by retracting the electrode, thus worsening the situation.

These shortcomings in simple voltage-dependent control have led in recent years to improved methods based on alternating random components of the arc signal[9]. These fluctuating components may be likened to the "hiss" of the normal carbon arc or welding arc in air and have led to the description of the newer control system as "hash", "noise" or "pulse-rate" control. It has been found that these fluctuations or pulses vary according to arc gap, both amplitude and frequency increasing as the electrode approaches the pool. These pulses are separated from the mains frequency ripple and used as an additional control system. Since such signals are also sensitive to the nature and temperature of the arc they offer a better solution to the correction of diffuse arc conditions than the simple voltage-dependent control.

An alternative design to the electric motor drive utilises the control signals to operate a variable displacement hydraulic pump in conjunction with a rotary hydraulic motor. A wide speed range is achieved, coupled with maximum speed of response for reversal of the electrode system. Again the system is operated by voltage-dependent or "hash" signals either alone or together and incorporates other refinements to optimise uniform melting conditions[39].

Another system operating hydraulically utilises the stinger rod itself as part of a hydraulic piston controlled by a sensitive two-way valve. In practice, the feed system is set to drive the electrode down at a rate slightly faster than is actually required to keep pace with the burn-off, thus causing the arc gap to gradually shorten. As this occurs the frequency of the fluctuations or pulses on the arc potential changes accordingly. When the conditions reach a previously selected state the downward travel is halted for a pre-set period of time. This "dwell" period is so short as to affect the arc gap insignificantly. In the event of an actual short circuit taking place the feed system rapidly withdraws the electrode to the appropriate distance required to clear the condition simultaneously measuring the extent of the withdrawal and returning the electrode to its former position. Advantages claimed for this particular system are the elimination of several moving parts and the fact that the basis of the hydraulic system already exists to power the furnace lift and swing movements[40].

Since the inception of consumable arc melting, considerable attention has been paid to the development of improved methods of arc control following up on the early realisation that uniformity of the heat balance conditions in the area of the arc and molten pool is all-important if the full metallurgical advantages of the process are to be gained. Particular importance must also be attached to stability and reliability under operational conditions. Failure of such a control system during a melt can be extremely costly and, in the case of reactive metals, very dangerous. The use of modern electronic design techniques, however, has considerably improved reliability, and simplified maintenance by the use of plug-in printed circuits.

In the event of an absolute failure, most systems provide simple manual control for emergency completion of the melt.

3.1.5 Arc stabilising and stirring

Because the arc is an electrical phenomenon comprising flowing electrons and ionised particles it may be influenced and deflected mechanically by magnetic fields, intentional or otherwise. From the very early stages of the development of consumable arc furnaces it has been customary to provide an electromagnetic coil around the crucible in order to generate a magnetic field coaxial with the axis of the ultimate ingot. The application of elementary electromagnetic laws relating field, current and force would tend to suggest that such a coaxial field would generate no resultant force on the arc or pool since the field would be in the same direction as the flow of (melting) current. In practice of course the current flow from the electrode tip through arc and molten pool to the crucible sides is far from coaxial with the coil field. Being quite divergent, it is capable of deflection in the direction of the resultant force.

In most designs the electromagnetic coil is contained within the water jacket, thus effecting both protection and adequate cooling. In order to minimise the power requirements the coil may be tapped at several places so that only the section surrounding the melting zone is utilised. As melting proceeds and the ingot builds up, the power is transferred to a higher section of the coil. Present-day coils are usually designed to generate a magnetic field of about 100–250 oersted.

The need for, and the effects of, electromagnetic stirring or stabilising may be examined under three headings:
(a) intentional stirring and agitation of the molten pool;
(b) the avoidance of undesirable rotation of the pool;
(c) alteration of the heat distribution within the melting zone.

Intentional stirring is of value during the primary melts of alloys produced from sponge or powder electrodes, *e.g.* titanium and zirconium. Intensive agitation ensures thorough mixing of the alloying material and any attendant undesirable metallurgical effects may be obviated in the subsequent remelt.

In the case of final melts, however, particularly of alloys which are prone to segregation effects, any rotation of the pool is thought to be a prime cause of some of the undesirable segregation features found in these ingots. As mentioned earlier, undesirable pool rotation can occur in most units owing to stray magnetic fields and asymmetrical current connections. It has been found possible to counteract this by the judicious application of a small neutralising field. An alternative method, however, is to utilise a periodically reversed field whereby the direct current supply to the coil is reversed every few seconds. By asymmetrical adjustment of the periods it is possible to compensate for the inherent rotation of the pool. The concept is also of importance in appreciating the third characteristic of stirring fields—that of heat distribution effects in the melting zone. A feature of this periodically reversed stirring is that although the arc is affected, having little or no mass, the actual molten pool is less affected on account of its relatively high inertia[9]. Thus it is possible to effect some measure of control over the arc and minimise misplaced arcs to the crucible wall without the attendant disadvantage of vortex effects deepening the actual molten pool. This is well illustrated by the following figures for pool depth measured on 110-mm-diam. experimental ingots, utilising uranium as a radioactive tracer[41]. In each case the effective stirring field was the same (130 oersted).

	Pool depth
(1) Unidirectional stirring	8.1 cm
(2) Stirring reversed at 4-second intervals	7.4 cm
(3) Stirring reversed at 1-second intervals	6.3 cm

3.1.6 Instrumentation

In the early days of vacuum arc melting—particularly of reactive metals—a permanent trace of the progress of the melt was considered essential, the reason being to aid enquiry following any accidental explosion. It was quickly realised, however, that such records are of great value in achieving optimum performance from the control system and in maintaining consistent standards from melt to melt.

Thus it is usual to provide for *recording* in addition to *indication* of the following melting parameters:
(1) Melting current
(2) Arc voltage
(3) Electrode travel (distance)
(4) Ultimate vacuum.

Figure 29 shows a typical control desk for a modern furnace. In addition to the recording instruments mentioned above, indicating meters are included to enable the operator to assess the satisfactory functioning of the furnace and to ensure safe operating conditions. These will generally include the following in order of importance:
(a) Crucible cooling water temperature and flow rate
(b) Fore vacuum (backing)
(c) Electrode travel rate
(d) Pressure gauge (for argon atmosphere melting)
(e) Elapsed time clock
(f) Stirring field current indicator
(g) Electrode, upper and lower limit warning signals, and sundry meters to indicate the level of the various control signals (*e.g.* "hash", or voltage difference).

References pp. 642–647

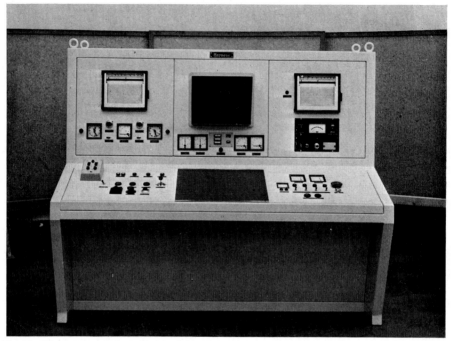

Fig. 29. View of a typical control desk for a vacuum arc plant showing furnace controls, instruments for indicating and recording essential melting parameters and the tele-optic viewing screen. (Courtesy Leybold–Heraeus)

Despite the improvements in present-day furnace control systems and the information provided to the operator by array of instruments already mentioned, direct viewing of the surface of the melt is still regarded as advantageous. In the smaller laboratory furnaces the melting controls are usually situated on the furnace and viewing may be direct through conveniently situated viewing ports. In the case of the larger installations, however, and particularly where melting of reactive metals is concerned, the melting console is situated some distance from the furnace and a remote viewing system is called for. One obvious method which springs to mind is of course closed-circuit television, but although this system is convenient and versatile much of the information on the state of the arc and molten pool is lost in a black and white picture, and since colour television is considerably more expensive recourse has been made to "piped" optical systems relying on prisms and lenses to project an image on to a ground glass screen at the control desk. A typical system is illustrated in Fig. 30. Two similar optical systems convey images from opposite sides of the pool and combine them into a composite projected image at the desk. In the melting of long ingots it may be necessary to refocus the system, as the melt proceeds, by remote-controlled mechanisms. With a well designed optical system a great deal of information is available to the melter. It is possible to see whether or not the melting conditions are rendering a full pool,

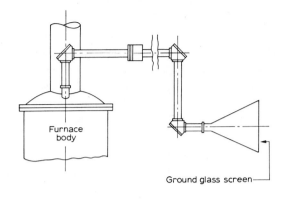
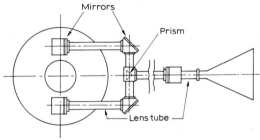

Fig. 30. Optical viewing system used for projecting an image of the arc and molten pool on to a ground glass screen at the control desk.

and hence good skin on the ultimate ingot and whether the arc is behaving normally or is displaced or diffused. The effect of any selected stirring field can be observed and perhaps most important of all the appearance of a green coloration is an obvious danger sign that the copper crucible is being damaged and that the melt should be terminated immediately.

With the increase in growth of vacuum arc melting some measure of automation has crept in, in an endeavour to minimise the labour content of melting by reducing the amount of supervision required. Once the optimum melting procedure for a particular type and size of melt has been established, *i.e.* from starting up through the main melting period and finally through the hot-topping schedule, it is possible to pre-programme the melt using punched cards or tape so that subsequent melts may proceed with minimum intervention from the melting staff. In the simplest cases the melting power and other parameters are merely adjusted appropriately according to a pre-set time cycle. This is adequate where size, shape and soundness of electrode are uniform and exactly reproducible but more often than not unavoidable differences in melting of the electrode result in variable yields. The variation may be further aggravated by differences in the amount of electrode dressing and cleaning required so that simple time cycle control is inadequate. What is really required is a continuous measurement of electrode *weight* so that the hot-topping cycle can be commenced when the correct amount of

References pp. 642–647

electrode material is left and the system is not misled by large cavities and areas of unsoundness in the top of the electrode.

Various methods have been devised for continuously monitoring electrode weight, the most promising being that which utilises a resistance wire strain gauge load-cell dynamometer mounted in the stinger rod assembly[42]. This load cell weighing system includes multi-range read-out covering the range 0–25 ton, with an accuracy of ± 25 kg below 250 kg and $\pm 1\%$ above.

3.1.7 Cooling systems and safety features

In view of the amount of heat generated during consumable arc melting, water-cooling facilities form a fairly important aspect of furnace design. The major portion of the heat is generated in the copper crucible where melting takes place but in addition to this, heat flow along the electrode by conduction may have to be removed by cooling the stinger rod. Also the furnace body may require cooling to prevent damage due to splashes from the molten pool. For cooling the body the furnace chamber may either be supplied with attached cooling coils or alternatively a double-walled construction may be used. During periods between melts when the furnace is open to the atmosphere, condensation may occur inside the chamber, and reaction take place with any volatile hygroscopic constituents which may be deposited on the furnace wall (*e.g.* magnesium chloride). In order to avoid this the circulating water to the body is normally warmed so that the furnace walls are maintained above ambient temperature.

It will be appreciated that the most critical aspect of cooling is in the removal of heat from the copper crucible in which melting takes place. Normally the water jacket is designed to accommodate the largest diameter crucible which is to be used with the particular furnace. Therefore a large clearance exists when smaller diameter crucibles are in use, and in order to maintain adequate surface flow velocity, matching water guide tubes may be used with each size of crucible. These give a clearance of one or two centimetres and minimise the possibility of water stagnation and steam bubble formation at the copper surface.

In the event of crucible penetration by a stray arc, cooling water may enter the melt, with the attendant generation of large volumes of steam. This causes the arc to be extinguished and, since modern furnaces are strengthened with a view to containing the resulting pressure rise, no further external damage results in the majority of cases. As a precaution, however, it is usual to site furnaces so that the crucible area is below ground level.

In the melting of reactive metals such as Ti and Zr the situation is particularly hazardous, however, since the reaction between molten Ti and water is exothermic, producing the oxide of titanium and gaseous hydrogen with a resulting heat of reaction of 39,000 cal/mol of titanium. Given sufficient quantities of molten metal and water, the amount of hydrogen generated may exceed the capacity of the pumping system and the pressure generated rises rapidly, eventually causing disruption of the water jacket and releasing hydrogen to the atmosphere. This may result in a second (oxy-hydrogen) explosion, with disastrous consequences to the whole of the surrounding area.

The melting of certain titanium-base alloys which contain volatile elements such as manganese as an essential alloying element presents a special problem. These alloys are usually melted under a low pressure of inert gas, and explosive pressures may be attained much more quickly since no vacuum pumps are in operation.

Some measure of protection against explosion may be effected by incorporating a pressure-release valve in the pumping line or furnace chamber. This consists of a large-diameter rubber-sealed valve spring-loaded from the outside. In principle, any potential explosive tendency generated within the furnace is released to atmosphere, the valve then returning to the closed position to obviate the admittance of air to the furnace. In view of the high velocity rates of explosive phenomena, however, their efficiency is under question and following a number of serious accidents during the early days of reactive metal melting it is now common practice to situate such furnaces behind concrete or metal blast walls and to operate them from a safe area.

In recent years, alternative methods of cooling which would obviate the use of water have been considered. Inert gas cooling is feasible and safe but has not found favour on account of the poor heat transfer characteristics. Another method which has now been accepted as a much safer, though more expensive, alternative to water cooling is the use of liquid sodium–potassium (NaK) alloy as a coolant.

At concentrations between 40 and 90% potassium the alloys are liquid at room temperature, have good heat conductivity and have similar fluidity to water. Though these alloys were initially considered to be too dangerous in themselves for such applications, experience has shown that, by the utilisation of properly designed systems, the liquid may be pumped, stored, cooled and circulated almost with the facility of water.

The entire crucible cooling system must, of course, be on a closed circuit, the liquid NaK being pumped through specially designed crucibles in which the cooling jacket and copper melting crucible are integral. Special valves are incorporated at the inlet and outlet to the crucible so that at the termination of the melt the crucible assembly may be removed from the furnace without the spillage of any coolant which remains in the assembly throughout the life of the crucible. At some distance remote from the furnace the NaK is pumped through a heat exchanger, which may be either water- or air-cooled depending upon the size of the installation.

The successful application of this type of cooling to a reactive metal consumable arc furnace has been described by Cooper and Dilling[43], and comprehensive "package" systems are now commercially available. Although considerably more expensive initially than the normal water-cooling system, a proportion of the extra cost is offset by the avoidance of the high cost and relative inconvenience of blast-proof siting of the furnace.

3.2 Special designs of furnace

While the vast majority of consumable arc melting operations are conducted in the standard type of furnace hitherto considered, certain rather specially designed furnaces with unconventional features have been produced for specific projects. Two such examples deserve mention.

Perhaps the earliest modification to the standard consumable arc furnace was the so-called "skull" melting unit in which the normal, deep, round crucible used for ingot production is replaced by a shallower water-cooled crucible capable of being tilted through 90° so that its contents may be utilised to make shaped castings by pouring into conventional moulds.

The term "skull melting" is derived from the fact that a thin shell or skull of metal remains behind in the crucible after pouring owing to the rapid chilling effect of the water-cooled crucible. Depending upon the thermal constants of the metal being melted and the power available, up to 90% of the available crucible contents can be poured, however, and the principle has obvious applications in the manufacture of castings in the reactive metals which do not lend themselves to melting in ceramic crucibles. In addition, the principle has been used to demonstrate the possibility of making investment castings in the more mundane materials such as nickel-base heat-resisting alloy, where freedom from exogenous inclusions is of paramount importance[44].

It is in the field of titanium and titanium alloy castings, however, that the method has been most widely adopted, and various mould-making techniques have been developed to resist the action of the molten titanium on pouring. Most techniques are based on the use of carbon in some form or other[45] or on methods of introducing carbon into normal ceramic moulds[46].

The skull melting principle is illustrated in Fig. 31. High melting powers are required in order to achieve a large quantity of molten metal as quickly as possible. When the required amount of metal has been melted from the electrode the power is switched off and the crucible tilted as rapidly as possible so as to minimise the loss of heat. It will be appreciated that little superheat of the metal is possible using a consumable electrode technique since, broadly speaking, higher melting power merely results in quicker melting rates, and metal temperatures in excess of 30–40°C above the melting point are difficult to achieve. One method of achieving some degree of superheat is to weld the required amount of metal in the form of an electrode on to a tungsten rod so that following the melting of the electrode material the melt is continued, using the tungsten rod as a non-consumable electrode over the molten pool.

Quite large skull furnaces have been built capable of pouring something like 300 kg of titanium and the technique has been extended to the making of castings

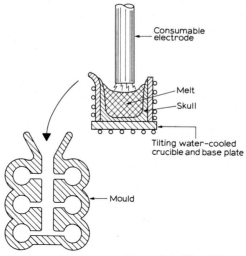

Fig. 31. Principle of the consumable arc "skull" melting technique.

in tungsten and molybdenum for aerospace applications. It seems likely, however, that in due time the electron-beam furnace will prove a more feasible route for such castings as development and increased usage render the economics more attractive.

A more recent innovation in furnace techniques has been the development of the alternating-current consumable arc furnace. Although the attractive possibilities of using a.c. had been evident since the early days of consumable melting, there were problems associated with the difficulty of maintaining an arc owing to the fact that the power passes through zero on each half-cycle and the arc is extinguished. This difficulty was overcome by the use of 3-phase system feeding three electrodes into a single crucible and by the continuous addition of an ionising agent (KCl) throughout the melt. The KCl is completely volatilised and, it is understood, leaves no trace of contamination in the resulting ingot.

The furnace (see Fig. 32) is basically similar to the normal consumable arc furnace but requires three electrode control systems. Each consumable electrode is sector-shaped as shown in Fig. 33 to minimise electrode length and travel requirements since these are a function of the ratio of total electrode cross-sectional area to ingot area. This unusual electrode shape presents some problems in manufacture and increases the cost of conditioning by way of surface preparation, but against this disadvantage and the extra cost of the control systems must be set several postulated advantages. The system offers better heat distribution and a shallower pool for a given melting current, and since no current flows through the actual melting crucible the danger of perforation is minimised. The improved heat distribution is of particular interest in the production of certain superalloys which are prone to segregation, and Cooper et al.[47] have described the successful

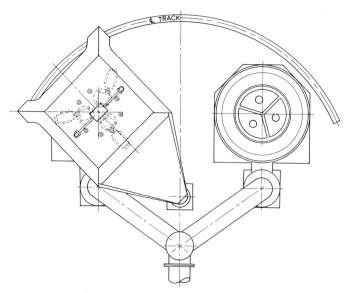

Fig. 32. Plan view of the 3-phase vacuum consumable arc furnace showing the 3 sector-shaped electrodes *in situ* in the loading position (right) and the head with the 3-electrode control system in position for melting on the left. (Courtesy Lectromelt)

Fig. 33. Sector-shaped electrodes required for melting in the 3-phase consumable arc furnace. (Courtesy Lectromelt)

experimental production of 700-mm-diam. ingots in these materials, using this three-electrode a.c. furnace.

The principle of three-electrode melting has also considerable potential as a means of producing slab ingots by an "in-line" arrangement of the electrodes.

3.3 Process techniques

3.3.1 Electrode fabrication and handling

The major prerequisite for any material to be vacuum arc melted is that it be electrically conducting in order to carry the necessary melting current. It must also be capable of being fabricated into the appropriate shape.

In the case of the reactive and refractory metals for which the process was primarily developed the raw material in the form of powder, sponge or granules is mixed with the appropriate alloys in the correct proportions and pressed in a closed-die press into half-round or half-octagonal compacts of suitable size. These compacts are then "stagger" welded together to form a long electrode, as shown in Fig. 34. Depending upon the reactivity of the particular metal, tungsten-inert-

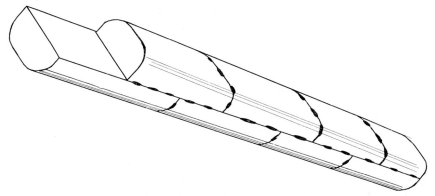

Fig. 34. Consumable electrode of reactive metal formed from welded compacts of pressed powder or sponge.

gas arc welding may be used or, where absence of contamination is all important (*e.g.* zirconium alloys), the welding may be carried out in a totally enclosed welding box under a completely inert atmosphere.

As an alternative to compacting the raw material by die pressing, isostatic pressing is also possible and has been described by Dorsett and Krebs[48] and also by Noesen[49]. Densities of the order of 70–90% of theoretical may be achieved by either method depending upon the electrode size and pressure available, and in the case of titanium and zirconium alloys adequate "green" strength is achieved for direct melting without further treatment. In other cases, however, some degree of semi-sintering treatment is required to give the electrode adequate mechanical strength and to improve its electrical conductivity. This may take place under vacuum or inert atmosphere or in many cases under hydrogen to effect some degree of pre-purification by reduction.

In most instances it is difficult to achieve the required degree of homogeneity and ingot soundness in one melting sequence when starting from compacted raw

References pp. 642–647

material electrodes and it is usual to carry out a second melt using the first ingot as an electrode and melting into a larger diameter crucible.

In the case of steels or nickel-base alloys the electrodes may be prepared by any of the usual fabrication techniques. There is little doubt that a forged electrode free from porosity or pipe is the ideal consumable electrode; expediency and economics, however, have led to the use of cast electrodes. These have proved completely successful provided that appropriate care is taken to minimise segregation and porosity. Any piped portion should normally be discarded and all surface oxidation should be removed by grinding or machining. While parallel-sided electrodes are desirable for uniformity of melting conditions over the whole vacuum arc ingot, difficulties in casting and stripping have led to the use of a slight taper on electrode moulds.

Depending upon the size and quality of the required electrode the casting may be by any of the well known air-melting techniques, *e.g.* acid or basic electric arc or by air induction melting. For the most exacting specifications the electrode may be cast by the vacuum induction process. This is considered essential where the alloy contains appreciable proportions of reactive constituents such as Ti, Zr and Al, and the rapid growth and widespread use of these iron and nickel base "superalloys" have led to the development of larger vacuum induction melting units so that wherever possible electrodes may be cast in one single piece. This obviates the necessity of joining small electrodes together end to end either by welding or screw-threading.

The electrode having been prepared by one of the foregoing means, it must then be introduced into the furnace and mounted on the stinger rod ready for striking the arc and commencing the melt. This is usually achieved by inserting the electrode into the empty copper crucible and centralising it by suitable wedges. The crucible is then offered to the furnace body, clamped in position and the evacuation cycle commenced. When adequate vacuum is attained the stinger rod complete with clamp holding a stub of similar material but of smaller diameter is lowered to strike an arc with the top of the electrode to form a weld. On completion of the weld the stinger rod now carrying the electrode is retracted into the furnace body and the crucible removed. Jigging devices or wedges are removed from the crucible and the starter plate of similar material placed on the crucible base. The opportunity may also be taken to examine the weld for soundness. The crucible is replaced in position and when the appropriate vacuum is achieved actual melting may commence. The ranges of melting supply requirements are illustrated in Fig. 35.

It will be appreciated that the method of welding the stub to the electrode described above occupies valuable furnace time and in the interests of economy many organisations carry out this operation independently of the furnace using external apparatus.

In the case of pressed raw material electrodes the lack of density means that

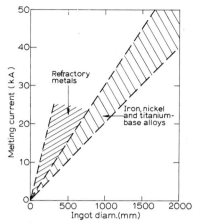

Fig. 35. Typical range of melting supply requirements for consumable arc melting.

comparatively long electrodes are required to produce a reasonable ingot length and to simplify matters the melting may be carried out from two separate electrodes. A typical sequence for the production of a titanium ingot is as follows:

(1) The first electrode is melted to produce the "A" ingot. This ingot is then stripped from the crucible.

(2) The second electrode is then melted to form the "B" ingot.

(3) At the end of this "B" melt the stub is welded on to the ingot and the "B" ingot retracted.

(4) The primary melt crucible is removed and replaced by the larger diameter remelt crucible with the "A" ingot centralised in position in the bottom.

(5) The "B" ingot is lowered and welded on to the "A" ingot. The whole is then retracted and the final remelt commenced using the "A" and "B" composite electrode.

A simpler, alternative, procedure may be employed where smaller ingots are required or where adequate furnace and crucible length is available. The "B" primary ingot may be melted on top of the "A" ingot and the stub finally welded on. The whole is then retracted and melted into the larger-diameter remelt crucible.

A novel furnace has been developed for the production of molybdenum[50] in which the whole process of electrode production and melting is carried out continuously in the one unit. Powder is continuously pressed into a round electrode die and fed down through electrical contact shoes, which introduce a heavy current through the material and effect a certain amount of sintering. Thus strengthened, the electrode passes into the actual melting furnace and is consumably arc melted in the normal manner.

3.3.2 Melting parameters

With the electrode suitably mounted on the stinger-rod and the furnace pumped down to the required degree of vacuum the actual melting process may

References pp. 642–647

be commenced. The electrode is fed down by the control gear and an arc is struck on the starter pad. This must be of similar material to the melt and is usually a 10–20-mm slice machined from the parent electrode. A few turnings of the parent material placed on the starter pad are also used to assist in generating the arc quickly.

High melting power is avoided initially until a molten pool has been formed to afford some degree of protection to the crucible bottom. The melting power may then be raised to a higher level than is intended for the main melting period, in order to offset the chill effect of the crucible base. Figure 36 illustrates a typical

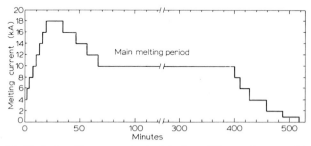

Fig. 36. Typical melting current schedule for a 500-mm-diam. steel ingot.

current schedule for a 500-mm-diam. steel ingot. As ingot build-up proceeds the cooling effect from the base is reduced to negligible proportions and the melting power is reduced to avoid the formation of too deep a molten pool. The power is then maintained constant over the major portion of the ingot until the end of the melt, when the power is reduced according to a predetermined "hot-top" programme in order to minimise shrinkage in the top of the ingot.

The choice of melting current is essentially a compromise and from the point of view of productivity should be as high as possible in order to minimise furnace time. On the other hand, however, metallurgical requirements demand a low melting power in order to maintain a shallow molten pool and minimise segregation effects. Once satisfactory melting conditions have been established for a particular alloy and ingot diameter the current requirements for larger ingots may usually be specified by simple ratio since to maintain constant power per unit area of ingot, I^2 must be held proportional to area, therefore current should be increased in proportion to diameter.

Again, since the arc is a constant voltage phenomenon the melt-off rate of material from the electrode is proportional to current and may conveniently be expressed in terms of weight per kiloampere-minute (k_1). Some typical values are shown in Table 5. These factors are of use in estimating total melting times and are of particular value in the calculation of hot-topping schedules, by indicating the amount of electrode required and hence the point at which hot-topping should be commenced.

A typical example is shown in Table 6, which lists the current–time schedule

TABLE 5

TYPICAL VALUES FOR SPECIFIC MELT-OFF RATES (k_1)

Material	lb/kA min	kg/kA min
Iron and nickel-base alloys	1.25–1.55	0.565–0.70
Niobium and molybdenum alloys	0.5–0.75	0.255–0.32
Titanium (sponge)	1.18	0.53
Titanium (remelt)	1.22	0.55
Copper	1.54	0.70

TABLE 6

TYPICAL HOT-TOPPING SCHEDULE AND CALCULATION FOR DETERMINING AMOUNT OF ELECTRODE REQUIRED (500-mm-diam. STEEL INGOT)

Current (kA)	Time (min)	Current quantity (kA min)	Calculated quantity of metal melted off at each stage* (kg)
8	5	40	25
6	8	48	30
4	15	60	38
3	15	45	27
2	15	30	19
1	15	15	9
Total	73	238	148

* Assuming specific melt-off rate (k_1) of 0.63 kg/kA min.

for the 500-mm steel ingot previously referred to in Fig. 36. The current and time values are arrived at largely by experience and the figures shown are known to give almost complete absence of shrinkage pipe in this particular ingot size following a main melting current of 10 kiloamperes. For each current step the appropriate amount of material required is calculated and the sum total is converted into terms of electrode length (assuming uniform density and cross section) and is used to determine the stage in the melt at which the hot-topping should be commenced in order to finish the melt with the minimum amount of electrode material remaining on the stub.

The choice of electrode diameter for a given ingot size is a compromise affected largely by technological considerations. A large annulus is desirable from the purification point of view by improving the gas conductance up the crucible, but too large a gap results in inadequate filling of the pool and poor ingot skins. For obvious reasons, however, the clearance between electrode and crucible must be larger than the arc gap between electrode and ingot in order to minimise the appearance of unwanted side arcs and a value of 15–30 mm is normal. Electrodes which have been produced by the vacuum induction process may warrant the minimum gap since less gas is evolved during arc melting.

In the case of the refractory metals, particularly tungsten, appreciably larger electrode–crucible separations have been found necessary and slight variations can have a profound effect on melt-off rate as shown by Noesen[50].

The electrode–crucible diameter ratio also influences furnace design indirectly by controlling the amount of stinger rod travel required and hence the furnace height.

Total travel required for a particular melt is given by the electrode length minus the resulting ingot length (neglecting arc gap). Hence:

$$\text{Total travel} = l\left(1 - \frac{d^2}{D^2}\right) \text{cm}$$

where l = electrode length, d = electrode diameter and D = ingot diameter in cm.

Electrode feed rate is also influenced by the ratio of ingot–electrode diameters and is shown by the expression:

$$\text{Electrode travel rate} = \frac{\left(1 - \frac{d^2}{D^2}\right)}{\rho \pi d^2} I k_1 \text{ cm min}^{-1}$$

where I = melting current in A, k_1 = specific melt-off rate and ρ = metal density.

On completion of the melt, air may be admitted to the furnace as soon as solidification is complete and ingot stripping commenced within a matter of minutes since the resulting surface oxidation of the ingot is immaterial if the ingot is to be further processed by hot working. In the case of primary melts of the reactive metals titanium and zirconium, however, the ingot must be allowed to cool down under vacuum in the furnace to avoid the oxidation and hence contamination of the subsequent remelt. The extent of this cooling is naturally dependent on the size of the ingot and the melting power used to produce it but is normally between 50% and 100% of the melting time.

3.4 Metallurgical effects

3.4.1 Purification

In the case of sponge and powder electrodes a certain amount of degassing may take place before actual melting, *i.e.* in the heated portions of the electrode adjacent to the arc. In the case of solid electrodes, however, degassing is restricted almost entirely to the arc zone and takes place on the face of the electrode, the surface of the bath and to a lesser extent during the passage of the metal from one to the other in the form of droplets.

The effectiveness of the process in removing volatile impurities will depend on the pressure and temperature prevailing in the arc and also on the residence time in the molten state. This latter function is dependent to a certain extent on the melting current used, but as shown by Sperner and Persson[51] and Bungardt and Vollmer[52], melt diameter is the major factor influencing the time exposed to vacuum and improved refining conditions can be expected with a larger ingot.

Pressure is usually measured at some point in the furnace body at a considerable distance from the actual melt and will normally be in the region of 10 to 1 millitorr or less but it will be appreciated that a considerable pressure difference can exist down the crucible, particularly in the early stages of the melt on account of the length of the conductance path. Various workers have endeavoured to measure the actual arc pressure[53,54] and it seems clear that the actual pressure existing in the centre of the arc zone may be several Torr when the normal indication in the body of the furnace is in the millitorr range.

The temperature existing in the arc is not easily measured. Cobine[55] has estimated that anode spot temperatures of the order of twice the melting point of the metal may exist but it is doubtful whether these conditions apply for the vacuum arc. Practical measurements made by Bungardt and Vollmer[52] indicate that even at relatively high melting currents little more than 30°C of superheat may be achieved in the pool when melting steel.

In the case of the high melting point refractory metals such as tungsten, however, the high arc temperature existing coupled with the low vapour pressure of the metal is of value in aiding purification. In general the vapour pressure of any

TABLE 7

BEHAVIOUR PREDICTIONS BASED ON RELATIVE VAPOUR PRESSURES OF METAL AND METAL OXIDE

Data at 2000 K

Should deoxidise	Should not deoxidise
$\dfrac{MoO}{Mo} = 10^{0.5}$	$\dfrac{Ti}{TiO} = 1$
$\dfrac{NbO}{Nb} = 10$	$\dfrac{V}{VO} = 10^2$
$\dfrac{WO}{W} = 10^2$	$\dfrac{Be}{BeO} = 10^3$
$\dfrac{ZrO}{Zr} = 10^2$	$\dfrac{Cr}{CrO} = 10^4$
$\dfrac{BO}{B} = 10^2$	$\dfrac{Mn}{MnO} = 10^5$
$\dfrac{ThO}{Th} = 10^3$	$\dfrac{Fe}{FeO} = 10^6$
$\dfrac{HfO}{Hf} = 10^4$	$\dfrac{Ni}{NiO} = 10^7$
$\dfrac{TaO}{Ta} = 10^4$	
$\dfrac{YO}{Y} = 10^5$	

impurities which will be present will be considerably higher than that of the base metal. Quite small quantities of the interstitial elements (C, N, O) are harmful to the properties of refractory metals and alloys, and purification by consumable vacuum arc melting is particularly effective for these metals.

Deoxidation by vaporisation of the metal oxide is possible with the higher melting point metals, where the vapour pressure of the monoxide is several orders of magnitude higher than that of the metal. Smith[56] has calculated the vapour pressure ratios of metals and their monoxides. These are listed in order of deoxidation favourability in Table 7.

Deoxidation is further enhanced by the presence of elements having a high affinity for oxygen, and it is common practice to introduce carbon into the electrode when melting the refractory metals. Thus:

$$W + WO_2 + 2C \rightarrow 2W + 2CO \uparrow$$

and the gaseous product of reaction is continuously pumped off.

Purification during consumable arc melting also occurs by simple removal of dissolved gases. Oxygen and nitrogen may occur in this form in certain alloys but by far the more important example occurs in the removal of hydrogen from steels and also titanium alloys.

In addition to the removal of undesirable gases and elements, detrimental loss of desirable alloying elements may occur where the alloying element has a high vapour pressure relative to that of the base. Examples occur in the case of titanium and zirconium from molybdenum and niobium-base alloys and more particularly in the case of manganese from steels. The appropriate amount lost is

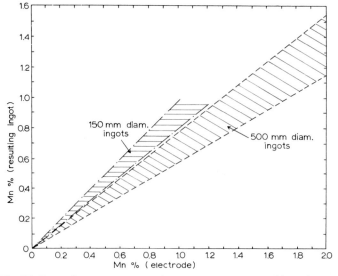

Fig. 37. Loss of manganese occurring on vacuum arc remelting of steels.

usually determined by experience and excess is then added to the electrode. Figure 37 illustrates the loss of manganese occurring on the remelting of steels. The two shaded areas indicate the approximate spread in manganese loss over a wide range of steel types for two ingot sizes. The major loss occurs with the larger ingot diameter because of the greater area presented to vacuum and the longer dwell time in the molten state.

3.4.2 Cleanness

In the case of steels and superalloys, one of the main purposes of the vacuum arc remelting stage is to achieve improved cleanness, and it is pertinent to note that the actual process of inclusion removal may be chemical or physical. Some of the less stable oxides and nitrides may for instance suffer dissociation at the temperatures and pressures existing in the arc. In the main, however, the most harmful inclusions present are the more stable ones such as Al_2O_3, TiN, Ti(CN), etc., together with the more complex compounds and spinels for which the conditions for dissociation are not attained. Carbon, if present, acts as a deoxidant, as evidenced by the fact that a small but measurable amount of carbon is lost during vacuum arc remelting.

Since sulphur and phosphorus are not removed by the vacuum arc process, alloys which require a low residual sulphide content must use selected raw materials for the production of the electrode.

The major factor in improving cleanness, however, is probably physical dispersion and flotation of inclusions. As material passes from the electrode to the melt in the form of droplets, the forces prevailing in the arc disperse the gross inclusions, resulting in a more homogeneous distribution of smaller innocuous ones throughout the ingot.

The following three examples of cleanness improvement[57] well illustrate the beneficial effects of vacuum arc remelting on steels for a variety of applications.

Following the early developments of the consumable arc process for reactive metal melting, one of the first applications of the technique to the remelting of steels was in the improvement of the 11% chromium martensitic steels widely used in the U.K. for the manufacture of turbine and compressor discs for aircraft gas turbine engines. These discs are subjected to rigorous examination by the immersion ultrasonic flaw detection method and in certain critical areas internal defects no larger than 0.8 mm may result in complete rejection of the component.

Figure 38 (a), (b) and (c) shows a comparison between the defects found by ultrasonic examination in 100 air melted discs and 100 vacuum arc remelted discs. The 100 vacuum melted discs have substantially less defects than the air melted components and were entirely free from defects greater than 1.2 mm equivalent diameter. Considering only the best 95 discs, the comparison with air melted material is even more favourable since no defects greater than 0.8 mm were observed. At 90% utilisation no defects greater than 0.4 mm were found in the

References pp. 642–647

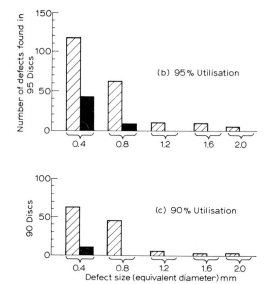

Fig. 38. Comparison of number of defects found by ultrasonic (immersion) inspection on air melted and vacuum melted turbine discs[57]. (a) 100% utilisation; (b) 95% utilisation; (c) 90% utilisation.

vacuum arc melted material. A further example concerning turbine wheels and compressor discs is quoted by Barraclough[58]; vacuum arc melting reduced the overall rejection rate to 0.26% compared with a computed rejection rate of 31.5% on air melted components examined to the same standard.

The improvement in vacuum arc remelted material is of particular value in engineering components which require a high degree of surface finish, as, for example, the steel rolls used for the production of metal foil. The out-cropping of gross inclusions at the surface during polishing is minimised and a relatively blemish-free product results.

A somewhat parallel case occurs in the production of high purity stainless steel for the fuel element cans for the British advanced gas-cooled nuclear reactor. Since the material is drawn down to very thin-walled tubes the inclusion specification is quite stringent. The alloy used is a 20% Cr–25% Ni steel stabilised with

TABLE 8

CLEANNESS OF DOUBLE VACUUM MELTED STAINLESS BAR MATERIAL

Cast No.	Number of inclusions (other than sulphides) found in 540 fields at magnification of ×130, in each of the following size ranges (mm)			
	<0.025	0.025–0.075	0.075–0.13	>0.13
1	16	0	0	0
2	32	1	0	0
3	34	0	0	0
4	20	0	0	0
5	16	0	0	0
6	28	2	0	0
Approx. comparison on air-melted stainless steel (20–12+Nb)	470	100	30	70

niobium, and in order to minimise cobalt and boron levels, the electrodes are prepared by vacuum induction melting from high grade raw materials. Table 8 summarises the total inclusion count (neglecting sulphides) on six production casts of this material. Longitudinal sections representing outside, mid-radius and centre are examined from forged 80-mm-diam. bar. Sixty fields on each of nine sections are examined, making a total of 540 fields from each cast. Four of the six casts under consideration had no inclusions in any of the 540 fields examined greater than 0.025 mm, and in the other two casts only one and two inclusions respectively were found in the next size range. In fact, in the 3,000 or so fields represented by the examination, no inclusion (other than sulphide) greater than 0.05 mm was detected and the largest sulphide inclusion detected was only 0.15 mm.

For comparative purposes an estimate is given of the cleanness of a basic–electric air melted 20% Cr–12% Ni (niobium-stabilised) stainless steel.

Perhaps the most important factor influencing ultimate cleanness following vacuum arc remelting is the inherent quality of the starting material, and this is well illustrated by the third example, which concerns the 1% carbon–chromium steel universally used for bearings. This particular alloy has been the subject of a great deal of process development over the last 15 years or so in endeavours to improve its cleanness and hence its operating life under stress. Improvements in normal air melting techniques made initial gains and later, application of bulk degassing further enhanced the properties.

In Table 9 the results of extensive examination of this material produced by two vacuum melting routes are compared with air melted material produced by the normal basic–electric arc process. The total area examined for each type of melting was 25 cm² and a complete list of the inclusions found in this area is given under four size ranges.

The air melted material had extensive inclusions greater than 0.5 mm in size, whereas the vacuum arc material remelted from air melted electrodes had only

References pp. 642–647

TABLE 9

INCLUSION FIGURES SHOWING COMPARATIVE CLEANNESS OF AIR MELTED, VACUUM ARC MELTED, AND DOUBLE VACUUM MELTED EN31 BALL BEARING STEEL

Process	Number of inclusions found in each size range (mm)				Average inclusion length, mm
	0.025–0.075	0.1–0.25	0.28–0.5	>0.5	
Air melted material (basic electric arc)	466	90	16	16	0.079
Vacuum arc melted material	126	3	0	0	0.033
Double vacuum melted material (vacuum induction followed by vacuum arc)	63	0	0	0	0.025

three inclusions in the size range 0.1–0.25 mm and none greater. Finally, the double vacuum melted material, made by vacuum arc remelting an electrode prepared by the vacuum induction process, had no inclusions greater than 0.075 mm, thus illustrating the progressive diminution of inclusion content as more refined vacuum techniques are employed.

It should be borne in mind of course that improvement must be paid for in increased processing costs and the ultimate choice is governed mainly by economics.

3.4.3 Segregation effects

In the vacuum arc melting process, the very high thermal gradients existing in the molten pool from the contra-effect of the arc power supplied at the surface of the melt and the rapid cooling effect from the base and sides of the water-cooled crucible, favour columnar crystallisation in the ingot. A typical structure is illustrated in the etched longitudinal ingot section shown in Fig. 39. At the base of the ingot, crystal growth is vertical on account of the rapid chill effect. As the ingot builds up, the cooling effect from the base is minimised and the cast structure takes on the familiar "pine-tree" appearance with crystal growth following the direction of heat flow and normal to the solidification front. In a longer ingot, this appearance occurs over the major portion of the ingot, finally terminating in a zone of mixed crystal structure and orientation due to the hot-topping process.

Columnar crystal growth is accentuated by high melting temperatures and by a high degree of purity in the material being melted owing to the deficiency of nucleants. It is therefore very pronounced in ingots of the refractory metals such as niobium, tungsten and molybdenum, purified by arc melting.

The larger the grain size, the smaller the total grain boundary area present, and the greater the concentration of residual impurities in these areas for a given impurity level. Grain boundary embrittlement and its attendant effect on subsequent working of the ingot are thus accentuated by large grain size, and attempts have been directed at methods of achieving finer grain structures during melting.

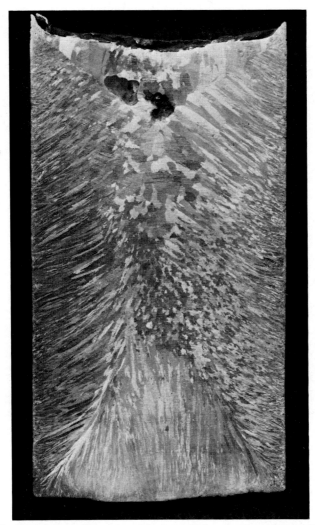

Fig. 39. Longitudinal section of a consumable arc melted austentic steel ingot (150-mm-diam.).

The addition of small quantities of alloying elements can have a grain refining effect, but if the purity of the metal is to be maintained, this course is not open. The use of vibrational energy has been tried with some success[59] but on account of practical difficulties has not been used on any scale commercially.

In the case of materials other than pure metals, particularly high alloy steels and superalloys, two forms of segregation are frequently encountered.

The first—aptly named "tree-ring" segregation—is manifest by a series of roughly concentric bands revealed on etching a transverse section of the material. It persists throughout the subsequent reduction of the ingot and is in fact accentuated in appearance in the resulting billet or bar, as shown in Fig. 40. The rings

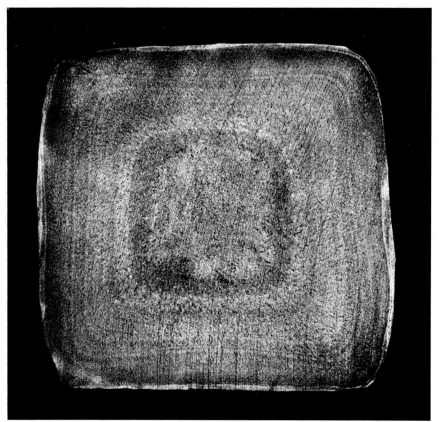

Fig. 40. Illustration of "tree-rings" in alloy steel (140 mm square billet produced from 500-mm-diam. ingot).

are in fact a cross-section of the consecutive pool markings and are caused by sudden changes in the rate of advance of the solidification front at the base of the molten pool. Such changes are caused by interruption in the steady flow of heat from the arc, due to the occurrence of diffuse arcs or random short circuits between electrode and pool, either of which may result in an abrupt change in heat flow. Attempts to check compositional gradients across these bands have not usually revealed any significant degree of chemical segregation, and the effect is largely one of differing grain size as illustrated by the micro section taken across such a zone in a 140-mm-square billet (Fig. 41). The sudden transition from large to very fine dendrite size is clearly evident.

The effect is accentuated by high melting currents due to the deeper pool formation and sharper angle of incidence made by the solidification front and the transverse plane. In general, however, the phenomenon is worse in appearance than effect and mechanical properties are not usually impaired.

A more serious defect which has beset high alloyed materials since the incep-

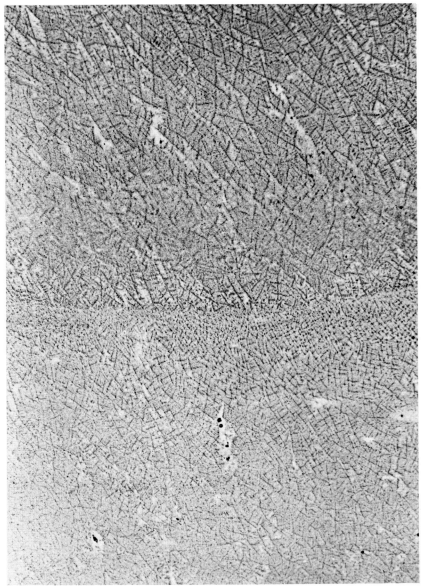

Fig. 41. Micro section (×7) across a "tree-ring" showing the sudden transition from large to very fine dendrite size (NiCr–steel).

tion of vacuum arc melting is that known as "spot" or "freckle" segregation. A typical example of "freckles" appearing in a transverse slice from a billet of vacuum arc remelted material is shown in Fig. 42. Again the effect is most marked following some degree of reduction of the ingot and appears most prevalent in the outer and mid-radius zones. Investigation shows that the "freckle" area is in fact

Fig. 42. Example of "freckle segregation" appearing in transverse slice from a billet of vacuum arc melted tool steel.

a terminal solidification product, a local area of interdendritic segregation high in alloy content.

The actual mechanism of formation of these spot segregations has been the subject of a great deal of speculation and investigation over the years. Gould[60] offers an explanation based on controlled solidification experiments on a laboratory scale. During freezing, the liquid ahead of the advancing solidification front becomes enriched and the composition changes to the point where a second immiscible phase is formed. Being less dense than the primary liquid the secondary phase tends to rise through the liquid melt and eventually solidifies—after the primary liquid—resulting in a discrete pocket of segregation. The phase rejected from solution may of course be gaseous and, once nucleated, a bubble of gas is formed by subsequent absorption of gaseous solute. The bubble rises through the melt, sweeping out a rod-like trace which is filled in by the liquid of terminal composition.

Bubbles of gas could result from release of hydrogen, oxygen or nitrogen from solution, but the more likely explanation seems to be the formation of CO as a result of deoxidation taking place within the melt at the low pressures prevailing. Hence the necessity of producing well "killed" electrodes by the addition of aluminium to the parent melt.

Liquid movement during freezing is known to accentuate segregation effects, and undesirable motion in the liquid pool due to unwanted asymmetrical electromagnetic disturbances is believed to be the major factor in the formation of freckles. As mentioned earlier, every endeavour is made during furnace design and in layout to ensure balanced current input leads and the absence of stray fields owing to the presence of large ferromagnetic masses near the crucible.

3.4.4 Improvements in properties

With the improvement in cleanness already discussed and the superior ingot structure found in vacuum arc remelted ingots, some degree of improvement in mechanical properties is to be expected, particularly in the transverse direction.

This improvement is particularly significant in the case of high-tensile materials. Not only is ductility improved but reproducibility from one portion of the ingot to another and from cast to cast is usually superior to that measured on ingots cast by normal methods—even following degassing treatment—as shown by Peter and Spitzer[61], who studied properties on forging ingots in a wide range of steels and alloys.

Barraclough[62] also quotes a striking example of improvement in both level and scatter of properties due to vacuum arc remelting of a martensitic chromium steel for turbine disc production. In addition, Fig. 43, due to DeVries, shows that the transverse ductility of vacuum arc remelted A-286 is approximately doubled when compared with an air-melted, conventionally cast ingot of the same alloy.

Extensive use of vacuum melting has been made for the improvement in fatigue life of 1% C, 1.3% Cr, ball-bearing steel and also in the melting of 18% tungsten high-speed steel which finds extensive use in the U.K. for bearings operating at higher temperatures. In the latter case, segregation problems have necessitated the use of smaller ingots, normally of less than 230 mm in diameter.

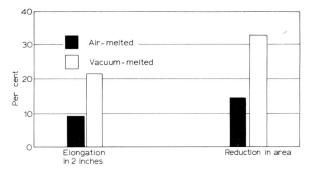

Fig. 43. A comparison of the transverse ductilities of air melted and consumable electrode vacuum arc melted A-286 superalloy.

References pp. 642–647

TABLE 10

COMPARISON OF THE FATIGUE STRENGTH
OF AIR AND VACUUM MELTED BEARING STEELS

Method of melting	Fatigue strength at 100×10^6 reversals (kgf/mm^2)	
	1% Carbon 1.3% chromium steel	18% Tungsten high speed steel
Basic electric arc	±60	±50
Acid open hearth	±69	—
Consumable electrode vacuum arc	±75*	±78*
	±90**	±85†
Vacuum induction	±82	±85

* Using basic electric arc electrodes
** Using acid open hearth electrodes
† Using acid high frequency electrodes.

Table 10 illustrates the improved fatigue strength achieved by vacuum melting and underlines the dependence of properties achieved after consumable arc melting on the origin of the electrode material[63].

The superior strength attributed to "acid" melting is maintained through the vacuum remelting process.

In recent years, considerable improvement has been achieved in the nickel-containing "maraging" steels owing to the application of improved vacuum processing. Single vacuum arc remelting of the air-melted product is popularly employed, but for the ultimate in properties, electrodes produced by the vacuum induction process in order to minimise impurity levels are favoured as shown in Table 11. Here again, ductility in the transverse direction is doubled in the case of the double vacuum melted material.

As mentioned in Section 2, removal of deleterious volatile impurities such as lead, bismuth and tellurium can be achieved under vacuum, resulting in improve-

TABLE 11

EFFECT OF MELTING TECHNIQUE ON TRANSVERSE TENSILE
PROPERTIES OF "MARAGING" STEEL (18% Ni)
(500-mm-diam. ingots forged to 130-mm-square billet)

Material	0.2% Proof-stress (kgf/mm^2)	U.T.S. (kgf/mm^2)	Elongation on $4\sqrt{A}$ (%)	Reduction of area (%)	Izod mkg
Single vacuum arc melted (mean of 10 casts)	172	182	5.5	12.9	0.78
Double vacuum melted (vacuum induction–vacuum arc) (mean of 42 casts)	172	183	9.2	30.7	1.83

ment in creep properties of high temperature alloys. It is doubtful whether adequate removal is achieved in the case of vacuum arc remelting, however, because of the limited time that the metal resides in the molten state—a factor which again underlines the importance of vacuum induction melting for electrode preparation in the case of these superalloys. Nevertheless, such removal of detrimental impurities as does occur is considered to be a contributory factor to improving the hot-workability of superalloys produced by the consumable arc process.

4. Electron Beam Melting

4.1 Introduction

By definition, electron beam melting is the process which utilizes the energy of highly accelerated electrons to produce thermal energy generated in the material itself for melting and refining it. The melting is a consequence of transfer of the kinetic energy of the electrons to the lattice of the material on impact, leading to local temperature rise and, eventually, to melting. While it is often difficult to establish the actual beginning of a process, this is not the case for electron beam melting. Accidental fusion of cathodes did occur, it is true, on many occasions at the end of the 19th century, but these incidents, while related, cannot be considered as consciously dealing with the invention. The conceptual birth of this process must have taken place during or about 1905, and its inventor is M. von Pirani. The 1907 United States patent[64] bearing his name appears to be the recorded beginning of this process.

This patent is amazingly advanced in terms of the ideas expressed in it (see Fig. 44). In principle, at least, it contains many of the concepts related to the process as used in industry today. We see in it the early work-accelerated Temescal gun system; in essence, the type of installation which was first applied to commercial melting and refining. We see the self-accelerated guns which superseded them, as well as the multiple self-accelerated devices which appear to be preferred by some of today's leading electron beam melting equipment producers. Three additional early electron beam melting installations and attempts to carry out the process are known. They are the work of Trombe[65] of France, Tiede[66] of Germany and Hultgren et al.[67] of the United States. The descriptions given in these three references appear to be of modest laboratory installations. Despite its obvious and considerable advantages, the technique did not become a useful tool until almost half a century later.

Invention of a device, a tool and a process at an early date does not necessarily assure its industrial adaptation, economic success or growth. It is the need for these devices, tools or processes which is really responsible for their introduction and which leads to the eventual development of technologies based on them. This,

References pp. 642–647

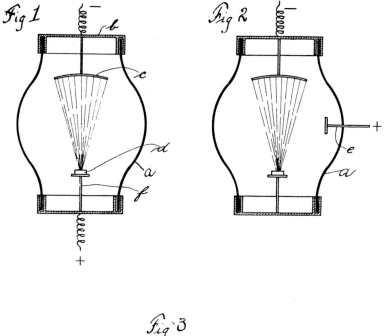

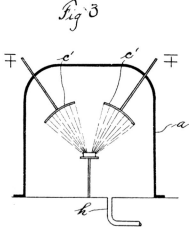

Fig. 44. Marcello von Pirani's sketches of various types of electron beam melting furnaces (1907)

indeed, was the case with electron beam melting. Though it was invented in 1907, nothing of significance took place until 1954, when Temescal activities got under way. The pressures brought about by new materials and problems related to their adaptation for use led to fresh reconsideration of the problem of electron beam melting.

It can be said that the first and most significant factor which catalyzed the growth of electron beam melting was the emergence of the nuclear and aerospace industries. The need for new developments in the processing of the reactive and refractory metals which were becoming absolute essentials for industrial growth in these areas was most acute. Vacuum technology, having received a major boost from early needs of the nuclear industries, was continuing to make important advances. Details concerning the fundamentals of vacuum and its production have already been discussed in Chapter II of this book and will not be considered here. The status of vacuum technology in the mid-fifties made it eligible for consideration for use with new economically feasible and reliable processing techniques in general, and electron beam melting in particular. The very large pumping requirements of this process could be met with greater ease by the new and higher capacity diffusion pumps being developed. Last, but not least, there was the realization that the electron beam itself had some truly outstanding advantages as a heat source for melting. Its cleanness, the lack of a real upper temperature limit, the realization that vacuum, its natural environment, is a most desirable atmosphere, were all important contributing factors.

While electron optics were well advanced, at that time the design and construction of high-power electron guns were in fact only in their beginning. The sound electron optics background, however, permitted rapid strides to be made here. The progress being made in the building of reliable heavy-duty, high-voltage power supplies for continued operation was also an important contributing factor; these supplies have improved further since.

The vacuum environment permits selective distillation of impurities. Unlike arc melting, electron beam melting allows the material to be fused and then maintained in a molten state for as long as is desired. In this way the removal of higher volatility impurities by direct evaporation is made possible. This important facet of the purification process combined with the water-cooled non-contaminating casting mould bring about the basic improvements and the higher purities ascribed to electron beam melting, as will be discussed in detail in the following sections.

The author has never been able to establish whether the Temescal group was aware of the early activities in the field[65-67], or if it was original thinking on their part without the realization of past happenings that led to the process. All that he knows is that the individuals who formed Temescal were, prior to the organization of this company, part of the Berkeley electron accelerator group. It is not inconceivable that accidental electrode fusion in connection with this work, which must have been certain to occur, led to the idea of using the electron beam as a source of thermal energy and, in turn, its application for melting and refining. According to H. R. Smith[68] of Temescal, apparently one of the initial problems resolved subsequent to the organization of Temescal and the construction of early melting systems was the consolidation of titanium, which was carried out by Smith and his associates in 1954. The few years which followed were most productive, for electron beam melting in general and Temescal in particular. In 1957[68] facilities capable of processing 80-mm-diam. titanium ingot up to 1.5 m in length were completed. About that time, C. Hunt, of the same organization, arranged to have

References pp. 642-647

a 25-mm ingot of tantalum melted for the author to be subsequently processed in tantalum foil for capacitors.

This coincides with the beginning of the almost explosive increase in the use of tantalum for this purpose. In 1958 capacities to melt an 80-mm-diam. tantalum ingot were available and in 1959 an 80-mm-diam. tungsten ingot[68] could be processed. The early papers on the subject began to appear in 1958. This relatively modest beginning has developed so rapidly that it is now the most important technique for processing the refractory and reactive materials and their alloys. It is also probably one of the most exciting techniques for the processing of superior alloy steels—potentially, at any rate; the economic aspect has yet to be tested.

4.2 Equipment design as manifested by commercially produced systems

Equipment design for electron beam melting has of necessity undergone rapid development since the early quasi-commercial application of the pioneering work-accelerated annular Temescal guns[69]. It is noteworthy that ten years after their initial commercial use virtually all work-accelerated guns have been replaced by comparatively trouble-free and more efficient self-accelerated guns.

"Work-accelerated" means that the voltage for accelerating the electrons is applied between the electron-emitting cathode and the material to be heated or melted. "Self-accelerated" means that the voltage is applied between this cathode and an auxiliary anode, suitably arranged in a relatively short distance from it, in order to achieve the desired beam-forming effect. The potential between the self-accelerated gun and the material to be heated or melted can, therefore, be kept very low.

The schematic diagram in Fig. 45 represents an electron beam melting system. In terms of its components and their functions, it has remained the same, though its scale has grown and its components characteristics have changed.

The early work-accelerated Temescal guns (Fig. 46) suffered substantially from arcing instability and operational difficulties. These were essentially brought about by the random pressure surges common in the process which occur when

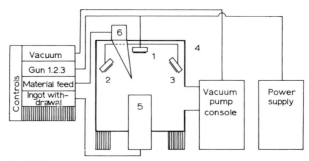

Fig. 45. Schematic of electron beam melting system. 1, 2, 3, electron guns; 4, work chamber; 5, mold and ingot withdrawal; 6, material feed.

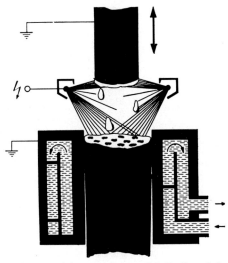

Fig. 46. A work-accelerated gun with coaxial material feeding (schematic) (after Heraeus).

sudden gas evolution takes place for a variety of reasons. While, as stated above, these guns are no longer used for large-scale melting, they continue to be used to good advantage in electron beam crystal growing and zone refining installations. (See Chapter III, Subsection 4.4 and this chapter, Section 5.)

They have been replaced by several types of self-accelerated guns and gun configurations, including annular self-accelerated guns, axial guns, axial differentially pumped guns, transverse guns and radial multi-filament guns. Single and/or multiple grouping of these guns powers industrial electron beam melting installa-

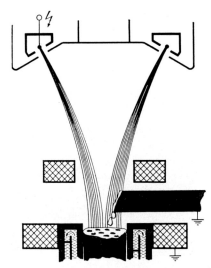

Fig. 47. Self-accelerated and focused gun with side feeding of consolidated material (schematic) (after Heraeus).

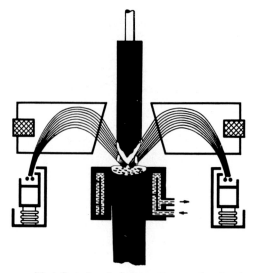

Fig. 48a. Transverse guns with deflected path for simultaneous bombardment of melt stock and molten pool (schematic) (after Heraeus).

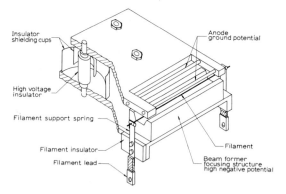

Fig. 48b. Transverse gun (schematic). (Courtesy Temescal)

Fig. 48c. Transverse gun (actual). (Courtesy Temescal)

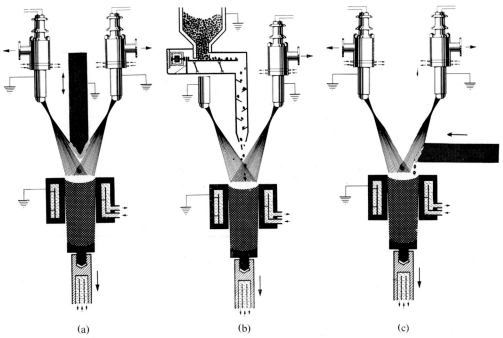

Fig. 49. Axial differentially pumped guns with three possible raw material feed schemes. (a) Drip melting of vertically fed consumable electrodes. (b) Bath melting of granulated material and scrap. (c) Drip melting of horizontally fed consumable electrodes. (Courtesy Heraeus)

Fig. 50a. Radial multi-filament gun, 150 kW rating. (Courtesy B. Paton, Paton Institute, Kiev)

tions today. These various types of installation are presented in Figs. 47–51. In these figures, the guns, as well as the geometry of systems and melting material feed provision, are presented. It is also of interest to refer to Fig. 52, which shows the type of power distribution associated with different number and gun geometry as observed in a Heraeus system, for example.

References pp. 642–647

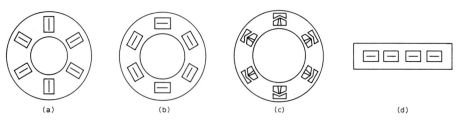

Fig. 50b. Possible alternate multi-filament configurations, three circular (a, b, c) and one linear (d) offering different beam configurations. (Courtesy B. Paton, Paton Institute, Kiev)

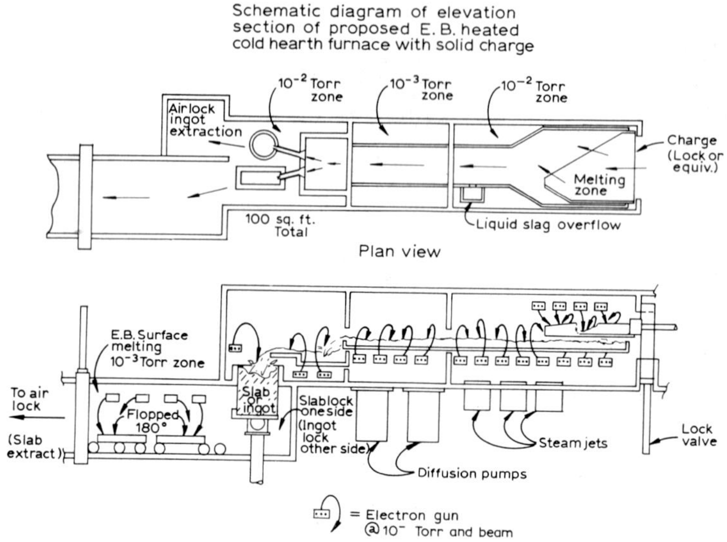

Fig. 51. Advanced concept Temescal melting approach utilizing transverse guns. (Courtesy Temescal) (None of information at hand permits decision as to the existence of an operational system such as this.)

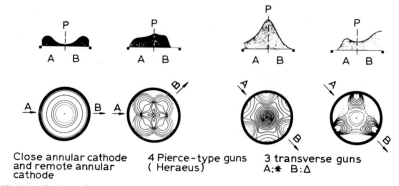

Fig. 52. Distribution of electron beam power on the pool surface where no consumable electrode is inserted in the beams. After electrode is inserted, power on pool is reduced in the centre of the pool. (Courtesy Heraeus)

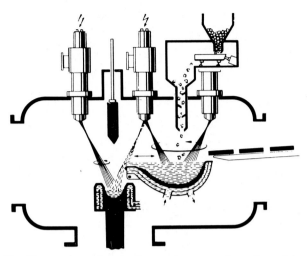

Fig. 53a. Electron beam furnace for continuous casting using axial guns.
(Courtesy Heraeus)

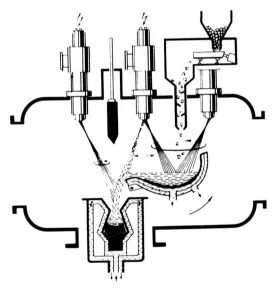

Fig. 53b. Electron beam furnace for mold casting using axial guns.
(Courtesy Heraeus)

In addition to basic melting systems, a schematic presentation of casting approaches is also given in Figs. 53(a), 53(b) and 54. Though little specific industrial recorded experience on electron beam casting has been reported in the literature[70-76], this certainly is an area of important growth and potential future developments.

The various corporations (see Table 12) which have entered the field of equip-

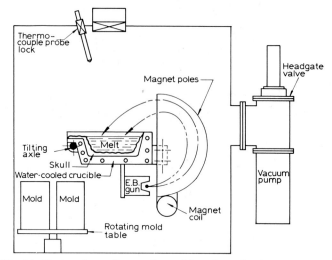

Fig. 54. Electron beam furnace for mold casting using transverse guns. (Courtesy Temescal)

ment manufacturers have developed their own systems for efficient melting and refining, but, in principle, only a few basic approaches exist in the ultimate systems manufactured by these suppliers. Inasmuch as this group is small, and these producers have relatively few basic systems on the market, the author believes that it will be best to discuss most, if not all, of the typical units. This survey of modern equipment is presented also because it is believed that much has been stabilized in the systems being built today. Except, perhaps, for power levels of a very high order (in excess of 3 megawatts), advances in the matter of controls and automation on provision systems produced by each organization and expressing their most forward thinking will essentially remain minimal.

Temescal, the pioneering electron beam melting equipment organization, which has recently become a subsidiary of Air Reduction Corporation, has un-

TABLE 12

ELECTRON BEAM MELTING EQUIPMENT PRODUCERS

Producer	Gun types utilized	Max. power installation constructed (kW)	Max. power installation planned (kW)
Airco–Temescal	Transverse	1400	3600
NRC	Axial	200	?
Heraeus	Axial	200	600
Alcatel	Axial	200	600
Paton Inst.	Radial heater	200	?
Moscow Power Inst.	Axial	?	?
von Ardenne–LEW	Axial	1700	3000
Japan Electro Optical	Axial	200	?

TABLE 13

ELECTRON BEAM MELTING FURNACES (TEMESCAL)

	Vacuum tank dimensions (cm)	Product dimensions (cm)	High vac. pump rate (liters/sec)	Installed electrical beam power (kW)	Number of electron guns	Type of system
A. Existing electron beam melting systems						
Super-alloy and refractory metal development	300×180×180	Ingots up to 20 diam. ×180 long	50,000	600	4	Semi-commercial scale pilot plant
Super-alloy and refractory metal production	75×360×180	Ingots up to 48 diam. ×210 long	150,000	1400	8	Commerical scale production plant
B. Electron beam melting systems now in construction or preliminary design stage						
Specialty alloy steel production	1200×300×300	Ingots up to 50 diam. ×300 lg.	300,000	2000	15	Commercial prod. plant
Refractory metal production	600×240×180	Ingots up to 30 diam. ×240 lg.	150,000	1000	6	Commercial prod. plant
C. Forecast of typical electron beam melting based on reasonable scale-up factors						
High strength and stainless prod. (100,000 ton per year)	1500×600×300	Ingots up to 60 diam. and slabs up to 15×125×600	1,000,000	9000	60	Commercial production
Mild steel production (1,000,000 ton per year)	3800×900×300	Molten metal delivered to atm. pressure continuous casting system	3,000,000	20,000	100	Commercial production

doubtedly been the most vocal and powerful driving force responsible for the acceptance and growth of the process. The melting equipment which it has manufactured or still has on its drawing boards has covered in the last decade the power span of 25 kW up to 36 megawatts. The melting installations built, designed or planned for the future are given in Table 13[77]. The initial work-accelerated annular guns[69] which pioneered the approach have been replaced by transverse guns, and the single-gun installations of yesterday are replaced by multi-gun installations containing as many as 25 self-accelerated transverse guns in the latest Temescal-developed melting throft installation concept, as already referred to in Fig. 51; and in principle at least, guns supplying power can be as numerous as 200. In all of its most recent installations, Temescal relies exclusively on the self-accelerated transverse guns. Further, Temescal has expressed the opinion that these guns are the most advantageous guns for larger power systems, a statement which is certainly not accepted by other producers in the field. In support of this opinion, the following facts are submitted. It is said that the guns are simple in principle, and quite rugged in operation, and that they can be upscaled to any power level

References pp. 642–647

desired without extensive additional development work. These guns are no doubt easier and less expensive to build than, for example, the axial guns used by several of Temescal's competitors, as a comparative examination will reveal, but the problem of selection of the most superior approach is by no means simple.

Another characteristic feature of Temescal's equipment aimed at improved operation is the location of guns completely outside the line of metal vapor flow, and in particular, below and off to the sides of the vaporization interfaces. With reference to Temescal's operational installations, the author has been unable to establish if the steel processing installations referred to in Table 13, and obviously aimed at the biggest potential business segment for electron beam melting equipment, *i.e.*, the melting of ferrous alloys, has been constructed. This type of installation and any similar power level system, however, are upscaled versions of components, details of which have been field tested and well proven in small installations by the Temescal group, and no major difficulties should be expected. In principle, at least, much larger installations are well within the range of technical reality but, to date, even though they have been proposed, they are apparently not yet an economic proposition. It should also be mentioned that electron beam melting is often combined with other vacuum melting processes in so-called duplex processing[78].

Moving from the overall aspect to specific details of components, comments are needed on the power supplies employed. Though essentially these comments refer to Temescal's installations[74,77,79], their validity is considered general and basically applies to any uses of power supplies for electron beam melting. First and foremost, whatever the principle of design, the power supply must be so designed as to assure trouble-free and reliable continuous duty operation. It should be immune to difficulties brought about by conditions, already mentioned, prevailing in the application of the process. Any process disruption leads to difficulties with the end product and this should be strongly guarded against.

Both constant voltage and constant current power supplies are known to have been used in electron beam melting operations. They can provide the stability and reliability needed by the process. Since the power supply is as important as the gun which it powers it should be discussed in some detail. This is perhaps done best by looking at the systems produced to date.

The total power of melting installations operational in the United States in 1968 approached about 6 megawatts. Here and elsewhere the 15-35 kV acceleration is the common acceleration voltage. Most of the installations are usually equipped with cathode emission controls, which are provided either by process control elements in a closed loop system or from externally programmed input.

An important advantage of the constant voltage supplies is the fact that any number of guns can be controlled with a single power supply. Inasmuch as Temescal makes extensive use of the multi-gun approach, it customarily utilizes this type of power supply. Up to three guns have been known to be operated with a constant current power supply, but this is not usually recommended. Figure 55 shows the characteristics of a constant current power supply operated in current-limited mode. Voltage level is maintained with a separate electron beam filament controller (Fig. 56 illustrates similar type characteristics in a constant voltage supply). Here, vacuum tubes are used as switching elements in series with the output.

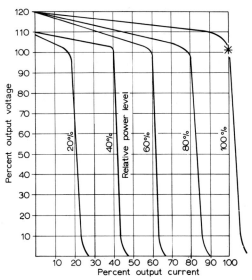

Fig. 55. Performance characteristics of constant current supply for electron beam melting installation. (Courtesy Temescal)

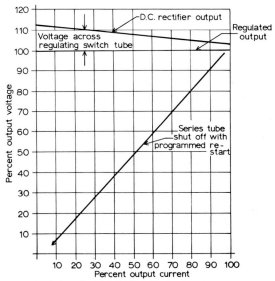

Fig. 56. Performance characteristics of a constant voltage triode cut-back electron beam melting power supply. (Courtesy Temescal)

References pp. 642–647

The notable feature of this type of supply is its ability to accommodate shorts or arcs which often take place in the operations through output control switches. In essence, it reduces power to essentially zero for several milliseconds. Power is resumed by programmed restart. Such fluctuations are beyond detection for most practical purposes as far as the operation is concerned, but they can be observed with an oscilloscope, if desired.

Figure 57 shows voltage current characteristics of a third type of power

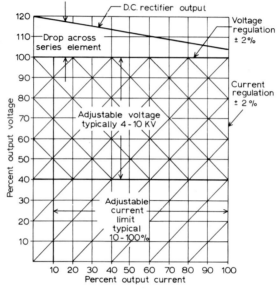

Fig. 57. Performance characteristics of a constant voltage triode system. (Courtesy Temescal)

supply. Here, output is also controlled by a tube in series, the difference from the previous supply being that the entire power is channelled through and is controlled by the tube. The voltage current characteristics of this type of supply show the wide range of power output which can be obtained in this manner. The external appearance of one of the commercially produced Temescal power supplies is shown in Fig. 58.

Fig. 58. 600 kW, 12-channel electron beam power supply. (Courtesy Temescal)

The pumping system for electron beam installation should have adequate capacity in order to assure a minimum of the better than 10^{-3} Torr levels needed for efficient and trouble-free operation. The rating of the pumps will depend upon power level of the furnace, desired feed-through rate and quality of the finished product, with specific values for melting systems to be presented in the course of this section.

NRC (of recent date NRC–Norton), a noted vacuum equipment supplier, is the only other United States organization which is at present active in the electron beam melting equipment field. NRC entered the field early[80] and has been also a pioneering organization in the U.S.A., though on a much more modest scale. The largest electron beam melting installation known to have originated from this company does not exceed the 200-kW power level, but it appears at the moment that there is only a 60-kW furnace[81] among the standard products of this company. As a producer of tantalum and niobium, however, NRC has pioneered extensively electron beam applications to the melting, refining and alloying of these and related refractory metals.

Fig. 59. 4-gun cluster, guns rated at 150 kW. (Courtesy Leybold–Heraeus)

References pp. 642–647

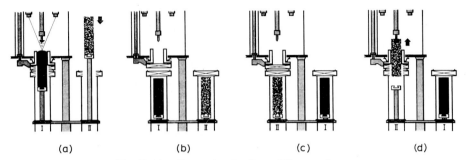

Fig. 60. Loading and unloading of Heraeus furnace.

(a) During the current melt, the ingot previously melted and meanwhile cooled can be unloaded. Subsequently, the starting block and the new consumable electrode can be loaded into the ingot container, and the closed chamber can be evacuated with a fore pump.

(b) Upon completion of the melt, the ingot is retracted into the container, and the valves on crucible and ingot container are closed under vacuum, so as to ensure a vacuum-tight seal. The space between the two valves is vented.

(c) By means of an oil hydraulic system passing through the tank, the crucible and valve are raised so that the two ingot retracting units can at the same time be interchanged. Subsequently, the valve and crucible are again let down to the other ingot retracting unit which has been moved to a position beneath them and the space between the two valves is evacuated.

(d) The pressure in the tank which was increased by opening the valves is reduced very quickly. The consumable electrode is moved into the crucible, and on the upper end a small molten pool is produced by electron bombardment. After dipping and solidification of the stub, the consumable electrode is lifted and, at the same time, the starting block is moved into the crucible. The next melt can be started.

Fig. 61. 250 kW furnace with potential of conversion of 600 kW.
(Courtesy Leybold–Heraeus)

In West Germany, Heraeus (now Leybold–Heraeus) has been an important producer of electron beam melting equipment. Though it lagged behind Temescal by a considerable margin in entering the field, this group has been quite active and has developed the differentially pumped gun cluster approach for melting[72,73,82] in contrast with von Ardenne's single differentially pumped gun approach[83]. Today, this group manufactures installations operational at power levels from 10 kW (small laboratory installation) up to several hundred kW. The largest installation to have the potential of 500 kW is powered by a cluster of four 150-kW rated guns (see Fig. 59).

Figure 60 shows approaches used to load melt stock and unload processed ingots, while Fig. 61 shows the actual furnace. In the first stage of development, this installation[82] is operational at the 260-kW production level and is provided by a 50,000 l/s diffusion pump with suitable mechanical pumping. Information on its performance capabilities on tantalum is given in Table 14. At full power level, *i.e.* when the four guns will be operational, two 50,000 l/s pumps will provide the needed vacuum. This system has a rotational charging system with two ingot chambers, as already illustrated. A weekly output of tantalum at the first stage of operation is shown in Fig. 62 and represents an approximate production rate of 1 ton weekly.

TABLE 14

RESULTS OF TANTALUM MELTING IN AN EBM FURNACE WITH 260 kW MELTING POWER AND 50,000 l/s PUMPING CAPACITY (HERAEUS)

Material	Ingot diam. (mm)	Melting speed (kg/h)	Total beam power (kW)	Average pressure (Torr)	Ingot yield % of matl. charged	Splattered and evap. regained % of matl. charged	Analysis (ppm) H	O	N	Brinell hardness (kgf/mm²)
Pressed bars from scrap 80×80 mm							50	11300	125	
1st Melt	100	25–30	180–200	$4-8 \times 10^{-4}$			3	1200	100	
2nd Melt	100	120–150	220–240	$5-8 \times 10^{-5}$	76	17	1–2	182	90	75–80
Pressed bars from powder 80×80 mm							146	2500	258	
1st Melt	100	35–45	180–200	$3-6 \times 10^{-4}$	80	12	5	200	70	
2nd Melt	100	130–150	220–240	$5-8 \times 10^{-5}$	80	12	1–2	115	43	70–75
Pressed bars from powder 80×80 mm							50	800	180	
1st Melt	100	60–70	180–200	$2-5 \times 10^{-4}$			2	150	50	
2nd Melt	100	130–150	220–240	$5-8 \times 10^{-5}$	85	8	1–2	100	22	70–75

References pp. 642–647

Fig. 62. Weekly tantalum production 50- and 100-mm-diam. ingots of the furnace illustrated in Fig. 61 (rate 1 ton weekly). (Courtesy Leybold–Heraeus)

Heraeus has probably produced the bulk of the electron beam melting installations in Western Europe as well as a few in the United States which were marketed by Heraeus–Engelhard. Heraeus, as a processor of tantalum, like NRC, has also been active in the application of electron beam melting to the processing of tantalum and its alloys.

Degussa, another vacuum equipment supplier in West Germany, was one of the first to produce electron beam melting equipment in Europe under a licence from Temescal; it used to build a 60-kW melting system[84] but is apparently no longer in this field.

In France, the picture regarding electron beam melting equipment producers has, at best, been somewhat confusing. Several companies, including Fours Cyclop, Sogev and Alcatel, have been active on different occasions and have produced installations alone or in conjunction with other firms since 1960, but no concrete and detailed facts are available about past activities here. The most recent information[85] indicates that Alcatel has become the main producer of electron beam melting equipment in France. It has recently built a 200-kW installation (see Fig. 63) and has scheduled for 1969 the production of a 500-kW melting furnace. It uses the single gun approach up to 200 kW and, though it is believed that the same approach will be used, no information as to the specific mode to power the 500-kW furnace is to hand.

Fig. 63. 200 kW melting furnace using single differentially pumped gun approach. (Courtesy Alcatel)

In Japan, the most active group appears to be the Japan Electron Optical Corporation, which is known to manufacture a series of commercial electron beam melters[86] with power levels of up to 200 kW. This company, according to the most recent information, produces three basic industrial furnaces, namely, the JEBM 200D rated at 200 kW (30 kV, 6.7 A), the JEBM 120D rated at 120 kW (30 kV, 4 A) and the JEBM 80D rated at 80 kW (25 kV, 3.2 A). It also produces two laboratory installations, the 30D rated at 30 kW (20 kV, 1.5 A) and the JEBM 10D rated at 10 kW (10 kV, 1.0 A).

In East Germany one of the early pioneering organizations was the Institut M. von Ardenne in Dresden[83,87,88]. This group continues its active involvement in the melting field to this very date, but only as a development group, and has transferred all its manufacturing operational and production rights[89] to the LEW Hans Beimler of Henningsdorf. The latter has been extremely active and probably has produced the majority of the electron beam melting installations used in Eastern Europe.

References pp. 642–647

Fig. 64. EMO 60—60 kW power level electron beam melting system.
(Courtesy M. von Ardenne)

It pioneered the differentially pumped gun principle[83] and today builds systems[90] in the 60 kW–1.7 megawatt range, utilizing a single differentially pumped gun to power the system. It is known that plans were made several years ago to build also a 6-megawatt single gun installation, but recent information[91] indicates that plans for this have been scrapped in favor of possibly multi-gun installations of comparable levels. Further, though two 1.7-megawatt installations were built, it appears from information on the performance of one of the furnaces that it is preferable to run the system at 1.2-megawatts level[92,93]. This group has capabilities for manufacture of small laboratory furnaces, but no such systems are in the line of LEW's productions. This group produces three basic units, namely, the EMO 60 (Fig. 64), the EMO 200 and the EMO 1700 (as already referred to), operating at the 1200-kW level (Figs. 65a and 65b). The performance capabilities of EMO 60 and EMO 250 are given in Tables 15a and 15b. No comparable data are available for the EMO 1700s, though one would believe that they might be somewhat more efficient than a multiple of EMO 200s. In terms of specifics for the case of steel where the ratio of ingot weights is available, the 690 kg output for a EMO 250 furnace comparable with the EMO 200 goes to 3500 kg for the 1200-kW unit, or as suggested, an approximate power-dependent multiple of the smaller type units. All of LEW's furnaces have programmed oscillation of the beam in the molten

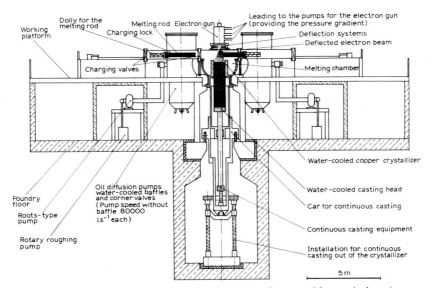

Fig. 65a. Vertical cross-section of EMO 1700. (Courtesy M. von Ardenne)

Fig. 65b. Pictorial view of EMO 1700. (Courtesy M. von Ardenne)

pool to ensure uniform temperature distribution—another special feature of these systems.

The Soviet Union has also been very active and is still the only country in which a book devoted exclusively to electron beam melting has been published[94]. Several centers, the more important ones being the Moscow Power Institute and the Paton Institute for Electrowelding, have pioneered the activities in the Soviet Union and numerous production facilities manufacture the prototypes developed there.

References pp. 642–647

TABLE 15a

SUMMARY OF TECHNICAL PARAMETERS ON THE EMO 60 ON PROCESSING OF DIFFERENT MATERIALS[90]

Material	Specific energy use (kWh/kg)	Diam of water-cooled mold (mm)	Ingot size		Melt time		Duration of a melt cycle		Yrly. prod. 3-shift operation		Cooling time to 400°C (h)
			a* (kg)	b* (kg)	a* (h)	b* (h)	a* (h)	b* (h)	a* (ton)	b* (ton)	
Cu	1.5	150	79	126	1.9	3.1	3.3	4.5	122	143	0.3
U	1.5	150	169	270	4.0	6.6	5.6	8.2	154	168	0.5(?)
Ni	1.5	130	59	94	1.4	2.3	2.8	3.7	107	130	0.3
Fe	1.5	130	52	83	1.2	2.0	2.6	3.4	102	124	0.3
Ti	2.5	110	21	34	0.83	1.4	2.53	3.1	42	56	0.6
Zr	3	100	25	40	1.2	1.9	2.9	3.6	44	57	0.6
Nb	10	80	21	34	3.3	5.4	4.9	7.0	22	25	0.5
Mo	10	80	25	41	3.9	6.6	5.5	8.2	23	25.5	0.5
Ta	15	50	16	26	3.8	6.3	5.4	7.9	15	16.8	0.5
W	30	25	—	—	—	—	—	—	—	—	—

* a Ingot length 500 mm.   b Ingot length 800 mm.

TABLE 15b

SUMMARY OF TECHNICAL PARAMETERS ON THE EMO 250 ON PROCESSING OF DIFFERENT MATERIALS[90]

Material	Diameter of water-cooled ingot mold (mm)	Ingot length (mm)	Ingot size (kg)	Specific energy consumption (kWh/kg)	Melt time (h)	Duration of a melting cycle (h)	Yrly. prod. 3-shift operation (ton)	Cooling time to 400°C (h)
Steel	280	1500	690	1.5	4.1	6.4	510	0.5
Ti	250	1500	315	2.2	2.7	7.5	200	3
Ti	250	1100	230	2.2	2.0	4.2	290	—
Nb	150	1200	260	10	10.2	14.0	88	2
Nb	150	800	170	10	6.6	8.4	96	—
Mo	150	1200	205	10	7.9	11.7	83	2
Mo	150	800	135	10	5.2	7.0	92	—
Ta	100	1200	150	12	7.0	10.3	69	1.5
Ta	100	800	100	12	4.8	6.6	72	—

The first group has developed systems in some ways reminiscent of the von Ardenne installations[95], i.e. large power, single, differentially pumped gun installations. Other axial guns for melting have been developed by Sushkin[96].

The Paton Institute, on the other hand, has developed its own rather unique design, the "radial gun" principle[97-100] already referred to (see Figs. 50a and 50b) which presents a 150-kW radial gun and some alternative geometry approaches for use of these guns. It is also believed that much of the early transverse gun work, later adopted by Temescal, is a derivation of the radial gun of the Soviet-type installations. Besides manufacturing its own installations in the Soviet Union,

Fig. 66. Soviet electron beam melting furnace, 150 kW, 15 kV, ingot size 200 mm diameter, 1.2 m long. (Courtesy B. Paton, Paton Institute, Kiev)

it has also imported numerous installations from East Germany. Figure 66 shows the Soviet electron beam installation.

While it is known that Italy, Holland, Belgium, England, Switzerland and Sweden are using electron beam melting installations, no commercial electron beam furnace building activities in these countries are known to the author.

4.3 Melting and purification in electron beam furnaces

As already stated, the electron beam melting and purification process owes its success to the ability to fuse any material known to undergo a solid/liquid transition and maintain the same in the fused state for as long as is necessary to effect the purification required. In this manner, related purification takes place, either by direct evaporation of impurities or the products of reaction occurring in the fused metal at elevated temperatures in accordance with prevailing laws of kinetics and thermodynamics. This, when carried out in a water-cooled non-contaminating copper hearth, led to the results ascribed to electron beam refining.

References pp. 642–647

In principle, at least, electron beam melting has no limits as regards either the nature of the material or the scale on which this process can be practised. In the Western world, except for France, where electron beam processing of ceramics is carried out, electron beam melting has almost exclusively been applied for the processing of metallic materials. Though fragmentary, information to hand indicates that users of the process in Eastern Europe apply the process for melting ceramics as well as metals.

There are, however, some limitations to the extent of purification possible[101]. At this stage of development of the process it appears, however, that the foremost limitations to its growth are based on economic factors. Several items must be examined in order to characterize the purification achievable in the electron beam melting and refining process:

(a) Vaporization of sub-oxides
(b) Evolution and removal of gases
(c) Carbon–oxygen reactions and metal–oxygen reactions
(d) Vaporization of metallic impurities and the interrelation between purification by evaporation and the economics of the process
(e) Floating of non-melting insoluble compounds
(f) Composition segregation, *i.e.*, zone solidification effects
(g) Potential sources of contamination.

(a) The ability to remove or not remove a specific lower oxide by the process of electron beam melting depends on the relative vapor pressures of an oxide to its respective metal. Table 7, Chapter V, Section 3 gives approximate values for these ratios. If the oxide has higher vapor pressure than the metal, we can readily remove it and the relative readiness depends on the ratios as given in this table. However, when this is not the case, we are in no position to remove the specific oxide in this manner and the relative difficulties of such removal have also been tabulated in order in this table (see also Chapter III, Subsection 4.2.2).

(b) The next item to be discussed is the removal of gases in general, and the interstitial O, H and N in particular. Typical data on removal of oxygen from tantalum[102] are given in Fig. 67, which presents the dependence of oxygen removal and, specifically, oxygen content in terms of melt time. Similarly, this type of data for oxygen and nitrogen in niobium[102] is given in Fig. 68. In the case of the interstitial gases, removal is essentially a function of time, which, in turn, affects the melt rate, the accepted interstitial levels and economic factors; *i.e.* cost to accomplish the same.

An alternative approach used for electron beam purification is the admixture of carbon to combine with oxygen, in particular, so that the products of this reaction can be removed. Typical data for tantalum[102] are given in Fig. 69.

The fourth active operative mechanism is the direct removal of higher volatility metallic elements from the metal to be melted and refined. The vapor pressure data in Fig. 4 a, b, Chapter III, allow for a rough assessment as to which specific

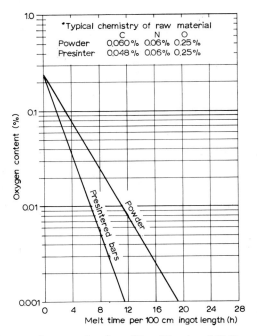

Fig. 67. Removal of oxygen from tantalum in melting. (Courtesy J. T. Perryman, Fansteel Metallurgical Corp.)

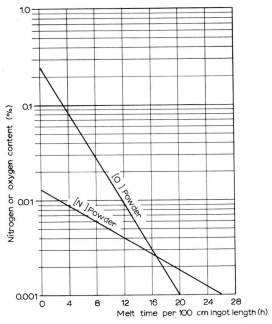

Fig. 68. Removal of oxygen and nitrogen from niobium. (Courtesy J. T. Perryman, Fansteel Metallurgical Corp.)

References pp. 642–647

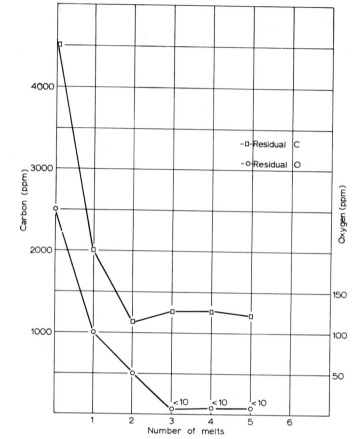

Fig. 69. Decarburization of tantalum by electron beam melting. (Courtesy J. T. Perryman, Fansteel Metallurgical Corp.)

metallic impurities can be removed from a given metal in electron beam processing and to what degree. Purification is feasible only if the ratio of impurity to solvent content in the vapor phase is much greater than in the melt, otherwise the loss of solvent becomes too high and the economy of the process goes down[101].

Actual rate of removal of higher volatility metallic impurities[103] can be seen in Figs. 70 and 71 showing the removal of chromium and iron, respectively, from zirconium at different melt rates.

While higher volatility impurities removal is an important part of the electron beam purification process, it also works against the process, or more specifically, against alloy manufacture by the process when some alloys containing these materials are to be prepared. Figure 72 shows such a case for the processing of Nb–10 Ti–5 Zr alloy[102]. Additional information on these effects can be found in a paper by Smith et al.[104].

Next of the factors contributing to the purification in the electron beam

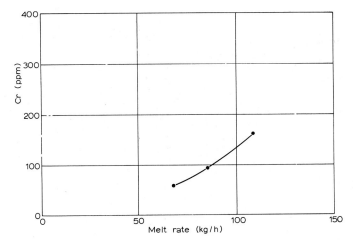

Fig. 70. Removal of chromium from zirconium. (Courtesy W. Aschoff, Wah–Chang Corp.) Material: zirconium; 360 ppm Cr in starting material; power input: 160 kW; pressure: approximately 10^{-4} Torr.

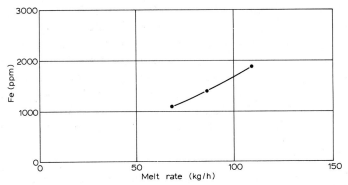

Fig. 71. Removal of iron from zirconium. (Courtesy W. Aschoff, Wah–Chang Corp.) Material: zirconium; 2600 ppm Fe in starting material; power input: 160 kW; pressure: approximately 10^{-4} Torr.

melting process is the floating to the surface of the melt of all insolubles in the melt materials and inert inclusions which are neither decomposed nor in any way affected in the processing. Understandably, this contribution is relatively small because of the usual cleanness of most materials which are so processed. This contribution, however, could be considerable when unusually contaminated materials and materials containing substantial amounts of non-metallic inclusions are being processed. Inasmuch as, besides melting and resolidifying of the molten metal, the condition of directional solidification exists, both composition segregation and a quasi zone refining effect take place in electron beam melting. Full documentation studies are not readily available[105].

The facts contributing to purification have been discussed above. In order to round off this discussion, reference must also be made to potential sources of

References pp. 642–647

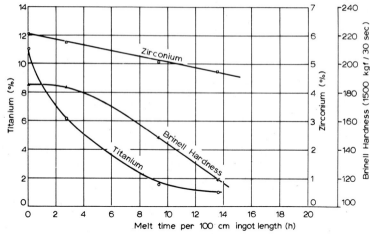

Fig. 72. Melt losses in a Nb–10Ti–5Zr alloy. (Courtesy J. T. Perryman, Fansteel Metallurgical Corp.)

contamination in the process of electron beam melting. A most recent and very careful study by ORNL*[106] focuses well on the different potential danger areas here. In this study, the nature of furnace walls and deposits on them was established as one of the two main sources of contamination, the second being the electron emitter cathode material. The latter can be particularly troublesome when transverse guns are used. Cleanness of furnace walls and judicious selection of emitter material, in cases where this is possible, are recommended. Specific data obtained in the course of this study are given in Table 16. As to the filament contamination,

TABLE 16

CONTAMINATION BY WALL AND EMITTER CATHODE[106]

Ingot No.	Melting conditions	Number of samples	Avg. impurity content (weight ppm)		
			Ta	W	Hf
—	As received powder	4	84	4	≤1
T	T-111 alloy emitter; uncleaned chamber walls	16	564	50	413
N	No emitter; uncleaned chamber walls	16	424	27	302
W	W emitter; uncleaned chamber walls	16	477	25	287
TC	T-111 alloy emitter; cleaned chamber	7	115	5	≤10

* Oak Ridge National Laboratory.

for example, when tantalum is processed, use of tantalum cathodes is feasible in order to avoid tungsten contamination with a tungsten cathode.

One facet of future development of the electron beam process is its potential for direct reduction. The author knows of only one instance[74] where this has been investigated and niobium exceeding nuclear grade requirements has been prepared in this manner.

4.4 Electron beam processed materials

Some electron beam processed materials will now be examined to establish the extent of utilization of the process and the type of material that reaches the market subsequent to electron beam melting and refining. In surveying the facts related to electron beam processed materials[107-111], it is appropriate to begin a description with the refractory and reactive metals, as it is through processing of these metals that the electron beam became established, and it is the processing of these materials and their alloys which provides to date the economic support for the process.

4.4.1 Tantalum

Tantalum was not one of the first metals to be electron beam processed on a large commercial scale, but since the initial successful melts, it has become exclusively so-processed. Table 17 shows relatively early data on tantalum processing, illustrating the important purification already effected at this stage of development of the process. It shows the type of analysis obtained in 1959, while Table 18 presents partial analysis data from 1962[72] and 1968[82]. These data, influenced by many variables besides the nine years span in processing, are not satisfactory for exact comparison, but they do indicate that as far as tantalum, in particular, is concerned and most probably the refractory metals in general, the processing of

TABLE 17

TANTALUM—ANALYTICAL DATA, 1959—TRACE ELEMENT CONTENT[104]

Element	Starting material (weight ppm)	After electron beam melting (weight ppm)
Al	250	50
C	36	20
Cu	50	50
Fe	100	10
H	10	1
Mo	100	25
N	30	10
Nb	100	25
Ni	100	10
O	82	6
Si	250	20
Ti	50	10

References pp. 642–647

TABLE 18
RESULTS OF TANTALUM MELTING IN 200 kW HERAEUS FURNACE[72]

Material	Ingot diam. (mm)	Melt. speed (kg/h)	Beam power (kW)	Pressure average (Torr)	Ingot yield % of matl. charged	Total loss by evap. % of matl. charged	Analysis (ppm) H	C	O	N	Brinell hardness (kgf/mm²)
Presintered bars							4	500	8900	100	
1st melt	40–50	6–8	80–100	1–2×10⁻⁴			1	95	1000	60	
2nd melt	60–80	10	120–160	5×10⁻⁵	80	8	1	44	30	26	89 53
Presintered bars							3	70	2100	90	
1st melt	40–50	6–8	80–100	1×10⁻⁴			1	45	65	40	
2nd melt	60–80	10	120–160	5×10⁻⁵	90	4	1	30	15	10	60 50
Pellets or granules							2	125	1300	150	
1st melt	40–50	6–8	80–100	1–2×10⁻⁴			1	65	70	20	
2nd melt	60–80	10	120–160	5×10⁻⁵	90	4	1	50	15	5	65 50
Scrap							1–10	500	5000	70–130	
1st melt	60–80	10	120–160	1×10⁻⁴	85	6	1	60	72	30	75

these materials by the electron beams had just about been established by the turn of the sixties. One must also reiterate that the process has capabilities of producing material of substantially higher purity. The degree of purification effected within the limits permitted by the process is a matter of economic justification and market requirements.

A glance at Table 17 reveals the substantial reduction of Al, Fe, Cu, Ni, Si and Ti, all elements with appreciably higher vapor pressures than that of tantalum. On the other hand, Mo and Nb are not as readily removed inasmuch as their vapor pressures are comparable with that of Ta. The removal of C and O can be ascribed to evaporation of Ta suboxide as well as CO which is formed by interaction of these two elements in the melt. As already stated, it is believed that H and N are removed from Ta as gaseous molecules.

The removal of the interstitials is, of course, of greatest importance because of the drastic effect which these elements have on the mechanical properties in general, and on the ductility in particular. Because of space considerations, no mechanical properties will be discussed here; these are readily available in monographs dealing with these materials.

4.4.2 Niobium

Another material which is also essentially electron beam processed is niobium (columbium). Data on an early study[108] are given in Table 19. More recent data[72] are given in Table 20. Here, purification is essentially equivalent to that of tantalum. Neither W nor Ta can be removed by means of electron beam processing and

TABLE 19
EFFECTS OF ELECTRON BEAM MELTING ON TRACE ELEMENTS IN NIOBIUM[108]

Element	Starting material	Impurity content % (atomic) Melt		
		1	2	3
Al	0.002	<0.002	<0.002	0.002
B	—	<0.0001	<0.0001	<0.0001
C	0.007	0.01	<0.003	<0.003
Nb	—	>99.8	>99.8	99.8
Cd	—	<0.0001	—	<0.0001
Cr	<0.005	<0.002	<0.002	<0.002
Cu	0.001	0.007	—	0.001
Fe	<0.003	<0.01	<0.01	<0.01
Mg	<0.0005	0.002	<0.002	<0.002
Mn	<0.002	<0.002	<0.002	<0.002
Mo	0.03	0.004	0.002	<0.002
Ni	0.007	0.0103	0.005	0.004
N	<0.002	<0.002	<0.002	<0.002
O	0.013	0.0131	0.0052	0.005
Pb	0.003	0.004	0.005	<0.002
Si	0.003	<0.01	<0.01	<0.01
Sn	—	<0.002	—	<0.002
Ta	<0.15	0.096	0.078	0.062
Ti	0.003	<0.015	<0.015	<0.015
V	—	<0.002	<0.002	<0.002
W	—	<0.03	<0.0287	<0.03
Zn	—	<0.002	—	<0.002
Zr	<0.001	<0.05	—	<0.05
Brinell hardness	—	55.7	59.3	51.0

thus in an application where low content of these elements is required, as for instance in nuclear technology, purification of these must be effected prior to melting. Levinson[109] indicated, for example, that one can remove from 0.02% to 0.007% carbon if the starting oxygen content is of the order of 0.15%. Once the oxygen content falls below 0.005%, removal of carbon becomes impossible, thus again indicating that oxygen addition is one of the possible ways of effecting carbon removal. The difference in macrostructure between arc melted and electron beam melted niobium is readily observed. Vickers hardness decreases from 160 to 52.

4.4.3 Molybdenum

Electron beam melting of molybdenum also tends to produce a high quality metal. In electron beam remelting[110] the amounts of N, O, C, Si, P, Fe and Cu are reduced and metallographic analysis indicates the virtual absence of inclusion. Though considerable purification is accomplished, a study by Smith and his associates[111] did not observe any improvement in its transition temperature. These workers ascribe this behavior to the fact that very minute trace element amounts

References pp. 642–647

TABLE 20

ANALYSIS AND PROCESSING DATA ON NIOBIUM AND Hf+2%Zr[72]

Material	Ingot diam. (mm)	Melt. speed (kg/h)	Beam power (kW)	Press. average (Torr)	Ingot yield % of matl. charged	Evap. loss % of matl. charged	Analysis (ppm) H	C	O	N	Brinell hardness (kgf/mm²)
Nb Presintered bars							11	460	8940	580	
1st melt	50–60	6–8	50–80	1×10^{-4}			1		580	64	75
2nd melt	60–80	8–10	70–100	5×10^{-5}	80	10	1	40	95	20	51
3rd melt	80–100	8–10	110–140	5×10^{-5}			1	30	18	15	45
Nb Presintered bars							2	80	1560	85	
1st melt	40–60	8	50–80	1×10^{-4}			1	55	105	30	60
2nd melt	60–80	8–10	70–100	5×10^{-5}	90	4	1	30	15	15	50
Nb Pellets or granules							4	259	2100	105	
1st melt	40–60	6–8	50–80	1×10^{-4}			1	50	80	40	55
2nd melt	60–80	8–10	70–100	5×10^{-5}	90	4	1	35	20	15	50
Nb Pellets with very high N-content							3	280	655	1085	
1st melt	40–60	5	50–80	$1\text{–}2\times10^{-4}$			1	70	155	460	
2nd melt	60–80	6–8	70–100	5×10^{-5}	90	7	1	55	21	45	55
3rd melt	80–100	8	110–140	5×10^{-5}			1	40	20	35	55
Nb Scrap							2	30	100	30	
1st melt	60–80	8–10	80–100	1×10^{-4}	92	3	1	25	30	30	55
Hf+2% Zr sponge (Ugine)							2		1550	70	
1st melt	40–70	6–8	40–70	8×10^{-5}			2		390	55	
2nd melt	60–80	8–10	60–80	5×10^{-5}	90	5	2		170	25	160

already affect the properties. There further appears to be evidence at hand that even 2 ppm O, 34 ppm C and 8 ppm N have a deleterious effect on the properties. In another study[112], a reduction of Vickers hardness by electron beam melting from 160 to 130 is reported. Though effective purification of molybdenum can be obtained, it is believed that this metal is not customarily electron beam melted because of the economics, quality relation and market needs of the same.

4.4.4 Tungsten

The behavior of electron beam melted tungsten is very much analogous to that of molybdenum. Though substantial purification[112] results, the actual ductility-related properties are not improved. A 1962 study by Gruber et al.[72] indicates the type of purification which is brought about in tungsten and molybdenum and its effect on their respective hardness (see Table 21).

TABLE 21

RESULTS OF TUNGSTEN AND MOLYBDENUM MELTING (HERAEUS)

Material	Ingot diam. (mm)	Melt. speed (kg/h)	Beam power (kW)	Pressure (average) (Torr)	Ingot yield % of matl. charged	Total loss by evap. % of matl. charged	Analysis (ppm) H	C	O	N	Brinell hardness (kgf/mm²)
W Presintered bars							1	70	4100	30	
1st melt	40–50	4	120–160	1×10^{-4}			1	45	115	11	220–250
2nd melt	40–50	4	120–160	6×10^{-5}	90	6	1	30	5	2	200–210
Mo Pellets							1.5	42	60	10	
1st melt	40–60	6–8	50–90	$1-2 \times 10^{-4}$			1	35	12	8	168
2nd melt	60–80	8–10	70–100	5×10^{-5}	85–90	6	1	20	3	3	145
Mo Presintered bars							2	170	810	51	
1st melt	40–60	6–8	50–90	$1-2 \times 10^{-4}$			1	64	105	15	
2nd melt	60–80	8–10	70–100	5×10^{-5}	86	6	1	25	6	3	140–150

4.4.5 Zirconium, hafnium

Zirconium and hafnium, on the other hand, are two of the reactive metals which are customarily electron beam processed. Data on results obtained on processing of these metals[113-116] are given in Tables 22 and 23, indicating not only the substantial purification possible, but also the advantage of electron beam melting related to consumable electrode, vacuum arc or vacuum induction melting in the case of hafnium.

TABLE 22

CHEMICAL ANALYSIS OF HAFNIUM PROCESSED BY ARC AND ELECTRON BEAM MELTING (IMPURITIES %)[113]

Ref.	No.	Al	Fe	Si	Ti	Ni	O	Cu	Cr
Arc	(115)	0.006	0.032	0.007	0.002	0.0028	0.09	—	—
EB	(115)	0.0025	0.005	0.007	0.001	0.0027	0.04	—	—
Arc	(116)	0.012	0.044	—	0.007	0.005	0.090	0.0085	0.0155
EB	(116)	0.005	0.001	—	0.002	0.006	0.040	0.007	0.001

References pp. 642–647

TABLE 23

IMPURITIES AND HARDNESS OF DIFFERENTLY PROCESSED ZIRCONIUM
(IMPURITIES %)[113,114]

Material	C	N	O	H	Brinell hardness
Sponge	0.007	0.0008	0.043	0.0012	91
Iodide crystal bar ingot	0.004	0.002	0.014	0.0004	70
Electron beam melted	0.003	0.0008	0.0145	0.0002	58

Process	C	O	N	H
Arc melted	0.0082	0.044	0.0085	0.0175
Slow rate electron beam melted	0.0149	0.036	0.0026	0.0022
Intermediate rate electron beam melted	0.0114	0.014	0.0015	0.0027
Fast rate electron beam melted	0.0105	0.017	0.0027	0.0009

4.4.6 Titanium

Data on electron beam processed titanium from a variety of sources[117] are given in Table 24, which contains results of analysis for iron in addition to the interstitials. While on the basis of economics it appears that today the bulk of titanium is vacuum arc processed, a considerable amount of this metal is apparently also electron beam melted, especially for scrap recovery[82].

TABLE 24

ELECTRON BEAM AND ARC PROCESSED TITANIUM (IMPURITIES %)[117]

Sponge type	Melting process	O	N	C	Fe	H
Dow	Arc	0.055	0.001	0.04	0.10	0.0041
	EB	0.057	0.008	0.04	0.03	0.0006
DuPont	Arc	0.070	0.009	0.04	0.15	0.0047
	EB	0.066	0.010	0.02	0.16	0.0009
Electromet	Arc	0.099	0.010	0.03	0.05	0.0108
	EB	0.125	0.007	0.12	0.04	0.0007
Toho	Arc	0.075	0.005	0.01	0.18	0.0049
	EB	0.068	0.008	0.03	0.06	0.0016
M.S.T.	Arc	0.021	0.004	0.02	0.01	0.2250
	EB	0.030	0.004	0.02	0.03	0.0010

4.4.7 Other non-ferrous metals and alloys

While virtually any element can be electron beam melted, this is not customarily done. The performance at an economic price is a powerful factor in technological development. However, studies on the electron beam processing of Be[118], Co[109], V[119], U[119], Cu[120] and Ni[121] have been conducted and the last two elements, for example, are commercially supplied in the Soviet Union, as shown in the analysis given in Table 25[120,121]. No comparable processing and commercial availability of these materials is found in the United States.

TABLE 25

ELECTRON BEAM MELTED NICKEL AND COPPER (IMPURITIES %)

	O	N	Fe	Ni	Pb	Sn	Zn	Bi	Sb	As
Cu (starting matl.)	0.0014	0.0006	0.005	0.002	0.001	0.0002	0.001	0.0002	0.0007	0.0007
EB processed	0.0001	0.00001	0.0007	0.0007	0.0007	N.D.	N.D.	N.D.	N.D.	N.D.
	C	O	H	Cu	Fe	Si	Mn	Mg	Pb	Zn
Ni starting brickettes	0.05	0.0021	0.002	0.04	0.034	0.096	0.002	0.05	0.0006	0.0005
EB processed	0.04	0.0008	0.0001	0.02	0.018	0.0017	N.D.	N.D.	0.0008	N.D.

N.D. = Not detected.

In addition to pure metals, electron beam melting is applied to the manufacture of a number of refractory metal alloys, specific examples being the 10% W–90% Ta, the 9.5% W–2.5% Hf, balance Ta, the 1% Zr, balance Nb, the FS–85, i.e., 10% W–27% Ta–1% Zr and balance Nb and a considerable number of Nb–Ta, Mo and W alloys containing about 65% Nb, 30% Ta with Mo and Hf in the 1–10% range and a number of the DOD* Ti alloys containing Ti, Al, V, among other elements.

In the preparation of alloys via the electron beam process, the problem of vapor pressure and effect of composition on processing has to be reconsidered. While evaporation of impurities is one of the basic electron beam refining functions, it is at the same time a leading cause of difficulties in maintaining alloy composition when alloys of elements with diverse vapor pressures are to be produced. Estimation of rates of removal and of the melting rate, depending on surplus addition for maintenance of composition control, assumes a certain degree of importance in electron beam alloy processing.

4.4.8 Steel

The refractory and reactive elements and their alloys, however, are not all the material that is electron beam processed today. Steel, in its many ramifications and varieties, has not remained outside the consideration of those interested in wider acceptance of the electron beam melting process and the benefits of the same. As a matter of fact, the eventual acceptance, or lack thereof, in the ferrous field remains as the most challenging aspect of the future of the process. To date, little concrete information on the properties of electron beam processed steels has appeared in the United States literature, though overseas publications show evidence of much work in West Germany[122-128], East Germany[129-134] and the Soviet Union[135].

Although steel processing on a larger, but still experimental, scale began at the turn of the sixties, only in recent studies have sufficient data been revealed to call serious attention to the process.

---

* DOD = Department of Defense.

References pp. 642–647

Now is the time for those in both the steel industry and electron beam equipment production to evaluate the merit and determine the potential of electron beam melting in the iron and steel industry that can be developed within the scope of the existing economics of the process. To be sure, it appears that economics are still against the process, assuming an equal product price for electron beam and vacuum arc processed metals, but this could be a matter of scale. The author believes that at power levels of an order of higher magnitude, *e.g.*, in the 3–10 megawatt furnaces, economics could become much more favorable.

A contingency factor which has not been considered to date is the obtaining of higher prices for materials with the known improved properties brought about by electron beam processing than for materials processed by alternative means. It appears that such properties are indicated from work published to date. These must, however, become sufficiently well documented and established beyond doubt to permit the evaluation of a realistic upcharge by virtue of these improved properties. From information at hand to date, it appears that electron beam processed steel is sold at the price of vacuum processed steel. This is not making possible any additional profit margins to be used to offset higher processing costs, although preliminary evidence indicates properties possibly superior to those of related vacuum processed steels. If the possibility of charging more for the superior product steel is ignored, and reliance simply placed on upscaling, it can be seen from Table 26[77] that quantity alone could bring the process to economic level. In view of the enormous size of the capital investment (see Table 27[77]) and the fact that a considerable uncertainty does exist in the operational parameters when one resorts to such an operation scale, the alternative approach, *i.e.*, establishing the

TABLE 26

APPROXIMATE DISTRIBUTION OF OPERATING COSTS FOR A TYPICAL FAMILY OF SMALL MEDIUM AND LARGE-SCALE ELECTRON BEAM PROCESSING SYSTEMS IN $ PER YEAR AND TON[77]

	2000 kW (Specialty steel ingot production) (15,000 ton/year)	10000 kW (Stainless steel ingot, ingot and slab production, extra-low-carbon grade) (100,000 ton/year)	20000 kW (Super-degassing of molten steel) (1,000,000 ton/year)
	Dollars per year		
Labor and overhead	350,000	500,000	600,000
Utilities	200,000	850,000	1,000,000
Maintenance	150,000	450,000	700,000
Totals	700,000	1,800,000	2,300,000
	Dollars per ton		
Labor and overhead	23.33	5.00	0.60
Utilities	13.33	8.50	1.00
Maintenance	10.00	4.50	0.70
Totals	46.67	18.00	2.30

TABLE 27

ESTIMATED APPROXIMATE CAPITAL COSTS (INSTALLED) FOR UNITS IN TABLE 13[77]

	2000 kW	10,000 kW	20,000 kW
Electron beam systems, $	1,300,000	5,000,000	9,000,000
Vacuum systems, $	500,000	1,600,000	3,000,000
Materials handling systems, $	700,000	1,800,000	3,000,000
Totals, $	2,500,000	8,400,000	15,000,000
Total, $ per ton	167	84	15

existence of superior qualities (if this is indeed so) and using this as a leverage to obtain the higher price needed, is preferred as it eliminates the element of gamble from the return on investment. Another justification might be provided by improved yields in overall processing.

Let us now see in concrete terms what has apparently been done with electron beams in the processing of steels. Actually, comparison among steels is even more difficult than the similar task related to reactive and refractory metals because of the wide range of compositions to which steels are manufactured. No such attempt will be made; instead, some specific data obtained on certain steels will be examined. Results from one of the early investigations of steel processing are given[135] in Table 28. This information reveals three basic factors: these are the composition modification brought about by the loss of high vapor pressure alloying elements, the removal of the interstitial gases and the substantial reduction of the nonmetallic inclusions. Although not as detailed, but containing additional information, Table 29 reveals a partial analysis as well as processing information on several specialty steels[72]. Here, the substantial interstitial content reduction is given and, by analogy, even though not reported, the analysis changes and reduction in non-metallic inclusions of consequence which must have accompanied the melting.

In a more recent study by Yefimenko[135], the Soviet steel, Shkhl 5, in a comparison between the normally produced and that produced by electron beam melting, shows very much the same type of information. The composition of this precision ball bearing steel before and after electron beam processing is given in Table 30, and the considerable improvement of interstitial content and the same type of purification effects as in the study by von Ardenne et al.[129] can be noted. The non-metallic inclusions are also considerably reduced and this apparently is reflected in a minor change in the density of the steel: i.e., for the solutionized condition, an increase from 7.814 to 7.819, and in the quenched martensite state, from 7.792 to 7.802. Both of these studies, as well as the other referred to, claim superior properties for these steels which are best manifested in heavy duty performance. The type of improvement in fatigue reported for Soviet steel, Shkhl 5, can be seen in Fig. 73.

Several additional steels produced by VEB Edelstahlwerk Freital[132], i.e., the steels identified as IUR 100CR6, 2UR 100CR6 and 3 UR 100CR Mn6, are

References pp. 642–647

TABLE 28

CHEMICAL COMPOSITION OF STEEL BEFORE AND AFTER ELECTRON BEAM PROCESSING[135]

Material	Alloying element (%)										Tests	Gas content	
	C	Si	Mn	P	S	Cu	Cr	Mo	V	Al	N		O (%*)	H (cm³/100g)

Material	C	Si	Mn	P	S	Cu	Cr	Mo	V	Al	N	Tests	O (%*)	H (cm³/100g)
Starting electrode	0.632	0.26	0.74	0.029	0.016	0.15	0.17	—	—	0.14	0.009	Avg.	0.008**	0.7
Cast ingot	0.31	0.26	0.30	0.029	0.014	0.09	0.15	—	—	0.09	0.005	Top	0.002	0.5
												Middle	0.003	0.3
												Bottom	0.003	0.3
Starting electrode	0.20	0.46	0.32	0.025	0.025	0.14	12.64	0.76	0.43	0.015	0.009	Avg.	0.009**	0.7
Cast ingot	0.20	0.46	0.12	0.012	0.021	0.07	12.25	0.76	0.43	0.015	0.006	Top	0.005	0.7
												Middle	0.004	0.7
												Bottom	0.005	0.9

Non-metallic inclusions in steel type 35[135]

Condition	Composition (%)				Quantity of non-metallic inclusions (%)	Oxygen content of steel (O%)
	SiO₂	Al₂O₃	MnO	FeO		
Starting electrode	68.5	17.4	11.6	2.5	0.016	0.008
	69.0	16.4	10.4	3.9	0.015	0.007
Cast ingot†	71.0	14.2	6.6	8.2	0.004	0.002
	77.0	13.4	5.6	4.0	0.004	0.002

* Obtained by hot extraction.
** Obtained by equivalent calculations.
† The oxide content is based on analysis with 10% experimental error.

TABLE 29

RESULTS OF MELTING STEEL AND Ni-BASE ALLOYS—1962 (HERAEUS)[72]

Material	Ingot diam. (mm)	Melt. speed (kg/h)	Beam power (kW)	Press. (avg.) (Torr)	Ingot yield % of matl. charged	Total loss by evap. % of matl. charged	Analysis (ppm)		
							H	O	N
High strength structural steel CNOV44, vacuum arc remelted							2	40–60	90–120
1st melt	80–100	30–90	50–80	1×10^{-4}	92–96	2	1–2	10–20	50–60
Stainless steel Wironit 600, vacuum arc remelted							1	30–40	100–120
1st melt	80–100	30–90	60–90	1×10^{-4}	90–94	4	1	10–20	70–80
Nimonic 105, vacuum arc remelted							1–10	180–200	80–110
1st melt	80–100	30–40	60–70	1×10^{-4}	92	3	1–2	10–20	15–25

TABLE 30

COMPOSITION OF PRECISION BALL BEARING STEEL BEFORE AND AFTER ELECTRON BEAM PROCESSING[135]

Processing mode	Elements (%)									
	C	Mn	Si	S	P	Cr	Ni	Cu	O	N	H
Standard processing	1.05	0.28	0.28	0.015	0.015	1.50	0.11	0.06	0.0040	0.0070	0.0001
EB processing	1.03	0.04	0.28	0.008	0.008	1.41	0.012	0.01	0.0007	0.0013	0.00014

claimed by the manufacturer to give 100 to 150 times improvement of life in bearing application and 30–50% increase in fatigue strength at a 30% increase fatigue limit. For the most realistic evaluation of the significance of this information, direct comparison with the properties of the same steels vacuum melted by an alternative process will be needed. These are not available and, as a consequence, the true significance of this performance improvement will have to be taken with some reservation.

A four-year investigation at Stahlwerke Südwestfalen AG in Hüttental–Geisweid was reported by Hentrich[122]. The following steels, 1% C, 1.5% Cr bearing material, Cr, Mo and CR Mo 1885, as well as most stainless and acid-resistant steels, were processed on a semi-commercial scale. The cost of these operations remains higher, and some comments referred to above pertaining to cost of the

References pp. 642–647

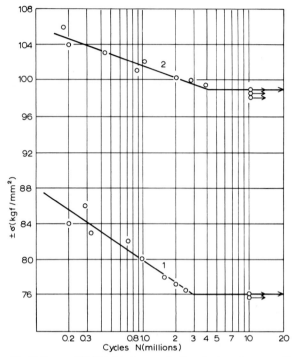

Fig. 73. Fatigue of Shkhl 5 before and after electron beam melting.

product and processing are certainly valid. No cost information on these materials is available.

Some important advantages of the electron beam processing of steels were established by Hentrich et al.[127,136] and are reiterated here:

(1) A wide range of the physical configuration of the feed stock, including differently shaped ingots, scrap iron of various dimensions and lumpy or granular materials, can be used.

(2) It makes it possible to produce ingots of different sections having the same degree of purity with the same melting speed in kg/h.

(3) Within the limits of available power the energy in kWh/kg required to melt can be controlled independently of the energy needed for refining. This includes the ability to obtain at slow melting rates very pure materials, as well as to work at melting rates comparable with vacuum arc melting, for example, and obtain a product of comparable quality.

(4) The charge cross-section can be equal to or considerably greater than the section of the finished product.

(5) Both circular and rectangular cross-section ingots can be produced.

(6) The versatility of electron beam heating requires strict control of the melting parameters in order to assure product uniformity. In general, industrial

scale operations tend to show considerable improvement in comparison with experimental systems.

A recent, detailed study of the corrosion properties of electron beam processed steels by Drews[128] for the first time provides the potential for the cost upcharge referred to above. It has been established that electron beam melted 18/8 Cr–Ni steel is comparable in corrosion resistance with higher alloy steels.

When such improvements become established above and beyond doubt alongside mechanical property improvements and the improved flexibility of the process, as already referred to, little doubt of the growth of electron beam melting as a tool for steel processing will remain.

4.4.9 Ceramics

Not only metallic materials, however, are subject to successful electron beam melting. Although to a limited extent only, Accary et al.[137,138] of the French Atomic Energy Commission have pioneered the electron beam melting of ceramics of widely diverse properties. Ceramics covering the property range from Al_2O_3 to ZrO_2 to those of uranium, thorium and zirconium carbides of different composition have been so prepared. Further, glasses have also been subjected to successful melting, although no specific information is available on the subject.

5. Crystal Growing

5.1 Introduction

Crystal growth and the deformation behavior of crystalline solids constitute two of the most frequently studied areas of the field of metallurgy, materials science and engineering. The ability to grow crystals of reproducible purity, orientation and structural detail is a prerequisite to any systematic study of properties (mechanical, physical, electrical, magnetic). In considering metals and alloys as a grouping within the spectrum of crystalline solids, mechanical properties (*i.e.*, the response of the metal to a particular stress environment) and deformation characteristics are of prime importance; this is a consequence of the unique combination of strength, ductility and toughness exhibited by most metals and alloys.

In this section, consideration is given to current crystal growing practice with emphasis on high melting point and reactive metals and alloys. These constitute a majority of the materials for which processing involves the techniques of vacuum metallurgy. In practice, the methods of crystal growing frequently involve *in situ* zone refining operations. Although consideration has to be given to crystal purity, primary interest is centered on the growth technique since zone melting and zone refining are dealt with elsewhere in this book (see Chap. III, Section 4). It will be seen that the electron beam floating zone technique is the most widely used method of crystal growth for high-melting-point and reactive metals and alloys; the essential

References pp. 642–647

details of this technique are described along with recent modifications and improvements. Other recent techniques having general applicability are considered.

In the area of crystal growing[139-146] excellent review articles and texts exist. The objective of this survey is to update these reference sources. This is best accomplished by presenting much of the new material in tabular form, with an associated comprehensive bibliography.

5.2 Methods of crystal growing

5.2.1 General considerations

Two major classifications of crystal growing exist: techniques involving containment of the material, and techniques in which the material is self-contained by surface tension (*i.e.* floating zone crystal growing). The former approach is relatively simple and has been utilized extensively for many years in the growth of crystals of low melting point materials. Since there have not been any radical departures or developments in procedure over the last two or three years, only a brief summary of the pertinent details is included, and the bibliography updated. In many instances, crystals of low melting point metals and alloys are grown in a controlled atmosphere, as opposed to high vacuum.

Extensive use has recently been made of the floating zone technique for crystal growth; in particular refractory and reactive metals and alloys having melting temperatures $\gtrsim 1500°C$ are amenable to this approach. It is therefore appropriate to discuss this technique in some detail and to illustrate its potential, keeping in mind that most crystal growing facilities produce material for metallurgical research, as opposed to large scale production requirements.

5.2.2 Crystal growing in a container

Apart from the obvious role of the container in the shaping of the crystal, a number of physical and chemical properties of the container material are important. In most cases, the container must be inert in relation to the metal or alloy being purified. Wetting of the container by the melt is undesirable since this leads to sticking and possible fracture of the container. The coefficients of thermal expansion of the container and charge should be carefully matched. In order to minimize heat transfer problems, the thermal conductance of the container material should be comparable with or less than that of the charge.

Fortunately, these guideline criteria allow for a wide range of possible container materials, although the choice becomes more limited as the temperature and/or reactivity of the molten charge increases. Commonly used materials include: plastics, metals, glass, silica, mullite, alumina, zirconia, beryllia, silicon nitride and graphite. Reactivity between the molten charge and the container usually limits melting temperatures to $\sim 1500°C$. Having decided upon the required crystal shape (bars of circular or square cross section are most common), all that is required is a heat source giving the required thermal gradient along the charge. The main difference between crystal growing per se, and crystal growing concurrent with zone purification, lies in the volume of charge actually molten at any given time. In zone-refining a short length of molten zone is passed through a relatively long solid charge to bring about purification, whereas in crystal growth it is usual to have all or most of the material molten in the container prior to solidification. In general traversal rates for zone refining will be lower than those used for crystal growth without significant *in situ* purification. Other considerations such as the method of specimen traversal, container materials and properties, and the means to obtain a vacuum or inert atmosphere, will be similar for crystal growing and zone refining.

When reaction between the molten charge and the surrounding air can occur, recourse is made to vacuum or an inert atmosphere. The simplest approach consists of enclosing the material and container in a suitable ampoule (or simply enclosing the material in the ampoule) and sealing off the assembly under vacuum or with the required pressure of inert gas. In some cases, a dynamic vacuum or circulating inert gas cover is satisfactory.

These principles have been utilized in the growth of crystals of aluminum[147], copper[148], lithium[149], lead alloys[150], tin[151,152], zinc[151], gold[147] and silver[147]. In these studies, the interrelationship between crystal growing and purification by zone refining is again emphasized.

The growth of crystals in a water cooled container is of importance since the technique can be applied to high melting point and/or reactive metals. The usual container material is copper and the heating source a d.c. arc[153] or induction heater[154–156]. Inert gas atmospheres, hydrogen or vacuum can be used. In general the material will be in the form of a high purity polycrystalline aggregate and may require subsequent annealing to increase the grain size. The method is particularly suited to titanium, zirconium, iron, cobalt, and all the body-centered cubic refractory metals[153–156].

5.2.3 Crystal growing by the floating zone technique

In floating zone crystal growing, a molten zone of the material is held in place by its own surface tension between two collinear solid rods of the same material. The technique was originally developed by Kech and Golay[157], Emeis[158] and Theuerer[159] in order to prepare high purity single crystals of silicon. The main advantages of the method are: avoidance of crucible contamination; the material usually solidifies behind the molten zone to form a single crystal along the entire length of the specimen rod; if the crystal is grown under vacuum, the original material is usually purified by evaporation of volatile impurities, outgassing, and in many cases segregation of impurities, which are preferentially soluble in the molten zone, at one end of the specimen.

(*a*) *Induction heating*. A number of heating sources have been utilized, including

Joule heating[141], dielectric heating[141], solar energy[141], arc-image[160,161], radiation[158], induction[157–159] and electron bombardment. Of these sources, induction heating and electron bombardment heating are the most common. With induction heating, it is possible to work under high vacuum, in a reducing or oxidizing atmosphere, or at a positive pressure of inert gas. The latter capability makes induction heating particularly attractive for the growth of crystals of the reactive metals titanium[162–164], zirconium[162], beryllium[165], iron[162,166–168], nickel[162,163,169] and yttrium[170].

It is usual to hold the coil stationary, and to move the specimen through the coil by a mechanical feed mechanism. In high vacuum operations, it is usual to have the coil surrounding the specimen inside the working chamber; this minimizes the space between the inside diameter of the coils and the outside diameter of the crystal, so that good coupling and a sharp focusing of the heat zone are possible. However, care must be exercised in order to prevent shorting of the coil turns by condensation of metal evaporated from the specimen[171]. For crystal growing in an inert atmosphere at a small positive pressure, it is convenient to suspend the specimen inside a quartz tube with the induction coil on the outside of the tube. By having a double-walled tube, continuous cooling is possible. A number of floating zone crystal growing (and zone refining) units utilizing induction heating have been described in detail in the literature[141,157,159,162,163,168,170–175]. Single crystals of lengths of ~ 250 mm and up to ~ 25 mm diameter may be grown by the floating zone induction heating technique. It is relatively easy to grow crystals of a specific orientation by initiating the molten zone within a seed crystal placed in good thermal contact with the specimen rod at the appropriate angle.

(b) *Electron beam heating.* In the last five years there has been an intensive and continuing study of the deformation behavior of the body-centered cubic refractory metals and alloys. Crystals suitable for the property evaluations have, in most cases, been grown by a floating zone technique involving electron beam heating of the material in vacuum. This form of heating was first used successfully by Calverley *et al.*[176]; subsequently, numerous modifications and improvements have been introduced[177–190]. All aspects of the electron beam floating zone technique have been reviewed by Lawley[182] and Schadler[183]. Control of the dimensions of the molten zone is easier than with induction heating since the electrons can be focused on the specimen.

A concise description is given of the unit developed by the author and Dr. H. L. Prekel[190]. This unit has been used extensively in the preparation of refractory metals, iron and titanium, and can be considered representative of the many designs currently in operation. The cross-section (schematic) is illustrated in Fig. 74. A dynamic vacuum $\sim 5 \times 10^{-6}$ Torr is obtained in the bell jar by means of a 140 mm diam. oil diffusion pump fitted with a refrigerated trap. Since the pumps have to work against leaks, outgassing from the surfaces within the bell jar and outgassing of the specimen, it is necessary to maintain a pumping speed (at the pump) of

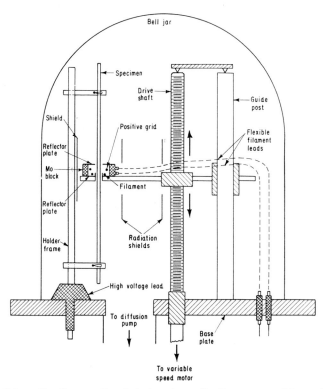

Fig. 74. Schematic of conventional electron beam floating zone crystal growing unit.

~1000 l/s. The specimen is held stationary and the heater traversed up or down; most units operate in this fashion since vibration of the specimen is minimized, and the working volume is approximately one-half of that required in the case of a moving specimen arrangement. The unit can accommodate rods of up to ~230 mm in length and ~6 mm diam.

The most important component of the crystal growing unit is the electron beam heater, and two basic designs are commonly used. In the simplest arrangement, the specimen is held at a positive potential of ~4000 volts, with the circular tungsten filament (a single turn loop of ~20 mm diam. made from 0.5-mm-diam. wire) at ground potential. Focusing of the electron beam is accomplished through flat plates of tantalum parallel to but above and below the plane of the filament. Automatic control of the temperature in the molten zone is obtained by passing the beam current through a resistor and comparing the associated voltage drop with a reference voltage. The out of balance error signal is then amplified and in turn controls the heater voltage to the filament. A typical temperature gradient in the rod near the molten zone is $\sim 900°C\ cm^{-1}$.

The main disadvantage of this design lies in the fact that there is a direct "line of sight" between the specimen and the filament (Fig. 74), and cross-con-

References pp. 642–647

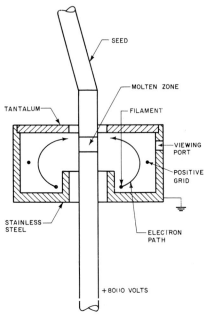

Fig. 75. Filament cage configuration for floating zone crystal growth.

tamination takes place. Since tungsten is always used as the filament material, crystals will show a pick-up of this element. More serious, however, are the problems associated with the coating of the filament by material evaporated from the surface of the specimen; this leads to a continuing loss in emission current which has to be compensated for by a continuous increase in filament current. In the case of materials having a high vapor pressure at the melting point, the filament current can become so high that the filament burns through, or the control unit becomes unstable. A characteristic feature of crystals grown in this way will be the non-uniformity of the diameter along the length of the crystal.

In order to prevent cross-contamination, a number of filament cage designs have been developed[181,185] in which the electrons take a curved path to the specimen. The unit described here uses a filament cage design similar to that of Brownsword and Farr[185]. The filament is housed in a stainless steel cage (Fig. 75) and is screened from the specimen by the inner tube, thereby forcing the electrons to make a curved trajectory to the specimen rod. In the absence of the positive grid (Fig. 75), accelerating voltages in excess of 9000 volts are required to melt 4-mm-diam. molybdenum, since a space charge builds up in the vicinity of the filament, limiting the beam current. By introducing a single turn grid ring (at a positive potential of ~ 120 volts) emission currents sufficient to promote melting are obtained at $\gtrsim 8000$ volts with correspondingly lower filament currents.

Use of a filament cage arrangement of this form is definitely recommended if crystals of consistently uniform diameter are required. In the particular case of molybdenum, typical tolerances on the diameter over a 150 mm length of crystal were 3 ± 0.025 mm. This uniformity is a direct result of the reduction in filament contamination, and a consequent stability in the temperature control circuit. Other advantages include: a substantial increase in filament life with no appreciable change in filament characteristics, the ability to view the complete molten zone without obstruction by the filament, and a relaxation on the problem of centering the specimen relative to the filament.

To illustrate the extensive use made of the electron beam floating zone technique in the growth of single crystals of refractory metals and alloys, and iron, a summary has been prepared (Table 31). The table pertains to work appearing in the open literature within the last three years. For a similar summary covering the earlier aspects of crystal growing, the reader should consult references 139, 144, 145 and 182. Chromium and vanadium are not amenable to electron beam floating zone crystal growth because of the high rate of evaporation during melting. For these elements, it is usual to enclose the charge in a quartz tube under a positive pressure of inert gas and to revert to induction heating.

TABLE 31

GROWTH OF CRYSTALS OF HIGH MELTING POINT METALS AND ALLOYS
(E.B.F.Z. = electron beam floating zone)

Material	Method	Form	References
Mo	E.B.F.Z.	Rod*	191–197
	E.B.F.Z. plus hydrogen anneal	Rod	198, 199
Ta	E.B.F.Z.	Rod	195, 196, 200, 201
Ta–Nb; Ta–W	E.B.F.Z.	Rod	201, 202
Ta–Mo	E.B.F.Z.	Rod	202
Ta–Re	E.B.F.Z.	Rod	202, 203
V	Float-zone induction heating in argon or helium	Rod	204
W	E.B.F.Z.	Rod	191, 205, 206
	Electron-beam melting	Button**	207
	High-vacuum annealing	Sheet**	208
W–Re; W–Ta	Electron-beam melting plus vacuum annealing	Ingot**	209
Cr	Zone induction annealing in argon	Rod	210
Nb	E.B.F.Z.	Rod	187, 188, 195, 211
	Electron beam melting	Ingot**	212
	E.B.F.Z. plus high vacuum annealing	Rod	213–215
Nb–Mo	Electron beam melting plus E.B.F.Z.	Rod	216
Fe	Strain-anneal	Sheet	217–219
Fe–V	Zone melting in hydrogen/argon mixture	Rod	220

* In general, the rod will be in the form of a large single crystal.
** These forms are generally polycrystalline.

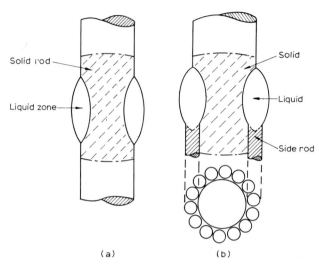

Fig. 76. Schematic representation of "skin melting" and "skin melting" of a "fascia"[221].

(c) *The fascia technique for float-zone crystals.* In the electron beam floating zone technique, limits on the zone size, in relation to the rod diameter, are imposed by the level of surface tension forces, the thermal conductivity and the density of the molten metal. As a general rule, the length of a stable molten zone will be between one-half and one rod diameter. When melting is carried out such that a solid core of metal is maintained along the axis of the rod, then this is termed "skin melting". The core gives support to the circumferential molten zone, and also acts as a seed crystal (Fig. 76a).

The concept of skin melting has led to the "fascia" technique whereby the diameter of a float-zoned crystal can be increased. Specifically, a bundle of rods (or wires) is bound around an existing crystal, and skin melted to fuse the rods onto the central core (Fig. 76b). Promising results, in terms of obtaining true float zone crystals larger than those conventionally manufactured, have been reported by Bucklow and Wilson[221] for tantalum.

(d) *Hollow cathode float zone crystals.* The hollow cathode provides a relatively new way in which to prepare float zone crystals. The geometry of the hollow cathode technique is similar to that of conventional electron beam zone melting; unique features come about through the manner in which the molten zone is formed, namely by highly localized heating within a hollow cathode gas discharge[222].

A large d.c. voltage is applied to a gas at a reduced pressure to develop the discharge. The cathode geometry (Fig. 77) is such that the discharge or glow region is completely confined within the cathode hollow. "Run-away" electrons from the cathode wall will converge at the focal point of the hollow, namely the specimen rod. The important feature of the technique is that electron bombardment heating takes place within the hollow cathode glow area, and since the glow is a plasma,

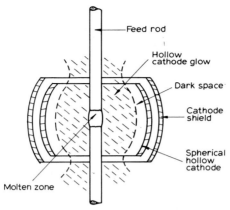

Fig. 77. Geometry of a hollow cathode floating zone crystal grower[222].

any space charge developed at the rod will be neutralized. Application of the method to non-conducting materials is now under investigation. The advantages of the method may be summarized as: high heating efficiency comparable with that of the standard electron beam technique; ability to operate in almost any gas at pressures up to ~3 Torr; elimination of the need for a thermionic emitter as the source of electrons; application to crystal growth of refractory metallic and non-metallic (dielectric) materials; possibilities of controlled levels of purification or doping via the gas; inherently stable operation having a high degree of control.

5.2.4 Growth of bicrystals

In order to study the role of grain boundaries in phenomena such as liquid metal embrittlement, boundary energies, dislocation–grain boundary interactions and dislocation–dislocation interactions, it is convenient to work with bicrystals having a controlled orientation between grains. Bicrystals of the high melting point metals are particularly difficult to process. Hook and Hirth[223] were able to prepare controlled orientation bicrystals of silicon–iron by diffusion bonding or pressure welding. Diffusion bonding of single crystal seeds of niobium[224] gave rise to voids and oxide particles at the boundary.

Brehm and Gregg[225] have recently succeeded in producing bicrystals of niobium in which the boundaries are straight, symmetrical and free from voids or

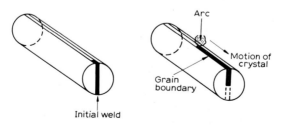

Fig. 78. Bicrystal preparation procedure[225].

oxide particles. On subsequent annealing, the boundaries were stable in that no recrystallization, polygonization or boundary migration was observed. Basically, the method involves the joining of two single crystal seeds each in the shape of a half round, Fig. 78; initially the two seeds having the required misorientation are arc welded at one end. The bicrystal is then grown in a water-cooled travelling hearth arc melter under a gettered argon atmosphere. The boundary is formed vertically in the arc melter, but does not run the complete depth of the seed. In this way, the newly formed boundary is effectively pinned in the center of the bicrystal, resulting in a straight, symmetrical geometry.

REFERENCES

Sections 1–3
1 W. ROHN, *Ger. Pat. 345 161*, 1918.
2 W. ROHN, *AIME Tech. Pub. 470*, 1932.
3 W. W. DYRKACZ, R. S. DE VRIES AND R. K. PITLER, *Symp. on Arcs in Inert Atmospheres and Vacuum*, Electrochem. Soc., Wiley, 1956.
4 H. C. CHILD AND G. T. HARRIS, *J. Iron Steel Inst. (London)*, Dec. (1958) 414.
5 R. SCHLATTER AND A. SIMKOVICH, *Blast Furnace Steel Plant*, Nov. (1966) 1029.
6 P. E. BUSBY, *Trans. Vacuum Met. Conf., 1964*, Am. Vacuum Soc., Boston, Mass., 1965, p. 380.
7 J. D. NISBET, *Trans. Vacuum Met. Conf., 1967*, Am. Vacuum Soc., New York, 1968, p. 15.
8 F. SPERNER AND H. C. CHILD, *Trans. Vacuum Met. Conf., 1967*, Am. Vacuum Soc., New York, 1968, p. 73.
9 R. SCHLATTER, *Trans. Vacuum Met. Conf., 1968*, Am. Vacuum Soc., New York, 1968, p. 333.
10 L. S. TAYLOR, *Foundry Trade J., 117* (1964) 587.
11 Vacuum Engineering (Scotland) Ltd., private communication (basis of Brit. Pat. Appl.).
12 R. SCHLATTER AND A. SIMKOVICH, *Trans. Vacuum Met. Conf., 1966*, Am. Vacuum Soc., Boston, Mass., 1967, pp. 338 and 458.
13 D. R. WOOD AND R. M. COOK, *Metallurgia*, March (1963) 109.
14 H. C. CHILD, P. M. GRAY AND H. H. SCHOLEFIELD, *Symp. on User Experience of Large Scale Industrial Plant, March 1961*, Inst. Mech. Engrs., London.
15 O. WINKLER, *Met. Rev., 5* (1960).
16 A. V. BRADSHAW AND E. D. RICHARDSON, *I.S.I. Special Rept. 92*, 1965, p. 24.
17 T. E. PERRY, *I.S.I. Special Rept. 92*, 1965, p. 105.
18 S. W. GRAHAM AND B. B. ARGENT, *J. Iron Steel Inst. (London), 205* (1967) 1066.
19 G. H. J. BENNETT, H. T. PROTHEROE AND R. G. WARD, *J. Iron Steel Inst. (London), 195* (1960) 174.
20 Y. ICHINOSE, *Nippon Kinzoku Gakkaishi, 29* (1965) 294.
21 G. E. CARLSON AND E. J. DELGROSSO, *U.S. At. Energy Comm. Rept., 1965 (PWAC-441)*.
22 A. SIMKOVICH, *J. Metals, 118* (1966) 505.
23 K. BUNGARDT AND H. SYCHROVSKY, *Stahl Eisen, 76* (1956) 1040.
24 S. E. VOLKOV, B. V. LINCHEVSKII, A. Y. POLYAKOV AND A. M. SAMARIN, *Stal* (in English), Feb. (1965) 115.
25 R. G. WARD AND R. HALL, *J. Iron Steel Inst. (London), 185* (1960) 75.
26 W. E. COUPETTE, *Vakuum-Tech., 14* (1965) 157.
27 P. P. TURILLON, *Trans. Vacuum Met. Conf.*, Am. Vacuum Soc., 1963, p. 88.
28 W. E. DUCKWORTH AND B. APPLEBY, *BISRA Rept. MG/C/107/62*, 1962.
29 W. J. BOESCH AND J. F. RADAWICH, in R. F. BUNSHAH (Ed.), *Trans. Vacuum Met. Conf., 1960*, Interscience, New York, 1961.

30 R. C. Leinbach and H. J. Hamjian, *J. Metals*, Feb. (1966) 219.
31 H. C. Child, *J. Met. Club*, University of Strathclyde, 1961–1962.
32 G. T. Harris and H. C. Child, *Journées Intern. Appl. Cobalt*, Cobalt Inform. Centre, Brussels, 1964.
33 H. Vollmer, *Giesserei, 56* (1969) 318.
34 W. A. Faust and G. W. Lake, *Trans. Vacuum Met. Conf., 1968*, Am. Vacuum Soc., New York, 1968, p. 403.
35 H. Richardson, *Elec. Rev.*, 7th Aug. (1964) 210.
36 G. N. Okorokov et al., *Stal* (in English), 23 (1963) 23.
37 E. J. Borrebach, in R. F. Bunshah (Ed.), *Vacuum Metallurgy*, Reinhold, New York, 1958, p. 121.
38 H. Scheidig, in M. A. Cocca (Ed.), *Trans. Vacuum Met. Conf., 1964*, Am. Vacuum Soc., Boston, Mass., 1965, p. 399.
39 *Brit. Pat. 1048436*, 2 April 1964.
40 Anon., *33 Mag. of Metals producing Industry*, July 1965, p. 72.
41 G. E. Oldfield, *Congr. intern. sur les applications des techniques du vide à la métallurgie, Strasbourg, 1967*, p. 208.
42 P. J. Wooding, (Consarc Corp.) private communication.
43 D. E. Cooper and E. D. Dilling, *J. Metals*, Feb. (1960) 1.
44 A. Dunlop, *Foundry Trade J.*, June 2 (1966) 7.
45 H. W. Antes, J. T. Norton and R. E. Edelman, *Mod. Castings, 33* (1958) 69.
46 A. Dunlop, *8th Ann. Conf. Brit. Investment Casters Tech. Assoc.*, May 1967.
47 D. E. Cooper, K. S. Snow, J. C. Fulton and J. W. Tommaney, *J. Metals*, (1965) 1368.
48 L. S. Dorsett and T. M. Krebs, in R. F. Bunshah (Ed.), *Trans. Vacuum Met. Conf.*, Interscience, New York, 1960, p. 3.
49 S. J. Noesen, *Columbium Metallurgy, A.I.M.E.*, Vol 171, 1947, p. 416.
50 S. J. Noesen, *J. Metals*, Nov. (1960) 842.
51 F. Sperner and G. Persson, *Stahl Eisen, 82* (1962) 401.
52 K. Bungardt and H. Vollmer, *Stahl Eisen, 82* (1962) 1099.
53 H. Gruber, in R. F. Bunshah (Ed.), *Vacuum Metallurgy*, Reinhold, New York, 1958, p. 138.
54 S. J. Noesen, *Vacuum Symp. Trans., 1957*, Am. Vacuum Soc., Pergamon, London, 1958, p. 150.
55 J. D. Cobine and E. E. Burger, *J. Appl. Phys., 26* (1955) 895.
56 H. R. Smith Jr., in R. F. Bunshah (Ed.), *Vacuum Metallurgy*, Reinhold, New York, 1958, p. 221.
57 H. C. Child and G. E. Oldfield, *Iron and Steel Inst. Special Rept. No. 77, 1963*, p. 73.
58 K. C. Barraclough, *Steel Times*, March 20 (1964) 376.
59 D. R. Carnahan et al., *Trans. Vacuum Met. Conf., 1959*, New York Univ. Press, New York, 1960, p. 49.
60 G. C. Gould, *Trans. AIME, 233* (1965) 1345.
61 W. Peter and H. Spitzer, *Stahl Eisen, 86* (1966) 1383.
62 K. C. Barraclough, *Iron & Steel (London), 35* (1962) 412.
63 H. C. Child and F. Sperner, *C.I.A.V.I.M. Conf.*, Strasbourg, Nov. 1967; *Le Vide, 23* (1968) 191.

Section 4
64 H. von Pirani, *U.S. Pat. 848,600*, 1907.
65 F. Trombe, *Bull. Soc. Chim. France*, (1934) 262.
66 E. Tiede, *Chem. Ber., 46* (1913) 2229.
67 R. Hultgren and M. H. Pakkala, *J. Appl. Phys., 11* (1940) 643.
68 H. R. Smith, Jr., in R. Bakish (Ed.), *Introduction to Electron Beam Technology*, Wiley, New York, 1964, p. 168.
69 H. R. Smith, Jr., C. d'A Hunt and C. W. Hanks, *Trans. 5th Natl. Symp. Vacuum Technology, 1958*, Pergamon, Oxford, 1959, p. 164; also *J. Metals*, Feb. 1959.
70 C. Hayashi, C. Kashu, M. Kosani and J. Hiruta in R. Bakish (Ed.), *Proc. 1st Intern. Conf. Electron and Ion Beam Science and Technology, 1964*, Wiley, New York, 1965, p. 415.

71 H. GRUBER, *Z. Metallk.*, *52* (1961) 291.
72 H. GRUBER, W. DIETRICH, E. ERBEN, J. HEIMERL AND H. STEPHAN in R. F. BUNSHAH (Ed.), *Trans. Vacuum Met. Conf., New York, 1962*, Am. Vacuum Soc., Boston, Mass., 1963, p. 68.
73 G. GRUBER, H. STEPHAN, W. DIETRICH and R. LESSER, *Trans. 8th Natl. Vacuum Symp. and 2nd Intern. Congr., 1961*, Vol. 2, Pergamon, London, 1962, p. 722.
74 H. R. SMITH, JR., in R. F. BUNSHAH (Ed.), *Trans. Vacuum Met. Conf., New York, 1962*, Am. Vacuum Soc., Boston, Mass., 1963, p. 95.
75 C. W. DEAN, R. E. MCDONALD AND C. F. LEITTEN, in R. F. PEASE (Ed.), *Proc. 9th Symp. on Electron, Ion and Laser Beam Technology, Berkeley, 1966*, San Francisco Press, 1967, p. 262.
76 H. STEPHAN, F. SPERNER AND W. DIETRICH, *Trans. Vacuum Met, Conf., 1968*, Am. Vacuum Soc., New York, 1968, p. 389.
77 Personal communication.
78 R. BAKISH AND G. BARBER, in I. E. CAMPBELL AND E. M. SHERWOOD (Eds.), *High-Temperature Materials and Technology*, Wiley, New York, 1967, p. 557.
79 H. R. SMITH, JR AND C. d'A. HUNT, in R. BAKISH (Ed.), *Proc. 2nd Intern. Conf. Electron and Ion Beams in Science and Technology, 1966*, Gordon and Breach, New York, 1969, p. 277.
80 E. S. CANDIDUS, M. H. HABLANIAN AND H. A. STEINHERZ, *Trans. 6th Natl. Symp. Vacuum Technology, 1959*, Am. Vacuum Soc., Pergamon, Oxford, 1960, p. 185.
81 M. H. HABLANIAN, personal communication.
82 W. DIETRICH AND K. STEPHAN, *Proc. Fourth Intern. Vacuum Congr., Manchester, 1968*, Pergamon, Oxford, p. 458.
83 M. VON ARDENNE AND S. SCHILLER, *Kernenergie*, *3* (1960) 507.
84 W. SCHEIBE, *Metall*, *14* (1960) 401.
85 Personal communication.
86 JEOL Product Description Literature.
87 M. VON ARDENNE, *Tabellen zur Angewandten Physik*, Band 1, DVW, Berlin, 1962.
88 M. VON ARDENNE, in R. BAKISH (Ed.), *Proc. 1st Intern. Conf. on Electron and Ion Beam Science and Technology, 1964*, Wiley, New York, 1965, p. 370–405.
89 S. SCHILLER, personal communication.
90 M. VON ARDENNE, S. SCHILLER AND P. LENK, *Kernenergie*, *11* (1968) 81.
91 S. SCHILLER, personal communication.
92 M. VON ARDENNE, S. SCHILLER, P. LENK, H. FIEDLER, G. SCHARF AND H. SCHOENBERG, in R. BAKISH (Ed.), *Proc. 2nd Intern. Conf. on Electron and Ion Beams in Science and Technology, 1966*, Gordon and Breach, New York, 1969, p. 323.
93 H. SCHOENBERG AND S. SCHILLER, *Neue Hütte*, *12* (1967) 146.
94 G. F. ZABORONOK, T. I. ZELENZOV, A. C. RONJIN AND B. G. SOKOLOV, *Electron Beam Melting*, Metallurgia, Moscow, 1965.
95 M. J. SMELYANSKI, V. A. BOYAROSHINOV, K. D. GUTERMAN, D. G. TKACHEV AND CISHEVSKY, *Vacuum Arc and Electron Beam Furnaces*, Metaloizdat, 1962.
96 H. G. SUSHKIN, *Electron Beam Melting*, Metallurgia, Moscow, 1965.
97 B. MOVCHAN AND A. ZLOTIN, *Avtomat. Svarka*, 1963, No. 3, 120.
98 B. E. PATON AND B. MOVCHAN, in R. BAKISH (Ed.), *Proc. 1st Intern. Conf. Electron and Ion Beams in Science and Technology, 1964*, Wiley, New York, 1965, p. 406.
99 Electron beam melting, *Econ. J.*, 1962, No. 37 (Russian).
100 B. MOVCHAN AND A. TICHONOVSKII, *Avtomat. Svarka*, 1963, No. 4.
101 T. KRAUS AND O. WINKLER, in R. BAKISH (Ed.), *Introduction to Electron Beam Technology*, Wiley, New York, 1964, p. 145.
102 J. T. PERRYMAN, Fansteel Metal. Corp., 1965, personal communication.
103 W. ASHOFF, Wah Chang Metallurgical, 1966, personal communication.
104 H. R. SMITH, JR., C. d'A. HUNT AND C. W. HANKS, in W. R. CLOUGH (Ed.), *Reactive Metals*, (Met. Soc. Conf., 1958), Vol. 2, Interscience, New York, 1959, p. 131.
105 H. FIEDLER, G. SCHARF AND F. ESSER, *Vacuum*, *19* (1969) 205.
106 R. E. READ, C. W. DEAN, R. E. MCDONALD AND J. F. EMERY, in R. BAKISH (Ed.), *Proc. 3rd Intern. Conf. on Electron and Ion Beams in Science and Technology, 1968;* also *ORNL-TM-2208*, July 1968.
107 E. S. CANDIDUS AND J. C. SIMONS, JR., *Trans. 5th Natl. Symp. on Vacuum Technology*,

Am. Vacuum Soc. 1958, Pergamon, Oxford, 1959, p. 86.
108 D. MAYKUTH AND R. JAFFEE, *Problems in Contemporary Metallurgy*, 1960, No. 6.
109 D. LEVINSON, in W. R. CLOUGH (Ed.), *Reactive Metals*, (Met. Soc. Conf., 1958), Vol. 2, Interscience, New York, 1959, p. 123.
110 L. VEROT AND A. TORRESTIER, *Compt. Rend.*, No. 4 (1960) 255.
111 H. R. SMITH, JR., J. Y. K. HUM, A. DONLEVY AND C. D'A. HUNT, *J. Less-Common Metals, 2* (1960) 69.
112 G. OGIERMANN AND W. SCHEIBE, *Metall, 15* (1961) 3.
113 L. BANGERT AND K. HENNEMANN, *Metall, 14* (1960) 704.
114 W. ASCHOFF AND E. F. BAROCH, *J. Metals, 14* (1962) 204.
115 Production and purification of hafnium, *J. Metals, 12* (1960) No. 1., p. 25.
116 C. E. ARMANTROUT AND H. KATO, in W. L. CLOUGH (Ed.), *Reactive Metals* (Met. Soc. Conf., 1958), Vol. 2, Interscience, New York, 1959, p. 429.
117 C. B. DITTMAR AND S. ABKOWITZ, in R. F. BUNSHAH (Ed.), *Trans. Vacuum Met. Conf., New York, 1959*, New York Univ. Press, 1960, p. 109.
118 Personal communication.
119 D. PECKNER, Electron beam melting of materials, *Design Eng., 53* (1961) No. 3.
120 *Copper*, Paton Institute for Electrowelding, Kiev, 1965.
121 *Nickel*, Paton Institute for Electrowelding, Kiev, 1965.
122 R. HENTRICH, The installation of an electron beam melting furnace in a stainless steel plant, *Public. No. 136*, Stahlwerke Südwestfalen AG, D 593, Hüttental-Geisweid, 1968.
123 A. R. RANDAK AND R. HENTRICH, Manufacture and properties of vacuum handled and vacuum molten steels, *Tech. Rept. 26/63*, Stahlwerke Südwestfalen AG, D 593, Hüttental Geisweid, 1968.
124 R. HENTRICH AND G. OGIERMANN, *Radex Rundschau, 5* (1965) 623.
125 A. R. RANDAK AND J. KURZOJA, *Stahl Eisen, 86* (1966) 1017.
126 W. WESSLING, *Werkstatt Betrieb, 98* (1965) 33.
127 R. HENTRICH, *VIth Intern. Congr. on Electro-Heat, Brighton, May 1968*. Union Intern. d'Electrothermie, Paris, N. 136.
128 T. DREWS, The electrochemical behavior of vacuum molten chromium–nickel steels, *Thesis*, Techn. Univ. Berlin, 1966.
129 M. VON ARDENNE, S. SCHILLER, W. KUNTSCHER, H. THIEL AND L. MEYER, *Neue Hütte, 6* (1961) 198.
130 K. KÖHLER, H. THIEL AND H. FÖRSTER, *Neue Hütte, 6* (1961) 4.
131 H. THIEL, H. MISCHLER AND H. FÖRSTER, *Neue Hütte, 8* (1963) 313.
132 H. FIEDLER, *Neue Hütte, 9* (1964) 129.
133 H. FIEDLER, personal communication.
134 G. SCHARF AND D. RUMBERG, *Freiberger Forschungsh., B122* (1966) 89.
135 YU. M. YEFIMENKO *et al.*, *Physico Chemical Mechanics of Materials*, Vol. 1, No. 4 (1965) 470.
136 R. HENTRICH AND J. FISCHHUBER, *CIAVIM Intern. Vacuum Met. Congr., Strasbourg, 1968*, Panel discussion on vacuum melting.
137 A. ACCARY, A. TREILOU AND J. TROUVÉ, in R. BAKISH (Ed.), *Proc. 1st Intern. Conf. Electron and Ion Beams in Science and Technology, 1964*, Wiley, New York, 1965, p. 439.
138 A. ACCARY AND J. TROUVÉ, *Proc. Intern. At. Energy Assoc., Prague, July, 1963*, Vol. 2.

Section 5
139 J. J. GILMAN (Ed.), *The Art and Science of Crystal Growing*, Wiley, New York, 1963.
140 E. A. D. WHITE, *Brit. J. Appl. Phys., 16* (1965) 1415.
141 N. L. PARR, *Zone Refining and Allied Techniques*, Newnes, London, 1960.
142 J. H. WERNICK, *Ultra High Purity Metals*, ASM, Cleveland, Ohio, 1962, p. 55.
143 W. G. PFANN, *Zone Melting*, Wiley, New York, 2nd edn., 1966.
144 A. LAWLEY, in R. F. BUNSHAH (Ed.), *Techniques in Metals Research*, Interscience, New York, 1968.
145 R. BAKISH (Ed.), *Proceedings First Intern. Conf. on Ion and Electron Beam Technology*, John Wiley, New York, 1965.

146 Intern. Conf. on Crystal Growing, *J. Phys. Chem. Solids*, Suppl. No. 1, 1967.
147 R. Schaefer, Y. Nakada and B. Ramaswami, *Trans. AIME*, 230 (1964) 605.
148 E. D. Tolmie, *J. Sci. Instr.*, 37 (1960) 175.
149 I. G. D'Yakov and I. R. Khvedchuk, *Phys. Metals Metallog.*, 17 (1964) 139.
150 W. C. Johnston and W. A. Tiller, *Trans. AIME*, 221 (1961) 331.
151 P. J. Schlicta, *J. Sci. Instr.*, 39 (1962) 392.
152 A. F. Armington and G. H. Moates in M. S. Brooks and J. K. Kennedy (Eds.), *Ultrapurification of Semiconductor Materials*, Macmillan, New York, 1962, p. 502.
153 G. A. Geach and F. O. Jones, *J. Less-Common Metals*, 1 (1959) 56.
154 A. Berghezan and E. Bull Simonsen, *Trans. AIME*, 221 (1961) 1029.
155 E. Bull Simonsen, *J. Iron Steel Inst. (London)*, 200 (1962) 193.
156 V. G. Epifanov and A. G. Lesnik, *Akad. Nauk Ukr. SSR*, 20 (1964) 185.
157 P. H. Kech and M. J. Golay, *Phys. Rev.*, 89 (1953) 1297.
158 R. Emeis, *Z. Naturforsch.*, 9a (1954) 67.
159 H. C. Theuerer, *Trans. AIME*, 206 (1956) 1316.
160 R. P. Poplawsky and J. E. Thomas, Jr., *Rev. Sci. Instr.*, 31 (1960) 1303.
161 R. E. De La Rue and F. A. Halden, *Rev. Sci. Instr.*, 31 (1960) 35.
162 R. L. Smith and J. L. Rutherford, *J. Metals*, 9 (1957) 478.
163 J. H. Wernick, D. Dorsi and J. J. Byrnes, *J. Electrochem. Soc.*, 106 (1959) 245.
164 E. J. Darnell, *Trans. AIME*, 212 (1958) 356.
165 G. J. London and M. Herman, *Conf. Intern. sur la Métallurgie du Beryllium*, Grenoble, Presses Universitaires de France, Paris, 1966, p. 21.
166 T. Ooka, H. Mimura, S. Yano and S. Soeda, *Proc. Japan Acad.*, 39 (1963) 294.
167 E. J. Koepel and B. Park, A new method of using the zone refining technique, *ASM Tech. Rept.* 17-3-64 (1964).
168 W. M. Williams, G. B. Craig and W. C. Winegard, *Can. Mining Met. Bull.*, 55 (1962) 35.
169 V. J. Albano and R. R. Soden, *J. Electrochem. Soc.*, 113 (1966) 511.
170 W. C. Necker, in R. F. Bunshah (Ed.), *Trans. Vacuum Met. Conf.*, Interscience, New York, 1961, p. 289.
171 E. Buehler, *Trans. AIME*, 212 (1958) 694.
172 R. W. Warren, *Rev. Sci. Instr.*, 33 (1962) 1378.
173 B. F. Oliver, *Trans. AIME*, 227 (1963) 960.
174 S. J. Silverman, *J. Electrochem. Soc.*, 108 (1961) 585.
175 J. L. Rutherford, R. L. Smith, M. Herman and G. E. Spangler, in *New Physical and Chemical Properties of Metals of Very High Purity*, Centre Nat. de la Recherche Sci., Gordon and Breach, New York, 1965, p. 345.
176 A. Calverley, M. Davis and R. F. Lever, *J. Sci. Instr.*, 34 (1957) 142.
177 F. E. Birbeck and A. Calverley, *J. Sci. Instr.*, 36 (1959) 460.
178 R. G. Carlson, *J. Electrochem. Soc.*, 106 (1959) 49.
179 A. Lawley, *Electronics*, 32 (1959) 39.
180 H. W. Schadler, *Trans. AIME*, 218 (1960) 649.
181 M. Cole, C. Fisher and I. A. Bucklow, *Brit. J. Appl. Phys.*, 12 (1961) 577.
182 A. Lawley, in R. Bakish (Ed.,), *Introduction to Electron Beam Technology*, Wiley, New York, 1962, p. 184.
183 H. W. Schadler, in J. J. Gilman (Ed.), *The Art and Science of Crystal Growing*, Wiley, New York, 1963, p. 343.
184 H. G. Sell and W. M. Grimes, *Rev. Sci. Instr.*, 35 (1964) 64.
185 R. Brownsword and J. P. G. Farr, *J. Sci. Instr.*, 41 (1964) 350.
186 L. C. Skinner and R. M. Rose, in R. Bakish (Ed.,), *Proc. 2nd Intern. Conf. on Electron and Ion Beam Science and Technology, 1966*, Gordon and Breach, New York, 1969, p. 207.
187 R. E. Reed, in R. Bakish (Ed.), *Proc. 2nd Intern. Conf. on Electron and Ion Beam Science and Technology, 1966*, Gordon and Breach, New York, 1969, p. 225.
188 E. B. Bas and H. Stevens, in R. Bakish (Ed.), *Proc. 2nd Intern. Conf. on Electron and Ion Beam Science and Technology, 1966*, Gordon and Breach, New York, 1969, p. 167.
189 H. L. Prekel and A. Lawley, in R. Bakish (Ed.), *Proc. 2nd Intern. Conf. on Electron and Ion Beam Science and Technology, 1966*, Gordon and Breach, New York, 1969, p. 189.

REFERENCES

190 H. L. Prekel, A study of the mechanism of plastic flow in high purity molybdenum, *PhD. Thesis*, Faculty of Science, University of Pretoria, Pretoria, South Africa, 1967.
191 L. Kaun, A. Luft, J. Richter and D. Schulze, *Phys. Status Solidi*, 26 (1968) 485.
192 F. Guiu and P. L. Pratt, *Phys. Status Solidi*, 15 (1966) 539.
193 L. D. Whitmire and F. R. Brotzen, *Trans. AIME*, 239 (1967) 824.
194 S. Feuerstein and L. Rice, *Trans. AIME*, 236 (1966) 1674.
195 P. J. Sherwood, F. Guiu, H. C. Kim and P. L. Pratt, *Can. J. Phys.*, 45 (1967) 1075.
196 S. S. Lau, S. Ranji, A. K. Mukherjee, G. Thomas and J. E. Dorn, *Acta Met.*, 15 (1967) 237.
197 H. W. Loesch, Jr. and F. R. Brotzen, *J. Less-Common Metals*, 13 (1967) 565.
198 D. F. Stein, *Can. J. Phys.*, 45 (1967) 1063.
199 R. G. Davies and A. Gilbert, *Acta Met.*, 15 (1967) 665.
200 J. F. Byron and D. Hull, *J. Less-Common Metals*, 13 (1967) 71.
201 R. J. Arsenault and A. Lawley, *Phil. Mag.*, 15 (1967) 549.
202 T. E. Mitchell and P. L. Raffo, *Can. J. Phys.*, 45 (1967) 1047.
203 P. L. Raffo and T. E. Mitchell, *Trans. AIME*, 242 (1968) 907.
204 E. S. Greiner and D. M. Boulin, *Trans. AIME*, 239 (1967) 965.
205 A. S. Argon and S. R. Maloof, *Acta Met.*, 14 (1449) 1966.
206 D. R. Hay, R. K. Skogerboe and E. Scala, *J. Less-Common Metals*, 15 (1968) 121.
207 G. D. Rieck, G. H. G. Vaessen and D. L. Vogel, *Trans. AIME*, 242 (1968) 575.
208 K. Farrell, A. C. Schaffhauser and J. O. Stiegler, *J. Less-Common Metals*, 13 (1967) 141.
209 J. R. Stephens, *Trans. AIME*, 242 (1968) 634.
210 C. N. Reid, A. Gilbert and G. T. Hahn, *Trans. AIME*, 239 (1967) 467.
211 E. Votava, *Acta Met.*, 16 (1968) 285.
212 B. J. Shaw, *J. Less-Common Metals*, 13 (1967) 294.
213 G. Taylor and J. W. Christian, *Phil. Mag.*, 15 (1967) 873.
214 D. K. Bowen, J. W. Christian and G. Taylor, *Can. J. Phys.*, 45 (1967) 903.
215 R. A. Foxall, M. S. Duesbery and P. B. Hirsch, *Can. J. Phys.*, 45 (1967) 607.
216 I. Milne and R. E. Smallman, *Trans. AIME*, 242 (1968) 120.
217 A. S. Keh and Y. Nakada, *Can. J. Phys.*, 45 (1967) 1101.
218 T. Yamashita and Y. Taneda, *Japan. J. Appl. Phys.*, 2 (1963) 266.
219 T. Takeuchi, *Trans. Nat. Res. Inst. Metals (Tokyo)*, 8 (1966) 1.
220 H. G. Suzuki, M. Tanino and K. Aoki, *Japan. J. Appl. Phys.*, 5 (1966) 879.
221 I. A. Bucklow and J. F. Wilson, *J. Sci. Instr.*, 1, series 2 (1968) 363.
222 W. Class, *Trans. First Conf. and School on Purification by Zone Refining*, Materials Research Corporation, Orangeburg, New York, 1967, p. 64.
223 R. E. Hook and J. P. Hirth, *Acta Met.*, 15 (1967) 535.
224 M. R. Achter, *Naval Res. Lab. Mem. Rept.*, 1718, 1966.
225 W. F. Brehm and J. L. Gregg, *J. Less-Common Metals*, 14 (1968) 463.

Chapter VI

Determination of the Gas Content of Metals by Vacuum Degassing Methods

E. LASSNER

1. Introduction

First of all it should be made clear what the term "gas content" of a metal implies. Although it is in general use nowadays, it was not, perhaps, a fortunate choice; in fact it is rather misleading, as it is used to indicate the total content of oxygen, nitrogen and hydrogen in a metal, without any indication of the form in which these elements are present. Gaseous inclusions in the form of gas bubbles (pinholes, shrinkage cavities)—if present at all—represent by far the smallest percentage of the gas content. The major portion is either combined as a separate phase (oxides, nitrides) or dissolved as interstitials. Usually both types are present, and this shows clearly that in fact the gas content of metals is of a rather complex nature.

All forms in which the three elements, oxygen, nitrogen and hydrogen, may occur markedly affect the mechanical properties of the metals even in very small

TABLE 1

METHODS FOR THE DETERMINATION OF OXYGEN, NITROGEN AND HYDROGEN IN METALS

Method	Used for the determination of		
	O	N	H
Vacuum extraction	×	×	×
Inert gas fusion	×	×	×
Emission spectrographic analysis	×	×	
Activation analysis	×		
Isotope dilution analysis	×	×	×
X-Ray structural analysis	×	×	
Mass spectroscopy of solids	×	×	
Determination of internal friction	×	×	
Determination of nuclear magnetic resonance			×
Degassing by ion bombardment			×
Wet chemical oxide isolation	×		
Hydrogen reduction method	×		
Chlorination or bromination method	×		
Sulphur method	×		
Kjeldahl method		×	
Combustion in a stream of oxygen			×

References pp. 668–672

concentrations; the gas content of most metals is below 0.1%. As the effects are usually unfavourable, every effort is made nowadays to keep the gas content of metals low, which requires careful, continuous checking during every stage of production.

For the determination of the elements oxygen, nitrogen and hydrogen in metals a variety of techniques is employed using physical and chemical methods or a combination of these. A survey of the methods in current use is shown in Table 1.

Vacuum and *inert-gas extraction methods* will be dealt with in detail later on. *Emission spectrographic analysis* is well known and does not require detailed discussion. In *activation analysis* the sample is exposed to a standardized neutron bombardment, and the concentration of an active isotope formed is determined by gamma spectrography. In *isotope dilution analysis* the sample is heated in contact with an isotope of the element to be determined to a temperature at which rapid isotope exchange takes place. After equilibrium has been obtained the isotope ratio in the atmosphere of the apparatus is determined with a mass spectrometer. *X-Ray structural analysis* is in certain cases a very valuable means for the determination of gases in metals. Provided the concentration of the gases is not too low, oxide or nitride phases may be detected. Another possibility is the measurement of the lattice parameter of the metal matrix, as this parameter may be affected by dissolved elements such as oxygen or nitrogen. The extent of the change of the lattice parameter makes possible the derivation of the concentration of the elements. *Mass spectroscopy* of solids analyses the total bulk sample and not, as explained under isotope dilution analysis, a gas mixture only. *Measurement of the internal friction* (damping measurement) for determination of the gas content is applicable in rare cases only, although it has been very successful in the determination of nitrogen in tantalum[1]. *Nuclear magnetic resonance* and degassing by *ion bombardment* allow the determination of hydrogen, but are of no importance for routine analysis.

All other methods, listed in Table 1, are purely chemical and most of them have been known for a long time. The *wet chemical isolation* of oxides is divided into two sub-groups, as it may be carried out by use of selective solvents (*e.g.* bromine/methanol) or by anodic dissolution of the sample in an electrolyte under definite conditions (voltage and current density). As the name of this method implies, only the oxygen in the form of oxides is determined. The *hydrogen reduction method* is usually applied for the determination of higher oxygen contents only. Furthermore, its use is restricted to metals whose oxides are reducible by hydrogen. *Bromination and chlorination* methods are frequently employed for oxide determinations in metals which form easily volatile halides, particularly titanium and zirconium. In the *sulphur method* the sample is treated with sulphur at 1000°C; at this temperature the total oxygen reacts with the sulphur, forming sulphur dioxide, which is determined. Owing to the generality of its application the

Kjeldahl method is even nowadays the method most frequently used for the determination of nitrogen in metallic materials.

The choice of the most suitable method from this great variety of techniques is governed by the following considerations:

(1) The nature of the sample. Every method is not applicable to all metals.

(2) The analytical requirements. For instance, it is a matter of importance whether the total quantity of an element should be determined or only that portion that is present in the combined form. The wet chemical method evaluates only combined oxygen, and not that in solution.

(3) The apparatus available. Some of the methods mentioned require extremely expensive equipment. Mass spectrometry and activation analysis are examples of this type. Neutron sources and the auxiliary equipment required for this process are still not standard items in an analytical laboratory.

(4) The time factor. Especially for routine analyses it is very important that they be completed in a very short time. From this point of view the physical methods are far superior to the chemical processes.

(5) The accuracy required, as this varies for the different methods.

(6) The personnel available. Owing to the universal shortage of skilled labour, the employment of physical methods is here also to be preferred as these may easily be automated and therefore be operated by unskilled labour.

By combining two methods it is sometimes possible to discriminate between combined and dissolved oxygen. In this case the total oxygen and the oxides are determined.

The vacuum extraction method and the inert-gas extraction method are now the methods most frequently used for the analysis of gases in metals, in particular for the total quantities of oxygen, nitrogen and hydrogen. The principal difference between these two methods is small; both techniques are a combination of chemical and physical processes. The inert-gas method operates in a stream of rare gases; the rare gas carries the gas extracted from the sample to the analyser. In the vacuum extraction technique the sample is degassed in a high vacuum of 10^{-3} to 5×10^{-6} Torr, the pumped-off gases are collected and analysed. In both methods the sample is usually heated in a graphite crucible. The total oxygen reacts with the carbon to form carbon monoxide, whereas nitrogen and hydrogen are given off in the elementary form. Even if the nitrogen is present in the form of nitrides in the sample, quantitative decomposition can always be guaranteed in the presence of carbon[2]. The degassing of the samples in both techniques may take place in the liquid (molten) or in the solid states. The decision as to which method should be employed is governed by the nature of the sample and by the element to be determined.

Only the vacuum degassing method falls within the scope of the present book. This method is the one most used for the determination of the gas content of metals, especially in Europe.

References pp. 668–672

After the principles of vacuum extraction analysis and the apparatus employed for this technique have been discussed, the sources of error and in connection with it the accuracy and precision of such determinations will be considered. This will be followed by a systematic compilation of the relevant literature in order of metals and the three elements to be determined. It should be pointed out that this literature compilation is not claimed to be complete. The main object of this compilation is to include all work of fundamental interest and importance. References published before 1950 are included in exceptional cases only.

The only book solely concerned with the determination of the gas content of metals was published in 1959 and all theories and analytical methods known up to that time are critically reviewed in it[3]. Besides this several review papers have been published which may be useful for reference purposes[4-10].

2. Technology of the Vacuum Extraction Analysis

In vacuum extraction analysis the sample to be analysed is heated in a high vacuum. Heating is usually carried out in a graphite crucible; only for the determination of hydrogen is a quartz crucible used, in cases, for instance, where hydrogen alone is to be determined. The operating temperature varies according to the method employed between 600° and 2200°C. The sample may be degassed in the solid or the molten state; this mode of operation determines the nature of the various techniques. The first mode of operation—degassing in the solid state—is also called *diffusion extraction*; the second—degassing in the molten state—*fusion extraction*. Diffusion extraction is most frequently used for the determination of hydrogen, the main reason for its use being the high diffusion capability of hydrogen. This means that the operation may be carried out at relatively low temperatures; it has also the advantage that hardly any other gases, such as carbon monoxide or nitrogen, are given off at the same time.

The diffusion method, however, may also be used for the determination of oxygen if the samples are in the form of powder and if the melting point of the powder is higher than the working temperature of the extraction procedure; in other words, for use with high-melting-point metals. The best procedure for introducing the sample into the crucible is the so-called graphite ampoule method which was first mentioned by Zakharov[11] and which has been modified and perfected more recently by Paesold[12]. For this method the powder samples are sealed into graphite ampoules and then dropped into the crucible. In this way any losses which might occur during the introduction of the sample, or during degassing, are avoided, *e.g.* the ejection of powder particles by gas bubbles if the flux technique or a carrier melt is used.

Diffusion extraction has also been suggested by several authors for the determination of oxygen in compact, bulk samples of metals of high-melting point[13-18], at a working temperature of 2000°C. However, Lassner[19,20] showed that this procedure is of doubtful value and should therefore be rejected since diffusion-inhibiting elements, which are frequently present in metals as impurities, may cause marked discrepancies. This method gives usable results only if the metals are of high purity and if the kind and quantity of the impurities are exactly known.

The most frequently applied variant of the vacuum extraction method is fusion extraction. In this technique the sample is dropped either directly into the graphite crucible at a temperature above the melting point of the sample or into a carrier melt of a special metal which is already in the crucible. This carrier melt is also known as a bath. The carrier melt helps with all samples of melting point below the operating temperature to dilute the metal to be analysed and to prevent heavy losses by evaporation. The carrier melt may also increase the activity of the elements to be analysed. In special cases the type and quantity of the graphite precipitation which always occurs may be modified and controlled by use of a special carrier melt. The phenomenon of graphite precipitation will be dealt with in detail later. For all metals with a melting point higher than the operating temperature the bath serves as a quasi-solvent. A low-melting-point eutectic may be formed.

A modification of the carrier melt method is the so-called flux method. In this modification the sample, together with a certain quantity of a carrier metal, usually wrapped in a foil or sealed in an ampoule made of the carrier metal, is dropped into the crucible. A condition for this technique is a uniformly low gas content of the carrier metal of which the foils or ampoules are made.

A schematic survey of the various vacuum extraction methods is shown in Table 2. As can be seen a variety of metals can be used as carrier melts and even alloys of these metals have been employed. However, with regard to the metals most frequently used, the ferrous metals take the first place, followed by platinum which is preferred for the analysis of titanium, zirconium, rare-earth metals, uranium, etc. When platinum is used as a carrier metal, the employment of the flux technique is preferred owing to the high price of platinum, as it requires less carrier metal than does the bath technique.

Regarding the best working procedure for a given metal, the literature on the subject contains quite conflicting statements. Almost all proved working techniques have been both recommended and rejected.

The causes for these discrepancies are that when describing optimum working conditions it is very often overlooked that not only the kind of metal plays an important role, but also the apparatus employed. Besides this, the material to be analysed is frequently not sufficiently characterized. There are hardly any indications in the reports on the homogeneity and structure of the samples, the degree of purity or the level of impurities. All these factors may affect the degassing of the sample to a considerable extent.

Characteristics of the whole research in this direction until recently have been a purely empirical approach and often a not very sound or convincing interpretation of single results.

References pp. 668–672

TABLE 2
METHODS OF VACUUM EXTRACTION ANALYSIS

In the solid state	Degassing of the metals	
	In the liquid state	
	Using a carrier melt (Fe, Ni, Co, Sn, Au, Pt)	Without a carrier melt
Suitable for:	Suitable for:	Suitable for:
Compact samples (for hydrogen determination only)	Compact samples of m.p. above 2200°C or showing a getter effect (for hydrogen, oxygen and nitrogen determination)	Compact samples of m.p. below 2200°C without getter effect (for hydrogen and oxygen determination)
Powder samples (for hydrogen or oxygen determination)	Modification: Flux method	Powder samples (for hydrogen and oxygen determination)
Particularly suitable with graphite ampoule method	Suitable for: Compact and powder samples, particularly if platinum is used as a carrier melt, but also with ferrous metals	Particularly suitable with graphite ampoule method

Kallmann et al.[21] have very clearly stigmatized this state of affairs when they said in a recently published report: "Unfortunately, the determination of the interstitial elements, particularly of oxygen, is still surrounded by considerable mystery. In the vacuum fusion technique, the preference of one or the other bath or flux or for a particular temperature appears often to be based more on wishful thinking than on valid experimental data which can be reproduced by others".

Only in the last few years have attempts been made to clarify the mechanism of vacuum extraction analysis by systematic experiments and theoretical computations. A number of papers[22-25] have been published on the degassing kinetics of metal melts. Thermodynamic calculations of the reducibility of oxides and of the decomposition temperatures of hydrides and nitrides could be carried out on the basis of the many known numerical data[3,26-30]. Two recent publications which are concerned with systematic studies on the best method for clarification of a definite problem and which at the same time take account of the theoretical fundamentals should be especially mentioned here[19,31].

Far fewer data are available on the diffusion of oxygen and nitrogen in the solid state, particularly on the mutual interference of elements which diffuse simultaneously, despite the fact that such data would be of particular interest for the research on diffusion extraction. Moreover, in most theoretical publications the application of the model case to practice is omitted. Particular lack of data on the equilibria of the following multicomponent system is felt:

Carbon ↔ carrier metal ↔ sample metal ↔ oxygen or nitrogen

and

Carbon ↔ sample metal ↔ oxygen or nitrogen

Although a large number of data are already known it is usually still impossible to indicate immediately the optimum conditions for the analysis of any given material. For the determination of these conditions and for evidence that the analyses are correct the following criteria are usually required:

(1) Analysis of reference samples
(2) Comparison with other analytical methods.

Neither procedure is entirely satisfactory. It is hardly ever possible to produce samples of exactly known gas content which have properties analogous to those of the sample to be analysed. This particularly concerns the bonding and distribution of oxygen, nitrogen and hydrogen and the microstructure and level of impurities. When comparing different methods of analysis it should be realized that the physical methods of measurement in particular, such as neutron activation analysis and spectroscopic analysis, always require calibration. Such calibration is carried out on reference samples, and the reference standards are often calibrated by the vacuum extraction technique. In this way a cyclic series of faulty results can occur.

3. Apparatus

Every vacuum extraction apparatus consists of three principal parts:

Furnace
Pumping system
Analyser.

According to the list shown in Table 3, these parts may be quite different. In the literature a large variety of apparatus has been described[3,9,31-49]. A large

TABLE 3

VACUUM EXTRACTION APPARATUS

Furnace	Pumping system	Analyser
Graphite resistance heating	Backing pump: Rotary vane pump Töpler pump	Volumetric Manometric Infra-red absorption
High-frequency heating	Diaphragm pump	(CO, CO_2) Thermal conductivity (H_2)
	High-vacuum pump: Mercury diffusion pump Oil diffusion pump	Conductimetric (CO_2) Coulometric (CO_2) Gas chromatographic Mass spectrometric

References pp. 668–672

proportion of the apparatus is either of the do-it-yourself type, or has been assembled from different parts. These and the first apparatus available commercially were mainly constructed of glass or fused quartz. The individual parts of such an assembly were either connected by ground glass joints, or were welded together to an integral part. The great risk of breakage of this type of construction and the difficulty of carrying out repairs were the driving forces for the development of all-metal equipment. The ruggedness of such apparatus is of great importance, particularly for laboratories in industry which are mainly concerned with routine analyses.

Not only the equipment as such, but also the methods of analysis of the gas mixture evolved, are being steadily improved and refined. Although in the beginning micro-Orsat analysis and manometric measurement and freeze-out of individual components in cooled traps were regarded as adequate, gas chromatography[38,40,50], measurement of infra-red absorption[43], measurement of thermal conductivity[43], and in special cases even mass spectroscopy[37,51], are used nowadays for the analysis of the gas mixture. These methods are much more expensive, but the results obtained are more easily reproducible; they also offer the advantages of requiring very short times for completion and the possibility of automation.

According to the present state of knowledge, the following requirements should be placed on a universal vacuum extraction apparatus which may be used for all purposes.

3.1 Furnaces

The equipment should be provided with both a resistance furnace, using a graphite heater, and a high-frequency induction furnace. The resistance furnace has the greater capacity and is more rugged, but shows a higher, although constant, blank value. It is therefore particularly suited for routine analysis of samples of high and medium gas content and permits the attainment of a working temperature of up to 2000°C. The high-frequency induction furnace, owing to its lower blank value, is more suited for the analysis of samples of low gas content. It should be mentioned that the constancy of its blank value is not as good as that of the

TABLE 4

EFFECT OF PUMPING SPEED OF THE VACUUM PUMP ON THE RESULTS OF THE DETERMINATION OF HYDROGEN IN FERRO-MANGANESE

Pumping speed (l/s)	Hydrogen content* (ml/100 g)	Degassing time (min)
150	35	8
45	32	40
12	25	40

* Mean of six determinations.

resistance furnace, particularly if relatively low frequencies are employed, these being preferable for obtaining agitation of the melt. Induction heating always offers advantages if really high temperatures (above 2000°C) are required, as it is possible to reach 2500°C and even higher temperatures. Each furnace should be provided with a lock system for the introduction of the samples during operation. This is of particular importance for apparatus which is used for routine analysis.

For reasons that will be explained later, provision should be made for the crucible to be emptied or exchanged without the apparatus being exposed to the atmosphere, or having to be cooled down.

3.2 Pumping system

The pumping system should be arranged for the attainment of a vacuum of about 10^{-6} Torr. The speed of the high-vacuum pump should be of the order of 150–200 l/s. An oil diffusion pump best meets these requirements.

3.3 Analyser

The analyser system should have a sensitivity as high as possible with respect to the gases to be determined and at the same time it should guarantee a good reproducibility of the results. The threshold of response should be at least 0.5 μg both for oxygen and for nitrogen, and 0.02 μg for hydrogen. A gas chromatograph best meets these requirements. Recently, however, it has been reported that the above-mentioned thresholds in determination are also reached by measurement of the infra-red absorption and the thermal conductivity[30].

The analyser should also be provided with a measuring device for calibration gases. This is required for calibration and control of the analyser as well as for the measurement of getter effects in the furnace vessel.

The following two case histories show that the requirements specified above are not at all unjustified.

The results of a collective study instigated by the Chemical Committee of the Gesellschaft Deutscher Metallhütten- und Bergleute e.V. showed clearly that installations having too low a pumping speed of the high vacuum pump always gave deficiencies in the determination of the hydrogen content of ferro-manganese[52]. The results of these analyses are shown in Table 4 and may be interpreted as follows:

During prolonged dwell times of the hydrogen in the furnace—in accord with a low pumping speed of the pump—this gas is absorbed by the graphite parts, and is released only very slowly. This phenomenon has been observed at a temperature of about 1100°C, but not at higher temperatures.

Similar effects were observed during other getter processes, and it generally holds good: the higher the pumping speed of the high vacuum pump the lower is any getter effect. The reason for this is easily seen if it is realized that the probability of the collision of a gas molecule with the gettering substance decreases

References pp. 668–672

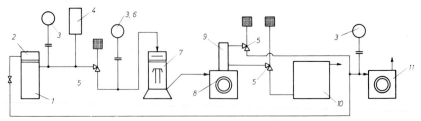

Fig. 1. Exhalograph EA-1, block diagram.
1 Resistance furnace or rotary crucible furnace; 2 Sample lock; 3 Medium vacuum gauge head; 4 Calibration device for admission of calibration gases; 5 Magnetic valve; 6 High vacuum gauge head; 7 Oil diffusion pump; 8 Gas collector pump; 9 Analyser with thermal conductivity gauge and U-tube total pressure manometer with electric reading; 10 Infra-red analyser; 11 Backing pump.

with decreasing dwell time. Besides a high pumping speed of the high-vacuum pump, adequately high conductance between furnace vessel and pump is necessary so that the pumping speed may become effective.

Another collective study has shown that a working temperature of about 2200°C is best for the determination of oxygen in molybdenum[53]. The sample is melted at this temperature without the use of a carrier melt as the Mo_2C–C eutectic, *i.e.* without the many sources of error inherently involved in the bath technique. Such a temperature can hardly be maintained over a prolonged period of time by graphite resistance heating, as the graphite begins to evaporate to a marked extent. In this case preference is given to induction heating.

The range of apparatus commercially available today is small. In principle there are only two which meet the requirement set out earlier. These are the Exhalograph EA-1 of Balzers A.G., and the Evolograph VH-8 of Heraeus and

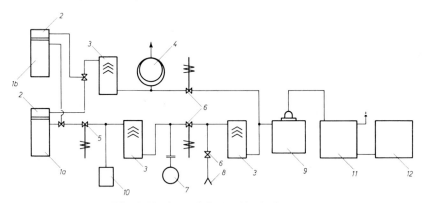

Fig. 2. Evolograph VH-8, block diagram.
1a Resistance furnace; 1b Induction furnace; 2 Sample lock; 3 Diffusion pump; 4 Backing pump; 5 Furnace isolation valve; 6 High vacuum valve; 7 Vacuum gauge; 8 Oven used in connection with hydrogen determination; 9 Gas collector pump; 10 Argon thrust valve; 11 Gas chromatograph; 12 Compensation strip chart recorder.

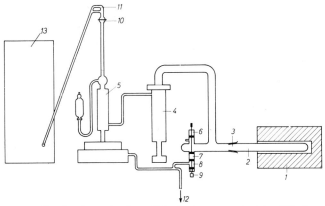

Fig. 3. Apparatus for the determination of the hydrogen content by diffusion extraction.
1 Resistance tube furnace; 2 Quartz tube; 3 Cooled joint; 4 Mercury diffusion pump; 5 Automatic Töpler pump; 6 Piston-type vacuum lock; 7 Sample introduction; 8 Closing lid; 9 Valve; 10 Stop-cock at burette; 11 Carrier gas admittance; 12 Lead to oil backing pump; 13 Gas chromatograph.

Feichtinger. The respective schematic diagrams of the two pieces of apparatus are shown in Figs. 1 and 2.

The Exhalograph EA-1 is extremely well suited for routine analyses. A detailed description of this apparatus has been published in the literature[43]. Its particular suitability for routine operation in the iron and steel industry is provided by the very short time required for an analysis (2–3 min). Moreover, the Exhalograph is the only vacuum apparatus which makes possible the exact, routine determination of nitrogen in steel as it is equipped with a rotary crucible which allows the crucible to be emptied after each analysis[31].

The Evolograph VH-8 should be regarded rather as a scientific tool or as an apparatus for special purposes because of its greater versatility (two different furnaces, temperature programming, gas chromatograph). A detailed description of this equipment has been given in the literature by Sperner and Koch[41].

A modern, do-it-yourself, high-efficiency micro-apparatus has recently been described by Lilburne[54].

For the determination of hydrogen alone in iron and steel much more simple apparatus can be used. A schematic diagram of such a piece of apparatus is shown in Fig. 3. A detailed description of it is given in the *Handbuch für das Eisenhütten-Laboratorium*[55].

4. Sources of Error

As was mentioned earlier, the vacuum extraction technique involves a number of potential sources of error. According to the nature of the origin these sources of error may be classified under three main headings.

References pp. 668–672

4.1 Sources arising from the sample

Both sample taking and sample preparation play a decisive role in the determination of the gas content of metals. During sample taking from the melt, which has been described in a number of papers[55-65], attention must be paid to the possible escape of gases in the process. This type of sample taking is mainly used in the steel industry. A number of techniques and devices have been developed for this purpose; they allow a sample to be taken, the gas content of which corresponds to that of the melt in the furnace[56].

However, improper sample taking from solid bodies may also easily lead to faulty results. Inhomogeneities are inherent, particularly in large pieces. The sample-taking technique from bulk steel parts has been described in detail[55]. In sintered bodies of, for instance, refractory metals, the gas content differs in the surface zone from that of the interior according to the sintering technique[66].

Sample preparation too may be the cause of faulty results. Metals susceptible to oxidation should not be subjected to heavy mechanical working, because surface oxidation may occur. Even with samples of steels—which are not particularly susceptible to oxidation—incorrect lathe turning may lead to an increase in the level of the oxygen at the surface[67]. Original surfaces must, however, be removed before analysis. This may be done either by mechanical means or by chemical or electrochemical processes. Which of these methods is chosen depends mainly on the nature of the metal. Certain metals pick up hydrogen and oxygen during treatment with acids, others do not. Hard metals show, for instance during filing, an increase in the hydrogen level; for other metals this technique is the best[20]. A number of further papers dealing with the very important subject of sample preparation have been published[55,58,64,67-71]. On no account should the sample be subjected to fine pulverization as is customary for other analytical purposes, although even here there are exceptions[52,72]. For brittle material, crushing to small pieces is a method frequently used which does not involve a great risk of gas pick-up.

Even the size of the crystallites of the sample material may be of importance in the analysis of refractory metals if a carrier melt is employed. Fine-grain material dissolves very much more quickly in the carrier melt than coarse-grain material[68]. Care should therefore be taken to prolong the degassing time for coarse-grained material accordingly.

4.2 Errors caused by the carrier melt or other factors in the furnace vessel

One factor is of vital importance, *viz.*, that every carrier melt dissolves graphite from the crucible wall, which is precipitated at the cooler parts of the bath as a result of the temperature gradient. These graphite precipitates are of the lamellar type, and cause a marked increase in the viscosity of the melt which, in turn, results in a slower and in most cases also in incomplete liberation of the gases[31]. The following techniques may be used as a remedy for this shortcoming:

(a) To keep the working temperature as low as possible, as the dissolution of the graphite decreases with decreasing temperature. On no account should the temperature of pure iron melts exceed 1700°C.

(b) To choose for the carrier melt a metal or alloy of lower solubility of carbon than iron, for instance, platinum or a nickel–iron alloy of composition 80 wt.% nickel and 20 wt.% iron. But even here the above-mentioned effect of graphite precipitation occurs, although more slowly, so that the bath may be used either for a longer time or at a higher temperature. For instance, platinum allows the use of a working temperature of about 1900°C, the 80:20 nickel–iron alloy one of 1850°C.

(c) To add a metal to the carrier melt which lowers the solubility of carbon, for instance, tin to nickel–iron carrier melts[68].

(d) To dilute the carrier melt by occasional addition of fresh carrier metal. This is, in combination with the methods described under (a) and (b), probably the most frequently employed working technique.

(e) To use a fresh carrier melt for each analysis. This technique involves either a large expenditure in time or requires a special device which allows the crucible to be emptied after each analysis, such as the rotary crucible furnace developed by Balzers A.G.[31].

(f) To change the precipitation of the graphite from the lamellar form to the spheroidal type by addition of cerium to the carrier melt. Spheroidal graphite increases the viscosity of the carrier melt very much less than does lamellar graphite. This technique has been used by Gerhardt, Kraus and Frohberg[31] in special applications, for instance, for the determination of nitrogen in steel.

(g) To use bath agiation. Lilburne[54] recently reported that by application of medium range frequencies (5 kHz) for induction heating a marked agitation effect could be observed in the melt, which facilitates degassing. This agitation or stirring effect has been postulated for high-frequency heating earlier, but this was subsequently refuted.

The requirement is often found in the literature that the concentration of the metal to be analysed should not exceed a certain level in the carrier melt. This requirement could have several causes:

(a) The sample metal may cause an increase in the solubility of one of the elements to be determined in the melt, for instance, titanium for oxygen.

(b) The metal to be analysed may show a marked vapour pressure at the operating temperature, and the metal vapour may have a gettering effect on one of the gases to be determined, for instance, manganese for carbon monoxide.

(c) The metal to be analysed may form a carbide of high melting point which may be difficult to dissolve in the carrier melt, so that it is precipitated from the melt, thus causing an increase in viscosity.

(d) The metal to be analysed displaces the equilibrium sample metal ↔ oxygen ↔ carbon monoxide in the direction such that a fine disperse phase, such as

References pp. 668–672

Al_2O_3 or SiO_2, is precipitated.

All these phenomena make an inspection of the melt in the crucible imperative. This inspection should be continuous by direct visual observation.

4.3 Errors caused by the apparatus

Potential errors which might be introduced by the use of pumps of too low a pumping speed were discussed in detail earlier.

Most apparatus requires calibration of the analyser system. As is known, even calibrations themselves may form a source of errors if they are applied incorrectly or at too long intervals. Frequent checking of the calibration is therefore necessary.

Another factor which must not be overlooked is the blank value of the apparatus which is inherent. This blank value is caused by continuous gas evolution from the apparatus. The gas evolution has its origin to a lesser extent in leakages, and to the major extent in the greatly retarded degassing process of the graphite parts of the furnace. The blank values differ for the various types of apparatus according to the design of the furnace (quantity of the graphite parts). The difference in blank values between resistance and induction furnaces has already been discussed. Besides this, the magnitude of the blank value depends on the working technique. Employment of a carrier melt always involves a higher blank value than working with a "dry" crucible. Moreover, the blank value is not constant over a period of time when working with carrier melts.

When apparatus with a lock system is used, small quantities of gas are admitted into the furnace vessel with the sample; these quantities are also regarded as adding to the blanks. In order to keep this part of the blank value as constant as possible, the operation of the lock should be made strictly reproducible.

To obtain a result as exact as possible, continuous recording of the blank value is imperative. The blank value should be determined before and after each analysis. Care should also be taken to keep this blank value as low as possible by outgassing the furnace before the analyses so that the absolute values of its fluctuations also remain small. It should be remembered that the limit of determination is in turn determined by the range of scatter of the blank values. The furnace should therefore be outgassed for a sufficiently long time at a temperature 200 °C higher than the working temperature.

This discussion of the various sources of error has been recorded in order to show that a vacuum extraction analysis requires considerably more vigilance than is usual for other instrumental methods. For this reason efforts were made a long time ago to devise optimum and strictly standardized working methods for the gas analysis of the various individual metals. In most cases, however, it has been overlooked that these conditions depend not only on the type of metal but also on the specific properties of the sample, such as microstructure, surface quality, type and level of impurities, etc. As has been mentioned earlier, certain conditions for the analysis, such as, for instance, the time required for degassing, may depend on the characteristics of the apparatus. Such data apply to each individual apparatus, and are not valid in every case. Many contradictory statements made on this subject can probably be attributed to these causes.

5. Accuracy and Precision

The correct working of the analyser system can usually be checked very simply with calibration gases, but the accuracy of the analysis cannot. Unfortunately, the extent to which inaccuracies are introduced during the analysis by the various sources of errors described earlier can also not be predicted. A method which is frequently employed for the control of the accuracy of the technique used is the analysis of standard or reference samples. Accurate standard samples, however, are obtainable for a very restricted number of metals only. It is not, for instance, permissible to check the correctness of the determination of oxygen in a titanium sample with a correct result obtained with a standard sample of steel. For the analysis of a particular metal only standard samples of the same metal should be used. In many cases, therefore, standard samples have to be "home-made". This procedure is, however, rather doubtful as it is only possible in exceptional cases to introduce the gases in such a form and with the same distribution as are met

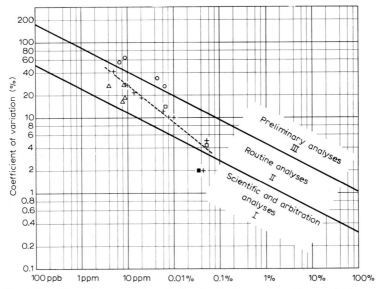

Fig. 4. Precision of oxygen determination by the fusion extraction method.
○ Comparative studies of several laboratories[73] on refractory metals and alloys
△ Comparative studies of several laboratories[21] on refractory metals
□ Literature references on analysis of steel
■ Plain carbon steel
+ Own investigations on various metals

References pp. 668–672

with in the samples in the actual analysis.

Comparison of results obtained with the same sample material, but by different methods, is therefore often used in testing the accuracy of any one method[21,73]. Caution should, however, be exercised in using samples of low gas content (below 10 ppm) for such a comparison, as the gas adsorption on the surface then attains a similar order of magnitude as the gas contained in the interior. Identical treatment of the samples, assuming that a high proportion of adsorbed gas is liberated easily, may give rise to *apparent* good agreement in the results.

The precision of the determination of oxygen in metals by vacuum degassing is shown in the diagram Fig. 4 and the values of precision results from the literature for various gas levels and various metals are compiled in Table 5; the data are listed separately for oxygen, hydrogen and nitrogen. In Fig. 4 the coefficient of variation (standard deviation in relative percentages) is plotted in relation to the oxygen level of the sample. A logarithmic division was used for both parameters.

TABLE 5

REPRODUCIBILITY OF VACUUM EXTRACTION ANALYSIS

Metal	Content (ppm)	Coefficient of variation (%)	Group after Püschel[74]	Reference
Oxygen				
Steel	900	5	II	39
	400	14	III	39
	380	3	I	75
	270	4	I	75
	200	16	II	39
	115	2–5	I	76
	116	5	I	51
	80	25	III	39
	55	6–12	I–II	51
Copper	1.27	10	I	51
	0.38	13	I	51
	0.28	25	I	70
	0.24	18	I	51
	0.10	9	I	70
	0.07	19	I	51
	0.04	26	I	70
Titanium	2100	3	II	77
	1300	4	II	35
	1100	6	II	77
	1000	5	II	39
	580	2	I	35
	310	11	II	77
Niobium	190	1	I	13
	160	5	I	78
	76	10	II	78
	63	4	I	78

TABLE 5 (continued)

Metal		Content (ppm)	Coefficient of variation (%)	Group after Püschel[74]	Reference
	Tantalum	6	28	II	79
		7	17	II	79
		94	3	I	78
		104	10	II	78
		120	3	I	78
	Chromium	2800	4	II	80
		1000	3	II	80
		500	3	I	80
	Molybdenum	40	2–12	I–II	76, 79
		32	6	I	51
		7.5	21–36	II	15
		2–3	8–42	I–II	76, 78
	Tungsten	3	21–46	II	81
Hydrogen					
	Steel	1.24	6	I	82
	Copper	0.41	15	I	70
		0.02	15–75		70
	Tantalum	7.3	11	I	20
	Molybdenum	5	12	I	83
		2.6	9	I	15
		0.58	43	II	15
		0.40	14–31	I	20
	Tungsten	0.50	30	I	20
		0.16	34	I	15
Nitrogen					
	Steel	164	3	I	2
		100	7	II	2
		31	10	II	51
	Molybdenum	9.7	5	I	83

Moreover, a classification of the analyses into three groups according to precision was introduced as proposed by Püschel[74]. In this diagram results obtained by the present author on bulk samples of refractory metals, as well as some taken from the literature, are shown.

It may clearly be seen from both Fig. 4 and Table 5 that precision may vary to a marked extent for the same gas content and even with the same metal. One reason for this discrepancy could be attributed to the fact that samples of different homogeneity have been used, another reason being the dependence of each measurement on the analyser. Precision is therefore not constant if different pieces of apparatus are employed for the measurements. For instance, considering all the data of Table 5, 8.2% lie in Group III, 42.8% in Group II, and 49% in Group I.

References pp. 668–672

If, on the other hand, the same calculations are carried out on results obtained with one type of apparatus (Exhalograph EA-1), only 7.3% are in Group II but 92.7% in Group I[84].

The dependence of the precision on the type of material analysed may also be seen from Table 5. Despite the use of different types of apparatus and an extremely low oxygen level, all analyses of copper lie in Group I. It is also evident that the determination of hydrogen can always be carried out with better precision than that of oxygen or nitrogen.

For the analysis of powder and bulk samples at oxygen levels higher than 100 ppm, reproducibility is often lower than might be expected from the diagram. Such materials are usually inhomogeneous with respect to oxygen. If, therefore, the oxygen content should exceed some 100 ppm, or even reach the range of some tenths percent, coefficients of variation of less than 10% are rarely obtainable.

6. Literature Survey

In Table 6 the metals are arranged in groups according to the Periodic Table. The most important publications on the determination of the elements hydrogen, nitrogen and oxygen in metals are quoted in the form of the literature reference number for the three elements separately under each metal.

TABLE 6
LITERATURE SURVEY ARRANGED ACCORDING TO METALS AND TO ELEMENTS TO BE ANALYSED

Periodic Table Group No.	Metal	Hydrogen	Nitrogen	Oxygen
I	Sodium			85, 86
	Potassium	87	87	87
	Copper	32, 70, 87–89	32, 87	32, 70, 81, 87–93
	Silver	32, 87	32, 87	32, 87
II	Beryllium	87–89, 94	87, 94	87–89, 94–102
	Magnesium	103		
	Calcium			104
III	Yttrium	87, 94	87, 94	87, 94
	Lanthanum		105	105
	Rare earths			106
	Aluminium	116, 117		91, 118
	Gallium			119
	Indium			119
IV	Titanium	32, 50, 87–89, 120–122	32, 50, 87, 107	14, 17, 32, 35, 39, 50, 77, 87–89, 97, 107, 120–130
	Germanium			133
	Zirconium	49, 88, 89, 108, 131	49, 107, 108	
	Thorium	49	49, 88, 89	49, 88, 89, 107
V	Vanadium	89, 107		89, 107, 134
	Niobium	33, 50, 87, 89, 135	6, 33, 41, 50, 66, 87, 136–140	6, 13, 33, 41, 50, 66, 78, 87, 89, 101, 135–147
	Tantalum	15, 20, 50, 88, 89, 94, 148	6, 15, 41, 50, 66, 94, 137–140	1, 4, 6, 15, 41, 50, 66, 78, 79, 88, 89, 94, 101, 130, 133, 137–141, 144, 148, 149
VI	Chromium	32, 88, 89, 150	32, 151	32, 80, 88, 89, 91, 150–153
	Molybdenum	15, 20, 32, 33, 87	15, 32, 33, 41, 66, 87, 107, 137, 138, 140, 144, 151, 154–158	4, 6, 15, 16, 19, 20, 32, 33, 41, 51, 53, 66, 68, 78, 79, 87, 97, 107, 123, 124, 133, 137, 138, 140, 141, 144, 151, 154–157, 159–163
	Tungsten	15, 20, 33, 83, 87, 89, 164	15, 33, 41, 66, 83, 87, 137, 140, 154, 164	4, 15, 33, 41, 66, 79, 87, 89, 91, 124, 137, 138, 141, 144, 154, 161, 164, 165
	Uranium	49, 88, 89, 94, 108–112	49, 94, 107, 108	49, 88, 89, 94, 107–111, 113–115
VII	Rhenium	74, 183		
VIII	Iron and steel	31, 32, 37, 40, 44, 56, 60, 82, 88, 89, 166–176	32, 37, 40, 46, 58, 69, 75, 166, 168, 172, 173, 176–178	32, 37, 39, 40, 42, 43, 46, 51, 57–65, 71, 75, 88, 89, 166, 168, 172–174, 179–182
	Ferrous alloys	48, 52, 72, 150, 184	48	48, 72, 150
	Cobalt	150		91, 124, 150
	Nickel	150, 173	173	91, 133, 150, 173
	Platinum	87, 108	87, 108	87, 108

References pp. 668–672

REFERENCES

1 F. Schlät and K. Friedrich, *Tagung "Stickstoff in Metallen"*, Freiberg, April, 1964.
2 A. Gerhardt, *Dissertation*, Fakultät für Bergbau und Hüttenwesen der Technischen Universität Berlin, 1965.
3 Z. M. Turovtseva and L. L. Kunin, *Analysis of Gases in Metals*, Consultants Bureau, New York, 1961.
4 E. Lassner and E. Wölfel, *Mikrochim. Acta*, (1960) 394.
5 Z. M. Turovtseva, *Tr. Komis. po Analit. Khim. Akad. Nauk SSSR, Inst. Geokhim. i. Analit. Khim.*, 40 (1960) 82.
6 M. W. Mallett, *Rept. DMIC-Memo-49*, 1960.
7 W. Fischer, *Z. Chem.*, 1 (1961) 354.
8 L. Pape and R. Platzer, *Comm. Energie At. (France), Ser. Bibl. No. 23*, (1962) 85.
9 G. A. Gokcen and E. S. Tankins, *J. Metals*, 14 (1962) 584.
10 K. Friedrich and E. Lassner, *J. Less-Common Metals*, 13 (1967) 156.
11 E. L. Zakharov, *Byul. Tsentr. Inst. Inform. Chernoi Met.*, (1957).
12 G. Paesold, *Tagungsber. 2. Balzers Kundenkolloquium, Balzers, Liechtenstein, 1966*, p. 22; *Tagungsber. 3. Balzers Kundenkolloquium, Balzers, Liechtenstein, 1967*, pp. 23, 32.
13 W. R. Hansen and M. W. Mallett, *Anal. Chem.*, 29 (1957) 1868.
14 H. G. Lange von Stockmeier, *Arch. Eisenhüttenw.*, 29 (1958) 95.
15 J. E. Fagel, R. F. Witbeck and H. A. Smith, *Anal. Chem.*, 31 (1959) 1115.
16 K. Friedrich, *Acta Chim. Acad. Sci. Hung.*, 28 (1961) 187.
17 T. A. Sullivan, B. J. Boyle, A. J. Mackie and R. A. Plott, *U.S. Bur. Mines Rept. Invest. No. 5834, 1961*, p. 30.
18 K. Friedrich and G. Ehrlich, *Reinststoffe in Wissenschaft und Technik*, Akademie Verlag, Berlin, 1963, p. 521.
19 E. Lassner, in E. Rexer (Ed.), *Reinststoffprobleme*, Vol. 2, *Reinststoffanalytik*, Akademie-Verlag, Berlin, 1966, p. 689.
20 E. Lassner, *Tagungsber. 2. Balzers Kundenkolloquium, Balzers, Liechtenstein, 1966*, p. 23.
21 S. Kallmann, R. Liu and H. Oberthin, *Tech. Rept. AFML-TR-65-194*.
22 T. Kraus, *Tagungsber. 2. Balzers Kundenkolloquium, Balzers, Liechtenstein, 1966*, p. 15.
23 T. Kraus, *Schweiz. Arch.*, 11 (1962) 452.
24 T. Kraus, *Balzers Hochvakuum Fachbericht 3*, April, 1965.
25 T. Kraus and O. Winkler, *Iron Steel Inst. (London) Spec. Rept.*, 92 (1965) 45.
26 Y. A. Klyachko and E. M. Chistyakova, *Zavodsk. Lab.*, 26 (1960) 1335.
27 L. L. Kunin, *Zh. Analit. Khim.*, 20 (1965) 822.
28 G. W. Goward, *Anal. Chem.*, 37 (1965) 117 R.
29 L. L. Kunin, *Zh. Analit. Khim.*, 20 (1965) 822.
30 T. Kraus, personal communication.
31 A. Gerhardt, T. Kraus and M. G. Frohberg, *Z. Anal. Chem.*, 218 (1966) 192.
32 D. L. Guernsey and R. H. Franklin, *ASTM Spec. Tech. Publ.*, 222, 1957, p. 3.
33 G. V. Mikhailova, Z. M. Turovtseva and R. Sh. Khalitov, *Zh. Analit. Khim.*, 12 (1957) 338.
34 P. D. Blake, *J. Iron Steel Inst. (London)*, 188 (1958) 261.
35 S. J. Bennett and L. C. Covington, *Anal. Chem.*, 30 (1958) 363.
36 N. A. Gokcen, *Trans. AIME*, 212 (1958) 93.
37 J. F. Martin, J. E. Friedline, L. M. Melnick and G. E. Pelissier, *Trans. AIME*, 212 (1958) 514.
38 C. Baque and L. Champeix, *Comm. Energie At. (France) Rappt. No. 1386*, 1958.
39 L. C. Covington and S. J. Bennett, *Anal. Chem.*, 32 (1960) 1334.
40 P. Tyou and A. Hans, *Rev. Met.*, 2 (1961) 187.
41 F. Sperner and K. H. Koch, *Metall*, 18 (1964) 701.
42 K. Teske and A. Hundrieser, *Neue Hütte*, 8 (1963) 370.
43 T. Kraus, *Arch. Eisenhüttenw.*, 33 (1962) 527.
44 S. Bergenfelt and C. A. Akerblom, *Jernkontorets Ann.*, 146 (1962) 461.
45 T. Kraus, *Z. Anal. Chem.*, 209 (1965) 206.

46 A. GERHARDT, *Tagungsber. 2. Balzers Kundenkolloquium, Balzers, Liechtenstein, 1966*, pp. 6, 9.
47 T. KRAUS, *Tagungsber. 2. Balzers Kundenkolloquium, Balzers, Liechtenstein, 1966*, p. 11.
48 A. FUCHS, H. REINHARD, J. NIEBUHR AND R. BECK, *Arch. Eisenhüttenw.*, 34 (1963) 361.
49 *U.K. At. Energy Authority Prod. Group, PG Rept. 381*, 5, 1962, p. 16.
50 R. LESSER AND H. GRUBER, *Z. Metallk.*, 51 (1961) 195.
51 M. L. ASPINAL, *Analyst*, 91 (1966) 33.
52 Chemiker-Ausschuss der G.D.M.B. e.V., *Erzmetall*, 20 (1966) 419.
53 K. FRIEDRICH AND E. LASSNER, *J. Less-Common Metals*, 13 (1967) 156.
54 M. T. LILBURNE, *Talanta*, 14 (1967) 1029.
55 Chemikerausschuss des VDE, *Handbuch für das Eisenhüttenlaboratorium*, Verlag Stahleisen m.b.H., Düsseldorf, 1966.
56 W. SCHWARZ AND H. ZITTER, *Tagungsber. 1. Balzers Kundenkolloquium, Balzers, Liechtenstein, 1965*, p. 2.
57 Z. NOVOTNY, *Tagungsber. 1. Balzers Kundenkolloquium, Balzers, Liechtenstein, 1965*, p. 2.
58 K. STYBLO, *Tagungsber. 1. Balzers Kundenkolloquium, Balzers, Liechtenstein, 1965*, p. 3.
59 J. MOCEK, *Tagungsber. 1. Balzers Kundenkolloquium, Balzers, Liechtenstein, 1965*, p. 4.
60 F. V. GLEESON, *Tagungsber. 1. Balzers Kundenkolloquium, Balzers, Liechtenstein, 1965*, p. 6.
61 W. SCHWARZ, *Tagungsber. 2. Balzers Kundenkolloquium, Balzers, Liechtenstein, 1966*, p. 32.
62 M. KOVARIK, *Tagungsber. 2. Balzers Kundenkolloquium, Balzers, Liechtenstein, 1966*, p. 35.
63 V. MAYER, *Tagungsber. 2. Balzers Kundenkolloquium, Balzers, Liechtenstein, 1966*, p. 37.
64 M. BAJ, *Tagungsber. 2. Balzers Kundenkolloquium, Balzers, Liechtenstein, 1966*, p. 38.
65 Z. NOVOTNY AND K. STYBLO, *Tagungsber. 2. Balzers Kundenkolloquium, Balzers, Liechtenstein, 1966*, p. 39.
66 E. WÖLFEL AND E. LASSNER, *Planseeber. Pulvermet.*, 9 (1961) 162.
67 J. BRUCH AND A. WUTSCHEL, *Balzers-Hochvakuum-Fachber. 1, February*, 1965.
68 E. LASSNER, *Metallk. Koll., St. Christoph, 1965*, and *Tagungsber. 1. Balzers Kundenkolloquium, Balzers, Liechtenstein, 1965*, p. 8.
69 R. SUAREZ-ACOSTA, *Tagungsber. 1. Balzers Kundenkolloquium, Balzers, Liechtenstein, 1965*, p. 7.
70 L. HEIMONEN, *Tagungsber. 1. Balzers Kundenkolloquium, Balzers, Liechtenstein, 1965*, p. 7.
71 J. GROCHOL, P. KOMAREK AND A. KLIMSA, *Tagungsber. 2. Balzers, Kundenkolloquium, Balzers, Liechtenstein, 1966*, p. 27.
72 Chemiker-Ausschuss der G.D.M.B. e.V., *Erzmetall*, 19 (1966) 526.
73 Materials Advisory Board Subpanel on Analysis Methods of the Refractory Metals, Sheet Rolling Panel, *MAB-217-M*, February 1966.
74 R. PÜSCHEL, *Mikrochim. Acta*, in press.
75 V. A. FASSEL, F. M. EVENS AND C. C. HILL, *Anal. Chem.*, 36 (1964) 2115.
76 P. D. DONOVAN, J. L. EVANS AND G. H. BUSH, *Analyst*, 88 (1963) 771.
77 CH. VENKATESWARLU AND M. W. MALLETT, *Talanta*, 5 (1960) 283; *Anal. Chem.*, 32 (1960) 1888.
78 M. W. MALLETT, D. F. KOHLER, R. B. IDEN AND G. G. KOEHL, *Batt. Mem. Inst. Tech. Rept. WAL TR 823/5*, 1962.
79 G. EHRLICH AND K. FRIEDRICH, *Neue Hütte*, 8 (1963) 27.
80 A. LENCH, G. S. MARTIN AND G. CUMMING, *Anal. Chem.*, 36 (1964) 337.
81 W. F. HARRIS AND W. M. HICKAM, *Anal. Chem.*, 31 (1959) 281.
82 C. C. CARSON, *Anal. Chem.*, 32 (1960) 936.
83 YU. A. KLYACHKO, T. A. IZMANOVA AND E. M. CHISTYAKOVA, *Zavodsk. Lab.*, 29 (1963) 923.
84 *Prospectus DN 1126*, Balzers AG, Balzers, Liechtenstein.
85 N. F. LITVINOVA, V. I. MALYSHEV AND Z. M. TUROVTSEVA, *Tr. Komis. po Analit. Khim. Akad. Nauk SSSR, Inst. Geokhim. i. Analit. Khim.*, 10 (1960) 97.
86 J. A. J. WALKER AND H. SEED, *Analyst*, 90 (1965) 19.
87 Z. M. TUROVTSEVA AND N. F. LITVINOVA, *Proc. U.N. Intern. Conf. Peaceful Uses At. Energy, Geneva*, 28 (1958) 593.
88 E. BOOTH, A. PARKER AND F. J. BRYANT, *Analyst*, 82 (1957) 50.

89 A. Parker, *Iron Steel Inst. (London) Spec. Rept. No. 68*, 1960, p. 64.
90 W. F. Harris and J. Easha, *Trans. AIME*, 221 (1961) 1264.
91 H. A. Sloman, *J. Inst. Metals*, 71 (1945) 391.
92 L. Heimonen, *Tagungsber. 2. Balzers Kundenkolloquium, Balzers, Liechtenstein, 1966*, p. 24.
93 J. Mocek, *Tagungsber. 2. Balzers Kundenkolloquium, Balzers, Liechtenstein, 1966*, p. 25.
94 Z. M. Turovtseva, N. F. Litvinova, N. M. Vasileva and K. G. Semenyuk, *Tr. Komis. Analit. Khim. Akad. Nauk SSSR, Inst. Geokhim. i. Analit. Khim.*, 10 (1960) 109.
95 J. N. Gregory and D. Mapper, *Analyst*, 80 (1955) 230.
96 E. Booth and A. Parker, *Analyst*, 83 (1958) 241.
97 J. E. Still, *Iron Steel Inst. (London) Spec. Rept. No. 68*, 1960, p. 43.
98 E. Booth and A. Parker, *Analyst*, 84 (1959) 546.
99 A. Parker, *Iron Steel Inst. (London) Spec. Rept. No. 68*, 1960, p. 64.
100 N. F. Litvinova and Z. M. Turovtseva, *Tr. Komis. Analit. Khim. Akad. Nauk, SSSR, Inst. Geokhim. i Analit. Khim.*, 12 (1960) 341.
101 J. Bril, F. Dugain and M. Occhiminuti, *Bull. Soc. Chim. France*, (1965) 562.
102 M. R. Everett and G. E. Thompson, *Analyst*, 87 (1962) 515.
103 R. Berry and J. A. J. Walker, *Analyst*, 88 (1963) 280.
104 D. T. Peterson and V. G. Fattore, *Anal. Chem.*, 34 (1962) 579.
105 D. T. Peterson and D. J. Beernstsen, *Anal. Chem.*, 29 (1957) 254.
106 G. G. Glavin and Y. A. Karpov, *Zavodsk. Lab.*, 30 (1964) 306.
107 H. A. Sloman, C. A. Harvey and O. Kubaschewski, *J. Inst. Metals*, 80 (1952) 391.
108 R. E. Taylor, *Anal. Chim. Acta*, 21 (1959) 549.
110 Y. Morimoto and A. Hiramaya, *Bunseki Kagaku*, 11 (1962) 163.
111 Y. Morimoto and T. Ashizawa, *Bunseki Kagaku*, 10 (1961) 1383.
112 *U.K. At. Energy Authority, Prod. Group, PG Rept. 412(s)*, 1962, p. 8.
113 C. B. Griffith, W. A. Albrecht and M. W. Mallett, *Rept. BMI-1033*, 1955.
114 H. E. Rölling, E. Trommer and A. Minenko, *Acta Chim. Acad. Sci. Hung.*, 32 (1962) 160.
115 J. N. Gregory, D. Mapper and J. A. Woodward, *Analyst*, 78 (1953) 414.
116 V. A. Danilkin, K. M. Konstantinov and G. I. Bulatova, *Zavodsk. Lab.*, 27 (1961) 259.
117 Y. A. Klyachko, L. L. Kunin and E. M. Chistyakova, *Sb. Tr. Tsentr. Nauchn.-Issled. Inst. Chernoi Met.*, 24 (1962) 42.
118 L. Kopa, *Hutnicke Listy*, 14 (1959) 322.
119 N. M. Vasileva, N. F. Litvinova and Z. M. Turovtseva, *Zh. Analit. Khim.*, 18 (1963) 250.
120 H. Goto, S. Suzuki and A. Onuma, *Sci. Rep. Res. Inst. Tohoku Univ., Ser. A*, 8 (1956) 24.
121 R. M. Fowler, *ASTM Spec. Tech. Publ.*, 204, 1957, p. 197.
122 Z. B. Turovtseva and R. Sh. Khalitov, *Zh. Analit. Khim.*, 12 (1957) 720.
123 R. S. McDonald, J. E. Fagel and E. W. Balis, *Anal. Chem.*, 27 (1955) 1632.
124 D. H. Wilkins and J. F. Fleischer, *Anal. Chim. Acta*, 15 (1956) 334.
125 W. R. Hansen, M. W. Mallett and M. J. Trzeciak, *Anal. Chem.*, 31 (1959) 1237.
126 K. Shimasaki, S. Shindo and K. Hosada, *Bunseki Kagaku*, 9 (1960) 489, 494.
127 Y. A. Klyachko and E. M. Chistyakova, *Tr. Komis. Analit. Khim. Akad. Nauk SSSR, Inst. Geokhim. i. Analit. Khim.*, 12 (1960) 126.
128 D. I. Walter, *Anal. Chem.*, 22 (1950) 297.
129 G. Derge, *J. Metals*, 1 (1949) 31.
130 W. M. Albrecht and M. W. Mallett, *Anal. Chem.*, 26 (1954) 401.
131 F. K. Heumann, *Reactor Technol. Quart. Rept. No. 6, Chemistry, KAPL-2000-3*, October 1958, p. 7.
132 J. K. Stanley, J. V. Hoene and G. Wiener, *Anal. Chem.*, 23 (1951) 377.
133 A. L. Beach and W. G. Guldner, *ASTM Spec. Tech. Publ.*, 222, 1957, p. 15.
134 C. V. Banks, *Ann. Summary Res. Rept. in Chemistry*, July 1960–June 1961, p. 67.
135 Anon. *PWAC-340*, June 30, 1961, p. 34.
136 W. H. Harris, *NP-6689*, 1957.
137 E. S. Tankins, *J. Metals*, 16 (1964) 109.
138 R. Lesser, *Trans. 8th Natl. Vacuum Symp./2nd Intern. Vacuum Congr., 1961*, Pergamon, Oxford, 1962, Vol. 2, p. 782.
139 E. Fromm and H. Jehn, in E. Rexer (Ed.), *Reinststoffprobleme*, Vol. 2, *Reinststoffanalytik*,

Akademie Verlag, Berlin, 1966, p. 681.
140 J. A. Klyachko, T. A. Izmanova and E. M. Chistiakova, *Zavodsk. Lab.*, 29 (1963) 1425.
141 E. Lassner, in F. Benesovsky (Ed.), *Metalle für die Raumfahrt*, 4th Plansee Seminar, 1964, Springer-Verlag, Vienna, 1965, p. 511.
142 J. A. Karpov, G. G. Glavin, O. V. Savjalov and R. V. Ivanova, *Zavodsk. Lab.*, 31 (1965) 1190.
143 H. E. McCoy, *Rept. CF-60-8-25*, 1960.
144 R. Lesser, *Reinststoffe in Wissenschaft und Technik*, Akademie Verlag, Berlin, 1963, p. 511.
145 M. J. Leadbetter and B. B. Argent, *J. Less-Common Metals*, 3 (1961) 19.
146 W. F. Harris, in B. W. Gonser and E. M. Sherwood (Eds.), *Technology of Columbium*, Wiley, New York, 1958.
147 F. Burns, *Rept. CONF-313-11*, see *Nucl. Sci. Abstr.*, 18 (1964) 2685.
148 N. Oda, N. Katayama and K. Endo, *J. Japan Inst. Metals (Sendai)*, 25 (1961) 693.
149 S. Suzuki, *Japan Analyst*, 11 (1962) 356.
150 O. Ettrich, H. Taxhet and W. Thomich, *Arch. Eisenhüttenw.*, 35 (1964) 613.
151 Yu. A. Klyachko, E. M. Chistyakova and L. L. Kunin, *Tr. Komis. po Analit. Khim. Akad. Nauk SSSR, Inst. Geokhim. i Analit. Khim.*, 12 (1960) 281.
152 Chemiker-Ausschuss der G.D.M.B. e.V., *Erzmetall*, 20 (1966) 466.
153 W. S. Horton and J. Brady, *Anal. Chem.*, 25 (1953) 1891.
154 M. W. Mallett, *Trans. Am. Soc. Metals*, 41 (1949) 870.
155 M. W. Mallett and C. B. Griffith, *Trans. Am. Soc. Metals*, 41 (1949) 870.
156 G. A. Picklo, *Rept. NRL Progr.*, 1961.
157 L. E. Olds and G. W. P. Rengstorff, *J. Metals*, 8 (1956) 150.
158 K. Muramatsu, *J. Vacuum Soc. Japan*, 3 (1960) 341, 416.
159 J. Niebuhr, in F. Benesovky (Ed.), *Hochschmelzende Metalle*, 3rd Plansee Seminar, Reutte/Tyrol, 1958, Springer-Verlag, Vienna, 1959, p. 313.
160 M. W. Mallett and W. R. Hansen, in J. J. Harwood (Ed.), *The Metal Molybdenum*, Am. Soc. Metals, Cleveland, Ohio, 1958, p. 365.
161 E. Lassner, *Tagung "Gase in Metallen"*, Darmstadt, Feb. 1964.
162 R. K. McGeary, J. K. Stanley and T. D. Yensen, *Trans. Am. Soc. Metals*, 42 (1950) 900.
163 M. W. Mallett and C. B. Griffith, *Trans. Am. Soc. Metals*, 46 (1954) 375.
164 Yu. A. Klyachko, T. A. Izmanova and E. M. Chistyakova, *Sb. Tr. Tsentr. Nauchn. Issled. Inst. Chernoi Met.*, 31 (1963) 133.
165 H. A. Sloman, *Metallurgia*, 32 (1945) 223.
166 M. Signora and F. Baldi, *Chim. Ind. (Milan)*, 37 (1955) 794.
167 K. Sachs and M. Odgers, *J. Iron Steel Inst. (London)*, 196 (1960) 406.
168 Y. A. Klyachko and E. M. Chistyakova, *Zavodsk. Lab.*, 27 (1961) 135.
169 G. Timo and U. Lodi, *Metallurg. Ital.*, 55 (1963) 560.
170 H. Zitter and W. Schwarz, *Arch. Eisenhüttenw.*, 35 (1964) 109.
171 F. R. Coe, N. Jenkins and D. H. Parker, *Proc. SAC Conf.*, W. Heffer and Sons Ltd., Cambridge, 1965, p. 516.
172 M. Hanin and E. Jaudon, *Rev. Met.*, 62 (1965) 37.
173 N. N. Timoshenko, T. A. Izmanova and E. M. Chistyakova, *Zavodsk. Lab.*, 31 (1965) 1068.
174 L. C. Pasztor and D. E. Wood, *Talanta*, 13 (1966) 389.
175 H. Severus-Laubenfeld, J. Steiger and R. Stahel, *Z. Anal. Chem.*, 218 (1966) 241.
176 R. Suarez-Acosta, *Tagungsber. 2. Balzers Kundenkolloquium*, Balzers, Liechtenstein, 1966, p. 26.
177 Y. A. Klyachko, L. L. Kunin and E. M. Chistyakova, *Sb. Tr. Tsentr. Nauchn.-Issled. Inst. Chernoi Met.*, 19 (1960) 127.
178 S. M. Gnuchev, *Sb. Tr. Tsentr. Nauchn.-Issled. Inst. Chernoi Met.*, 19 (1960) 132.
179 H. Goto, S. Suzuki, J. Kimura and A. Onuma, *Sci. Rept. Res. Inst. Tohoku Univ., Ser. A*, 11 (1959) 271.
180 T. Kraus, M. G. Frohberg and A. Gerhardt, *Arch. Eisenhüttenw.*, 35 (1964) 39.
181 K. Furuya and H. Kamada, *Bunseki Kagaku*, 14 (1965) 544.
182 P. H. Scholes and E. W. Gill, *Tagungsber. 2. Balzers Kundenkolloquium*, Balzers, Liechten-

stein, *1966*, p. 28.
183 E. M. Savitskii, G. E. Cuprikov and G. G. Glavin, *Zavodsk. Lab.*, *28* (1962) 957.
184 N. Christensen and K. Gjermundsen, *J. Iron Steel Inst.*, *190* (1958) 248.

Chapter VII

Vacuum Sintering

R. KIEFFER, G. PAESOLD AND O. WINKLER

A large number of metals and alloys give rise to difficulties if sintered in a protective atmosphere. These difficulties can be overcome only if the sintering process is carried out under a high vacuum. In this way also substantial improvements in quality may be obtained.

However, the whole sintering process is not always carried out in a vacuum. In many cases a pre-sintering stage under reduced pressure in a reducing atmosphere, such as of hydrogen, precedes the sintering process proper and only the final high-temperature or dense sintering gives rise to the high-quality, dense products which are required as finished parts or as blanks for further fabrication.

Although the vacuum sintering of metals is at present in direct competition with the various methods of high-vacuum melting, the powder metallurgical process has in many cases fully maintained its position. Sintering is usually the easiest way of obtaining a ductile, dense body from those metals which are produced in the form of powder, sponge or fine crystals owing to the method of their preparation, or which pose serious problems to the metallurgist as a result of their extremely high melting points.

Vacuum sintering is now the established process for the production of tantalum, niobium, tungsten, molybdenum and beryllium—to name only the principal metals—as well as for the production of hard metals, cemented carbides and compound metals*. Mention should also be made of the vacuum-sintered, hard-magnetic materials of the type of the Alnico sintered magnets which have maintained their place in electrical engineering along with cast magnets and sintered ferrites as a result of their excellent mechanical and magnetic properties.

1. Phenomena during Sintering

It is outside the scope of this book to discuss in detail current and preferred theories on sintering. For a more thorough study of the subject the critical reviews by Roberts[1], Geach[2], Schwarzkopf[3], Thümmler[4], Schreiner[5] and Kieffer and Benesovsky[6] should be consulted; attention is also drawn to the bibliography by Michaelson[7] as well as to the work of Fischmeister and Exner[8].

The results of research in the field of solid-state physics are reflected in the

---

* (a) Hard metals include refractory hard materials like carbides, nitrides, borides and silicides of transition metals. (b) Cemented carbides refer to cobalt-bonded metallic carbides like WC, TiC, TaC, NbC and HfC, etc.

References pp. 718–719

development of more recent concepts of the sintering mechanism. The first theoretical treatments of the powder metallurgists aroused the interest of chemists, especially of physical chemists. The theories of sintering of J. A. Hedvall, and in particular those of G. F. Hüttig, are therefore characterized by physical–chemical reasoning. Initially, sintering was regarded as a chemical reaction, though with the recognition that surface tension forces play a major role.

Even when, about 1950, the pure physicists took an interest in the problems of sintering, no complete clarification was then arrived at. The results led rather to the recognition that the problems involved are far more complex than was originally thought.

In principle, two classes may be distinguished in the sintering of powder materials:
(1) The sintering of homogeneous systems (final product homogeneous)
(2) The sintering of heterogeneous systems (final product homogeneous or heterogeneous).

1.1 Sintering of homogeneous systems

The sintering of homogeneous systems represents quantitatively the main portion of the production of the sintering industry. Most metal sintering firms and a proportion of the cemented carbide industry belong to this group.

In the sintering of homogeneous powders two overlapping sintering phases may be distinguished. The first phase involves the formation and growth of bridges, *i.e.* of contact regions between adjacent powder particles. The growth of these contact zones in the first stage of the sintering process results in an increased cohesion of the powder particles in the green compact and also in a very substantial increase in the electrical conductivity of the compact. This sudden increase in conductivity takes place before the compact begins to shrink. Only during the second phase of the sintering process do increases occur in the density of the compact; the pore volume is reduced and, under favourable conditions, the pores disappear almost entirely. The driving force in both cases is the change in surface energy (surface tension) of the powder particles. In other words, the energy required for sintering is supplied from the reduction of the large surface areas of the powders or from the replacement of contact zones of high free energy by those of lower energy, such as healed lattice networks and grain boundaries. Calculations showed that surface tension forces are large enough to initiate the sintering process. However, the existence of mechanisms which cause a displacement of atoms is required, *i.e.* a materials transport, leading to a strengthening of the powder compact. The following five mechanisms are possible in the sintering of homogeneous compacts:
(1) Evaporation and condensation
(2) Surface and grain-boundary diffusion
(3) Lattice diffusion

(4) Viscous flow (Newtonian flow)
(5) Plastic flow (Bingham flow).

During the last few years numerous workers have made attempts to determine with the aid of model tests which of the five mechanisms is dominant at a given sintering temperature and which is rate-determining. Typical of this way of working were the model tests by Kuczynski[9]. In these the changes in structure which take place during the sintering of wire or balls on to plane metal surfaces were observed. The increase in the radius of the contact bridges, *i.e.* the growth of the contact areas at a certain sintering temperature with time, was used as a measure of sintering. By comparing the measured with calculated data, descriptions of the sintering mechanism present could be given. According to Kuczynski, lattice diffusion (volume diffusion) is the rate-determining mechanism for the growth of the contact zones in the first phase of sintering. Whether, in the second phase of sintering, which involves shrinkage of the compact, diffusion or flow is rate-determining has still not been firmly established. The evaporation–condensation mechanism contributes apparently only to the rounding-off of the pores, but does not affect the shrinkage process. According to Kieffer and Benesovsky[6] traces of impurities and, following this, eutectic films formed by them at the grain boundaries, as well as the fluid-like high mobility of the surface atoms of the powder particles, play a major role in the various flow mechanisms. These phenomena lead to the conclusion that in practice ideal homogeneous systems exist in very rare cases only. In fact they form the transition to the sintering of heterogeneous systems with or without the presence of a liquid phase described in the following subsection.

1.2 Sintering of heterogeneous systems

The sintering of magnetic materials, hard metals, mixtures as well as solid solutions of hard metals, and hard metals cemented with auxiliary metals is usually a process of sintering of multi-component systems.

Kieffer and his co-workers[10–12] distinguish in principle between four types of sintering according to whether the components during sintering, with or without

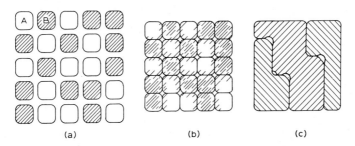

Fig. 1. Formation of a homogeneous sintered body from two completely miscible components.

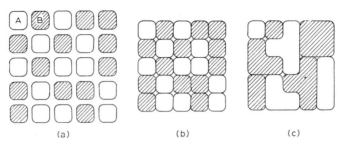

Fig. 2. Formation of a heterogeneous sintered body from two immiscible components.

the presence of a liquid phase, give rise to homogeneous or heterogeneous sintered bodies or alloys.

The processes which from the crystal-structural point of view occur during sintering may be discussed in the terms used by Kieffer and Benesovsky[6] employing a highly simplified model (Figs. 1–4).

I. Complete miscibility of the phases A and B or formation of a homogeneous intermediate phase, which after complete homogenizing in the solid state shows a single-phase structure (see Fig. 1), for instance:

Fe–Ni–Mo (Hastelloy)
Fe–Ni, W–Mo, Nb–Ta, W–25% Re,
Mo–35% Re, V+C → VC, Ti+2B → TiB$_2$,
Mo+2Si → MoSi$_2$.

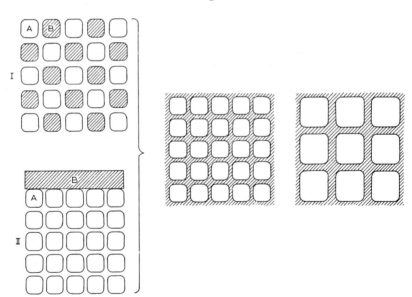

Fig. 3. Formation of a heterogeneous sintered body from two components, one of which is liquid at the temperature of sintering (I = Eutectic sintering; II = Infiltration by superposition). The crystals (white phase) may have more or less pronounced bridges and contact areas.

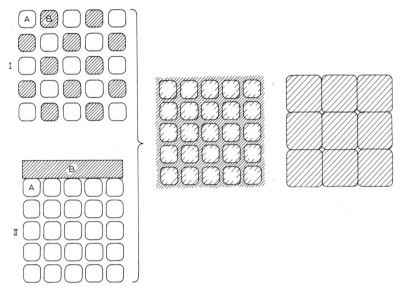

Fig. 4. Formation of a homogeneous sintered body from two components, one of which is liquid at the temperature of sintering (I = Eutectic sintering; II = Infiltration by superposition).

II. Immiscible crystal types A and B give after sintering or pressure sintering in the absence of a liquid phase a sintered body of conspicuously heavy grain growth. (In the case that small mutual solubility exists, the pure crystal types A and B are replaced by the solid-solution crystal types A' and B' (see Fig. 2).) Examples: Fe–Cu, W–Ag, Ni–Ag, Cu–graphite, Al–Al_2O_3, Ni–ThO_2, Mo–ZrO_2, Be–BeO, TiC–ZrB_2, $MoSi_2$–Al_2O_3, cermets.

III. In Fig. 3 the case is depicted in which component B becomes liquid—probably even dissolves a little of component A with increasing temperature—and covers component A with a molten film. Component B may also be introduced by infiltration. Favoured by the liquid phase B grain growth already takes place at relatively low temperatures. Examples: W–Cu, W–Ag, Ta–Cu, W–Ni–Cu, W–Ni–Fe, Mo–Ag, WC–Co, WC–TiC–TaC–Co, TiC–VC–Ni–Fe and so on.

IV. This is the classic case, in which the liquid component B is fully or partly absorbed by crystal type A under formation of a solid solution or of an intermetallic compound (see Fig. 4). Examples: Cu–Sn–bronze, Alnico–magnets, TiC–Mo_2C–Ni–hard metals with less than 3% nickel, Cr (<10%)–Ni–Cu alloys, $3Nb + Sn \rightarrow Nb_3Sn$.

1.3 Fundamentals of the reactions

The fundamentals of the reactions taking place in the solid state have been the subject of an extensive literature[5,13-16], which may be briefly reviewed here.

Diffusion phenomena, such as surface diffusion, lattice diffusion or lattice

References pp. 718–719

self-diffusion, play such a predominant role in the sintering of dispersed substances that "powder metallurgy" could equally well be called "diffusion metallurgy"[13]. Taking into account grain properties, such as grain size, grain shape, surface area, deformation and the conditions of production (pressing and sintering conditions), the following concepts in general terms may be suggested:

The sintering of single-phase systems, such as homogeneous metals without additions, involves, with increasing temperatures, phenomena which are mainly caused by adhesion and exchange mechanisms or crystallization processes in which the course of sintering according to Hüttig[17] is characterized by definite temperature ranges.

How the combination of two metals or of hard metals takes place is shown in a schematic form according to Hüttig[17] in Fig. 5.

If, as depicted in Stage I, the surfaces of two particles, *i.e.* the lattice planes, face each other, the fields of forces from the two lattices will extend into the space between them and will therefore cause a mutual attraction, Stage II. The structure of the left-hand crystal A may be assumed to be more stable than the structure of the crystal B. If the temperature is raised, the more stable lattice A will grow at the expense of the crystal B until finally this crystal has been completely consumed. In Stage III the effect of surface diffusion, which has already begun at a quarter of the absolute melting temperature, is depicted. Atoms diffusing across the surface are caught between the crystals and then fixed. The gap acts, to use an expression coined by Sauerwald[18], like a trap for surface atoms. The result of this diffusion is a mutual growth of both crystals towards each other and the combination to a single unit. Stage IV finally shows the effect of lattice self-diffusion. Self-diffusion within the crystal lattice differs from self-diffusion on the surface in that the migration of the atoms takes place not only across the crystal surface but through the whole cross-section of the crystals. Lattice self-diffusion, in contrast with surface self-diffusion, becomes effective only at higher temperatures, but may,

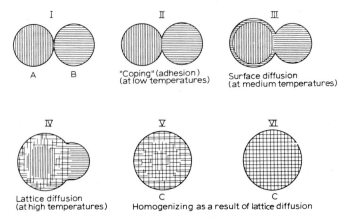

Fig. 5. Unification of two crystals in the solid state, according to Hüttig.

with further increasing temperatures, increase to such an extent that it far exceeds the material transport by surface diffusion. For practical purposes we may assume the diffusion process of crystals A and B as a firm coalescence in the first place, followed by a material transport towards the interior and across the surface resulting in the formation of a new unit C (Stages V, VI) which is larger than A by the addition of the volume of B.

A marked effect on the sintering process and consequently on the quality of finished sintered products is exercised by gases taken up by sorption during production, storage and further processing of the powders. These quantities of sorbed gases may be very considerable, particularly for fine powders owing to their large free surface[19], and they may also lead through chemical reactions to very high gas contents of the final products. In principle, a distinction should be made between purely physical adsorption and the chemisorption, where it is assumed that chemisorption is always preceded by physical adsorption[20]. The heat of adsorption amounts to only one-tenth or one-fifteenth of the heat of reaction in chemical combination. The heat of sorption determines, on the other hand, the energy required for the removal of the sorbed gases. Desorption may be obtained either purely physically by lowering the partial pressure or by raising the temperature (in practice a combination of these is usually employed) or by the initiation of a chemical reaction. It would be expected that desorption, particularly of vapours of high boiling point, such as water or hydrocarbons, would be greatly accelerated by application of reduced pressure, not only by a displacement of the equilibria but also by the improved diffusion in the capillaries of the sintered body. In order to prevent chemical reactions from taking place, the process should be carried out at the lowest possible temperature, *i.e.* the temperature should be raised very slowly indeed. This recommendation should, however, not be generalized, as absorption from gases in the furnace atmosphere may outweigh the desorption from such reactive elements as tantalum, niobium, zirconium or titanium, which act as getters even at very low temperatures. In this case it is necessary to traverse the lower temperature region as quickly as possible[21].

The desorption of combined gases which have a low heat of absorption, such as hydrogen, is effected quite easily, as investigations have shown that almost all metal–hydrogen compounds can be decomposed at 600°C.

The heats of formation of several nitrides, for instance those of iron, chromium, molybdenum and tungsten, are so low that degassing is still possible by dissociation of these compounds at relatively low temperatures; but this does not apply to the metals titanium, zirconium, hafnium, vanadium, niobium and tantalum*.

The removal of oxygen is difficult to achieve even from copper, a relatively noble element, as the dissociation pressure of CuO is only 10^{-4} Torr at 1000°C. A degassing with respect to oxygen is only successful if the sintering temperature

---

* See also Chapter IV, Section 2: "Vacuum degassing in the solid state".

References pp. 718–719

can be raised to very high levels, or when highly volatile oxides exist as in the case of molybdenum, tungsten, tantalum, niobium or rhenium*.

In general, only the application of a chemical reaction achieves the removal of oxygen. If hydrogen is employed as deoxidation agent, the use of a vacuum does not offer any advantages, as the efficiency of the deoxidation process depends solely on the partial pressure of the water vapour. It is therefore better to carry out the reduction by hydrogen at ambient pressure and then to effect the high-temperature sintering in a vacuum. It would be expected that advantages would be obtained, despite the higher costs, in those cases in which a deoxidation with carbon is not possible. The most frequently applied method of deoxidation is that by the chemical reaction with carbon. Here the vacuum treatment offers great advantages. The reaction equilibrium

$$Me_xO + C \rightarrow xMe + CO \uparrow$$

is displaced to the right by the continuous removal of the carbon monoxide in such a manner that: (i) the reduction takes place at lower temperatures, and (ii) lower oxygen contents are at the same time attained at lower carbon levels.

The reactions usually take place at the surface of the individual grains, either by contact of an oxide phase with carbon or by diffusion of the carbon through the grains towards the oxide phase. A great number of further reactions may participate during the deoxidation process; these appear to a greater or lesser extent according to the type of material to be sintered. In this connection the role of carbon monoxide in the sintering of cemented carbides should be mentioned, which finally led to the necessity for the control of the carbon monoxide content of the sintering atmosphere.

Brownlee et al.[22] showed that, depending on the furnace pressure, a loss of carbon or a carbon pick-up of the cemented carbides may result. In high vacuum the reduction of oxide films by the carbon contained in the carbides may lead to a deficiency. If higher furnace pressures prevail and the carbon monoxide partial pressure becomes too high, carbon monoxide acts as a carrier for carbon from the graphite trays and crucible to the tips[21]. According to Brownlee et al., at 1200°C with a partial pressure between 0.15 and 0.3 Torr mild carburizing conditions already exist which become stronger with rising partial pressure. Here it should not be overlooked that when carbon monoxide is formed the pressure inside the crucible which is normally connected with the furnace chamber only via a sight tube may well be much higher than the pressure measured at the pumping port. Not only desorption mechanisms and chemical reactions are active during sintering in a vacuum, but also the evaporation of non-gaseous components. Both the self-purification from metallic impurities and the increased density of vacuum-sintered products are caused to a major extent by processes of this type.

---

* See footnote on p. 679.

Transport of evaporated substance can, however, take place only if a partial pressure gradient exists. This may be created in various ways, for instance, by the presence of a cool surface, or—if only a small temperature difference exists—by absorption of the vapour by this surface, so that its activity, *i.e.* its re-evaporation rate, is small.

A sufficiently large free path at the surface of the sintered body and a large partial pressure gradient are therefore presuppositions for a high evaporation rate of the impurities. The optimum free path is attained as soon as the total pressure falls to less than 0.1 Torr.

If, on the other hand, evaporation should be reduced, the easiest way to obtain this is to shorten the mean free path by admittance of an inert gas. This is usual practice, for instance, in the sintering of cemented carbides above 1000°C in order to prevent the loss of auxiliary metal. (In order to avoid the above-mentioned carbon pick-up care should be taken that the carbon monoxide partial pressure is kept sufficiently low, *e.g.* by changing the inert gas atmosphere from time to time.) Studies of the evaporation rates of molten metal surfaces showed that this rate is markedly lowered only if the total gas pressure exceeds about 1 Torr[23,24*]. This, however, applies only to evaporation from the surface proper. In the interior of a sintered body evaporation is only impaired if the mean free path of the evaporating metal atoms is substantially shorter than their distance from the neighbouring surface. Taking into account the small distances between the individual grains in the sintered body, it would be expected that the pressure of the inert gas could be increased by several orders of magnitude during the sintering process, before there is any noticeable effect on the exchange processes connected with this evaporation, which contribute to a large extent to the densification of sintered structure. In other words, the inert gas pressure is not critical to the sintering process. Of much more significance is the surface quality from the physical and chemical points of view, which depends closely on the production conditions and pretreatment of the starting materials.

2. Production of Sintered Materials

Historically, the high-vacuum sintering of the Group VA metals represented the beginning of a new technical era. W. v. Bolton[25] produced, at the turn of the century, more or less ductile loops of tantalum wire by high-vacuum sintering of an extruded oxide–carbon mixture plasticized by addition of wax. This so-called "reaction sintering" became technologically increasingly successful—in the sense of increasing melting points of the metals—from vanadium through niobium to tantalum[26].

While the high-vacuum sintering process still dominates the production of

---

** See also Chapter I, Section 3.

high-purity, fine-grain tantalum today, the competition of the various high-vacuum melting processes becomes more and more noticeable for niobium and niobium alloys.

Molybdenum and tungsten are sintered preferably under a hydrogen atmosphere, but although the bulk of these metals is still produced by sintering, here too the high-vacuum melting methods slowly gain ground, particularly for the production of molybdenum and molybdenum alloys with titanium and zirconium additions.

Beryllium, which became of major importance with the establishment of nuclear technology, is produced exclusively by pressure sintering or vacuum hot pressing. The dense slabs are then manufactured to semi or final products.

In the cemented carbide industry, which has now existed for a long time, high-vacuum sintering methods were introduced only reluctantly to WC–Co alloys, in contrast with TiC–TaC–(NbC, HfC)-containing WC–Co alloys, which had been vacuum-sintered since 1931. Recently, vacuum sintering began to be employed to an ever-increasing extent also in the case of WC–Co alloys, mainly for economic reasons. The applicability of a vacuum to the sintering of compound materials had long ago been pointed out.

Hydrogen- or vacuum-sintered Alnico magnets, which were of great importance when first introduced, have lost their share in the market because melting processes have been improved, and new oxide magnets (ferrites) have been developed, so that today sintered magnets find application only as small special magnets or as magnets with sintered-on pole pieces or of complex shape[27].

2.1 Tantalum

The process of "direct production of metal semis from oxides" by v. Bolton no longer plays any role in the manufacture of bulk tantalum metal. Even in his own time this process could be successful only for fine tantalum wire with the ideal possibility of self-purification by direct resistance heating. This process has, however, seen a revival on a technological scale for the production of niobium and tantalum raw metal from an oxide–carbide reduction mixture[28,29]*.

The sinter-technological processing of tantalum has frequently been described in the literature. More or less detailed accounts of industrial working methods employed by various firms exist[30–35].

Allen[35] reviewed the effect of the composition of the starting material on the course of the technological sintering process.

The metal powder produced by one of the two main processes—reduction of K_2TaF_6 by sodium or fused-salt electrolysis of Ta_2O_5- and K_2TaF_6-containing alkali metal halides—is compacted to rods in steel dies with the aid of powerful hydraulic presses (1000–3000 tonf). Particular attention should be paid to grain

---

* See also Chapter III, Subsection 3.1.1.

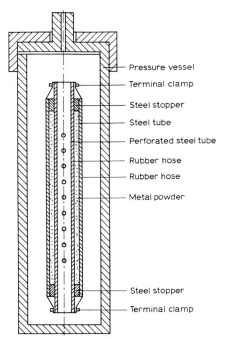

Fig. 6. Hydrostatic press for the cylindrical compacting of green tubular bodies from metal powders, according to Skaupy.

size distribution, as green compacts with a high proportion of fines (fine sponge or powder produced from scrap by hydriding and intensive grinding and milling) tend to swelling and poor degassing owing to too rapid a densification during the initial period of sintering. The production of sintered tubular compacts for the manufacture of tubes or sheet is preferably carried out in hydrostatic presses which were described by Skaupy[36] as far back as 1930 (Fig. 6).

If the direct sintering process is employed it has been found advantageous to presinter large rods, plates (60 mm × 60 mm or 30 mm × 100 mm) or tubes before clamping them into the terminals in the sintering bells in a vacuum at 1000–1700°C. Increasing the compacting pressure would also lead to a higher strength of the green bodies, but might in certain circumstances greatly impair the degassing of large rods.

A typical diagram for the sintering of a pair of flat tantalum rods (32 mm × 16 mm × 350 mm) according to Titterington[32] is shown in Fig. 7. The sintering temperatures as estimated by Kieffer and Braun[26] are plotted at the side of this diagram. The heating up of the tantalum and also of niobium rods is carried out in several stages of heating in order to accomplish the evaporation of impurities in such a way that on the one hand no swelling of the rods occurs owing to too heavy a gas evolution, and on the other hand no sintering to high density takes place before all impurities which are volatile in that temperature range have been evaporated.

References pp. 718–719

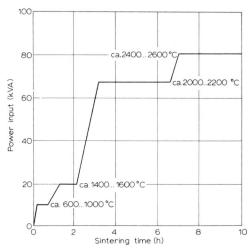

Fig. 7. Sintering diagram of flat tantalum rods, according to Titterington and Simpson.

In the lower temperature stages (between about 400° and 800°C) hydrogen escapes as well as traces of alkali metals which have, for example, remained from a reduction of the powder by sodium metal. At higher temperatures, between about 800° and 1200°C, traces of alkali metal salts and difluorides are given off. At about 1400–1600°C the oxide–carbide reaction begins, which is rather energetic between 2000° and 2200°C. Above 2200°C the lower oxides distil off; above this temperature the nitrides decompose, giving off elementary nitrogen*. At the final sintering temperature of 2500–2700°C substantial quantities of tantalum metal evaporate. The high loss in weight during sintering at high temperature (1–3 %) is therefore made up mainly of metal oxide and metal vapour, to a lesser extent of nitrogen, carbon monoxide, silicon (probably in the form of silicon monoxide) and of traces of the iron group metals.

The self-purification of tantalum rods takes place very easily indeed owing to the high sintering temperatures which may be applied to tantalum and therefore owing to the high vapour pressures and decomposition pressures of the impurities.

If after high-temperature sintering the tantalum rods should not show the required density of more than 92 %, preferably more than 95 %, of the theoretical density and a hardness below 100 HV, it is recommended that the sintered rods be subjected to a working of 10–20 % by pressing or rolling and subsequently sintered again for 2–6 hours at the highest temperature in a vacuum. This "duplex sintering" is also established practice for obtaining a material of extreme softness and ductility[26,37]. During the second sintering operation nitrogen and oxygen are still further removed, but the carbon level is lowered only if sufficient oxygen is

---

* See also Chapter IV, Section 2.

present. Grain growth by recrystallization is often very considerable—it depends on the amount of the intermediate working—and resulting coarse grains remain in the final product. The power consumption for sintering one kg of high-density tantalum metal is around 50 kWh.

The major portion of the available sintered tantalum metal is still produced by the Coolidge process[26,33,38], *i.e.* sintering by direct passage of current, despite the unavoidable losses at the ends of the rods, although there is now a certain trend to produce large-size bodies and tubes by the indirect sintering technique as no waste is then involved.

Increasing sizes of ingot (100–1000 kg) lead logically to the high-vacuum melting processes which today usually consist of a combination of vacuum arc melting and electron-beam melting[26,33,38].

The property of tantalum to form a dense oxide barrier layer on the surface by anodic oxidation—adjustable with respect to thickness—is utilized for the manufacture of high-grade electrical capacitors[26]. For this purpose, cylindrical compacts of pure tantalum powder of a volume of less than 1 cm^3 are produced, which are sintered at temperatures of about 1850–2150°C.

A porous structure is welcome in the sintered bodies in this case so that the internal surface which determines the capacity is large and that subsequent impregnation of the sintered body with a suitable electrolyte is possible. The larger the required capacity per unit volume, the smaller should be the grain size of the tantalum and the lower is then the sintering temperature which is required for obtaining sufficient conduction between the individual grains. As no marked purification effect can be expected at these low temperatures, maintenance of good vacuum conditions during the whole sintering cycle is of vital importance for obtaining a good and uniform quality. The temperature distribution in the sintering zone should for the same reasons be very even.

These requirements are best met by the employment of tantalum strip furnaces with radiation shields of tantalum; these furnaces should be equipped with a high-vacuum pumping system of relatively high pumping speed at low pressures (see Figs. 16–17).

2.2 Niobium

Niobium, closely related to tantalum, as its position in the Periodic Table indicates, has been produced on a commercial scale by sintering since about 1930. Its field of application lies where its excellent chemical resistance, its high melting point and its low density, compared with tantalum, are displayed to full advantage[26]. In nuclear technology, niobium is gaining importance owing to its high-temperature properties and its small capture cross-section for thermal neutrons. In space technology, niobium and niobium alloys are employed for rocket motors after application of an oxidation-resistant coating.

Niobium powder, the starting material for all fabrication processes, was for

References pp. 718–719

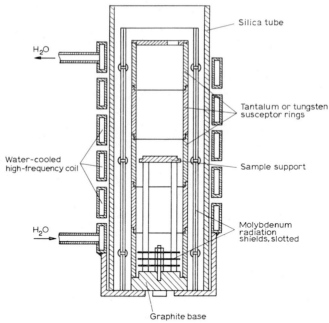

Fig. 8. Sintering furnace for niobium with susceptors made of tantalum or tungsten, according to O'Driscoll and Miller.

a long time produced by high-vacuum reaction sintering of niobium oxide and niobium carbide. Today, however, the alumino-thermic production with subsequent refining of crushed reguli in the electron-beam furnace is more common practice. Niobium powder is then obtained from the ingot by hydriding, crushing, milling and dehydriding of the powder[26].

The manufacture may be carried out by any of the possible sintering processes after compacting the powder into green bodies at a pressure of 7–10 tonf/cm². As the final sintering temperatures are 300–500 °C lower than those for tantalum,

TABLE 1

ANALYSIS OF NIOBIUM, BEFORE AND AFTER VACUUM SINTERING[38]

	Before sintering	After sintering at 2300°C for 45 min
C	0.18%	0.001–0.02%
O	0.71%	0.005–0.018%
N	0.09%	0.002–0.013%
Fe	0.04%	0.02%
Si	0.08%	<0.01%
Ti	0.05%	0.05%
Na	<0.01%	<0.002%
K	<0.01%	<0.002%
Ta	<0.3%	<0.3%

the indirect sintering process is now dominant for niobium, particularly as this process does not involve any waste in rod material (see Fig. 8)[38,39].

The lower degree of self-purification as a result of the lower sintering temperature has a great influence on the course of the sintering process for niobium. The sintering times are about two to four times those for tantalum. The effect of self-purification of large niobium rods from metallic and non-metallic impurities by sintering may be seen from Table 1[38]. This shows that a sintered rod of 99.9% niobium has been obtained from niobium of 98.75% purity. The tantalum content was disregarded in this comparison. An almost pore-free, massive metal may be obtained by an intermediate working operation with up to 20% deformation and subsequent sintering at high sintering temperatures.

High-vacuum sintering has also proved successful for the production of Group VA metal alloys, such as niobium–vanadium, niobium–molybdenum, tantalum–molybdenum and tantalum–tungsten–hafnium alloys[26].

2.3 Tungsten and molybdenum

Only a small portion of the total tungsten and molybdenum production is obtained by the high-vacuum sintering process, the reason being the inertness of these two metals to hydrogen and therefore the simplicity of the design of the protective-gas sintering furnaces.

The metal powders prepared by reduction of the oxides by hydrogen are compacted to rods in hydraulic presses at 2–6 tonf/cm^2 and, in case the green strength is low, presintered at 1100–1200°C in a hydrogen atmosphere. While the tungsten rods are almost exclusively sintered by the Coolidge process owing to the high final sintering temperatures, and only large parts of complex shape, for instance rocket nozzles, are produced by the indirect sintering technique, molybdenum is sintered by all possible processes. The final sintering temperatures are, for tungsten 2500–2800°C and for molybdenum 1700–2200°C[33,34].

For no other metals prepared by sintering methods do the previous history, *i.e.* the grain distribution and grain size, and the properties and method of production of the initial oxides, play such a vital role as they do for tungsten and molybdenum. Particular attention should, therefore, be paid to these characteristics, besides the impurities, in order to obtain a high-quality product.

Lutz and co-workers[40] showed that high-vacuum sintering of molybdenum gave results equally as good as those obtained by deoxidation by carbon, boron or silicon in the form of carbides, borides and silicides. Residual oxygen remaining after indirect sintering of molybdenum may be lowered to less than 10 ppm by volatilisation of molybdenum oxides.

2.4 Beryllium

Beryllium gained industrial importance only through the growth of nuclear technology, although it had been used for a long time in X-ray techniques because

References pp. 718–719

of its specific properties. For instance, beryllium shows a seventeenfold higher transparency to X-rays than does aluminium. The application of beryllium as moderator and reflector in nuclear reactors requires an extreme purity which finally led to a new designation of purity—reactor-pure.

While no particular obstacles to the production and manufacture of beryllium exist, the toxicity of beryllium vapour and dust presents a major difficulty. Very small quantities indeed may be fatal, and extreme safety precautions are therefore required.

The technological preparation[28,41,42] of beryllium may take place by magnesium reduction of purified beryllium fluoride or by electrolysis from beryllium chloride melts containing alkali metal halides.

In the first process the metal is drained off as a liquid from the reaction crucible and refined in vacuum induction furnaces. The metal is not further manufactured by melting and casting techniques, as a metal so prepared shows very low ductility and strength owing to the extremely coarse as-cast structure and to the precipitation of the residual impurities at grain boundaries. The cast ingot is machined on large lathes and the chips produced are ground and milled to powder. The manufacture to semi-finished products may be carried out by the following processes:
(1) Vacuum hot pressing
(2) Hot pressing or pressure sintering in an inert gas or in air
(3) Pressure-free vacuum sintering (without dense sintering)
(4) Cold or hot compacting of green bodies followed by sintering
(5) Cold or hot compacting of green bodies followed by extrusion or hot pressing.

The vacuum hot pressing process developed by Dodds[43] is widely used in technology. Using a pressure of 9–18 kgf/cm^2, a vacuum of about 5×10^{-3} Torr and temperatures of 1050–1100°C, fine-grain, dense pressings of almost theoretical density are obtained. Modern vacuum hot presses (Fig. 9) permit the production of billets of a total weight of up to 1 ton[28]. The conversion into the final products takes place by modern forming or machining operations.

2.5 Hard metals and cemented carbides

In the hard-metals and cemented carbide industry vacuum sintering plants are employed to an increasing extent for the production of multi-component carbide alloys. Whereas cobalt cemented titanium-, niobium-, tantalum-, and hafnium-carbide-containing alloys, which are sensitive to nitrogen and oxygen, have always been preferably sintered in vacuum furnaces[6,44-47], it was thought for a long time that for straight tungsten carbide–cobalt alloys better properties were obtained by sintering in hydrogen[48,49].

More recently[50], vacuum sintering of these alloys also has begun to be predominant, mainly for economic reasons. Moreover, better and more consistent strength values and a better surface quality by subsequent polishing are obtained

Fig. 9. Vacuum hot pressing furnace for producing "QMV" beryllium block up to 165 cm in diameter. "QMV" is a trademark of the Brush Beryllium Company. (Courtesy Brush Beryllium Company)

owing to the reduced final porosity. A prerequisite for obtaining these improvements is, however, the proper working of the vacuum plant with particular reference to any leakage.

Modern cemented carbides consist of fine-grain mixtures of tungsten carbide with 4–25% cobalt to which are added, for instance, 2–50% WC–TiC, WC–TaC, WC–TiC–TaC or WC–TiC–NbC(TaC)–HfC solid solutions[6].

During the wet milling of the components and drying of the wet slurry, part of the cobalt usually oxidizes, so that either a reducing post-treatment or presintering under hydrogen or with a small carbon addition is required. During the final sintering process a eutectic W(Ti, Ta)–C–Co phase is produced which causes the formation of a dense compound body consisting finally of one or more carbide phases and a network of cobalt (see Fig. 10).

For the final sintering at temperatures of 1400–1550°C either resistance or induction furnaces may be employed; the latter have slightly lower operating costs.

The induction-heated furnace was used for the first time by Ballhausen[51] in 1930 for the sintering of cemented carbides under hydrogen or in a vacuum. A graphite crucible served as susceptor (Fig. 11) into which the sintering bodies were placed on shelves.

Particular attention must be paid to the wax addition which is often used as an aid to compacting. This wax must be distilled off before the sintering process proper starts in order to prevent an unwanted pick-up of carbon by the sinterings from the decomposition products of the wax and also to prevent their condensa-

Fig. 10. Microstructure of cemented carbides (×2000).

tion in the vacuum plant. For this distillation an additional vacuum or hydrogen presintering furnace may be employed, but this prolongs the working cycle and also requires the transfer of the presintered parts into the high-temperature sinter-

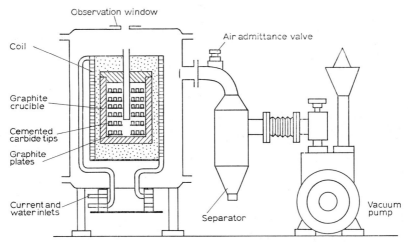

Fig. 11. High-frequency vacuum furnace for the sintering of cemented carbides, according to Ballhausen.

ing plant. There are nowadays plants in which both the dewaxing process and the high-temperature sintering can be carried out in one operation. The evaporating wax is either blown out from the furnace space by an inert gas and burnt at the exhaust nozzle or it is pumped off by a separate pumping system in a low vacuum. This pumping system is connected to the sintering crucible by a heated extraction pipe (see Fig. 34). The wax vapours, therefore, cannot enter the rest of the vacuum plant, so that no reactions with the wax vapours can take place during the subsequent sintering period. Another important advantage of the one-operation dewaxing process is that the sintering of tool tips with the cutting edge pressed to the finished shape is greatly facilitated, so that these tool tips may be produced much more cheaply. The transfer of presintered cemented carbide tool tips with the final cutting edge pressed from a dewaxing and presintering plant to the final sintering furnace can easily lead to a high proportion of rejects because the edges are rather delicate in the presintered state.

Figure 12 shows the typical course of temperature and pressure with time during sintering of cemented carbides. Here in the last stages of the sintering process an argon partial pressure >0.1 Torr was maintained in order to avoid evaporation of the cobalt binder.

Whereas in Fig. 12a the starting material was already dewaxed and cooling after sintering took place in a stationary atmosphere, in Fig. 12b the dewaxing process is included by pumping off the evaporating wax. By lifting the charge in a cooling tower after sintering and raising the pressure (see Subsection 3.4), the cooling down period is shortened so that the total cycle time is about the same in both cases.

Vacuum hot pressing (see Fig. 35) which is usual practice in the cemented carbide industry, for instance, for the production of pore-free drawing dies and

References pp. 718–719

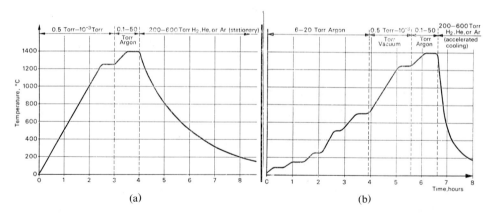

Fig. 12. (a) Typical course of temperature and pressure with time during sintering of a charge of dewaxed cemented carbides in a vacuum induction furnace with a working volume of 26 litres. (b) The same with preceding dewaxing in the same cycle (stepwise rise of temperature). After sintering, the cooling-down time was reduced by recirculation of the cooling gas.

rolls as well as in the experimental preparation of dense hard metals, reduces the sintering times from one-half to one-tenth of the former. But it is considerably more expensive than conventional sintering owing to the large requirements of graphite, electric power and labour.

2.6 Compound metals

The production of compound metals is closely related to the tungsten and cemented carbide industry and is nearly as old as that industry. It was suggested as early as 1922 in a patent[52] to join carbide skeletons by infiltration with a liquid auxiliary metal. Although this technique was not used for industrial application for a long time, it finally gained importance for tungsten/copper and tungsten/silver contact materials in the electrical industry[5] as well as for iron/copper machine tool parts[53,54].

Tungsten/silver infiltration alloys, which show good machinability, play an important role nowadays as substitutes for tungsten in rocket nozzles. Large quantities of these so-called "sweat-cooling alloys" are used for military rockets[55].

According to Kieffer and Benesovsky[56] three principal methods for the infiltration process are possible: (1) infiltration by dipping (capillary infiltration); (2) infiltration by full immersion; (3) infiltration by superposition (see Fig. 13). Combinations of these methods are, of course, also possible.

For obtaining completely pore-free bodies it is advisable to carry out the infiltration process in a vacuum. Since the cleanness of the surface of the sintered body exercises a major influence on the wettability and the surface tension of the infiltration metal, both processes—reducing presintering and infiltration—may be carried out in one and the same furnace without exposing the sintered body to a

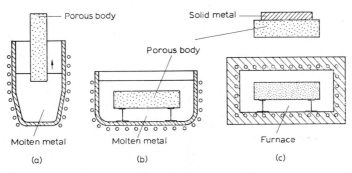

Fig. 13. Types of infiltration: (a) Infiltration by dipping (capillary infiltration); (b) Infiltration by full immersion; (c) Infiltration by superposition, according to Kieffer and Benesovsky.

reactive gas between the two operations. The combination of sintering in a protective atmosphere and infiltration in a vacuum can be carried out without any difficulty.

2.7 Alnico sintered magnets

As has been mentioned earlier, sintered magnets have to a large extent been replaced by improved cast Alnico magnets and by Ba(Sr)–ferrite magnets. However, owing to their good magnetic properties, their excellent grinding characteristics and the advantage of being produced in complex shapes, sintered magnets have held their own for certain applications. A particular advantage is the possibility of providing these sintered magnets with soft magnetic, machinable pole pieces, and of producing combination materials of very close tolerances without subsequent difficult machining operations[27].

In the production of sintered Alnico magnets, mixtures of iron, nickel and cobalt powder, together with a crushed aluminium–iron (titanium, copper, cobalt) master alloy, are compacted and heated in a vacuum to such a temperature that a eutectic melt of the master alloy is formed which combines with the matrix to form a homogeneous solid solution.

Sintered magnets have a finer grain than cast magnets but usually contain a little finely dispersed Al_2O_3 from their previous treatment[27].

2.8 Stainless steel

Recently, more and more components of compacted stainless steel powder are sintered in vacuum[57,58]. In processing these components, first the stearate lubricant used for pressing the green compacts is removed in a separate furnace by heating to 400–500°C. The compacts are placed on graphite trays with an intermediate layer of a stable oxide like alumina and then sintered in vacuum at a pressure of 0.1 Torr at about 1150°C. After cooling down, the parts are coined to the final dimensions and the sintering process repeated. Compared with sintering

References pp. 718–719

under hydrogen or dissociated ammonia atmosphere, corrosion resistance and ductility of stainless steel are improved.

By using heater elements and trays made of graphite, reducing conditions exist during sintering, and a lower pressure which could lead to chromium losses by evaporation is therefore not necessary during heat treatment.

Naeser and Wessel[59] also reported about the application of vacuum sintering as a first step for the fabrication of semi-finished products (bars, tubes) of stainless steel by hot extrusion.

With the development of an economic process for the production of steel powder[60] this process may become rather important, especially because new possibilities exist for influencing the structure and the technological qualities of the finished product by addition of small amounts of non-metallic material and by chemical reactions (decarburization, deoxidation, nitriding) during sintering.

Semi-finished products of chromium–nickel–steel with exceptionally low carbon content between 20 and 90 ppm could be produced by adjusting the carbon additions to the oxygen content of the powder. During sintering at 1200–1250°C at 10^{-1} Torr to 10^{-2} Torr oxygen and carbon are easily removed as carbon monoxide. The still porous ingots were encapsulated and hot-extruded at ambient pressure, giving a material with the expected high corrosion resistance.

3. High-vacuum Sintering Furnaces

The high state of development of the sintering processes and of the sintered materials was to a large extent made possible by the perfecting of the sintering furnaces.

For industrial use three main types of heating of these furnaces have emerged:
(a) Direct resistance heating (Coolidge process).
(b) The numerous variants of the resistance furnace with indirect heating of the sintering material.
(c) The induction furnace with either direct or indirect induction heating.

The general vacuum-technological problems, which are common to all types of sintering furnace, are dealt with in Chapter III. The specific characteristics of the individual type of vacuum sintering furnace will, however, be discussed here.

The type of heating of a furnace depends on the type of material to be sintered, on the size of the bodies to be sintered, on the temperature required and, finally, but not of least importance, on the economy of the process.

3.1 Direct-resistance heating

This method of heating has been used for a long time in protective-atmosphere furnaces for the production of tungsten and molybdenum bars. The application of this method to vacuum sintering involved some difficulty at the beginning because of the flexible, vacuum-tight lead-ins for the current, but this difficulty has for quite a time been overcome.

This so-called Coolidge process is restricted to rods or slabs of a cross-section that is constant over the entire length and to materials which have an adequate electrical conductivity. Since the Joule heat is generated in the compacted material itself, the Coolidge technique is particularly suited for the production of very high-melting-point metals.

A particular advantage of this technique is the simplicity of the sintering bells used and consequently their reliability and availability. Proved installations for the

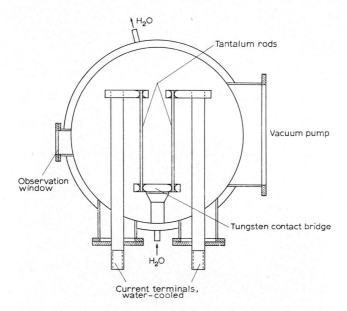

Fig. 14. Principle of the lay-out of a two-rod high-vacuum sintering bell. (Courtesy Metallwerk Plansee)[26]

simultaneous sintering of two rods[26] are depicted in Figs. 14 and 15. Any difficulties in supporting the rods arising from the use of high temperatures are almost completely absent. The ends of the rods are clamped into the current terminals and no other supports are necessary. These current terminals—usually one fixed and the other movable—are water-cooled to prevent welding to the rod. This, however, is also the major shortcoming of this technique, as the end pieces of the rods do not attain the required sintering temperature and are therefore not properly

References pp. 718–719

Fig. 15. Large sintering bells for tantalum, front view and side view. (Courtesy Metallwerk Plansee)[26]

sintered. The end pieces must therefore be removed before any working operation can be carried out on the rod or slab. The end pieces usually amount to about 1/5–1/8 of the length of the rod.

Another problem is the uneven temperature distribution over the cross-section of the rod despite constant distribution of the energy, as a temperature gradient is created from the centre of the rod to its surface by the high radiation loss and the consequent high heat flow. This effect may become so pronounced that the density of the material varies over the cross-section; it also may cause a marked difference in the distribution of the impurity elements over the cross-section.

From the economic point of view direct-resistance heating is a particularly advantageous technique as the application of high inputs at mains frequencies makes possible high heating rates and because temperatures very close to the melting points of the sintered materials may be attained.

Equally important is the rapid cooling of the rod which, at temperatures below 1000°C when the radiation falls to rather low values, occurs mainly through the extraction of heat through the water-cooled clamping jaws. Metals which are particularly reactive, such as tantalum or niobium, should be removed from the furnace only when completely cold.

Fig. 16. Tantalum heater assembly installed in a sintering furnace with special charge holder and radiation shield assembly. The upper radiation shields are withdrawn. (Courtesy Balzers A.G.)

References pp. 718–719

Fig. 17. 3-Phase heater unit made from tungsten or tantalum. Working space: ϕ 120 × 250 mm. Temperature maximum: 2400°C. (Courtesy Balzers A.G.)

The largest installations for the sintering by direct-resistance heating accommodate rods of a length of about 1 m and a cross-section of about 50 mm × 50 mm or 35 mm × 120 mm, respectively. A restriction in the size of the sintering rods is imposed only by the problems of core superheating[26].

3.2 Indirect-resistance heating

In this type of furnace design, charge and heat source are separate and heating

Fig. 18. High-vacuum furnace for sintering (brazing, annealing and heat treatment) resistance-heated. Furnace chamber open, with tantalum heater. Working space: ϕ 50, 120 mm. Temperature maximum: 2400°C. (Courtesy Balzers A.G.)

takes place by radiation. Type, cross-section and shape of the sintered bodies play a minor role in this case, and materials without electrical conductivity can be sintered equally as well as the metals. The requirements placed on the resistance material for the heater are that it should be suitable for the sintering temperature employed and for the conditions of the vacuum and that it should not lead to contamination of the sintered material. The working temperatures of most high-temperature resistance materials, with the exception of graphite and tantalum, are restricted by their tendency to recrystallization and embrittlement which lead sooner or later to failure.

Problems in supporting and insulation presented the designers with difficulties. The design was therefore changed particularly for high-temperature vacuum sintering furnaces from the cage-type heater (Figs. 21 and 22) to the self-supporting three-phase or single-phase strip furnaces (Figs. 16, 17, 18, 19 and 23).

References pp. 718–719

Fig. 19. 1-Phase tantalum heater with radiation shield assembly. Outer casing and current lead-ins water-cooled. Working space: ϕ 50, 120 mm. Temperature maximum: 2400°C. (Courtesy Balzers A.G.)

A design suitable for particularly high working temperatures is that employed by Sylvania Electric Products, which is shown in Fig. 20. It is a mains-frequency resistance furnace, the heating element of which consists of a wire mesh of doped tungsten. The elasticity and mobility of the individual wires ensure a long service life despite the unavoidable recrystallization and embrittlement caused by the high working temperatures. Even fracture of individual wires has no adverse effect on current and temperature distribution. Only at extreme high temperatures of about 3000°C is the service life restricted owing to deformation by electrodynamic

TABLE 2

UPPER LIMITING TEMPERATURE OF HEATING ELEMENT DURING PERMANENT OPERATION IN VACUUM

Heating element	Upper limiting temperature (°C)
Chromium–nickel (80/20)	~1000
Kanthal A	~1100
Molybdenum	1700
Tantalum	2500
Tungsten	2800
Graphite	<2500

Fig. 20. Tungsten mesh resistance-heated high-vacuum furnace. (Courtesy Degussa)

forces and by evaporation.

The vapour pressure of the resistance-heater material at the ultimate working pressure plays an important role also. For instance, the vapour pressure of graphite above 2200°C may be detrimental for the charge and can also lead above 2500°C to a rapid change in the cross-section of the conductors, particularly at hot spots. (This evaporation can only be suppressed by the use of an inert gas.)

The charge, on the other hand, can damage or even destroy the resistance heater by evaporating off constituents or reactive gases and therefore affect the furnace as an economic proposition. As an example of this phenomenon, carbide formation may be mentioned, which for tantalum heaters starts at temperatures even as low as 1000°C and is for tungsten heaters critical at 1400°C. Proven heating element materials and their limiting temperatures are listed in Table 2.

References pp. 718–719

3.2.1 Sintering furnaces with metallic heating elements

If only small heated volumes are required, as, for instance, in the laboratory, the tube furnace is adequate for a wide range of temperatures. The tube, either bent from sheet or seamless drawn and made of molybdenum, tantalum or tungsten, is clamped at its ends in water-cooled terminals, one of which should be able to absorb the thermal expansion of the tube.

For industrial production purposes, *i.e.* when large heated volumes are required and when metallic heating elements must be used, cage-type heating elements which consist of one or of several parallel wires are more suitable than tube resistors for temperatures of up to 1700°C. Cage-type heating elements are less liable to fracture and may also be operated with a considerably higher nominal potential (up to about 60 V) because of their higher resistance[61-63].

The heater forms a cylindrical heating space. Its stability with respect to its cylindrical form is maintained by several wire rings around its circumference. The wire rings carry insulating tubes of pure Al_2O_3 on to which the heating-element wire is bound by a wire of the same material as that of the heating element. In order to ensure a good contact at the terminals, these are reinforced by a large number of parallel wires. Figures 21 and 22 show such a heating element for three-phase delta connection with their radiation shields in both the dismounted and the assembled state, as used in furnaces for sintered magnets.

The heating element is suspended freely on the three terminals and can therefore expand in the downward direction.

In furnaces with horizontally arranged cage-type heating elements (the heating zone in this case usually has a rectangular cross-section) the support of the heating

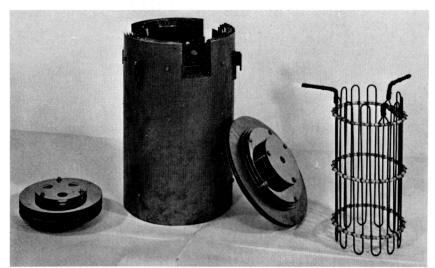

Fig. 21. Molybdenum wire heater assembly for resistance heating. Working space: ϕ 120×250 mm. Temperature maximum 1700°C. Radiation shield assembly consisting of: lid, base and cylinder. (Courtesy Balzers A.G.)

Fig. 22. High-vacuum furnace for sintering (brazing, annealing and heat treatment) resistance-heated. Pneumatically operated lid swung aside. Molybdenum wire heater assembly. Working space: ϕ 380×500 mm. Temperature maximum: 1700°C. With current lead-ins, heater and radiation shield assembly. A water-cooled copper cylinder surrounds the radiation cylinder for rapid cooling by circulation of gas. Front left and bottom right, the thermocouples for temperature measurement are shown in the working position. (Courtesy Balzers A.G.)

element as well as that of the charge is usually rather intricate. If the shape of the charge and the charging conditions permit, the vertical arrangement of the heating elements is therefore to be preferred.

If very high temperatures are to be attained with metallic heating elements, as, for instance, during the sintering of tantalum capacitors, strip-like heating elements are employed, and for still higher temperatures the considerably more expensive tungsten wire mesh heating elements (Fig. 20). Strip-like heating elements may be constructed for one-, two-, or three-phase current just as can cage-like heating elements. Figure 17 depicts a three-phase arrangement in star connection as an example. This arrangement requires for the same nominal rating and the same thickness of the sheet a lower heating current intensity than does a strip-type element, as shown in Fig. 23, particularly if this element is designed for three-phase operation. In order to keep the electrical load of the terminals within reasonable limits, sheet of thinner gauge must be used, which, of course, adversely affects both the stability of the shape of the sheet and its service life. With larger dimensions and high sintering temperatures a gradual deformation or warping of the heating element and the inner radiation shields takes place and may finally cause

References pp. 718–719

short-circuits. If too thin a gauge is used for the sheet, the heating element usually melts away, but if a thicker gauge is employed, in most cases only a hole is burnt through the sheet so that it may be used further even without interruption.

An accurate calculation of the power input for obtaining a certain temperature in relation to the number and type of radiation shields and to the geometry of the furnace is rather difficult. The furnace manufacturers base their calculations as a rule on experience. Also, the required power input does not usually remain constant during the life of the radiation shields.

If the thermal conduction in the direction of the heat flow is negligibly small, the required power input for maintaining a certain temperature follows quite accurately the Stefan–Boltzmann law, *i.e.* it is proportional to T^4, where T is the absolute temperature of the sintering zone. For furnaces of large heating zone diameter at which the surface area of the outer radiation shields is not materially larger than that of the inner radiation shields, this power input is, as a first approximation, proportional to the surface area of the heating space and inversely proportional to the quantity $n+1$, where n is the number of radiation shields arranged one behind the other.

Since the emission coefficient, e, for metallic materials is always lower than unity, somewhat higher power inputs are obtained in a calculation in which $e = 1$ is used than heating practice really requires. It should, however, be taken into account that the emission coefficient gradually increases with the occurrence of recrystallization and the consequent roughening of the surface of the radiation shields as well as with possible deposition from vapours. This is usually one of the main reasons for the increase in the power required. Another reason is in many cases that not all radiation shields are manufactured from refractory metals, *i.e.* molybdenum, tungsten or tantalum, in order to reduce the costs of these shields. Frequently those radiation shields the maximum temperature of which will remain below 1000°C are made of stainless steel which, as a result of its high chromium content, has the property of combining with any oxygen that might be present in the residual gas atmosphere, and this surface oxidation leads very quickly to an increase in the emissivity coefficient, up to nearly unity, and consequently to higher radiation losses. This process may be prevented by the use of nickel or iron sheets, but their lower high-temperature strength must be tolerated.

3.2.2 Sintering furnaces with graphite or carbon heating elements

The pick-up of gas from the atmosphere by heating elements and insulation materials made of graphite or carbon is extremely large because of the porous structure and the large surface area which is even greater for carbon than for graphite. Consequently, the quantities of gas given off during heating are also very large. These materials will therefore be employed instead of metallic materials only if the requirements imposed on the quality of the vacuum are not too strict, and if the presence of carbon is not detrimental to the charge. However, heating

Fig. 23. Resistance-heated insert surrounded by a radiation screen assembly and tungsten or tantalum Ω heating element for temperature max. of 2400°C. (Courtesy Balzers A.G.)

elements and heat insulation units made of graphite or carbon have the advantage over metallic ones that their manufacture is simple and relatively inexpensive and that they are, moreover, mechanically more stable and keep their shape.

The heating element may be constructed like a cage from rods with links of the same material. Large area heating elements woven from graphite filaments may also be used, if eventual oxidation can be avoided. For the working spaces necessary, for instance, in the vacuum sintering of cemented carbides (10–50 l), tube-like heating elements give a more compact, space-saving construction. In most cases slotted graphite tube heating elements of a wall thickness of 10–20 mm with three-phase star connection are used (see Fig. 24).

Because of the heat extraction by the water-cooled current terminals only 2/3 to 3/4 of the total length of the heating tube according to the required homogeneity in temperature of the sintering zone is available as useful working space. In the sintering of cemented carbides a uniformity in temperature of ± 10 °C, and in special cases of ± 5 °C, is required at a working temperature of about 1400°C. To comply with these requirements, a sufficiently thick heat-insulating layer is provided at the upper and lower ends of the heating zone. Additionally,

References pp. 718–719

Fig. 24. Three-phase graphite-tube resistance furnace. (Courtesy Balzers A.G.)

the wall section of the upper end of the heating element may be reduced in order to change its specific-power consumption, or the section size of the insulating layer over the length of the useful heating space varied accordingly.

Previously, graphite granules were preferred as material for insulation. The granules were filled into the space formed by two concentric tubes surrounding the heating element. The heat insulation at the top and bottom was provided by inserts consisting of thin-walled graphite containers filled with graphite granules. Nowadays graphite or carbon felt is more and more used for this purpose. This type of insulation requires less space, keeps its shape and may easily be adapted to the required insulation factor. It is also more easily assembled and removed.

The upper temperature limit of sintering furnaces with heating elements of graphite is around 2200°C for long-time operation, as above this temperature carbon begins to evaporate to a marked degree. If, however, frequent change of the heating elements is permissible, operating temperatures of up to 2500°C are possible. Special furnaces of this type were built, for instance, for sintering 50-kg tungsten ingots; a slight surface carburization of the ingot can, however, hardly be avoided in this case.

In the calculation of the holding power requirement for graphite or carbon tube furnaces using a heat insulation of graphite granules or graphite or carbon felt, the thermal conduction in the direction of the heat flow must not be neglected, in contrast with furnaces equipped with metallic radiation shields. In most cases the required holding power input follows the Stefan–Boltzmann law only above about 1000°C, and then more closely the higher the temperature is.

3.2.3 Retort type furnaces

Vacuum retort furnaces are employed in sintering technology only if the maximum temperature required does not exceed about 1100°C. The charge is contained in a cylindrical retort made of heat-resistant steel or a nickel alloy. A bell-type furnace is then placed over the retort. The pumping system is connected to the cold open end of the retort.

The retort must be pressure-relieved for working temperatures above about 950°C. The bell-type furnace is in this case sealed gas-tight with the retort and evacuated by a separate roughing pump.

The schematic design of such furnaces is depicted in Figs. 25 and 26. These furnaces are employed in sintering technology, *e.g.* for dewaxing, *i.e.* for the

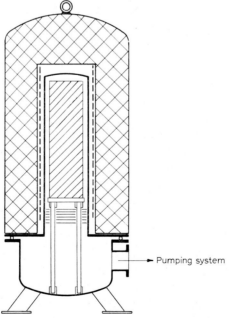

Fig. 25. Vacuum retort furnace.

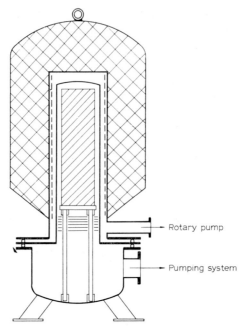

Fig. 26. Vacuum retort furnace, pressure-released.

removal of the compacting lubricant additions from green compacts, and for the subsequent presintering, for instance, in the manufacture of cemented carbides. A presintering operation of the carbide tips is always necessary if shaped parts, such as stamping or drawing dies or complex wear-resistant parts, are manufactured, ed, as these can only be brought into the required shape by a working process in the presintered state.

3.3 Induction furnaces

The induction furnace has gained more and more in importance in recent times. The simple construction of the furnace and its economy are attractive and contribute to its increasing application in sintering technology.

Similarly to resistance heating, direct and indirect heating of the compacts are also possible in induction heating. In direct heating only the compacts to be sintered absorb energy, and the length of the heating zone corresponds to the length of the compact.

Whereas in direct resistance heating there is the risk of melting the core, in induction heating there is a possibility of a premature formation of a densely sintered surface layer which may prevent the degassing of the interior of the compact. This occurrence is inherent in induction heating as the current density is greatest in the surface zone; it is based on the so-called skin effect. This risk of premature surface sintering may be avoided by proper selection of the frequency employed and by a slow increase in temperature. Conditions for the application

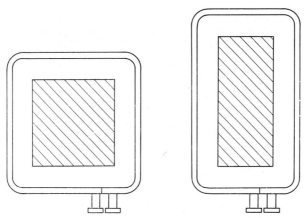

Fig. 27. Induction coil for rods of square and oblong cross-sections.

of this method are a uniform cross-section over the total length of the compact, so that a uniform power absorption is guaranteed, and a diameter of the rod that is at least twice as great as the depth of penetration of the induced current.

The view that induction heating is restricted to cylindrical compacts no longer applies. By use of a suitable shape of the induction coils (Fig. 27) and by selection of a suitable frequency, overheating of the edges of the compacts as a result of too high a power concentration may be prevented.

The higher radiation loss at the ends of the ingot may be counteracted by end-compensated coils (Fig. 28). The energy density transmitted to the ends of the compact may be increased either by tighter electrical coupling or by a higher density of the winding.

For indirect induction heating a closed, electrically conducting cylinder which surrounds the compacts to be sintered is required. This cylinder is called the susceptor and such materials as graphite in the form of crucibles (Fig. 29) or

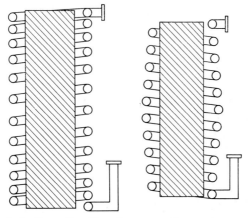

Fig. 28. Induction coils with end compensation, various approaches.

References pp. 718–719

Fig. 29. Medium-frequency high-vacuum sintering furnace with graphite susceptor. (Courtesy Balzers A.G.)

tantalum or tungsten in the form of rings (Fig. 8) may be used. Susceptors have certain advantages with respect to energy absorption, but require nevertheless a higher power input because of their larger emitting surface area. The section size of the wall should not be much smaller than the depth of penetration of the inducing current, so that on the one hand adequate energy absorption of the susceptor is ensured[35] and on the other hand the residual field in the heating space remains low so that no additional energy absorption and overheating of the parts to be sintered occur.

The cylindrical heating space must be closed at the top and the bottom by sufficiently thick closing plates which are also induction-heated and which therefore contribute to the creation of a homogeneous temperature field.

In most cases the true pressure conditions in the sintering zone are given far too little attention. Since the vacuum gauge head can only be attached to the vacuum vessel, no direct indication is obtained on the conditions prevailing in the sintering zone. A centre chimney or a ring-type slot in the crucible lid must therefore be provided for lowering the conductance to the outer vacuum space for the gases liberated so that they may be pumped off.

3.4 Vacuum sintering installations for the production of cemented carbides

It cannot be claimed that induction furnaces predominate over the resistance type in the cemented carbide industry, despite the fact that the induction furnaces are economically superior. Although the capital outlay is high owing to the relatively expensive medium-frequency generators, this is soon balanced by the considerably lower operating costs. Modern generators show efficiencies as high as 75–80%. Further, if the restriction of the heating zone to the effective length of the charge to be sintered is taken into consideration, as well as the absence of heat losses through water-cooled current terminals and the much lower radiation losses because of the relatively smaller diameter of the heating zone, with equal thermal insulation a substantially lower power input is required for induction furnaces and particularly so the higher the temperatures are. Another point in favour of the induction furnace is that the quantity of graphite to be heated is much smaller and so consequently are the quantities of gas liberated.

Experience shows that for operating temperatures above 1700°C and using graphite tube heating elements, even the capital outlay for induction furnace installations is lower than for resistance-type furnaces. Another advantage is in the operating of the furnace at temperatures above 2000°C, when the evaporation of carbon becomes noticeable, as a minor change in the cross-section of the susceptor has no marked effect on power input and temperature distribution.

Figure 30 shows the set-up of an induction coil of a sintering furnace for a working volume of 50 litres. The coil is arranged inside a double-walled, water-cooled vacuum vessel, the lid of which is turned sideways. At the left-hand side

Fig. 30. High-vacuum furnace for sintering (brazing, annealing and heat treatment) induction-heated. With the lid swung back, with induction coil, but no crucible. Part of the support of the insert can be seen under the coil. On the left over the pumping station port is the water-cooled screen. (Courtesy Balzers A.G.)

References pp. 718–719

the connecting duct for the high-vacuum pumping system may be seen. The furnace coil is coated with a thin layer of an electrically insulating material on the individual windings, so that no electrically conducting bridges can be formed between the windings when the crucible and the thermal insulation are installed. This coating may be made of plastics or of ceramics. Figure 29 shows the furnace ready for operation. The upper insulating layer has been removed so that the uppermost layer of the carbide tips can be seen.

The length of time required for cooling down the charge in the furnace is of particular importance for the economy of any vacuum heating or annealing plant. In most cases no high requirements are placed on the homogeneity of the temperature distribution in the heating zone during this period so that a cooling rate as high as possible is no disadvantage.

The sintered parts should in many cases be exposed to the atmosphere only

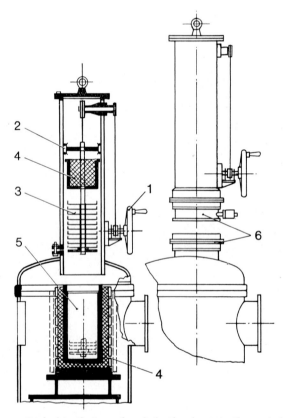

Fig. 31. Cooling tower attached to the top of an induction heat-treating and sintering furnace.
1 Lifting device
2 Side guides for the charge carrier
3 Charge carrier
4 Heat insulation
5 Working space
6 Two gate valves

at temperatures below 150°C in order to prevent staining or discoloration. The cooling period in the vacuum below 1000°C, when the heat extraction by radiation becomes very low, therefore determines the cycle time for a charge.

There are three ways of shortening this period:

(1) Transfer of the charge from the heating space into a cooling zone in which the charge may radiate towards cold walls. If this cooling zone is arranged above the vacuum vessel (see Fig. 31) and if it can be cut off by valves from the furnace space, there is the possibility of transferring the charge immediately after completion of the sintering operation and of introducing a new charge, which is waiting in a second pre-evacuated cooling tower, into the furnace. The time required for cooling does not in this case need to be considered in the total time for a working cycle. The working space remains all the time under vacuum; this

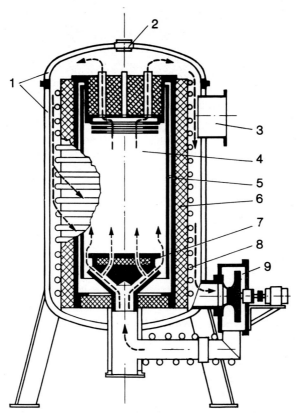

Fig. 32. Gas-circulation cooling in a resistance heat-treating furnace, with graphite heating element.

1 Water-cooled furnace shell
2 Observation window
3 Port for vacuum pumping system
4 Working space
5 Heating element
6 Heat insulation
7 Charge carrier
8 Internal cooler
9 Fan

References pp. 718–719

Fig. 33. High-vacuum furnace for sintering (brazing, annealing and heat treatment) resistance-heated. Rear side, giving a better view of the gas circulation fan for rapid cooling and pneumatically-operated lid-lifting mechanism. (Courtesy Balzers A.G.)

means that cooling down the whole furnace is not necessary, and the heating-up time of the next charge will also be shorter.

(2) Removal of the heat insulation which surrounds the crucible. This method is rarely applied.

(3) Admittance of a protective gas of high thermal conductivity, such as helium or hydrogen, so that an additional heat transport by thermal conduction and convection becomes feasible at a temperature below 1000°C. The cooling rate may be still further increased by recirculating the gas with the aid of a blower. An arrangement of this type is shown in Figs. 32 and 33 for a resistance furnace, but it may equally well be employed in induction furnaces.

Some brief indications of the possibility of designing a continuous vacuum sintering plant or at least an automatic process cycle may now follow. Plants have been developed to technological reliability which permits a semi-continuous cycle (dewaxing, presintering, high-temperature sintering, gas admittance for cooling of the charge, and opening to the atmosphere) by a system of control instruments and devices over a time interval of 72 hours.

The difficulties in continuous sintering furnaces lie equally in the design and construction of a reliable lock system and in the materials selection for the transport of the charge. Existing designs are limited so far to semi-continuous cycles,

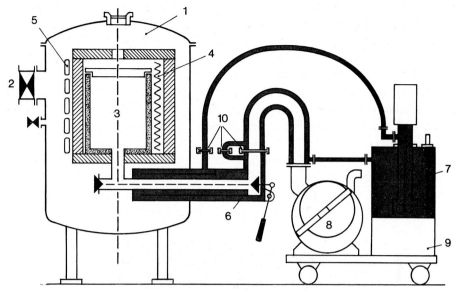

Fig. 34. Dewaxing installation for hard-metal plants. (Courtesy Balzers A.G.)
1 Vacuum vessel
2 Port for vacuum pumping system
3 Working space
4 Resistance-heated arrangement
5 Induction-heated arrangement
6 Heated dewaxing duct
7 Heater for oil with oil circulation pump
8 Dewaxing pump
9 Movable part
10 Quick couplings

i.e. while one charge is being sintered in the heating zone, the second is cooling in a cooling tower and the third is withdrawn through a lock system and the carrier again loaded with a new charge. The use of such equipment is, however, only justified if working round the clock is intended and if sufficiently large quantities of similar materials, *i.e.* with the same temperature cycle, are to be processed.

Beyond the stage of the semi-continuous furnaces no installations seem to have been developed which are in industrial operation, and it might well be that no genuine demand exists for such installations as the same object, *viz.* the saving in manpower, reproducible working conditions and continuous, automatic process control probably equally economically and with a lower capital investment, can be achieved with several furnaces for individual charges. There is also the advantage that the risk of breakdown is much smaller with these smaller units.

Earlier it has been shown how the cooling time and probably even the heating-up time may be shortened in such furnaces. Of equal importance is also the incorporation of the dewaxing process in the process cycle, at least for such products as tool tips which may be pressed into their final shape. Figure 34 shows the diagram of a sintering furnace in which the wax vapour may be pumped off by a separate backing pump, either through a condenser or directly through a heated gas duct which is connected to the sintering crucible, so that no wax vapour can penetrate in the vacuum space outside the heating zone. As soon as the charge is

References pp. 718–719

ready for sintering, this duct is closed and the large pumping system switched on.

Reliable and constant temperature control of the charge is also required for an automatic process programme. Nowadays there are sheaths for thermocouples available, which are sufficiently impervious and resistant even against carbon-containing gases so that contamination of the Pt/Pt–Rh thermocouples can be avoided and their accuracy maintained over longer periods (1000 hours). As thus the conditions for a close adherence to a predetermined temperature programme are secured, the only manual operation involved is that of charging and emptying the sintering plants.

Just as with the type of heating, no unified opinion exists among the various cemented carbide producers on the choice of the pumping system. There are installations which are equipped with a high-vacuum pumping system, consisting of an oil-diffusion pump (1500–5000 l/s) with the necessary backing pump, or with a

Fig. 35. Laboratory-type vacuum pressure sintering furnace. (Courtesy Degussa)

rotary pumping system, which consists usually of a Roots-type vacuum blower (500–2000 m^3/h) with an oil rotary pump (100–200 m^3/h) as the backing pump.

If a diffusion pump is employed, it is usually in operation only during the heating-up period up to the presintering stage, as the quantity of gas liberated at high temperatures is too high. During the high-temperature sintering, when the quantities of gas released have become less, an argon partial pressure of several Torr is often admitted in order to suppress the evaporation of cobalt. The advantage of the employment of a high-vacuum pump is that if a low pressure can be attained in the plant this indicates that the plant is leak-free. (This task of indicating leaks

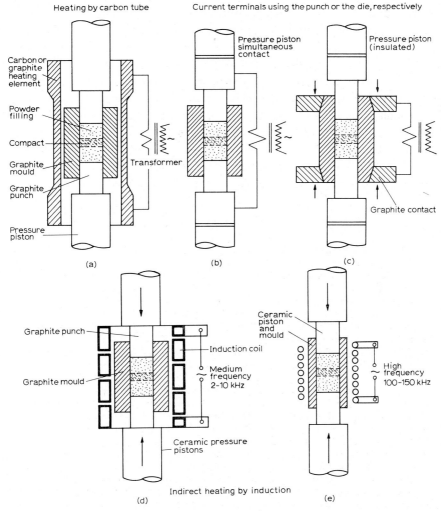

Fig. 36. Principles of pressure sintering installations, according to Ballhausen: (a), heating by carbon tube; (b) and (c), current terminals using the punch or the die, respectively; (d) and (e), indirect heating by induction.

References pp. 718–719

is even more reliably carried out by a mass spectrometer, which, however, is not less expensive.) By this indication the high vacuum diffusion pump becomes a means of process control. Whether or not the quality of the sintered products is improved by the better vacuum conditions during the heating-up stage has so far not convincingly been established. Several producers believe that TaC- and TiC-bearing grades of superior quality could be achieved.

3.5 Pressure sintering furnaces

Vacuum-pressure sintering furnaces which are very special installations should also be mentioned here. The material in such equipment is subjected to the simultaneous application of high temperature, mechanical pressure and vacuum and makes possible the manufacture of sintered products of a density which is surpassed only by casting. The powder-like material is filled into suitable moulds, usually made of graphite, and subjected to a mechanical pressure of 70–150 kgf/mm^2 at temperatures of up to 2500°C in a vacuum. The mould is heated either directly by passing an electric current through it or indirectly by induction (Figs. 35 and 36). Such installations are mainly used for the dense sintering of gas-sensitive high-temperature materials, such as carbides, nitrides, borides and the like. The densities attained are very high and are usually greater than 95% of the theoretical density combined with the advantage of a high purity. The weight loss of the graphite resistors and graphite moulds is extremely low in vacuum operation.

REFERENCES

1. J. P. ROBERTS, *Metallurgia*, 42 (1950) 123.
2. G. A. GEACH, *Progr. Met. Phys.*, 4 (1953) 174.
3. P. SCHWARZKOPF, *Powder Metallurgy*, Macmillan, New York, 1947; *Powder Met. Bull.*, 5 (1950) 4.
4. F. THÜMMLER, *Fortschritte der Pulvermetallurgie*, Akademie-Verlag, Berlin, 1963.
5. H. SCHREINER, *Kontaktwerkstoffe*, Springer-Verlag, Berlin, 1964.
6. R. KIEFFER AND F. BENESOVSKY, (a) *Hartstoffe*, Springer-Verlag, Berlin, 1963; (b) *Hartmetalle*, Springer-Verlag, Berlin, 1965.
7. H. B. MICHAELSON, *The Theories of the Sintering Process: A Guide to the Literature (1931–1951)*, Sylvania Electric Products, Bayside, N.Y., 1951.
8. H. FISCHMEISTER AND E. EXNER, *Metall*, 18 (1964); 19 (1965) 113, 941.
9. G. C. KUCZYNSKI, *Trans. AIME*, 185 (1949) 169; *Acta Met.*, 4 (1956) 58; 8 (1960) 209.
10. R. KIEFFER, in K. WANKE (Ed.), *Einführung in die Pulvermetallurgie*, Stiasny, Graz, 1949.
11. R. KIEFFER, in *The Physics of Powder Metallurgy*, McGraw-Hill, New York, 1951.
12. R. KIEFFER AND F. BENESOVSKY, *Berg-Hüttenmänn. Monatsh.*, 94 (1949) 284.
13. W. SEITH, *Diffusion in Metallen*, Springer-Verlag, Berlin, 1939.
14. W. E. KINGSTON, *The Physics of Powder Metallurgy*, McGraw-Hill, New York, 1951.
15. C. G. GOETZEL, *Treatise on Powder Metallurgy*, Vol. II., Interscience, New York, 1950.
16. F. THÜMMLER, *Fortschritte der Pulvermetallurgie*, Akademie-Verlag, Berlin, 1963.
17. G. F. HÜTTIG, *Kolloid-Z.*, 97 (1941) 227, 281; 98 (1942) 6,263; 99 (1942) 262.
18. F. SAUERWALD, *Z. Anorg. Allgem. Chem.*, 122 (1922) 277; *Kolloid-Z.*, 104 (1943) 144.
19. H. A. SLOMAN, *Symp. Powder Metallurgy, London, 1954*, Iron Steel Inst. (London) Spec. Rept. No. 58, 1956, p. 44.
20. A. WHEELER, *Structure and Properties of Solid Surfaces*, Chicago University Press, 1952.

21 O. Winkler, *Powder Met.*, *1/2* (1958) 114.
22 L. D. Brownlee, R. Edwards and T. Raine, *Symp. Powder Metallurgy, London, 1954*, Iron Steel Inst. (London) Spec. Rept. No. 58, 1956, p. 143.
23 L. Schumann-Horn, A. Mager and W. Deisinger, *Z. Metallk.*, *47* (1956) 145.
24 B. Ilschner and J. Humbert, *Z. Metallk.*, *51* (1960) 626.
25 W. v. Bolton, *Z. Elektrochem.*, *11/3* (1905) 45, *13/15* (1907) 145.
26 R. Kieffer and H. Braun, *Vanadin, Niob, Tantal*, Springer-Verlag, Vienna, 1948.
27 R. Kieffer and W. Hotop, *Sintereisen und Sinterstahl*, Springer-Verlag, Vienna, 1948.
28 C. A. Hampel, *Rare Metals Handbook*, 2nd edn. Reinhold, New York, 1961.
29 (a) R. Kieffer, H. Bach and H. Lutz, *Metall*, *1* (1967) 19.
   (b) R. Kieffer, F. Lihl and E. Effenberger, *Z. Metallk.*, *60* (1969) 94.
30 Siemens u. Halske A.G., *D.R.P. 397,641*, 1922.
31 C. W. Balke, *Ind. Eng. Chem.*, *27/10* (1935) 1166; *Trans. Electrochem. Soc.*, *85* (1944) 89; *Chem. Ind. (London)*, *6* (1948) 83.
32 (a) R. Titterington and A. G. Simpson, *Symp. Powder Metallurgy, London, 1954*, Iron Steel Inst. (London) Spec. Rept. 58, 1956, p. 11.
   (b) G. L. Miller, in F. Benesovsky (Ed.), *Hochschmelzende Metalle*, 3. Plansee Seminar, Reutte/Tyrol, June 1958, Springer-Verlag, Vienna, 1959, p. 306.
33 R. Kieffer and F. Benesovsky, *Metall*, *13* (1959) 379, 652.
34 R. Kieffer and K. Sedlatschek, *Österr. Chemiker-Ztg.*, *61* (1960) 217.
35 B. C. Allen, *DMIC Memorandum 90*, 1961.
36 F. Skaupy, *Metallkeramik*, 1st ed., Verlag Chemie, Berlin, 1930; 4th edn., Verlag Chemie, Weinheim-Bergstr., 1950.
37 G. F. Cox, *Metal Ind. (London)*, (1960) 186, 207, 231.
38 G. L. Miller, *Tantalum und Niobium*, Butterworths, London, 1959, p. 279.
39 W. G. O'Driscoll and G. L. Miller, *Symp. Metallurgy of Niobium, 1957; J. Inst. Metals*, *85* (1956/57) 367, 379.
40 H. Lutz, F. Benesovsky and R. Kieffer, *J. Less-Common Metals*, *16* (1968) 249.
41 A. R. Kaufmann and E. Gordon, *Metal Progr.*, (1947) 387.
42 A. R. Kaufmann, P. Gordon and D. W. Lillie, *Trans. Am. Soc. Metals*, *42* (1950) 785.
43 H. W. Dodds, *U.S.P. 2,818,339*, Dec. 31, 1957.
44 E. Zscherpe, *Z. Ver. Deut. Ing.*, *97* (1915) 1167.
45 M. Donovan, *Powder Met.*, *1/2* (1958) 104.
46 F. G. Cox, *Murex, Ltd. Rev.*, *2* (22) (1960) 41.
47 K. Yoshida, *Kikai to Kogu*, *5* (6) (1961) 93.
48 J. T. Norton, *J. Metals*, *8* (1956) 49.
49 J. Hinnüber and W. Kinna, *Tech. Mitt. Krupp*, *19* (1961) 130.
50 J. J. Löffler, K. H. McKee and R. J. Stuligross, *Progr. Powder Met.*, *19* (1963) 174.
51 C. Ballhausen, *Stahl Eisen*, *71* (1951) 1090.
52 H. Baumhauer, *D.R.P. 443,911* (1922).
53 G. Zapf, *Ind.-Anz.*, *78* (1956) 348.
54 H. Silbereisen, *Fertigungstechnik*, *8* (1958) 163.
55 R. Kieffer and G. Jangg, *Chem. Ing.-Tech.*, *39* (1967) 43.
56 R. Kieffer and F. Benesovsky, *Berg.-Hüttenmänn. Monatsh.*, *94* (1949) 284.
57 Anon., *Metal Progr.*, Sept. (1968) 92.
58 J. E. Anderson, *Trans. Vacuum Met. Conf., New York, 1968*, Am. Vacuum Soc., New York, 1968, p. 595.
59 G. Naeser and O. Wessel, *Arch. Eisenhüttenw.*, *40* (1969) 257.
60 M. Michalke and W. Scholz, *2nd Europ. Symp. on Powder Met., Stuttgart, 1968*, Rept. No. 4.1.
61 B. Natter, Molybdän als Ofenbauwerkstoff, (Discussion), *Hochschmelzende Metalle*, 3. Plansee-Seminar, Reutte/Tyrol, June 1958, Springer-Verlag, Vienna, 1959, p. 132.
62 R. Kieffer and K. Sedlatschek, *Molybdän-Metall als Sonderwerkstoff, Molybdän-Dienst Nr. 12*, Düsseldorf, 1961; *Bänder, Bleche, Rohre*, 1962, April, p. 173.
63 R. Kieffer and F. Benesovsky, *Metallurgia*, *58* (1958) 119.

Chapter VIII

Vacuum Heat Treatment

S. S. WHITE

1. The Vacuum Atmosphere

Although the principle of the elimination of all atmosphere as a means of preventing oxidation has long been understood, the practical development and general use of vacuum heat treating techniques is relatively recent. During the last ten years, the advantages of the vacuum method particularly in heat treating, hardening and certain brazing operations have been realized by industry and have led to the wider development and use of the vacuum type furnace. Some of these advantages are relative freedom of a potentially contaminating atmosphere, removal of undesirable gases from products, frequently lower installation and operating costs, and adaptability to operating temperatures as high as 3000°C.

Vacuum as an atmosphere usually connotes space relatively free from matter. The pressures that are usually encountered in most vacuum metallurgical processes cover the range of 1 to 10^{-6} Torr. With this in mind, we may consider the two outstanding features of a high vacuum atmosphere that are of particular interest in vacuum heat treatment. The first feature is that vacuum provides an atmosphere of rather low chemical activity, being almost inert because of the relatively low gas constituents. Because of this, the usual reactions that take place (gas–solid heat treatment reactions) such as carburization, decarburization, reduction or oxidation will not be so likely to occur. The surface chemistry remains unchanged and the surface finish is protected. The second feature of interest with a vacuum atmosphere is that it provides conditions under which the removal of gases from alloys and metals or the decomposition of metal oxides and other gas–metal compounds can occur. In any application of vacuum heat treatment, either or both of these features may be at work and theoretically all problems connected with the generation of suitable protective atmospheres are circumvented. Yet both features must be considered in the light of the fact that a vacuum atmosphere is not entirely free from matter, but does contain some residual gases. The extent of changes of the surface properties of the metal which may occur depends on the partial pressure of these impurities. In Chapter IV, Section 2 and Chapter XI, Section 5 these reactions are discussed in more detail.

Normally in vacuum heat treatment the removal of nonmetal impurities is of less importance than the avoidance of detrimental surface reactions, like oxygen pick-up or decarburization. Therefore pressure and composition of the residual atmosphere must be kept under close control and adjusted to the type of material to be heat treated. (Because of economical reasons it is, for example, not feasible to use expensive vacuum systems with a low base pressure for heat treatment of materials like high speed steels, which are less critical than alloys with reactive components or refractory metals.)

In spite of the fact that in the interaction of metals with gases diffusion in the body from and to the surface is in general the slowest step and therefore rate-determining for the bulk of the material, in heat treating only the rates of phase boundary reactions have to be considered. Only the changes which occur at the surface are of importance, and we may look at this problem as if we had to deal only with a thin metal foil.

Of the four steps which lead to the absorption of gases or nonmetals,
(a) transport of the molecule to the surface
(b) adsorption
(c) dissociation of the molecule
(d) diffusion into the solid
— if we neglect the last one — the third step, dissociation, is in most cases rate-determining in this sequence. The same is valid for the association of the gas atoms in the reverse direction. The dissociation reaction of the diatomic molecules O_2, H_2 and N_2 leads to the dissolution of these gases and by equilibration with gas mixtures like CO_2–CO, H_2O–H_2, carbon, oxygen or hydrogen may be exchanged even simultaneously. Equilibrium exists if the activity of the nonmetals in the solid phase and that in the gas phase are equal. If this is not the case a reaction takes place[1].

If there is a surplus in the gas phase, it depends on the activity and the chemical affinitiy to the metal whether the nonmetal is dissolved or a compound, *e.g.* an oxide, is formed, which with an excess of oxygen may grow to a scale.

In order to understand the significance of the reduction in oxidation rate of a metallic surface when heated in vacuum rather than air with an oxygen partial pressure of 0.21 atmosphere, it is important to consider the appropriate theories of metallic oxidation.

Every scale must pass through a thin-film stage at the beginning of oxidation. Especially at low temperatures, the films may continue to be thin for a long time. Experimental work has shown that many metals oxidize rapidly at the beginning when exposed to room temperatures. But, after only a few minutes, the oxidation rate drops to low or negligible values, a stable film of thickness 20–50 Angstrom units being formed[2].

The region of thin film formation having a limiting thickness may extend to temperatures above room temperature, the maximum temperature varying from metal to metal. At intermediate temperatures which are still below heat treating temperatures, neither the low temperature asymptotic oxidation mechanism nor the high temperature oxidation mode prevails. In this intermediate temperature range, various relationships have been found to apply. Cabrera and Mott[3] explain at least some of the logarithmic, cubic and parabolic relationships that have been observed.

When metals are heated to within their normal heat treating temperature ranges, the lower temperature oxidation modes are overshadowed in terms of rate and total thickness by a parabolic time law based on a diffusion-controlled process. This high temperature oxidation mode is adequately covered by Wagner's theory[4].

Wagner's theory of oxidation of metals is confined to surface reactions that obey the parabolic time law and for which a true diffusion process of ions in the oxide lattice is rate-determining. By considering the effect of oxygen concentration on the mechanism of this theory, the importance of relative degrees of partial pressure of the different species in the vacuum atmosphere (including reducing gases) may be realized.

In order to maintain electrical neutrality, ionic diffusion must be accounted for by a simultaneous diffusion of electrons in the same direction as the cations. The rate of diffusion in a metallic lattice is determined by the movement of the cations between vacant lattice sites which is much slower than the movement of electrons. To have a directed diffusional movement through an oxidation layer, a concentration gradient must exist as shown in Fig. 1. According to Wagner[4] an expression exists for the rate of thickening of the oxidation layer in terms of the specific conductivity of the film (Z), the transport numbers of the cations, anions and electrons (τ_c, τ_a, τ_e), and the free energy decrease of the oxidation reaction (E_0).

Kubaschewski and Hopkins[2] present the derivation of special forms of

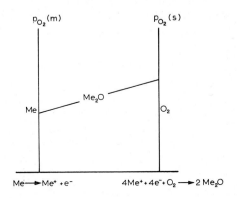

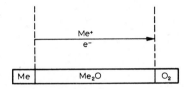

Fig. 1. Schematic representation of diffusion in the system Me–Me$_2$O–O$_2$.

References pp. 744

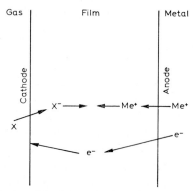

Fig. 2. Electrochemical model of the oxidation of a metal Me by oxygen gas X_2.

Wagner's equation as given by Hoar and Price[5] and Jost[6]. This fairly simple derivation gives a very clear picture of the basic processes. In these analyses, the transport of ions and electrons through the oxidation layer may be viewed as a flow of current through a cell, with the film providing the electrolyte (due to ionic transport) and the external circuit (due to conduction) (see Fig. 2). The e.m.f. of the cell is the change in free energy E_0 of the reaction, if polarization is ignored. The cathodic reaction at the film–gas interface is represented by

$$X + e^- = X^- \tag{1}$$

or

$$O_2 + 4e^- = 2 \cdot O^= . \tag{2}$$

The anodic reaction takes place at the metal–film interface and corresponds to

$$Me = Me^+ + e^- . \tag{3}$$

The total resistance (electrolytic plus electronic) of an area A (cm^2) of film of thickness $\Delta\varepsilon$ (cm) is

$$R = (1/Z \text{ electrolytic} + 1/Z \text{ electronic}) \Delta\varepsilon/A$$
$$= \left(\frac{\tau_a + \tau_c + \tau_e}{(\tau_a + \tau_c)Z} + \frac{\tau_a + \tau_c + \tau_e}{\tau_e Z} \right) \Delta\varepsilon/A = \frac{\Delta\varepsilon}{(\tau_a + \tau_c)\tau_e Z A} \tag{4}$$

(Z in Ω^{-1}cm^{-1}).

If the film area consists of n gram-equivalents and is formed in t seconds, Faraday's and Ohm's laws yield (assuming rate of film growth is equivalent to the current I in amperes) with E_0 in volts

$$dn/dt = I/F = E_0/FR \tag{5}$$

(F = Faraday constant)

$$\therefore dn/dt = \frac{(\tau_c + \tau_a)\tau_e Z A E_0}{96500 \, \Delta\varepsilon} \tag{6}$$

which is a simple form of the equation derived by Wagner[4]. dn/dt is proportional to the rate constant K_p in the parabolic law of film growth

$$dn/dt = (\tfrac{1}{2})V_{eq}(z/M)^2 K_p \tag{7}$$

which would hold for oxide Me_2O with V_{eq} = volume of 1 gram-equivalent of the oxide, z = valency of oxygen ion and M = atomic weight of oxygen. (Note that the theory only requires a constant concentration difference across the film, not a uniform concentration gradient.)

So far in this discussion, the variation of Z with the pressure (p_x) of the negative component has not been considered. For our purposes (and indeed whenever serious calculations of surface contamination are made), this cannot be neglected. Therefore, the fundamental equation (7) must be expanded.

The mathematical treatment of this problem by Kubaschewski and Hopkins[2] leads to the equation

$$\frac{dn}{dt} = \frac{8.9 \times 10^{-10}}{\Delta\varepsilon \cdot z}(\tau_a + \tau_c)\tau_e Z_o AnT(p_x^{1/n}(s) - p_x^{1/n}(m)) \tag{8}$$

where Z_o = conductivity at a pressure p_x = 1 atm of the negative component of the surface compound (oxide), n = constant (for example, n = 7–8 for Cu_2O; -4 to -6 for ZnO), z = valency factor, $p_x(s)$ = pressure of the gas X_x at the outside surface of the oxide film, i.e. the gas pressure, $p_x(m)$ = pressure at the film–metal interface, i.e. the dissociation pressure of Me_mX_n.

Equation (8) points out the extreme importance of oxygen partial pressure [$p_x(s)$] on the oxidation rate [dn/dt] and hence of the total film thickness. When $p_x(s)$ is essentially zero, there is, of course, no film growth rate. In fact, $p_x(s)$ must be greater than the dissociation pressure of the oxide [$p_x(m)$] (both raised to the $1/n$ value) for noticeable film growth rate. As the partial pressure of oxygen increases, growth rate increases at first very fast and more slowly at higher pressures.

It is for these reasons that vacuum systems are designed to operate at the lowest pressure attainable, consistent with justifiable equipment costs.

The magnitude of the oxygen pressure required in order to avoid any oxidation can be derived from Fig. 20 in Chapter IV. From this we can conclude that the oxygen activity or partial pressure in the residual atmosphere over the common metals has to be extremely low. Dissociation of oxide scales formed during or before heat treatment is not likely to occur at the vacuum conditions which are normally achieved in technical equipment.

If experience shows that very often a metal surface can be maintained almost free from oxide at pressures several decades higher than those suggested by these theoretical figures, there are different explanations. The most likely mechanism is one of solution and diffusion of oxygen into the body of the metal at a rate greater than that of the formation of oxide at the surface, or mechanisms such as local nucleation of oxides rather than continuous film formation.

References pp. 744

Another possibility is the reaction of oxygen with reducing agents like carbon or hydrogen either in the solid or in the residual atmosphere.

The oxidizing effects of gases other than oxygen (such as carbon dioxide and carbon monoxide) should also be considered. For example, at steel austenitizing temperatures, carbon dioxide reacts with carbon on a steel surface as follows:

$$\underline{C} + CO_2 = 2CO \tag{9}$$

where $\underline{C}$ = carbon dissolved in austenite.

The reaction continues until all carbon dioxide is consumed, or all dissolved carbon depleted from the steel surface. If all the carbon has been removed from the surface and an excess of carbon dioxide remains, the following reactions take place:

$$Fe + CO_2 = FeO + CO \tag{10}$$

$$3FeO + CO_2 = Fe_3O_4 + CO. \tag{11}$$

Above 554°C, FeO is the stable oxide; Fe_2O_3 is stable below this temperature.

Equations (9), (10) and (11) proceed to equilibrium at a rate that decreases with time and with decreasing temperature and pressure of the system.

The CO/CO_2 partial pressure ratio, which can be tolerated in order to avoid oxidation of various pure metals depending on the temperature, is also shown in Fig. 20, Chapter IV.

Water vapour too may act as an oxidizing medium. As in the case of CO_2, it is not the partial pressure of the oxidizing gas *per se*, but the H_2/H_2O partial pressure ratio (see also Chapter I) which decides whether reduction or oxidation predominates. Figure 3 shows the partial pressure ratios which are in equilibrium with different metal oxides[7] (see also Fig. 20, Chapter IV). From this we conclude

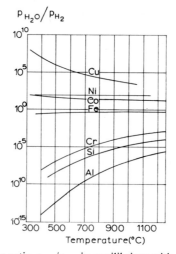

Fig. 3. Partial pressure ratio p_{H_2O}/p_{H_2} in equilibrium with different metal oxides.

that at least with metals of high oxygen affinity the pressure ratio is less critical the higher the temperature.

Residual hydrogen can act as a reducing or decarburizing agent. The extent of decarburizing depends upon carbon content, temperature, time and moisture content. At temperatures below 800°C, decarburizing effects due to hydrogen are negligible. Decarburizing increases significantly above this temperature, with water vapor greatly enhancing this effect since it is a supply of atomic hydrogen (as well as oxygen). The dissolved carbon in steel combines with such hydrogen as follows:

$$\underline{C} + 4H = CH_4. \quad (12)$$

Even relatively dry hydrogen will decarburize steel according to the following relationship:

$$\underline{C} + 2H_2 = CH_4 \quad (13)$$

with the equilibrium constant

$$K = \frac{p_{CH_4}}{p_{H_2}^2}. \quad (14)$$

It should be remembered that the decarburizing effect of hydrogen is increased with increasing carbon content, whether the hydrogen is wet or dry. The partial pressure ratio $p_{CH_4}/p_{H_2}^2$ which is in equilibrium with steel of different carbon content is shown in Fig. 4[7].

Water vapor also combines with dissolved carbon in steel as follows:

$$\underline{C} + H_2O = CO + H_2 \quad (15)$$

In consideration of eqn. (13) the decarburizing effect of water vapor desorbing from internal surfaces of the vacuum equipment may therefore be partly or fully

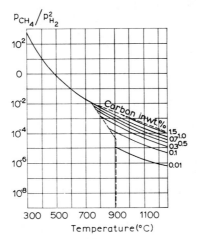

Fig. 4. Partial pressure ratio $p_{CH_4}/p_{H_2}^2$ in equilibrium with steel of different carbon content.

References pp. 744

compensated by a hydrocarbon partial pressure originating from the pumping system, *e.g.* if only oil-sealed mechanical pumps are used, provided that there are no leaks in the vacuum plant.

While there are many advantages to be gained by using high vacuum, there are certain disadvantages which should also be considered. Figures 4a and 4b, Chapter III, show vapor pressure *versus* temperature curves for a number of metals, from which it may be seen that there is considerable variation between one metal and another. This means that at very low pressures there is a possibility of selective volatilization (or sublimation) of the more volatile constituents of an alloy, with a subsequent depletion in the surface layers of this particular constituent. It is possible, for example, to convert brass into copper sponge by vacuum heat treatment.

Another example is the depletion of chromium and aluminum present as alloying constituents in iron- or nickel-based alloys. Fortunately, the picture is not quite as bad as it might appear from these curves, as the vapor pressure of a metal present in a dilute form in an alloy will certainly be less than that of the pure metal. Furthermore, there are frequently thin oxide films, which reduce metal vaporization tendencies.

If the vapor pressure of volatile constituents becomes too high at elevated temperatures, surface depletion can only be avoided by back-filling or continuous bleeding in of an inert gas like argon at a pressure where the mean free path of the evaporating atoms becomes so small that most of them are scattered back to the surface (see also Chapter I, Subsection 3.2 and Chapter III, Section 4). By raising the pressure to about 10 Torr the evaporation rate is reduced by several orders of magnitude.

Another disadvantage is the absence of a conductive gas, especially at lower temperatures, where the heat transmission by radiation is very low. A less homogeneous temperature distribution in the heating zone is therefore to be expected. Cooling in vacuum at lower temperatures is also retarded for this reason. Finally, a leak in the vacuum plant may be more detrimental than the incorrect composition of a protective atmosphere[8].

2. Types of Furnaces

The major proportion of furnace types used for vacuum heat treatment work fall into two categories. These are as follows:
1. The hot wall type which is externally heated.
2. The cold wall type which is internally heated.

2.1 Hot wall furnaces
The hot wall furnace is similar in construction to a protective atmosphere

furnace. In this unit the charge is placed in a metal retort, which is in itself enclosed within resistance heating elements placed on the furnace walls. Instead of feeding a particular protective or reducing atmosphere to the retort, the space is evacuated. When the retort diameter is large and the operating temperature high, it is standard procedure to evacuate the furnace casing as well (termed "double pumping") in order to avoid the risk of collapsing the retort. Generally, hot wall furnaces for vacuum use are limited to a maximum operating temperature similar to those limiting the temperature of standard atmosphere resistance furnaces. For industrial hot wall furnaces, the retort may be constructed from a variety of suitable materials and thicknesses, depending on the temperature used. At higher temperatures, the limiting factor is the strength of the retort material. Below 300°C, plain carbon steels are quite adequate for retort materials. Steels containing 3–9% chromium, known as heat resisting chromium steels, are useful when operating temperatures are as high as 500°C. Stainless steels of any category may be used for operating temperatures up to 850°C. However, stabilized austenitic stainless steel (types 321 or 347), or low carbon stainless steel (types 304-L or 316-L), are preferred since these stainless types are less prone to weld cracking during fabrication and resultant furnace leaks. For temperatures up to 1050°C, Inconel is used. With "double pumping" even 1150°C may be reached. Hot wall furnaces are made in many conventional types such as tube furnaces (horizontal, bell, etc.). In most cases, the metal container is cylindrical (see also Figs 25 and 26, Chapter VII).

Perhaps the greatest commercial use of hot wall furnaces is for the bright annealing of copper. Typical installations have capacities in excess of 100 tons per week, consisting of two or more furnaces of the elevator bell type. Each furnace base is loaded, transferred to a furnace station and connected by flexible hose to a mechanical pump capable of producing 10^{-1} Torr vacuum, although it has been shown that a pressure of 2 Torr produces satisfactory results. The bell is evacuated and raised into the furnace. To assure removal of lubricant from the wire, the vacuum pump is run continuously through the heating cycle. Times vary from 4 to 5 hours for a temperature between 200°C and 300°C, after which the charge is removed from the furnace and allowed to cool.

2.2 Cold wall furnaces

The cold wall furnace is of a very different design and is engineered to reach temperatures considerably higher than the useful upper limiting temperatures of the materials used in the outer shell. The cold wall is the outside of the furnace itself (*i.e.*, the tank) and is water cooled. The furnace heating system and any required insulation is built into the inner part of the tank. This type of construction allows a much wider choice of materials for heating elements which are usually protected during their life by an appropriate vacuum or a proper atmosphere when these furnaces are back-filled. Heating elements such as molybdenum, tanta-

References pp. 744

lum, tungsten, graphite rods or graphite cloth are used and insulation is provided in the form of radiation shields made of stainless steel or other materials such as molybdenum, graphite, carbon felt or even ceramics. When non-metallic shields are used, a poorer vacuum usually results owing to the outgassing of such materials unless the system is designed to handle this extra load.

With the most commonly used heating elements, progressively higher temperatures may be reached using molybdenum, graphite, tantalum or tungsten, in that order. Molybdenum and graphite are the most popular materials used for heating elements since they are fabricated with ease compared with tungsten. They are also considerably less expensive than tungsten. Tantalum is easy to fabricate but is quite expensive. Temperatures up to 1600°C are obtainable using molybdenum wires supported on alumina and up to 1700°C is obtainable when molybdenum elements are directly connected to molybdenum terminals. Tungsten elements are usually obtained in the form of cylindrical sheet or wire mesh assemblies and furnaces of this type are made with operating temperatures of up to 3000°C (see also Chapter VII).

Cold wall furnaces come in many sizes. A typical cold wall furnace suitable for laboratory experiments and high quality production is shown in Fig. 5. The water-cooled stainless steel horizontal cylindrical chamber (tank) used in this furnace is shown in Fig. 6. The front door is flanged and fitted with an access door sealed with an O-ring for loading and unloading. Four clamps precompress

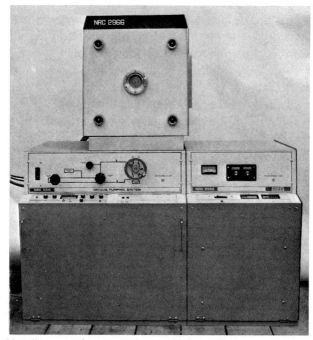

Fig. 5. Cold wall vacuum furnace. (Courtesy of Norton Vacuum Equipment Division)

TYPES OF FURNACES

Fig. 6. Water-cooled stainless steel tank and jacket for a cold wall vacuum furnace. (Courtesy of Norton Vacuum Equipment Division)

the O-ring and secure the door when closed. The rear of the chamber is flanged for ease of maintenance and accommodates thermocouple glands and test accessory feed-throughs. Other ports for additional accessories as well as the large flanged port for connection to the vacuum pumping system may also be seen in Fig. 6. A viewing port (see Fig. 5) is provided for the optical determination of specimen temperature and condition. A manually operated spring return shutter

Fig. 7. Molybdenum rod heating element for cold wall vacuum furnace. (Courtesy of Norton Vacuum Equipment Division)

References pp. 744

is provided to minimize the deposition of metallic vapors on the Pyrex window. Brackets are provided for mounting the heater and shield assemblies.

A heater and shield assembly for this unit is shown in Fig. 7. Although the water-cooled jacket construction allows almost any type of suitable heating element to be used, the one shown is of molybdenum rod. The heating element is graded along its length to provide a greater concentration of power at the ends where the rate of heat loss is highest. Molybdenum shielding surrounds the heating element and molybdenum plate radiation shields reduce radiation loss to the front opening door and to the rear of the chamber. The furnace hearth is supported on molybdenum pins from the bottom of the chamber.

A similar furnace is shown in Fig. 8. The versatility of such a furnace allows it to be a high temperature horizontal or vertical furnace, or to be used for induction melting, inert gas welding, arc melting and casting, and even a system for

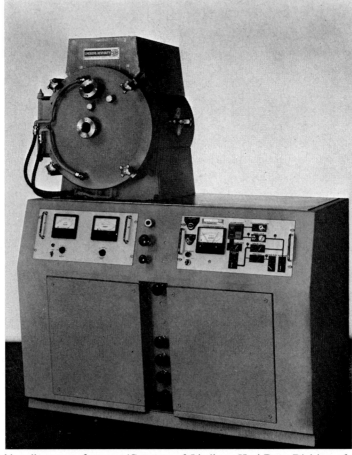

Fig. 8. Cold wall vacuum furnace. (Courtesy of Lindberg Hevi-Duty Division of Sola Basic Industries)

Fig. 9. Cold wall vacuum furnace. (Courtesy of Lindberg Hevi-Duty Division of Sola Basic Industries)

vacuum-to-liquid quenching. In its role as a horizontal cold-wall resistance furnace, it houses a 150 × 150 × 400 mm long work zone with tungsten rod heating elements. Four separate tungsten elements connect to water-cooled solid copper electrodes.

Fig. 10. Cold wall vacuum furnace with loading and unloading dolly. (Courtesy of Lindberg Hevi-Duty Division of Sola Basic Industries)

Fig. 11. Cold wall vacuum furnace. (Courtesy of Lindberg Hevi-Duty Division of Sola Basic Industries)

Molybdenum and stainless steel are used for shielding. When the unit is converted to a vertical vacuum and inert atmosphere furnace, tantalum mesh elements allow a maximum temperature of 2200°C to be reached. With tungsten mesh elements and refractory metal shields, 3000°C can be reached in the 100 mm diameter × 180 mm high working area.

A typical cold wall furnace used for industrial bright hardening, annealing, brazing, sintering, stress relieving, tempering, normalizing and degassing is shown in Figs. 9 and 10. Figure 9 shows the external view of the unit and Fig. 10 shows a loading operating into the 60 cm wide × 90 cm deep × 40 cm high hot zone. The heat in this particular furnace is supplied by 12 graphite rod heating elements (6 top and 6 bottom). Shielding is accomplished by molybdenum or Inconel shields, backed up with 8 cm of asbestos fiber insulation retained by an expanded metal stainless cage. A uniform temperature of 1425°C maximum is attained within

60 minutes using a 120 kW power supply. 1300°C maximum is attained within 60 minutes using an 85 kW power supply.

A somewhat larger cold wall vacuum furnace is shown in Fig. 11. This unit is capable of operating continuously at 1300°C practical maximum temperature with an internal pressure of 10^{-4} Torr, and a clear usable work space of 1.2 m diameter × 1.2 m high. The unit can handle loads of up to 340 kg/m² of hearth area, or 500 kg gross. Graphite heating elements are used with molybdenum heat shields backed up by high temperature thermal insulation (as with the furnace previously described). This 500 kg furnace utilizes a 20 in. diffusion pump of 10000 l/s, backed up by a 1000 m³/h mechanical pump. A total of 275 kW of power to the graphite pipe heating elements is required to maintain uniform temperature. Larger furnaces have been built and more complex units tested experimentally.

Most modern furnaces for heat treatment have forced argon or nitrogen cooling with heat exchangers (closed loop gas cooling) for fast cooling of the work after heating. With this device not only shorter cycle times, but also heat treatment of superalloys (solution treatment and precipitation hardening) and hardening of the so-called air hardening steels, are possible[9,10]. Naturally, a high degree of purity of the quench gas is necessary if a bright finish with reactive alloys is to be maintained (<3 vol. ppm O_2, <1 vol. ppm H_2O)[9]. In this way a load of 500 kg can be cooled from 1250°C to 400°C in 15 min[11].

For the heat treatment of carbon, stainless and alloy tool steels even faster cooling with oil quenching is in some cases necessary. Such furnaces with an integrated[12] or additional quenching chamber, separated by a large valve, have also been developed. Units with capacities of up to 700 kg with quenching facilities are being built[13].

No doubt the future will see units capable of faster pump-down times as well as larger hearth size and probably also furnaces for continuous vacuum processing.

2.3 Other furnace types

Although hot and cold wall resistance furnaces embrace the majority of vacuum furnaces employed in industrial heat treating processes, a number of other furnace types can be used. For example, gas fired furnaces are employed. With hot wall furnaces, there is little reason why they cannot be gas fired, especially for low temperature application, either by direct heating or by heating from radiant tubes.

One interesting promise for the future of vacuum heat treatment furnaces seems to be the use of electron beam furnaces. The advantages of electronic bombardment heating for vacuum melting of metals are well known; its value as a means of heating for high vacuum heat treatment is just being realized. In comparison with resistance heated furnaces, an electron bombardment furnace is more efficient and compact, and such a unit is already operating at the Temescal Division

References pp. 744

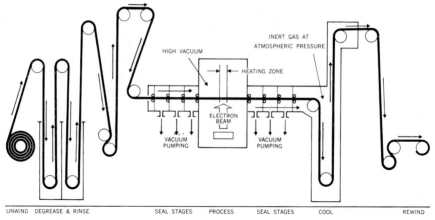

Fig. 12. Continuous vacuum heat treating annealing of strip by the electron beam process.

of Airco in Berkeley, California. This unit can handle strip of 600 mm width. Another, smaller, unit handling 400 mm width strip is located at the Youngstown Sheet and Tube Company. Both units are long horizontal chambers equipped with four pumped vacuum seals at each end to allow the strip to enter and leave a central chamber continuously (see Fig. 12). Beam guns in the center chambers convert high voltage, direct current into electronic energy as the strip passes.

The unit at Berkeley can have up to several million watts of power consumed in its main chamber, although 600,000 watts is the most common mode of operation. Densities of up to 540 W/cm^2 are reached.

3. Heat Treatment Processes Carried Out in Vacuum

3.1 Hardening

3.1.1 Steels

Any air hardening steel is capable of being hardened in a vacuum furnace after an appropriate gas quench. According to Reynoldson[9,14] vacuum hardening is done for the following reasons:

(a) In a vacuum hardening furnace, there is no movement of hot components in any critical state in the heat treatment (this means less distortion of the parts).

(b) A very bright finish is obtained. This alone justifies the use of vacuum for the heat treatment of tools and dies.

(c) There is usually neither carburization nor decarburization.

(d) Absence of salt from heat treating salt baths in blind holes or cavities avoids corrosion.

(e) Little reworking of expensive dies is necessary.

(f) Once a technique has been determined, it is quite reproducible.

Vacuum hardening involves putting the tools or dies—in clean condition—in suitable fixtures in the vacuum furnace. The furnace is evacuated to the required pressure for the work pieces. Preheating, when desired, is begun, then there is possibly a soaking period, after which the work (charge) is raised to the appropriate hardening temperature and, if necessary, soaked at temperature. During this entire cycle, the vacuum pumps are used to maintain low pressure. After this, the pumps are stopped, the furnace backfilled with an appropriate atmosphere (quickly) and often a fan started. The fan allows the gas to circulate through the work until the temperature of the work pieces has fallen to the required level. Cooling must also be carried out quickly to produce the desired structure and hardness. The rate of cooling is a function of the mass and volume of the gas employed, as well as the rate of circulation through the charge (and any heat exchangers used) as well as the efficiency of the fan (if used) and heat exchanger. The important characteristics of such gases are density, specific heat and thermal conductivity. Although hydrogen, because of its high thermal conductivity, provides a high cooling rate at all pressure levels, pure nitrogen or argon is usually adequate, less dangerous and sometimes less damaging metallurgically. Similarly, helium, also noted for high thermal conductivity, can be used when lower pressures of the order of 100 Torr are utilized and helium costs are acceptable. The lower pressures still permit the benefits of the high heat transfer.

Most tool and die steels require tempering after hardening. In the case of high speed steels, this is termed secondary hardening and is performed at 500–560°C for periods of 2–3 hours, this treatment being repeated 2 or 3 times for some steels. These tempering processes are carried out at temperatures below the lower limit (700°C) specified for vacuum hardening furnaces (usually cold wall type) and for very long times. Thus, it is not economical to use vacuum hardening furnaces for tempering and secondary hardening treatments[9].

Typical steps involved in the hardening of various steels are given below according to AISI and SAE nomenclature as described by Becket and Reynoldson[9]:

A series—Air hardening medium alloy cold work tool steels

    (a) Preheating is done over the range of 750–800°C.

    (b) Hardening is then accomplished by gas quenching from 850° to 980°C. (Note—the vacuum required during preheating and hardening is 10^{-3} Torr.)

    (c) The charge is cooled to 100°C from the hardening temperature in 20–25 minutes.

    (d) Tempering is done in the range of 150–530°C in a separate atmosphere furnace, yielding a final hardness of 52–60$^R{}_C$.

D series—High-carbon, high chromium cold work steels

    (a) Preheating is done over the range 800–820°C.

    (b) Hardening is then accomplished by gas quenching from 950° to 1025°C. (Note—the vacuum required during preheating and hardening is 10^{-2} Torr.)

References pp. 744

(c) The charge is cooled from the hardening temperature to 50°C in 25–30 minutes.

(d) Tempering is done in the range of 200–350°C in a separate atmosphere furnace, yielding a final hardness of 54–61R_C.

H series—Hot work tool steels containing 5.0–7.0% chromium

(a) Preheating is done over the range of 800–820°C.

(b) Hardening is then accomplished by gas quenching from 900° to 1060°C except for H15 steel. This steel is hardened by quenching from 1150° to 1250°C. (Note—the vacuum required during preheating and hardening is 10^{-2} Torr.)

(c) The charge is cooled from the hardening temperature to 100°C in 25–30 minutes.

(d) Tempering is done in the range of 540–680°C in a separate atmosphere furnace, yielding a final hardness of 36–58R_C.

H series—Hot work tool steels containing 9.0–18.0% tungsten

(a) Preheating is done over the range of 810–870°C.

(b) Hardening is then accomplished by gas quenching from 990° to 1275°C. (Note—the vacuum required during preheating and hardening is 10^{-2} Torr.)

(c) The charge is cooled from the hardening temperature to 100°C in 25–30 minutes.

(d) Tempering is done in the range of 540–680°C in a separate atmosphere furnace, yielding a final hardness of 35–58R_C.

H series—Hot work tool steel containing 5.0–8.0% molybdenum

(a) Preheating is done over the range of 730–840°C.

(b) Hardening is then accomplished by gas quenching from 1090° to 1230°C. (Note—the vacuum required during preheating and hardening is 10^{-2} Torr.)

(c) The charge is cooled from the hardening temperature to 55°C in 25 minutes.

(d) Tempering is done in the range of 500–650°C in separate atmosphere furnaces, yielding a final hardness of 45–58R_C.

T series—Tungsten high speed steels

(a) Preheating is done over the range of 810–970°C and soaked there for 15–20 minutes.

(b) Hardening is then accomplished by gas quenching from 1180° to 1320°C. (Note—the vacuum required during preheating and hardening is 10^{-2} Torr.)

(c) The charge is cooled from the hardening temperature to 200°C in less than 18 minutes.

(d) Secondary hardening in the range 550–570°C for 1 or 2 hours is done and repeated 2 or 3 times (all in a separate atmosphere furnace—vacuum purged and fed with an inert gas or nitrogen). Final hardness is 61–64R_C.

M series—Molybdenum high speed steels

(a) Preheating is done over the range of 800–850°C and soaked there for 15–30 minutes.

(b) Hardening is then accomplished by gas quenching from 1190° to 1240°C. (Note—the vacuum required during preheating and hardening is 10^{-2} Torr.)

(c) The charge is cooled from the hardening temperature to 200°C in less than 18 minutes.

(d) Secondary hardening in the range of 550–570°C for 1 or 2 hours is done and repeated 2 or 3 times (all in a separate atmosphere furnace). Final hardness is 60–64$^R{}_C$.

3.1.2 Specialty metals

The solution treating of Nimonic alloys, Hastelloys, Udimet alloys, Waspaloy, and Rene alloys is also carried out in vacuum with gas quenching as is the solution treating and precipitation hardening of stainless steels. Brazing followed by hardening is also a process which should become more important industrially in the future[10,15].

3.2 Vacuum annealing

As with the previous processes described, the use of cold wall furnaces greatly extends the limits and possibilities of application of these processes. The most important of these are annealing, brazing and sintering. Brazing and sintering are covered in Chapters IX and VII.

3.2.1 Steels and refractory metals

The biggest commercial use for vacuum annealing applies to the more costly exotic metals and alloys. General information on the annealing temperatures and pressures required are given in Table 1.

TABLE 1

TEMPERATURES AND PRESSURES REQUIRED FOR VACUUM ANNEALING

Metal	Annealing temperature	Vacuum range
Ferritic stainless steel (S.S.)	630–830°C	10^{-3} Torr
Martensitic S.S.	830–900°C	10^{-3} Torr
Austenitic		
(Unstabilized) S.S.	1010–1120°C	10^{-3} Torr
(Stabilized) S.S.	950–1120°C	10^{-4}–10^{-5} Torr
A Series tool and die steel	730–870°C	10^{-3} Torr
D Series tool and die steel	870–900°C	10^{-2} Torr
H Series tool and die steel	815–900°C	10^{-2} Torr
T Series tool and die steel	870–900°C	10^{-2} Torr
Molybdenum	1000–1100°C	10^{-3}–10^{-6} Torr
Tungsten	1400°C upwards	10^{-3}–10^{-4} Torr
Titanium	700–750°C	10^{-2}–10^{-4} Torr
Zirconium	900–1000°C	10^{-2}–10^{-4} Torr
Uranium	600–700°C	1.0– 10^{-1} Torr

3.2.2 Soft ferromagnetic alloys

Electromagnetic devices are frequently adapted for electromechanical energy conversion in data processing machines. The main requirements in such applications are fast response, high efficiency and reliable operation. The material used in these devices are in the magnetically soft category. According to the ASM Metals Handbook[16] some of the outstanding characteristics required of magnetically soft materials are: low hysteresis loss, low eddy-current loss, high permeability and high saturation value. These requirements are best met by iron-base alloys. The effect of the metallurgical structure on magnetic properties is well known. Becker[17,18] indicates that the microstructure actually controls the magnetic domain configuration, which is responsible for the magnetic properties. Among structure-sensitive properties are permeability and coercive force[19]. Magnetic properties are also affected by composition, impurities, strain and crystal structure[19].

Once a particular alloy is selected for an application, composition and crystal structure are fixed. Optimum magnetic conditions are thus best chosen for this alloy by heat treatment. The treatment is designed to minimize residual stresses, while preventing impurities from entering the material, and even removing existing impurities. Thus, the heat treatment usually consists of annealing in a strongly reducing atmosphere followed by slow cooling to at least below the gamma-alpha transformation temperature. For best results, high annealing temperatures are maintained to obtain the largest practical grain size since this improves some magnetic properties[20,21].

All reducing furnace atmospheres contain a substantial amount of hydrogen. Such atmospheres are carriers of contaminating impurities. Vacuum heat treatment is a proper substitute for neutral furnace atmospheres; and it can even be used in place of reducing atmospheres. It has been found that a diffusion pumped pressure of 2×10^{-4} Torr can yield results equivalent to those produced by heating in a very carefully purified hydrogen environment[21]. This development should, in time, prove to be as important as the application of vacuum to the annealing and/or heat treating of the more exotic metals and other requirements of the aerospace industry.

3.2.3 Copper

The advantages of vacuum annealing have been recognized in the copper wire industry and, for much the same reasons, in the steel wire industry. In particular, no gases come in contact with the vaporized lubricant and inter-turn adhesion of fine wires annealed on spools is prevented[22]. Tubing for gasoline and refrigeration is annealed and cleaned internally and externally simultaneously; no further processing is required[22,23]. Even capillary tubing of 0.6 mm diameter can be degreased and cleaned in coil form in lengths (at the present time) of up to 500 m. It is, however, advisable to use a water-soluble lubricant in the last drawing stage and rinse the tubing prior to vacuum annealing.

3.2.4 Oxidizing annealing (blueing)

Usually a uniform blue color is the target in the controlled oxidation of mild steel wire and strip. Theoretically, this should be possible if clean material is exposed to air at a temperature of approximately 300°C. But, if there are temperature differences throughout the charge, this will lead to large differences in color and the material may become mottled. Even differences of 10 °C are considered too high.

Clean oxygen is required for oxidation. The introduction of air into a standard non-vacuum furnace hinders a uniform flooding of the surface as the large mass of nitrogen prevents some areas from receiving an adequate oxygen supply at the proper temperature. It is not even sufficient to introduce clean oxygen if it does not enter under high pressure in order to flush out quickly the previous gaseous atmosphere. However, in vacuum blueing, clean oxygen floods into an evacuated container, penetrating with a pressure greater than that in the chamber, easily and uniformly filling all spaces and covering all exposed surfaces.

Two basically different methods have been developed for the blueing of steel wire and strip in a vacuum furnace[24,25]. In the first method, the charge is simply heated to 300°C and oxygen flooded into the system. For this reason, it is usually desirable to choose lubricants (for any cold work prior to blueing) that vaporize under high vacuum below 300°C. Should the vaporization temperature be higher, the temperature of the material must be increased to above the blueing temperature. This causes a time delay, since the entire charge must be cooled to 300°C after the lubricants have been expelled at some higher temperature. However, even such a procedure produces bright material[24].

Special requirements for vacuum blueing arise when dealing with certain products such as flower-wires, packaging tapes, deep-drawn strip, chain links, etc. If the cold-worked material has to be annealed simultaneously, 700°C in vacuum is reached, the material cooled, then blued. Indeed, degreasing, annealing and blueing can be carried out in vacuum during the same thermal cycle.

The second vacuum blueing method takes advantage of the fact that considerably less oxygen is required for satisfactory coloring at temperatures above 300°C. Indeed, at 300°C, excessive quantities of oxygen can be used without any deleterious effect on the surface color. At higher temperatures, when the oxygen partial pressure is controlled to produce the proper shade of blue, cost savings are realized because less oxygen is required.

Herdieckerhoff[24] found that appreciation of this point led to practical results. If a steel was to be simultaneously annealed and blued, he proceeded as follows:

The material was placed in a vacuum chamber which was then pumped to a low pressure (high vacuum). Relatively pure nitrogen was introduced which could

References pp. 744

escape through a one-way valve as the charge was heated to 700°C. The furnace was then placed in a cooling pit. When the temperature cooled to approximately 400–500°C, the partially created vacuum was brought up to one atmosphere with clean oxygen. The inert atmosphere acted as a throttle against admission of excess oxygen and also assured that oxygen was distributed uniformly throughout the many interstices of various charges such as wire or loosely packed strip or pressed parts. Very uniform blueing resulted[24].

3.2.5 Annealing to a mirror brightness

When strip or wire is annealed to a mirror brightness in a gaseous protective atmosphere furnace, the results are excellent in many cases. Intensive investigation of the surface, however, shows that the annealed material is often covered with a very slight—often invisible—film (oxide or residual lubricants) and so does not have a truly clean metallic surface (one which has only a slight thickness of oxide). By vacuum annealing the material, all lubricant residues adhering to the material are removed (provided that the lubricants are chosen so as to evaporate at the annealing temperature). Thus, not only an extra bright but also a well annealed product results. As an example, the inlay wire for wire glass may be cited. If this wire carries with it even traces of grease or moisture on its surface, the mesh inlay made of this wire triggers small explosions that tend to form gas inclusions (bubbles) along the wire and at wire crossings[25]. With vacuum annealed wire, such formation does not occur.

3.2.6 Annealing for uniform strength and elongation values

Uniformity of annealing is especially important in wire to be used in weaving as well as in deep drawing from strip. More uniform strength and equal elongation is attained in vacuum heat treating. No other annealing method has so far been found that can produce the uniform drawing values obtained from vacuum annealed stock[20].

Equally important as obtaining uniform structure is the ability to avoid surface decarburization. If the anneal is carried out under a protective gaseous atmosphere, danger arises from two factors. If the protective atmosphere contains more than 1% of CO_2, surface decarburization will take place owing to the decomposition of CO_2 into CO and O, the free oxygen combining with the carbon of the steel. A protective atmosphere must, therefore, be provided in which the CO_2 content is as small as possible. Surface decarburization can also be produced by even a slight moisture content in the protective atmosphere. Both dangers can be partly overcome by precisely controlled composition of the protective atmosphere and careful drying. Surface decarburization is avoided with a high degree of safety if the whole annealing process is carried out in the right temperature range under vacuum. Numerous trials on steel in vacuum reported by Herdieckerhoff[23,27] show that no decarburization of steel surfaces occurs at 850°C and very little even at 1000°C despite these seemingly high temperatures under the condition that the surface was free from rust and other oxides.

It should also be noted that vacuum annealing offers another possibility—the intentional decarburization of a surface. In this way, a surface may be made soft and ductile (and in a uniform manner). Experiments on wires have shown that, after a heating under vacuum to 750°C and the flooding of the vacuum chamber with moist hydrogen, twenty minutes of exposure produced uniform decarburization to a depth of 0.1 mm[27].

3.2.7 Annealing to improve galvanizing or tinning operations

If a material is to be galvanized or tinned, a dry surface prior to the application of the metallic film is of great importance. Usually galvanizing or tinning is preceded by pickling in order to produce a dry, clean and degreased surface. With vacuum annealed material, this is unnecessary, provided that a water-soluble drawing lubricant is used before the annealing. This innovation is especially interesting for the coating of fine wire in high-speed galvanizing plants. If the charge (especially if in wire form) is drawn through a stearate solution, a simple dipping in weak hydrochloric acid adequately cleans the wire since calcium changes to calcium chloride in the hydrochloric acid bath. Good results have been reported with wire drawn to about 2.5 mm in soap and then given a drawing in weak HCl to about 2.2 mm[25]. Vacuum annealed wire shows a particularly brilliant surface after galvanizing, which is more attractive and durable than that of pickled wire[24,25,27] and is presumably caused by the superior surface of the wire prior to coating.

4. Conclusion

Section 3 described many of the processes and techniques associated with vacuum heat treatment. Other areas of interest such as elimination of different pickling operations, protection of various surfaces during heating, removal of unsightly rusty surfaces, etc. exist at either planning or development stages. The growth of the vacuum heat treatment field is such that history will no doubt add many more developments, bigger and more versatile equipment, etc. in the years to come. All of these new ideas and paths will, however, be based on the fact that a vacuum keeps clean metallic surfaces clean, and can clean dirty or contaminated metallic surfaces. Indeed, that has been the entire message of this chapter.

References pp. 744

REFERENCES

1 H. J. GRABKE, Kinetics of phase boundary reactions between gases and metals. *34th Meeting on Reactions between Gases and Solids, Dayton, Ohio, October, 1969.*
2 O. KUBASCHEWSKI AND B. E. HOPKINS, *Oxidation of Metals and Alloys*, Butterworths, London, 1953.
3 N. CABRERA AND N. F. MOTT, *Rept. Progr. Phys.*, *12* (1948–49) 163.
4 C. WAGNER, *Z. Physik. Chem.*, *B 21* (1933) 25.
5 T. P. HOAR AND L. E. PRICE, *Trans. Faraday Soc.*, *34* (1938) 867.
6 W. JOST, *Diffusion u. Chem. Reaktion in Festen Stoffen*, Leipzig, 1937.
7 H. E. MÖBIUS AND F. KRALL, *Metall*, *23* (1969) 314.
8 D. F. MALLOY, SR., in R. B. BARROW AND A. SIMKOVICH (Eds.), *Trans. Intern. Vacuum Met. Conf., 1968*, Am. Vacuum Soc., New York, 1968, p. 626.
9 F. J. BECKET AND R. W. REYNOLDSON, *VIth Intern. Congr. on Electro-Heat, Brighton, May 1968, Paper No. 325*, Union Intern. d'Electrothermie, Paris.
10 ANON., *Metal Progr.*, *95*(9) (1968) 89.
11 ANON., *Métallurgie*, *101* (1969) 216.
12 ANON., *Metal Progr.*, *95*(9) (1968) 91.
13 N. K. KOEBEL, *Metal Progr.*, *96*(10) (1969) 136.
14 R. W. REYNOLDSON, *Intern. Symp. for Metallurgy and Heat Treatment of Metals, Warsaw, 1967*, D. 5.
15 J. VANDERSLUIS, *Metal Progr.*, *95*(9) (1968) 96.
16 *ASM Metals Handbook*, 8th edn. Am. Soc. Metals, 1961, p. 785.
17 J. J. BECKER, *Met. Rev.*, *7* (1962) 371.
18 J. J. BECKER, *Trans. Am. Soc. Metals*, *68* (1959).
19 R. M. BOSORTH, *Ferromagnetism*, Van Nostrand, New York, 1951.
20 J. K. STANLEY, *Trans. Am. Soc. Metals*, *72* (1963).
21 G. KOVES, in M. A. COCCA (Ed.), *Trans. Vacuum Met. Conf., 1964*, Am. Vacuum Soc., Boston, Mass., 1965, p. 189.
22 ANON., *Wire Wire Prod.*, Sept. (1964) 1331.
23 W. HERDIECKERHOFF, *The Wire Industry*, *35* (1960) 593.
24 W. HERDIECKERHOFF, *Wire Wire Prod.*, *41* (1966) 257.
25 W. HERDIECKERHOFF, *Wire Wire Prod.*, *36* (1961) 470.
26 ANON., *Metal Progr.*, *95*(9) (1968) p. 95.
27 W. HERDIECKERHOFF, *Wire Wire Prod.*, *41* (1966) 892.

Chapter IX

Joining

Section 1

R. BAKISH

Section 2

S. S. WHITE

Section 3

R. BAKISH

1. Electron Beam Welding

1.1 Introduction

The idea of using electron beams as "working tools" was probably first conceived by M. von Ardenne in 1938[1], yet the thought of using them for welding did not appear until the mid-1950's[2,3]. The reason for this is that neither the need for accomplishing this nor adequate tools to perform this operation were really at hand much before. The need for new materials and their processing requirements, initially for the nuclear power industry and later for the aerospace industry, created the environment necessary for the invention of electron beam welding and cutting. The steady advances of vacuum technology at that time made the commercial success of this process possible. The names of Stohr[2], Weyman[3] and Steigerwald[4] should be remembered by those who derive benefits from this metal joining and cutting technique. Under the aegis of the nuclear industry, Stohr and Weyman realized the potential of electron beams as joining tools, and electron beam joining was born. Steigerwald, on the other hand, pioneered electron beam cutting.

1.2 The process and its modifications

Electron beam welding by definition is the fusion welding process which utilizes the electron beam as a heat source to effect fusion.

The full and exact details of the mechanism operational in the process have not yet been resolved to the satisfaction and agreement of all those active in the field, though several detailed studies of the problem have been made[5-18] and

---

R. Bakish wishes to thank all those who generously provided pictorial and reference material used in this chapter.

References pp. 800–802

much additional work has been performed[19]. The record of these investigations is valuable and is recommended to those who wish to obtain better understanding and become intimately familiar with the subject. It might be added that because electron beam welding is already an accepted commercial process capable of solving a wide range of problems the pressure truly to resolve all the details of the mechanism is virtually non-existent. The essence of what is most widely accepted as the prevailing state of affairs can be explained on the basis of classical electron penetration alone. It is proposed that the impacting high velocity electrons penetrate slightly below the surface of the metal, where on collision with the work piece they release the bulk of their kinetic energy and transfer it initially to the lattice electrons and then to the total lattice. This energy transfer leads to an increase in lattice vibrations and, in turn, brings about a substantial rise in temperature, leading to local melting and evaporation. The displacement of the molten material by this vapor, which is of considerably lower density than the solid, permits additional penetration of the beam. The interaction of the beam with this vapor leads also to formation of plasma in the area. The beam, the plasma and their interaction in the material lead to the well-known high penetration and, in turn, to the favorable depth-to-width ratio of the weld zone, usually associated

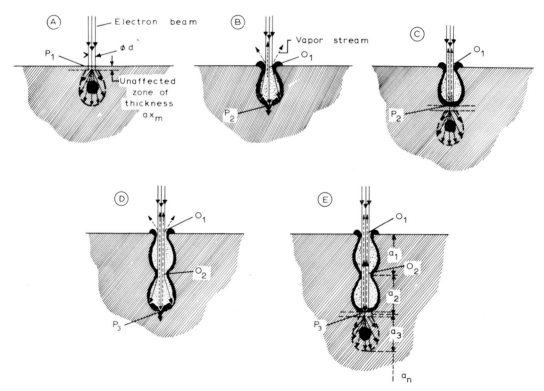

Fig. 1a. Schematic of electron beam penetration (after H. Schwartz).

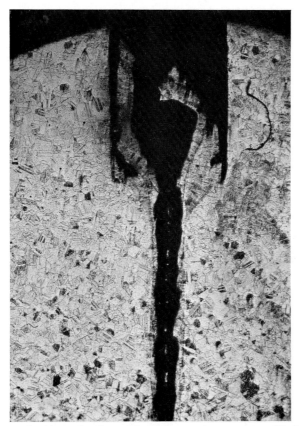

Fig. 1b. Experimental evidence for electron beam penetration as proposed in Fig. 1a. (courtesy Muelemans[11])

with the electron beam welding process. (See Fig. 1a for schematic representation of mechanism, Fig. 1b for experimental evidence for this type of step-wise penetration and Fig. 2a for a schematic presentation of all items which must be considered in a detailed analysis.) As the beam is displaced the molten metal in the cavity is moved up opposite to the direction of the beam displacement producing the seam.

At this point, it should be pointed out that under no circumstances would every electron beam generated tend to produce the "signature" usually associated with electron beam welding, namely, the high depth-to-width ratios. (See Fig. 2b, which illustrates the highest depth-to-width ratio known to the writer.) This matter has been the subject of much legal deliberation in connection with patents in the United States. Hashimoto and Matsuda[10] illustrated this at an early date. Recent work of Muelemans[11] confirms this convincingly and presents the complete dependence of fusion zone configuration and the deep penetration welding phenomenon on the power density values of the beam. Donovan[12] and Steffens and Wilkens[13] have also shown this effect. Deep penetration is primarily not dependent

References pp. 800–802

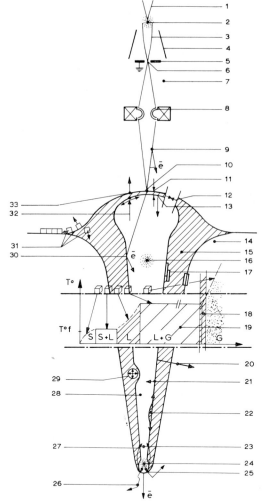

Fig. 2a. Schematic of all factors which must be considered in detailed analysis of beam penetration. (courtesy Muelemans[11])

Points 1 to 8 are the details of electron gun used; points 9 to 33 the various physical aspects of the electron penetration in a metal.

1. Filament.
2. The thermionic emission.
3. Curvature of the electrons paths, due to the electrostatic field.
4. High tension Wehnelt cathode.
5. Zero potential anode.
6. First electrostatic focus.
7. The pressure in the housing is $\pm 10^{-4}$ Torr.
8. Magnetic lens.
9. Electron/molecule collision; the mean free path, independent of acceleration voltage, is six times that of the molecule hit.

10. Focus of the magnetic lens.
11. Thin, liquid, spherical calotte, transparent for the electron beam.
12. Surface tension responsible for the closing of the channel.
13. Thermal screen for the high temperature radiating from the impact at the bottom of the channel.
14. Solid substrate of the target metal.
15. Liquid substrate of the target metal.
16. Gas phase (complex plasma).
17. Surface diffusion; the energies playing a part in the surface diffusion are smaller than those of volume diffusion; the resistance to displacement in viscous media is proportional to the deflection speed and not, as in the elastic theory, to the deflection itself.
18. Change of the density of the target metal resulting in a significant increase of the mean free path of the electrons.
19. Schematic diagram temperature–heat of a metal considered as pure; the allotropic changes and the exact values of the specific heats have not been taken into account. The diagram shows the thermodynamic state of the volume elements from a cross section during the electron impact. However, it shows that the liquid and the gaseous metal hold a fairly large amount of heat in a very small volume, that condensation phenomena may occur and that the gases expand and carry very hot liquid particles when the spherical calotte ruptures on account of overpressure.
20. Thermal gradient.
21. The pressure in the channel results from the pressure of the high temperature plasma and from the initial reaction force of the evaporating and reevaporating atoms.
22. Liquid macroparticles from projections at the direct impact.
23. The energy of the electrons is lessened by the energy lost in the collisions in the liquid film and the collisions in the plasma and the liquid walls.
24. Very high temperature zone.
25. Direct impact electrons–ions of the crystalline network; loss of metallic cohesion. Greater probability of energy transfer.
26. The electron paths in the solid metal are determined by the probability functions.
27. When the liquid spherical calotte is ruptured and the beam repenetrates to the bottom of the channel, a film forms by an obstruction effect related with the surface tension of the liquid. This new film rapidly ascends to the surface where it ruptures owing to the new overpressure.
28. The fastest electrons of the Maxwellian speed distribution reach the bottom of the channel.
29. The gases trapped in the base metal expand in the liquid phase and explode in the channel. They are mainly responsible for the irregular shape of the melting bath. They sometimes block the electron beam leading to a spherical melting bath at that point.
30. The collisions electrons–ions of the liquid phase of the channel give rise to quantum forces of repulse nature and with fast variation. Since the angle of incidence of the collision is small and since the frequency and the amplitude of ions oscillation are important, one may assume that there is a minimum of electron–ion energy transfer.
31. The oxides are dissolved in the liquid bath or evaporated when they are more volatile than the base metals.
32. Impact of liquid micro and macroparticles on the inner walls of the liquid film. Sometimes this phenomenon can be very well observed by direct optical means.
33. Evaporation in the vacuum of the liquid film. When the film tends to thicken, a larger part of the beam energy is lost because of a greater mass evaporation, the liquid remaining bonded by the capillary forces.

on the system which generates the beam or on the voltage to which the beam electrons are accelerated. This was initially suggested by Hablanian[14] and Franco-Fereira[15] very early in the development of electron beam welding machines. It should be remembered, however, that by virtue of space charge problems higher

References pp. 800–802

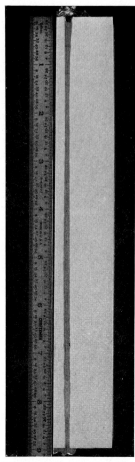

Fig. 2b. High depth-to-width ratio electron beam weld with 5083 aluminum alloy, 230 mm thick, obtained at 50 cm/min, using 525 mA at 58 kV. (courtesy J. Solomon, Sciaky)

voltages simplify beam focusing and make it easier to achieve the necessary high power density beams at lower total power.

Figures 3 a–d illustrate changes in nugget geometry as the current increases from 2 to 5 mA with corresponding increases in power density. Here a stainless steel strip was placed on top of a copper strip in order to show more clearly the penetration in metallographic examination. These runs were carried out at a constant acceleration voltage of 150 kV and in one continuous operation[11]. On the basis of these and other studies already referred to, it can be stated with a good degree of confidence that beam power densities of the order of 10^6–10^8 W cm^{-2} are needed for welding and even greater power densities for drilling and cutting.

It is important to state here that use of this process is no longer confined to the extravagant "off-the-beaten-track" research projects hitherto associated with it. It is rather a recently introduced joining method which is already well entrenched,

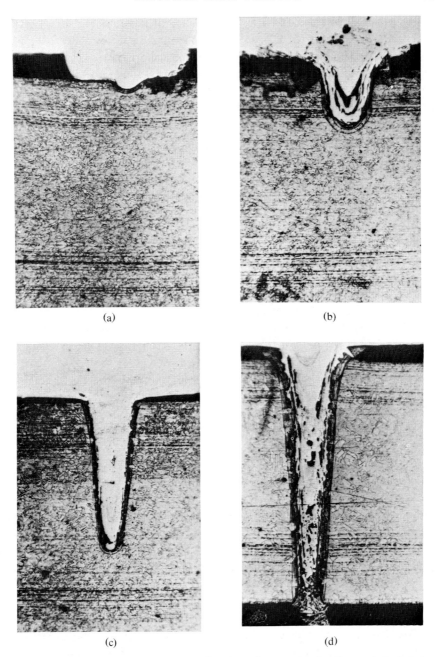

Fig. 3a-d. Changes in nugget geometry as function of power density. Respectively 2, 3, 4 and 5 mA at 150 kV. (courtesy Muelemans[11])

offering great advantages and showing promise of considerable future growth. There is virtual certainty of this because of the many features which have firmly established electron beams in welding in the very short time since its discovery. The ultimate level of acceptance here will simply be a function of equipment cost and productivity.

The important features of the process are:
(1) high power density
(2) low heat input
(3) very high purity environment (*i.e.*, vacuum or controlled purity inert gas)
(4) superior control
(5) high speed (250–600 cm/min, and possibly higher, depending on material and thickness).

As a consequence, we can among other things:
(1) obtain fusion welds with varied nugget geometry and fusion zone width including the highest known depth-to-width ratios
(2) obtain low distortion joints and introduce minimum thermal damage
(3) obtain minimum heat-affected zones
(4) join thin to thick section materials
(5) join very heavy cross-sections (up to 230 mm in some materials)
(6) process all materials which benefit from these superior features
(7) adopt them to automatic and programmed control
(8) effect cost savings in many joining problems and, in particular, in joining of heavy sections
(9) join dissimilar materials
(10) weld after finish machining
(11) increase welding accuracy
(12) shorten welding times
(13) obtain superior joint properties in many materials.

Note, however, that with all these advantages which they certainly offer for welding, electron beams are not a "cure all" and they definitely operate within the confines imposed on materials by the laws of physical thermodynamics and metallurgy. It should never be accepted that two materials deemed unweldable on the basis of metallurgical thermodynamics can be welded with electron beams. The fusion of two metals to each other by means of the electron beam, if it can be accomplished at all, implies that it is a weld. A weld, by definition, must have structural and functional integrity.

The electron beam machines today are capable of welding an extremely wide range of materials in an equally broad range of thicknesses in a single pass. The premise which brought the utilizing of electron heating to the welding of nuclear materials was the fact that vacuum processing prevented the interstitial embrittlement which plagues these materials. This, indeed, was the earliest recognised virtue of the electron beam welding process prior to the realization of its other capabilities. Now it has become a much more diverse and powerful tool than it was in the days of its discovery. Today, electron beam welding can be used in three operational modes: (a) in vacuum (10^{-4} Torr or better), the original mode, to be referred to as hard vacuum mode; (b) in air at reduced pressures (10^{-1}–3×10^{-1} Torr), the soft vacuum mode; and (c) in air or under inert gas cover at one atmosphere, the so-called IAEBW. The mechanism of beam penetration already discussed prevails in all modes, with the available beam power density at the work piece governing the geometry of the fusion zone.

1.2.1 Hard vacuum mode electron beam welding

The hard vacuum mode utilizes guns with acceleration voltages in the 20–150 kV range and power levels from 2 to 30 kW. It is the most powerful of the three modes referred to above. It is capable of resolving the broadest range of tasks, materials and a thickness range of from several thousand angstroms in thin film electronics interconnections up to 230 mm in a single pass in some special cases[20]. Review of early work carried out is to be found in Bakish[21], Michler and Monroe[22] and a British Welding Research Association publication[23]. Further activities up to 1964 are described by Miller and Takenaka[24] and Bakish and White[25]. For those interested, most, if not all, of the information available on the subject will be found in papers by Hetherington[26], Bakish[19,21,25,27–31], Morley[32,33], ElKareh[34], Haddad[35], Funk[36], Pease[37], Eichhorn, Koch and Matting[38] and Nazarenko[16], and in the Welding Journal.

Although in principle the process is capable of doing the job, some problems remain in the joining of heavy cross-sections, as will be seen in the discussions below. Not only is the technique capable of welding metallic materials, but ceramics too have successfully been joined[39] and ceramic components manufactured after special precautions have been taken and special heat-up and cooling provisions made[40].

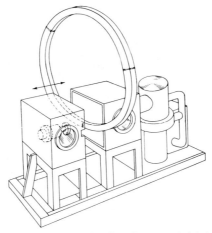

Fig. 4. Split vacuum chamber. (courtesy Sciaky)

This mode has an important practical limitation, *i.e.*, the size of object which it can process, as a consequence of the fact that all such work must be in a vacuum chamber. Although these chambers have grown tremendously in size—one can easily maneuver an automobile in some of the bigger ones or have a 12 m linear travel in others—it is unlikely that housings large enough to accommodate a deep diving submarine or a sizeable rocket casing, for example, will ever be built. One of the approaches utilized here to offset this to some extent has been the building of vacuum chambers in which the object to be welded becomes an integral part of the walls of the vessel as seen in Fig. 4[41]. In addition to this, movable chambers which can be attached to the work to be welded have also been considered. Initial work here was done by Anderson[42] of North American Aviation, and this idea is under renewed consideration in a joint contract between Sciaky and North American Aviation under U.S. Air Force sponsorship. The French Atomic Energy Commission, a pioneer and an ardent supporter of electron beam welding, has also carried out research in this field, and introduced the local chamber (Fig. 5).

Fig. 5. Local chamber, system for joining of nuclear fuel elements. (courtesy J. Stohr, Saclay)

Electron beam welding has led to major advances in joining technology and solved joining problems deemed impossible or difficult to resolve. The hard vacuum mode has no peers when it comes to welding of the refractory and reactive metals. It is the technique which is invariably specified when high reliability nuclear reactor applications are considered. As a matter of fact, this is the area in which the first electron beam welding specifications were developed by the U.S. Navy.

High quality equipment is manufactured commercially and can be obtained without difficulty, and a very long list of case histories with jobs well done establishes the worth of electron beam welding performance here.

This mode of electron beam welding, in addition to the size limitation referred to above, has at least one problem area (as revealed in a recent study by Groves and Gerken[43]) which needs much work before complete understanding and solu-

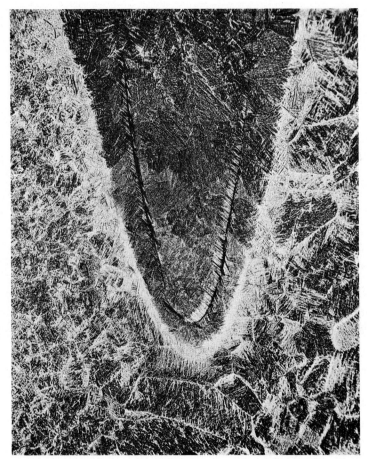

Fig. 6. Root of 2-pass weld in Ti–6Al–4V 45 mm plate, 22 kV, 300 mA, 65 cm/min, 2 passes, 1 each side. (courtesy J. Gerken, TRW, and with the permission of the U.S. Air Force)

tion are achieved. Reference is made to electron beam welding defects encountered when some heavy cross-section materials—*i.e.*, sections of 25 mm thickness and beyond—are joined. Figures 6–9 show the more important type of defects which have been observed in the course of this study. They are a good example of the type of problems that a new user could conceivably encounter. While speculations

References pp. 800–802

regarding the causes are given in the legends to Figs. 6–8, no good explanation can be advanced today for the appearance and geometries of the defects shown in Fig. 9. This problem could have serious consequences in pressure vessel welding operations in the aerospace, chemical and nuclear industries. The exact nature of the defects, the reason for their occurrence or means to avoid them still apparently remain beyond the understanding and competence of most, if not all, of those utilizing the technique.

Fig. 7 Cold shut at bottom of electron beam weld in 25 mm Ti–6Al–4V Plate 23, kV, 330 mA, 45 cm/min, 1 pass. Probably caused by insufficient beam power. (courtesy J. Gerken, TRW, and with the permission of the U.S. Air Force)

Additional difficulties are those of the occurrence of spikes (see Fig. 10) and wavy contours of the fusion zone (see Fig. 11) when certain materials are processed[44]. Both these problem areas are domains where valuable contributions to the state of the art can be made and are recommended for detailed studies. The successful resolution of these problems will give additional impetus to the already importantly entrenched technology.

This sphere is one in which the electron beam equipment producers have shown a sad degree of neglect. The writer is aware of no published account of any work in this direction. This criticism of the equipment manufacturers should by no means be considered to detract from their effort in regard to the equipment itself.

Inasmuch as, except for standard photomicrographs, no detailed studies considering quality of fusion zone, heat-affected zone and transition zone in either the soft vacuum or the IAEBW modes have appeared in the literature, no comments

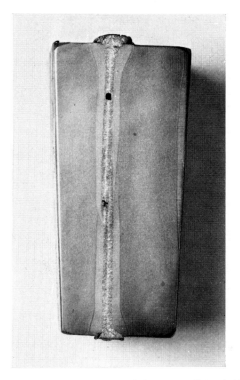

Fig. 8. Porosity in electron beam weld in HP-9Ni-4 Co-0.25C Steel 38 mm thick (bead on plate), 50 kV, 320 mA, 130 cm/min, 2 passes. Probably caused by too high speed to allow gas to escape. (courtesy J. Gerken, TRW, and with the permission of the U.S. Air Force)

on these can be made. It is suspected that difficulties not unlike those observed in the high vacuum mode can be expected. These will be modified by the unknown effect of ambient pressure and power density changes which will prevail as a consequence of the respective operational mode.

1.2.2 Soft vacuum mode electron beam welding

The second mode of electron beam welding is referred to as either partial vacuum or soft vacuum mode; we shall use the latter. This mode became practicable because of the realization that the impurity level of rarefied air is comparable with that of accepted commercial inert atmospheres (see Table 1). 4×10^{-2} Torr

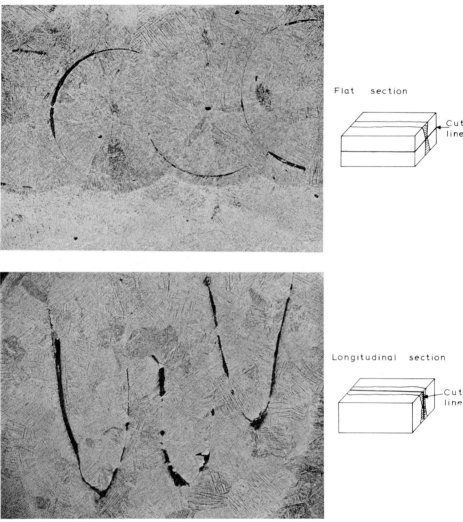

Fig. 9. Electron beam welds in Ti–6Al–4V, 38 mm thick, 22 kV, 300 mA, 70 cm/min, 2 passes, 1 each side. Defect and schematic of sites of this occurrence. (courtesy J. Gerken, TRW)

air pressure, for example, is comparable with the contamination level of the best commercially available argon.

The satisfactory contamination levels from the standpoint of pressures are those which can be generated with mechanical pumps. These also permit limited scattering of the beam and the maintainance of good beam power densities. (See Fig. 12 for pressure effects on weld penetration.) Here, the working distance which in the hard mode can be up to 50 cm is reduced to a reasonable depth of constant

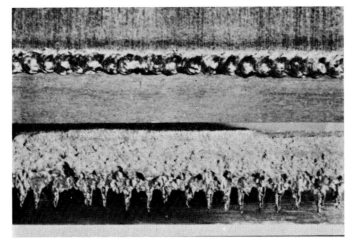

Fig. 10. Spikes in high strength aluminum alloys (surface view top) and cross-sectional view (bottom)). (courtesy Hicken and Booco, Dow)

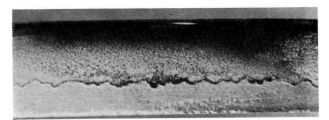

Fig. 11. Wave-like contour in uranium weld. (courtesy Hicken and Booco, Dow)

TABLE 1

Air pressure (Torr)	Contamination level (ppm)	Argon purity grade	Contamination level (ppm)
10^{-4}	0.132	99.995	50
10^{-3}	1.32	99.95	500
10^{-2}	13.2		
10^{-1}	132		

References pp. 800–802

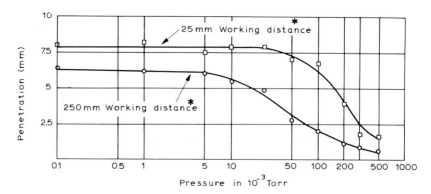

Voltage 100 kV
Current 9 mA
Material AISI 304

*Note: Beam path length at indicated pressure equal to working distance plus 400 mm

Fig. 12. Effects of pressure on weld penetration. (courtesy J. Meier, Hamilton Standard)

Fig. 13. Underwater electron beam welds (courtesy B. Schumacher, Westinghouse)

power in the vicinity of 20 cm. While this will be discussed subsequently, it is important to note that in IAEBW working distances of only a fraction of a centimeter at best appear possible.

The soft vacuum mode electron beam welding was announced virtually simultaneously[45,46] by the two important United States producers in the Spring of 1966. The technique's main virtue, as already stated, is the fact that its pumping requirements can be met without diffusion pumps, and in terms of commercial and economical processing this is of major importance. This mode appears ideally suited for cases where a large number of parts are to be manufactured and a score of repetitive production tasks undertaken. Although the workpiece is at the higher pressure the environment in the beam generating components of the system remains hard vacuum. The automotive industry appears to be especially interested in this mode of electron beam welding.

Specimen (all 3 mm thick)		*Electron beam*		*Welding speed*	*Remarks*
Type	*#*	*(kV)*	*(mA)*	*(cm/min)*	
Copper (OFHC)	UW61*	125	50	75	Lap weld; X-ray shows little porosity
Copper (OFHC)	UW59	120	44	75	Butt weld; X-ray shows porosity
Copper (OFHC)	UW58	120	42	75	Butt weld; not fully fused
304 Stainless steel	UW49	100	33	130	Butt weld; top 2.3 mm, root 1.3 mm wide, half of length free from porosity, other half porous
304 Stainless steel	UW48	100	32	130	Butt weld with machined edges; top 2.5 mm, root 0.8 mm wide; slight porosity
304 Stainless steel	UW45	100	32	130	Butt weld (rear side shown), top 2.5 mm, root 1.5 mm wide, edges shear cut but not machined X-Ray shows no porosity
Carbon steel	UW68	105	42	130	Lap weld; X-ray shows porosity
304 Stainless steel	UW51	100	38	130	Lap weld; X-ray shows no porosity
304 Stainless steel	UW3	103	40	100	Bead on plate, top 1.3 mm, root 0.8 mm X-Ray shows 25 mm section free from porosity

*Upper left corner, next one below, etc. All specimens were 50 mm long; difference between gun orifice and work piece was 3 mm in all the above cases.

References pp. 800–802

1.2.3 Inert atmosphere electron beam welding (IAEBW)

One of the foremost problems preventing immediate wider acceptance of electron beam welding has been the confines of the vacuum chamber. The IAEBW welding in principle at least removes this restriction for the process. But this benefit in process simplification is not obtained without cost. This mode utilizes electron beams generated under equivalent conditions to those in the other two modes already discussed, but makes use of their energy only after it brings them at atmospheric pressure through a series of differentially pumped stages. Details of the physical phenomena and property changes associated with the bringing of an electron beam in air at atmospheric pressure have been the subject of extensive studies; these have been reviewed by Schumacher[47].

Scattering of the beam electrons by molecules of different gases through which the beam passes is responsible for the loss of much of the beam power, and the loss is dependent on the scattering cross-section of these molecules. Commercial systems must be so designed as to permit the preservation of sufficient power density for welding after the beam has passed through the specific medium. This necessitates operating voltages in the 150–200 kV range and reduction of working distance for the equipment to a value not exceeding 1 cm as indicated above. This reduction of working distance presents some serious limitations upon hardware configuration and, in turn, appreciably reduces the potential scope of this mode of joining. Yet, for some applications, such as for welded tube manufacture, to mention just one example, it appears to have exciting possibilities. Although it is referred to as inert atmosphere electron beam welding, it can also weld in air or under water[48] (see Fig. 13).

This technique as initially developed by Hamilton Standard necessitated considerable pumping capacity which, when combined with the high voltage, led to an installation which was exceedingly costly. Westinghouse has entered this field as a competitor[48-51] and there are some indications of possible cost reduction. Full details are not as yet available. It is interesting to note that the first mention[52] of the potentialities of the process was made as early as 1953. Some evidence of the presence of porosity in joints approaching 13 mm thickness is indicated in at least one study[53], but much more experience is needed before firm statements can be made about the ultimate quality possibilities of IAEBW-processed materials in sections approaching and exceeding this thickness.

1.3 Equipment

All electron beam joining systems fit the block diagram given in Fig. 14, though they differ in details, as well as in the specifications of the individual components which make up the system, consisting of gun and power supply with respective control, vacuum chamber with pumping provision and controls and work handling mechanism and controls. In the soft vacuum mode, for example, the welding chamber is replaced by a single work chamber with a provision for

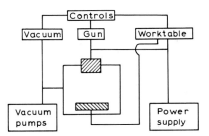

Fig. 14. Block diagram of an electron beam welder.

rapid loading and unloading, and this chamber is typically prepared for each and every type of part which will be mass produced. With the IAEBW, on the other hand, no chamber is present but, instead, the beam reaches the atmosphere through a series of differentially pumped stages. The exact number and design vary with the different equipment producers.

1.3.1 The electron gun

The electron gun is the most important component in an electron beam welding system. In principle, the electron gun is an electron optical device which is capable of generating, accelerating and, to some extent, even focusing electrons. It consists of two main components: (1) the beam-generating components—i.e., the cathode and related heating auxiliaries—and (2), the field-shaping components which cause the electrons emitted by the cathode to assume their proper configuration of a useful beam.

The cathode may be self-heated or indirectly heated by bombardment. It can be of hairpin shape, or shaped as a solid rod or a pill, and is habitually made of a variety of electron emitters in accordance with its performance and environment requirements. In welding systems tungsten and tantalum are mostly used, with occasional incidence of lanthanum hexaboride. The electron-emitting properties of the cathode are explained by quantum dynamics. The criterion for electron

TABLE 2[54]

Substance	Work function
Barium	2.1
Barium on barium oxide	1.0
Barium boride BaB_6	3.45
Barium carbonate	1.0–1.5
Lanthanum boride LaB_6	2.66
Molybdenum	4.15–4.44
Rhenium	4.75–5.1
Tantalum	4.03–4.19
Thorium boride ThB_6	2.92
Thoriated tungsten	2.5–2.6
Tungsten	4.25

References pp. 800–802

emission is the work function and it refers to the potential energy barrier which must be overcome in order to produce free electrons. Table 2[54] gives some values of the work function of some of the more important emitter materials. These electrons are then removed through the application of suitable fields. If, for example, the applied field is of sufficient strength to withdraw all available electrons, the saturation current density at T degrees K is given by:

$$I = AT^2 e^{-b_0/T} * \text{ A/cm}^2$$

The terms A and b_0 are constants and functions of the material with an approximate theoretical value for A of about 120 A/cm^{2}K^2, although recent research indicates values of 60 for tungsten, 55 for molybdenum and 37 for tantalum.

The gun of practical design customarily draws less than the saturation current in what is conventionally referred to as space charge limited operation. This mode of operation is independent of the temperature of the cathode. The actual current which flows between parallel plates is given by Child's law stating:

$$I_{sp} = 2.33 \times 10^{-6} \frac{V^{\frac{3}{2}}}{D^2} \text{ A/cm}^2$$

where D is spacing between electrodes and V the acceleration voltage. The 3/2 power of the acceleration voltage is valid for virtually all electrode configurations provided that proper constants are introduced.

While these two equations tend to dominate the fundamental aspects of gun design, the actual design of a superior gun system is a matter of considerable difficulty. Here, all matters from materials of construction selection to precise definition of the environment where the gun will be operational and the exact configuration of the beam needed for a specific application have to be most meticulously analyzed in order to produce a satisfactory beam generating system. The exact knowledge of the working environment is most important as contamination reduces drastically the electron emission properties of most materials.

There is a considerable variation in electron beam generating systems which could conceivably be used in electron beam welding installations. However, it appears that because of performance characteristics necessary, two main types of guns are mainly encountered; these are various modifications of the Pierce type guns and the telefocus or Steigerwald type guns. The Pierce type system[55] uses either directly or indirectly heated cathodes, while in the telefocus type the electrons are mainly supplied by directly heated filament cathodes. The beam emitted from either of the systems is either single or multiple stage focused by electromagnetic or electrostatic lenses. Many of the systems manufactured today have provision for oscillation deflection; some can even program the beam to be displaced along its desired trajectory in a variety of wave forms.

---

* Richardson's equation modified by Dushman.

1.3.2 Commercial welding systems

The power supplies for the miscellany of electron guns are habitually high quality filtered and stabilized supplies. The specific degree of filtering and stabilization depends very much on the individual system and its requirements.

The three modes of electron beam welding as stated in the foregoing are carried out by three types of equipment, though the remarks pertaining to electron guns given above apply equally well to all three. Ample literature on electron optics[56,57] and gun design[55,58] is available. Those interested in vacuum systems and vacuum system design should see Chapter II of this volume. With this information, and with time, money and engineering capabilities at hand, virtually anyone could produce an electron beam welding system. To those primarily interested in joining, the commercial equipment is recommended. A broad range of electron beam welding systems of various qualities is produced in at least eight countries (see Table 3). This table illustrates the more important types of equipment that appear to be manufactured by the various concerns.

Because of the fact that all the necessary information is not readily available and because occasional equipment revisions inevitably take place, the table is not as accurate and complete as the writer would have liked to make it. It does, however, contain the most important facts and it should remain reliable for some time to come. The available machines, regardless of mode of welding to be utilized,

Fig. 15. Large chamber, Hamilton Standard machine. (courtesy J. Meier, Hamilton Standard)

TABLE 3

Producer	Country of origin	Acc. volt. (kV)	Power (kW)	Gun	Gun type*	Operational modes	Observation modes**
Brad Thompson Industries	U.S.A.	30–60	2–5	Movable	P	HV	O
Hamilton Standard	U.S.A.	60–150	2–15	Fixed and movable	S	HV, SV, IAEBW	C
Westinghouse	U.S.A.	150	6–15	Fixed	NA	IAEBW	O
Sciaky	U.S.A.	30–60	6–30	Movable	P	HV, SV	C and O
Thompson Welder	U.S.A.	30–60	6–15	Movable	P	HV	O
N.R.C.	U.S.A.	30	5	Fixed	P	HV	O
Metropolitan Vickers	U.K.	150	5	Fixed	S	HV	C
Torvac Ltd.	U.K.	30–60	1–10	Movable	P	HV	C and O
Hawker Siddeley	U.K.	150	2–15	Fixed and movable	S	HV, SV, IAEBW	C
Alcatel	France	30–60	5	Fixed	P	HV, SV	O
Sciaky	France	30–60	6–30	Movable	P	HV, SV	C and O
Leybold-Heraeus	W. Germany	150	6–15	Fixed and movable	S	HV, SV, IAEBW	C
Steigerwald Strahltechnik	W. Germany	150	10	Fixed	S		C
Balzers AG	Liechtenstein	30	5	Fixed	P	HV	O
Inst. M. von Ardenne	E. Germany	20–125	5	Fixed	S and P	HV ⎫ Possibly	C
Paton Institute	U.S.S.R.	30–100	Up to 10	Fixed	S and P	HV ⎬ others	C and O
Moscow Power Institute	U.S.S.R.	30	?	?	P	HV ⎭	O
Japan Electro Optical	Japan	30–60	3–15	Fixed	S	HV	C

* P = Pierce; S = Steigerwald, Telefocus.
** C = Coaxial with beam; O = Oblique.

Fig. 16. Large chamber, Sciaky machine. (courtesy J. Solomon, Sciaky)

fall into two main categories: stock items of various configurations and custom systems. The latter are designed with the aim of solving a specific joining problem or serving the user's individual needs.

Let us now look briefly at typical equipment as produced by some of the suppliers mentioned. By virtue of the fact that the number of electron beam machines in use in the world must have passed the 1000 mark, at best only extreme selectivity is possible. Figure 15 shows typical large chamber standard items of Hamilton Standard obtainable in a wide range of work chambers, and Fig. 16 illustrates Sciaky's machine, which can likewise be obtained with considerable variation in chamber sizes, with or without seam tracking and beam deflection, with various degrees of programming and table motions. The Hamilton Standard's system uses Steigerwald's telefocus guns, while Sciaky employs modified Pierce variety guns. Most of the Hamilton Standard systems use stationary guns, while Sciaky likes the movable approach. Both systems can be obtained with powers of up to 30 kW and are operational in the 60–150 kV and 30–60 kV ranges respectively.

Figure 17 shows a Balzers system utilizing a bolt cathode gun developed by Bas[59], with provision for ion deflection aimed at extending cathode life. This is a 30 kV system. Here the separate evacuation of the gun chamber as with the soft vacuum mode was applied for the first time. In Fig. 18 the soft vacuum principle is illustrated schematically.

References pp. 800–802

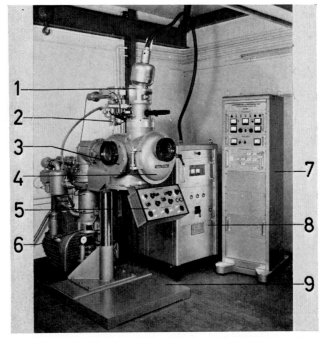

Fig. 17. Balzers system 30 kV. (1) Gun, (2) Gun diffusion pump, (3) Observation port, (4) Access door and port, (5) Controls, (6) Diffusion pump, (7) Vacuum controls, (8) Power supply, (9) Stand. (courtesy O. Winkler, Balzers)

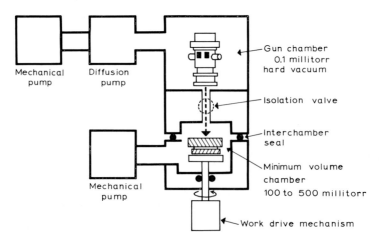

Fig. 18. Schematic of soft vacuum welding machine.

Figure 19 shows a custom-designed soft vacuum Sciaky system for the manufacture of hack saw bands, while Fig. 20 depicts a twin Hamilton Standard soft vacuum system.

A cross-section of an experimental approach utilized by the French Atomic

Fig. 19. Sciaky custom hack saw blade stock soft vacuum system. (courtesy J. Solomon, Sciaky)

Fig.20. Twin soft vacuum system. (courtesy J. Meier, Hamilton Standard)

References pp. 800–802

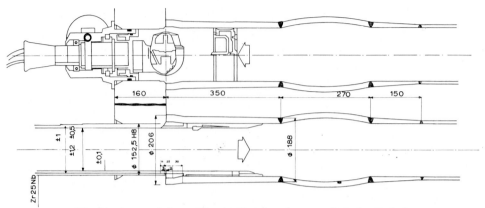

Fig. 21. Approach for welding inside tubes. (courtesy J. Stohr, Saclay)

Fig. 22. Soviet system for manufacture of electronic components. (courtesy N. A. Olshansky, Moscow Power Institute)

Energy Commission for welding inside tubes can be seen in Fig. 21. Figure 22, on the other hand, depicts a relatively early 1963, but unusual, Soviet system with provision for a revolving type loading, while Fig. 23 shows a large size chamber system of a more recent entry into the electron beam welding equipment field.

The Westinghouse systems for IAEBW are seen in Figs. 24a and 24b, while Figs. 25a and 25b show the portable Westinghouse space welder and its block diagram.

Fig. 23. Large chamber 60 kV 30 kW welding system. (courtesy A. Shannon and H. Lander, Thomson Electron Beam Corp.)

1.3.3 X-Ray hazards

All electron beam generating systems operating at voltages exceeding 22 kV, which is the accelerating potential for the generation of the Mo Kα-line, generate X-rays. The higher the acceleration voltage, the higher the intensity of the X-rays emitted. Their properties are understood and the necessary precautions to be taken are known[60,61]. Industrial health and plant safety regulations have placed the burden of safeguarding users of this equipment on the manufacturer. Depending upon the intensity of the X-rays generated and the mode of electron beam welding system which is being considered, mild steel, stainless steel, lead lined steel and concrete have been and are used as a protection with the miscellany of production electron beam installations.

1.4 Industrial applications and related topics

Electron beam welding, which was born in the sphere of the nuclear power industry, has long since transcended its bounds. The aerospace industry was the next to be entered with the technique now about to become accepted as a production tool by automotive and metal working industries. Up to 1962 great potential was envisaged for electron beam in microelectronics. Although applications have developed here, even the comparatively low heat input of the electron beam relative

References pp. 800–802

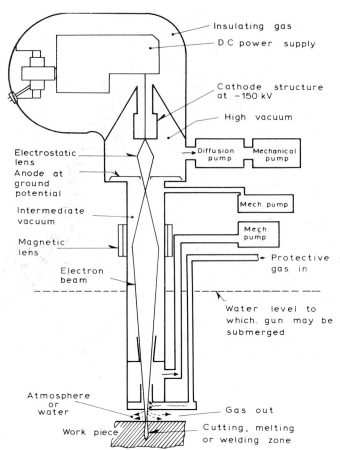

Fig. 24a. IAEBW welder, schematic. (courtesy B. Schumacher, Westinghouse)

to various welding techniques is most devastating when integrated circuits are concerned. Electron beams remain as a powerful tool in component hermetic packaging, while only thermocompressing bonding appears to be satisfactory for integrated circuits. Electron beam welding will continue to be used in the making of interconnections, but the future here is not particularly promising as a growth area.

A comprehensive review of the most voluminous work with different materials is outside the space limitation of this chapter. Suffice it to say that virtually any material or combination of materials deemed weldable by laws of physical metallurgy can be welded by electron beams. The actual welding can be carried out much faster than TIG or MIG processing and this will be accomplished without the need for filler metal. The experience accumulated to date indicates that almost invariably electron beam produced joints will have properties superior to those produced by other commercial joining techniques. In the ultimate, the decision as to whether any of the electron beam modes or alternative joining techniques is to

Fig. 24b. IAEBW welder. (courtesy B. Schumacher, Westinghouse)

be adopted in a production facility must depend on economics. Giving examples of problems tackled or properties of materials joined is also useless as no selection made could fully cover and comprehensively illustrate the vast scope of applicability of this powerful joining tool. With regard to thickness of sections to be joined, it can in general be stated that, on the basis of information available today[43], the heavier the cross-section the more favorable electron beam welding economics become. Although, as already stated, there remain some real problems with the joining of heavy section materials, this area of application has some exciting expansion in sight.

In terms of jigs and fixtures, those accepted for TIG and MIG and, in many cases, lesser fixturing will be sufficient for electron beam welding. Besides those suggested by the equipment manufacturers, some limited information on jigs and fixtures can be found in the literature.

One field of application which perhaps transcended industrial boundaries at an early stage is that of repair welding. By virtue of the unique capabilities of the process substantial savings have been and continue to be made by the repair welding of complex and expensive parts. No other welding technique has ever been able to tackle repair welding with the breadth of materials and configurations that electron beam welding has accomplished. The most recent adoption of elec-

Fig. 25a. Portable battery operated space welder. (courtesy B. Schumacher, Westinghouse)

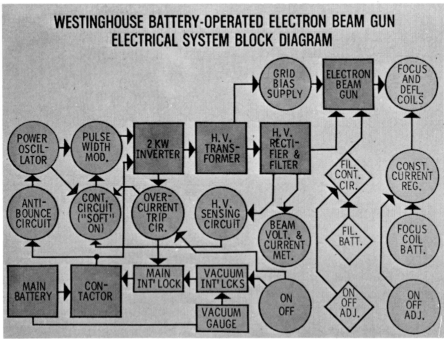

Fig. 25b. Electrical system block diagram for same. (courtesy B. Schumacher, Westinghouse)

tron beam repair welding is in the maintenance and repair facilities of the commercial airlines. Several of the larger carriers have acquired facilities for this purpose to perform a wide range of tasks, including restoration of jet engine blades.

1.5 Conclusions

In conclusion, it can be stated that electron beams have made a most important inroad into the field of heat sources for materials joining in the last decade. The three pressure-dependent operational modes are experiencing an increased acceptance by an ever widening group of users bridging a range of tasks beginning with those of great sophistication and terminating with simple low-cost high-multiplicity commercial tasks.

Large chambers, more automation and higher total power systems are becoming increasingly common. Of the three modes, it appears that the hard vacuum mode is the prevailing one, while the soft vacuum mode is the fastest growing at the time of writing, even though it is much less versatile and powerful than the hard mode. While this is taking place, considerable effort is in progress to utilize this as the tool for in-space welding operations and both self-contained battery-powered systems and systems to be fed with power supplies transported by space vehicle have become available. The United States and the Soviet Union are apparently equally active in this sphere, as would be any nation with comparable space effort. With the entry of Westinghouse into the IAEBW field the belief in the future growth and potential of the process originally expressed by Hamilton Standard has been impressively reinforced and evidence of increased activities is apparent.

Diversity of opinions exists pertaining to the growth and ultimate scope of acceptance of electron beam welding as represented by its different operational modes, but all concerned appear extremely optimistic. The only thing which can be stated with certainty is that the merit of the process and the breadth of its applicability cannot be underestimated. The key to the ultimate growth here is in the hands of equipment producers and their pricing. The novelty of electron beams in welding has worn off; from now on, growth will be on the basis of the well proven rule of the best service for a price.

2. Vacuum Brazing

2.1 Introduction

Brazing is preferred to welding when complicated assemblies with great variations in thickness have to be joined or when components with insufficient accessibility of the joints must be processed. Brazing is also preferred for joining materials which are difficult to weld, like some types of superalloys, some refractory metals and metal combinations which are incompatible when welded.

References pp. 800–802

Vacuum brazing is a commercial method for joining metals without flux at temperatures below their melting points or incipient melting temperatures. In this discussion the term brazing will often be used, but the reader should bear in mind that brazing and soldering are identical processes differing in but one detail—and that is an arbitrary one. Brazing takes place below the melting point of the parts or base metals to be joined by virtue of having a third metal which becomes molten placed between the parts that are to be joined. Since this "filler" metal becomes molten during the joining process it can form an excellent mechanical fit, even on an atomic level, with the base metals. It merely remains for any nonmetallic barrier on the base metals to be removed in order to have the filler material form an excellent metallurgical bond. This chemical cleaning action is commonly done by fluxes or furnace atmospheres. In the case of vacuum brazing formation of dense barriers is suppressed thus enabling complete dissolution in the parent metal or filler, in some cases also chemical reduction or dissociation of residual oxides.

Soldering and brazing are the same processes. They are given different names because, by international agreement, when the type of joining described takes place at 427°C and above it is called brazing. Below 427°C, the process is called soldering. In this context it should be recognized that all silver solders fall into the brazing range of temperatures. Silver soldering is unfortunately a misnomer with which the industry has lived for some sixty years. (Although it is of little interest in vacuum brazing, the reader will note that at approximately 40–180°C, the joining process may be termed epoxying and at 40°C and below the process may be termed glueing.)

The advantages of vacuum brazing are due primarily to its clean brazing environment. Since there is but small chance for contamination, vacuum brazed joints tend to be cleaner, stronger and less corrosion-prone than conventionally brazed joints[62].

Vacuum brazing allows joining of dissimilar metals and even metal–ceramic combinations. With refractory metals, it is one of the best ways to produce sound joints in critical applications. It is useful for joining in enclosed areas, or areas that cannot be reached by protective inert gases in conventional brazing[63,64].

No fluxes are needed with vacuum brazing and no post-cleaning. Corrosion phenomena at the joints are therefore avoided. In addition, entrapped gases can be eliminated and the materials to be joined will be degassed. Vacuum brazing often makes surfaces appear brighter by removing impurities. Also, since braze metals flow faster in vacuum, cycle time can be shortened and output rates increased[62,64].

Vacuum brazing is popular for aerospace applications. It has been used to join tantalum alloys and niobium alloys; superalloy aircraft parts joined by vacuum brazing include Haynes 25 and A-286.

Vacuum brazing is suitable for making refractory metal–ceramic seals which will be required in future ion, thermionic and magnetohydrodynamic propulsion systems.

A significant potential commercial use for vacuum brazing is for aluminum alloys. Other uses with many metals and non-metals embrace food, drug and chemical processing equipment, and hospital and laboratory hardware.

Vacuum brazing is also used for certain specialized joining applications. One of the most important of these is termed Vacuum-Braze-Cladding. In this variation of vacuum brazing, a high strength integral metallurgical bond is achieved over large areas by taking advantage of the relaxation of plates at elevated temperatures[64,65]. The vacuum atmosphere is used to keep the surfaces to be joined free from oxidation and causes atmospheric pressure to force the surfaces into contact. In these operations the alloy cladding sheet and backing plate are both in their final thicknesses before brazing, since the pressures involved are not sufficiently high to reduce thickness and the parts that are made are often too large or too complex to warrant further size reduction. The assembly, which may consist of one or two clad plates produced in a "sandwich", is sealed by welding around the edges. A vacuum pump is connected to the inside of the sandwich allowing external atmospheric pressure to produce the force for surface contact. The entire operation takes place in a furnace, the gaseous atmosphere of which protects the sandwich material from serious oxidation as required.

Vacuum brazing techniques may be employed for cladding large plates, pipes, forgings and large diameter rings. It is also used in the fabrication of channel plates for heat exchangers, curved plates, press platens and other cooling purposes[64,65].

2.2 Characteristics of vacuum brazing

Vacuum brazing compares with other high purity brazing atmospheres by having some of the following advantages and disadvantages:

I. The vacuum atmosphere eliminates the necessity of purifying supplied atmospheres, since it removes essentially all the gases from the brazing area. Most commercial vacuum brazing takes place at pressures between 10^{-4} and 5×10^{-1} Torr and above. The pressure is determined by the composition of the residual gas atmosphere, the oxidation potential of materials being brazed, the filler metals being used, the area of the brazing surface and the heating rate.

II. The low pressure obtained during the joining process also removes entrapped gases as well as volatile impurities from the metals. As is the case in vacuum melting, the properties of the base metals are often improved by having them undergo a vacuum brazing cycle. This characteristic may be a disadvantage in those cases where the base metals or the filler volatilize at brazing temperatures and the low surrounding pressure.

III. In vacuum, base metal oxides of low stability may dissociate at brazing temperatures. The proper temperature–pressure relationships depend upon oxide dissociation equilibria (see Chapter IV, Fig. 20). Similar data exist for the dissociation of hydrides and nitrides but these seldom need to be considered since they usually dissociate at lower temperatures and pressures significantly higher than those required for corresponding oxide dissociation.

IV. The difficulties of contamination of brazing interfaces due to base metal expulsion of gases are essentially nonexistent in vacuum brazing. Although gas removal rate is dependent upon temperature, pressure and time at temperature, any occluded gases are removed from the brazing surfaces immediately upon evolution.

V. One relatively important disadvantage of vacuum brazing centers upon the fact that when refractory metals are being joined, the leak rate and cleanness of the equipment become very important. Even though the vacuum pumping rate may be high, alloys containing 1% or more of titanium or zirconium, for example, must be joined in equipment having the lowest practically obtainable leak rate[64].

2.3 Wetting and spreading phenomena

2.3.1 Influence of surface tensions

The spreading of a liquid droplet on a solid surface is determined by the contact angle θ. In Fig. 26 the surface tensions between solid surface and liquid

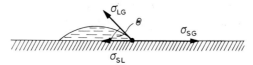

Fig. 26. Equilibrium between surface tensions at the circumference of a liquid droplet on a solid surface.

(σ_{SL}), liquid and gas atmosphere (σ_{LG}) and between solid and gas atmosphere (σ_{SG}) are in equilibrium if

$$\sigma_{SG} = \sigma_{SL} + \sigma_{LG} \cos \theta$$

therefore $\cos \theta = \dfrac{\sigma_{SG} - \sigma_{SL}}{\sigma_{LG}}$;

spreading should take place if $\cos \theta \to 1$ or

$$\sigma_{LG} < \sigma_{SG} - \sigma_{SL}.$$

Studies on the wetting and spreading of a liquid droplet on a solid surface have been made, e.g. by Beatson[66], Wall and Milner[67] and Lapujoulade[68]. These experiences show that wetting is promoted if solubility between liquid and base metal exists, but that it is not an absolute conditon. That is shown by the wetting in systems of extremely low solubility, e.g. Fe with Pb or Ag, Mo with Cu or Ag, graphite with Cu[67,69].

2.3.2 Influence of surface films

Much more important are the existence or nonexistence and the structure of oxide films covering the base metal or the filler metal[67]. Even after a careful cleaning in air, the surface of reactive metals or alloys containing reactive components like Cr, Al, etc. is still covered with a chemisorbed surface film newly formed, which during heating up in vacuum is getting even thicker by reaction with the residual atmosphere, depending on its composition. This can only be retarded or avoided by maintaining very low pressures from the beginning or by the application of protective metal coatings.

These oxide films are not wetted by the filler metal and can only be penetrated if imperfections or surface defects exist or by disruption of the films. Normally they have to be removed by chemical reactions. These include dissolution in the base metal or filler at elevated temperature or reduction, either by reducing agents like carbon in the base metal or Ti, B, Be or Si in the filler or by a partial pressure of reducing gases like hydrogen or hydrocarbons in the residual atmosphere. The chemisorbed oxide film on stainless steel which creates passivation is, for example, quickly reduced above 1000°C by its carbon content[70].

The relationship between reduction temperature and the H_2/H_2O partial pressure ratio for oxides is represented in Chapter IV, Fig. 20. This figure also shows their dissociation pressures. For most metals used in industry, with the exception of Cu, they are so low that they are out of the range of industrial vacuum technology.

The influence of the composition of the residual gas atmosphere on the wetting and spreading of fillers was demonstrated by Wall and Milner[67] in measuring the maximum pressure for satisfactory brazing of Cu, Ni and Fe with different pumping systems. The results are shown in Table 4. With a system pumped with

TABLE 4

VARIATION OF CRITICAL PRESSURE AT 800°C WITH PUMPING CONDITIONS FOR THREE METALS[67]

Conditions	Critical pressure at 800°C for:		
	Copper	Nickel	Iron
Evacuated with rotary pump only and heated to temp.	>0.5 Torr	0.2 ± 0.1 Torr	0.1 Torr
Evacuated with rotary and diffusion pumps to 10^{-3} Torr, heated to temp., air leaked in	0.015 ± 0.005 Torr	0.010 ± 0.005 Torr	0.005–0.010 Torr
System twice evacuated and purged with O_2, evacuated to 10^{-5} Torr, heated to temp., O_2 leaked in	0.005 Torr	0.003 ± 0.002 Torr	0.001–0.005 Torr

References pp. 800–802

rotary oil pumps only the critical pressure was two orders of magnitude higher than with a system pumped with diffusion pumps where the pressure was adjusted by air leak-in. This must be attributed to the influence of the hydrocarbon partial pressure originating from the rotary pump in the first case.

In practice the building up of a barrier film depends on the balance between the incoming flux of oxygen (nitrogen in most cases has not to be considered) through leaks or originating from water vapour desorbed inside the vacuum chamber and the rate of absorption or reduction at the base metal surface[67]. Metals like Ti, Ta and Nb with a high solubility for oxygen will retain a clean surface at elevated temperature at relatively high oxygen-partial pressures. Beatson[66] found, for example, that the critical pressure below which Cu was wetting Ti at 1000°C was surprisingly high (10^{-1} Torr). Another possibility to shift this balance in a favourable direction is the application of metal coatings, *e.g.* by electroplating with Ni or by vapour deposition of 3–5 μm Ti or Zr[69,71]. With Ti it could be shown by Adams[69] that even loosely adhering coatings have a pronounced effect on the wettability and spreading of fillers even with difficult-to-wet materials at temperatures above 800°C. Such a Ti film seems not only to prevent further oxidation of the base metal but also by dissolving in the filler to activate the filler metal. It was shown that the filler is tunnelling the Ti film, and because wetting of materials free from oxide films like WC, graphite and BN also is promoted, it is supposed that a favorable contact angle between the Ti film and the filler is formed which assists in pulling the filler across the surface between Ti film and base metal by mechanical forces.

Many films of oxides like alumina are so stable that a chemical reduction or dissolution is not feasible. Wetting in this case is still possible if defects in the film or destruction by mechanical forces allow for a contact between filler and base metal with subsequent peeling off of the oxide film, when the filler is spreading underneath it. In experiments with Cu–Al, Ni–Al and Fe–Al alloys with 0.5–6% Al, Wall and Milner[67] showed that during annealing in vacuum (10^{-4}–10^{-3} Torr) an alumina film is formed which creates an oxidation resistance in air. But anyway wetting was possible by different filler metals, whereby the spreading occurred along the metal–oxide interface. This penetration did not take place when the film was formed in air or in vacuum above a certain critical pressure or in a hydrogen atmosphere. They conclude therefore that the structures of the oxide films are different.

Spreading was also observed with a thick oxide film on the filler metal alone (Al on Cu, Ag, Ti) on an oxide-free base metal surface because of disruption of the film during melting.

2.4 Metallurgical reactions between filler and base metal

The amount of alloying between filler and base metal depends very much on their mutual solubility given by the phase diagram. Viscosity of the filler and brazing temperature are secondary factors. If the solubility is limited, there is only a slight erosion in the entrance to the capillary observed until the filler is saturated. A typical example is brazing of steels with Cu or Ag–Cu alloys. If the mutual solubility is very large, heavy erosion occurs. At the same time the melting point of the filler may rise and also its viscosity. Under these conditions the filler does not reach the remote parts of the capillary and the braze remains incomplete. That can happen, for example, in brazing Cu with Ag–Cu-eutectic[68] or with pure Zn on Al[67]. Therefore a limited solubility is preferred.

If the bulk of the filler is able to form intermetallic compounds with the base metal, similar conditions exist as in the case of very large solubility. Because most intermetallic compounds are hard and brittle and lead to low joint strength, only a very limited formation of such compounds is tolerable, *e.g.* by alloying additions which are needed for lowering the melting point of the filler or for reactive brazing.

Increasing temperature normally increases the spread and also the solubility. It should only be applied in a system with limited solubility, because higher temperatures favor erosion rather than spreading[67]. Only when spreading is fast are temperature and solubility less important.

On the other hand there are some cases where high solubility between filler and base metal is desirable in order to raise the melting point and strength of the joint by a subsequent heat treatment for interdiffusion. This is important, for example, for the production of high strength structures of refractory metals with high service temperatures. If filler materials with high melting point and corresponding high strength are used, erosion at brazing is more likely to occur than with fillers of lower melting point and a corresponding lower service temperature. But with a post-braze diffusion treatment, the so-called diffusion sink at temperatures below their melting point, which may be raised after some interdiffusion has taken place, melting point and service temperature can be raised again by several 100°C. This method has been applied, for instance, in the brazing of niobium honeycomb structures[72].

2.5 Metallurgical failure mechanisms in vacuum brazing

Outgassing and decomposition of oxides, nitrides and hydrides can lead to brazing imperfections and braze failure. Overlooking differences in coefficients of thermal expansion can likewise be the trigger of a failure mechanism. However, almost all of the brazing alloys are prone to "hot-shortness"—the most common metallurgical cause of braze failure.

"Hot-shortness" is a condition in which a seemingly single phased alloy is allowed to appear in a very poorly mixed condition owing to the slow cooling between the liquidus and solidus of the filler. Although a single phased alloy is usually a very well mixed material, slow cooling between the temperature limits bracketing the onset and completion of solid precipitation causes an unmixing to occur. This unmixing is very serious since the center of each grain will be rich

References pp. 800–802

in a high melting point constituent and hence strong. Each grain boundary will be composed of material high in the percentage of lower melting constituents and hence weak. In addition to this composition and strength gradient from grain core to extremity, the center of each grain is formed first upon cooling. Hence, strong material is formed upon cooling and shrinks while weaker hot material is formed in the grain boundary around the grain. The net result of this grain shrinkage is a high stress upon the grain boundaries—often too high a stress for the grain boundaries to support without rupture.

The standard way of minimizing this problem is to use the fastest possible cooling rates from the brazing temperature consistent with minimizing any distortion in the parts. Fast cooling is very difficult in vacuum brazing operations and the reader should, therefore, be even more on guard to use whatever design means he can to minimize the chances of hot-shortness. For example, the use of a gas for back-filling with eventual forced circulation will help achieve a higher cooling rate than obtainable in vacuum. As with any purging or partial vacuum brazing system, such a gas should be pure and dry and may even be highly reducing hydrogen, although argon or helium are adequate (even dry nitrogen is adequate for many operations).

2.6 The strength of brazed joints

Regardless of whether atmosphere or vacuum brazing is considered, the influence of joint thickness upon strength must be taken into account. This is all the more necessary because of a general ignorance of the importance of the part that joint thickness plays in braze strength.

In addition to the increased ductility and often slight increase in tensile strength that might be realized through vacuum brazing, a brazed joint usually has a tensile strength considerably higher than the ultimate tensile strength of the filler material.

Moffatt and Wulff[73] have shown that the strength of a brazed joint is always higher than the ultimate tensile strength of a filler material except for very thick joints. Figure 27 is representative of their work in which they used a carbon steel of 50 kgf/mm^2 yield strength for base metals and unalloyed silver of 14 kgf/mm^2 for filler material. As may be seen in Fig. 27, the strength of their brazed joints always lay between the limits of base metal yield strength and filler metal ultimate tensile strength. The reason for this high strength occurring in a brazed joint—even though the fracture occurs in the filler material—is that the tri-axiality of stresses occurs in the filler material when loaded in tension.

A very thick fillet might be represented by a large block of filler material. Taking silver once more as the example, it would be found that a bar of silver placed in a tensile machine and pulled to rupture would fracture at, say, 14 kgf/mm^2. This fracture would have been accompanied by a considerable "necking" of the specimen in the area of the fracture. Necking implies a thinning of the

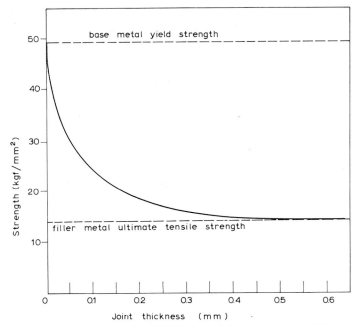

Fig. 27. The effect of joint thickness upon butt-brazed joint strength under ideal or laboratory conditions. The curved line is the resultant brazed joint strength.

specimen and a loss in dimension of the diameter of the specimen if a round tensile bar had been used in the tensile testing machine.

With a very thin fillet the situation is different. In this case, rather than consider a large block of silver being pulled in tension, consider the case of a very thin disc of silver metallurgically attached on both its flat faces to a strong carbon steel. When such a composite bar is loaded in tension (the flat faces of the silver disc being perpendicular to the axis of tension), the silver would like to neck as it did in the case of the large silver block. It would like to begin this necking action at a strength somewhat below its ultimate tensile strength. But it cannot do this since in order to neck the filler must decrease in diameter. The filler cannot decrease in diameter until the material to which it is metallurgically bonded also begins to decrease in diameter because the filler is so thin. The base material will not decrease in diameter until its yield point had been reached (by definition). Therefore, we should expect that an infinitely thin brazed joint has essentially the properties of the base material at the yield strength of the base material. The important point to remember here is that the thin disc of filler wants to decrease in diameter but cannot. Therefore a very high stress, *i.e.* a triaxial stress, is imposed upon the filler, forcing it to exist in non-fractured form at strengths significantly above its ultimate tensile strength and approaching the yield strength of the base metal.

Although Moffatt and Wulff[73] published their work with strength plotted as

References pp. 800–802

a function of joint thickness, it should be noted that their data could also have been plotted with strength as a function of thickness/diameter of tensile bar. In this way the absolute thickness of the joint can be seen to be of somewhat less importance than the ratio of joint thickness to total brazed area. Thus, we might expect that a brazed area of three square centimeters would be a thick joint if the gap between the base metals was originally set with a 0.25 mm thick shim of filler material. For a brazed area of several dm² such a joint thickness would represent a thin joint and would be expected to be higher in strength.

Figure 27 represents an ideal condition. The curve shown therein is seldom realized in real assemblies. Usually, one obtains data from real assemblies more in line with the curve shown in Fig. 28. The loss in strength of the brazed joint

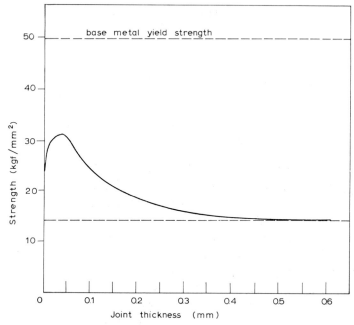

Fig. 28. The effect of joint thickness upon butt-brazed joint strength under most real or practical conditions. The curved line is the resultant brazed joint strength.

at small joint thicknesses is due to the existence of imperfections on the interfaces of the base metals as well as in the filler material. In real joints care is seldom taken to remove "capillary dams" from the interfaces of the base metals. Capillary dams are any deviation from perfectly smooth surfaces. For example, ridges from the two interfaces may touch and prevent the filler from flowing over them and completely filling the joint. Similarly, deep grooves retard the rate of filler movement. Practically speaking, the ridges and grooves from standard machining operations begin to have a significant effect upon joint strength at joint thicknesses of 0.02 mm and less. In addition to these defects, porosity from both the filler materials and base metal is difficult to remove at joint thicknesses of less than 0.05 mm. This is true even in high vacuum brazing, particularly in the case of large brazed areas.

2.7 Materials that can be vacuum brazed

2.7.1 Beryllium

Beryllium is not easy to braze in conventional atmosphere or vacuum because of its reactivity. However, vacuum brazing of beryllium has been carried out with simple shapes[62,74].

Aluminum and silver brazing fillers are common, but titanium alloys also are used[75]. Normal brazing periods cannot be used with silver or silver–copper eutectic alloys (heat must be just sufficient for the brazing alloy to melt and flow properly) since brittle intermetallics may form between filler constituents and beryllium. Temperatures of 780° to 900°C are satisfactory with silver or silver–copper alloys although the low side of this range is preferred. High percentage copper alloys also are satisfactory as fillers when joining is kept below 650°C.

The Ag–7.5 Cu alloy exhibits better wetting characteristics than does pure copper but joint strength is reduced in prolonged brazing cycles. However, nickel additions to the sterling alloy improve joint strength.

The use of titanium as a filler involves deposition of titanium as a film[75]. Indeed, vacuum metallizing with either titanium or zirconium is acceptable. These metallic films have been found to promote complete spreading and wetting not only for beryllium, but for all combinations of liquid metal and solid surfaces studied to date, including many which exhibit total non-wetting behavior without the metallic films. The development has proved most useful in the brazing of such difficult-to-wet materials as beryllium, graphite and the so-called superalloys (for use at elevated temperature)[71]. Even for more ordinary materials, the use of auxiliary wetting promotors widens the scope of filler metal selection by removing wetting as a criterion. The systems studied to date are limited to brazing temper-

TABLE 5

Filler flow temperatures	*Filler metals*
780–900°C	88Al–12Si, 92.5Ag–7.5Cu, 72Ag–28Al, 72Ag–28Cu, 94Ti–6Be, Ti–Cu–In, 92.5Ag–7.5Cu+Ni, titanium film, zirconium film.

atures above 850°C, but lower temperatures, even soldering temperatures below 420°C, should be reached with further research.

Table 5 is a partial summary of the joining temperatures and fillers suitable for vacuum brazing of beryllium.

References pp. 800–802

2.7.2 Aluminum and aluminum alloys

Cleanness is essential for producing sound joints in aluminum and its alloys. For example, no more than 12 hours should elapse between cleaning and brazing (without special precautions being taken to prevent recontamination of joint surfaces)[62,76]. A vapor degrease, followed by alkaline cleaning, water rinsing, nitric acid dipping, nitric/hydrofluoric acid dipping, water rinsing and oven drying at 110°C, is a good procedure. Assembly tolerances, brazing procedures and joint design are similar to flux-type dip or furnace brazing. Brazing time normally is only 1–2 min. However, longer time is required for the special heat-transfer characteristics in vacuum and for "pump down" time to achieve the desired vacuum. Brazing temperature is 590°C ± 5°C and vacuum is about 1×10^{-6} Torr[62,76]. To date, the aluminum alloys that have been found joinable by vacuum or fluxless brazing include 2002, 3003, 6061 and 2219. Alloy 718 (88Al–12Si) in wire and foil is the most popular brazing alloy[76,77].

2.7.3 Titanium and its alloys

Titanium alloys such as Ti–6A–14V, Ti–5Al–2.5 Sn and Ti–8Al–1Mo–1V usually are vacuum brazed with silver, silver–aluminum and silver–copper alloys. Brazing alloys such as Ag–Cu–Ge, Ti–15Cu–15Ni and Ti–43Zr–12Ni–2Be alloys are also promising[62,76,78].

Titanium vacuum brazing with silver alloys is limited because of the relatively poor corrosion resistance of the joint (when compared with titanium), especially in salt environments. However, there are corrosion-resistant brazing alloys, especially developed for titanium brazing. Several of these alloys have been produced in powder form. They are all brazed at temperatures below 980°C, and they do not react with titanium[79].

The most popular of these is a mixture of titanium, zirconium, beryllium and nickel. Other alloys are titanium–zirconium–beryllium, titanium–zirconium–beryllium–aluminum and titanium–zirconium–beryllium–cobalt[79].

Choice of the alloy depends on the types of titanium alloys to be vacuum brazed because the melting and beta-transition temperatures of titanium alloys

TABLE 6

Filler flow temperatures	Base metals	Filler metals
800–1250°C	Ti–6Al–4V Ti–8Al–1Mo–1V Ti–5Al–2.5Sn	Ag–Al, Ag–Cu–Ge, Ag, Ti–43Zr–12Ni–2Be, Ag–Cu, Ti–15Cu–15Ni

vary. When these brazing alloys are used, brazing temperatures range from 800°C to 940°C[79].

Table 6 is a partial summary of the joining temperatures and fillers suitable for vacuum brazing of titanium alloys:

2.7.4 Superalloys

With few exceptions, precious metal alloys and nickel alloys are the most common filler alloys for high temperature, hot strength alloys, such as titanium and iron-, nickel- and cobalt-base superalloys. Most of the standard grades, such as A-286, TD nickel, Hastelloy X, Inconel 718, Udimet 700 and Haynes 25 can be vacuum brazed with several brazing alloys, usually silver- or nickel-base, or alloys containing relatively high nickel content. Suitable brazing alloys also are available for the more recently developed dispersion strengthened nickel alloys.

On the basis of shear strength of vacuum brazed Inconel 718 joints and the effect of brazing temperature on the base metal strength, the optimum vacuum brazing alloy seems to be 82Au–18Ni with a flow point of 980°C[80]. Two other alloys—Au–8Pd–22Ni (flow point 1050°C) and Au–16.5Cu–2Ni (flow point 940°C)—can also be used where structural strength is not a consideration[81]. It is also advisable to plate all surfaces to be brazed with approx. 0.01 mm nickel to ensure good filler metal flow[80].

Table 7 is a partial summary of the joining temperatures and fillers suitable for vacuum brazing of superalloys.

TABLE 7

Filler flow temperatures	Base metals	Filler metals
900–1260°C	A–286	Ni–Cr–Fe–Ti, Ag–Cu
	Hastelloy X	Ni–Mn–Si–Cu, Au–Pd–Ni, Ni–Cr–Si, Ag–Cu
	Haynes 25	Au–Pd–Ni, Co–Ni–Cr–Si–W–B, Ni–Cr–Si, Pd–Ni–Cr, Ag–Cu
	Inconel 700	Ni–Cr–Mo–Co–Ti–Al
	Inconel 702	Ni–Cr–Al
	Inconel 718	Au–Ni–Cu, Mn–Ni–Co–B, Cu–Mn–Co–Ni, Au–Ni, Au–Pd–Ni, Ag–Cu
	Inconel 901	Ni–Cr–Mo–Fe–Ti
	Inconel X	Ni–Cr–Fe–Ti
	Inconel W	Ni–Cr–Fe–Ti
	J-1300	Ni–Cr–Mo–Ti
	J-1500	Ni–Cr–Mo–Fe–Co–Ti
	J-1570	Ni–Cr–Co–Ti
	L-605	Ni–Cr–Mo–Fe–Co–B
	N-155	Ni–Cr–Mo–Fe–Co–W
	Ni–Cr–Si	Ni–Cr–Fe–Si
	Ni–Si–B	Ni–Si–B
	R-235	Ni–Cr–Mo–Fe–Co–Ti–Al
	TD Nickel	Ni–Mo–Cr–W–Si, Pd–Cr–Ni, Ag–Cu
	TD–Ni–Cr	Ni–Fe–Cr–Si–Mo–Co
	Udimet 700	Ni–Cr–Mn–Si, Pd–Ni–Cr–Si, Pd–Ni–Cr, Ag–Cu
	17-7 Ph	Ni–Cr–Si

Note: In all cases, Ti is 4% or less, Al is 3% or less, Mo is usually in the 3–5% range. Nickel is usually in the 30–80% range. Exact braze compositions are not given since the optimum percentages of the alloying constituents do not appear to be known at the present time.

2.7.5 Refractory metals

Niobium, molybdenum, tantalum, tungsten and their alloys can be successfully vacuum brazed[62,76,82].

Excellent wettability has been reported with usable joint properties up to 1650°C when the diffusion sink process is used to give a higher remelt temperature[82]. Titanium-base brazing alloys are the most commonly used for niobium, tantalum and molybdenum. Three hafnium alloys, Hf–7Mo, Hf–19Ta–2.5 Mo and Hf–40Ta, also are popular for vacuum brazing tantalum. A tungsten alloy and platinum–boron are used for tungsten.

Table 8 is a partial summary of the joining temperatures and fillers suitable for vacuum brazing of refractory metals.

TABLE 8

Filler flow temperatures	Base metals	Filler metals
980–1800°C	Niobium (D-36, D-43, Cb-752)	Ti–8.5Si, Ti–33Cr, 66Ti–30V–4Be, 48Ti–48Zr–4Be, 75Zr–19Cb–6Be, Ti–13V–11Cr–1Al
980–1480°C	Molybdenum (TZM)	Ti–8.5Si, 62Ti–25Cr–13Ni
1650–2200°C	Tantalum	34Ti–33Zr–33V, Ti–30V, Hf–7Mo, Hf–40Ta
1100–1900°C	Tungsten	90W–6Ni–4Fe, Pt–3.5B

2.7.6 Non-metals, dissimilar metals and combinations thereof

Vacuum brazing of ceramics and graphite is difficult for several reasons. Firstly, ceramics are inherently difficult to wet (most conventional brazing metals dewet). Secondly, differences in thermal expansion (between ceramics and brazing alloys, or between ceramics and base metals in dissimilar joints) can cause cracking. In addition, since ceramics are poor heat conductors, it takes longer to get to temperature.

To overcome these problems, ceramics are premetallized to cause wetting. After this, they can be vacuum brazed with copper, silver–copper and gold–nickel brazing alloys. In addition, a recent 49Ti–49Cu–2Be brazing alloy exhibits excellent flow and produces crack-free joints in Al_2O_3, BeO and MgO ceramics[62].

This Ti–Cu–Be alloy, and a similar 4%-Be grade, show promise also for graphite brazing primarily because their major constituents are strong carbide formers. However, Au–Ni–Ta alloys containing less than 30% (by weight) tantalum and 35Au–35Ni–30Mo alloy also are being evaluated for graphite-to-graphite and graphite-to-metal combinations[62].

Many materials, although metallurgically dissimilar, are compatible for vacuum brazing. Typical examples include beryllium–stainless steel, beryllium–titanium, titanium–niobium, titanium–ceramics and molybdenum–graphite.

Titanium and precious metal alloys are the brazing alloys mostly used, although Ag–Cu alloys are sometimes analyzed[83]. Titanium brazing alloys are not limited to joints involving titanium-base alloys, but also are used for combinations such as beryllium oxide–pyrolytic graphite, beryllium–stainless steel, molybdenum–graphite and other dissimilar materials.

Table 9 is a partial listing of the joining temperatures and fillers suitable for vacuum brazing of non-metals, dissimilar metals and combinations thereof.

TABLE 9

Filler flow temperatures (°C)	Base materials	Filler metals
760–980	Metallized ceramics on graphite	Ag–Cu, Cu, Au–Ni
1010	Al_2O_3, BeO, MgO	49Ti–49Cu–2Be, 48Ti–48Cu–4Be
1010	Graphite	Au–Ni–Ta, Au–Ni–Mo
1010	Al_2O_3–BeO/Mo or tantalum	48Ti–48Zr–4Be(wt.%)
780	Ceramics–Kovar	72Ag–28Cu
1010–1050	Graphite–molybdenum	48Ti–48Zr–4Be(wt.%)
1390–1480	BeO–pyrolytic graphite	93Ti–7Ni, 93Ti–7Fe, 53Ti–47Cr
1050	Titanium–ceramics	49Ti–49Cu–2Be(wt.%)
940–1010	Stainless steel–beryllium	49Ti–49Cu–2Be(wt.%), Ag, Ag–Li, Ag–Cu
1180	Stainless steel–niobium	Co–21Ni–21Cr–8Si
760–1010	Stainless steel–titanium	48Ti–48Zr–4Be(wt.%), 81.1Pd–14.3Ag–4.6Si
890–930	Titanium–beryllium	92.7Ag–7.2Cu–0.2Li, Ag
930–1300	Titanium–niobium	Ti–Cd–Be, Ti–Fe–V, Ti–V–Be
970–1050	Titanium–copper	Ag, 49Ti–49Cu–2Be(wt.%)

2.8 Vacuum brazing furnaces

For low temperature brazing, *e.g.* for brazing aluminum or aluminum-alloys, hot wall furnaces are more economical than cold wall furnaces. The material to be brazed is placed inside a steel retort and heated from outside (see Chapter VII, Figs. 25 and 26). If the retort is pressure-released, the maximum temperature which can be attained is about 1150°C. For higher brazing temperatures and if more stringent vacuum conditions are required cold wall furnaces are necessary.

In principles the vacuum furnaces described in Chapters VII and VIII used for sintering and heat treatment can also be used for brazing. But in general the base pressure and the composition of the residual gases are here more critical. Therefore, furnaces with higher pumping capacities and lower outgassing rates of the surfaces and structural components inside the vacuum chamber have to be selected, especially for brazing reactive and refractory metals and alloys.

The thermal inertia of the furnace should also be as low as possible. Starting from the soaking temperature the desired brazing temperature should be reached quickly in order to avoid premature alloying of the filler and base metal before spreading. Also, as described in Subsection 2.5, fast cooling from the liquidus

References pp. 800–802

is important in order to avoid detrimental segregation in the filler.

Optimum conditions also with respect to the cycle time can only be achieved by (1) avoiding heavy thermal insulation and (2) avoiding materials for heating elements and thermal insulation with high gas absorption in the loading phase. When materials with high affinity to oxygen or carbon have to be brazed heating elements and heat shields made from refractory metals like molybdenum or tantalum are more suitable than graphite elements or thermal insulations made from carbon or graphite felt or porous refractory oxides. Graphite and carbon elements or insulations should only be used in brazing composites containing graphite, for carbon steels, copper and copper alloys or base metals or fillers with high vapor pressure, which could cause refractory metals to deteriorate.

2.9 Conclusion

It has been the main purpose of this Section to describe the broad outlines of vacuum brazing in its different forms. No attempt has been made to delve into the several areas of metallurgical bond or joint imperfections to any considerable extent. The many textbooks available dealing with furnace brazing adequately cover these fields. Rather, an attempt has been made to concentrate on explaining those phenomena and presenting those data associated with vacuum brazing with which the engineer or operator must become familiar in order to truly appreciate optimum utilization of vacuum brazing. Newer innovations such as maintaining vacuum while loading or unloading, continuous vacuum brazing and quenching in vacuum (without breaking the vacuum) will no doubt make vacuum brazing an even more versatile and fruitful joining technique.

3. Solid State Bonding

In the foregoing sections of this chapter two joining processes have been discussed, which in the last few years have already found rather extensive industrial application. A distinguishing feature of these processes is that joining occurs via a liquid phase.

Recently, joining by solid state bonding in a vacuum atmosphere has been considered for the union of reactive and refractory metals. This makes it possible to avoid the formation of embrittled structures and grain coarsening by recrystallization or alloying, resulting in joints of higher bond strength than would be obtained by other methods. Sometimes this method is selected because of design considerations, *e.g.* in order to save expensive materials in less critical areas. Another reason for proceeding in this manner is the need to use the assembly at temperatures above those which are feasible with brazed joints or because other joining processes are not applicable.

Of the numerous possibilities which seem to be promising two processes which might find industrial application have emerged: diffusion bonding by roll-bonding, and hot pressing and vapor deposition welding.

3.1 Joining by diffusion bonding

The maintenance of a clean surface unoxidised and free from various contaminants in a vacuum atmosphere also at elevated temperature promotes interdiffusion of metals and in turn leads to better bonding and improved products in systems where no brittle intermetallics are formed at the interface. The influence of contaminants on metallic adhesion is one of the most important factors in the processes described here. This problem is discussed in detail in Chapter XI, Section 4, especially for systems at ambient temperature. Fortunately, at the higher temperatures which are normally used in diffusion bonding, the requirements with respect to partial pressure of reactive gases and cleanness are not so severe. Also the diffusion coefficients are much higher and deformation under compression forces is facilitated because of the lower yield point. Oxide films and other contaminants are therefore much more easily dispersed between the two metallic surfaces, leading to a high-strength joint.

3.1.1 Joining by roll bonding

While the advantages which can be brought about by vacuum processing are important, the very factors which benefit the operations are also potentially severe handicaps to the performance of these operations. Reference is made to the tremendous problems that components in highly stressed rotating machinery, as for example rolling mills or drawing benches, can present when one considers these for operation at an elevated temperature and in vacuum environment. Under conditions of this sort, all surfaces free from oxides and conventional or special lubricants will be subject to severe galling problems and this will include all interfaces between materials to be rolled, rolling rolls, drawing auxiliaries and related equipment. In other words, vacuum metal working at elevated temperatures necessitates the employment of special equipment and special techniques if it is to be brought to successful completion.

Because of these problems, inert atmosphere processing, the so-called infab (inert atmosphere fabrication) type facilities, pioneered by U.S. Navy and Universal Cyclops Corporation at the turn of the 60's, has had important preference in relation to vacuum processing in the United States.

Alternatively, material has been vacuum sealed at room temperature in suitable combinations of geometry and external vacuum closure and the composite thus prepared subjected to further rolling in air. Removal of the shell material at the completion of the processing presents problems in terms of surface contamination and surface appearance, not to mention the problems in sealing metal packets and the cost involved in processing of a type such as this. In many cases, however, the concern is only for the internal interface and no outside wraps are

References pp. 800–802

used; this will be referred to later.

While the writer has been unable to locate any references in the U.S. and European literature pertaining to actual vacuum rolling, several articles on this subject have been found in the Soviet literature[84-91].

Work by Ulyanov and his associates[91] indicates the well known fact that vacuum rolling permits maintenance of high purity contact surfaces. Some indication that on vacuum processing dissociation of surface films takes place has been provided by Wagner[92]. A special high vacuum two-high rolling mill has been designed by Amonenko and his associates[93].

In another publication[94], Amonenko et al. report on vacuum rolling of bimetallic couples with Cu–Ni, Ti–steel, Cu–Nb, Cu–steel being under investigation. These couples, according to published accounts, are apparently virtually impossible to fabricate by conventional techniques. On vacuum processing, substantial improvement in the strength of the metal–metal interface is observed and some improvement in the percent reduction is effected.

Actually, the influence of percent reduction, residual pressure in processing chamber, rolling temperature and surface condition on the bond strength was investigated as part of this study.

Rolling was carried out at 0.3 m/s and it was found that 15–30% reduction in vacuum produces a bond strength equivalent to the strength of the weaker of the two elements. Reduction of up to 70% was needed to effect the same strength results in air, and needless to say the couples containing reactive and refractory metals cannot be processed in air. Figure 29 shows the relationship between metal bond strength and reduction for several couples at pressures below 10^{-4} Torr and at different temperatures. In the roll bonding of couples which

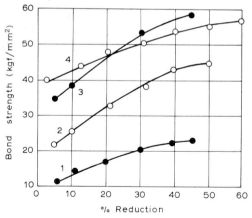

Fig. 29. Relationship of the metal bond strength to percent reduction[93].
    (1) Cu–steel 3 at 1000°C
    (2) Ti–steel 3 at 1050°C
    (3) Mo–Nb at 1200°C
    (4) 1 Kh 18 N9T steel–steel 3 at 1200°C.

form solid solutions, as for example Cu–Ni, Mo–Ti, Mo–Nb, etc., the bond strength increases as the temperature of processing increases. In other cases, where low-strength eutectics or brittle intermetallic compounds would form at the interface at elevated temperatures, bimetal production is also possible at lower temperatures.

The residual chamber pressure is of foremost importance in vacuum roll bonding. At pressures of 10^{-1}–10^{-2} Torr with Ta, Nb and other chemically active metals surface oxidation at the interface leads to low bond strength. To be specific, when Mo–Nb is rolled at this pressure and 1200°C with 30% reduction the bond strength is 5–8 kgf/mm². At a pressure of 2×10^{-5} Torr under the same rolling conditions a bond strength of 32 kgf/mm² was produced, corresponding to an almost 400% improvement. When reactive metals are bonded at high pressures, e.g. 10^{-2} Torr, fracture always occurs at the interface.

More details about the influence of the rolling temperature, pressure and relative reduction have been given by Aleksandrov et al.[95] in the production of nickel–copper bimetal by vacuum rolling. Here copper and pure nickel in 10 mm thickness vacuum annealed at 700°C and 900°C have been packrolled as 20 × 40 × 100 mm packets at 750–1050°C with 5–52% reduction. Figures 30a, b, c show the effect of the above-mentioned parameters on the bond strength. (In Fig. 30b the temperature of rolling was 950°C with 30% reduction.)

Metallographic investigations showed no transition zone in the as-rolled condition. Maximum bond strength was achieved with 30% reduction at 1×10^{-4} Torr between 1050° and 950°C. There was no significant influence of the quality of surface finish before roll bonding.

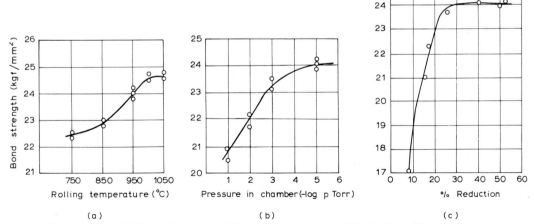

Fig. 30. Effect of different factors of rolling on bond strength of Ni–Cu bimetal[95].
        (a) Rolling temperature
        (b) Residual air pressure p in chamber
        (c) Relative reduction.

References pp. 800–802

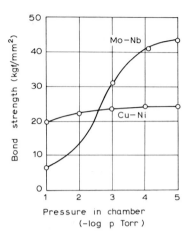

Fig. 31. Relationship of the bond strengths of bimetals Cu–Ni (950°C) and Mo–Nb (1200°C) to residual gas pressure p during heating and rolling with 30% reduction.

Also, the pressure conditions are not so critical as in roll bonding of metals of higher reactivity. This is shown by comparison with the results of Amonenko et al.[93] with Mo–Nb in Fig. 31. At up to 18% reduction failure on loading occurs at the interface, whereas at higher degrees of reduction the bond strength in the interface is higher than the strength of copper.

The experiments showed that vacuum rolling in one pass is sufficient to achieve the necessary bond strength, and that further reduction by rolling in air at ambient temperature is feasible without impairing the quality of the material. The production of Ni–Cu bimetal in this way for the tube manufacturing industry (for saving nickel and raising its heat conductivity) is therefore considered as an economic solution.

In another investigation Krivonosov et al.[96] report on producing a Nb–steel bimetal by vacuum rolling. The starting materials were 2 mm of Nb and 9–12 mm of steel. Rolling was done at 4×10^{-5} Torr pressure with a speed of 0.19 m/s and a reduction from 10 to 40% per pass in the range 900° to 1200°C. The actual packet was 50×140 mm. As in the study just reported above, an increase of temperature of rolling here also improved joint strength, the comparative value at 30% reduction being 10 kgf/mm^2 at 900–1000°C, while at 1100–1200°C rolling temperature and the same reduction a strength of 21–23 kgf/mm^2 was obtained.

Percent reduction is equally important and for example processing at 1100°C and 10% reduction produces 10 kgf/mm^2 strength while 30–40% reduction improves the strength up to 25 kgf/mm^2. It is an interesting fact that reduction beyond this tends to lead only to a slight additional increase in strength. In this instance a clear boundary between the two metals was visible. Evidence of decarburization on the steel part of the interface was also observed.

In general it can be stated that in all bimetallic couples phase diagram and metallurgical thermodynamics considerations determine the ultimate properties.

This actually proves to be one of the main reasons tending to limit the usefulness of vacuum processing in particular in view of the recent and rapidly advancing use of explosive bonding leading to joints of non-equilibrium structure. The explosive forming is a potential competitor to vacuum rolling but not to the other metal working processes. Another possibility which limits the application of vacuum rolling of bimetallic couples is the processing of bimetallic couples which have been electron beam welded in vacuum to protect the interface and prepare them in this manner for subsequent rolling in air.

3.1.2 Joining by hot pressing

The essence of all cladding and laminating processes, except those in which blanks are prepared by fusion welding or explosive forming where we are dealing with non-equilibrium configurations, is a rapid solid state joining process.

As considerable benefits including potential weight savings and better stress distribution can be envisaged for solid bonding or welding, this rather ancient approach is now being reexamined. While it is not part of the substance of this volume, it is pertinent to our discussions here and will briefly be mentioned.

A good way to start perhaps is to refer to the comprehensive review of Milner and Rowe[97] and their conclusion regarding the existence of two basic temperature ranges for solid state joining. The first range takes place at temperatures just over one-half of the melting points as expressed in terms of degrees Kelvin, the second range being below this temperature (see also Chapter XI, Section 4). In the high-temperature range bonding apparently takes place, provided that oxidation is suppressed, which will be the case in the vacuum environments considered in this volume. In the low-temperature range, bonding is not observed unless the metal is deformed and relative movement takes place at the interface.

In a more recent publication, Albom[98] reviews solid state bonding as a potential technique for the joining of reactive metals (Be, Ti, Zr) and refractory metals (Nb, Mo, Ta, W). He cites numerous advantages including the possibility of elimination of brittle intermetallics and even elimination of heat-treating cycles. It is further claimed that these benefits could be developed for a wide range of industrial domains including the aircraft, nuclear, electronics and aerospace industries. In order to carry out this process the parts to be joined have to be brought into intimate contact by applying external pressure and heated to a specific temperature and kept under these conditions for a specific time depending on the materials to be joined. During this treatment, diffusion of the metal atoms takes place across the contact surfaces. King and Owczarski[99,100] studied the different parameters of this process with titanium and proposed a tentative mechanism for the coalescence of two solid surfaces, and this approach has been utilized in exploration work in connection with several aerospace components prepared for the NASA Marshal Space Flight Center[101].

The proposed model consists of three stages, in which different metallurgical

References pp. 800–802

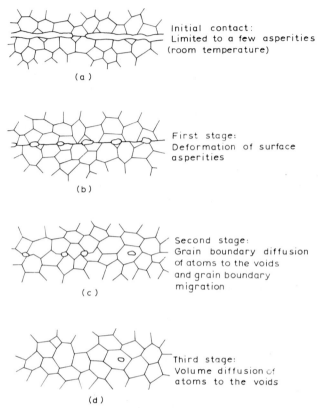

Fig. 32. The stages of bonding titanium according to King and Owczarski[100].

mechanisms are dominant in the bonding process. Figure 32 shows the initial conditions at room temperature where the real contact is limited to a few asperities. The contact area is growing by the deformation of these asperities in the first stage during heating to the bonding temperature selected. With decreasing strength of the asperities there is a rapid growth of the contact area, which on the other hand leads to a reduction of the stress applied by external pressure and the deformation is slowed down. Because of the original surface roughness voids remain in the interface (Fig. 32b). At the same time the areas of contact become grain boundaries. In the second stage the pores are shrinking and the grain boundaries migrate into the two halves of the composite (Fig. 32c). In the third stage the remaining voids become engulfed by further diffusional processes (Fig. 32d).

In the proposed model the first stage is dominated by the deformation mechanisms and the others by diffusional processes and therefore application of pressure should be necessary only in the first stage. Metallographic examinations showed that this assumption is correct in principle but that further application of pressure in the second and third stages leads to lower void content.

The influence of surface roughness is demonstrated in Fig. 33. It shows that

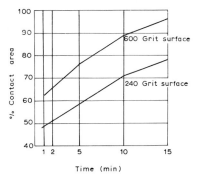

Fig. 33. Plot of contact area growth vs. time at 840°C with titanium for two initial surface roughnesses[100].

with rougher initial surface the growth of contact areas is impeded and also larger voids are produced which extend the later stages. Therefore, surface preparation is of great significance for good results. For optimum conditions metallographic polish and consecutive cleaning with acetone is recommended[98]. Electropolishing and etching are detrimental in most cases because of the formation of residual films on the surface. The requirements concerning cleanness are normally higher

Fig. 34. 15-Ton diffusion bonding vacuum hot press with water-cooled hydraulic pistons. Cold-wall furnace with refractory metal heating elements and heat shields, 30 kW. Maximum temperature 1600°C. (courtesy Vacuum Industries, Inc., Somerville, Mass.)

References pp. 800–802

than with brazing and also the tolerances are stricter in order to get the necessary good fit of the joints and maximum contact surface.

Heating may be done in a furnace by external heating of an evacuated retort containing the support tooling and the material to be joined or by induction resistance heating inside a cold wall vacuum or protective gas furnace. Also self-resistance heating may be applied.

Figure 34 shows a vacuum diffusion bonding hot press for temperatures up to 1600°C with a maximum 15 ton compacting force at 10^{-5} Torr. It is a resistance-heated cold-wall furnace with an external press frame supporting the chamber and a double acting ram.

The pressures applied are normally below 100 kgf/cm^2 and the maximum temperatures necessary correspond to about one-half the absolute melting point of the material to be processed. But in some cases higher pressures and temperatures are also necessary[98].

The fixtures have to fulfil the following conditions:

1. Sufficient strength should be obtained at the bonding temperatures (therefore refractory metals or ceramics have to be used normally).

2. The coefficient of expansion should be matched to the material to be processed, so that sufficient pressure is maintained during the heating cycle.

3. Reactions between fixtures and the parts to be joined should be avoided.

Albom[98] describes the process parameters necessary in diffusion bonding of reactive and refractory metals as well as of dissimilar materials. Additional information on experience gained in the joining of refractory metal compounds, also with oxides, by vacuum hot pressing is given by Kelley *et al.*[102].

3.2 Chemical vapor deposition welding

In order to complete the discussion related to solid state bonding one should refer also to a newcomer here in which vacuum technology plays an important role, as will be shown later. This is the so-called chemical vapor deposition or CVD welding which is the name used to describe adaptation to welding of the well established process of deposition of metallic materials through the thermal decomposition of suitably selected metallic or organometallic compounds. The use of this approach for the actual joining of materials is apparently of very recent origin and Macedo[103] demonstrated that this process is capable of making possible tungsten tensiles of up to 95 kgf/mm^2. This method makes possible at least in principle the production of weldments with mechanical, chemical and physical properties equal to or exceeding those of the substrate, provided that suitable compounds exist.

As in most cases decomposition temperatures are rather low in relation to temperatures encountered in fusion welding processes, no detrimental effects of the type associated with recrystallization embrittlement, heat-affected zone and composition changes will take place. In his studies Wilson[104] has investigated material systems for the joining, each to itself, of tungsten, aluminum alloys and 321 stainless steel and established in principle the feasibility of joining Be to itself and also Ti–8Mo–1V–1Co in like manner.

In this type of work cleanness of the substrate surface, the decomposition temperature, pressure of the operation, the flow rates involved and purity of the carrier gases are very critical. The problem of joint configuration is a new one and as a rule special equipment must be designed. Suitable catalysts which have been developed for some cases can make an important economic contribution to the process. At this juncture considerable additional work is required prior to commercial applications. According to Wilson, it is claimed that the process when fully developed will be suitable for particularly difficult joining tasks, as for example the joining of foamed metals, the joining of metallic to non-metallic materials, the joining of fiber-reinforced composite materials and the preparation of coated joints.

The following steps are recommended for the process[104]:

Substrate preparation:
1. Machine (bevel) specimens, to provide joint geometry.
2. Clean to provide a metallurgically acceptable bond surface.
3. Jig to assure correct joint alignment and weld gap.
4. Install thermocouple.
5. Place assembly into CVD system, in alignment with induction heating coils.

Preparation of CVD systems:
1. Remove atmospheric contaminants. A high vacuum bakeout procedure is preferred. Inert gas purge-flushing can be used as an alternative method.
2. Flush process lines with CVD vapors into the waste traps, bypassing the nozzle and specimens.
3. Reduce specimen surface oxides, at temperatures below the recrystallization temperature, with thermal-hydrogen gas.
4. Remove absorbed gases and moisture ("degas") by vacuum-heat treatments, typically 430°C at 10^{-6} Torr, for 15 min.
5. Apply first CVD ("flash") deposit, 0.7–2 μm thick.
6. Degas (repeat step 4) to remove absorbed CVD gaseous byproducts.
7. Apply CVD deposit to achieve the full weld nugget.
8. Terminate CVD welding process, cool specimen to room temperature, release vacuum, remove welded specimen, immediately close and evacuate system (to preserve cleanness of CVD system), and dispose of chemical waste.

This shows the many precautions which have to be taken in order to get good results.

References pp. 800–802

REFERENCES

1. M. von Ardenne, *Naturwissenschaften*, 26 (1938) 562.
2. J. A. Stohr, *Fuel Elem. Conf.*, Paris. Nov. *1958*, TID, 7546, Book 1, p. 917.
3. W. L. Weyman and W. I. Steinkamp, *Rept. HW 55667*, General Electric, 1958.
4. K. H. Steigerwald, Materialbearbeitung mit Elektronen, *Physik. Verhandlungen, 4* (1953) 123.
5. E. Bas, Materialbearbeitung mit Elektronen im Hochvakuum, *Advances in Vacuum Technology*, Pergamon, London, 1960.
6. N. A. Olshansky, On the peculiarities of electron heating of metals, *Avtomat. Svarka*, 5 (1962) 19.
7. N. A. Olshansky, in R. Bakish (ed.), *Proc. 1st Intern. Conf. on Electron and Ion Beam Science and Technology*, Wiley, New York, 1965.
8. H. Schwartz, in R. Bakish (Ed.), *Proc. 1st Intern. Conf. on Electron and Ion Beam Science and Technology*, Wiley, New York, 1965.
9. W. E. Meyer, W. Scheffels and K. H. Steigerwald, in El Kareh (Ed.), *Proc. Symp. Electron Beam and Laser Welding*, Univ. of Pennsylvania, 1965.
10. T. Hashimoto and F. Matsuda, *Trans. Natl. Res. Inst. Metals (Tokyo)*, 7 (1965).
11. M. Muelemans, in R. Bakish (Ed.), *Proc. 2nd Intern. Conf. on Electron and Ion Beams in Science and Technology, 1966*, Gordon and Breach, New York, 1969, p. 357.
12. R. F. Donovan, in A. El Kareh (Ed.), *Proc. Symp. Electron Beam and Laser Welding* Univ. of Pennsylvania, 1965.
13. H. D. Steffens and G. Wilkens, *Electron Beam Welding*, Vol. 1, Deutscher Verband für Schweisstechnik, Düsseldorf, 1967, p. 189.
14. M. Hablanian, in R. Bakish (Ed.), *Proc. 2nd Symp. on Electron Beam Technology, Boston, 1960*, p. 102.
15. E. Franco-Fereira, in R. Bakish (Ed.), *Proc. 2nd Symp. on Electron Beam Technology, Boston, 1960*, p. 122.
16. O. K. Nazarenko, *Electron Beam Welding*, Scientific World, Kiev, 1965, p. 6.
17. T. Hashimoto and F. Matsuda, *Trans. Nat. Res. Inst. Metals (Tokyo)*, 7, 1965.
18. H. Schwartz, *J. Appl. Phys.*, 33 (1962) 3464.
19. R. Bakish (Ed.), *Proc. 3rd Intern. Conf. on Electron and Ion Beams in Science and Technology, 1968*, Electrochem. Soc., New York, 1968.
20. J. Solomon, private communication, 1966.
21. R. Bakish (Ed.), *Introduction to Electron Beam Technology*, Wiley, New York, 1962.
22. H. W. Michler and R. E. Monroe, *DMIC Rept. 174*, Batelle Memorial Inst., 1962.
23. *Electron Beam Welding Bibliography*, British Welding Research Association, London, 1963.
24. K. J. Miller and T. T. Takenaka, *Electron Beam Welding in Aerospace Manufacturing*, Welding Research Council, 1964.
25. R. Bakish and S. S. White, *Handbook on Electron Beam Welding*, Wiley, New York, 1964.
26. J. Hetherington (Ed.), *1st Symp. on Electron Beam Technology, Boston, 1959*.
27. R. Bakish (Ed.), *Proc. 2nd Symp. on Electron Beam Technology, Boston, 1960*.
28. R. Bakish (Ed.), *Proc. 3rd Symp. on Electron Beam Technology, Boston, 1961*.
29. R. Bakish (Ed.), *Proc. 4th Symp. on Electron Beam Technology, Boston, 1962*.
30. R. Bakish (Ed.), *Proc. 1st Intern. Conf. on Electron and Ion Beam Science and Technology*, Wiley, New York, 1965.
31. R. Bakish (Ed.), *Proc. 2nd Intern. Conf. on Electron and Ion Beams in Science and Technology, 1966*, Gordon and Breach, New York, 1969.
32. J. Morley (Ed.), *Proc. 5th Symp. on Electron Beam Technology, Boston, 1963*.
33. J. Morley (Ed.), *Proc. 6th Symp. on Electron Beam Technology, Boston, 1964*.
34. A. El Kareh (Ed.), *Proc. Symp. on Electron Beam and Laser Welding, University of Pennsylvania, 1965*.
35. G. I. Haddad (Ed.), *Proc. Symp. on Electron Beam and Lasers, University of Michigan, 1966*.
36. E. Funk, *Electron Beam Welding Symp.*, Ohio State University, Columbus, Ohio, 1967.
37. F. Pease, *Proc. Symp. on Electron, Ion and Laser Beam*, IEEE, Catalogue No. F 79, San Francisco, 1967.

38 F. Eichhorn, H. Koch and A. Matting, *Electron Beam Welding, Vol. 1*, Deutscher Verband für Schweisstechn., Düsseldorf, 1967, (in German).
39 Hamilton Standard, sales literature.
40 N. A. Olshansky, in J. Morley (Ed.), *Proc. 5th Symp. on Electron Beam Technology*, Boston, 1963.
41 J. R. King, *Amer. Mach./Metalworking Manuf.*, April 27, 1964.
42 J. V. Anderson, in R. Bakish (Ed.), *Proc. 4th Symp. on Electron Beam Technology, Boston, 1962* and *Mid Western Welding Conf.*, I.I.T., Chicago, Ill., 1964.
43 M. T. Groves and J. M. Gerken, *Eng. Rept. 6847* based on *Contract AF 33 (615)-1626*, also in J. Morley (Ed.), *Proc. 5th Symp. on Electron Beam Technology*, Boston, 1963.
44 K. Hicken, in R. Bakish (Ed.), *Proc. 3rd Intern. Conf. on Electron and Ion Beams in Science and Technology*, Electrochem. Soc., New York, 1968.
45 J. Meier, in J. Morley (Ed.), *Proc. 5th Symp. on Electron Beam Technology*, Boston, 1963.
46 J. L. Solomon and H. A. James, in J. Morley (Ed.), *Proc. 5th Symp. on Electron Beam Technology*, Boston, 1963.
47 B. Schumacher, in R. Bakish (Ed.), *Proc. 1st Intern. Conf. on Electron and Ion Beam Science and Technology*, Wiley, New York, 1965.
48 B. Schumacher, Electron beams as tools for underwater cutting and welding, in E. Funk, *Electron Beam Welding Symp.*, Ohio State University, Columbus, Ohio, 1967.
49 J. Lempert, J. F. Lowry, C. S. Williams and F. D. Seaman, in G. I. Haddad (Ed.), *Proc. Symp. on Electron Beam and Lasers*, University of Mich., 1966.
50 J. Lempert, *IEEE Electronic Welding Conf.*, Detroit, Mich., Nov. 15, 1966.
51 Anon., Westinghouse brochure on *Electron Beam Welding in Air*, 1967.
52 *Stuttgarter Nachrichten*, May 23, 1953.
53 L. G. Steman, Technical program summary, *Contract NASA 8-11929*, 1967.
54 F. Rosebury, *Handbook on Electron Tubes and Vacuum Techniques*, Addison-Wesley, Reading, Mass., 1966.
55 J. R. Pierce, *Theory and Design of Electron Beams*, Van Nostrand, New York, 1949.
56 V. E. Coslett, *Introduction to Electron Optics*, Oxford University Press, Oxford, 1946.
57 O. Klemperer, *Electron Optics*, Cambridge University Press, Cambridge, 1953.
58 V. P. Taranenko, *Electron Guns*, Technika, Kiev, 1964.
59 E. B. Bas, G. Cremosnik and H. Lerch, *Trans. 8th Natl. Symp. on Vacuum Technology and 2nd Intern. Congr., 1961*, Pergamon, London, 1962, p. 817.
60 F. X. Masse, in R. Bakish (Ed.), *Introduction to Electron Beam Technology*, Wiley, New York, 1962.
61 W. G. Reininger, in J. Morley (Ed.), *Proc. 5th Symp. on Electron Beam Technology*, Boston, 1963.
62 M. M. Schwartz, *Mater. Eng.*, 66 (1967) 76.
63 A. L. Phillips (Ed.), *Welding Handbook*, Am. Welding Soc., 5th edn., Sect. 3, 1964.
64 A. L. Phillips (Ed.), *Welding Handbook*, Am. Welding Soc., 5th edn., Sect. 5, 1967.
65 A. L. Phillips (Ed.), *Welding Handbook*, Am. Welding Soc., 4th edn., Sect. 5, 1961.
66 E. V. Beatson, *Brit. Welding J.*, 5 (1958) 137.
67 A. J. Wall and D. R. Milner, *J. Inst. Metals*, 90 (1961–62) 394.
68 J. Lapujoulade, *Vide*, 95 (1961) 207.
69 C. M. Adams, Jr., in B. B. Barrow and A. Simkovich, (Eds.), *Trans. Intern. Vacuum Met. Conf., 1968*, Am. Vacuum Soc., New York, 1968, p. 324.
70 T. N. Rhodin, *Compt. Rend., 6ième Réunion Soc. de Chimie et de Physique, 1956*, p. 296.
71 R. G. Gilliland, C. M. Adams, Jr. and S. Weiss, in E. L. Foster (Ed.), *Trans. Intern. Vacuum Met. Conf., 1967*, Am. Vacuum Soc., New York, 1968, p. 837.
72 A. H. Freedman and E. B. Mikus, *Welding J.*, 46, Welding Res. Suppl., (1966) 258-S.
73 W. G. Moffatt and J. Wulff, *Trans. A.I.M.E.*, 209 (1957) 442.
74 M. M. Schwartz, *Mater. Eng.*, 64 (1966) 47.
75 S. Weiss and C. M. Adams, Jr., *Welding J.*, 46, Welding Res. Suppl., (1967) 49-S.
76 M. M. Schwartz, F. B. Gurtner and P. K. Shutt, Jr., *Welding J.*, 46 (1967) 423.
77 C. S. Beuyukian, *Welding J.*, 47 (1968) 710.
78 M. M. Schwartz, *Ind. Heating*, July (1967) 1266.

79 ANON., *Prod. Eng.*, *38* (1967) 144.
80 R. S. KIRBY AND G. S. HANKS, *Welding J.*, *47*, *Welding Res. Suppl.*, (1968) 97-S.
81 E. G. HUSCHE AND G. S. HOPPIN III, *Welding J.*, *37*, *Welding Res. Suppl.*, (1958) 233-S.
82 E. H. WHITSON, *Metal Progr.*, *82* (1962) 93.
83 H. UDIN, E. R. FUNK AND J. WULFF, *Welding for Engineers*, Wiley, New York, 1954.
84 YA. B. GUREVICH AND A. M. ZUBKO, *Stal'*, No. 10 (1959).
85 I. M. PAVLOV et al., *Tr. Inst. Met. Akad. Nauk SSSR*, *9* (1961).
86 I. M. PAVLOV AND YU. M. SIGALOV, *Izv. Vuzov SSSR, Chernaya Met.*, Nr. 8, (1961).
87 YA. B. GUREVICH, *Sbornik Nr. 7, M.*, Metallurgizdat, 1962.
88 V. S. SMIRNOV et al., *Tr. Leningr. Politekhn. Inst.*, Nr. 238, 1964.
89 V. M. AMONENKO et al., *Tr. Ukr. Nauchn. Issled. Inst. Met.*, *10* (1964).
90 V. S. SMIRNOV et al., *Tr. Leningr. Politekhn. Inst.*, Nr. 238, 1964.
91 R. A. UL'YANOV, N. D. TARASOV AND S. F. KOVTUN, *Tsvetn. Metal.* Nr. 3 (1964).
92 E. R. WAGNER, *Electronics*, No. 7 (1934) 104.
93 V. M. AMONENKO, A. S. TRON' et al., *Stal'*, No. 10 (1960) 920.
94 V. M. AMONENKO, A. S. TRON', V. V. MUKHIN, N. D. RYBAL'CHENKO AND E. A. KOVALEVA, *Brutcher Transl.*, *UDK 669-419*, *4*; 621, 771.
95 A. A. ALEKSANDROV, A. S. TRON' AND N. D. RYBAL'CHENKO, *Leningr. Politekhn. Inst. Trans.* Nr. 263, 1966, pp. 42–47, *Trans. Nr. F.T.D.-MT-24-264-67* (Foreign Technol. Div,) AD 677247. WP-AFB, Ohio.
96 YU. KRIVONOSOV, F. E. DO'ZHENKOV, O. A. MYAKSHIN AND L. A. ZAKHAROV, *Brutcher Abstract Transl.*, *UDK 669-293*. 14-419.
97 D. R. MILNER AND G. W. ROWE, *Met. Rev.*, *7* (1962) 433.
98 M. J. ALBOM, *Welding J.*, *43* (1964) 491.
99 W. H. KING AND W. A. OWCZARSKI, *Welding J.*, *Welding Res. Suppl.*, *46* (1967) 289-S.
100 W. H. KING AND W. A. OWCZARSKI, *Welding J.*, *Welding Res. Suppl.*, *47* (1968) 444-S.
101 ANON., *Metal Progr.*, *96* (1968) 100.
102 J. E. KELLEY, D. H. SUMNER AND H. J. KELLY, *U.S. Bur. Mines Rept. of Invest. 7225*, Febr. 1969.
103 J. W. MACEDO, Feasibility study on the vapor deposition joining of tungsten, *Report 574*, Martin Marietta Corp., Sept. 1961.
104 R. WILSON, *Welding J.*, *Welding Res. Suppl.*, *47* (1968) 345-S.

Chapter X

Vacuum Coating

E. RITTER

The subject of this chapter is the production of thin metal films and coatings by vacuum processes. These include evaporation or sputtering of metals with subsequent condensation on various substrates, such as metals, glass, ceramics, semiconductors and plastics. The condensed metal films find various applications in such fields as metallurgy, optics, electrical engineering and for decorative uses. In accordance with the aims of the present book the metallurgical applications are dealt with in more detail than the others.

There are three methods available for the production of metal films and coatings by vacuum processes: vacuum evaporation, cathode sputtering and thermochemical vapour deposition. Of these the last one has not found industrial application up to now and therefore only the first two methods will be described here. Also the materials used, the technological conditions required, the apparatus employed and the applications of the films and coatings will be reviewed.

1. Vacuum Evaporation

1.1 Principle of the method

The substrate to be coated is carefully cleaned and introduced into a vessel that can be evacuated. After the vessel has been pumped to a high vacuum, the metal to be used as coating is heated to such a temperature that its vapour pressure attains about 10^{-1} Torr. The evaporated atoms fly from the evaporation source into the evacuated space in a straight line and condense on the substrate and all other places which are in line of sight of the evaporation source and which therefore are in the path of the flying atoms, provided that the substrate is at a sufficiently low temperature. After the completion of the coating the vacuum vessel is opened to the atmosphere and the substrate taken out. This is usually the end of the process.

If large numbers of substrates or long strips are to be coated, introduction into the vacuum and removal from it may be effected via vacuum locks. In this case the process may be carried out semicontinuously or continuously.

1.2 Substrate preparation

1.2.1 Cleaning

The deposition of coatings with good adhesion by vacuum processes requires a more thorough cleaning of the surface to be coated than that for processes of electrolytic or chemical deposition, as adhesion of vacuum-deposited coatings on the substrate depends mainly on the relatively weak interaction of the van der Waals forces. Only if a chemical reaction is also involved does better anchorage on the substrate occur.

Glass and ceramics are easily cleaned, but plastics often require predegassing by baking.

With metals, heavy soiling, for instance, rust, grease, dust and residues from working or fabrication, must be removed by brushing, rinsing in wetting solutions, vapour degreasing or ultrasonic treatment.

1.2.2 Glow discharge

For the removal of adsorbed water and for the elimination of residual organic impurities the surface to be coated may be exposed to a glow discharge in the vacuum chamber at a potential of several kilovolts. The jigs for glass and ceramics are usually made the anode in this process, whereas metals are made the cathode. In the former case, besides the cleaning action, sputtering of the cathode may also occur. The sputtered metal atoms strike the substrate with high energy and form a thin film on it which consists, according to whether the gas used for the glow discharge is inert or active, of the cathode material or one of its compounds, for instance, a metal oxide. The formation of this adherent film is probably the most important condition for good adhesion of the coating subsequently applied. In the other case, a removal of material from the substrate occurs by cathode sputtering. In this way the surface layer, which might have been heavily distorted by previous working or which may carry impurities, is removed and the lattice of the matrix with its binding forces exposed.

1.2.3 Heating

Baking of the substrate up to 300°C may also serve for cleaning. In many cases this is more effective than a glow discharge and also results in faster degassing. If the substrate is not properly degassed, dark-brown, loose films are produced when, for instance, steel is coated with aluminium[1].

1.2.4 Shape and jigging of the parts

The passage of material from the evaporation source to the part to be coated is analogous to the radiation of light from a light source of similar shape; only surfaces which can "see" the evaporation source are coated. Shapes which deviate from a plane must therefore be rotated during the coating operation. Rotation is also necessary if a part is to be coated on all surfaces. For obtaining an even thickness on the substrate, a certain minimum distance from the evaporation source is required, which varies according to the size of the part to be coated and the size of the evaporation source. This minimum distance may be reduced if the parts are rotated and the evaporation source is situated off the rotation axis. This reduces the requirements on the quality of the vacuum, at the same time improving the utilization of the coating material. Holland[2] and Bunshah and Juntz[3] consider in detailed studies the distribution of the film thickness obtained with different source to substrate configurations.

Normal incidence of the evaporated atoms gives the best adherence and particularly smooth coatings.

From the above statements it is evident that the coating of bore holes and blind holes by vacuum deposition is extremely difficult.

1.3 Vacuum required and the effect of residual gases

The vacuum to be obtained depends to a great extent on the vacuum equipment used and on the maximum permissible gas inclusion of the deposit.

In order to obtain adherent and dense films and coatings, the mean free path of the evaporated atoms must be equal to, or greater than, the distance from the evaporation source to the point of condensation on the substrate. For a vacuum of 5×10^{-4} Torr the mean free path is about 10 cm, and in a vacuum of 5×10^{-5} Torr it is about 100 cm.

The incorporation of gases into the deposited coating occurs to a rather low percentage by association of gas molecules with the evaporated atoms during their flight. In fact, incorporation mainly takes place during condensation on the surface of the substrate by reaction with impinging residual gas molecules.

The quantity of the gases incorporated depends therefore on the ratio of the numbers of metal atoms to the residual gas molecules arriving at the point of condensation per unit time, that is on the evaporation rate and the temperature of the surface[4]. The composition of the residual gases plays, of course, an important role[5]. Low pressure, high evaporation rate and elevated substrate temperature improve the purity of the deposits.

With complex shapes, on the other hand, the evaporation process is sometimes conducted at 2×10^{-2} to 3×10^{-2} Torr argon pressure in order to improve the throwing power[6] and to obtain a more uniform coating.

1.4 Methods of evaporation

In principle, all metals which are solid at room temperature may be evaporated and subsequently condensed. However, the technological difficulties met in both evaporation and condensation vary greatly[7]. The type of heating employed depends on the melting point, the evaporation temperature and the reactivity of the metal to be evaporated[8]. The methods most frequently used for heating are:

1.4.1 Direct heating

Directly heated crucibles are made of high-melting-point metals, such as tungsten, tantalum or molybdenum or of graphite or intermetallic compounds,

References pp. 818–820

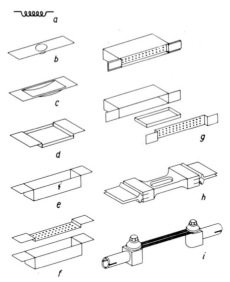

Fig. 1. Various evaporation sources using direct heating (according to Ross[9]).
(a) Tungsten helix; (b) and (c) Preformed tungsten boats; (d) to (f) Boats made of molybdenum, tantalum or tungsten sheet; (g) Box-type molybdenum boat for horizontal evaporation; (h) Graphite boat with flexible, cushioned support made of molybdenum sheet; (i) Parallel tungsten wires.

such as titanium–zirconium boride for the evaporation of aluminium. Evaporation sources made of iron may be used for metals of low melting point. In Fig. 1 some of the shapes most frequently used for evaporation sources made from the above-mentioned materials are shown[9,10]. Some of these can only be used for the evaporation of small quantities as required for optical and certain electrical applications. If evaporation of large quantities is required, the material may be fed either continuously or intermittently, in the form of wire, powder or granules[11].

1.4.2 Induction heating

In some cases, induction heating of the evaporant in crucibles made of metal, graphite or ceramics is employed instead of direct heating. This method does not impose such strict requirements on the crucibles material[12].

1.4.3 Heating by electron bombardment

Heating of the material by electron bombardment in a cooled crucible avoids almost completely any crucible reactions and allows concentration of high energies on small areas. Electron bombardment therefore also makes possible the evaporation of metals with the highest melting points, such as tungsten, tantalum, molybdenum, platinum and rhodium. This method is therefore particularly used for the large-scale evaporation of metals. In most cases, a water-cooled copper crucible is used for holding the material. But for aluminium a water-cooled steel trough with a mullite–zirconia ceramic lining[13] has proved adequate. In Fig. 2 several types

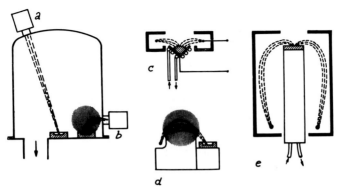

Fig. 2. Various devices for evaporation by electron beams (according to Zinsmeister[14]).
(a) Axial Pierce type gun; (b) and (d) Systems with magnetic focusing; (c) and (e) Ring systems with electrostatic focusing.

of electron beam evaporation devices are shown. They differ with respect to focusing and the arrangement of the cathode[14]. An electron beam evaporator of high output (75 kW) with magnetic deviation is shown in Fig. 3. This is used for the evaporation of large quantities of metal, for instance, for the production of protective coatings[15,16].

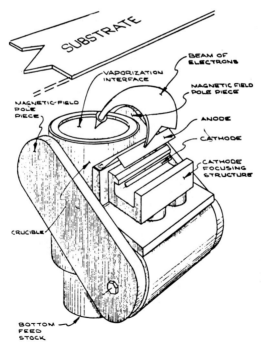

Fig. 3. Sketch of an electron-beam evaporation device of high output (according to H. R. Smith, Jr.[15]).

References pp. 818–820

1.4.4 Evaporation of alloys

Direct evaporation of alloys is difficult as the components evaporate according to their vapour pressure in the alloy. A possible method for almost complete avoidance of fractionating was found in the so-called flash evaporation. The alloy to be evaporated is fed continuously in minute quantities into an evaporator, which is kept at so high a temperature that the alloy evaporates instantaneously. Such an arrangement is shown in Fig. 4. The material is introduced into the evaporator in the form of granules of the alloy. Feeding is effected by a vibrating helix conveyor through a water-cooled tube. Another method of evaporation of alloys is multi-source evaporation, in which the individual components are simultaneously evaporated from separate evaporators. Alloy formation takes place during condensation[3].

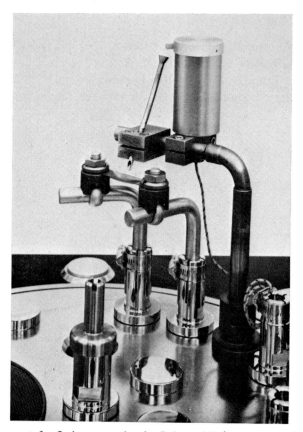

Fig. 4. Arrangement for flash evaporation by Balzers AG (by courtesy of Balzers AG).

1.5 Control of evaporation

In order to control the thickness of the coatings during evaporation, different methods may be applied, depending on the type of material used. For nonabsorbing dielectric films and for the production of semi-transparent metallic films, optical methods are applicable. The reflectance or transmittance of a test glass is continuously monitored. The measurement of transmittance is used for example in strip coating of plastics with metallic films in order to control the rate of evaporation.

Another method, introduced recently, is the measurement of the mass deposited on an oscillating quartz crystal[17,18]. The frequency shift which is proportional to the mass deposited during evaporation is again a measure for the film thickness. As films with high optical absorption and up to a thickness of some microns can be measured this method is especially suitable for metals.

The measurement of the desired physical characteristics is preferred only in the production of metallic films with a predetermined electrical resistance. Here a test glass with two electrodes is used which allows for the continuous recording of the change of resistance during film deposition.

With these three methods the test glass or oscillating quartz has to be arranged in such a way that the thickness of the film deposited is equal or at least proportional to the film thickness on the substrate to be coated.

In many cases, especially in strip coating, the rate of evaporation has to be controlled also. With metals, optical methods are only applicable for semi-transparent films. In metallizing plastics foils this may be done on a small strip parallel to the band to be coated, possibly by attenuation of the oncoming metal vapour stream if a thicker non-transparent coating is desired.

Another possibility for rate control is the measurement of the vapour density using the principle of a hot cathode high vacuum ionization gauge[19-22]. Here a portion of the vapour molecules is ionized by the impact of electrons from a separate electron emitter. The number of ions attracted to a collector is measured, and this ion current, which is proportional to the density of the vapour stream, can be used to control the evaporation rate. This method is also suitable for the control of several vapour sources, for the simultaneous evaporation of different materials in order to form alloys or metal oxide mixtures on the substrate. Here every vapour source has to be controlled by a separate rate meter and they have to be shielded against each other so that mixing occurs only near the substrate.

The third method is restricted to batch operation. Here the differential change of the frequency of the oscillating quartz crystal, already mentioned above, is used as a measure for the evaporation rate[23].

There are still other possibilities for rate monitoring which will not be considered here because they have not found application in industrial practice up to now.

1.6 Condensation

Nucleation, growth of the deposit and structure of the deposit depend on various parameters, as well as on the actual combination of substrate and coating, on rate of deposition, on the temperature of the substrate and on the residual gas pressure. The optimum conditions must be established for each case[24].

Mattox[25] and other authors[6] claim that the condensation and properties of metal films improve if the thermal evaporation is performed in the presence of a glow discharge, where the substrate is the cathode of a high-voltage system. A portion of the evaporated atoms is ionized and the ionized atoms are accelerated toward the substrate by the applied d.c. potential. Atoms and ions are deposited while the surface is continually bombarded with the gas ions of the glow discharge. The method is called "ion plating".

In the case of deposition of thick protective coatings on metals, it is usual practice to work with high vapour supersaturation, *i.e.* with high deposition rates. In order to obtain a certain degree of alloy formation with the substrate material, increased substrate temperatures are often used. For instance, if the substrate is heated to 450°C, the formation of the intermetallic compound Fe_2Al_5 at the interface between aluminium and steel was found during the coating of steel with aluminium[26]. However, good mechanical adherence of aluminium to steel was obtained even when the substrate temperatures were only higher than 150°C. At low temperatures and with substrates that are not specially cleaned in a vacuum binding via an oxide probably occurs. If such coatings are immersed in dilute hydrochloric acid, they spall. This may be avoided by alloy formation which takes place at higher substrate temperatures[26].

In another example Ti–Cu diffusion couples are formed by the deposition of titanium on copper at 700°C[27].

The substrate temperature also influences strongly the structure of the deposits[28].

Porosity depends to a large extent on the surface quality of the substrate[1]. Pores are formed preferentially at sharp edges, scratches and holes. Dust, surface roughness or gas occlusion may also lead to porous deposits. Minimum porosity of aluminium coatings on steel was obtained at deposition rates above 2000 Å/sec and substrate temperatures of 250°C. Coatings of a thickness of more than 1.7 μm usually show very little porosity[26].

Stresses may arise as a result of the different thermal expansion coefficients of substrate and coating, or as a result of oblique incidence of the atoms, or of lattice distortion, diffusion and gas occlusion. These stresses may, in the extreme case, lead to a loosening or even spalling of the coating, particularly at greater thicknesses.

1.7 Operation of the processes

1.7.1 Discontinuous practice

It is the usual practice to operate this process in the batch fashion, unless long strips of paper, plastics or steel, or large sheets in great numbers, are to be coated.

Commercial plants for the batch operation have vacuum vessels in the form of vertical or horizontal cylinders, and more recently also as cubes. The diameter of the vacuum vessel of production units varies between 50 and 200 cm.

A plant consists mainly of the pumping system, the vacuum vessel, the jig, the glow discharge device, the substrate heating unit, as well as of the electrical auxiliaries and indicators and recorders. A diagram of such a plant is shown in Fig. 5.

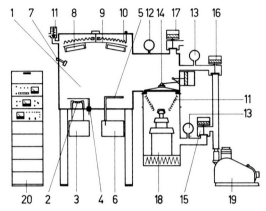

Fig. 5. Block diagram of a discontinuous coating plant.
1 Vacuum vessel; 2 Evaporation source; 3 Current supply to evaporation source; 4 Shutter; 5 Glow discharge electrode; 6 High-tension current supply for glow discharge unit; 7 Device for measuring evaporation rate and film thickness; 8 Substrate; 9 Rotary substrate support; 10 Heater for substrate; 11 Gas admittance valve; 12 High-vacuum gauge head; 13 Fore-pressure gauge head; 14 Plate valve; 15 Backing-line isolation valve; 16 By-pass valve; 17 Valve to atmosphere; 18 Diffusion pump; 19 Roughing pump; 20 Control unit.

1.7.2 Continuous practice

This, as mentioned above, is particularly well suited for the coating of long strips or large sheets. In Fig. 6 the diagram of a continuous plant for the coating of steel strip after Smith, Hunt and Edwards[29] is shown. In front of the vacuum plant a washing plant is situated. The cleaned strip passes through several air locks and is heated by electron bombardment. After being coated, for instance with aluminium using electron beam evaporators, it leaves the plant via several air lock chambers and is cooled in an inert atmosphere. Finally it is coiled.

Detailed descriptions of actual plants particularly designed for metallurgical application were given by Smith and Hunt[30] and by Reichelt, Dietrich and Hauff[13,31]. The plant described by Smith and Hunt (*loc. cit.*) can accept strip of thicknesses between 0.12 and 1.0 mm and of widths of up to 600 mm. The coating speed varies between 0.6 and 260 m/min. The nominal power supply is laid out for 1200 kW. The pump-down time from atmospheric pressure to the operating vacuum (below 10^{-4} Torr) is about 30 minutes.

References pp. 818–820

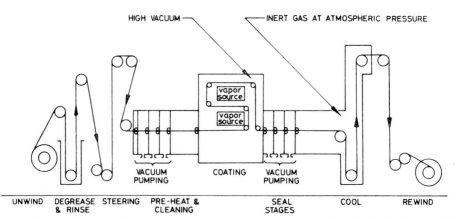

Fig. 6. Block diagram of a continuous plant for strip coating (according to Smith, Hunt and Edwards[29]).

1.8 Applications

Thin metal films and coatings find applications in almost all fields of technology. Some of the most important applications are:

1.8.1 Optical applications[2,32,33]

Vacuum deposited, metallic surface mirrors are almost exclusively used in optical instruments nowadays[34].

The most widely used metal is aluminium, as it shows a high reflectivity in the whole range of the spectrum from the ultraviolet to the infrared. It is also frequently combined with dielectric films for protection or to increase the reflectance. Such combinations find, *inter alia*, application for the control of the temperature of satellites[35].

1.8.2 Electrical engineering applications

The coating of paper or plastics strips with zinc or aluminium in continuous plants for foils of self-healing capacitors and the deposition of electrical contacts and metal resistors (NiCr) are carried out on a production scale.

Applications for microelectronics, active elements and memory units[36-38] are still in the development stage.

In microelectronics complete passive networks are to be produced by deposition of metals (Al, Cu, Ni, Au) and alloys (Ni–Cr) in combination with dielectric films (SiO).

In active elements, such as transistors, vacuum-deposited metal films play only the role of auxiliaries (contacts, conductors, electrodes).

As memory units for computers, magnetic films of Ni–Fe as well as superconductive films of tin, lead and niobium are used.

1.8.3 Decorative applications[39]

Plastics of various types and in a variety of shapes are vacuum-coated for numerous purposes. The metal most frequently used is aluminium. Wear and corrosion resistance may be improved by lacquering; this also provides coloration if desired.

1.8.4 Metallurgical applications

In contrast with the decorative applications where large quantities are produced, the metallurgical applications are in some instances still in the development stage.

(a) Aluminium and tin may be vacuum-deposited in thicknesses of 0.1–2 μm on steel strip by a continuous process. Such strip is to be used in the packaging industry, also for sign boards, motor car parts and toys. Pilot plants are in operation at various large steelworks[40].

It is estimated that the vacuum aluminizing of steel strip may be carried out under certain conditions at a third of the cost of electrolytic tinning of the same thickness[13].

A combination of a 3.7-μm layer of electroplated zinc with a 1.1-μm film of vacuum-deposited aluminium is said to show a corrosion resistance which far exceeds the resistance of the individual layers[11].

(b) Cadmium films at a thickness of between 5 and 12 μm are used for the protection against corrosion of struts, joining elements and structural parts of high-strength steels[40]. For the landing gear of aircraft, cadmium is deposited in thicknesses of up to 75 μm[41]. Vacuum-deposition avoids embrittlement as a result of hydrogen pick-up which occurs during electroplating[42].

Cadmium films can be used only up to temperatures of 260 °C. For this reason aluminium deposits are now being tested for rockets and Mach-2 aircraft parts[40]. Aluminium deposits of thicknesses up to 100 μm have proved so strong, elastic and pore-free that they protect the landing gear of aircraft up to temperatures of 480°C[6,41].

(c) Vacuum-deposited zirconium in a thickness of 2.5 μm protects beryllium from corrosion for more than 240 h in weathering tests according to MIL–E–5272B[40].

(d) Further products in an experimental stage are nickel-coated copper foil, ferro-chromium-coated stainless steel and stainless-steel-coated copper or steel strip[29].

(e) Foils of metals, such as beryllium, titanium or aluminium, may be produced by vacuum deposition and subsequent removal from the substrate[43,44].

(f) By deposition of an active metal film by vacuum metallizing with titanium or zirconium the spreading and wetting of the filler metal in brazing of difficult-to-wet materials is promoted[45].

References pp. 818–820

2. Cathode Sputtering

2.1 Principle and modifications of the process

2.1.1 Classic cathode sputtering

In a glow discharge at a pressure of between 1 and 5×10^{-2} Torr and a potential of 1–5 kV, the cathode is sputtered by positive ions. In Fig. 7 the arrangement for the classic cathode sputtering process is shown. A gas, such as argon,

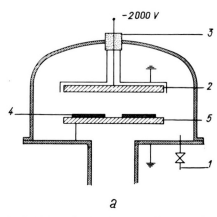

Fig. 7. Set-up for classical cathode sputtering.
1 Gas admittance valve; 2 Cathode; 3 High-tension lead-in; 4 Substrate; 5 Anode.

is admitted into the evacuated system by a valve holding the gas pressure constant at about 10^{-2}–10^{-1} Torr. It is the usual practice to choose an inert gas for the sputtering of metals and semiconductors. The cathode is then connected to a potential of 2000 V and the glow discharge initiated. The positively charged argon ions strike the negative cathode and sputter the individual atoms of the cathode. The sputtered atoms diffuse through the gas atmosphere and deposit on the substrate, which is placed opposite the cathode, as a film. As cathode sputtering is caused by a collision mechanism and not by evaporation, the speed of the sputtered atoms is about one order of magnitude and their kinetic energy therefore two orders of magnitude higher than that of evaporated atoms. This explains why it is possible to obtain coatings of excellent adherence by sputtering, even if the gas pressure is as high as 10^{-1} Torr. The mean sputtering rate is one atom per ion; sputtering is therefore a somewhat slower process than evaporation. Another shortcoming of the classic sputtering process is the risk of gas being occluded into the deposit. This is caused by the high gas pressure. The object of further development of the cathode sputtering process was therefore to reduce the inclusion of gas in the deposit and to increase the sputtering rate. (For further details on the physics and potential of cathode sputtering, see ref. 46.)

2.1.2 Modern methods of cathode sputtering

(a) During bias sputtering a small negative potential is placed on to the substrate so that the substrate is also being bombarded by positive ions. It is assumed that the surface of the growing film is continuously being cleaned of adsorbed gases in this way and therefore that films of improved purity are being obtained[47]. It is said that bias sputtering also allows internal coverage of slots and grooves to a certain degree[48].

(b) Sputtering with an asymmetric alternating current has a similar effect to bias sputtering. In this process deposition takes place only during each half wave, while the deposit is bombarded with ions during the other half-wave period. In this way too adsorbed gases are removed[49].

(c) In protective sputtering the inert gas used for sputtering is purified from impurities of active gases by the getter effect of the sputtered material before it enters the region of sputtering proper[50]. This effect is obtained by encapsulation of cathode and anode in a cage.

(d) For sputtering by an assisted gas discharge a hot cathode must be used as a source for the electrons. This process is therefore called triode or tetrode sputtering. Furthermore, if care is taken that the probability of ionization in the gas is sufficiently high by increase of the electron paths, a relatively high sputtering rate may be obtained even at working pressures below 10^{-3} Torr. This object may be realized in various ways, and an example of it is shown in Fig. 8. An arc-discharge between a hot cathode and an auxiliary anode is produced in a chamber above the proper working chamber. This chamber, which has a somewhat higher

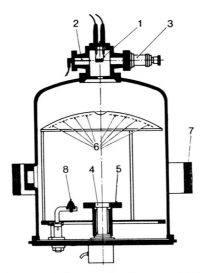

Fig. 8. Set-up for triode sputtering (Sputron of Balzers AG).
1 Hot cathode; 2 Auxiliary anode; 3 Gas admittance valve; 4 Cathode and target, water-cooled; 5 Ring-shaped anode, water-cooled; 6 Substrate; 7 Magnet coil; 8 Conventional evaporation source.

References pp. 818–820

pressure (about 10^{-2} Torr) than that of the working chamber, is connected with the working chamber by an orifice 10 mm in diameter. If a positive potential of about 50 V is applied to the lower annular anode, the plasma consisting of ions and electrons is drawn into the working chamber through the orifice, which acts as a pressure stage. The magnetic guiding field keeps the plasma together so that all ions are directed on to the target. If the target is connected to a negative potential of 500 to 2000 V, high sputtering rates are obtained with ion currents of several amperes.

If the pressure is low it is also possible to place the substrates at a greater distance from the target so that the coating of large areas may be carried out in correspondingly longer times.

(e) In high-frequency sputtering in most cases the same arrangement is used as in normal diode sputtering, but instead of d.c. a radio frequency field is used (mostly 13.5 or 27 MHz). By this method also dielectrics like glass and oxides may be sputtered. Recently it is also sometimes applied for sputtering of metals, especially for those which have a tendency to form oxides on the surface like aluminium and also in those cases where dielectrics and metals have to be sputtered in the same vacuum process.

2.2 Materials which may be deposited by cathode sputtering

With the aid of the sputtering methods described, nearly all metals may be sputtered. Even alloys may be sputtered, maintaining their composition quite well which is a particular advantage of this method[51]. Even if one component of the target material sputters more easily, an equilibrium is established after some time because the other component is enriched on the surface up to a point where the composition of the sputtered material becomes equal to the composition of the basic target material.

However, with heterogeneous material this is only true if the temperature of the target is sufficiently low so that no diffusion inside the target takes place and if the diffusion is also not enhanced by the ion bombardment[52].

Films of stainless steel of a thickness of some millimetres have, for example, been sputtered as protective coatings. It is interesting to note that the grain microstructure and the crystal structure of stainless steel are strongly dependent on the substrate temperature. Dahlgreen and McClanahan[53] found, in high rate sputtering of stainless steel 304 L, that between 360° and 400°C the microstructure of the deposit changed from columnar to equiaxed grains whereby the columnar grains were associated with the body-centered-cubic crystal structure (the structure which is theoretically the ferritic equilibrium structure below 400°C and normally does not exist because the austenitic condition is always undercooled) and the equiaxed grains showed the face-centered-cubic phase.

Composition and density of the deposit in the austenitic condition corresponded very closely to those in the target material. We come, therefore, to the conclusion that deposits made by sputtering in a clean atmosphere tend to grow under the equilibrium conditions, corresponding to the substrate temperature.

2.3. Sputtering practice

Many problems of cathode sputtering, such as substrate preparation, condensation, etc., are similar to those existing in vacuum deposition by evaporation. Even the design and construction of the plants do not show any fundamental differences. Usually before starting deposition the surface layer of the substrate is removed by reverse sputtering. This promotes the adherence of the material deposited afterwards. Also the target material is sometimes sputtered firstly under a shield in order to remove the surface layers eventually formed on the target during exposure to the atmosphere.

A great advantage of the sputtering technique compared with evaporation is the very simple adjustment of the desired thickness of the coating by control of the time only, provided that the electrical parameters and also the composition of the vacuum atmosphere are kept constant.

In order to get reproducible conditions from the very beginning of the process it is therefore necessary to evacuate the plant to low pressures in order to remove the water vapour adsorbed at the chamber walls. If the gas discharge would be started immediately after reaching the operating pressure and bleeding in the gas to be ionized the partial pressure of hydrogen (originating from water vapour) would be rather high and take over a large part of the ion current. Since the sputtering rate for hydrogen is much lower than for gases of higher molecular weight a continuous change of the deposition rate would be observed associated with the change of the hydrogen partial pressure in the atmosphere, and thickness control by time would therefore be impossible.

2.4. Applications

In comparing the potentialities of cathode sputtering with those of vacuum evaporation the advocates of sputtering very often create the impression that sputtering would overcome a lot of the limitations inherent in the evaporation process which in reality do not exist or may probably disappear in the near future. A comparison is therefore only possible by those who are knowledgeable in both fields.

Besides the easier control of deposition rates there are no doubt additional distinct advantages of sputtering over evaporation which are due to the higher energies of the sputtered atoms or molecules and to the fact that sputtering is not directly related to the vapour pressure of the material to be sputtered. The high energy of the sputtered particles offers in some cases particularly good bonding to the substrate, and therefore excellent adherence and the quasi independence of the vapour pressure make sputtering very suitable for refractory metals with very high melting points, such as tantalum and tungsten. On the other hand, it is a relatively slow process compared with vacuum evaporation.

References pp. 818–820

2.4.1 Electrical engineering applications

Sputtering has found wide applications in the electrical industry for the manufacturing of monolithic integrated circuits. It permits the deposition of insulators as well as conductors. With the so-called tantalum technique for example conductors, resistors and capacitors are produced on the base of sputtered tantalum or tantalum nitride films[54,55]. Sputtered nichromium films are also used for resistors. Also pinhole-free chromium films for the production of photomasks (which are used in the photo etching process in the production of hybrid circuits) are made in this way, but not exclusively, since vacuum evaporation is also successful.

2.4.2 Metallurgical applications

Here protection against corrosion is the main application. Sputtered coatings of titanium, tantalum, chromium, stainless steel and Hastelloy C are said to protect magnesium against corrosion by sea-water and therefore offer a solution to this problem[54]. Sputtered high purity chromium is used on the edges of razor blades in order to prevent corrosion and to extend their lifetime. Relatively thick stainless steel deposits, as already mentioned, are used for protection against corrosion in high temperature applications, for example on turbine blades. Sputtered films of precious metals and other corrosion-resistant materials are used for protection of a cheaper base material in the chemical industry in some cases.

It has also been suggested that aluminium, beryllium and titanium can be coated with chromium, tungsten or stainless steel in order to improve the wear resistance[54] and to apply thin films of highly abrasion resistant material in the form of alloys or compounds by sputtering on cutting tools for the same purpose.

The application of dry lubricants deposited as thin films by sputtering is of similar importance.

2.4.3 Decorative applications

Sputtered films on glass and plastics materials like Mylar are used for decorative effects. The materials which have been applied are gold, silver, brass, bronze, chromium and stainless steel[56].

REFERENCES

1 W. REICHELT, *Vakuum-Tech.*, 15 (1966) 1.
2 L. HOLLAND, *Vacuum Deposition of Thin Films*, Chapman and Hall, London, 1956.
3 R. F. BUNSHAH AND R. S. JUNTZ, in E. L. FOSTER (Ed.), *Trans. Vacuum Met. Conf., 1967*, Am. Vacuum Soc., New York, 1968, p. 799.
4 E. RITTER, in E. HAHN, P. B. BARNA AND J. PEISNER (Eds.), *Proc. Colloq. on Thin Films*, Budapest, 1967, p. 79.
5 H. L. CASWELL, in G. HASS AND R. THUN (Eds.), *Physics of Thin Films*, Vol. 1, Academic Press, New York, 1963, p. 1.
6 L. E. MCCRARY, J. F. CARPENTER AND A. A. KLEIN, in R. B. BARROW AND A. SIMKOVICH (Eds.), *Trans. Vacuum Met. Conf., New York, 1968*, Am. Vacuum Soc., New York, 1968, p. 221.

7 W. Espe, in M. Auwärter (Ed.), *Ergebnisse der Hochvakuumtechnik und der Physik dünner Schichten*, Wissenschaftliche Verlagsgesellschaft, Stuttgart, 1957, p. 67.
8 M- J. White, *Vacuum, 18* (1968) 651.
9 A. Ross, *Vakuum-Tech., 8* (1959) 1.
10 A. H. Plaisted, *Proc. 12th Ann. Tech. Conf., Detroit, Mich., 1969*, Soc. Vacuum Coaters, Cleveland, Ohio, 1969, p. 57.
11 J. M. Roblin, *Metal Progr., 89* (1966) 115.
12 G. F. Jacobus, R. P. Madden and L. R. Canfield, *J. Opt. Soc. Am., 53* (1963) 1084.
13 W. Reichelt, W. Dietrich and A. Hauff, *Metalloberfläche, 20* (1966) 474.
14 G. Zinsmeister, in L. Steipe (Ed.), *Mikroelektronik*, R. Oldenbourg Verlag, München, 1965, p. 92.
15 H. R. Smith, Jr., Materials Science and Technology for Advanced Applications, II, *ASM Golden Gate Metals Conf., 1962*, p. 569.
16 H. R. Smith, Jr., in E. L. Foster (Ed.), *Trans. Vacuum Met. Conf., New York, 1967*, Am. Vacuum Soc., New York, 1968, p. 719.
17 H. Pulker and E. Ritter, *Vakuum-Tech., 14* (1965) 91.
18 K. H. Behrndt, in G. Hass and R. Thun (Eds.), *Physics of Thin Films*, Vol. 3, Academic Press, New York, 1966, p. 1.
19 O. Haase, *Z. Naturforsch., 12* (1957) 941.
20 R. Olson and L. Hale, in R. B. Barrow and A. Simkovich (Eds.), *Trans. Vacuum Met. Conf., 1968*, Am. Vacuum Soc., New York, 1968, p. 263.
21 C. A. Prawdzik, *Proc. 12th Ann. Tech. Conf., Detroit, Mich., 1969*, Soc. Vacuum Coaters, Cleveland, Ohio, 1969, p. 64.
22 B. Zega, *Proc. 4th Intern. Vacuum Congr. Manchester, April 1968*, Inst. Phys. and Phys. Soc., London, p. 523.
23 G. Sauerbrey, *Z. Physik, 155* (1959) 206.
24 H. Mayer, *Physik Dünner Schichten*, Vol. 2, Wissenschaftliche Verlagsgesellschaft, Stuttgart, 1955.
25 D. M. Mattox, *Develop. Rept. No. SC-DR-281-63*, Sandia Corp., November 1963.
26 R. G. Meyers and R. P. Morgan, in M. A. Orehoski and R. F. Bunshah (Eds.), *Trans. Vacuum Met. Conf., 1966*, Am. Vacuum Soc., Boston, Mass., 1967, p. 271.
27 R. Baboian, C. D. Turk and H. L. Marcus, in R. B. Barrow and A. Simkovich (Eds.), *Trans. Vacuum Met. Conf., 1968*, Am. Vacuum Soc., New York, 1968, p. 203.
28 K. Kennedy, in R. B. Barrow and A. Simkovich (Eds.), *Trans. Vacuum Met. Conf., New York, 1968*, Am. Vacuum Soc., New York, 1968, p. 195.
29 H. R. Smith Jr., C. d'A. Hunt and R. Edwards, in M. A. Orehoski and R. F. Bunshah (Eds.), *Trans. Vacuum Met. Conf., New York, 1966*, Am. Vacuum Soc., Boston, Mass., 1967, p. 240.
30 H. R. Smith, Jr. and C. d'A. Hunt, in L. M. Bianchi (Ed.), *Trans. Vacuum Met. Conf., 1965*, Am. Vacuum Soc., Boston, Mass., 1966, p. 227.
31 W. Dietrich, A. Hauff and W. Reichelt, *Proc. 4th Intern. Vacuum Congr., Manchester, 1968*, Part 2, Inst. Phys. and Phys. Soc., London, p. 573.
32 O. S. Heavens, *Optical Properties of Thin Solid Films*, Butterworth, London, 1955.
33 H. Anders, *Dünne Schichten für die Optik*, Wissenschaftliche Verlagsgesellschaft, Stuttgart, 1965.
34 G. Hass, in R. Kingslake (Ed.), *Applied Optics and Optical Engineering*, Vol. 3, Academic Press, New York, 1965, p. 309.
35 L. F. Drummeter, Jr. and G. Hass, in G. Hass and R. Thun (Eds.), *Physics of Thin Films*, Vol. 2, Academic Press, New York, 1964, p. 305.
36 R. Thun, W. N. Caroll, Ch. J. Kraus, J. Riseman and E. S. Wajda, in E. Keonjion (Ed.), *Microelectronics*, McGraw-Hill, New York, 1963, p. 173.
37 L. Holland (Ed.), *Thin Film Microelectronics*, Chapman and Hall, London, 1965.
38 A. Lewicki, *Einführung in die Mikroelektronik*, R. Oldenbourg-Verlag, München, 1966.
39 F. J. Horn and H. B. Hebble, Jr., in C. Powell, J. H. Oxley and J. M. Blocher, Jr. (Eds.), *Vapor Deposition*, Wiley, New York, 1966, p. 579.

40 C. A. Krier, in C. Powell, J. H. Oxley and J. M. Blocher, Jr. (Eds.), *Vapor Deposition*, Wiley, New York, 1966, p. 512.
41 J. E. Sanford, *Amer. Mach./Metalworking Manuf.*, March 1 (1965) 51.
42 D. J. Fishlock, *Electroplating Metal Finishing*, June (1959) 221.
43 P. J. Clough and R. W. Steeves, in R. F. Bunshah (Ed.), *Trans. Vacuum Met. Conf., New York, 1963*, Am. Vacuum Soc., Boston, Mass., 1963, p. 167.
44 R. F. Bunshah and R. S. Juntz, in M. A. Orehoski and R. F. Bunshah (Eds.), *Trans. Vacuum Met. Conf., New York, 1966*, Am. Vacuum Soc., Boston, Mass., 1967, p. 209.
45 C. M. Adams, Jr., in R. B. Barrow and A. Simkovich (Eds.), *Trans. Vacuum Met. Conf., New York, 1968*, Am. Vacuum Soc., New York, 1968, p. 279.
46 G. Wehner, *Sci. Technol.*, Sept. 1968, p. 32.
47 L. J. Maissel and P. M. Schaible, *J. Appl. Phys.*, 36 (1965) 237.
48 J. M. Seemann, *Vacuum*, 17 (1967) 129.
49 R. Frerichs, *J. Appl. Phys.*, 33 (1962) 1898.
50 H. C. Theuerer and J. J. Hauser, *J. Appl. Phys.*, 35 (1964) 554.
51 R. C. Krutenat, R. F. Wielonski and F. M. Lawrence, in R. B. Barrow and A. Simkovich (Eds.), *Trans. Vacuum Met. Conf., New York, 1968*, Am. Vacuum Soc., New York, 1968, p. 235.
52 G. S. Anderson, *J. Appl. Phys.*, 40 (1969) 2884.
53 S. D. Dahlgreen and E. D. McClanahan, *Symp. on the deposition of thin films by sputtering*, Univ. of Rochester, Rochester, N.Y., Sept. 1969.
54 J. M. Seemann, in L. M. Bianchi (Ed.), *Trans. Vacuum Met. Conf., New York, 1965*, Am. Vacuum Soc., Boston, Mass., 1966, p. 269.
55 W. S. Shirley, in R. B. Barrow and A. Simkovich (Eds.), *Trans. Vacuum Met. Conf., New York, 1968*, Am. Vacuum Soc., New York, 1968, p. 251.
56 V. Hoffmann, *Proc. 12th Ann. Tech. Conf., Detroit, Mich., 1969*, Soc. of Vacuum Coaters, Cleveland, Ohio, 1969, p. 1.

Chapter XI

Effect of High Vacuum on some Important Properties of Metals and Alloys

D. T. BOURGETTE, D. V. KELLER AND R. L. STEPHENSON

1. Introduction

This chapter has been written to aid those responsible for conducting research in high vacuum metallurgy. Major emphasis is directed toward materials problems associated with application in space or ultrahigh vacuum rather than equipment or equipment design. An attempt is made to define the more important metallurgical problems and show how they are compounded when metals and alloys are used in vacuum.

Our discussion is oriented from the space applications viewpoint because countless hours of testing have been and will be undertaken to advance our knowledge of environmental effects on space vehicle systems and components. The future success of space exploration depends to a great extent on the development of lightweight, efficient power plants to supply both the thrust power and electrical needs of space vehicles. Design studies and experiment have shown that high thrust performances are achievable with both chemical and nuclear power, while electrical needs can be satisfied by nuclear reactors and isotope cells. Within this broad area of application, materials will frequently be required to operate at the limits of their chemical and mechanical capabilities for periods of several years.

It is wrong to think that there exists a class of materials that are characterized by the title "Space Materials". Many commonplace alloys are presently being used successfully in space and ultrahigh vacuum. On the other hand, there are certain specific applications, such as liquid metal and nuclear fuel containment, component reliability, vacuum heat transfer (radiators) effects, bearing and metal fatigue, metallic adhesion, and boundary or surface lubrication that all present metallurgical problems that are more serious in an unconventional environment such as space. The principal interactions between the space or vacuum environment and metallic materials occur at the metal surfaces; however, considerable changes in bulk properties can also result. Therefore, this chapter describes the space environment in terms of altitude, composition, pressure, density and temperature, which in turn are compared with space simulation in vacuum chambers. High temperature problems such as the effects of vacuum on creep and fatigue

References pp. 858–861

properties are delineated. Lower temperature problems such as metallic adhesion and lubrication are also discussed.

2. The Space and Vacuum Environment

Surface reactions that may result from exposure to a space environment will be easier to understand if the characteristics of space are first summarized and compared with vacuum chamber simulation. Pressure, particle density, composition, and temperature are the major characteristics of space that are of interest to us. They vary with altitude, latitude, and seasons. For the purpose of this discussion, variations due to solar activity and seasons will be neglected.

The principal characteristics of space are summarized as a function of altitude in Table 1. An increase in altitude is accompanied by a decrease in particle density, pressure, and molecular weight, while the composition tends toward the lighter elements. Above 1600 km, the space composition is 100% atomic hydrogen. Later measurements[1,2] have shown that two transition regions exist—oxygen ions to helium ions and helium ions to hydrogen ions—rather than a single transition from oxygen to hydrogen ions. At an altitude of 800 to 1300 km the oxygen ion to helium ion ratio is unity. The ions that form the greatest concentration in the lower ionosphere are N_2^+, O_2^+ and O^+. The N_2^+ ions dissociate and recombine very rapidly at low pressures; therefore their concentration remains small (Table 1).

TABLE 1

CHARACTERISTICS OF SPACE ENVIRONMENTS[a] AT VARIOUS ALTITUDES

Altitude (km)	(miles)	Molecular weight	Particles (No./cm³)	Density (g/cm³)	Pressure (Torr)	Constituents and proportions
0	0	29	2.5×10^{19}	1.22×10^{-3}	760	21% O_2–79% N_2
500	300[b]	16.6	1.44×10^{8}	4.10×10^{-15}	2.17×10^{-8}	88% O–5% N–7% N_2
500	300[c]	15.9	1.20×10^{7}	3.10×10^{-16}	1.24×10^{-9}	91% O–8% N–1% N_2
1600	1000[b]	8.3	1.00×10^{4}	1.35×10^{-19}	5.10×10^{-12}	46% O–8% N–46% H
1600	1000[c]	1.0	3.7×10^{4}	6.3×10^{-20}	3.87×10^{-12}	100% H
Interplanetary space[d]		1.0	30	5.02×10^{-24}	5×10^{-16}	100% H

[a] F. S. Johnson, *Satellite Environment Handbook*, Stanford University Press, Stanford, Calif., 1961.
[b] At maximum sunspot activity.
[c] At minimum sunspot activity.
[d] Values from NRC literature.

The kinetic temperature is a function of altitude to about 400 km. Above this altitude, the environment is isothermal and has temperatures ranging from 525° to 1225°C, depending on the time of year and sunspot activity. The particle density varies significantly with altitude in accordance with hydrostatic laws until interplanetary space is reached where it will vary with the distance from the sun. Therefore, the average particle energy is greater than that in vacuum chambers at 25°C.

3. Space Simulation

Complete environmental simulation would include facilities for simulating various pressures, atmospheric composition, thermal and solar radiation, particle radiation, electromagnetic radiation, magnetic and gravitational fields, and space vehicle environments. It is emphasized that these characteristics are simulated and not duplicated. The effects that must be simulated depend on the object being tested and the purpose of the test because there is a considerable difference between an attempt to duplicate the spatial environment and the simulation of the particular effects that are essential to performance or reliability testing. Space-simulation testing has been applied to such diverse areas as the study of thermal balance and vehicle skin temperature, thermal life testing, arc and glow discharge, electromagnetic phenomena, evaporation or changes in surface characteristics, metallic adhesion or cold welding, friction and lubrication, metallurgical studies, creep and fatigue behavior, and reliability and interaction testing of vehicle components. Because many fine articles have been written[3-9] about space-simulation techniques and associated experimentation and because many of these techniques and simulation effects are beyond the scope of this text, we shall concern ourselves only with those simulation characteristics which are pertinent to metallurgical testing in small laboratory vacuum systems.

In vacuum chamber simulation, the primary attempt is usually directed toward achieving particle densities that are comparable with those of selected altitudes, using pressure as an indicator. In using this approach, we fail to attain identical particle energies. For example, at 1600 km (1000 miles), the kinetic pressure 5.16×10^{-9} dyn/cm^2 (3.87×10^{-12} Torr) and kinetic temperature of 1300 K correspond to an average kinetic particle energy of 2.69×10^{-13} erg or an energy density of 9.95×10^{-9} erg/cm^3. However, if we attain the same pressure (3.87×10^{-12} Torr) in a vacuum chamber at an ambient temperature of 300 K, the average particle energy will be 6.21×10^{-14} erg and the energy density 2.29×10^{-9} erg/cm^3. We fail to create identical energy densities in vacuum chambers.

The vacuum characteristics actually achieved in a simulator depend on the

TABLE 2

PRESSURE, GAS SPECIES AND FLUX DENSITY FOR A SPACE SIMULATOR[a]

Pressure (Torr)	Probable gas species	Flux density (molecules cm^{-2} s^{-1})
1×10^{-6}	H$_2$O	5.0×10^{14}
1×10^{-8}	H$_2$O, N$_2$, H$_2$, CO	5.0×10^{12}
1×10^{-10}	H$_2$, CO, CH$_4$	1.4×10^{11}
1×10^{-12}	H$_2$, CO, He	1.4×10^{9}

[a] D. H. HOLKEBOER, F. PAGANO, D. W. JONES AND D. J. SANTELER, *Vacuum Engineering*, Boston Technical Publishers, 1967, p. 39.

nature of the gas load that originates from the test item and walls of the chamber. Gas species present in vacuum chambers are different from those in space, as seen by comparing Table 2 with Table 1. Bonis[1] reports that hydrocarbons heavier than mass 40 are also present in certain vacuum chambers at working pressures of 10^{-6} to 10^{-10} Torr. Molecular composition is therefore not duplicated in working vacuum chambers.

Present technology will allow us routinely to achieve working pressures of 10^{-8} to 10^{-12} Torr in laboratory vacuum chambers. When these pressures are compared with pressures in space (10^{-12} to 10^{-16} Torr), it is evident that limitations are automatically imposed on many types of experiments when testing components for space application. Such areas of research are the achieving and maintaining of ultraclean surfaces for adhesion studies or the investigation of gas–metal reactions, corrosion, and boundary lubrication.

The factors governing the kinetics of reactions that occur between low-pressure gases and metal surfaces and result in contamination or purification can be deduced from a model based on the kinetic theory of gases[10]. The mass incident rate, Q_i, of a gas at a surface in a vacuum is given by $5.833 \times 10^{-2} \, P(M/T)^{\frac{1}{2}}$. Let it be assumed that a gas molecule in a vacuum system incident on the metal surface either is absorbed or rebounds. The rate of the surface reaction is then given by

$$Q_a = \alpha Q_i = \alpha(5.833 \times 10^{-2} \, P(M/T)^{\frac{1}{2}}), \tag{1}$$

where

Q_a = absorption rate, g/cm^2 s,
Q_i = incident rate, g/cm^2 s,
P = gas pressure, Torr,
M = molecular weight of the gas, and
T = absolute temperature, K.

The parameter α is dimensionless and is the probability that the impinging gas will become absorbed; hence, this factor is called the sticking probability. Equation (1) simply states that the surface contamination or purification rate of a metal by a gas is proportional to the sticking probability and the test pressure. Further, the extent of the contamination or purification is proportional to the reaction time. (The effect of temperature will not be discussed.) Finally, space acts as an infinite sink for gas molecules leaving the surface of any object, while in vacuum chamber simulation the molecules may return to the surfaces of the test piece after collision with the walls of the chamber.

These differences between space and space simulation may contribute to differences in surface behavior involving metal evaporation, self-welding or metallic adhesion, thin-film lubrication, and impurity adsorption. Therefore, simulation techniques may lead to questionable predictions.

4. Metallic Adhesion

4.1 Introduction

A potentially serious surface reaction that can occur in ultrahigh vacuum is the adhesion or "cold welding" of clean metallic surfaces. In many applications such surfaces are designed to move over one another at bearing pressures ranging from a few grams to several hundred kilograms. The nuclear reactor, isotopic fuel cells, electronic instrumentation, and propulsion engines operating in space at temperatures ranging from 0° to 1000°C all afford opportunity for contacting surfaces to "weld" together.

When two solid metallic surfaces are brought into intimate contact with each other, interactions occur along the interface that result in an attractive force between the bodies. The magnitude of this force will vary, depending on the respective surface conditions and the metals involved, and may range from insignificantly minute attractions to one greater than the magnitude of the respective metallic cohesive forces. Atomic interactions are responsible for the attractive forces and the formation of the quasiequilibrium interface. Therefore, the study of metallic adhesion phenomena, in general, reduces to the investigation of the effects of such variables as surface contamination, temperature, dissimilar metals, normal and tangential loading forces, and crystallographic orientation on the strength of the interface and the kinetics of its formation.

In most investigations, the analysis of metal adhesion data is always reduced to a discussion of the strength characteristics of the interface tested in shear or in tension. The role of contaminants in inhibiting the formation of a mechanically stable junction between metallic surfaces is not known in terms of monolayers and surface chemical interactions. An understanding of these phenomena is now beginning to emerge from ultrahigh vacuum adhesion studies[11-15] and the extent of the open literature directed toward examining metallic adhesion, in general, has reached significant proportions as indicated by recent reviews[16-22].

A critical examination of these presentations, however, immediately exposes an interesting situation. It appears as if each experimenter or technique, since each school seems to have a unique experimental approach, produces data and often complete interpretations which do not appear simply consistent with those conclusions of his colleagues. This situation is immediately evident when one reviews the data and conclusions of Sikorski[19] who studied the adhesion of metals using "in air" experiments, and those of Buckley[23] who generally uses ultrahigh vacuum techniques. The conclusions of each, for the most part, are similar. The experimental procedures, however, are so radically different that one hesitates to establish a line of consistency between them. The purpose of the following discussion is to examine the pertinent variables of the phenomena of metallic adhesion in a most general fashion, correlate these parameters with current investigative work, and to establish a set of boundary conditions on future analyses of similar data. The

References pp. 858–861

experimental results which follow provide one experimental attack which holds considerable promise in the identification of some of the variables which will be cited herein.

Two metallic surfaces brought into physical contact are usually said to experience "metallic adhesion" if an observable net tensile load is required to separate the joined system[22]. The magnitude of metallic adhesion is dependent on the physical and chemical properties of the metals[15,24–26], the nature and extent of loading[16] and the characteristics of the contaminant layers present on all but atomically clean metal surfaces[26]. Generally, the contacting process involves the elastic and plastic deformation of surface asperities, deformation of the bulk substrate and the rupturing and dispersal of contaminant surface films[27]. If the contaminant barrier can be sufficiently dispersed, the ensuing metal–metal contact along the interface results in a welded junction, the tensile strength of which may approach that of the bulk metal[22]. The conclusion that similar metal couples weld under near-zero normal loads, provided that both surfaces are atomically clean, has been well accepted in adhesion literature[22,28] and would be predicted from ultrahigh vacuum epitaxy studies using low-energy electron diffraction equipment[29], adsorption studies[30] and other investigations.

Adhesion studies which have involved deliberate gaseous contamination (*e.g.*, *cf.* Gilbreath[31]) from a fraction of a monolayer to ambient atmospheric conditions present an analytical problem which is most complex. Very simply, the mechanical compressive forces producing physical contact through asperity deformation with or without subsequent bulk substrate deformation can act to disperse the contaminant barrier into an ineffective state, which permits metal–metal contact regions to be established, which in turn resist tensile fracture on unloading. The disruptive mechanical forces acting within the interfacial zone—or more generally, the mechanical work imparted to the interface—is only one of several energy transfer mechanisms which can provide contaminant barrier dispersal. For example, increased thermal energy could cause evaporation or dissolution of the contaminant layer, or shock wave energy either from explosive impact or an ultrasonic source could also act as energy inputs which could promote contaminant dispersal along the interface. Since the contaminant layer is developed by the mechanisms of adsorption and surface creep or bulk diffusion to the surface, the energy inputs to the interface which causes disdersal may also enhance the rate of contaminant film development. In the consideration of metallic adhesion, therefore, the system must be limited to a degree of contamination which does not exceed that of a nominally clean surface exposed to ambient conditions. Specifically lubricated systems will not be considered as the analysis becomes more complex. Of the various modes of energy inputs to the interface only normal compressive loading at room temperature will be considered.

The description of metallic adhesion phenomena in real systems under bulk compressive loads corresponding to less than a 10% deformation of the massive coupled system requires a clear description of the microtopography of each of the two free surfaces before contact. The description is necessary to provide a definition of the *real* area of contact relative to the massive system geometry. This has been presented recently by Greenwood and Williamson[27] as a distribution function, the exact form of which depended on the prior history of the surface. The macro-radii of curvature of the surfaces must also be considered. As has been suggested by many authors (*cf.* a recent review by Bowden and Tabor[32]) a reasonable surface roughness model consists of a large diameter sphere contacting a flat or second sphere upon which are superimposed asperities the size and shape of which are dependent on the surface finishing techniques utilized before contact. For example, metallographic polishing techniques on the harder metals may result in a hill-and-valley contour in which the hill–valley depth is less than a micron and the peak-to-peak distance is in the range of 10 microns. The consequences which result when two such nominally flat surfaces are brought into physical contact under normal load have been reviewed by Greenwood and Williamson[27], Greenwood[33] and Kragelsky *et al.*[34]. The generally accepted model for surfaces in contact under a specific load is that the highest of the asperities, which can be represented by a Gaussian distribution of heights, will yield until a sufficient number of asperities have been deformed to accept the impressed load. On account of the very small size of the asperities, such deformation on a microscale will occur well before the onset of what is classically considered bulk plastic deformation. Since the uniqueness of the surface asperity configuration is retained until rather high compressive forces are realized[35] (*e.g.*, some[36] have suggested the range of at least 10% bulk deformation for flat surfaces) the real area of the interfacial system will consist of islands of various sizes surrounded by regions of non-contact. The real area of physical contact and the nominal area of contact are, therefore, quite different for all but the most severely loaded systems. Since surface mass transport in the form of plastic deformations must be involved during loading, the variation of real area with load time (*e.g.*, creep) will also be involved in the expansion of the real contact area[37]. The real area of contact will then be a function of the nature of the metal, impressed load, time and temperature; this has been substantiated by hardness measurements[38] and electrical contact studies[39]. Without question, the most important aspect of the study of metallic adhesion is the definition of the real area of contact with respect to its magnitude and constitution since the fracture strength of this adhesion junction, the only measure of metallic adhesion stability, is dependent on the *real* stresses developed within this *real* area during the unloading process.

4.2 Adhesion strength data

As is indicated in a review of the recent literature[16–22] most adhesion strength data which have been presented in the literature have involved only reference to the fracture load per unit of nominal area of contact. Let us, therefore, consider this aspect in more detail.

The process of adhesion may be considered as being comprised of two steps: two free surfaces are brought into physical contact and subjected to a compressive load; and then the applied load to the system is removed, possibly to some tensile load representing a nominal adhesion junction strength. The entire process is directly dependent on the nature and extent of the real area of contact and the fracture stresses developed therein.

Numerous suggestions based on macro-observations have been presented which relate the real area of contact (A) to the impressed load (W)[33,36,37]. In most general form this can be given as

$$A = (k)_x W^x, \qquad (2)$$

where k and x are related to the particular deformation process involved in expanding the load-supporting area as the load is increased or the time is extended at a fixed load (creep). Thus, the value of k is directly related to x through the process. Under lightly loaded conditions (e.g., less than the bulk compressive yield point of the material involved in contact) such an area expansion process will involve a number of individual asperities which will have a distribution in size and position along the contacting interface as well as a relationship to massive geometrical effects such as the overall relative radii of curvature of the two macroscopic systems. Consequently, as the load is impressed, the loading conditions on each individual asperity and relatively between adjacent asperities will be unique—that is, at equilibrium some asperity contact points may have been subjected to heavy plastic deformation while others may have only experienced a low-level elastic contact. A more complete general expression for the real area, therefore, ought to be a summation of the contributions from each asperity in the contact system with regard to each asperity (i) in the interface system and the respective position (j) of that asperity.

$$A_{ij} = \sum_{ij}^{n} (k_{ij})_x W^x{}_{ij}. \qquad (3)$$

Explicit in this equation are two necessary assumptions which appear reasonable but which have not been justified experimentally. First, it is assumed that each asperity deformation is a unit process (i.e., not related to the adjacent asperity) and as such follows a simple power law of deformation similar to that observed in macrosystems. Equation (2), therefore, is representative of one unit process and not generated through an averaging process of significantly different microprocesses. The second assumption which is necessary and yet unproven is that the representative equation is constant throughout an asperity deformation process irrespective of the percent deformation which is experienced by that unit process. Since geometrically reproducible surfaces cannot be generated in dimensions below microinches on real surfaces, it is unlikely that proofs will be presented in the immediate future. Consequently, we must rely on macroscale observations to provide a possible path for interpretation. As an example of the problem facing the analyst, let us consider a simple hardness experiment in which the indenter is assumed to represent an asperity unit process. If the load (W) is sufficient to cause general plastic transport under the indenter, the projected area (A_p) has been shown[40] to be approximately

$$A_p = \frac{W^m}{3Y}, \qquad (4)$$

where (m) is a material constant very nearly equal to one and (Y) the yield point of the material. The reasonably valid assumption necessary for this macro-approximation, but not necessarily valid for a similar microprocess, is that surface contaminants will not affect the plastic flow process. Such is not the case on two accounts: first, the real area supporting the load represents only a fraction of the apparent projected area because of the effect of asperities as pointed out by Williamson[35]. Second, the very flow processes occurring along the interface of the indenter which are necessary to expand the area are most sensitive to the lubricative properties of the contaminants which aid or restrict the material flow along the interface. The phenomenon has been clearly demonstrated during the observation of the sensitivity of hardness measurements to surface lubricants[41]. Gane et al.[42] also have shown that our knowledge of the mechanical properties of metal surfaces on a microscale is not satisfactory.

Although the presentation of eqn. (4) rests on some rather nebulous assumptions regarding the behavior of the individual asperity, it does bring forth the recognition that physical contact behavior is the result of a multitude of such interactions with plastic deformations ranging from near 100% to those near forceless contact. More specifically, the interface system has been placed in a rather complex state of stress which may per unit volume be resolved into two components: the applied stress (σ^a) and the residual stress (σ^r). As the flow stress in a unit volume of the material is exceeded that unit volume will deform plastically. In an implicit manner the model suggests that an absolute correlation of adhesion data with atomic properties, structure of the material, or defect mechanics requires a rather adventurous extrapolation, if any but the most gross generalizations are involved.

The instant that any fraction of the compressive load is removed from the system, each unit of area supporting that load will be subjected to a new stress relative to the fraction of the applied load removed from that unit area and also the availability of residual stresses adjacent to the unit area under consideration. If such a unit area is exposed to a tensile stress which exceeds some critical fracture stress (σ^c) the unit area will separate (i.e., permit crack propagation) which in turn will relieve a portion of the accumulated stresses. The condition for fracture per unit area can be presented as

$$\sigma^c < \sigma^a + \sigma^r. \qquad (5)$$

References pp. 858–861

Under relatively light contact loads (*i.e.*, very small bulk deformations), much of the real contact area will be subjected to rather severe stress concentrations of nearly infinite sharpness on account of the presence of voids along the interface. A careful stress analysis of the system must contend with this factor[43]. The magnitude of the critical fracture stress is related directly to the physical properties of the material through which the crack must propagate, and as a consequence is extremely sensitive to the structure and temperature of this phase, as was emphasized by Gilman[44]. For example, the critical fracture strength of a pure metallic junction can be compared with that of a clean grain boundary within the bulk metal while critical fracture stress of a junction completely contaminated with an organic oil ought to be compared with that of the organic material and not with that of a metal. A more extreme situation can be envisioned in the case of very lightly loaded regions along the interface between two glass plates in which the adsorbed water is not entirely dissipated in the compression process. In conclusion, the unit area resisting fracture can vary from some value approaching the bulk strength of the metal involved in the metallic couple to near zero, depending on the interfacial material and the degree of its dispersion. Furthermore, fracture of a small unit area can occur even though the overall system is still in a state of compression as long as the corresponding applied load is less than the maximum load experienced by the system during the compression mode of the adhesion process. For example, the addition or deletion of applied load only effects (σ^a) in a unit asperity process, or microunit volume adjacent to the interface. Under certain circumstances, the residual stresses, which to a degree are independent of the applied stresses, could effect a high tensile stress in a microunit volume even though other regions of the contact area are bearing the compressive load. Such was clearly identified by Bowden and Tabor[16] in their discussions of "released elastic stresses" during hardness measurements.

Again, the most important parameter of the process is the definition of the unit area over which the critical fracture stress must operate and again some rather extreme simplifying assumptions in the model must be made since we must consider the real contact area of a one asperity contact to be homogeneous in σ^c even though it is clear that this need not necessarily be the case for any except the ideally clean metallic adhesion system. If we make the further simplifying assumption that the stress state is unique and homogeneous within each asperity contact region, then we can represent the second half of the adhesion cycle as an equation based on $F = \sigma A$, where the force (F) on the interface of a one asperity contact is given by the nominal stress (σ) per unit real area (A) such that fracture ensues when $\sigma > \sigma^c$. In order that all of the asperities involved in one adhesion interface are considered, a summation can again be applied for the total force (F_T):

$$F_T = \sum_{ij}^{n} \sigma_i A_{ij}, \tag{6}$$

where σ_i is the effective stress developed on the i-th asperity junction with a real area A_{ij} and the total force represents the effects of n junctions. The fracture of the i-th junction will occur when some critical stress (σ^c) is exceeded in that microvolume of the junction which will permit a crack to move, thus releasing the accumulated applied (σ^a) and the residual stresses (σ^r) as indicated in eqn. (5). The necessity for studying the fracture process on a single asperity basis becomes evident, if one considers that the real area of contact is made up of contact points in varying degrees of deformation and further that the contaminant dispersal effect resulting in a metallic adhesion bond strength between these two points can be presented as a function of the contaminant as well as the percent deformation to which the metallic system has been subjected.

4.3 Bulk dispersal mechanisms

Since there is no direct evidence on precisely how an asperity undergoes gross deformation in a surface system while subjected to compressive loading and it is through just such a process that the contaminant barriers to adhesion are removed, the examination of bulk dispersal mechanisms ought to provide some insight. The roll-bonding studies by Milner et al.[18] serve as a simple example. Such adhesion studies are significant only if we presume that similar processes could be operative at the scale of asperities. The Milner experiments involved the rolling of two slabs of metal in air to some degree of bulk deformation and then testing the interface bond in shear. In this case our assumptions are probably more nearly correct since for the most part the real area of contact is expanded under conditions of constant availability of contaminants and chemical reaction rates tending to disperse the oxide contaminant layer. A portion of the voluminous data developed by Milner from roll-bonding studies of various metal couples is presented in Fig. 1. The numerous data points delineating these curves in the original data were left

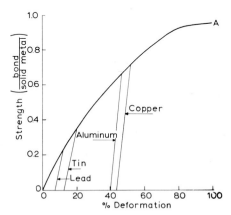

Fig. 1. Roll-bonding studies of various metals in air at room temperature[13]. The data are presented as the strength ratio of the interface bond formed by roll bonding to the solid vs. percent deformation.

References pp. 858–861

out here for convenience. The curves illustrate several significant points regarding one possible mechanism for the dispersion of oxide films between the two metal surfaces. First, let us consider the case of aluminum in which the variables of temperature, rolling speed, and surface structure are held constant. The curve indicates that a threshold of about 40% bulk deformation is required before any bond strength is observed. Between 40 and 45% deformation, the dispersion rate of the oxide, as well as that of the adsorbed gas, is quite rapid, as is indicated by the increase in the shear strength of the system (*i.e.*, a sizeable fraction of the real strength of aluminum). At deformations greater than 45%, the oxide dispersal process seems to follow a limiting curve which is representative for the other metals shown. The interfacial strength compares favorably with the bulk metal strength above 80% deformation. It is interesting to compare the aluminum curve with that of lead since the deformation threshold for lead is only 8% deformation, yet lead encounters the same limiting curve (*cf.* Fig. 1, O–A) as that experienced by Sn, Al and Cu. What is suggested by this set of curves is that after the brittle oxide layer is fractured[45] (*i.e.*, deformation threshold, which is dependent on the substrate material-oxide characteristics) a limiting rate process of contaminant dispersal is attained which is dependent on the degree and type of deformation and independent of the material which is involved. Since these systems were prepared in a similar manner (wire brushing and severe rolling), one might suspect that the limiting oxide dispersal is a function of asperity interaction (light loads—less than 20% deformation) and metal flow patterns along the interface under the severe rolling conditions. In comparing these data with the normal loading interface contact model under discussion, it is unrealistic to carry this analogy too far since in a simple contact process extrusion-type flow (*i.e.*, parallel to interface) would not be expected to such a severe extent. Furthermore, during roll bonding the interfacial area is grossly expanded, whereas in normal adhesion relative motion in the interface is quite small. Milner *et al.* have clearly examined other models of energy input (*e.g.*, the deformation threshold decreases with increasing temperature and extension of the duration of exposure to roll pressure). They have also examined the effects of limiting contaminants (*e.g.*, the deformation threshold of aluminum was reduced to about 2% by brushing in medium range vacuum). The softer metals indicated a lower deformation threshold except for magnesium which did not respond to the simple analysis as presented for Fig. 1 since the limiting curve was very low.

Although roll-bonding studies do not simply represent the state of affairs in a normal contact problem, they do clearly demonstrate the *contaminant dispersal effect* that has been interjected into the contact fracture argument. Vacuum adhesion studies presented by Hordon[46] in Fig. 2 were obtained by wire brushing two small flat plates of the respective metals in very high vacuum (better than 10^{-10} Torr), subjecting the plates to near-normal loading, and then testing the welded system in tension. The data are shown as the relative strength of the inter-

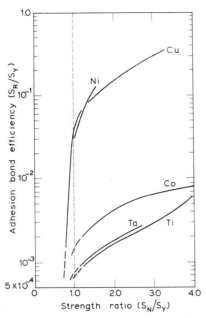

Fig. 2. Variation of the relative strength of polycrystalline metals with the degree of compression[42]. The ratio of the bond fracture stress (S_R) to the yield stress (S_Y) is plotted vs. the reduced compressive stress (S_N/S_Y)

face bond (S_R) to the yield strength of the material (S_Y) based on the nominal area of contact which is compared with the normal loading force (S_N) ratio with S_Y. The general character of the curves is precisely what would be predicted. For example, the natural surface roughness of the samples ensures asperity interaction which will provide an exceedingly small real area of contact until at least a few percent plastic deformation is attained (*i.e.*, a nominal load in excess of the yield point—1.0 on the abscissa of Fig. 2). It is evident that normal loading does not provide the rapid oxide dispersal which accompanied roll-bonding experiments as indicated by the lack of a simple symmetrical limiting curve. More severe interfacial dispersing is, however, observed in the softer metals Ni and Cu when compared with the harder metals Co, Ta, and Ti. Hordon also observed that by increasing the ambient temperature the bond strength at a fixed load was also increased. The amount of contamination present on the wire-brushed metal surfaces in ultrahigh vacuum (Hordon) was impossible to ascertain; however, the degree was certainly considerably less than that present in the roll-bonding experiments. Another important unknown in the analysis of the adhesion system is emphasized since no technique has been successfully applied to ascertain the exact amount, or character, of the contaminant phase available for the interruption of adhesion during the study of interface bond strength relative to fraction of deformation.

References pp. 858–861

4.4 Interfacial forces and material transfer during tensile tests

Since the real area of contact formed in the contact zone during the compressive mode (eqn. (3)) of the adhesion test is identical with that operated upon during the removal of the compressive force (eqn. (6)) and possibly the same as that during the tensile test of the adhesion junction, the total interfacial force may be written as

$$F_T = \sum_{ij}^{n} \sigma_i(k_{ij})_x W^x{ij}, \qquad (7)$$

where k must be evaluated under the conditions of x which is due to the deformation process involved at the i-th asperity. For example, if the i-th asperity is undergoing simple plastic deformation in compression, we might assume ($k = 1/3Y$) and $x = 1$; however, if the i-th asperity is under tension $x = 0$ and $k = W_{max}/3Y$ might be assumed as a first approximation until the critical stress (σ^c) in the i-th asperity is exceeded. Under any circumstances when the compressive load is reduced infinitesimally portions of the system may be exposed to a tensile stress even though the entire system is still considered as being under a compressive load. If the tensile stress experienced by the asperity contact area exceeds the critical fracture stress (σ^c) of the interface, a crack will propagate through that region to relieve the internal stress but will stop when the balance ($\sigma^c = \sigma^a + \sigma^r$) is achieved.

The path of the crack will, of course, follow the path of least resistance which will couple a minimization of molecular bond strengths with a maximization of regional tensile stress. The chemical composition, therefore, of the free surfaces resulting from fracture cannot simply represent the precontact surfaces since material transfer is expected in all cases. For example, in the case of severely oxidized metal surfaces, adhesion should be expected between some of the oxide particles in contact; however, on fracture the path of least tensile force resistance may not include such an adhesion junction. Material transfer would result. A situation quite similar to this state of affairs was clearly described by Bowden and Tabor[32] in their discussion of "released elastic stresses" during normal hardness measurements. Johnson and Keller[15,24] also reported a similar phenomenon in adhesion studies between similar and dissimilar couples under contaminated conditions.

If a very weak boundary exists all over the interface (*e.g.*, σ^c is very small) a plot of the variation of contact area with applied load from maximum load to zero load should very nearly superimpose on the loading curve provided that no massive plastic flow of either system has been effected. Plastic flow would provide a larger real area of contact on unloading than was available on loading, depending on the magnitude of residual elastic stresses in that region.

Next, let us consider the real area of contact developed between two atomically clean surfaces such that each contact point becomes a welded junction, a case which closely resembles a clean grain boundary[28]. Since compressive loading prior to a tensile test tends only to slightly distort the tensile stress–strain diagram of the metal under consideration, one would predict[28] that the strength of each asperity adhesion junction would be approximately that of the tensile strength of that metal based on that *real* area of contact. The relationship of the junction strength to impressed load is only through the asperity deformation necessary to expand the contact area. If, for example, atomically clean and flat surfaces were brought into intimate contact without an impressed load, the junction strength would still be the tensile strength of the metal still based on the real contact area, which in this ideal case would be the nominal area.

Another important aspect for the consideration of eqn. (7) lies in the fact that the distribution and the degree of contaminant dispersal is a function of the degree of deformation. As a consequence, the critical fracture stress (σ^c) will vary with the contact point area, depending on the amount and type of contaminant present at that point and the degree of dispersal experienced by that point during the compressive mode. Studies directed toward the evaluation of specific contaminants and their ability to interrupt the adhesion process ought therefore to be conducted in a system in which rigorous control is maintained over all secondary impurities, surface roughness, and loading variables such as contact time, temperature and rate. One suggested configuration[25] was to evaluate eqn. (7) under atomically clean conditions at various maximum loads and then compare these values with those observed under one specifically contaminated condition, maintaining all of the other variables constant in the test system, *e.g.*,

$$\eta = \frac{[\sum \sigma_i A_{ij}] \text{ contaminated}}{[\sum \sigma_i A_{ij}] \text{ atomically clean}}. \qquad (8)$$

Such an approach and the assumption that the only change in the system is (σ^c) permit a rather simple analysis. Extensive details of the value of this assumption and a detailed analysis have been presented by Westwood[47].

4.5 Coefficient of adhesion

The coefficient of adhesion (α) was developed[11] as the ratio of the fracture load of a nominal adhesion junction to that compressive load utilized in the formation of the interface. The implicit assumption is that, on an average, the area supporting the load is identical with that which resists a tensile force to fracture the system; however, according to a more careful examination this is only the case when absolutely no contamination exists between two metal surfaces. The presence of only a fraction of a monolayer of contamination on either surface immediately invokes the necessity to sum the varying degrees of asperity deformation necessary to generate the real area or the application of a compressive force which will generate complete dispersal at all points. In equation form we can use the maximum force in compression,

$$F_{max} = \sum_{ij}^{n} \sigma_i [(k_{ij})_x W^x ij]_{max}, \qquad (9)$$

as the load to form the junction and eqn. (7) that to cause fracture. The coefficient of adhesion (α), thus, takes form

$$\alpha = \frac{F_{total}}{F_{max}} = \frac{\sum_{ij}^{n} \sigma_i(k_{ij})_x W^x{}_{ij}}{\sum_{ij}^{n} \sigma_i[(k_{ij})_x W^x{}_{ij}]_{max}}. \tag{10}$$

The coefficient of adhesion (α), therefore, may vary from zero to infinity depending on the conditions of the experiment. For example, if atomically clean, flat surfaces are brought into forceless contact the denominator approaches zero, and if any force of attraction exists between the two bodies, the numerator has a finite number and α approaches infinity. On the other hand, a perfect lubricant reduces the numerator to zero at any load or (α) approaches zero. Wide variations in α for the same metal system tested by different experimenters under approximately the same conditions are common (*e.g.*, in the case of copper *cf.* Buckley[48] for large α values and Ham[49] for small α values). The definition of an α for each i-th contact or an average $\bar{\alpha}$ also appears to be a fruitless path because of the difficulty in ascertaining either the precise degree of contamination of the i-th contact or the total amount of contaminant dispersion energy available to the system necessary for the dissipation of the contaminant layer which is preventing the two asperities from welding together, and establishing the α for that asperity.

In the utilization of various α values, as reported in the literature, it would appear that for the purposes of comparing data produced between ultra-clean *versus* specifically contaminated surfaces which are produced by one investigator utilizing the same technique for each experiment such as is done in the experiments by Gilbreath[31], one could assume a degree of qualitative relationship between the different values of α on similar metal couples at corresponding loads without much error. Interrelating data from different physical systems of study as suggested by Rittenhouse[50] or those between dissimilar metal couples as used occasionally by Buckley[48], however, should be considered dangerous since F_{max} and σ^c are strongly dependent on the test temperature relative to the absolute melting temperature, crystal structure, cohesive strength, etc., of the bulk materials as well as the nature of the contaminants and surface roughness in the system. The equation which is used when a comparison of this type is made follows the form:

$$(\alpha) \text{ contaminated} = \frac{\left[\frac{\sum \sigma_i A_{ij}}{\sum \sigma_i (A_{ij})_{max}}\right]_{contaminated}}{\left[\frac{\sum \sigma_i A_{ij}}{\sum \sigma_i (A_{ij})_{max}}\right]_{clean}}, \tag{11}$$

where one must assume that the $[\Sigma \sigma_i(A_{ij})_{max}]$ values are equivalent and may be cancelled. In relating two different metals, however, the maximum values cannot be related since the deformation mechanisms providing the real area of contact are different because of the different modes of deformation as are the rates of contaminant dispersal, etc.

Clearly the mechanism of the dispersal rate of the contaminant barrier to metallic adhesion, and its balance with film growth rate, is the key to the overall analytical problem; and until quantitative studies which are initiated with a known degree and type of surface contaminant are undertaken, correlation of data produced by one investigator is not likely to agree in detail with that of another. Under compressive loads below the yield point of the material, several authors[17,20,21] have shown that a monolayer of certain contaminants[15,24] emanating either from the vapor or by diffusion from the bulk[26] can reduce the adhesion strength to zero. The detail with which the original metal surface (*i.e.*, prior to specific contamination) must be defined is established through this limitation.

The lack of clear definition of the amount and type of contaminant layer present on a metallic system prior to study has also inhibited our ability to gain any insight into possible cross-correlations between different modes of energy inputs for contaminant dispersal. For example, the correlation of adhesion data produced by the normal load contact methods[26] cannot be precisely correlated with a normal contact plus some fraction of tangential motion or the comparison of normal contact at some temperature (T) with that at ($T+500°C$). If both of these examples could be clearly resolved, a significant step toward the understanding of the adhesion theory of friction could be achieved.

List of Symbols

A = Real area of contact
A_p = Projected area of contact
F = Force on asperity
F_T = Total force on contact area (A)
k = Constant dependent on deformation process
m = Materials constant
n = Number of asperity contacts
S_N = Nominal compressive stress
S_R = Nominal fracture stress
S_Y = Yield stress
W = Impressed normal load
W_{max} = Load at maximum compressive stress in adhesion cycle
x = Constant dependent on deformation process
Y = Yield point
α = Adhesion coefficient
σ = Total stress

References pp. 858–861

σ^a = Applied stress
σ^c = Critical fracture stress
σ^r = Residual stress

5. Mechanical Properties in Vacuum

5.1 Interaction of materials with residual contaminants

The interaction of materials in vacuum, or more precisely their interaction with residual contaminants at very low partial pressures, has enjoyed intense interest in the past decade. This is partly because the quest for higher operating temperatures has culminated in the attempt to utilize the more reactive higher melting point metals (*e.g.*, Nb, Ta, Mo, and W). Interest is further intensified by the fact that the vacuum of space is being contemplated as the operating environment for some devices. Hence, while a few vacuum effects at lower temperatures have been known for some time, most of our discussion will involve the refractory metals at temperatures of the order of 1000°C or higher.

For the sake of discussion it seems helpful to divide the field into two parts: bulk effects which involve the interior of the material and surface effects which at most involve a superficial reacted layer or conceivably only an adsorbed layer of gases. This chapter attempts to systematize what is known about these interactions and the conditions under which one might expect to encounter them, then discuss examples of how such interactions influence various mechanical properties.

The composition of the residual impurities in a vacuum system is quite inindividualized. It is a function of the type of system, its geometry, its materials of construction and their histories, and its state of repair. An unbaked system might exhibit a high partial pressure of water vapor, a poorly trapped system using a diffusion pump might exhibit a high hydrocarbon content, or a leak might result in a high nitrogen content. Hence, every situation must be analysed individually.

To begin with, equilibrium must obviously call for a reaction. It becomes clear that this condition is usually met when one realizes that, for the materials and conditions to be discussed, an effect might result from a reaction proceeding in either direction. In the example of a reactive metal forming a compound with a gas, a specific pressure is expected to be in equilibrium with the compound at any given temperature. If a higher pressure is maintained (and if time permits) the metal will be converted entirely to compound. Conversely, a lower pressure will ultimately deplete the compound in the system. Further, at concentrations below the solubility limit one would expect the equilibrium pressure to vary with the square of the concentration in accordance with Sieverts' law. Unless equilibrium is precisely met for each gaseous species in the system, the metal would tend to gain or lose solute in an attempt to establish equilibrium. Beyond the simple cases involving only the equilibrium between the metal or alloy and the partial pressure of gaseous elements one must also consider the possibility of competing reactions. The preferential reaction of carbon with oxygen has been shown[51] to result in the decarburization of D-43 (Nb–10% W–1% Zr–0.1% C). A knowledge of such multicomponent equilibria can also be used to advantage. Perkins[52] was able to prevent decarburization of TZM (Mo–0.5% Ti–0.08% Zr) by the intentional introduction of CH_4 during vacuum annealing while Inouye[53] was successful in minimizing oxygen contamination of TZM using controlled partial pressures of CH_4. Under some conditions, sublimation of metal–gas compounds proceeds at a significant rate relative to their formation (*e.g.*, the oxides of Mo and W).

In addition to establishing the thermodynamic requirement for a reaction one must consider whether, under the conditions in question, kinetics permit the reaction to proceed at a significant rate relative to the time scale of the experiment or application. The mass of gas molecules per unit time (Q_i) impinging on the surface of a metal at a given pressure and temperature is given by

$$Q_i = 5.833 \times 10^{-2} \, P \, (M/T)^{1/2} \qquad (12)$$

where P = pressure in Torr, M = molecular weight in grams, T = solute temperature in K.

Comparison with experimentally determined reaction rates indicates that while the reaction rate is proportional to this quantity, only a small fraction of the molecules arriving at the metal surface actually react; the majority ultimately are desorbed and reflected immediately from the surface. A factor (α) is therefore introduced representing the probability that an impinging atom will contribute to the accumulation of atoms at the surface. Hence, the actual reaction rate (Q_a) is given by eqn. (13):

$$Q_a = \alpha Q_i. \qquad (13)$$

Values of α are typically 0.1 or less. Typical curves of α *versus* temperature are shown in Fig. 3. The points at which the factor (α) becomes zero correspond to

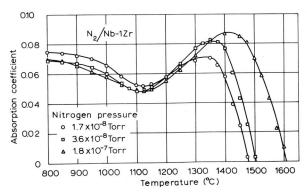

Fig. 3. Absorption coefficient α *vs.* temperature for the reaction of nitrogen with Nb–1% Zr at selected pressures, after Inouye[54]. (Courtesy of Plenum Publishing Corporation, New York)

References pp. 858–861

the equilibrium temperature for the particular pressures. While the actual processes involved in governing the value of α at low temperature are not known, it seems clear that near equilibrium the process is dominated by the reverse reaction as a dynamic equilibrium is approached. In fact, it has been shown that at temperatures near equilibrium α can be computed from the relationship

$$\alpha = \alpha_0(1 - C_m/C_e), \tag{14}$$

where

C_m = mean gas concentration in the metal,
C_e = equilibrium concentration in the metal,
α_0 = sticking probability at zero concentration.

Incidentally, at concentrations greater than equilibrium, negative values of α are useful in treating the degassing reaction. The foregoing considerations are reviewed in detail by Inouye[54].

In computing the rate of incidence of material on the surface, the effect of chamber geometry must be considered. Many experiments are performed with the specimen in a hot zone which is tightly enclosed in a radiation shield. The conductance between the hot zone and the vacuum chamber may be quite small, making it very difficult to determine the pressure actually seen by the specimen.

A further condition for the observation of a bulk effect is that diffusion rates must permit the reacting species to permeate a significant fraction of the cross section. For a gas entering into solid solution in a metal, a useful treatment is given by Darken[55]. Figure 4 plots $(C-C_0)/(C_s-C_0)$ versus $x/(Dt)^{\frac{1}{2}}$, where C is the instantaneous concentration at any point x distance from the surface at time t, C_0 is the initial concentration in the metal, C_s is the surface concentration which is assumed to be equal to the final or equilibrium concentration, and D is the diffusivity at the temperature in question. This curve has the form of an error

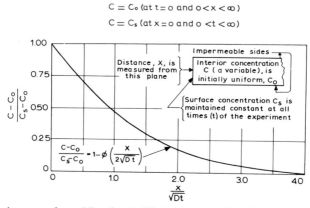

Fig. 4. Penetration curve for unidirectional diffusion in a semi-infinite medium, after Darken[55]. (Courtesy of American Society for Metals)

function. For example, half of the compositional change called for by equilibrium will have taken place at a point x distance below the surface when value of the quantity $x/(Dt)^{\frac{1}{2}}$ reaches slightly less than one (0.9538). If the diffusivity is not a function of composition, this curve may also be used for diffusion outward during degassing. (Curves are also given for estimating the time required for a material of known diffusivity to come to equilibrium with a diffusing gas.) The foregoing treatment assumes that the surface of a metal specimen, initially of uniform composition, is brought to a new concentration, C_s, instantaneously and maintained at this concentration throughout the time of the experiment.

It is immediately clear that the rate of contamination can be limited either by the number of atoms impinging and sticking, eqn. (13), or by diffusion. It can be shown that the flux, J, required to satisfy the assumptions of the diffusion treatment will be given by

$$J = -D \frac{C_s - C_0}{(\pi D t)^{\frac{1}{2}}}, \qquad (15)$$

where
D = diffusivity,
C_0 = initial concentration,
t = time.

At short times, the flux required to satisfy the error function form of the penetration curve is very high; hence the assumption of the foregoing treatment will not be satisfied. Under these conditions the reaction rate, given by eqn. (13), will be limiting. If the flux computed from eqn. (15) falls below that available by eqn. (13), within a time which is short relative to the total time of the experiment, one can assume that the surface is very close to the equilibrium concentration and can apply Fig. 4 in estimating the contamination.

The internal oxidation of a reactive solute in a less reactive solvent is treated by Rapp[56]. Where no scale interferes with the absorption at the metal surface and where the diffusion rate of the reactive solute metal is negligible compared with that of the desolving gas he shows that the distance (ξ) the precipitation front has proceeded inward from the surface in a given time is given by

$$\xi \simeq \left[\frac{2 N_o^{(s)} D_0 t}{N_b^{(o)} v} \right]^{\frac{1}{2}}, \qquad (16)$$

where
$N_o^{(s)}$ = mole fraction oxygen at the surface,
$N_b^{(o)}$ = mole fraction of reactive solute in the bulk alloy,
D_0 = diffusivity of the gas in the base metal at the temperature in question,
t = time,
v = number of gas ions per solute atom in the compound.

Many refractory alloys contain reactive elements such as Zr or Hf; hence, this

References pp. 858–861

equation would be expected to be useful in predicting the fraction of any given cross section affected by contamination for relatively simple thermal histories.

The implications of the appearance of time in the expression for ξ seem noteworthy. Since the rate decreases with time, one would expect particle size to vary with distance from the surface. Further, the assumption that the diffusion rate of the solute metal is negligible may be valid during the earlier stages of contamination and invalid for longer times; hence, the composition could vary through the cross section.

Finally, contaminants must be present in a form which affects the property in question. For example, in the case of high temperature creep, precipitates must be of such size and distribution as to affect dislocation motion, or else they will not significantly affect creep rates. Figure 5 shows the effect of pretest annealing

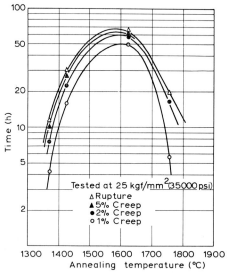

Fig. 5. Effect of pretest annealing temperature on the creep–rupture properties of FS-85 Alloy at 980°C.

temperature on the creep rupture properties of FS-85 (Nb–27% Ta–10% W–1% Zr) tested at 25 kgf/mm² (35,000 psi) and 980°C. Chemical analysis indicates that the compositions of these specimens did not change significantly during annealing and testing. Although the grain size changed with annealing temperature, it is quite likely that the solution and reprecipitation of interstitial compounds produced a microstructure more effective in inhibiting dislocation motion. Chang[57] concludes that Mo alloys are strengthened by dispersions of TiC and ZrC. From electron microscopy studies Perkins[52] suggests that mechanical properties of niobium-modified TZM (Mo–1.5% Nb–5% Ti–0.8% Zr–0.2% C) can be altered by thermomechanical history. A profound effect of fabrication variables on the long-time creep–rupture properties of Nb-TZM has been demonstrated[58]. In short, it

MECHANICAL PROPERTIES IN VACUUM

is generally accepted that thermomechanical history can affect the mechanical properties of refractory metal alloys as profoundly as many substantial changes in composition.

The requirements for the observation of a surface effect depend on the mechanism of the effect. Some theories suggest that adsorbed layers of gas are responsible for effects. A table of times for the formation of a monolayer at various pressures is given in Chapter II. If, on the other hand, a layer of actual reaction product is required, one must decide (1) does equilibrium call for a reaction and (2) do kinetics permit the development of such a layer. In this context, diffusion into the metal must be viewed as a competing reaction along with sublimation and reactions with other gaseous species.

5.2 Hardness

The work of Gebhardt[59] illustrates the effect of residual contaminants on the hardness of reactive metals. Gebhardt outgassed Ta wires at 2500°C and 5×10^{-6} Torr, then exposed them to pressures of 2×10^{-2} Torr at 700°, 900° and 1000°C. Their hardness is shown as a function of oxygen content in Fig. 6. The hardness increased with oxygen content and is apparently independent of engassing temperatures up to the solubility limit. At the solubility limit the curve shows a discontinuity and the hardness is relatively insensitive beyond this point. (Resistivity and lattice parameter measurements confirm the solubility limit at these temperatures.) One can usually assume that the presence of interstitials in refractory metal solid solution will cause an increase in hardness. Conversely, their loss from the material will yield a decrease in the hardness. The prediction of hardness changes simply

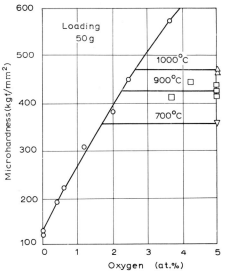

Fig. 6. Microhardness *vs.* oxygen content for tantalum, after Gebhardt and Seghezzi[59]. (Courtesy of Metallwerk Plansee)

References pp. 858–861

involves deciding, from free energy data, the direction of the reaction, and, from reaction and diffusion kinetics, the extent of the cross section affected. In fact, with the aid of published hardness *versus* interstitial concentration data, such predictions can be quantitative. Loria[60] suggests that the hardness of V can be predicted from its interstitial content by the following equation:

$$DPH = 47 + 900\,(\%N) + 780\,(\%C) + 600\,(\%O).$$

Situations in which precipitates are involved are frequently of greater practical interest but are unfortunately more complex. The hardness is profoundly affected by the particle size and distribution of the precipitate. These are, in turn, affected by any of several variables.

A solid solution, in equilibrium at one temperature, may become supersaturated upon cooling to another temperature. A hardness change may thus be produced via an aging reaction. Such a situation was studied by de Lamotte, Huang and Altstetter[61] in the Nb—N system. Niobium wires were outgassed at 2100°C and 10^{-7} Torr, then equilibrated with known amounts of nitrogen. The wires were quenched by admitting He to the system. Figure 7 shows the hardness changes observed after aging at various temperatures. It is clear that the hardness is a function of aging time and temperature.

Mukherjee and Martin[62] allowed Mo–1% Ti specimens to react with nitrogen at 1100°, 1300°, and 1500°C. Hardness traverses of these specimens are shown in Fig. 8. At 1100° and 1300°C, a hardness increase accompanied the penetration of the reaction front into the alloy. At 1500°C, however, an unexpected peak appears in the hardness. The authors suggest that as the progress of the reaction front slows down (as predicted by eqn. (16)) more time is allowed for the diffusion of Ti toward the reacting nitrogen. The peak may therefore coincide with the achieve-

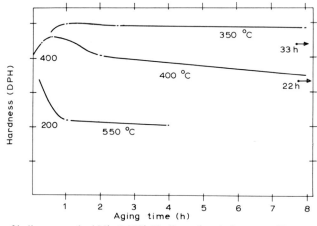

Fig. 7. Aging of helium quenched Nb–0.35% N alloy, after de Lamotte, Huang and Altstetter[61]. (Courtesy of the American Institute of Mining, Metallurgical and Petroleum Engineers)

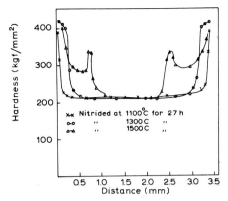

Fig. 8. Reichert microhardness traverse of nitrided Mo–1% Ti alloy, after Mukherjee and Martin[62]. (Courtesy of Elsevier Publishing Company, Amsterdam)

ment of an optimum composition. The results of a study of the Mo–Zr–N system[63] lend further credence to this hypothesis. The features of the hardness traverse can be displaced inward by nitriding in ammonia rather than nitrogen. The use of ammonia resulted in more rapid penetration because of more rapid dissociation at the specimen surfaces. Strictly speaking the last-named two investigations cannot be regarded as vacuum reactions since the nitrogen was added at atmospheric pressure, but it is reasonable to infer that similar effects are possible whenever a contaminant is introduced into a refractory metal above the solubility limit.

Another such complexity is illustrated by the data of Cortes and Field[64] who studied the Nb–C system by adding carbon to the melt. Figure 9 gives the hardness of as-cast niobium as a function of carbon content. The hardness rises to a maximum, after which it actually decreases with increasing carbon content. The authors suggest that the hardness peak may be related to a pronounced substructure forma-

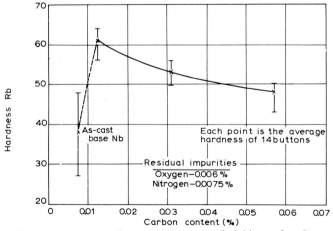

Fig. 9. Effect of carbon content on the as-cast hardness of niobium, after Cortes and Field[64]. (Courtesy of Elsevier Publishing Company, Amsterdam)

References pp. 858–861

tion in this range of carbon contents. The existence of this peak was qualitatively corroborated by Begley and Lewis[65].

5.3 Ductility and ductile to brittle transition temperature

Great concern arises over the high ductile to brittle transition temperatures frequently observed in refractory metals. This behavior is of practical significance whenever a device, once exposed to high temperatures, is exposed to ambient temperatures.

Ductile to brittle transition temperatures of 400°C and higher are frequently observed in some metals. Much of the problem is intimately related to interstitial impurities, exactly those contaminants which metals are expected to gain or lose at high temperatures in vacuum. The data of Lomis and Carlson [66], shown in Fig. 10, illustrate the pronounced effect of such interstitial elements on the transition temperature of vanadium.

In examining the possibility of a bulk effect in the metal resulting from residual contaminants in a vacuum environment, it is useful to compare the Group VA metals (V, Nb, and Ta) with the Group VIA metals (Cr, Mo and W). This field has been reviewed by Hahn, Gilbert and Jaffee[67] who advance several propositions for further discussion and investigation, not meaning to suggest that they have been proven for all cases. They first point out that available data indicate that the

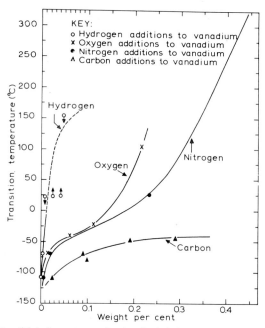

Fig. 10. Effect of interstitial elements on the ductile–brittle transition temperature of unalloyed vanadium tested in bending, after Lomis and Carlson[66]. (Courtesy of John Wiley and Sons, Inc., New York)

Group VA metals show much higher solubility for interstitials than do the Group VIA metals. Therefore, commercially pure Group VA metals are single phase while commercially pure Group VIA metals are generally supersaturated, two-phase alloys with respect to interstitial solutes. Hahn et al.[67] contend that it is reasonable to expect the Group VA metals to be capable of retaining three to four orders of magnitude greater concentrations of oxygen, nitrogen, hydrogen and carbon in solution than the Group VIA metals. These authors further suggest that low temperature embrittlement is primarily caused by the presence of interstitial precipitates and that all of these metals are inherently ductile when sufficiently pure. On the basis of these proposals, the tolerance of the VA metals would be expected to be high. Low transition temperatures are achieved for these metals with relative ease. Figure 11 indicates that with the attainment of somewhat greater purities, low transition temperatures are observed for Cr[68]. Lawley, Van den Sype, and Maddin[69] have shown that zone refining of Mo produces ductile behavior at very low temperatures. The zone refining of W resulted in only a slight improvement in the transition temperature[70], but owing to the extremely restricted solubility it is unlikely that a solid solution was ever achieved.

It is possible that the order of efficacy of the interstitial elements shown in Fig. 10 is general (*i.e.*, that hydrogen is always the most embrittling, followed by oxygen and nitrogen with carbon being the least embrittling).

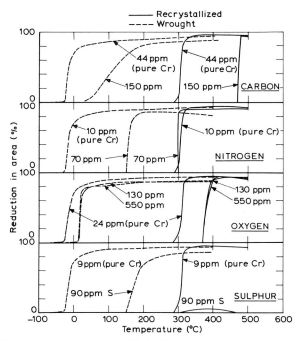

Fig. 11. Effect of C, N, O and S on the ductile–brittle transition temperature of chromium tested in tension, after Allen, Maykuth and Jaffee[68].

References pp. 858–861

Fractographic[71-74] evidence indicates that precipitates may result in an embrittling behavior by segregating at grain boundaries where they initiate fracture. This contention is further supported by the observation that fine grain materials and worked structures are more tolerant of interstitials than are coarse-grained recrystallized structures[68,74]. On the other hand, recent work by Tsuya and Aritomi[75] has injected further complexity into this subject. They show that carburizing under specific conditions can actually increase the ductility of coarse-grained Mo, even though numerous carbides are observable in the grain boundaries.

Substitutional alloying elements have varied and profound effects on the transition temperatures of refractory metals. The addition of small concentrations of reactive elements (*e.g.* Hf, Y, Zr, and Ti) which act as scavengers ties up the interstitials in more innocuous distributions and thereby improve low temperature ductility[76]. Although more complex, the improvement in the transition temperature of tungsten effected by large additions of Re[77] is noteworthy. The fact that this subject is more complex than indicated by these examples is illustrated in Figs. 12 and 13[76,78]. Some elements lower the transition temperature throughout the range studied; others raise it throughout the range studied; still others lower it to a minimum beyond which they raise it. Finally, the effects of surface condition

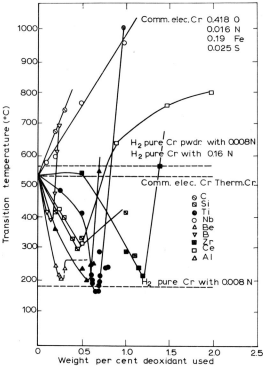

Fig. 12. Effect of substitutional alloying on the bend transition temperature of chromium, after Klopp, Holden and Jaffee[77].

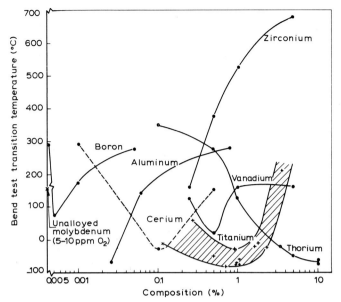

Fig. 13. Effect of small binary additions on the bend transition temperature of cast molybdenum, after Olds and Rengstorff[76]. (Courtesy of the American Institute of Mining, Metallurgical and Petroleum Engineers)

have been shown to change bend transition properties[79].

From the foregoing considerations it seems clear that in applying the criteria of equilibria, reaction kinetics, and diffusion kinetics, one must also bear in mind the tolerance of the solvent metal or alloy, the relative efficacy of the contaminant, the structure of the material (grain size, cold worked, etc.), state of stress, and the surface condition. Great care must also be taken in the interpretation of any ductility data. A small amount of hydrogen, for example, may enhance the effect of a much larger amount of oxygen, or the effects of grain size, surface condition, or substitutional impurities may affect the results.

With regard to surface effects, Stephens[80] has shown that an oxide film, formed under specific conditions, can have an effect on ductility identical with that of surface removal by electropolishing. Hence, the environment may determine what type of oxide, if any, is formed and thus alter the ductility without the occurrence of bulk diffusion.

5.4 Tensile properties

The presence of interstitial impurities in solid solution in a refractory metal is usually assumed to raise its tensile strength. The data of Tottle[81] for niobium (Fig. 14) are typical. Vaughan and Rose[82] studied this same metal system at several temperatures and their data, illustrated in Fig. 15, show that this same strengthening trend is dependent on temperature. The maximum effect is observed at approximately 400°C with a lesser effect resulting at higher temperatures. Schmidt et al.[83],

References pp. 858–861

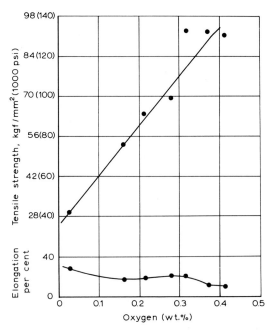

Fig. 14. Effect of oxygen content on the room-temperature strength and ductility of commercial-purity niobium, after Tottle[81].

by treating both tensile and creep data on the same Larson–Miller plot, show a decreasing effect of oxygen and nitrogen on the strength of tantalum with increasing temperature. Tantalum–carbon alloys show a strength difference which persists to higher temperatures. They suggest that this is due to dispersion hardening (Fig. 16).

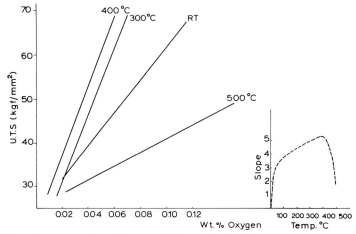

Fig. 15. Effect of oxygen on the ultimate tensile strength of niobium, after Vaughan and Rose[82]. (Courtesy of U.K.A.E.A.)

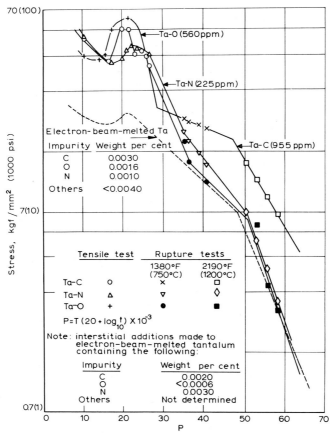

Fig. 16. Larson–Miller plot comparing rupture properties of recrystallized electron-beam-melted Ta, Ta–C, Ta–N and Ta–O alloys after Schmidt[83].

This behavior, in turn, suggests that systems involving precipitates are more complex, depending on particle size and distribution as well as composition. Evans[84], studying the effect of nitrogen on the room temperature yield strength of niobium, showed that nitrogen in solution affected the lower yield stress, but that neither of two precipitates (obtained by different cooling rates) produced any significant change in the lower yield point. He suggested that both dispersions were too coarse to produce an optimum effect.

The possible permutations of pressure, temperature, and stress are obviously tremendously varied. There are many indications that strength of a given alloy is profoundly affected by thermal history and reaction rate. In some cases, contaminants are added at moderate temperatures followed by a high temperature homogenizing anneal. In other cases, interstitial elements are added during melting. Michael and Gentry[85] show that these two methods do not produce equivalent tensile strengths in at least one case. The effects on mechanical properties may be greatly dependent on such minor differences; hence, the ability to predict me-

References pp. 858–861

chanical behavior subsequent to a specific exposure must await a thorough understanding of the mechanism controlling the property change.

The role of metal surfaces in determining mechanical behavior is profound in some cases. Kramer and Demer[86] reviewed this field in detail in 1961. The field is sufficiently broad that a complete discussion will not be undertaken in this text; however, a few examples of observed effects may be helpful. In general, investigators have explained their results in terms of an oxide surface layer which inhibits the egress of dislocations from the metal. This causes a "pile-up" of dislocations at the metal surface, resulting in a stress field which opposes the approach of subsequent dislocations.

Kramer and Podlasech[87] find that under a vacuum of 10^{-8} Torr, the flow stress at 6% strain is only 50% of the flow stress at atmospheric pressure for aluminum single crystals. It has also been shown[88] for molybdenum that 95% of the resistance to flow at $-80°C$ is attributable to an oxide surface layer. Other studies indicate that removal of the surface layer reduces the rate of work hardening in aluminum single crystals[89] and changes the apparent activation energy for plastic deformation in single crystals of Al, Cu and Au and also in polycrystalline aluminum[90].

Appropriate conditions of temperature and low pressure can result in removal of oxide films found on most metals by dissociation, sublimation, or dissolution in the metal. In view of the magnitude of the observations described, it is clear that knowledge of the role of surface layers in determining the mechanical behavior of the particular metal is necessary under such conditions.

5.5 Creep properties

The energy-producing devices contemplated for the atomic age call for unprecedented service temperatures. Temperatures above 1000°C focus attention on the refractory metals and their superior high temperature creep properties. Tottle[81] in 1955 observed that oxygen increased the creep strength of niobium at 600° to 700°C. It was not his purpose, however, to treat the partial pressure of oxygen as an experimental variable; his response to this discovery was an attempt to minimize oxygen contamination and hence produce valid creep data. Until very recently the primary concern has been to ascertain the inherent creep properties of refractory metals and alloys (*i.e.*, properties unclouded by compositional changes). Considerable evidence, however, has been accumulated concerning the effect of controlled additions of interstitials made before creep testing. Stoop and Shahinian[91] studied the effect of oxygen on the creep of niobium by adding oxygen at low temperatures, then homogenizing the specimens at higher temperatures. Their results, illustrated in Fig. 17, show that increasing oxygen concentration progressively increased the rupture time and decreased the secondary creep rate. Mukherjee and Martin[92] added nitrogen to Mo–Ti and Mo–Zr specimens and subsequently creep tested them. (These alloys definitely contained precipitates.) The nitriding increased rup-

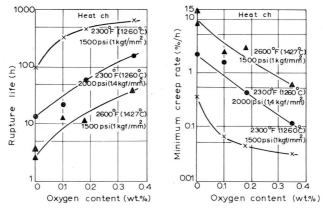

Fig. 17. Influence of oxygen on the creep properties of niobium. (a) Rupture life and (b) creep rate, after Stoop and Shahinian[91].

ture lives and reduced creep rates in both alloys. Michael and Gentry[85] report the complex dependence of secondary creep rate on oxygen content for Nb and Ta which is shown in Fig. 18.

These results indicate the need for caution in determining the creep properties of such materials (although they do not necessarily reflect the effect of the simultaneous exposure to stress and residual contaminants at low pressures). If inflated creep strengths were determined in poor vacua, then used in the design of devices

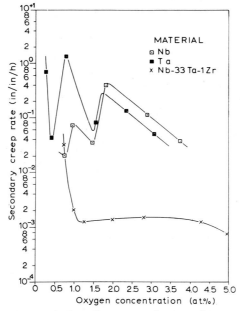

Fig. 18. Influence of oxygen content on the compressive secondary creep rate of Nb, Ta and Nb–Ta–Zr alloy at 1200°C and 7 kgf/mm² (10,000 psi, after Michael and Gentry[85]. (Courtesy of Pergamon Press, Inc., New York)

References pp. 858–861

for operation in the hard vacuum of space, for example, serious errors could result. Accordingly, a study was undertaken by the TRW Equipment Laboratories under the sponsorship of the National Aeronautics and Space Administration in which great care was taken to minimize compositional changes. The results of their creep studies at 10^{-9} Torr are summarized in the literature[93]. Other investigators[94-96] have also studied the creep of refractory metal alloys in this pressure range. The results of the determinations of the creep–rupture properties from 10 to 1000 h of 15 alloys in the pressure range from 10^{-7} to 10^{-8} Torr were compared with these data where possible[97]. Considering the potential variations in the properties of different heats of the same nominal composition and due to different heat treatments (Fig. 5), the data seem in reasonable agreement. It therefore seems reasonable to assume that the technology for determining valid creep properties is now in hand.

It is of practical and theoretical interest, however, to treat vacuum as an experimental variable (*i.e.*, systematically to study the simultaneous effect of stress, temperature and partial pressure of the various reactants in a vacuum system). Recently, attention has been turned to this problem. It has been shown[98] that tantalum creep specimens, protected by a foil wrapping, are weaker than unprotected specimens in creep at 2000°C when tested in argon*. Chemical analysis verified that the foil protected the specimens from interstitial contamination. Titran and Hall[96] creep tested FS-85 (Nb–27% Ta–10% W–1% Zr) at pressures of 10^{-6} to 10^{-7} Torr, then extrapolated their results using the Manson–Haferd parameter. The authors then performed long-time creep tests at 10^{-8} to 10^{-9} Torr to test the accuracy of these predictions. Their results are illustrated in Fig. 19. Although the results are reasonably close to the predicted values, Titran and Hall are alarmed by the fact that all of the ultrahigh vacuum data fall below the predicted values. On the other hand, using values read from the curves of Titran and Hall, the authors employed a computer program to determine the parameter constants and subsequently construct a master curve**. This prediction, also included in Fig. 19, not only comes closer to the actual values but no longer falls above all the experimental points. Roche[99] studied the effects of pressure (10^{-5} to 10^{-7} Torr) on the bending creep behavior of Nb–0.6% Zr. He found that the deflection rate of cantilevered beam specimens was accelerated in the early stages of testing at the higher pressures but retarded at longer testing times. Since the precipitation of interstitial compounds involves a volume expansion, he suggests that such precipitation causes the observed increase in the initial creep rates. The

---

* This statement embodies the implicit assumption that a given partial pressure of oxygen will yield the same effect in argon as in vacuum. In many cases this assumption is probably justified although it is always desirable to verify it experimentally. In other cases it is not justified (*e.g.*, where sublimation is a factor).

** The authors wish to acknowledge with thanks the use of a computer program written by R. W. Swindeman for this purpose.

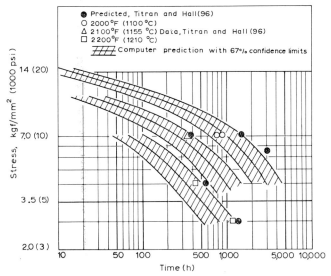

Fig. 19. Time to 1% creep vs. stress for FS-85.

decrease in creep rates at longer times may be due to dispersion or solution strengthening, he suggests.

Inouye[100] has studied the creep of tungsten and W–50% Re at 1650°C in various partial pressures of oxygen. He finds that creep is accelerated at the higher pressures. In this temperature range, sublimation of oxides proceeds at a significant rate relative to their formation. Measurements of the weight loss indicate that if the loss of material were uniform, the cross-sectional area would only be reduced by about 0.1%. Although the work is not complete, metallographic examination revealed that the material loss was not uniform; hence, sublimation of volatile oxides is still being considered as a plausible cause of this effect.

The effect of a vacuum environment on the creep properties of nickel-base alloys has received considerable attention. Effects have been observed at times and temperatures where significant bulk diffusion is not possible; hence, they are regarded as surface or near-surface phenomena. Shahinian and Achter show that both Nichrome V[101] and nickel[102] are stronger in air than in vacuum at low stresses (long times) but stronger in vacuum at high stresses (short times). They tentatively explain their observation in terms of two competing processes. "Oxidation strengthens the material but surface adsorption of gases reduces the surface energy and facilitates the propagation of cracks. At low strain rates and high temperatures, conditions conducive to extensive oxidation, the metals are strengthened in air while at high strain rates and low temperatures the competing process becomes controlling and they are stronger in vacuum".

5.6 Fatigue

The field of fatigue is quite complex and its literature is extensive. It is not

References pp. 858–861

possible to treat even the effect of vacuum on fatigue in detail here.

Vacuum usually extends the fatigue life of metals. As the pressure is reduced, fatigue life is usually seen to be weakly dependent on pressure until a rather well-defined critical pressure is reached at which the life increases abruptly. Upon further decreasing of the pressure, this effect saturates and the life is again found to be weakly dependent on pressure.

Microcracks are usually nucleated during the first 10% of the fatigue life. This seems to be true in both vacuum and air[103,104]. The effect of vacuum on fatigue life seems, therefore, to be connected with crack propagation as opposed to crack nucleation. Graphic evidence of the effect of vacuum on crack propagation is provided by the fractograph in Fig. 20. This specimen, of 2024–T3 aluminum, was cycled alternately in air and vacuum[105]. Striations were formed in air

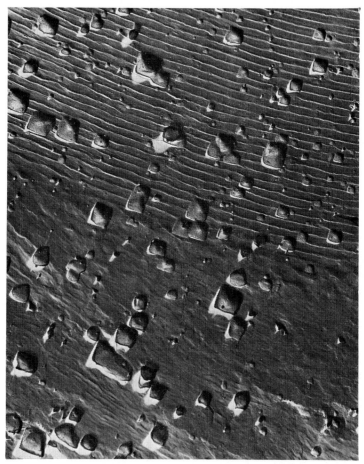

Fig. 20. Electron fractograph of aluminum fatigue specimen (etch pits were induced chemically to determine orientations), after Pelloux and Walner[105]. (Courtesy of the American Society for Metals)

while smooth fracture surfaces were formed in vacuum. (The etch pits were introduced to determine the crystallographic orientation of the fracture plane.) Let us discuss this phenomenon in terms of three theories, realizing that this may oversimplify a complex effect.

It has been suggested[106] that if cracks, opened during the tension half of the cycle, remain uncontaminated by gases they can "reweld" during the compression half of the cycle. Therefore, in vacuum, crack propagation is retarded in comparison with air. It was later suggested that an atmospheric corrosion layer, formed during the tension half of the cycle, weakened atomic bonds at the crack tip, thus enhancing crack propagation in air. Finally, it has been suggested[104,107] that a corrosion layer, present in air, inhibits the egress of dislocations from the surface and causes the pile-up of dislocations in a "surface debris layer" resulting in high local stresses favorable to crack propagation. The following facts seem to have relevance to these theories.

In some cases, oxygen and/or water vapor seem to have particular significance. Exposure to other gases does not seem to reduce fatigue life significantly in such cases. Bennit[108] applied cellophane tape to specimens prior to cycling them in a moist environment. Gas bubbles appeared under the tape after cycling. Although chemical analysis was inconclusive, the gas was probably hydrogen, suggesting that a chemical reaction was involved. Jacisin[109] found evidence of an oxide on fracture surfaces of fatigue specimens.

Hordon and Wright[110] cycled aluminum specimens between low and high tensile stresses with no compressive loading. A vacuum effect was observed in this case. If rewelding were important, one would expect purely tensile loading to diminish its effect.

At low pressures, an appreciable time is required for residual gases to cover the freshly opened surface of a crack. An increase in cyclic frequency would shorten the time of exposure of the crack to contaminants. One would therefore expect the effect of reduced pressure to be enhanced by high frequencies. Wright and Hordon[111] have found such a frequency effect and were able to correlate the critical pressure with the time to form a monolayer. Other investigators[104,107,109] contend that a frequency effect in the opposite direction is sometimes observed. They suggest that, if the surface debris layer theory is invoked, such cases could be explained by a time-dependent relaxation of the stresses in the layer. These investigators cite the profound effect of surface removal on tensile and creep properties, discussed previously, as additional evidence for the surface layer concept.

References pp. 858–861

List of Symbols

P	=	Pressure in Torr
M	=	Molecular weight in grams
T	=	Temperature, K
Q_i	=	Mass of gas molecules per unit time
α	=	Absorption coefficient (eqn. 13)
α_0	=	Sticking probability at zero concentration
C_m	=	Mean gas concentration
C_e	=	Equilibrium gas concentration
C	=	Instantaneous concentration
C_0	=	Initial concentration
C_s	=	Surface concentration
D	=	Diffusivity
x	=	Distance
t	=	Time
J	=	Flux
ξ	=	Penetration depth of internal oxidation
$N_o^{(s)}$	=	Mole fraction oxygen of the surface
$N_b^{(o)}$	=	Mole fraction reactive solute in the alloy
D_0	=	Diffusivity of oxygen in the base metal
v	=	Number of gas ions per solute atom in compound
DPH	=	Diamond pyramid hardness

REFERENCES

1 L. J. BONIS, in E. L. FOSTER (ed.), *Trans. Intern. Vacuum Met. Conf., 1967*, Am. Vacuum Soc., New York, 1968, p. 207.
2 R. E. BORDEAU AND F. J. BAUER, *Structure of the Upper Atmosphere Deduced from Charged Particle Measurements from the Rockets in the Explorer VIII Satellite*.
3 H. REICH, in W. G. MATHESON (ed.), *Trans. Fifth Nat. Symp. on Vacuum Technology, 1958*, Pergamon, 1959, p. 279.
4 B. V. WACHOLDER AND E. FAYER, *Proc. Inst. Environmental Sciences*, Inst. Environmental Sciences, Mt. Prospect, Illinois, 1960, p. 19.
5 D. J. SANTELER, in C. R. MEISSNER (ed.), *Trans. Sixth Nat. Symp. on Vacuum Technology, 1959*, Pergamon, 1960, p. 129.
6 J. C. SIMONS, *Proc. Inst. Environmental Sciences*, Inst. Environmental Sciences, Mt. Prospect, Illinois, 1960, p. 5.
7 A. LENGYEL, P. A. MARFONE AND D. J. SANTELER, *Trans. ASME*, Paper No. 61-AV-52, 1961.
8 H. P. KERFOOT, R. E. GAUMER AND S. SOKOLSKY, *Space/Aeronautics*, Vol. 38, No. 5, Pt. 2, October 1962, p. 15.
9 R. W. ROBERTS AND T. A. VANDERSLICE, in N. HOLONJAK, JR. (ed.), *Ultrahigh Vacuum and Its Applications*, Prentice-Hall, Englewood Cliffs, N. J., 1963.
10 S. DUSHMAN, *Scientific Foundations of Vacuum Technique*, 2nd edn., Wiley, New York, 1962, p. 14.
11 D. V. KELLER, *Wear*, 6 (1963) 353.
12 T. SPALVINS AND D. H. BUCKLEY, *Importance in lubrication of various interface types formed*

during vacuum deposition of thin-metallic films, NASA TM X-52305, 1967, NASA, Washington, D.C.
13 W. P. GILBREATH AND H. T. SUMSION, *Proc. AIAA 6th Structure and Materials Conf.*, Palm Springs, Calif., April 5–7, 1965.
14 D. H. BUCKLEY, *The influence of recrystallization and preferred orientation on the friction of copper in vacuum (10^{-11}) Torr*, NASA TN D-3794, 1967.
15 K. I. JOHNSON AND D. V. KELLER, JR., *J. Appl. Phys.*, 38 (1967) 1896.
16 F. P. BOWDEN AND D. TABOR, *Friction and Lubrication of Solids, Part II*, Clarendon Press, Oxford, 1964.
17 Adhesion or cold welding of materials in space environments, *ASTM Spec. Tech. Publ. No. 431*, 1968.
18 D. R. MILNER AND G. W. ROWE, *Metallurgical Rev.*, 7 (1962) 433.
19 M. E. SIKORSKI, *Wear*, 7 (1964) 144.
20 P. M. WINSLOW AND D. V. MCINTYRE, *J. Vacuum Sci. Tech.*, 3 (1966) 54.
21 T. H. BATZER AND R. BUNSHAH, *J. Vacuum Sci. Tech.*, 4 (1967) 115.
22 D. V. KELLER, JR., in E. L. FOSTER (ed.), *Trans. Intern. Vacuum Met. Conf.*, 1967, Am. Vacuum Soc., New York, 1968, p. 235.
23 D. H. BUCKLEY, *Influence of chemisorbed films on adhesion and friction of clean iron*, NASA TND-4775, September 1968 (also *cf.* bibliography included therein).
24 K. I. JOHNSON AND D. V. KELLER, JR., *J. Vacuum Sci. Tech.*, 4 (1967) 115.
25 D. V. KELLER, JR., The analysis of metallic adhesion data, NASA TM X-52305, 1967, NASA, Washington, D.C., p. 181.
26 T. MACNICHOLAS AND D. V. KELLER, JR. *J. Adhesion*, 1 (1969) 164.
27 J. A. GREENWOOD AND J. B. P. WILLIAMSON, *Proc. Roy. Soc. (London)*, A295 (1966) 300.
28 D. V. KELLER, JR., Invited discussion, *NASA Symp.*, San Antonio, Texas, November 1967, 42-1 to 42-12.
29 J. W. MAY, *Ind. Eng. Chem.*, 57 (1965) 19.
30 G. EHRLICH, Adsorption and surface structure, *Metal Surfaces*, Am. Soc. Metals, Metals Park, Ohio, 1963, p. 221.
31 W. P. GILBREATH, Definition and evaluation of parameters which influence the adhesion of metals, NASA TM X-52305, 1967, NASA, Washington, D.C., p. 128.
32 F. P. BOWDEN AND D. TABOR, *Brit. J. Appl. Phys.*, 17 (1966) 1521.
33 J. A. GREENWOOD, *Trans. ASME*, 89 (1967) 81.
34 I. V. KRAGELSKY AND N. B. DEMKIN, *Wear*, 3 (1960) 170.
35 J. B. P. WILLIAMSON, *Proc. Symp. on Interdisciplinary Approach to Friction and Wear*, San Antonio, Texas, November 1967, NASA.
36 G. W. ROWE, *Wear*, 7 (1964) 204.
37 A. G. AKINS AND D. TABOR, *J. Inst. Metals*, 94 (1966) 107.
38 A. G. AKINS, A. SILVERIO AND D. TABOR, *J. Inst. Metals*, 94 (1966) 369.
39 R. HOLM, *Electric Contacts*, Springer-Verlag, New York, 1967.
40 D. TABOR, *Hardness of Metals*, Clarendon Press, Oxford, 1951.
41 R. CHAIT, Ph. D. Dissertation, Department of Chemical Engineering and Metallurgy, Syracuse University, 1966.
42 N. GANE AND F. P. BOWDEN, *J. Appl. Phys.*, 39 (1968) 1432.
43 V. WEISS AND S. YUKAWA, A critical appraisal of fracture mechanics, *Fractured toughness testing and its application*, ASTM, Spec. Tech. Publ. No. 381, p. 1.
44 J. J. GILMAN, *Am. Soc. Metals, Trans. Quart.*, 59 (1966) 596.
45 J. R. OSIAS AND J. H. TRIPP, *Wear*, 9 (1966) 388; also private communication to J. B. Williamson.
46 M. J. HORDON, Adhesion of metals in high vacuum, NASA TM X-52305, 1967, NASA, Washington, D.C., p. 109.
47 A. R. C. WESTWOOD, *Phil. Mag.*, 8 (1963) 787; also 9 (1964) 199.
48 D. H. BUCKLEY, *Influence of crystal structure orientation and solubility on the adhesion and sliding of various metal single crystals in vacuum* (0.01 nTorr), NASA TM X-52305, 1967, NASA, Washington, D.C., p. 248.
49 J. L. HAM, *Trans. Am. Soc. Lubrication Eng.*, 6 (1963) 20.

50 J. B. Rittenhouse, *Space simulation testing of the adhesion of metals*, NASA TM X-52305, 1967, NASA, Washington, D.C., p. 88.
51 D. T. Bourgette, in L. M. Bianchi (ed.), *Trans. 8th Vacuum Met. Conf., 1965*, Am. Vacuum Soc., Boston, Mass., 1966, p. 57.
52 R. A. Perkins, *Effect of processing variables on the structure and properties of refractory metals*, AFML-TR-65-234, Part II, 1965.
53 H. Inouye, in R. I. Jaffee (ed.), The contamination of refractory metals in vacua below 10^{-6} Torr, *Refractory Metals and Alloys III: Applied Aspects, Part 2, Vol. 30*, Gordon and Breach, New York, 1963, p. 871.
54 H. Inouye, in I. Machlin, R. T. Begley and E. D. Weisert (eds.), Interactions of refractory metals with active gases in vacua and inert gas environments, *Refractory Metals and Alloys*, Plenum Press, New York, 1968.
55 L. S. Darken, Formal basis of diffusion theory, *Atom Movements*, Am. Soc. Metals, Cleveland, 1951.
56 R. A. Rapp, *Corrosion*, 21 (1965) 382.
57 W. H. Chang, *Trans. Am. Soc. Metals*, 57 (1964) 527.
58 R. L. Stephenson, The effect of fabrication variables on the creep-rupture properties of molybdenum base alloys, submitted to *Trans. AIME*.
59 E. Gebhardt and F. D. Seghezzi, in F. Benesovsky (ed.), New investigations of the Ta-O system, *3rd Plansee Seminar, Reutte, Austria*, p. 280.
60 E. A. Loria, *J. Less-Common Metals*, 10 (1966) 296.
61 E. de Lamotte, Y. Huang and C. Altstetter, *Trans. AIME*, 239 (1967) 1625.
62 A. K. Mukherjee and J. W. Martin, *J. Less-Common Metals*, 2 (1960) 392.
63 A. K. Mukherjee and J. W. Martin, *J. Less-Common Metals*, 3 (1961) 216.
64 F. R. Cortes and A. L. Field, *J. Less-Common Metals*, 4 (1962) 169.
65 R. T. Begley and A. I. Lewis, *Development of Nb base alloys*, Contract AF 33(616)-6258, First Quart. Progr. Rept., December 1959.
66 B. A. Lomis and O. N. Carlson, *Reactive Metals, Met. Soc. Conferences, Vol. 2*, Interscience, New York, 1959, p. 227.
67 G. T. Hahn, A. Gilbert and R. I. Jaffee, in M. Senchyshen and I. Parlmutter (eds.), The effects of solutes on the ductile to brittle transition in refractory metals, *Refractory Metals and Alloys II*, Interscience, New York, 1963.
68 B. C. Allen, D. J. Maykuth and R. I. Jaffee, *Influence of impurity elements, structure, and prestrain on tensile-transition temperature of chromium*, NASA TN D-837, April 1961.
69 A. Lawley, J. van den Sype and R. Maddin, *J. Inst. Metals*, 91 (1962) 23.
70 B. C. Allen, D. J. Maykuth and R. I. Jaffee, *J. Inst. Metals*, 90 (1961) 120.
71 A. C. Gilbert, C. N. Reid and G. T. Hahn, *J. Inst. Metals*, 92 (1963) 351.
72 H. Scott, W. A. Taebel and P. P. Lawthers, *The Metal Mo*, Am. Soc. Metals, Cleveland, 1958, p. 51.
73 J. W. Clark, *Physical metallurgy of welding tungsten and tungsten-base alloys*, ASDTR 61-584, 1962.
74 A. G. Ingram, *Further investigations of notch sensitivity of refractory metals*, ASD-TDR-62-1004, 1962.
75 K. Tsuya and N. Aritomi, *J. Less-Common Metals*, 15 (1968) 245.
76 L. E. Olds and G. W. P. Rengstorff, *Trans. AIME*, 209 (1957) 468.
77 W. D. Klopp, F. C. Holden and R. I. Jaffee, *Further studies on rhenium alloying effect in Mo, W and Cr*, Battelle Memorial Institute Technical Rept., Nonr-1512(00), 1960.
78 E. P. Abrahamson II and N. J. Grant, *Trans. Am. Soc. Metals*, 50 (1958) 705.
79 M. Semchyshen and R. Q. Barr, Mechanical properties of Mo and Mo base alloy sheet, *ASTM Spec. Tech. Publ. No. 272*, 1959, 12.
80 J. R. Stephens, *Effect of oxygen on mechanical properties of W*, NASA TND-1581, 1963.
81 C. R. Tottle, *J. Inst. Metals*, 85 (1957) 375.
82 H. G. Vaughan and R. G. Rose, *The tensile properties of Nb*, IGR-TN-C-583, 1958.
83 F. F. Schmidt et al., *Investigation of the properties of Ta and its alloys*, WADD TR 59-13, 1959.
84 P. R. V. Evans, *J. Less-Common Metals*, 4 (1962) 78.

REFERENCES

85 A. B. MICHAEL AND W. O. GENTRY, in N. E. PROMISEL (ed.), The trend and status of development of refractory metal alloys, *The Science and Technology of Selected Refractory Metal Alloys*, Pergamon, New York, 1964.
86 I. KRAMER AND L. J. DEMER, *Progr. Mater. Sci.*, 9 (1961).
87 I. R. KRAMER AND S. PODLASECH, *Acta Met.*, 11 (1963) 70.
88 I. R. KRAMER, *Trans. AIME*, 239 (1967) 520.
89 H. SHEN, S. E. PODLASECH AND I. R. KRAMER, *Trans. AIME* 233 (1965) 1933.
90 I. R. KRAMER, *Trans. AIME*, 230 (1964) 991.
91 J. STOOP AND P. SHAHINIAN, *Effect of oxygen on the creep-rupture properties of Nb*, NRL Rept. 6095, 1964.
92 A. K. MUKHERJEE AND J. W. MARTIN, *J. Less-Common Metals*, 5 (1963) 403.
93 J. C. SAWYER AND E. A. STEIGERWALD, *Generation of long time creep data on refractory alloys at elevated temperature*, ER-7203, June 1967.
94 R. H. TITRAN AND R. W. HALL, *Ultrahigh vacuum creep behavior of columbium and tantalum alloys at 1100°C (2000°F) and 1210°C (2200°F) for times greater than 1000 hours, NASA TN D-3222*, January 1966.
95 R. W. HALL AND R. H. TITRAN, in R. I. JAFFEE (ed.), Creep properties of columbium alloys in very high vacuum, *Refractory Metals and Alloys III, Applied Aspects*, Part 2, Vol. 30, Gordon and Breach, New York, 1963, p. 885.
96 R. H. TITRAN AND R. W. HALL, *High temperature creep behaviour of a columbium alloy, FS-85, NASA TN D-2885*, June 1965.
97 R. L. STEPHENSON, in R. B. BARROW AND A. SIMKOVICH (eds.), *Trans. Vacuum Met. Conf., 1968*, Am. Vacuum Soc., New York, 1968, p. 15.
98 *High-temperature materials program progress rept. No. 25*, Part A, GEMP-25A, July 1963.
99 T. K. ROCHE, in R. I. JAFFEE (ed.), Effect of degree of vacuum on the slow-bend creep behavior of columbium-0.6% zirconium at 1000°C, *Refractory Metals and Alloys III*, Part 2, 1963, p. 901.
100 H. INOUYE, in D. T. BOURGETTE AND A. L. SIMKOVICH (eds.), *Trans. Vac. Met. Conf., New York, 1969*, Am. Vac. Soc., New York, 1969, p. 91.
101 P. SHAHINIAN AND M. ACHTER, *Trans. Am. Soc. Metals*, 51 (1959) 244.
102 P. SHAHINIAN AND M. ACHTER, *Trans. AIME*, 215 (1959) 37.
103 N. J. WADSWORTH AND J. HUTCHINS, *Phil. Mag.*, 3 (1958) 1154.
104 H. SHEN AND I. R. KRAMER, in E. L. FOSTER (ed.), *Trans. Intern. Vacuum Met. Conf., 1967*, Am. Vacuum Soc., New York, 1968, p. 263.
105 R. M. V. PELLOUX AND H. WALNER, *Metal Progr.*, 94 (1968) 82.
106 N. THOMPSON, N. WADSWORTH AND N. LOUAT, *Phil. Mag.*, 1 (1956) 113.
107 H. SHEN, S. E. PODLASECH AND I. R. KRAMER, *Acta Met.*, 14 (1966) 341.
108 J. A. BENNIT, in J. A. BURKE, N. L. REED AND V. WEISS (eds.), Effect of reactions with the atmosphere during fatigue of metals, *Fatigue–An Interdisciplinary Approach*, Syracuse University Press, 1964.
109 J. M. JACISIN, *Trans. AIME*, 239 (1967) 821.
110 M. J. HORDON AND M. A. WRIGHT, *Mechanism of the atmosphere with the fatigue of metals, NASA CR-1165*, 1968.
111 M. A. WRIGHT AND M. J. HORDON, *Acta Met.*, 15 (1967) 430.

Conversion Factors of Several Important Metric Units into British Units

Physical Quantity	Metric Unit	British Unit
Length	cm	0.394 in
	m	3.28 ft
Area	cm^2	0.155 in^2
	m^2	10.76 ft^2
Volume	cm^3	0.0610 in^3
	dm^3	0.0353 ft^3
Mass	kg	2.205 lb
Force	kgf	2.205 lbf
Pressure	kgf/mm^2	1422 lbf/in^2 = 0.635 $tonf/in^2$
Density	g/cm^3	0.0361 lb/in^3
	kg/m^3	0.0624 lb/ft^3

Author Index

Abanin, D. D., 332$^{394,\ 401}$
Abdellatif, A., 333^{428}
Abiseva, R. U., 325^{144}
Abkowitz, S., 645^{117}
Abrahamson, E. P., II, 860^{78}
Abshev, D. N., 324^{97}
Accary, A., 330^{346}, 645$^{137,\ 138}$
Achard, J. C., 321^{20}
Achter, M., 861$^{101,\ 102}$
Achter, M. R., 647^{224}
Ackermann, B. R., 323^{64}
Adams, C. M., Jr., 801$^{69,\ 71,\ 75}$, 820^{45}
Afanasev, J. M., 509^{110}
Ageev, N. V., 332^{400}
Ailloud, P., 330$^{347,\ 350}$
Aitken, R., 507^{2}
Akerblom, C. A., 668^{44}
Akerman, K., 326^{181}, 335^{496}
Akins, A. G., 859$^{37,\ 38}$
Aksoy, A. M., 323^{79}, 512^{276}
Albano, V. J., 646^{169}
Albert, P., 323$^{72,\ 73}$, 324$^{117,\ 118,\ 119,\ 120}$, 330^{348}, 331^{351}
Albom, M. J., 802^{98}
Albrecht, W. A., 670^{113}
Albrecht, W. M., 670^{130}
Alcock, C. B., 92^{3}, 321^{8}
Aleksandrov, B. N., 327^{247}, 334$^{481,\ 484}$
Allen, B. C., 333^{461}, 719^{35}, 860$^{68,\ 70}$
Allred, W. P., 334^{475}
Altstetter, C., 860^{61}
Alyamovsky, C. I., 322^{54}
Améen, E., 508^{44}
Amonenko, V. M., 327^{244}, 329$^{287,\ 288,\ 294,\ 296,\ 312,\ 313}$, 335^{505}, 802$^{89,\ 93,\ 94}$
Anable, W. E., 331^{370}
Anders, H., 820^{33}
Anderson, G. S., 820^{52}
Anderson, J. E., 719^{58}
Anderson, J. V., 801^{42}
Anderson, P. D., 92^{2}
Andrieux, J. L., 326^{180}, 329^{284}
Ansheles, I. I., 510^{158}
Antes, H. W., 643^{45}

Aoki, K., 647^{220}
Appleby, B., 642^{28}
Ardenne, M. von, 644$^{83,\ 87,\ 88,\ 90,\ 92}$, 645^{129}, 800^{1}
Argent, B. B., 333$^{446,\ 447}$, 642^{18}, 671^{145}
Argon, A. S., 647^{205}
Aritomi, N., 860^{75}
Arkel, A. E. van, 331^{384}
Armand, M., 331^{355}
Armantrout, C. E., 645^{116}
Armington, A. F., 646^{152}
Arsenault, R. J., 647^{201}
Arzamastsev, Yu. S., 328^{255}
Asano, K., 512^{241}
Aschoff, W., 645^{114}
Ashizawa, T., 670^{111}
Ashoff, W., 644^{103}
Aspinal, M. L., 669^{51}
Auler, H., 327^{225}
Aurini, T. D., 327^{207}

Baboian, R., 819^{27}
Bach, H., 322^{39}, 719^{29a}
Badiali, M. A., 331^{381}
Bagdavadze, D. I., 324$^{126,\ 129}$, 325^{133}
Baj, M., 669^{64}
Bandel, G., 510^{150}
Bakhireva, L. D., 323^{95}, 328^{255}
Bakish, R., 331^{381}, 644^{78}, 645^{145}, 800$^{19,\ 21,\ 25,\ 27,\ 28,\ 29,\ 30,\ 31}$
Baldi, F., 671^{166}
Balis, E. W., 670^{123}
Balke, C. W., 719^{31}
Ballhausen, C., 719^{51}
Bandel, G., 510$^{151,\ 164}$
Bangert, L., 645^{113}
Banks, C. V., 670^{134}
Baque, C., 668^{38}
Baranov, J. A., 513^{324}
Barber, G., 644^{78}
Barber, J. B., 93^{55}
Baroch, E. F., 331^{378}, 645^{114}
Barr, R. Q., 860^{79}
Barraclough, K. C., 513^{291}, 643$^{58,\ 62}$

AUTHOR INDEX

Barrick, J. R., 513[333]
Barthel, J., 332[419], 333[431, 458]
Bas, E. B., 333[427, 428, 429], 646[188], 800[5, 59]
Bastien, P., 512[253]
Batzer, T. H., 859[21]
Bauer, F. J., 858[2]
Bauer, H., 92[23]
Baum, B. A., 330[340]
Baumann, S., 326[169]
Baumhauer, H., 719[52]
Bazin, J., 323[92]
Beach, A. L., 670[133]
Beatson, E. V., 801[66]
Beck, R., 669[48]
Becker, J. J., 744[17, 18]
Becket, F. J., 744[9]
Becvar, J., 514[337]
Beeley, P. R., 93[54]
Beernsten, D. J., 670[105]
Begley, R. T., 860[65]
Behrndt, K. H., 819[18]
Beljaev, V. S., 509[111]
Belk, J. A., 333[455]
Bellamy, R. G., 332[395]
Belvedere, C., 513[283]
Benesovsky, F., 323[88, 89], 718[6, 12], 719[33, 40, 56, 63]
Bennet, C. F., 508[34]
Bennett, G. H. J., 642[19]
Bennett, S. J., 668[35, 39]
Bennit, J. A., 861[108]
Benson, K. E., 334[483]
Berezin, A. B., 335[495]
Bergenfelt, S., 668[44]
Berghezan, A., 646[154]
Berlin, I. K., 323[67], 333[441]
Berry, R., 670[103]
Bertea, O., 331[373]
Berthel, K. H., 334[463]
Berthold, C. E., 325[135]
Berve, J., 513[305]
Besobrazov, S. V., 330[333, 334]
Bessemer, H., 508[24]
Betterton, J. O., Jr., 333[436], 508[27]
Beuyukian, C. s., 801[77]
Bezarashvili, S. M., 324[127]
Bezobrazov, S. V., 330[337]
Bibenina, G. A., 330[322]
Bibina, I. A., 330[323, 325]
Bieber, J. L., 335[506]
Bigge, H. C., 510[149]
Birbeck, F. E., 333[434], 646[177]
Blake, P. D., 668[34]
Blanc, B. L., 329[297]
Blocher, J. M., 331[390]
Block, F. E., 325[132], 326[178]

Blum, P. L., 323[87]
Boccone, G. B., 513[320]
Böhm, W., 507[14]
Boesch, W. J., 642[29]
Bogatina, K. G., 321[10], 330[323, 325]
Bogdandy, L. v., 93[39]
Bolton, W. v., 719[25]
Bonar, K. M., 509[83]
Bonnier, E., 328[261], 329[284], 334[489]
Bonis, L. J., 858[1]
Booth, E., 669[88], 670[96, 98]
Bordeau, R. E., 858[2]
Borrebach, E. J., 643[37]
Bosorth, R. M., 744[19]
Boulin, D. M., 647[204]
Bourgeois, H., 331[357]
Bourgette, D. T., 860[51]
Bowden, F. P., 859[16, 32, 42]
Bowen, D. K., 647[214]
Boyaroshinov, V. A., 644[95]
Boyle, B. J., 668[17]
Bradshaw, A. V., 92[25], 93[41], 326[193], 513[307], 642[16]
Brady, J., 671[153]
Braun, H., 322[43], 719[26]
Brehm, W. F., 647[225]
Bretschneider, O. R., 326[175]
Breuer, E., 507[18]
Brewer, L., 92[18], 330[343], 509[82]
Bril, J., 670[101]
Broom, R. F., 328[258]
Brotherton, T. D., 325[150]
Brotzen, F. R., 647[193, 197]
Brotzmann, K., 92[19], 508[61, 71, 72], 509[79], 511[210, 223], 513[314, 318]
Brownlee, L. D., 719[22]
Brownsword, R., 646[185]
Broxham, P. H., 510[156]
Bruch, J., 669[67]
Bryant, F. J., 669[88]
Buckley, D. H., 858[12], 859[14, 23, 48]
Bucklow, I. A., 646[181], 647[221]
Buehler, E., 333[456, 457], 646[171]
Buhre, T., 510[140, 141]
Buhrig, E., 332[421]
Buker, D. O., 323[75]
Buketov, A. E., 324[97]
Bulatova, G. I., 670[116]
Bull Simonsen, E., 646[154, 155]
Bungardt, K., 642[23], 643[52]
Bunshah, R. F., 326[195], 327[208, 209], 329[289, 298], 510[167], 810[3], 820[44], 859[21]
Burger, E. E., 643[55]
Burlakov, V. D., 326[205], 329[287]
Burnazyan, A. S., 324[103]
Burnet, G., 327[211]

AUTHOR INDEX

Burnett, G., 327^{218}
Burns, F., 671^{147}
Burtsev, V. T., 509^{112}
Busby, P. E., 642^6
Bush, G. H., 669^{76}
Bychkov, Y. F., 328^{274}
Byrnes, J. J., 333^{444}, 646^{163}
Byron, J. F., 647^{200}

Cabrera, N., 744^3
Cadoff, I., 329^{286}
Calderbank, P. H., 93^{40}
Calverley, A., $333^{433, 434, 442}$, $646^{177, 178}$
Campbell, C. S., 335^{519}
Campbell, I. E., 332^{397}
Campbell, T. T., 325^{132}, 326^{178}
Candidus, E. S., $644^{80, 107}$
Canfield, L. R., 819^{12}
Capuano, A., 511^{227}
Carlson, G. E., 642^{21}
Carlson, O. N., 326^{183}, $328^{269, 276}$, $331^{376, 387}$, 335^{501}, 860^{66}
Carlson, R. G., 334^{466}, 646^{178}
Carlsson, C. G., 508^{47}
Carnahan, D. R., 643^{59}
Caroll, W. N., 820^{36}
Carpenter, J. F., 819^6
Carson, C. C., 669^{82}
Caswell, H. L., 819^5
Čečulin, B. B., 514^{351}
Chadwick, C. G., 330^{338}
Chait, R., 859^{41}
Champeix, L., 668^{38}
Chang, W. H., 860^{57}
Charushnikova, G. V., 330^{334}
Charveriat, M., 328^{261}
Cheburkov, V. I., 328^{274}
Chedaille, J., 510^{166}
Chelokhsaev, I. S., 321^{18}
Chel'tsov, V. M., 326^{170}
Chernobrovkin, V. P., 509^{111}
Chernyaev, V.N., $327^{243, 245, 246}$
Child, H. C., $642^{4, 8, 14}$, $643^{31, 32, 57, 63}$
Chipman, J., 92^{11}, 93^{59}, 322^{29}
Chistyakova, E. M., 668^{26}, 669^{83}, $670^{117, 127, 140, 151, 164, 168, 173, 177}$
Chizhikov, D. M., 324^{102}
Christensen, N., 672^{184}
Christian, J. W., $647^{213, 214}$
Chuiko, N. M., 509^{118}
Church, C. D., 513^{326}
Chuveleva, N. P., $323^{60, 61}$
Cishevsky, 644^{95}
Cizeron, G., 335^{498}
Clark, J. W., 860^{73}
Clasen, H., 322^{24}

Class, W., 647^{222}
Clemmons, B. H., 325^{151}, 326^{176}
Clough, P. J., 820^{43}
Cobine, J. D., 643^{55}
Cochran, A. A., 330^{314}
Cochran, C. N., $335^{515, 517}$
Coe, F. R., 671^{171}
Cole, M., 646^{181}
Cole, O. N., 325^{150}
Colombier, L., 513^{284}
Concina, A., 513^{283}
Conrad, A. A., 513^{330}
Cook, R. M., 642^{13}
Cooper, D. E., $643^{43, 47}$
Cordes, J. F., 325^{158}
Corrigan, D. A., 92^{11}
Cortes, F. R., 860^{64}
Coslett, V. E., 801^{56}
Cotterill, P., 514^{350}
Coupette, W., $508^{23, 35, 70}$, 510^{152}, 513^{285}, 514^{345}, 642^{26}
Cousin, F., 511^{186}
Covington, L. C., $668^{35, 39}$
Cox, G. F., 719^{37}
Cox, F. G., 719^{46}
Craig, G. B., 333^{440}, 646^{168}
Cremosnik, G., 801^{59}
Cumming, G., 669^{80}
Cunnell, F. A., 328^{258}
Cuprikov, G. E., 672^{183}
Czakó, K. A., 325^{168}

Daane, A. H., $324^{122, 123, 125, 128}$, $328^{263, 265, 266, 269}$
Dabney, J. B., 514^{334}
Därmann, H. J., $511^{181, 221}$
Daeves, K., 508^{22}
Dahlgreen, S. D., 820^{53}
Dammer, A., $512^{236, 237}$
Danckwerts, P. V., 93^{27}
Danilkin, V. A., 670^{116}
Danner, G. E., 510^{168}, 512^{261}
Darbinyan, M. V., 324^{103}
Darby, C. W., 513^{330}
Darken, L. S., $92^{1, 12}$, 860^{55}
Darling, A. S., 322^{34}, 508^{64}
Darnell, E. J., 646^{164}
Davey, R., 327^{232}
Davey, T. R. A., $326^{197, 198}$
Davies, R. G., 647^{199}
Davis, K. G., 335^{492}
Davis, M., 333^{433}, 646^{176}
Davis, R. E., 325^{150}
Dautzenberg, N., 323^{82}
Decroly, C., $321^{19, 21, 22, 23}$, 322^{26}, 325^{143}, 328^{278}

Dean, C. W., 644[75, 106]
Dean, K. C., 325[151], 326[176]
Deisinger, W., 326[202], 719[23]
De Lamotte, E., 860[61]
De la Rue, R. E., 646[161]
Delgrosso, E. J., 642[21]
De Martini, R., 513[320]
Demer, L. J., 861[86]
Demkin, N. B., 859[34]
De Negri, A., 512[256]
Dennis, N. T., 143[4]
Dennison, D. H., 328[265]
Den Tandt, C., 328[278]
Derenbach, M., 93[50]
Derge, G., 670[129]
De Vries, R. S., 642[3]
Dewsnap, P., 511[177]
Diener, A., 511[228]
Dietrich, W., 644[72, 73, 76, 82], 819[13, 31]
Dilling, E. D., 643[43]
Dittmar, C. B., 645[117]
Ditzenberger, J. A., 328[256]
Djakonov, V. I., 509[104]
Dmitriev, V. N., 331[391]
Doane, D. V., 332[409]
Dodds, H. W., 719[43]
Dode, M., 322[33]
Doeling, R. F., 327[221]
Doerschuk, V. C., 508[29]
Domalski, H. H., 326[203], 508[67]
Donlevy, A., 645[111]
Donovan, M., 719[45]
Donovan, P. D., 669[76]
Donovan, R. F., 800[12]
Dorn, J. E., 647[196]
Dorsett, L. S., 643[48]
Dorsi, D., 333[444, 483], 646[163]
Douglas, R. W., 334[478]
Do'zhenkov, F. E., 802[96]
Dragomir, I., 513[309]
Drangel, I., 334[469]
Dressler, G., 335[493, 494]
Drevermann, A., 511[215]
Drews, T., 645[128]
Driole, J., 334[489]
Drowart, J., 515[385, 386]
Drummeter, L. F., Jr., 820[35]
Duckworth, W. E., 642[28]
Dürrschnabel, W., 514[348, 349, 376]
Duesbery, M. S., 647[215]
Duflot, J., 509[117]
Dugain, F., 670[101]
Dunlop, A., 643[44, 46]
Dushman, S., 92[15], 143[1], 858[10]
D'yachkov, P. N., 509[93]
D'yachkova, S. S., 509[93]

D'yakov, I. G., 334[468, 481, 486], 646[149]
Dyble, E., 510[168]
Dyrkacz, W. W., 642[3]

Easha, J., 670[90]
Eaton, N. F., 332[424]
Eberl, E., 321[4]
Eckart, F., 330[317, 319]
Eckstein, H., 327[217]
Eda, S., 324[115]
Edelman, E. R., 643[45]
Edmond, J. T., 328[258]
Edström, J. O., 512[280]
Edwards, A. R., 332[402]
Edwards, K. L., 329[292], 334[488]
Edwards, R., 719[22], 819[29]
Effenberger, E., 322[35, 42], 719[29b]
Effer, D., 329[310], 335[499]
Efremkin, V. V., 326[179], 328[255]
Egorov, S. M., 330[327]
Eichberger, R. L., 328[275a]
Eichhorn, F., 801[38]
Eichner, A., 509[85]
Ehrlich, G., 668[18], 669[79], 859[30]
Eketorp, S., 509[78], 513[295]
El Kareh, A., 800[34]
Elkins, D. A., 326[176]
Ellinger, J., 514[338]
Elliott, J. F., 92[5, 9], 93[33, 34], 321[6], 514[363]
Ellis, T. G., 331[377]
Emeis, R., 646[158]
Emery, J. F., 644[106]
Eminger, Z., 509[99], 512[268]
Emley, E. F., 325[160]
Ende, H. vom, 509[96, 105]
Endo, K., 671[148]
Engelbrecht, H., 93[57]
Enright, D. P., 328[257]
Epifanov, V. G., 646[156]
Erben, E., 331[354], 644[72]
Ershova, S. a., 327[243, 245]
Espe, W., 514[356], 819[7]
Esser, F., 644[105]
Esyutin, V. S., 321[14], 327[238, 239], 328[248], 330[315]
Etterich, O., 92[23]
Ettrich, O., 671[150]
Evans, E. Ll., 92[3], 321[8]
Evans, E. U., 514[362]
Evans, J. L., 669[76]
Evans, P. R. V., 860[84]
Evens, F. M., 669[75]
Everett, M. R., 670[102]
Exner, E., 718[8]

Fagel, J. E., 668[15], 670[123]
Fairgrieve, D. S., 331[353]

Falke, W. L., 330[314]
Farr, J. P. G., 646[185]
Farrell, K., 647[208]
Fassel, V. A., 669[75]
Fast, J. D., 514[360]
Fatkullin, O. C., 513[325]
Fattore, V. G., 670[104]
Faure, C., 326[172]
Faust, W. A., 643[34]
Fayer, E., 858[4]
Fedock, M. P., 92[22], 510[142]
Fedorov, T. F., 325[136], 328[271]
Ferriss, D. P., 333[448]
Feuerstein, S., 647[194]
Fiala, A., 512[265]
Fiedler, H., 644[92, 105], 645[132, 133]
Field, A. L., 860[64]
Finkel', V., 334[473]
Finkl, C. W., 509[94, 95]
Finlay, W. L., 511[198, 199, 208]
Fischer, W., 668[7]
Fischer, W.-A., 93[50, 53, 57, 58, 60], 508[60, 66]
Fischhuber, J., 645[136]
Fischmeister, H., 718[8]
Fisher, A. Ya., 327[229], 330[326, 327]
Fisher, C., 646[181]
Fishlock, D. J., 820[42]
Fitzgerald, E. J., 511[205]
Fleischer, H.-J., 92[23]
Fleischer, J. F., 670[124]
Flemings, M. C., 514[334]
Flux, J. H., 513[288]
Fölkl, O., 513[310]
Förster, H., 645[130, 131]
Fogleman, E. L., 511[224]
Forsblom, G. V., 324[111]
Forster, G. B., 512[233]
Fortner, J. W., 331[353]
Foster, L. M., 329[308], 335[515]
Fowler, R. M., 670[121]
Foxall, R. A., 647[215]
Frade, G., 330[329]
Frade, M. G., 330[328]
Franco-Fereira, E., 800[15]
Franklin, R. H., 668[32]
Freedman, A. H., 801[72]
Frerichs, R., 820[49]
Friedline, J. E., 668[37]
Friedrich, K., 668[1, 10, 18], 669[16, 53, 79]
Friedrichs, H. A., 327[206, 210]
Frohberg, M. G., 668[31], 671[180]
Fromm, E., 322[52, 53, 55, 56], 323[65], 331[358, 359, 360, 361, 365, 366], 514[352, 353, 361, 365, 366, 367, 368, 369, 370, 371, 372], 515[381], 670[139]
Fuchs, A., 669[48]
Fulton, J. C., 643[47]

Funk, E., 800[36]
Funk, E. R., 802[83]
Furuya, K., 671[181]

Galaktionova, N. A., 514[357]
Gane, N., 859[42]
Garber, H. J., 93[38]
Garnik, G. A., 509[88]
Gaumer, R. E., 858[8]
Geach, G. A., 646[153], 718[2]
Gebhardt, E., 322[56], 331[359, 360, 364, 365], 514[346, 347, 348, 349, 352, 353, 355, 361, 365, 370, 375, 376], 515[380, 381], 860[59]
Gedeon, T. G., 325[168]
Geil, W., 334[480, 482, 485]
Geiselman, D., 328[264]
Gel'd, P. V., 322[49, 50, 54], 323[59]
Gelles, S. H., 329[291]
Gentry, W. O., 861[85]
Gerasimov, Ya. I., 323[71]
Gerhardt, A., 668[2, 31], 669[46], 671[180]
Gerken, J. M., 801[43]
Gerlach, J., 332[406, 408]
Ghodsi, M., 321[19, 21, 22, 23], 322[26]
Gilbert, A., 647[199, 210], 860[67]
Gilbert, A. C., 860[71]
Gilbreath, W. P., 859[13, 31]
Gill, E. W., 671[182]
Gilliland, R. G., 801[71]
Gilman, J. J., 645[139], 859[44]
Ginsberg, H., 329[302]
Givord, J. P., 331[355]
Gjermundsen, K., 672[184]
Glavin, G. G., 670[106], 671[142], 672[183]
Gleeson, F. V., 669[60]
Gleiser, M., 92[5, 17], 321[6], 330[344], 514[363]
Gmelin-Durrer, 92[7]
Gnuchev, S. M., 671[178]
Goering, H. L., 328[259], 334[475], 335[497]
Goetzel, C. G., 718[15]
Gokcen, G. A., 668[9]
Gokcen, N. A., 668[36]
Golay, M. J., 646[157]
Goldfinger, P. 515[386]
Goodwin, H. B., 332[399]
Gordon, E., 719[41]
Gordon, P., 719[42]
Gordon, R. H., 508[25]
Goss, A. J., 334[479]
Gosse, G., 323[72, 73], 324[117, 118, 119, 120]
Goto, H., 670[120], 671[179]
Gottwald, H.-J., 331[371, 375]
Gould, G. C., 643[60]
Goward, G. W., 668[28]
Grabke, H. J., 744[1]
Graham, S. W., 642[18]

Granitzki, K., 509[76]
Grant, N. J., 860[78]
Gravenhorst, H., 513[305]
Gray, P. M., 642[14]
Green, J. I. T., 508[65]
Greenwood, J. A., 859[27, 33]
Gregg, J. L., 647[225]
Gregory, J. N., 670[95, 115]
Greiner, E. S., 647[204]
Gribov, A. I., 331[391]
Griffith, C. B., 92[22], 510[142], 670[113], 671[155, 163]
Grimes, W. M., 646[184]
Grinyuk, V. N., 329[296], 334[486]
Grjotheim, K., 335[514]
Grochol, J., 669[71]
Gromov, B. I., 328[274]
Gross, P., 329[304], 335[519]
Groves, M. T., 801[43]
Gruber, G., 644[73]
Gruber, H., 643[53], 644[71, 72], 669[50]
Günther, K. G., 335[508]
Guerin, F., 330[328]
Guernsey, D. L., 668[32]
Guiu, F., 647[192, 195]
Guldner, W. G., 670[133]
Gulyanitsky, B. S., 324[101, 102]
Gulyayeva, E. I., 330[320]
Gurevich, Ya. B., 802[84, 87]
Gurry, R. W., 92[12]
Gurtner, F. B., 801[76]
Guterman, K. D., 644[95]
Gvelesiani, G. G., 324[126, 127, 129, 130], 325[133], 326[189]

Haase, O., 819[19]
Habermann, C. E., 324[125], 328[263, 266]
Hablanian, M. H., 644[80, 81], 800[14]
Hackspill, L., 324[106]
Haddad, G. I., 800[35]
Haefling, J. A., 328[269], 335[501]
Häuser, M., 321[2]
Hahn, G. T., 647[210], 860[67, 71]
Halden, F. A., 646[161]
Hale, L., 819[20]
Hall, R., 642[25]
Hall, R. W., 861[94, 95, 96]
Ham, J. L., 859[49]
Hambleton, R. C., 93[55]
Hamjian, H. J., 643[30]
Hammarlund, E., 510[129]
Hampel, C. A., 323[68], 719[28]
Hanak, J. J., 324[128]
Hanawalt, J. D., 325[163]
Hanin, M., 671[172]
Hanks, C. W., 643[69], 644[104]
Hanks, G. S., 802[80]

Hans, A., 668[40]
Hansen, M., 514[359]
Hansen, W. R., 668[13], 670[125], 671[160]
Harada, R. H., 335[509]
Harders, F., 508[61], 511[179, 180, 190, 212]
Harman, T. C., 335[500]
Harris, G. T., 642[4], 643[32]
Harris, W. F., 669[81], 670[90], 671[146]
Harris, W. H., 670[136]
Harvey, C. A., 670[107]
Harwood, J. J., 331[383]
Hashimoto, T., 800[10, 17]
Hashimoto, Y., 323[69]
Hass, G., 820[34, 35]
Hauff, A., 819[13, 31]
Hauser, J. J., 820[50]
Hay, D. R., 334[465, 474], 647[206]
Hayashi, C., 325[164], 643[70]
Hayes, D. M., 508[65]
Hazelby, D., 335[512]
Heavens, O. S., 819[32]
Hebble, H. B., Jr., 820[39]
Heimerl, J., 644[72]
Heimgartner, R., 335[518]
Heimonen, L., 669[70], 670[92]
Heinkel, O., 323[65], 514[368]
Heise, M., 335[516]
Heiss, A., 322[33]
Heitmann, G., 328[253]
Helfand, E., 332[420]
Hellbrügge, H., 513[286]
Henke, G., 512[266]
Hennemann, K., 645[113]
Hentrich, R., 511[204], 513[282], 645[122, 123, 124, 127, 136]
Heppell, T. A., 143[4]
Heraeus, W. C., 514[364]
Herbert, W. M. P., 514[335]
Herkart, P. G., 328[260]
Herman, M., 329[295], 646[165, 175]
Herold, R., 331[355]
Herrmann, E., 329[300, 306]
Herstad, O., 335[514]
Hess, W., 512[266]
Hess, W. T., 329[293]
Hetherington, J., 800[26]
Heumann, F. K., 670[131]
Hewitt, J., 510[162]
Hickam, W. M., 669[81]
Hicken, K., 801[44]
Hierdieckerhoff, W., 744[23, 24, 25, 27]
Higbie, R., 93[26]
Hill, C. C., 669[75]
Hill, N. A., 332[395]
Hillmann, H., 332[417]
Himes, R. C., 334[475]

Hinnüber, J., 719[49]
Hiramaya, A., 670[110]
Hirase, M., 512[242]
Hirata, R., 512[235]
Hirsch, P. B., 647[215]
Hirschwald, W., 322[28], 329[305]
Hirth, J. P., 647[223]
Hiruta, J., 643[70]
Hitt, E. S., 508[26]
Hoar, T. P., 744[5]
Hobson, J. D., 510[160]
Hobson, J. P., 143[11]
Hoene, J. V., 670[132]
Hörbe, R., 321[7]
Hörz, G., 93[45], 331[362, 363, 364, 369], 514[348, 373, 375, 376, 377], 515[380, 382, 383, 384]
Höy-Petersen, N., 325[162]
Hoffmann, A., 93[58], 508[66]
Hoffmann, V., 820[56]
Hogarth, C. A., 335[511]
Holden, F. C., 334[478], 860[77]
Holland, L., 819[2], 820[37]
Hollwitz, R., 324[113]
Holm, R., 859[39]
Holst, R., 324[112]
Holtkamp, J., 332[405]
Hook, R. E., 647[223]
Hooper, E. W., 329[290]
Hopkins, B. E., 744[2]
Hoppin, G. S., III, 802[81]
Hordon, M. J., 329[293], 859[46], 861[110, 111]
Horn, F. J., 820[39]
Hornak, J. H., 513[292]
Hornak, J. N., 510[146]
Horton, W. S., 671[153]
Hosada, K., 670[126]
Hotop, W., 719[27]
Houis, J., 509[125]
Hoyle, G., 511[177], 512[246, 247, 250]
Hsu Chia-Lung, 508[62]
Huang, Y., 860[61]
Hüttig, G. F., 718[17]
Huffine, C. L., 335[503]
Hughes, W. T., 325[160]
Hull, D., 647[200]
Hulme, K. F., 329[309], 333[445]
Hultgren, R., 92[2], 643[67]
Hum, J. Y. K., 645[111]
Humbert, I., 93[47]
Humbert, J., 326[200], 719[24]
Hundrieser, A., 668[42]
Hunt, C. d'A., 643[69], 644[79, 104, 111], 819[29, 30]
Hunter, C. J., 513[292]
Hunter, C. W., 92[21]
Hurwitt, S., 333[449]
Husche, E. G., 802[81]

Hutchins, J., 861[103]
Hutchinson, F., 335[504]
Hyams, W. M., 511[202]

Ichinose, Y., 642[20]
Iden, R. B., 669[78]
Ikeda, Ta., 511[189]
Ikeda, T., 511[197]
Ikemi, T., 510[172]
Ilschner, B., 93[47], 326[200], 719[24]
Ingram, A. G., 860[74]
Inouye, H., 860[53, 54], 861[100]
Isabaev, S. M., 323[96], 324[97], 325[144]
Isakova, R. A., 321[12, 15, 16, 17, 18], 327[241], 330[315]
Isbell, W. T., 327[227]
Ivanov, V. E., 326[205], 329[287, 288, 294, 296]
Ivanova, R. V., 671[142]
Ivanovsky, G. F., 334[464]
Izhvanov, L. A., 324[124]
Izmanova, T. A., 669[83], 671[140, 164, 173]

Jacisin, J. M., 861[109]
Jackson, G. F. W., 510[130]
Jackson, G. W. S., 510[134]
Jackson, P. L., 511[202]
Jackson, W. J., 514[344]
Jacobus, G. F., 819[12]
Jaekel, G., 325[161]
Jaffee, R. I., 322[51], 332[399], 333[461], 334[478], 645[108], 860[67, 68, 70, 77]
Jakob, D., 322[56], 331[359], 514[352, 365], 515[381]
Jakoby, L., 325[167]
James, H. A., 801[46]
Jangg, G., 719[55]
Jansen, F., 508[21]
Jaudon, E., 671[172]
Jauer, H., 327[206, 210]
Jayne, G., 335[498]
Jehn, H., 322[55], 331[361], 514[371, 372], 670[139]
Jenkins, N., 671[171]
Jensen, R. V., 329[307]
Johnson, K. I., 859[15, 24]
Johnston, W. C., 646[150]
Joly, M. F., 322[37], 331[379]
Jones, F. O., 646[153]
Jost, W., 514[378], 744[6]
Juntz, R. S., 327[208, 209], 819[3], 820[44]

Kadarmetov, K. N., 330[334]
Kämpf, F., 321[3]
Kajikawa, K., 513[329]
Kallmann, S., 668[21]
Kamada, H., 671[181]
Karasaev, R. A., 322[36]
Karasev, R. A., 509[112]

Karpenko, G. V., 514[358]
Karpov, Y. A., 670[106], 671[142]
Karsanov, G. V., 323[80]
Kashin, V. I., 93[61], 322[36]
Kashu, C., 643[70]
Kashu, S., 325[164]
Katayama, N., 671[148]
Kato, H., 645[116]
Kato, R., 511[188, 194]
Kato, T., 514[340]
Kaufmann, A. R., 329[291], 719[41, 42]
Kaun, L., 647[191]
Kawamoto, K., 511[191]
Kech, P. H., 646[157]
Keen, N. J., 329[290]
Keh, A. S., 647[217]
Keil, H., 514[355]
Keller, D. V., 858[11]
Keller, D. V., Jr., 859[15, 22, 24, 25, 26, 28]
Kelley, J. E., 802[102]
Kelley, K. K., 92[2]
Kellogg, H. H., 92[16], 321[9]
Kelly, H. J., 802[102]
Kennedy, K., 819[28]
Kent, P. J. C., 335[519]
Kerfoot, H. P., 858[8]
Kermes, J., 514[338]
Kesler, G. H., 331[389]
Khalitov, R. Sh., 668[33], 670[122]
Khvedchuk, I. R., 646[149]
Kidyarov, B. I., 327[226]
Kieffer, R., 322[35, 39, 43], 323[77, 88], 718[6, 10, 11, 12], 719[26, 27, 29a, b, 33, 34, 40, 55, 56, 62, 63]
Kikuchi, T., 335[520]
Kim, G. V., 327[239]
Kim, G. W., 322[25]
Kim, H. C., 647[195]
Kimura, H., 323[57, 58], 331[368]
Kimura, J., 671[179]
King, J. A., 325[159]
King, J. R., 801[41]
King, W. H., 802[99, 100]
Kingston, W. E., 718[14]
Kinna, W., 719[49]
Kinsky, F., 509[99], 512[268]
Kirby, R. S., 802[80]
Kirshenbaum, N. W., 331[381]
Kishinevsky, M. C., 93[28]
Kiyote, U., 511[197]
Klein, A. A., 819[6]
Klemperer, O., 801[57]
Kletecka, Z., 509[99]
Klimsa, A., 669[71]
Klisiewicz, Z., 513[312, 313]
Klofáč, J., 332[422]
Klofach, I. I., 327[245]

Klopp, W. D., 322[51], 332[399], 860[77]
Klushin, D. N., 321[10], 330[323, 325, 326, 327]
Klyachko, Y. A., 668[26], 669[83], 670[117, 127], 671[140, 151, 164, 168, 177]
Knacke, O., 321[7], 322[28, 31], 326[199], 327[206, 210], 328[253], 329[305]
Knaggs, K., 510[156]
Kneip, G. D., Jr., 333[436]
Knighton, J. B., 324[110]
Knüppel, H., 508[61, 71], 509[119], 511[212, 215, 270], 513[318]
Kobrin, C. L., 510[127]
Koch, H., 801[38]
Koch, K. H., 668[41]
Koebel, N. K., 744[13]
Koehl, G. G., 669[78]
Köhler, K., 645[130]
Koepel, E. J., 646[167]
Köthemann, K. H., 93[60]
Kohler, D. F., 669[78]
Kolchin, P. O., 322[45, 46], 323[60, 61, 67]
Kolchin, O. P., 333[441]
Kollmann, W. C., 511[207]
Koloyartsev, V. L., 330[337]
Komarek, K., 329[286]
Komarek, P., 669[71]
Komarova, T. N., 321[11]
Komlev, G. A., 328[249]
Kondo, Y., 323[83]
Konstantinov, K. M., 670[116]
Koo, R. C., 333[454]
Kopa, L., 670[118]
Kopetsky, Ch. V., 334[471]
Kopineck, H. J., 512[267]
Kopytov, S. A., 328[249]
Kornegay, R. L., 332[420]
Kornelsen, E. V., 143[11]
Kosakevitch, P., 93[32, 35]
Kosani, M., 643[70]
Kosolapova, T. Ya., 323[70, 81]
Kovalev, Yu. T., 327[246]
Kovaleva, E. A., 802[94]
Kovarik, M., 669[62]
Koves, G., 744[21]
Kovtun, G. P., 334[468, 472, 473]
Kovtun, S. F., 800[91]
Kowal, R. F., 511[185]
Kozhevnikov, G. N., 323[93, 94, 95, 98, 107, 108], 326[179, 188]
Kozin, L. F., 328[251]
Kragelsky, I. V., 859[34]
Krahe, J., 322[31]
Krall, F., 744[7]
Kramer, I., 861[86]
Kramer, I. R., 861[87, 88, 89, 90, 104, 107]
Kramer, R. A., 329[308]

Kraus, Ch. J., 820[36]
Kraus, T., 93[29, 37, 46, 51], 326[190, 191], 512[244], 644[101], 668[22, 23, 24, 25, 30, 31, 43, 45], 669[47], 671[180]
Krebs, T. M., 643[48]
Krestovnikov, A. N., 323[71]
Kribbe, H., 325[158]
Krichevets, R. B., 330[334]
Krier, C. A., 820[40]
Kripyakevich, R. I., 514[358]
Krivonosov, Yu., 802[96]
Krone, K., 92[14], 331[371]
Krüger, J., 92[14], 93[44], 322[27, 40, 41], 331[371, 372]
Kruglykh, A. A., 327[244], 329[288, 313], 334[468, 472, 473], 335[505]
Krulla, R., 508[39]
Krumme, J.-P., 332[408]
Krupp, W. E., 328[270], 331[376]
Krutenat, R. C., 820[51]
Kubaschewski, O., 92[3], 321[8], 514[362], 670[107], 744[2]
Kuchar, L., 332[418]
Kuczynski, G. C., 718[9]
Küntscher, W., 509[98], 645[129]
Kunin, L. L., 668[3, 27, 29], 670[117], 671[151, 177]
Kunze, C., 328[275]
Kunzler, J. E., 333[435, 457]
Kurilo, Yu. P., 326[205]
Kurockin, K. T., 509[124]
Kurosawa, T., 324[115], 335[520]
Kurz, W., 509[114]
Kurzoja, J., 645[125]
Kuslitsky, A. B., 514[339]
Kutscher, H., 511[203, 217, 218]
Kuznetsova, V. P., 330[331]
Kvas, O. F., 323[76]

Lacombe, M. P., 330[328, 329], 335[498]
Lafferty, J., 143[1]
Laidler, K. J., 514[374]
Lake, E. F., 508[43]
Lake, G. W., 643[34]
Lakin, J. R., 508[73]
Lange, K. W., 93[42]
Langeron, J. P., 327[214, 215], 330[347, 350], 333[437, 439]
Lange von Stockmeier, H. G., 668[14]
Langmuir, I., 93[48]
Lapujoulade, J., 801[68]
Larner, D., 507[13]
Lassner, E., 668[4, 10, 19, 20], 669[53, 66, 68], 671[141, 161]
Lawley, A., 331[380], 332[415], 333[450, 451], 645[144], 646[179, 182, 189], 647[201], 860[69]
Lawrence, F. M., 820[51]
Lawthers, P. P., 860[72]

Lau, S. S., 647[196]
Leach, J. C. C., 511[226], 513[300, 321, 331]
Leadbetter, M. J., 333[447], 671[145]
Lebedev, V. V., 329[312]
Leck, J. H., 143[6]
Leferrer, V. F., 327[222]
Lehr, P., 323[72, 73], 324[117, 118, 119, 120], 330[348], 331[351]
Leinbach, R. C., 643[30]
Leitten, C. F., 644[75]
Le Moyne, A., 509[125]
Lempert, J., 801[49, 50]
Lench, A., 669[80]
Lengyel, A., 858[7]
Lenk, P., 644[90, 92]
Leont'yev, G. A., 332[410, 411]
Lerch, H., 801[59]
Lesnik, A. G., 646[156]
Lesser, R., 644[73], 669[50], 670[138], 671[144]
Levaux, J., 512[252]
Lever, F. M., 324[121]
Lever, R. F., 333[433], 646[176]
Levi, D. L., 335[519]
Levin, E. S., 509[108, 123, 124]
Levine, E. D., 329[291]
Levinson, D., 645[109]
Lewicki, A., 820[38]
Lewis, A. I., 860[65]
Lewis, K. G., 512[278]
Lihl, F., 322[35], 719[29b]
Lilburne, M. T., 669[54]
Lillie, D. W., 719[42]
Lincevsky, B. V., 509[110], 642[24]
Lindvall, B., 327[231]
Litvinova, N. F., 669[85, 87], 670[94, 100, 119]
Liu, R., 668[21]
Lochiel, A. C., 93[40]
Lodi, U., 671[169]
Löffler, J. J., 719[50]
Loesch, H. W., Jr., 647[197]
Logan, C. E., 508[32]
Lohwater, K., 327[217]
Lomis, B. A., 860[66]
London, G. J., 329[295], 646[165]
Longo, H.-E., 334[463]
Loonam, A. C., 331[388]
Loria, E. A., 860[61]
Losikova, M. A., 331[391]
Louat, N., 861[106]
Loukashenko, E. E., 326[198]
Lowry, J. F., 801[49]
Luft, A., 647[191]
Lukashenko, E. E., 330[331]
Lunn, B., 508[49]
Lupton, J. M., 511[219]
Lutz, H., 322[39], 719[29a, 40]

Lux, B., 509[114]
Lyubimov, V. D., 322[54]

Maas, H., 512[231, 234]
Macedo, J. W., 802[103]
Machlin, E. S., 93[30], 326[192]
Mackie, A. J., 668[17]
Maclean, I. K., 511[219, 229]
MacNicholas, T., 859[26]
Madden, R. P., 819[12]
Maddin, R., 333[451], 860[69]
Maekawa, S., 513[304]
Mager, A., 326[202], 719[23]
Maissel, L. J., 820[47]
Makarenko, G. N., 323[76]
Makunin, M. S., 322[36, 38]
Mal'cev, V. S., 325[144]
Malé, G., 328[262]
Malkin, Ya. Z., 327[223]
Mallett, M. W., 668[6, 13], 669[77, 78], 670[113, 125, 130], 671[154, 155, 160, 163]
Malloy, D. F., Sr., 744[8]
Maloof, S. R., 334[487], 647[205]
Mal'tsev, V. S., 323[96], 324[97]
Malyshev, V. I., 669[85]
Mamro, K., 513[316]
Mantey, P. G., 509[80]
Mapper, D., 670[95, 115]
Marchal, J., 326[172]
Marcus, H. L., 819[27]
Maré, L., 508[31]
Marfone, P. A., 858[7]
Markaryants, A. A., 512[254]
Martin, A. J., 329[292], 334[488]
Martin, E., 335[517]
Martin, G. S., 669[80]
Martin, J. F., 668[37]
Martin, J. W., 860[62, 63], 861[92]
Martin, R. L., 331[373]
Masse, F. X., 801[60]
Matewossjan, P. P., 513[317]
Matsuda, K., 511[187, 225], 512[241], 514[340], 800[10, 17]
Matsuoka, H., 511[188, 191, 194]
Matting, A., 801[38]
Mattox, D. M., 819[25]
Matuschkowitz, R. K., 509[86]
May, E., 507[6, 10]
May, J. W., 859[29]
Mayer, H., 819[24]
Mayer, V., 669[63]
Maykuth, D., 645[108]
Maykuth, D. J., 332[399], 333[461], 860[68, 70]
Mazanek, T., 513[316]
McClanahan, E. D., 820[53]
McCoy, H. E., 671[143]

McCrary, L. E., 819[6]
McCreary, W. J., 328[277]
McDonald, R. E., 644[75, 106]
McDonald, R. S., 670[123]
McGeary, R. K., 671[162]
McIntire, H. O., 509[84]
McIntyre, D. V., 859[20]
McKee, K. H., 719[50]
McKisson, R. L., 328[275a]
McLafferty, J. J., 322[44]
McManus, G. J., 511[213], 512[263]
McMillan, A. G., 510[161]
Meadowcroft, T. F., 513[308]
Meadowcroft, T. R., 93[34], 508[69]
Mechenov, P. I., 321[13]
Meckelburg, E., 324[131]
Meerson, G. A., 323[74]
Meier, J., 801[45]
Melchior, G., 92[24], 515[388]
Melin, A., 322[27]
Melnick, L. M., 668[37]
Menkov, A. A., 328[267]
Merrill, T. W., 323[75]
Meyer, L., 645[129]
Meyer, W. E., 800[9]
Meyerhofer, D., 329[307]
Meyers, R. G., 819[26]
Mgaloblishvili, N. P., 326[189]
Michael, A. B., 861[85]
Michaelson, H. B., 718[7]
Michalke, M., 719[60]
Michler, H. W., 800[22]
Miholich, N., 513[303]
Mikhailova, G. V., 668[33]
Mikulinsky, A. S., 323[95], 324[98, 107, 108], 325[148, 149], 326[179], 328[255]
Mikus, E. B., 801[72]
Miller, E., 329[286]
Miller, G. L., 719[32, 38, 39]
Miller, K. J., 800[24]
Miller, O. G., 322[25]
Mills, D., 333[440]
Milne, G. J. C., 333[446]
Milne, I., 647[216]
Milner, D. R., 801[67], 802[97], 859[18]
Mimura, H., 646[166]
Minenko, A., 670[114]
Mischler, H., 645[131]
Mitchell, A., 509[113]
Mitchell, T. E., 647[202, 203]
Mitchell, W. R., 334[487]
Moates, G. H., 646[152]
Mocek, J., 669[59], 670[93]
Möbius, H. E., 744[7]
Moffatt, W. G., 801[73]
Monroe, R. E., 800[22]

Moolenaar, R. J., 325[134]
Moore, J. H., 93[56]
Morgan, R. P., 819[26]
Morgan, S. W. K., 327[233]
Mori, H., 511[220]
Morikawa, H., 510[172, 174]
Morimoto, Y., 670[110, 111]
Morlevat, J. P., 323[87]
Morley, J., 800[32, 33]
Moroz, L. S., 514[351]
Moschel, W., 326[175]
Motloch, Z., 510[159], 512[269]
Mott, N. F., 744[3]
Movchan, B., 644[97, 98, 100]
Muelemans, M., 800[11]
Müller, D., 331[367], 333[453]
Müller, F., 322[31]
Müller, L., 326[196], 327[212], 328[254]
Mukherjee, A. K., 647[196], 860[62, 63], 861[92]
Mukhin, V. V., 802[94]
Mullendore, J. A., 334[487]
Mullin, J. B., 329[309], 333[445]
Mund, A., 513[281]
Muramatsu, K., 671[158]
Murray, G. T., 334[469]
Mussler, R. E., 326[178]
Myakshin, O. A., 802[96]
Mydlarz, I., 512[259]
Myers, H. C., 513[319]

Nadinskaya, O. V., 321[10]
Nadiradze, A. A., 324[130]
Naeser, G., 323[82], 719[59]
Naimanov, S., 321[12]
Nakada, Y., 646[147], 647[217]
Nakagawa, Y., 513[304]
Nakayama, N., 512[241]
Narita, K., 513[297, 298]
Natter, B., 719[61]
Nazarenko, O. K., 800[16]
Necker, W. C., 335[502], 646[170]
Nehrenberg, A. E., 512[271]
Nepper, M., 512[252]
Nesterov, V. N., 321[12, 15, 16, 17], 327[241], 330[315]
Neubauer, E. P., 323[90]
Nichols, I. L., 325[151]
Niebuhr, J., 322[47], 323[63], 669[48], 671[159]
Nijland, L. M., 330[318]
Nisbet, J. D., 642[7]
Nish, J. I., 332[402]
Nishimura, S., 323[83]
Noda, T., 324[109]
Noesen, S. J., 643[49, 50, 54]
Norton, J. T., 643[45], 719[48]
Novik, L. M., 509[89, 90, 102]
Novotny, Z., 669[57, 65]

Oba, K., 511[187, 225]
Oberthin, H., 668[21]
Obrebski, J., 512[260]
Occhiminuti, M., 670[101]
Ocepek, D., 321[4]
Ockenhouse, G. H., 513[315]
Oda, N., 671[148]
Odgers, M., 671[167]
Odoevsky, L. S., 323[80]
O'Driscoll, W. G., 719[39]
Oeters, F., 511[190, 196, 217, 218], 512[270]
Ogiermann, G., 645[112, 124]
Ogneva, E. Ya., 321[11]
Ohji, M., 93[42]
Oiks, G. N., 509[92, 101], 513[317, 324], 514[341]
Okorokov, G. N., 513[311], 643[36]
Oldfield, G. E., 643[41, 57]
Olds, L. E., 671[157], 860[76]
Olette, M., 93[52], 327[213, 216]
Oliver, B. F., 646[173]
Olshansky, N. A., 800[6, 7], 801[40]
Olson, R., 819[20]
Onaev, I. A., 321[14]
Onillon, M., 93[49], 327[216]
Onuma, A., 670[120], 671[179]
Ooka, T., 646[166]
O'Railly, R. F., 513[322]
Orehoski, M. A., 510[146], 513[302]
Orman, M., 325[142], 326[181]
Orr, R. L., 92[2]
Osias, J. R., 859[45]
Ott, F., 326[182]
Ott, J. B., 328[275a]
Owczarski, W. A., 802[99, 100]
Owen, C. V., 331[387]
Oziashvili, D. Sh., 325[140, 141, 145]

Paesold, G., 668[12]
Pakkala, M. H., 643[67]
Palmer, P. E., 324[125], 328[266]
Panushkin, V. T., 323[96], 324[97]
Papamantellos, D., 93[42]
Pape, L., 668[8]
Papirov, I. I., 329[296], 334[486]
Papirov, L. I., 327[244]
Park, A. J., 511[185]
Park, B., 646[167]
Parker, A., 669[88], 670[89, 96, 98, 99]
Parker, D. H., 671[171]
Parmee, J. L., 335[512]
Parr, N. L., 332[414], 645[141]
Pastor, H., 334[489]
Pasztor, L. C., 671[174]
Paton, B. E., 644[98]
Pattoret, A., 515[385]
Paurise, J. C., 509[107]

Pavlov, I. M., 802[85, 86]
Pavlov, V. A., 334[472], 335[505]
Pawlek, F., 332[406, 408]
Payne, G., 508[73]
Payne, J. B., 324[121]
Pazukhin, V. A., 326[198]
Pease, F., 800[37]
Peckner, D., 645[119]
Peebles, F. N., 93[38]
Pehlke, R. D., 93[33]
Pekarev, A. I., 334[471]
Pelissier, G. E., 668[37]
Pelloux, R. M. V., 861[105]
Pemsler, J. P., 329[291]
Perelyaev, V. A., 323[66]
Perkins, R. A., 860[52]
Perry, T. E., 510[128, 131, 132], 513[289], 642[17]
Perry, T. W., 513[290]
Perryman, J. T., 644[102]
Persson, G., 643[51]
Peter, W., 643[61]
Peterson, D. T., 328[270], 670[104, 105]
Petri, R., 333[458]
Petrov, K. M., 509[104]
Pfann, W. G., 332[412], 334[479], 645[143]
Phillips, A. L., 801[63, 64, 65]
Piazza, J. R., 322[32]
Picardi, P., 512[256]
Picklesimer, M. L., 333[438]
Picklo, G. A., 671[156]
Pidgeon, L. M., 325[159]
Pierce, J. R., 801[55]
Pierret, J. A., 322[44]
Pirani, H. von, 643[64]
Pirani, M., 143[2]
Pitler, R. K., 642[3]
Plaisted, A. H., 819[10]
Planiol, R., 321[5]
Platzer, N., 324[106]
Platzer, R., 668[8]
Pleteneva, N. B., 330[316, 321]
Pletschun, G., 509[100]
Plott, R. A., 668[17]
Podlasech, S., 861[87, 89, 107]
Pöcze, L., 513[310]
Pogulev, B. A., 509[97]
Polyakov, A. Yu., 322[36, 38], 508[62], 642[24]
Ponomarenko, Yu. G., 330[334]
Ponomarev, V. D., 323[96], 324[94], 325[144]
Poplawsky, R. P., 646[160]
Pottgiesser, C. H., 511[221, 222]
Povedskaya, L. G., 327[245, 246]
Power, B. D., 143[5]
Pozdeev, N. P., 330[334]
Pratt, P. L., 647[192, 195]
Prawdzik, C. A., 819[21]

Prekel, H. L., 646[189], 647[190]
Presnetsova, N. V., 333[441]
Preston, J. K., 511[227]
Preusch, C. D., 511[198, 199, 206, 207, 208, 209]
Price, L. E., 744[5]
Probst, H., 332[406]
Proft, R., 324[112, 113]
Protheroe, H. T., 642[19]
Püschel, R., 669[74]
Pugh, J. W., 334[467]
Pulker, H., 819[17]

Radawich, J. F., 642[29]
Raffo, P. L., 647[202, 203]
Raine, T., 719[22]
Raisky, H., 512[264]
Ramaswami, B., 646[147]
Ramsbottom, M. A., 512[250]
Randak, A., 513[282], 645[123, 125]
Ranji, S., 647[196]
Ransley, C. E., 325[160]
Rapp, R. A., 860[56]
Rassaerts, H., 323[77]
Rassenfoss, J. A., 509[86]
Rayleigh, Lord, 93[31]
Read, R. E., 644[106]
Redhead, P. A., 143[11]
Reed, R. E., 646[187]
Reeve, H. T., 508[46]
Reich, H., 858[3]
Reichel, H. H., 512[238, 240]
Reichelt, W., 819[1, 13, 31]
Reid, C. N., 647[210], 860[71]
Reifman, M. B., 331[391]
Reinhard, H., 669[48]
Reininger, W. G., 801[61]
Reinitzer, P., 322[28]
Remy, A., 331[393]
Rengstorff, G. W. P., 671[157], 860[76]
Renucci, L., 323[72], 324[117], 330[347, 348, 349, 350]
Rexer, E., 331[382], 333[452, 453], 334[462]
Reynoldson, R. W., 744[9, 14]
Rhodin, T. N., 801[70]
Rhys, D. W., 334[477]
Rice, L., 647[194]
Richardson, E. D., 642[16]
Richardson, F. D., 93[41], 513[307]
Richardson, H., 643[35]
Richter, J., 647[191]
Richter, S., 328[254]
Ridge, H. M., 508[40]
Rieck, G. D., 647[207]
Riedel, G., 513[327]
Riseman, J., 820[36]
Rittenhouse, J. B., 860[50]
Ritter, E., 819[4, 17]

Ritter, F., 325[161]
Robens, A., 509[116]
Roberts, J. P., 718[1]
Roberts, R. W., 143[8], 858[9]
Robinson, N. W., 143[9]
Roblin, J. M., 819[11]
Roche, T. K., 861[99]
Rölling, H. E., 670[114]
Rohn, W., 508[45], 642[1, 2]
Rolsten, R. F., 331[385], 332[403, 404]
Ronjin, A. C., 644[94]
Rose, A., 514[336]
Rose, R. G., 860[82]
Rose, R. M., 333[443, 448], 646[186]
Rosebury, F., 801[54]
Rosenblatt, G. M., 92[18], 330[343]
Rosi, F. D., 328[260], 329[307]
Ross, A., 819[9]
Rothenbacher, R., 514[347]
Rowe, G. W., 802[97], 859[18, 36]
Roy, U., 514[370]
Rozanov, A. N., 328[274]
Rozian, J. W., 508[36]
Rüttiger, K., 508[72], 509[80], 511[196, 217, 228]
Rumberg, D., 645[134]
Rumyantsev, Yu. V., 321[11]
Runge, H., 325[158]
Russel, A. S., 335[515, 517]
Rutherford, J. L., 646[162, 175]
Rutkovski, V. B., 509[118]
Rutsch, W., 93[39]
Ryabchikov, L. N., 329[296]
Rybal'chenko, N. D., 802[94, 95]

Sachs, K., 671[167]
Sage, A. M., 509[77], 510[136]
Salmond, R. L. P., 510[161]
Samarin, A. M., 93[61], 322[36, 38], 323[84, 88, 89, 90, 91, 102, 109], 513[306], 642[24]
Samsonov, G. V., 324[116]
Samsonov, G. W., 323[81]
Samuelsson, P. 510[129]
Sandler, R. A., 324[111]
Sanford, J. E., 820[41]
Santalov, F. A., 328[250]
Santeler, D. J., 858[5, 7]
Sasaki, Y., 323[57, 58], 331[368]
Sauerbrey, G., 819[23]
Sauerwald, F., 718[18]
Savitsky, E. M., 334[470, 471] 672[183]
Savjalov, O. V., 671[142]
Sawyer, J. C., 861[93]
Sazhin, N. P., 322[45]
Scaife, D. E., 332[398]
Scala, E., 334[465, 474], 647[206]
Scalise, V., 512[255, 256, 257, 258]

Schadler, H. W., 333[459], 646[180, 183]
Schäfer, H., 331[386]
Schaefer, R., 646[147]
Schaeffer, W. H., Jr., 508[74]
Schaffhauser, A. C., 647[208]
Schaible, P. M., 820[47]
Scharf, G., 644[92, 105], 645[134]
Scharfenberg, R., 333[431]
Schaufler, G., 326[185]
Scheffels, W., 800[9]
Scheibe, W., 644[84], 645[112]
Scheidig, H., 643[38]
Schempp, E. G., 511[192, 211, 216]
Schenck, H., 93[42], 326[203], 508[67], 513[293]
Scherrer, E. D., 512[239]
Schildknecht, H., 332[413]
Schiller, S., 644[83, 89, 90, 91, 92, 93], 645[129]
Schissel, P. O., 515[387]
Schlät, F., 668[1]
Schlatter, R., 642[5, 9, 12]
Schlaubitz, K., 333[453], 334[463]
Schlechten, A. W., 327[220, 221]
Schlicta, P. J., 646[151]
Schmidt, F. A., 328[269, 270], 331[376, 377], 335[501]
Schmidt, F. F., 860[83]
Schmidt, M., 92[23]
Schmidt, P. H., 328[272]
Schneider, A., 325[157, 158]
Schöberl, A., 512[264]
Schoenberg, H., 644[92, 93]
Scholefield, H. H., 642[14]
Scholes, P. H., 671[182]
Scholz, W., 323[82], 719[60]
Scholz, W. G., 332[409]
Schreiner, H., 335[508], 718[5]
Schreyer, J. M., 335[506]
Schriempf, J. T., 334[476]
Schröder, W., 334[490]
Schulte, E., 512[267, 274]
Schulze, D., 647[191]
Schumacher, B., 801[47, 48]
Schumann-Horn, L., 326[202], 719[23]
Schwartz, H., 800[8, 18], 801[62, 74, 76, 78]
Schwarz, W., 669[56, 61], 671[170]
Schwarzfischer, P., 510[144, 173, 175, 176]
Schwarzkopf, P., 718[3]
Scott, H., 860[72]
Seagle, S. R., 331[373]
Seaman, F. D., 801[49]
Sedlatschek, K., 719[34, 62]
Seed, H., 669[86]
Seemann, J. M., 820[48, 54]
Seghezzi, F. D., 860[60]
Seghezzi, H. D., 331[360], 514[346, 349, 353, 355]
Seith, W., 718[13]
Sehgal, V. D., 509[113]

Seliger, H., 325[165], 328[279], 330[330]
Sell, H. G., 646[184]
Semchyschen, M., 331[383], 860[79]
Semenyuk, K. G., 670[94]
Sergeev, A. B., 330[334]
Sergeeva, V. M., 329[311]
Sergienko, V. Ya., 327[223, 228]
Severus-Laubenfeld, H., 671[175]
Sevin, R., 511[214]
Shahinian, P., 861[91, 101, 102]
Shakhov, A. S., 323[71]
Shalimov, A. G., 513[311]
Shamray, F. I., 325[136], 328[271]
Shapiro, Z. M., 331[392]
Shapoval, B. I., 329[312]
Sharp, J. D., 93[43], 510[155, 163], 512[272]
Shaw, B. J., 647[212]
Shchedrovitsky, Y. S., 330[333]
Shen, H., 861[89, 104, 107]
Shendyapin, A. S., 321[17]
Shengeliya, O. V., 325[138, 139, 140, 141, 153, 154, 155, 156]
Sherwood, E. M., 332[397]
Sherwood, P. J., 647[195]
Shesternin, P. S., 327[229], 330[326]
Shih, C. H., 327[220]
Shimasaki, K., 670[126]
Shindo, S., 670[126]
Shirley, W. S., 820[55]
Shokol, A. A., 328[251]
Shtrum, E. L., 329[311]
Shutt, P. K., Jr., 801[76]
Shveikin, G. P., 322[49, 50, 54], 323[59, 66]
Sibert, M. E., 322[48]
Sickbert, A., 510[144, 171, 173, 176]
Sieckmann, W., 511[193, 195, 200, 216]
Sigalov, Yu. M., 802[86]
Signora, M., 671[166]
Sikorski, M. E., 859[19]
Silbereisen, H., 719[54]
Silverio, A., 859[38]
Silverman, S. J., 646[174]
Simkovich, A., 92[10, 20], 642[5, 12, 22]
Simons, J. C., 644[107], 858[6]
Simpson, A. G., 719[32]
Simpson, W. S., 507[9]
Sims, C. T., 322[51]
Sinelnikov, K. D., 329[287, 294]
Sinel'nikova, V. S., 324[116]
Sinnot, M. J., 322[32]
Sipeiko, I. E., 325[148, 149]
Sivet, M., 510[166]
Skala, J., 513[332]
Skaupy, F., 719[36]
Skinner, L. C., 646[186]
Skogerboe, R. K., 334[465], 647[206]

Sloman, H. A., 670[91, 107], 671[165], 718[19]
Smallman, R. E., 647[216]
Smelyanski, M. J., 644[95]
Smirnov, J. D., 512[254]
Smirnov, M. P., 327[223, 228, 240], 330[321, 322]
Smirnov, V. S., 802[88, 90]
Smith, H. A., 668[15]
Smith, H. R., Jr., 330[345], 331[356], 643[56, 68, 69], 644[74, 79, 104], 645[111], 819[15, 16, 29, 30]
Smith, J. F., 328[276]
Smith, R. D., 513[328]
Smith, R. L., 646[162, 175]
Smithells, C. J., 514[354], 515[379]
Smols, S., 515[385]
Snape, E., 93[54]
Snow, K. S., 643[47]
Soden, R. R., 646[169]
Soeda, S., 646[166]
Soisson, D. J., 322[44]
Sokolov, B. G., 644[94]
Sokolov, G. A., 509[92, 101], 514[342]
Sokolsky, S., 858[8]
Solomon, J., 800[20]
Solomon, J. L., 801[46]
Sorby, W., 511[226], 513[331]
Sorkin, P. Ya., 509[103]
Sorokin, P. Ya., 330[336]
Soshuikova, L. A., 330[320]
Southern, L., 508[37]
Spaeth, G., 322[53], 514[369]
Spalvins, T., 858[12]
Spangler, G. E., 646[175]
Sparwald, V., 329[302]
Spedding, F. H., 324[123, 128], 328[265, 269], 335[501]
Speith, K. G., 509[96, 105, 122]
Spendlove, M. J., 326[201]
Speransky, W., 509[100]
Speransky, W. G., 509[87]
Sperner, F., 331[354], 513[294, 296], 642[8], 643[51, 63], 644[76], 668[41]
Sperner, H., 331[352]
Spindler, H., 332[396]
Spitzer, H., 643[61]
Spriet, B., 335[507]
Stahel, R., 671[175]
Stahl Johannessen, K., 335[514]
Stankiewicz, M., 512[260]
Stanley, J. K., 670[132], 671[162], 744[20]
Starovoitov, Yu. A., 514[339]
Starratt, F. W., 512[230]
Statsenko, V. I., 332[401]
Steeves, R. W., 820[43]
Steffens, H. D., 800[13]
Steiger, J., 671[175]
Steigerwald, E. A., 861[93]
Steigerwald, K. H., 800[4, 9]

Stein, D. F., 647[198]
Steinberg, M. A., 322[48]
Steinhauer, O., 509[122]
Steinherz, H. A., 143[3], 644[80]
Steinkamp, W. I., 800[3]
Steman, L. G., 801[53]
Stennenberg, R. K., 324[110]
Stepanov, Yu. D., 335[495]
Stephan, H., 644[72, 73, 76]
Stephan, K., 644[82]
Stephens, J. R., 647[209], 860[80]
Stephenson, R. L., 860[58], 861[97]
Stevens, H., 333[427], 646[188]
Stickler, H., 510[139], 513[301]
Stiegler, J. O., 647[208]
Still, J. E., 670[97]
Stirling, A., 321[1]
Stohr, J. A., 800[2]
Stoll, J. H., 510[147]
Stoop, J., 861[91]
Strain, R. V., 323[90, 91]
Stranski, I. N., 93[39], 326[199]
Strauss, A. J., 335[509]
Strelez, C. L., 324[101]
Strohmeier, T., 507[15]
Stroup, P. T., 329[299]
Struthers, J. D., 328[256]
Stuligross, R. J., 719[50]
Styblo, K., 669[58, 65]
Suarez-Acosta, R., 669[69], 671[176]
Suchi, M., 324[113]
Sullivan, T. A., 668[17]
Sully, A. H., 323[78]
Sumarokov, N. V., 322[45]
Sumarokova, N. V., 323[61]
Sumner, D. H., 802[102]
Sumsion, H. T., 859[13]
Sushkin, H. G., 644[96]
Suzuki, H. G., 647[220]
Suzuki, S., 670[120], 671[149, 179]
Sychrovsky, H., 642[23]
Sype, J. van den, 333[451], 860[69]
Szarowicz, T., 325[142]
Szöke, L., 510[169]
Szulyovszky, A., 325[168], 326[168], 330[324]

Tabatadze, L. S., 325[155, 156]
Tabor, D., 859[16, 32, 37, 38, 40]
Taebel, W. A., 860[72]
Taiz, A. J., 324[101]
Tajc, A. J., 326[177]
Takenaka, T. T., 800[24]
Takeuchi, S., 324[115]
Takeuchi, T., 647[219]
Taneda, Y., 647[218]
Tanenbaum, M., 334[479]

Tangen, T. A., 324[99]
Tanino, M., 647[220]
Tanizawa, K., 511[220]
Tankins, E. S., 668[9], 670[137]
Tao Tszu-Tsun, 334[470]
Taranenko, V. P. 801[58]
Tarasov, N. D., 802[91]
Tarby, S. K., 92[21]
Tarkhov, N. G., 327[223, 228]
Taussig, E., 507[4, 5, 7, 8]
Taxhet, H., 671[150]
Taylor, G., 512[261], 647[213, 214]
Taylor, K. C., 510[154]
Taylor, L. S., 642[10]
Taylor, R. E., 670[108]
Taziev, Zh. Sh., 327[239], 328[248]
Tedmon, C. S., Jr., 333[443]
Teghtsoonian, E., 335[492]
Temple, D. A., 327[233, 236]
Teske, K., 668[42]
Teworte, W., 328[252]
Tezuka, M., 323[69], 324[115]
Theuerer, H. C., 335[510], 646[159], 820[50]
Thiel, H., 645[129, 130, 131]
Thiel, W., 324[114]
Thielmann, H., 512[231]
Thien Chi, N., 331[393]
Tholander, H. 507[1]
Thomas, F., 508[41]
Thomas, G., 647[196]
Thomas, J. E., Jr., 646[160]
Thomich, W., 671[150]
Thompson, G. E., 670[102]
Thompson, N., 861[106]
Thomson, W. K., 508[30]
Thorn, R. J., 323[64]
Thorpe, A., 510[130, 134]
Thümmler, F., 718[4, 16]
Thun, R., 820[36]
Tiberg, L., 508[63], 510[165]
Tiberg, M., 510[140, 141]
Tichonovskii, A., 644[100]
Tiede, E., 643[66]
Tikhinsky, G. F., 329[288, 294, 296, 313], 334[486]
Tiller, W. A., 646[150]
Timmons, G. A., 332[409]
Timo, G., 671[169]
Timoshenko, N. N., 671[173]
Tirkina, A. N., 323[80]
Titran, R. H., 861[94, 95, 96]
Titterington, R., 719[32]
Tix, A., 508[50, 54, 56, 57], 510[150, 151, 152, 153], 513[286]
Tkachev, D. G., 644[95]
Tolmie, E. D., 646[148]
Tommaney, J. W., 643[47]

Tonomura, K., 323[83]
Torrestier, A., 645[110]
Tottle, C. R., 860[81]
Trapeznikov, V. A., 332[400]
Treilou, A., 645[137]
Trendelenburg, E. A., 143[10]
Treppschuh, H., 93[60]
Tripp, J. H., 859[45]
Trombe, F., 328[262], 643[65]
Trommer, E., 670[114]
Tron', A. S., 802[93, 94, 95]
Trotter, H. G., 511[229]
Trouvé, J., 330[346], 645[137, 138]
Trulson, O. C., 515[387]
Trzeciak, M. J., 670[125]
Tsaregorodtsev, I. D., 326[170]
Tseft, A. L., 321[12, 15, 18], 328[248]
Tsuya, K., 860[75]
Tulin, N. A., 330[334]
Turillon, P. P., 642[27]
Turk, C. D., 819[27]
Turner, T. H., 509[81]
Turovtseva, Z. M., 668[3, 5, 33], 669[85, 87], 670[94, 100, 119, 122]
Tusche, J., 321[3]
Tyou, P., 668[40]

Vaessen, G. H. G., 647[207]
Vanderscheuren, E., 333[430]
Vanderslice, T. A., 143[8], 858[9]
Vandersluis, J., 744[15]
Vasileva, N. M., 670[94, 119]
Vasynkova, A. A., 328[249]
Vasyutinsky, B. M., 329[312]
Vatsadze, D. V., 325[155, 156]
Vaughan, H. G., 860[82]
Veigel, N. D., 331[390], 332[397]
Venable, R. H., 509[75]
Venkateswarlu, Ch., 669[77]
Veprek-Siska, J., 328[282]
Verge, J., 509[117], 513[287]
Vergnolle, J., 331[393]
Verkin, B. I., 327[247]
Verö, J., 512[277]
Verot, L., 645[110]
Vershinin, N. P., 324[124]
Vertman, A. A., 323[84]
Vest, R. W., 328[276]
Vignial, P., 326[180]
Vogel, D. L., 647[207]
Volkov, S. E., 642[24]
Volkovich, A. V., 328[249]
Vollmer, H., 330[335], 643[33, 52]
Vorob'ev, V. V., 328[268]
Votava, E., 333[430], 647[211]
Vurm, K., 327[230]

Wacholder, B. V., 858[4]
Wacker, A., 507[16]
Wadsworth, N. J., 861[103, 106]
Wagner, C., 92[8], 744[4]
Wagner, E. R., 802[92]
Wahlster, M., 512[236, 238, 240]
Wain, H. L., 332[402]
Wainright, T., 508[38]
Wajda, E. S., 820[36]
Wakefield, G., 328[265]
Waldron, F. B., 508[33]
Walker, J. A. J., 669[86], 670[103]
Wall, A. J., 801[67]
Walldén, S., 327[231]
Walner, H., 861[105]
Walsh, W. J., 327[211]
Walter, D. I., 670[128]
Wang, C. T., 331[378]
Ward, R. G., 326[204, 207], 508[68], 642[19, 25]
Warner, N. A., 93[36], 326[194]
Warren, R. W., 646[172]
Watanabe, H., 512[241]
Watanabe, O., 323[69]
Wehner, G., 328[279], 820[46]
Weinig, S., 334[469]
Weisberg, L. R., 328[260]
Weiss, P., 324[100], 329[303]
Weiss, S., 801[71, 75]
Weiss, V., 859[43]
Werner, J. E., 513[315]
Wernick, J. H., 332[416], 333[435, 444], 334[483], 645[142], 646[163]
Wessel, O., 719[59]
Wessling, W., 645[126]
Westerheide, D. E., 327[218]
Westwood, A. R. C., 859[47]
Wever, F., 93[57]
Weyman, W. L., 800[3]
Wicher, A., 509[119], 514[336]
Wicher, H. A., 511[222]
Wieland, K., 335[516]
Wielonski, R. F., 820[51]
Wiener, G., 670[132]
Wiesner, U., 331[367]
Wilhelm, H. A., 323[90], 326[183], 331[377]
Wilkens, G., 800[13]
Wilkins, D. H., 670[124]
Willardson, R. K., 328[259], 335[497]
Williams, C. E., 508[28]
Williams, C. S., 801[49]
Williams, E. L., 335[506]
Williams, J. M., 335[503]
Williams, K. C., 327[232]
Williams, W. M., 646[168]
Williamson, J. B. P., 859[27, 35]
Willners, H., 508[44]

Wilson, J. C., 333[438]
Wilson, J. F., 647[221]
Wilson, L. H., 510[157]
Wilson, R., 802[104]
Wilson, W., 509[106]
Wiltna, H. W., 511[224]
Winegard, W. C., 646[168]
Winkler, O., 92[13], 93[46, 51], 326[190], 512[244], 642[15], 644[101], 668[25], 718[21]
Winslow, P. M., 859[20]
Winter, H., 507[20]
Winterhager, H., 92[14], 322[27, 40, 41], 332[405]
Witbeck, R. F., 668[15]
Witzke, W. R., 333[460]
Wheeler, A., 718[20]
Whelan, J. M., 328[256]
White, E. A. D., 645[140]
White, M. J., 819[8]
White, S. S., 800[25]
Whitmire, L. D., 647[193]
Whitson, E. H., 802[82]
Wölfel, E., 668[4], 669[66]
Wojcik, A., 512[262]
Wood, D. E., 671[174]
Wood, D. R., 642[13]
Wooding, P. J., 511[195, 200], 643[42]
Woodward, J. A., 670[115]
Worrell, W. L., 322[29, 30]
Wright, M. A., 861[110, 111]
Wulff, G., 333[428]
Wulff, J., 333[448], 801[73], 802[83]
Wutschel, A., 669[67]
Wylie, A. W., 332[398]

Yagihashi, T., 335[520]
Yamashita, T., 647[218]
Yano, S., 646[166]

Yarwood, J., 143[2]
Yefimenko, Yu. M., 645[135]
Yemelyanov, V. S., 332[394, 401, 411]
Yensen, T. D., 671[162]
Yevstyukhin, A. I., 332[394, 401, 411]
Yoshida, K., 719[47]
Yukawa, S., 859[43]
Yukhtanov, D. M., 330[321]
Yuki, S., 513[329]

Zaboronok, G. F., 644[94]
Zagorskaya, T. N., 334[464]
Zakharov, E. L., 668[11]
Zakharov, L. A., 802[96]
Zakowa, H., 512[273]
Zapf, G., 719[53]
Zedler, E., 331[367], 333[453]
Zega, B., 819[22]
Zelenzov, T. I., 644[94]
Zernov, V. B., 327[245]
Zhak, G., 330[339]
Zhukovetsky, V. V., 326[184, 187], 335[513]
Ziegenbalg, S., 324[114], 334[480]
Zinsmeister, G., 819[14]
Zitter, H., 669[56], 671[170]
Zlatnik, I., 513[323]
Zlatnikova, J., 513[323]
Zlotin, A., 644[97]
Zscherpe, E., 719[44]
Zubarev, A. G., 510[145]
Zubko, A. M., 802[84]
Zuev, I. M., 514[342, 343]
Zverev, B. F., 510[158]
Zviadadze, G. N., 325[138, 139, 140, 141, 145, 153, 154, 155, 156]
Zyryanov, G. G., 330[331]

Subject Index

Activity, 3, 7, 8
Activity coefficient, 7, 8, 9, 10, 14
Adhesion,
  coefficient of, 835
  in vacuum, 832
Adhesion strength in vacuum, 827
Alcan process, 254
Alkali metals,
  carbothermic reduction of, 181
  refining of, 243
  vacuum distillation of, 243
Alkaline-earth metals,
  carbothermic reduction of, 182
  metallothermic reduction of, 198
  refining of, 244
  vacuum distillation of, 244
Alnico sintered magnets, 693
"Alphatron" ionization gauge, 134
Aluminium,
  vacuum brazing of, 786
  zone-melting of, 317
Aluminium alloys,
  refining of, 253
  vacuum brazing of, 786
  vacuum distillation of, 253
Analysis, vacuum extraction, 649
Antimony, sublimation of, 239
Arc furnace, vacuum consumable, 555
Arc melting, vacuum, 553
Arkel–de Boer process, 279
Arsenic, sublimation of, 238
ASEA–SKF process, 377

Backstreaming, 107
Baffles, 115
Barium, vacuum distillation of, 247
Beryllium,
  metallothermic reduction of, 209
  refining of, 248
  vacuum brazing of, 785
  vacuum distillation of, 248
  vacuum sintering of, 687
  zone-melting of, 314
Bicrystals, growth of, 641

Bismuth,
  vacuum distillation of, 237
  zone-melting of, 313
Blue brittleness of steel, 392
Boron, zone-melting of, 317
Boundary layer, 32, 36
  concentration gradient in, 34
  friction, 40
  Nernst, 37
  thickness of, 36, 37
Brazing, vacuum, *See* Vacuum brazing
Brick-metal reaction, 28
Bridge breaker, 533
Bubble formation,
  mechanism of, 44
  in stream degassing, 45
Bubble nucleation, rate of, 43
Bubble nuclei,
  excess of, 52
  formation of, 42
Bubble radius, 42, 44
  critical, 42
Bubble volume *vs.* height, 59
Bubbles,
  ascending,
    growth of, 47
    partial pressure of hydrogen and nitrogen in, 56
  capillary pressure of, 42, 50
  CO,
    gas content of, 49
    partial pressure in, 51
  degassing through, 54
  diffusion boundary layer around, 47
  flow line net work around, 47
  gas content of, 50
  gas take-up of, 50, 57, 58
  growth of, 44, 45
  growth rate of, 51
  hydrostatic pressure of, 42
  minimum pressure for, 47
  nucleation of, 43
  rate of ascent of, 47, 49, 53
  size of, 51

SUBJECT INDEX

source of, 43
stability of, 42, 45

Cadmium, vacuum distillation of, 235
Calcium,
  carbothermic reduction of, 185
  metallothermic reduction of, 209
  vacuum distillation of, 247
Calcium production, 245
Carbides, preparation of, 175
Carbon monoxide reaction in iron, 349
Carbothermic reduction,
  of the alkali-metals, 181
  of the alkaline-earth metals, 182
  of calcium, 185
  of chromium, 176
  of magnesium, 183
  of molybdenum, 177
  of niobium, 164
  of tantalum, 172
  of titanium, 174
  of tungsten, 177
  of uranium, 179
  of vanadium, 161
  of zirconium, 175
Carbothermic reduction processes, 161
Cathode sputtering, 814
  applications of, 817
  materials deposited by, 816
Cemented carbides, vacuum sintering of, 688, 710
Ceramics, electron beam melting of, 633
Chemical potential, 7
Chemical transport reactions in a vacuum, 279
Chemical vapor deposition welding, 798
Chromium,
  carbothermic reduction of, 176
  refining of, by the iodide process, 281
Circulation degassing process, 409
Cleanness,
  of degassed steel, 447
  effect of vacuum degassing on, 456
  of ladle degassed steel, 377
  of steel, 403, 419
Cobalt, zone-melting of, 316
Coefficient of adhesion, 835
Cold cathode ionization gauge, 135
Cold wall furnaces, 729
Compound metals, vacuum sintering of, 692
Concentration change, time dependence of, 37
Condensation coefficient, 66
Conductance for molecular flow, 97
Contact angle, 43
Contact time, 34, 40, 41
Contaminant dispersal in roll bonding, 831
Consumable electrode, 575

Convection, 32
  in the gas space, 214
  in the melt, 214
Convection currents, 40
  natural, 39
Convective flow, 32
Coolidge process, 695
Copper, zone-melting of, 317
Creep properties, effect of residual contaminants on, 852
Crucible,
  decomposition of, 85
  decomposition rate of, 87
  effect of the structure of, 83
  equilibrium between melt and, 85
  gas permeability of, 86
  moisture in, 84
  reaction with melt, 84
  reactive impurities in, 84
  sintered, 86
  wettability of, 88
Crucible reaction, 82, 87
Cryosorption pumps, 123
Crystal growing, 633
  in a container, 634
  by electron beam heating, 636
  floating zone, 635, 637
  methods of, 634
Cycling and circulation degassing, 343
Cycling and circulation degassing processes, 396

Decarburization,
  of ferro-chromium, 262
  of steels, 29
Degassing, 14
  bubble-free, 40
  concentration change in the melt during, 71
  course of, 53, 70
  in crucibles and ladles, 70
  dust produced during, 351
  from falling metal drops, 61
  by gas bubbles, 42
  kinetics of, 483
  ladle, ordinary, 363
  pressure change during, 72
  in the receiving ladle, 63
  by scavenging gas, 54
  in the stream degassing process, 63
  stream droplet, 383
  tap, 387
  two-stage, 433
Degassing processes,
  appraisal of, 441
  cycling and circulation, 396
  effect of, on the properties of steel, 541

hydrogen removal in, 444
  incorporation of, in the production process, 449
  oxygen removal in, 445
  solid-state, 467
  vacuum-metallurgical potentialities of, 448
Degassing rate, 72
Degassing time, 53
Deoxidation,
  by carbon, 16, 18
  by deoxidizing elements, 17
  by evaporation of suboxides, 18
  by hydrogen, 18
  in stream degassing, 385
  via suboxides, 268
Desulphurization in a vacuum, 370
DH degassing process, 343, 396
  fall in temperature with, 399
  refractory linings for, 409
DH vacuum treatment of steel, oxygen elimination during, 402
Diffusion,
  in the gas space, 215
  in the melt, 214
  in solid state vacuum degassing, 487
Diffusion bonding, 791
Diffusion bonding vacuum hot press, 797
Diffusion boundary layer, 32, 34, 39
  around bubbles, 47
  of inductively stirred melt, 38
  thickness of, 37
Diffusion coefficient of gases in metals, 489
Dissociation pressure, 5
Distillation,
  of mixed binary phases, 221
  multi-stage, 234
  rate of, 216
  of silver crust, 222
  thin-film, 221
  in vacuum, 213
    separation of metals by, 213
Distillation metallurgy, 150
Distillation rate of metals, 217
Distribution coefficient, 287
  effective, 288
Ductile to brittle transition temperature, effect of residual contaminants on, 846
Ductility, effect of residual contaminants on, 846
Dust,
  composition of, in vacuum degassing, 352
  during degassing, 351
Dust filters in vacuum degassing, 353
Dwell time at the surface, 40

Ejectors, water-operated, 104
Ejector pump, vapour, 115

Electron beam furnaces, melting and purification in, 615
Electron beam melting, 593
  power supplies for, 604
Electron beam melting furnaces, 603
Electron beam melting process techniques, 609
Electron beam melting systems, basic designs used for, 596
Electron beam welding, 745
  hard vacuum mode, 753
  inert atmosphere, 762
  soft vacuum mode, 757
Electron beam welding equipment, 762
Electron beam welding guns, 763
Electron beam zone-refining, 295
Energy, free, 4
Enthalpy, 3, 4
  of formation, 7
  free, 4, 5
Entropy, 4
Equilibrium,
  dynamic, 16
  heterogeneous, 4
Equilibrium constant, 3, 6
  temperature function of, 6
Equilibrium pressure, 13, 14, 15, 22, 42
Equilibrium vapour pressure, 5
Evaporation,
  of gas molecules in solid state vacuum degassing, 494
  heat of, 22
  of manganese, 351
  of metallic elements, 74
  of metals, 215
  rate of, 74
  of refractory materials, 360
  in solid state vacuum degassing, 490
  of suboxides, 15, 18
    deoxidation by, 18
  during zone melting, 292
Evaporation coefficient, 80
Evaporation loss,
  of the basis metal during degassing, 76
  during zone melting, 293
Evaporation rate,
  maximum, 216
  of metals, 217
Extractive metallurgy, 145

Fascia technique, 640
Fatigue, effect of residual contaminants on, 855
Ferro-chromium,
  decarburization of, 262
  vacuum decarburization of, 262
Fick's first law, 34
Filtration in metallurgical processes, 119

Finkl–Mohr process, 377
Floating zone crystal growing, 637
Flow line network, 37, 38, 39
Flow model,
    of Kraus, 38
    of Machlin, 37
Flow velocity, 36, 37
Flux density, 64
Foreign layer, effect of, on the surface, 41
Furnaces for vacuum heat treatment, 728

Gallium, vacuum distillation of, 257
Gas ballast, 106
Gas content,
    after RH vacuum degassing, 417
    determination of, 649
    of DH degassed steel, 404
Gas evolution from refractories, 28
Gas sources, 101
    in metallurgical process plant, 95
Gases,
    diffusion coefficient of, in metals, 489
    influence of, on properties of metals, 464
    saturation concentration of, in metals, 465
    solubility of, 15
        in metals, 465
Gauge,
    ionization,
        "Alphatron", 134
        cold cathode, 135
        hot cathode, 135
        mechanical, 131
        McLeod, 132
        partial pressure, 137
        Penning, 135
        Pirani, 133
        thermocouple, 133
        vacuum, 130
Gero degassing process, principle of, 436
Gold, winning of, 159
Group III–V compounds, zone-melting of, 318

Hafnium,
    deoxidation via suboxides, 269
    electron beam melting of, 625
    metallothermic reduction of the halides of, 185
    refining of, 270
        by the iodide process, 282
Halogen leak detector, 141
Hard metals, vacuum sintering of, 688
Hard vacuum mode electron beam welding, 753
Hardening in a vacuum, 736
Hardness, effect of residual contaminants on, 843
Heat of formation, 6

Heat loss,
    in ladle degaassing, 368
    in vacuum degassing, 353
Heat transport, 215
Heat treatment, 721
Helium leak detector, 142
Henry's activity, 9
Henry's constant, 8
Henry's law, 8
Hot cathode ionization gauge, 135
Hot topping, 534, 579
Hot wall furnaces, 728
Hydrogen, solubility of, in iron, 348
Hydrogen content after stream degassing, 384
Hydrogen elimination,
    during RH vacuum degassing, 414
    during tap degassing of steels, 391
Hydrogen removal,
    in ladle degassing, 371
    in transfer degassing of steel, 424
    in vacuum ingot casting, 429
    in various degassing processes, 444

Ideal solution, 8
Imperial Smelting process, 225, 228, 232
Indium,
    vacuum distillation of, 257
    zone-melting of, 317
Induction melting, vacuum, 519
Inert atmosphere electron beam welding, 762
Ingot degassing, 436
    segregation in, 438
Interaction coefficient, 9, 10
Interaction of materials with residual gases, 838
Interaction of materials in vacuum, 838
Iodide process, 279
Iron,
    carbon-containing, oxygen saturation of, 349
    carbon monoxide reaction in, 349
    solubility of hydrogen in, 348
    solubility of nitrogen in, 348
Iron–hydrogen system, 470
Isbell de-zincing process, 229

Joining,
    in a vacuum, 745
    by vacuum diffusion bonding, 791
    by vacuum hot pressing, 795
    by vacuum roll bonding, 791

Kinetics of degassing, 483
    in the liquid state, 30
Knacke constitution diagram, 161
Knapsack process, 202
Knudsen flow, 96

Ladle degassing, 343
  with additional heating, 377
  ASEA–SKF, 379
  Finkl–Mohr, 378
  heat losses during, 368
  hydrogen removal in, 371
  with induction stirring, 374
  with injection of a purging gas, 371
  ordinary, 363
  oxide inclusions after, 368
  reduction of the hydrogen level during, 367
  refractory lining for, 380
  residual pressure during, 365
  stopper rods for, 832
  temperature losses in, 370
Ladle degassing processes, 363
Ladle to ladle degassing, 387
Ladle nozzle for stream degassing, 386
Ladles, refractory lining of, 356
Langmuir's hypothesis, 75
Lead,
  distillation of impurities from, 233
  refining of, 22
  vacuum de-zincing of, 229
  volatilization of, 157
  zone-melting of, 313
Lead bullion, vacuum de-zincing of, 224
Leak detector,
  halogen, 141
  helium, 142
Leak-finding, 138
Leak-proving, 138
Leak rate, 99
Leak-sensing devices, 140
Leakage, 98
  virtual, 99

Magnesium,
  carbothermic reduction of, 183
  metallothermic reduction of, 198
Magnesium processes, 201
Magnetherm process, 202
Manganese,
  evaporation of, 351
  loss of, 351
Manometer, "U" tube, 132
Mass action, law of, 2
Mass spectrometers, 137
  analytical, 142
  leak-test, 142
Mass transfer, 31, 39
  between melt and bubble, 55
  from the melt into the gas phase, 63
  pressure dependence of, 55
Mass transfer coefficient, 31, 33, 53
  determination of, 73
  overall, 65
Mass transport, 31, 38, 47, 49, 60, 216
  through the interface, 33
  in the liquid phase, 31
Material exchange, 32
Material transfer, 37
Materials transport by evaporation, 220
McLeod gauge, 132
Mean free path, 96, 216
Mechanical properties in vacuum, 838
Metal–gas binary systems, 467
Metal distillation, methods of, 221
Metals, refining of, 145
Metallic adhesion in vacuum, 825
Metallurgy, extractive, 145
Metallothermic reduction, 160
  of alkali metals, 193
  of alkaline-earth metals, 198
  of beryllium, 209
  of calcium, 209
  of the halides of hafnium, 185
  of the halides of vanadium, 185
  of magnesium, 198
  of the oxides of hafnium, 189
  of the oxides of titanium, 189
  of the oxides of zirconium, 189
  of plutonium, 186
  of rare-earth metals, 191
  of thorium, 186
  of uranium, 186
Metallothermic reduction processes, 185
Molecular flow, 96
Molybdenum,
  carbothermic reduction of, 177
  electron beam melting of, 623
  refining of, 278, 285
  vacuum sintering of, 687
  zone-melting of, 303
Mould degassing, 345
Mould degassing processes, 425

Nickel, zone-melting of, 316
Nickel base alloy, lead content of, 535
Niobium,
  carbothermic reduction of, 164
  degassing of, 498
  electron beam melting of, 622
  isothermal degassing of, 495
  liberation of nitrogen from, 497
  liberation of oxygen from, 501
  oxygen concentration in, during degassing, 503
  production of, 169
  refining of, 271
    by the iodide process, 281
  removal of oxygen and nitrogen from, 617

vacuum sintering of, 685
zone-melting of, 299
Niobium–nitrogen system, 471
Niobium–oxygen system, 476
Niobium alloys, production of, 171
Nitrides, decomposition pressure of, 475
Nitrogen, solubility of, in iron, 348
Nitrogen content after stream degassing, 385
Nitrogen elimination during tap degassing of steel, 391
Non-ferrous alloy scrap, vacuum distillation of, 260
Non-ferrous metals and alloys, electron beam melting of, 626
Non-metals, dissimilar metals and combinations thereof, vacuum brazing of, 788
Nucleation,
  of bubbles, 43
  heterogeneous, 43, 45
  homogeneous, 43
Nucleation frequency, 47, 52

Oxide inclusions,
  in DH-degassed steel, 403
  after ladle degassing, 368
  in RH degassed steel, 419
  in steel, 349, 393
Oxides,
  decomposition pressure of, 474
  vapour pressures of, 477
Oxygen concentration
  in niobium during degassing, 503
  in tantalum during degassing, 503
Oxygen content of steels during tap degassing, 389
Oxygen elimination,
  during DH vacuum treatment of steel, 402
  during RH vacuum degassing, 416
Oxygen liberation of vanadium, 504
Oxygen removal,
  in vacuum ingot casting, 429
  in various degassing processes, 445
Oxygen saturation of carbon-containing iron, 349
Outgassing, 99
Outgassing data, 100
Outgassing rates, 101

Partial pressure gauge, 137
Partition coefficient, 82
Penning gauge, 135
Phase transition, 64, 74
Pipe resistance, 96
Pirani gauge, 133
Platinum-group metals, zone-melting of, 310
Plutonium, zone-melting of, 314

Poiseuille equation, 96
Poiseuille flow, 96
Port Pirie plant, 229
Pourbaix–Ellingham diagram, 161
Precision casting furnaces, 552
Pressure,
  critical, 217
  of metals and alloys, 217
Pressure-measuring devices, 130
Pressure sintering furnaces, 718
Pump fluids, 116
Pumps, 102
  cryosorption, 123
  diffusion, 115
  ejector, 104, 111
    vapour, 115
  mercury diffusion, 119
  oil-sealed rotary, 105
  oil-vapour diffusion, 115
  oil-vapour ejector, 115
  Roots, 108
  rotary blade, 105
  rotary piston, 105
  sputter-ion, 121
  sublimation, 120
  turbomolecular, 125
  for U.H.V., 127
  vapour, 115
  water ring, 102
Pump-down curves, 98
Pumping of vapours, 106
Pumping systems, 102, 536
  choice of, 350
Pumping speed, effective, 96
Purging gas, injection of, in ladle degassing, 371

Rail steel, ladle-degassed, 366
Raoult's law, 8
Rare-earth metals,
  refining of, 241
  vacuum distillation of, 241
  zone-melting of, 314
Reaction, brick-metal, 28
  of iron melt with oxides, 27
  of melt with oxide-refractory materials, 28
  between melts, refractories and slags, 24, 82
Reducing agents, 160
Reduction process, 15
  carbothermic, 161
  metallothermic, 185
  vacuum-metallurgical, 160
Reduction,
  by carbon, 16, 27
  of crucible materials with iron melts, 28
  by hydrogen, 18

metallothermic, 160
Refining,
  of alkali metals, 243
  of alkaline-earth metals, 244
  of aluminium alloys, 253
  of beryllium, 248
  of hafnium, 270
  of lead, 22
  of metals, 145
    by the iodide process, 280
  of molybdenum, 278, 285
  of niobium, 271
  of rare-earth metals, 241
  selective, 29
  of tantalum, 271
  of thorium, 278
  of tungsten, 278
  in a vacuum, 29
  of vanadium, 271
  of zirconium, 270
  by zone melting, 286
Refractories, resistance of, to liquid steel, 357
Refractory linings,
  for DH degassing process, 409
  of ladles, 356, 380
  of ladles used in stream degassing, 386
  for RH degassing process, 413
  tensile strength of, 362
  thermal shock resistance of, 362
  of the vacuum ladles, 380
Refractory materials, evaporation of, 360
Refractory metals, vacuum brazing of, 788
Residual gas pressure, effect of, 217
Residual gases, interaction of materials with, 838
Residual pressure during ladle degassing, 365
Resistance furnaces, vacuum melting in, 518
Retort process, 202
RH degassing process, 343, 409
  degassing rate in, 411
  refractory linings for, 413
RH vacuum degassing,
  gas content after, 417
  hydrogen elimination during, 414
  oxygen elimination during, 416
  temperature fall during, 413
Rigid-flow model, 37
Roll bonding, 791, 831
  contaminant dispersal in, 831
Roots pump, 108

Scavenging gas, 54
  CO as, 58
  degassing by, 54
  degassing rates with, 57
  purging coefficient of, 56

uptake of gas by, 60
Selective refining, 29
Selenium, vacuum distillation of, 259
Sieverts' law, 12, 347
Silicon, zone-melting of, 318
Silver crust,
  distillation of, 222
  vacuum distillation of, 223
Sintering furnace, 691, 702
Slag reactions, 89
  resistance to, 359
"Skull" melting, 572
Soft vacuum mode electron beam welding, 757
Solid state bonding, 790
Solubility,
  of gases, 11, 14, 15
    in metals, 465
  of hydrogen in iron, 348
  of nitrogen in iron, 348
Solubility coefficient, 69
  Ostwald's, 52
Space simulation, 823
Space and vacuum environment, 822
Sputter-ion pumps, 121
Stainless steel, vacuum sintering of, 693
Standard state, 4
State variables, 3, 4
  partial values of, 7
Stationary states in solid state vacuum degassing, 481
Steam ejectors, 111, 114, 536
Steel,
  blue brittleness of, 392
  cleanness of, 403, 419
  degassed, cleanness of, 447
  DH-degassed,
    gas content of, 404
    oxide inclusions in, 403
  effect of degassing processes on the properties of, 541
  electron beam melting of, 627
  hydrogen elimination during tap degassing of, 391
  hydrogen removal in transfer degassing of, 424
  influence of vacuum treatment on grain size of, 459
  ladle degassed, cleanness of, 377
  maraging, 592
  nitrogen elimination during tap degassing of, 391
  oxide inclusions in, 349, 393
  oxygen content of, during tap degassing, 389
  oxygen elimination during DH vacuum treatment of, 402

RH degassed, oxide inclusions in, 419
vacuum arc melted ball bearing, 592
vacuum processed ball-bearing, 455
Sticking coefficient, 66
Stream degassing, 343, 383
  deoxidation in, 385
  from falling drops, 62
  hydrogen content after, 384
  ladle nozzle for, 386
  nitrogen content after, 385
  refractory lining of ladles used in, 386
  temperature loss during, 386
Stream degassing processes, 382
Stream droplet degassing, 383
Strontium, vacuum distillation of, 247
Sublimation pumps, 120
Suboxide partial pressure, 19
Suboxides, 19, 160
  deoxidation via, 268
  evaporation of, 15, 18
  vapour pressure of, 21
Superalloys, vacuum brazing of, 787
Surface-active substances, 41, 42

Tantalum,
  carbothermic reduction of, 172
  decarburization of, 618
  degassing of, 499
  electron beam melting of, 621
  isothermal degassing of, 495
  liberation of nitrogen from, 497
  liberation of oxygen from, 501
  metallothermic reduction of the halides, 185
  oxygen concentration in, during degassing, 503
  production of, 171
  refining of, 271
    by the iodide process, 281
  removal of oxygen from, 617
  vacuum sintering of, 682
  zone-melting of, 299
Tantalum–carbon–oxygen system, 478
Tantalum–nitrogen system, 472
Tantalum–oxygen system, 476
Tantalum melting in an EBM furnace, 609
Tap degassing, 387
  oxygen content of steels during, 389
  principle of, 387
  of steels,
    hydrogen elimination during, 391
    nitrogen elimination during, 391
Tellurium, vacuum distillation of, 259
Temperature loss,
  in continuous vacuum degassing, 423
  in ladle degassing, 370
  during stream degassing, 386

Tensile properties, effect of residual contaminants on, 849
Therm-I-Vac process, 393
  principle of, 394
Thermocouple gauge, 133
Thermodynamic relationships, 3
Thermodynamic equilibria,
  of metal–gas systems, 467
  in solid-state metal–gas systems, 466
Thermodynamics of solutions, 7
Thermo-Flow degassing process, 421
Thermo-Flow process, 343
Thorium,
  refining of, 278
    by the iodide process, 281
Titanium,
  carbothermic reduction of, 174
  deoxidation via suboxides, 269
  electron beam melting of, 626
  metallothermic reduction of the halides, 185
  refining of, by the iodide process, 282
  zone-melting of, 316
Titanium and its alloys, vacuum brazing of, 786
Titanium–hydrogen system, 470
Tin,
  vacuum distillation of, 235
  zone-melting of, 313
Transfer degassing, 343, 422
  of steel, hydrogen removal in, 424
Transport,
  within the diffusion boundary layer, 64
  in the gas phase, 64
  in the gas space, 492
Transport coefficient, 33, 40, 41, 42
Transport phenomenon, superposition of, 216
Transport resistance, 67
  in the gas phase, 70
  at the phase transition, 67
Traps, 107
Tungsten, carbothermic reduction of, 177
  electron beam melting of, 624
  refining of, 278
  vacuum sintering of, 687
  zone-melting of, 303
Turbomolecular pumps, 125

Ultra-high vacuum, 126
Ultra-high vacuum systems, 128
Uranium,
  carbothermic reduction of, 179
  refining of, by the iodide process, 281
  zone-melting of, 314
Uranium carbide, production of, 179

Vacher–Hamilton equilibrium, 16

Vacuum,
    adhesion strength in, 827
    chemical transport reactions in, 279
    desulphurization in, 370
    distillation in, 213
    interaction of materials in, 838
    mechanical properties in, 838
    metallic adhesion in, 825
    zone melting in, 286, 292
Vacuum adhesion, 832
Vacuum annealing, 739
Vacuum arc furnace, 3-phase, 574
Vacuum arc melting, 553
    arc stabilising and stirring in, 566
    cleanness obtained by, 583
    electrode fabrication and handling, 575
    electrode position control, 563
    improvements in properties in, 591
    melting parameters in, 577
    metallurgical effects in, 580
    power supplies for, 558
    pumping installations for, 557
    segregation effects in, 586
Vacuum atmosphere in heat treatment, 721
Vacuum brazing, 775
    of aluminium and aluminium alloys, 786
    of beryllium, 785
    metallurgical failure mechanisms in, 781
    metallurgical reactions in, 780
    of non-metals, dissimilar metals and combinations thereof, 788
    of refractory metals, 788
    of superalloys, 787
    of titanium and its alloys, 786
    wetting and spreading phenomena in, 778
Vacuum brazing furnaces, 789
Vacuum casting, 425
    using an intermediate ladle, 434
    without an outer vacuum chamber, 434
Vacuum coating, 803
Vacuum coating plant, 811
Vacuum consumable arc furnace, 555
Vacuum de-zincing of lead bullion, 224
Vacuum decarburization of ferro-chromium, 262
Vacuum degassing,
    composition of dust in, 352
    continuous, temperature losses in, 423
    dust filters in, 353
    effect of, on cleanness, 456
    heat loss in, 353
    historical survey, 337
    ingot, 436
    in the liquid state, 337
    refractory lining for, 356
    solid state, 463
        diffusion in, 487
        elementary steps of, 484
        evaporation in, 490
        evaporation of gas molecules in, 494
        stationary states in, 481
Vacuum diffusion bonding, joining by, 791
Vacuum distillation,
    of alkali metals, 243
    of alkaline earth metals, 244
    of aluminium alloys, 253
    of barium, 247
    of beryllium, 248
    of calcium, 247
    of gallium, 257
    of heavy non-ferrous alloy scrap, 260
    of indium, 257
    of metals, 218
    of rare-earth metals, 241
    of selenium, 259
    of silver crust, 223
    of strontium, 247
    of tellurium, 259
Vacuum distillation processes in the metallurgy of cadmium, zinc and tin, 234
Vacuum environment,
    interaction between metals and, 821
    space and, 822
Vacuum evaporation, 803
    applications of, 812
    control of, 808
    methods of, 805
Vacuum evaporation processes, 810
Vacuum extraction analysis, 649
    accuracy and precision in, 663
    apparatus for, 655
    sources of error in, 659
    technology of, 652
Vacuum filtration, 148
Vacuum gauges, 130, 536
Vacuum grinding, 149
Vacuum heat treatment, 721
Vacuum heat treatment furnaces, 728
Vacuum heat treatment processes, 736
Vacuum hot press, diffusion bonding, 797
Vacuum hot pressing, 691, 795
    joining by, 795
Vacuum induction melting, 519
    casting techniques in, 543
    cleanness obtained by, 550
    crucibles used for, 539
    melting and refining in, 542
    metallurgical effects in, 545
    power supplies for, 537
    pumping installations for, 536
    refractory components used in, 540
Vacuum induction melting furnaces, basic

SUBJECT INDEX

designs for, 521
Vacuum induction furnaces, for sintering, 708
Vacuum ingot casting, 425
  hydrogen removal in, 429
  oxygen removal in, 429
Vacuum ladles, refractory lining of, 380
Vacuum lifter process, 399
Vacuum melting, 517
  in resistance furnaces, 518
Vacuum-metallurgical potentialities of various degassing processes, 448
Vacuum-metallurgical process stages, 146
Vacuum-metallurgical reduction processes, 160
Vacuum precision casting, 553
Vacuum press, 148
Vacuum pressure sintering furnace, 716
Vacuum processed ball-bearing steel, 455
Vacuum pumps, 102, 536, 558
Vacuum pumping systems, 102
Vacuum rectification, 234
Vacuum refining of metals, 212
Vacuum roll bonding, joining by, 791
Vacuum rolling, 791
Vacuum sintering, 673
  of beryllium, 687
  of cemented carbides, 688, 710
  of compound metals, 692
  fundamentals of the reactions in, 677
  of hard metals, 688
  of molybdenum, 687
  of niobium, 685
  production of materials in, 681
  of stainless steel, 693
  of tantalum, 682
  of tungsten, 687
Vacuum sintering furnaces, 694
Vacuum sintering plants, 688
Vacuum stream degassing, 382
Vacuum sublimation, 150
Vacuum treatment,
  influence of, on grain size of steel, 459
  of liquid steel, 347
Vacuum valves, 118
Vacuum zone-melting apparatus, 294
Vanadium,
  carbothermic reduction of, 161
  metallothermic reduction of the halides of, 185
  oxygen liberation of, 504
  refining of, 271
    by the iodide process, 281
  zone-melting of, 299
Vapour pressure, 14, 23
  of bromides, 154
  of chlorides, 153
  of fluorides, 153

  of iodides, 154
  of metal, 22, 151
  of metals, 219
  of oxides, 155, 477
  of sulphides, 155
  of suboxides, 21
Vapour pressure ratios of metals and their monoxides, 582
Vaporization, rate of, 74
VAW process, 202
Volatilization,
  of lead, 157
  of zinc, 157
Volatilization metallurgy, 150

Water ring pumps, 102
Welding by vapour deposition, 798

Zinc,
  extraction of, 158
  kinetics of the distillation of, 225
  production of, 225, 230
  vacuum distillation of, 235
  volatilization of, 157, 158
Zirconium,
  carbothermic reduction of, 175
  deoxidation via suboxides, 269
  electron beam melting of, 625
  metallothermic reduction of the halides, 185
  refining of, 270
    by the iodide process, 282
  removal of chromium from, 619
  removal of iron from, 619
  zone-melting of, 298
Zirconium–hydrogen system, 470
Zone-melting,
  of aluminium, 317
  of beryllium, 314
  of bismuth, 313
  of boron, 317
  of cobalt, 316
  of copper, 317
  effects of, 298
  evaporation during, 292
  of Group III–V compounds, 318
  horizontal, 297
  of indium, 317
  induction, crucible-free, 296
  of lead, 313
  of molybdenum, 303
  of nickel, 316
  of niobium, 299
  of the platinum-group metals, 310
  of plutonium, 314
  of rare-earth metals, 314
  refining by, 286

of silicon, 318
of tantalum, 299
theory of, 290
of tin, 313
of titanium, 316
of tungsten, 303
of uranium, 314

in a vacuum, 286, 292
of vanadium, 299
of zirconium, 298
Zone-melting equipment, 294
Zone-melting process, 287
 principle of, 291
Zone-refining, electron-beam, 295